BMA

Neuroethics

Neuroethics
Anticipating the Future

Editor

Judy Illes

Associate Editor

Sharmin Hossain

UNIVERSITY PRESS

Great Clarendon Street, Oxford, OX2 6DP,
United Kingdom

Oxford University Press is a department of the University of Oxford.
It furthers the University's objective of excellence in research, scholarship,
and education by publishing worldwide. Oxford is a registered trade mark of
Oxford University Press in the UK and in certain other countries

Published in the United States of America by Oxford University Press
198 Madison Avenue, New York, NY 10016, United States of America

British Library Cataloguing in Publication Data
Data available

Library of Congress Control Number: 2016963308

ISBN 978-0-19-878683-2

Printed and bound by
CPI Group (UK) Ltd, Croydon, CR0 4YY

Foreword: Forays into neuroethics

It is early 2001, I had just been appointed as the inaugural Scientific Director of the Institute of Neuroscience, Mental Health and Addiction (INMHA) of the Canadian Institutes of Health Research (CIHR). The CIHR was created to usher in a new generation of thinking, action, and innovation about basic and translational research, for junior and more senior scientists and scholars alike.

One of the most exciting changes brought about by the creation of CIHR and its 13 institutes was the Institute Advisory Board structure and mandate enabling and encouraging the exploration of novel research avenues and frontiers without the traditional boundaries. For those of us in brain research, mental health, and addiction, the work of my inaugural Board and its vision were transformational. We nearly immediately departed from conventional ideas about funding in neuroscience to support a New Emerging Team (NET) program in 2002–2003, with topics such as the impacts of early life events on the developmental course of brain disorders, regenerative medicine, and understanding the placebo effect. This was merely our first call. The second followed directly from my own experiences and leadership of over 20 years in my laboratories at the Douglas Mental Health Institute affiliated with McGill University in Montreal. I was highly aware of ethical issues and challenges to optimal care for the mentally ill, their involvement in various types of clinical trials, and the high incidence of co-morbidity between mental health and addiction. Accordingly, I was keen to promote innovative research on ethical issues related to the broad mandate of INMHA. But neuroethics?

I first heard about neuroethics in the early years of the CIHR. I asked some of my colleagues and few ethicists locally and internationally about it. Most basically said that "It is a fad." They saw no reason to create this specialty or subfield. In their minds, research on ethics was complicated enough; neuroethics would only complicate the complicated further. In retrospect, by most measures, they were wrong. Neuroethics may have added a layer of complex inquiry to both neuroscience and biomedical ethics, but the sole fact that *Neuroethics: Anticipating the Future* (2017) today follows *Neuroethics: Defining the Issue in Theory, Practice and Policy* (2006), speaks for the need, force, and what was to come.

INMHA launched two calls to create NET teams on Neuroethics (awarded to researchers at Dalhousie University in Halifax, Nova Scotia) in addition to a partnered Chair in Neuroethics to recruit an expert from outside Canada. The University of British Columbia won the call, and book Editor Dr. Judy Illes was recruited back to Canada from the United States as the chair holder. This was great coup for all in Canada and for the field as it would ground neuroethics with the highest nationally named position and center yet.

Research in neuroethics by many scholars authoring chapters in this book and others has moved the field forward significantly over the past decade on topics as varied as incidental findings in brain imaging, forensic psychiatry, improved design of clinical trials in the fields of mental illnesses, addiction and incurable neurodegenerative diseases, neurotechnology, cognitive enhancements, consciousness, and brain death, to name a few. Of course, cutting-edge research in neuroethics is still urgently needed on these topics.

Novel areas have also emerged over the last few years to which neuroethics must lead with evidence-based, best-way-forward guidelines for experts and the civil society. Take for example, artificial intelligence and the interface between robots and the mind: How can the discourse of neuroethics help society to take up related challenges in a sensible, balanced manner? Where will the novel concept of a cognitive footprint take us? Will gene editing (CRISPR-Casp9-type) edit out incurable brain conditions or undesirable behavioral traits and states—and who will decide? How will the role and impact of various cultures and the environment on values relevant to mental health and addiction evolve? Will wearable technology, social media, digital e-health, and open access to large public and private sets of data bring solutions to global attitudes and demands of societies toward brain diseases and various forms of addiction? And, link to these considerations: cybersecurity and open access to large data sets.

Neuroethics has matured. It feels unstoppable not only because of the energy brought to it by its current and future leaders, but by the pressing issues awaiting discovery, definition, awareness, and answers. *Neuroethics: Anticipating the Future* is a signature of this momentum.

Rémi Quirion
Professor of Psychiatry, McGill University
Distinguished Researcher, Douglas Mental Health Institute
Quebec Chief Scientist

Preface

Part 1: Looking to the future

—Judy Illes and Sharmin Hossain

This publication of *Neuroethics: Anticipating the Future* marks both an anniversary of one of the first edited neuroethics volumes for the field (Illes 2006), as well as of the founding of the International Neuroethics Society, the major professional organization for the field. Like the publication of the first book volume that took the groundwork laid out at the Mapping the Field neuroethics meeting in San Francisco, California—now 15 years ago, and another anniversary—and gave it content anchor points in theory, practice, and policy, this one also signals milestones. The authors contributing to the original volume wrote to answer the question: *Where should the field go?* In this new volume, contributing authors answer the question: *Where must the field go next?* With significant progress over more than a decade, this is a critical undertaking with both legitimacy and gravitas.

In preparing this volume, the three sections on neurotechnology, healthcare, and policy emerged naturally from the contributions we received. The prescient Joseph J. Fins predicted the neurotechnology focus in his epilogue to the *Oxford Handbook of Neuroethics* (Illes & Sahakian 2011). Healthcare and public good are unwavering themes for the field, involving ethical challenges around brain future of youth, aging and dementia, brain injury and consciousness, stem cells and neurologic conditions, and global health and cross-cultural justice, among others. With one foot within and one foot outside the walls of the hospital and academy, neuroethics in the wild brings considerations of social and regulatory challenges to a leading role in new discourse about the environment, brain, and mental health, in part in a global rapprochement between neurology and psychiatry, internationalization and intersections of brain and security, and criminal justice.

Among the mix of authors are neuroethicists in the pipeline for leadership who are playing a pivotal role in charting the course of the field for decades to come. Like those who are more senior, they hail from diverse disciplines, spanning the humanities, public health and public policy, and science and medicine. Their diversity is reflected not only in their disciplines, but also in their cultural backgrounds—different lenses that bring focus and richness to any area of ethical inquiry.

Ten years ago, the International Neuroethics Society provided a platform for a small neuroethics community to engage; today it offers hundreds of faculty, trainees, and collaborators from around the globe expanded opportunities for networking, showcasing research, and career development, as well as a place to call a professional home. Over the past year, Sharmin Hossain also participated in a further initiative focused on the next generation of ethics and policy leaders: the Trainee Policy and Advocacy Committee (T-PAC) formed under the wing of Kids Brain Health Network (formerly NeuroDevNet, Inc.), a trans-Canada Network of Centres of Excellence. T-PAC aims to engage research trainees to advocate for evidence-informed policy making to benefit children living with neuro-developmental disorders, and serves as a practical example of initiatives that neuroethics scholars in training can and, we would argue, should undertake to collaborate toward a unified mission to bring about positive changes to people and society.

Throughout the volume, contributing authors bring both deep experience and fresh insights to critical issues in neuroethics, and they bring advanced scholarship, accounta-bility, and transparency to the profoundly interlocking themes of human rights and pro-tections, freedoms, and autonomy. Echoing Hillary Clinton's nod to the African proverb "It takes a village to raise a child" (Clinton 2006), we say that it takes a community to make a field. Much good work has been done in the 10 years past; and, as the chapters in this volume describe, there is much to do in those to come.

Preface

Part 2: A brief look back

—Judy Illes

Learning from lessons of the past

In the firm belief that to move forward it is essential to learn lessons from the past, the second part of this Preface is a brief look at the history of modern neuroethics. As I reflect on this task, one of the most frequently asked questions comes to mind: *Why do we, citizens and beneficiaries of advances in the neuroscience, need neuroethics?* And, formulated slightly differently: *How is neuroethics different than bioethics?* My regular answer is: neuroexceptionalism. Unlike the knee or the elbow, the liver or pancreas (but, with new knowledge about the microbiome, perhaps like the gut (Mayer et al. 2014)), the brain connects each of us to who we are. In health and in disease, it links us to our identity, relations, autonomy, personality, capacity to make decisions, and much more. Each is intricately intertwined with the other like the threads of a rope embedded in the context in which we thrive, derive, and survive. It has allowed scholars in the field to frame and reframe ethical concerns explicitly associated with the brain sciences, and to similarly consider governance and advisory frameworks carefully attuned to relevant issues, helping to define both good and ethically defensible neuroscience.

Reflections

I reflect on probably the most quoted phrase for the field, and I give it another nod here: Adina Roskies' "Ethics of neuroscience; neuroscience of ethics" (Roskies 2002). I have never quite agreed with this dichotomy as the latter cannot exist except within the context of the former, but no matter: even if the phrase is not perfectly dichotomous, its simplicity gave us a welcome and forceful catch phrase to propel neuroethics forward.

I continue with other reflections. The topic that never dies: enhancement, whether for the competitive boost or improved mental health (Illes 2016), and more recently for moral fortitude (Persson & Savulescu 2008). If the use of drugs for the moral application could give even a modicum of relief from its terror-riddled world, I would be all for it. Topics that should never die: addiction and mental health, aging and dementia, consciousness, justice and human rights, consent, dignity, pain, regenerative medicine, agency.

A topic that took on a life of its own: the management of unexpected (incidental, study off-target) findings in brain imaging research. As much as the question about how to handle incidental findings of possible health significance was not new to genetics, and was introduced originally by the US National Bioethics Commission in 1999 (National

Bioethics Advisory Commission 1999), the potency of the question took the neuroimaging community by surprise. Lively debates since 2005 and ongoing today (Illes et al. 2006, 2008; Wolf et al. 2008) have yielded well-balanced policies and guidance for brain anatomies. And, while ample questions still remain (Gibson et al. Chapter 3, this volume), new ones about the interpretation of the functional data in the context of functional outliers that may be medically or societally significant (Scott et al. 2012) are arising. They require urgent attention before solutions to inevitable challenges have to become reactive rather than anticipatory and thoughtfully proactive (Foster et al. 2003).

Allow me to move to other enduring imperatives: I would place education and mentorship of interested researchers and clinicians of all ranks and disciplines, and engagement of the community outside the Academy at the top of the list of priorities. Neuroethics teachers and mentors must support the move from didactic, unidirectional schooling of foundational theories and axioms to modern case- and problem-based teaching. The Internet has fueled a craving for autonomy in learning; classroom teaching should now elevate do-it-yourself learning in this digital age, rather than the historical reverse where book learning elevated lessons imparted from the more learned.

Similarly, there has not been a time in the history of biomedicine when the importance of including the voice of research participants and patients in the design, execution, and dissemination of research through methods such as participatory action and collaborative research (Stevenson et al. 2016) has been as recognized if not mandated. These interactions have brought a new level of meaningfulness to the specific translational goals of the biosciences (Illes et al. 2011), human subjects protection (Eijkholt et al. 2012), rights (Human Rights for Neuroethics: Reflection or Refraction, International Neuroethics Society Meeting, Washington, DC, November 2014), and public engagement (Robillard et al. 2015).

This discussion brings me now to one of the most cited text in neuroethics, *Mapping the Field* (Marcus 2002), the edited volume created from the transcripts of the landmark neuroethics meeting in San Francisco in that same year. The field has many thanks to offer to the Dana Foundation whose support is unfailing even 15 years later, and the vision for neuroethics of the late William Safire (said to have been first articulated in a conversation with Zack Hall in a taxi in New York City). Deep gratitude is also due to investment of the late William Stubing of the Greenwall Foundation whose support was foundational (e.g., Illes & Raffin 2002; Farah & Wolpe 2004), and also continues to date. Beyond these neuroethics funding visionaries, I count more than two dozen other private and government sponsors of research to whom the field owes thanks. Their financial contributions to the advancement of critical thinking about ethics, society, policy, and law alongside advances in the neurosciences, have enabled the field to gain traction, credibility, and sustainability. As Professor of Medicine and Director of the Usher Institute of Population Health Sciences and Informatics, Dr. Andrew Morris said at the International Association of Bioethics meeting in Edinburgh, Scotland in June 2016 (a meeting at which neuroethics was only minimally represented), ethics must be (and I paraphrase) "a first thought; not an after thought … in confluence to improve human and public health."

Perhaps it is both the alignment of neuroethics with neuro, and its practical pro-positioning in science that best define the field as it is today more than any particular the-oretical construct. Indeed, neuroethics seems well defined by what we, as scholars in the field do with it and the interdisciplinary, often pragmatic approach we take (Racine 2010). That is not to say that the theoreticians among us have not played a key role in integrat-ing philosophy fundamentally into the field, but still, as University of British Columbia Sociologist Professor D. Ralph Matthews and I concur about neuroethics at present: "We are what we study" (loosely translated from "You are what you eat"). To this end, per-haps we are raising more questions than answers, but that is not a failure, just a signal of dimensionality.

Measurables

There is little doubt that this is the century of the brain. As University of Chicago bio-ethicist Mark Siegler and I have discussed, technical advances in the last 75 years exceed those of 2000–3000 years that precede it. With each advance in the brain sciences, ques-tions about ethics arise. Think about end-of-life decision-making and today's new laws; predicting Alzheimer's disease in the absence of cure, or predicting adult-onset diseases in newborns and children; stem cells for autism or cerebral palsy; gene therapy; psychi-atric neurosurgery for mental health disorders; technologically assisted communication for the minimally conscious and other diagnostic and therapeutic advances for the spec-trum of mild to severe central nervous system injuries; magnetic resonance imaging in the courtroom; commercialization of health technology (some that preys on the most vulnerable in our society); big data; environment and culture; and manipulating genes to edit out disease and edit in desirable behavioral traits. The hype is phenomenal; pub-lic interest growing and unprecedented (Beaulieu 2000; Whiteley 2012). Are we doing enough? Are we progressing fast enough, meaningfully, transparently, interactively, and effectively enough?

The answer is, on the most part, yes, judging from the numbers of books and edited vol-umes, and international efforts in public engagement (Illes et al. 2005) ever since the early days of neuroethics. For example, shortly after the 2002 meeting, the Dana Foundation's journal *Cerebrum* published a special issue on neuroethics in Fall 2004. A special issue of the journal *Brain and Cognition* was published in 2005 devoted to ethical issues in advanced neuroimaging. The *American Journal of Bioethics* published its own special issue on neuroethics in 2005 and today, with a booming impact factor of more than 6, is accepting target articles in the area. I created the *AJoB–Neuroscience* series and served as its inaugural editor. It is now in Paul Wolpe's capable hands as the official journal of the International Neuroethics Society (INS). There are numerous relevant publications by the US Presidential Commission on the Study of Bioethical Issues (PCSBI; from 2008 to 2016, more than 50 neuroscientists and neuroethics scholars testified to the Commission). Journal papers, trainees in a diverse range of disciplines and presenters at the annual meeting of the INS and elsewhere, faculty in tenured, tenure-line, research stream, and private and government sector positions, programs of one type or another (41 worldwide at last count), media coverage abound.

The first-ever formal strategic plan of the INS, the major professional society for neuroethics, speaks clearly to accomplishments of the past and aspirations for the future with four priorities. With some editorial paraphrasing, the priorities are to:

1. position the INS, with its multidisciplinary membership and leadership, as an authoritative body on matters pertaining to neuroethics

2. promote and support research, scholarship, and education in all aspects of neuroethics

3. develop sustainable funding for the profession

4. grow in number and size.

The strategic achievement of the goals assigned to each of these priorities (http://www.neuroethicssociety.org/strategic) will bring the membership closer to citable policy or legal changes where it has yet to show its colors and truly define a legacy. The Neuroethics Response Task Force (now Emerging Issues Task Force) of the INS that I struck in 2016, initially chaired by Board of Directors member Mark Frankel, and engagement of the broadest range of the profession's diverse membership in rank, discipline, gender, and geographic background, will further enable this goal and the positioning of the INS as the authoritative go-to body in neuroethics.

Anniversaries

With this historical perspective, I also celebrate one of the concurrent past anniversaries of neuroethics—the 10th—the meeting of a small group neuroscientists and bioethicists in Asilomar, California in Spring 2006, to discuss a plan for the now established professional society. Early names were Neuroethics Interdisciplinary Association, competing with Society for Interdisciplinary Neuroethics. It would draw upon knowledge gained in prior meetings: besides the 2002 meeting (the 15th anniversary), in September 2003, the American Association for the Advancement of Science (AAAS) sponsored a meeting on neuroscience and law and, in 2005, another on neuroethics and religion. In May 2005, the Library of Congress sponsored a neuroethics meeting called *Hard Science—Hard Choices*. These are just to name a few.

We were a small group (Figure 0.1) in a small conference room—the rest of the rustic retreat center on the magnificent Pacific Ocean was occupied by a gathering of nearly 1000 church members seeking answers to questions of their own. In many ways, both the content and interpersonal dynamics of that 2-day meeting foreshadowed the many faces of neuroethics today: collaborations and entanglements; warmth and tensions; rigor and fluidity; curiosity and resistance; friends, no fig leafs. In that setting, we created a space for the liberal multiculturalism for the neuroethics that we live today. Culture is a hybrid of internal and external principles. Principles, like ideas, travel. I believe that we got it right then and that we continue to evolve productively. One aspect of the evolution, in fact, was a name-change from the originally settled-upon

Figure 0.1 Participants at the Asilomar Neuroethics Society planning meeting 2006. Kneeling from left to right: Martha Farah, Patricia Churchland. Standing from left to right: Anjan Chaterjee, Turhan Canli, Allison Mackey, Elizabeth Phelps, Hank Greely, Judy Illes, Laurie Zoloth, Mike Gazzaniga, Barbara Sahakian, Paul Wolpe. Not shown: Steve Hyman and Stephen Morse.
Reproduced courtesy of Judy Illes

Neuroethics Society to the International Neuroethics Society signaling a genuine commitment to a global involvement for the field.

With what historical events do we share 10th and 15th anniversaries? Here are a few examples for 2002 and 2006 that coincide with our own:

2002:

- George W. Bush signs into law the No Child Left Behind Act.
- The Netherlands legalizes euthanasia, becoming the first nation in the world to do so.
- The United States is re-elected to the United Nations Commission on Human Rights, 1 year after losing the seat it had held for 50 years.
- Michael Jackson dangles son Prince Michael II off a balcony from a Berlin hotel room (a zany and tragic example of a mental health challenge).

2006:

- The United Nations General Assembly votes to establish the UN Human Rights Council.
- Element 111, roentgenium (Rg) is officially named.

- Canadian researchers make two biomedical breakthroughs by stopping the build-up of toxic plaque in mice with Alzheimer's disease using a sugar-like substance known as scyllo-cyclohexanehexol; and, stopping a mutant gene from being split apart effectively preventing the onset of degenerative symptoms associated Huntington's disease.

Values and understanding

What constitutes knowledge, values, and understanding in neuroethics? I would argue it is both ways of knowing and ways of doing. Both contribute to good judgment, as do operational tools such as decision trees, logic models, checklists, and case studies. Questions are hard when values differ, and understandings perhaps even harder as they are further shaped by situational context. Neuroethics must continue to contribute to the marketplace of ideas or our work will become nothing more than a collection of rhetorical exercises. Similarly, time will tell whether investments in ethics and big neuroscience will inspire wave changes or only ripples of incremental impact in a sea of activity.

To conclude, I return to the first question I raised in its most simplistic form: *Why neuroethics*? Many scholars have spent time over the years in verbal and written discourse upholding this challenge (de Vries 2005, 2006; Wilfond & Ravitsky 2005; Parens & Johnston 2007). Much like competition in business, the existential question yields benefit by powering a critical reflexivity needed for the undeterred to stay sharp. We can be gratified that this healthy debate across our areas of specialty did not lead to a modern day *Tale of Two Cities* (Dickens 2012), with an irreparable collision of neuroethics revolutionaries toward older discipline aristocracy. Indeed, we are a collectively kind community focused on human health and well-being.

To conclude, as Chief of the Neuroethics Studies Program at Georgetown University, James Giordano has said, "No new neuroscience without neuroethics." And, as I said to the esteemed French neuroscientist Jean Pierre Changeux from the Collège de France, Institut Pasteur in a heated neuroethics meeting in Heidelburg, Germany in 2006, "It doesn't matter what we call it, it only matters that we do it" (Illes 2006).

References

Beaulieu, A.E. (2000). The brain at the end of the rainbow: the promises of brain scans in the research field and in the media. In: Marchessault, J. and Sawchuk K. (Eds.) *Wild Science: Reading Feminism, Medicine and the Media*, pp.39–54. London: Routledge.

Clinton, H.R. (2006). *It Takes a Village*. New York: Simon and Schuster.

De Vries, R.D. (2005). Framing neuroethics: a sociological assessment of the neuroethical imagination. *American Journal of Bioethics*, 5, 25–27.

De Vries, R.D. (2006, November). Firing the neuroethical imagination. In: *7th EMBL/EMBO Joint Conference Programme, Genes/Brain, Mind and Behavior*, Heidelberg, Germany.

Dickens, C. (2012). *A Tale of Two Cities*. New York: Vintage.

Eijkholt, M., **Anderson, J.A.**, and **Illes, J.** (2012). Picturing neuroscience research through a human rights lens: imaging first-episode schizophrenic treatment-naïve individuals. *International Journal of Law and Psychiatry*, **35**, 146–152.

Farah, M.J. and **Wolpe, P.R.** (2004). Monitoring and manipulating the human brain: new neuroscience technologies and their ethical implications. *Hastings Center Report*, **34**, 35–45.

Fins, J.J. (2011). Epilogue. In: Illes, J. and Sahakian, B. (Eds.). *The Oxford Handbook of Neuroethics*, pp.895–907. Oxford: Oxford University Press.

Foster, K.R., **Wolpe, P.R.**, and **Caplan, A.L.** (2003). Bioethics and the brain. *IEEE Spectrum*, **40**, 34–39.

Illes, J. (Ed.) (2006). *Neuroethics: Defining the Issues in Theory, Practice, and Policy*. Oxford: Oxford University Press.

Illes, J. (2006, November). From genetics to neuroethics: is imaging "visualizing" human thought? In: *7th EMBL/EMBO Joint Conference Programme, Genes/Brain, Mind and Behavior*, Heidelberg, Germany.

Illes, J. (2016). A feast of thinking on the naturalization of enhancement neurotechnology. In: Jotterand, F. and Dubljević, V. (Eds.) *Cognitive Enhancement: Ethical and Policy Implications in International Perspectives*, pp.346–350. Oxford: Oxford University Press.

Illes, J. and **Raffin, T.A.** (2002). Neuroethics: a new discipline is emerging in the study of brain and cognition. *Brain and Cognition*, **50**, 341–344.

Illes, J., **Blakemore, C.**, **Hansson, M.G.**, et al. (2005). International perspectives on engaging the public in neuroethics. *Nature Reviews Neuroscience*, **6**, 977–982.

Illes, J., **Kirschen, M.P.**, **Edwards, E.**, et al. (2006). Ethics: incidental findings in brain imaging research. *Science*, **311**, 783–784.

Illes, J., **Kirschen, M.P.**, **Edwards, E.**, et al. (2008). Practical approaches to incidental findings in brain imaging research. *Neurology*, **70**, 384–390.

Illes, J., **Reimer, J.C.**, and **Kwon, B.K.** (2011). Stem cell clinical trials for spinal cord injury: readiness, reluctance, redefinition. *Stem Cell Reviews and Reports*, **7**, 997–1005.

Mayer, E.A., **Knight, R.**, **Mazmanian, S.K.**, **Cryan, J.F.**, and **Tillisch, K.** (2014). Gut microbes and the brain: paradigm shift in neuroscience. *Journal of Neuroscience*, **34**, 15490–15496.

Marcus, S.J. (2002). *Neuroethics: Mapping the Field, Conference Proceedings*. New York: Dana Press.

National Bioethics Advisory Commission (1999). *Research Involving Human Biological Materials: Ethical Issues and Policy Guidance*. Available from: https://bioethicsarchive.georgetown.edu/nbac/hbm.pdf [accessed August 4, 2016.

Parens, E. and **Johnston, J.** (2007). Does it make sense to speak of neuroethics? *EMBO Reports*, **8**, S61–64.

Persson, I. and **Savulescu, J.** (2008). The perils of cognitive enhancement and the urgent imperative to enhance the moral character of humanity. *Journal of Applied Philosophy*, **25**, 162–177.

Racine, E. (2010). *Pragmatic Neuroethics: Improving Treatment and Understanding of the Mind-Brain*. Cambridge, MA: MIT Press.

Robillard, J.M., **Cabral, E.**, **Hennessey, C.**, **Kwon, B.K.**, and **Illes, J.** (2015). Fueling hope: stem cells in social media. *Stem Cell Reviews and Reports*, **11**, 540–546.

Roskies, A. (2002). Neuroethics for the new millennium. *Neuron*, **35**, 21–23.

Scott, N.A., **Murphy, T.H.**, and **Illes, J.** (2012). Incidental findings in neuroimaging research: a framework for anticipating the next frontier. *Journal of Empirical Research on Human Research Ethics*, **7**, 53–57.

Stevenson, S., Beattie, B.L., Bruce, L., and Illes, J. (2016). When culture informs neuroscience: considerations for community-based neurogenetics research and clinical care in a First Nation Community with early onset familial Alzheimer disease. In: Chiao, J.Y. (Ed.) *Cultural Neuroscience*, pp.171–181. Oxford: Oxford University Press.

Whiteley, L. (2012). Resisting the revelatory scanner? Critical engagements with fMRI in popular media. *BioSocieties*, 7, 245–272.

Wilfond, B.S. and Ravitsky, V. (2005). On the proliferation of bioethics sub-disciplines: do we really need "genethics" and "neuroethics"? *American Journal of Bioethics*, 5, 20–21.

Wolf, S.M., Lawrenz, F.P., Nelson, C.A., et al. (2008). Managing incidental findings in human subjects research: analysis and recommendations. *Journal of Law, Medicine & Ethics*, 36, 219–48.

Acknowledgments

We are deeply grateful for research funding and support over the past 15 years from numerous organizations: the Canadian Institutes of Health Research, US National Institutes of Health, Canada Research Chairs Program, Canada Foundation for Innovation, British Columbia Knowledge Development Fund, Vancouver Coastal Health Research Institute, Genome BC, Genome Canada, Stem Cell Network, Kids Brain Health Network (formerly NeuroDevNet, Inc.), Canadian Dementia Knowledge Translation Network, Vancouver Foundation, Djavad Mowafaghian Centre for Brain Health at the University of British Columbia (UBC), W. Maurice Young Centre for Applied Ethics at UBC, Peter Wall Institute of Advanced Studies, the Foundation for Ethics and Technology, Institute of Mental Health, American Academy of Neurology, Dana Foundation, Greenwall Foundation, North Growth Foundation, the Office for Science and Technology of the Embassy of France to Canada, and the British Consulate based in Vancouver, BC, and others. We extend our many thanks also to the team at the National Core for Neuroethics at UBC, colleagues, collaborators, trainees within and outside our university, province, and country, and family and friends who share and live the vision for neuroethics with us.

The opinions expressed in the chapters are solely those of the contributing authors, and do not necessarily reflect the views of the editors or research sponsors.

Contents

List of contributors *xxiii*

Part I **Neurotechnology: Today and tomorrow**

1 When emerging biomedical technologies converge or collide *3*
Debra J.H. Mathews

2 Emerging neuroimaging technologies: Toward future personalized
diagnostics, prognosis, targeted intervention, and ethical challenges *15*
*Urs Ribary, Alex L. MacKay, Alexander Rauscher, Christine M. Tipper,
Deborah E. Giaschi, Todd S. Woodward, Vesna Sossi, Sam M. Doesburg,
Lawrence M. Ward, Anthony Herdman, Ghassan Hamarneh, Brian G. Booth,
and Alexander Moiseev*

3 Incidental findings: Current ethical debates and future challenges in
advanced neuroimaging *54*
Lorna M. Gibson, Cathie L.M. Sudlow, and Joanna M. Wardlaw

4 Vulnerability, youth, and homelessness: Ethical considerations on the
roles of technology in the lives of adolescents and young adults *70*
Niranjan S. Karnik

5 The neuroethical future of wearable and mobile health technology *80*
Karola V. Kreitmair and Mildred K. Cho

6 Technologies of the extended mind: Defining the issues *108*
Peter B. Reiner and Saskia K. Nagel

7 Neuromodulation ethics: Preparing for brain–computer interface
medicine *123*
Eran Klein

8 Integrating ethics into neurotechnology research and development: The
US National Institutes of Health BRAIN Initiative® *144*
Khara M. Ramos and Walter J. Koroshetz

Part II **Neuroethics at the frontline of healthcare**

9 What do new neuroscience discoveries in children mean for their open
future? *159*
Cheryl D. Lew

10 Neuroprognostication after severe brain injury in children: Science
fiction or plausible reality? *180*
Sarah S. Welsh, Geneviève Du Pont-Thibodeau, and Matthew P. Kirschen

11 No pain no gain: A neuroethical place for hypnosis in invasive intervention *197*
Elvira V. Lang

12 Placebo beyond controls: The neuroscience and ethics of navigating a new understanding of placebo therapy *214*
Karen S. Rommelfanger

13 Ethical challenges of modern psychiatric neurosurgery *235*
Sabine Müller

14 At the crossroads of civic engagement and evidence-based medicine: Lessons learned from the chronic cerebrospinal venous insufficiency experience *264*
Shelly Benjaminy and Anthony Traboulsee

15 Ethical dilemmas in neurodegenerative disease: Respecting patients at the twilight of agency *274*
Agnieszka Jaworska

16 Anticipating a therapeutically elusive neurodegenerative condition: Ethical considerations for the preclinical detection of Alzheimer's disease *294*
Hervé Chneiweiss

17 When bright lines blur: Deconstructing distinctions between disorders of consciousness *305*
David B. Fischer and Robert D. Truog

18 Brain death and the definition of death *336*
James L. Bernat

Part III **Social, legal, and regulatory frameworks: Lessons of the past guide policy for the future**

19 Minors and incompetent adults: A tale of two populations *369*
Vasiliki Rahimzadeh, Karine Sénécal, Erika Kleiderman, and Bartha M. Knoppers

20 Behavioral and brain-based research on free moral agency: Threatening or empowering? *388*
Eric Racine and Veljko Dubljević

21 Cognitive enhancement of today may be the normal of tomorrow *411*
Fabrice Jotterand

22 Environmental neuroethics: Setting the foundations *426*
Laura Y. Cabrera

23 First Nations and environmental neuroethics: Perspectives on brain health from a world of change *455*
Jordan Tesluk, Judy Illes, and Ralph Matthews

24 The neurobiology of addiction as a window on voluntary control of behavior and moral responsibility *477*
Steven E. Hyman

25 Looking to the future: Clinical and policy implications of the brain disease model of addiction *497*
Adrian Carter and Wayne Hall

26 Concussion, neuroethics, and sport: Policies of the past do not suffice for the future *515*
Brad Partridge and Wayne Hall

27 Security threat versus aggregated truths: Ethical issues in the use of neuroscience and neurotechnology for national security *531*
Michal N. Tennison, James Giordano, and Jonathan D. Moreno

28 Communicating about the brain in the digital era *554*
Julie M. Robillard and Emily Wight

29 The impact of neuroscience in the law: How perceptions of control and responsibility affect the definition of disability *570*
Jennifer A. Chandler

30 Neuroethics and global mental health: Establishing a dialogue *591*
Dan J. Stein and James Giordano

Part IV **Epilogue**

31. Neuroethics and neurotechnology: Instrumentality and human rights *603*
Joseph J. Fins

Author Index *615*
Subject Index *633*

List of contributors

Shelly Benjaminy is a doctoral candidate at the National Core for Neuroethics, University of British Columbia, and a Clinical Ethics Fellow at Providence Health Care in Vancouver, BC. Her research interests lie at the intersection of ethics and novel biotechnologies that present the potential for new knowledge, therapeutics, and improved clinical care.

James L. Bernat is the Louis and Ruth Frank Professor of Neuroscience and Professor of Neurology and Medicine at the Geisel School of Medicine at Dartmouth. His research interests are in brain death, the definition and criterion of death, and chronic disorders of consciousness. He is the author of *Ethical Issues in Neurology*, 3rd edition (Lippincott Williams & Wilkins, 2008).

Brian G. Booth is a postdoctoral researcher at the University of Antwerp. His research interests are in medical image analysis with a focus on using machine learning techniques to identify patterns of abnormality in medical images. He received his PhD in computing science from Simon Fraser University where his thesis focused on diffusion magnetic resonance imaging analysis techniques specialized for the preterm infant brain.

Laura Y. Cabrera is Associate Professor of Neuroethics at Michigan State University. Her research focuses on the exploration of attitudes, perceptions, and values of the general public toward neurotechnologies, the normative implications of using neurotechnologies for medical and nonmedical purposes, and on the ethical and social implications of environmental changes for brain and mental health. She received a BSc in Electrical and Communication Engineering from the Instituto Tecnológico de Estudios Superiores de Monterrey (ITESM) in Mexico City, a MA in Applied Ethics from Linköping University in Sweden, and a PhD in Applied Ethics from Charles Sturt University in Australia.

Adrian Carter is Senior Research Fellow and Head of the Neuroethics and Policy Group, Monash Institute of Cognitive and Clinical Neurosciences, Monash University. His research interests are the impact of neuroscience on our understanding and treatment of addictive behaviors, including agency, identity, and moral responsibility; coercion; and the use of emerging technologies, such as deep brain stimulation and brain imaging, to treat addiction. Dr. Carter has been an advisor to the World Health Organization, the European Monitoring Centre for Drugs and Drug Addiction, the Australian Ministerial Council on Drugs Strategy, and United Nations Office on Drugs and Crime.

Jennifer A. Chandler is a Professor of Law and holds the Bertram Loeb Research Chair at the University of Ottawa. She researches and writes about the legal and ethical aspects of advances in biomedical science and technology, with particular interest in neuroethics, organ donation, and regenerative medicine. Recently she has written on the legal implications of advances in neurological therapies and neuroimaging technologies, the use of neuroscientific evidence in criminal law, regulatory policy related to medical practices such as organ donation and transplantation, and the ethics and law of scientific inquiry.

Hervé Chneiweiss is a Research Director at CNRS, neuroscientist, and neurologist. His research interests are the mechanisms linking the plasticity of the phenotype of the main population of cells of the nervous system, glial cells, and the development of brain tumors. He is Director of the research center Neuroscience Paris Seine-IBPS at Pierre and Marie Curie University in Paris, France, Chairman of the Ethics Committee of INSERM, and a member of the National Consultative Ethics Committee and of the International Bioethics Committee of UNESCO.

Mildred K. Cho is a Professor in the Departments of Medicine and Pediatrics at the Stanford University School of Medicine and Associate Director of the Stanford Center for Biomedical Ethics. Her research interests are in the ethics of research and translation of emerging biotechnologies.

Sam M. Doesburg is the Callum Frost Professor of Autism and Associate Professor of Biomedical Physiology and Kinesiology at Simon Fraser University. He completed his PhD in Neuroscience at the University of British Columbia (UBC) in 2008, and a postdoctoral fellowship at the UBC Department of Pediatrics. His research focuses on how brain network communication develops throughout childhood and adolescence and contributes to cognitive development and difficulties in clinical child populations.

Geneviève Du Pont-Thibodeau is an Assistant Professor of Pediatric Critical Care Medicine at Sainte-Justine University Hospital, University of Montreal. Her research interests include transfusion medicine as well as the neuromonitoring, neuroprotection and neuroprognostication of critically ill children with acute brain injury.

Veljko Dubljević is an Assistant Professor of Philosophy and affiliate of the Science, Technology and Society program at North Carolina State University. Before arriving at NC State, he spent 3 years as a Postdoctoral Fellow at the Neuroethics Research Unit at IRCM and McGill University in Montreal, Canada. His research focuses on ethics of neuroscience and technology, and the cognitive neuroscience of ethics. He coedited the volume *Cognitive Enhancement* (Oxford University Press, 2016) with Fabrice Jotterand, and is the inaugural managing editor and coeditor for the book series *Advances in Neuroethics* (Springer)

Joseph J. Fins is the E. William Davis, Jr. MD Professor of Medical Ethics, Professor of Medicine, Professor of Medical Ethics in Neurology, Professor of Medicine in Psychiatry and Chief of the Division of Medical Ethics at Weill Cornell Medical College, Co-Director of the Consortium for the Advanced Study of Brain Injury (CASBI) at Weill Cornell and Rockefeller University, and the Solomon Center Distinguished Scholar in Medicine, Bioethics and the Law at Yale Law School. His most recent book is *Rights Come to Mind: Brain Injury, Ethics, and The Struggle for Consciousness* published by Cambridge University Press.

David B. Fischer is a Resident Physician at the Harvard Neurology Residency Program at Brigham and Women's Hospital and Massachusetts General Hospital. His research interests are the neurobiology of consciousness, disorders of consciousness, and brain stimulation.

Deborah E. Giaschi is a Professor of Ophthalmology and Visual Sciences at the University of British Columbia. Her research interests are the typical and atypical development of visual perception and reading ability. She uses behavioral and MRI techniques to assess children with amblyopia or dyslexia.

Lorna M. Gibson is a Wellcome Trust Clinical Research Training Fellow at the University of Edinburgh and Honorary Specialty Trainee in Clinical Radiology in the National Health Service. Her research interests are in brain imaging, epidemiology, and research ethics. Her fellowship is focused on the epidemiology and impact on participants and health services of potentially serious incidental findings on brain and body magnetic resonance imaging in the UK Biobank cohort study.

James Giordano is a Professor in the Departments of Neurology and Biochemistry, and Chief of the Neuroethics Studies Program of the Pellegrino Center for Clinical Bioethics at Georgetown University Medical Center, Washington DC, USA. He serves as Senior Science Advisory Fellow of the Strategic Multilevel Assessment Group of the Joint Staff at The Pentagon, and is an appointed member of the Neuroethics, Legal and Social Issues Advisory Panel of the Defense Advanced Research Projects Agency (DARPA).

Wayne Hall is Professor and Director of the Centre for Youth Substance Abuse Research, University of Queensland, and Professor of the National Addiction Centre, Kings College London. His research interests are in drug use and addiction, epidemiology, mental health, ethics, and public health policy. He was formerly an NHMRC Australia Fellow at the University of Queensland Centre for Clinical Research and the University of Queensland Brain Institute (2009–2014), Professor of Public Health Policy, School of Population Health, University of Queensland (2005–2009), and Director of the Office of Public Policy and Ethics at the Institute for Molecular Bioscience (2001–2005), University of Queensland.

Ghassan Hamarneh is a Professor of Computing Science at Simon Fraser University (SFU). His research interests are in medical image analysis, and computer vision, optimization, and machine learning for biomedical imaging applications. Dr. Hamarneh is a Senior Member of the Institute of Electrical and Electronics Engineers (IEEE) and a Senior Member of the Association for Computing Machinery. He is the co-founder and director of the Medical Image Analysis Lab at SFU and a founding member of the IEEE Engineering in Medicine and Biology Chapter—Vancouver Section.

Anthony Herdman is an Associate Professor of Audiology at the University of British Columbia. His research interests are in the understanding of the brain dynamics involved in auditory and visual perceptions, attention, and language, and in advancing neuroimaging methodology to measure brain dynamics.

Sharmin Hossain is a Postdoctoral Research Fellow at the National Core for Neuroethics at the University of British Columbia (UBC) and at Kids Brain Health Network (formerly NeuroDevNet, Inc.), a Canadian Network of Centres of Excellence. She received a PhD in Neuroscience from UBC with a thesis focused on revealing the dynamic growth patterns of immature neurons in the brain. Her current research focuses on neurodevelopmental disorders, ethics, and policy.

Steven E. Hyman is Harvard University Distinguished Service Professor of Stem Cell and Regenerative Biology, a core faculty member of the Broad Institute of Harvard and MIT, and Director of the Stanley Center for Psychiatric Research at the Broad Institute. From 2001 to 2011, he served as Provost of Harvard University and from 1996 to 2001, director of the US National Institute of Mental Health. He is former Editor of the *Annual Review of Neuroscience* and served formerly as President of the Society for Neuroscience and the International Neuroethics Society.

Judy Illes is Professor of Neurology and Canada Research Chair in Neuroethics at the University of British Columbia, in Vancouver, Canada. She is a pioneer of the field of neuroethics, and has made ground-making contributions to a broad range of ethical, social, and policy challenges at the intersection of biomedical ethics and neuroscience. She is a Fellow of the Royal Society of Canada, Canadian Academy of Health Sciences, and the American Association for the Advancement of Science. She is the editor of the original neuroethics volume *Neuroethics: Defining the Issues in Theory, Practice, and Policy* (Oxford University Press, 2006).

Agnieszka Jaworska is Associate Professor of Philosophy at the University of California, Riverside. Her research interests are at the intersection of ethical theory, medical ethics, and moral psychology. Her current book in preparation concerns the ethics of treatment of individuals whose status as moral agents and persons seems compromised or uncertain, such as Alzheimer's patients, addicts, psychopaths, and young children. It is part of a larger project on the nature of the specifically human agency and the role of the capacity to care in moral psychology. She has published in *Ethics, Philosophy and Public Affairs*, and *Philosophy and Phenomenological Research*.

Fabrice Jotterand is Associate Professor of Bioethics and Medical Humanities and Director of the Graduate Program in Bioethics at the Center for Bioethics and Medical Humanities, Medical College of Wisconsin, Milwaukee, and Senior Researcher in Bioethics/Neuroethics at the University of Basel. His scholarship and research interests focus on issues including moral enhancement, neurotechnologies and human identity, the use of neurotechnologies in psychiatry, medical professionalism, and moral and political philosophy. He is the co-editor of *Cognitive Enhancement* (Oxford University Press, 2016, with Veljko Dubljeviić) and the co-editor of the book series *Advances in Neuroethics* (Springer).

Niranjan S. Karnik is the Cynthia Oudejans Harris, MD Professor of Psychiatry at the Rush University Medical Center in Chicago. In the Department of Psychiatry, he serves as the Director of the Section of Population Behavioral Health and Medical Director of the Road Home Program: Center for Veterans and their Families. He is also an Associate Faculty Member of the Maclean Center for Clinical Medical Ethics at the University of Chicago. His research interests are focused on homeless and underserved youth, mobile technologies, and the mental health needs of veterans and their families.

Matthew P. Kirschen is an Assistant Professor of Anesthesiology and Critical Care Medicine, Pediatrics and Neurology at the University of Pennsylvania Perelman School of Medicine and The Children's Hospital of Philadelphia. His research interests are neuromonitoring of critically ill children with acute brain injury, early detection of brain injury in critically ill children, neuroprognostication after acute brain injury in children, and death by neurologic criteria determination in children.

Erika Kleiderman is an Academic Associate at the Centre of Genomics and Policy at McGill University. Her research interests are the ethical, legal, and social implications surrounding minors, access to data and genetic information, and the regulation of stem cells, gene editing technologies, and regenerative medicine. She has also been involved in the development of controlled data and biosamples access documentation for longitudinal biobanks.

Eran Klein is a neurologist and philosopher at the Oregon Health and Science University, Portland VA Medical Center, and Department of Philosophy and the Center for Sensorimotor Neural Engineering at the University of Washington. His interests reside at the intersection of neurology, neurotechnology, and philosophy.

Bartha M. Knoppers is a Professor in the Department of Human Genetics in the Faculty of Medicine at McGill University, and Director of the Centre of Genomics and Policy. She is a Canadian lawyer and is an expert on the ethical aspects of genetics, genomics, and biotechnology. She holds a Canada Research Chair in Law and Medicine and is an Officer of the Order of Canada and Quebec.

Walter J. Koroshetz is the Director of the National Institute of Neurological Disorders and Stroke (NINDS). He works to advance the mission of the Institute, to improve fundamental knowledge about the brain and the nervous system, and to use that knowledge to reduce the burden of neurological disorders. He joined NINDS as the Deputy Director in 2007. Before coming to NIH, Dr. Koroshetz was a Harvard Professor of Neurology, Vice Chair of Neurology at the Massachusetts General Hospital, Director of Stroke and Neurointensive Care, and a member of the Huntington's disease unit. His research activities spanned basic neurobiology to clinical trials.

Karola V. Kreitmair is a Clinical Ethics Fellow at the Stanford Center for Biomedical Ethics. Her research interests include ethical aspects of mobile health and wearable technology, developments in citizen science, biohacking, and gaming, as well as questions surrounding justice and stewardship with respect to the minimally conscious state diagnosis.

Elvira V. Lang is a former Associate Professor of Radiology at Harvard Medical School and current CEO of Hypnalgesics, LLC d/b/a Comfort Talk®. Her research interests are use of nonpharmacologic analgesia interventions and hypnoidal language during invasive medical procedures which she tested in large-scale clinical trials. She further developed and also rigorously tested training in these methods, expanding their use to anxiety provoking medical examinations such as magnetic resonance imaging. Her recent interests examine patient/provider interactions and their impact on health outcomes, patient throughout and satisfaction, as well as overall impact on operational efficiency of healthcare systems.

Cheryl D. Lew is a Clinical Professor of Pediatrics (Clinician-Educator) at the University of Southern California Keck School of Medicine and Children's Hospital Los Angeles. Her research interests within biomedical ethics center on the unique moral status of children and the role of emotional investment in determining ethical values frameworks which drive clinical decision-making. In the biomedical clinical milieu, Dr. Lew is a specialist in pediatric rare lung diseases and is interested in the promises as well as pitfalls of regenerative medicine in ameliorating these diseases.

Alex L. MacKay is a Professor at the University of British Columbia with a joint appointment in the Department of Radiology and the Department of Physics and Astronomy. He is the Director of the UBC Magnetic Resonance Imaging Research Centre which hosts magnetic resonance projects in humans and in animal models for human disease. Dr. MacKay's research area is the application of magnetic resonance techniques to study brain pathology in neurodegenerative disorders. His research group developed a myelin water imaging technique that measures myelin content in vivo in brain and spine.

Debra J. H. Mathews is the Assistant Director of Science Programs in the Johns Hopkins Berman Institute of Bioethics and an Associate Professor in the Department of Pediatrics, Johns Hopkins University School of Medicine. Her research interests are around ethics and policy issues raised by emerging biotechnologies, with particular focus on neuroscience, genetics, stem cell science, and synthetic biology.

Ralph Matthews is Professor of Sociology at the University of British Columbia (UBC). His research focuses on resource management, climate change, and socioeconomic development. He has done research with First Nation indigenous groups in Canada on well-being, health, education, resource management, and environmental policy. With Judy Illes, he initiated the Environmental Neuroethics area of research inquiry. He is Past-President of the Canadian Sociological Association and Past-President of the International Sociological Association's Research Committee on Science and Technology. In 2014 he was named a UBC Senior Killam Research Fellow.

Alexander Moiseev is a Senior Scientist at the Behavioural and Cognitive Neuroscience Institute, Simon Fraser University, Vancouver, Canada. His research interests are in neuroimaging, data analysis, statistics and signal processing in MEG and EEG, bioelectromagnetic inverse solutions including traditional and multisource beamformers and other spatial filters, machine learning, and brain–computer interfaces.

Jonathan D. Moreno is the David and Lyn Silfen University Professor at the University of Pennsylvania where he is a Penn Integrates Knowledge (PIK) professor. At Penn he is also Professor of Medical Ethics and Health Policy, of History and Sociology of Science, and of Philosophy. His books include *The Body Politic*, which was named a Best Book of 2011 by *Kirkus Reviews, Mind Wars* (Bellevue Literary Press, 2012), and *Undue Risk* (Routledge, 2000). Moreno also frequently contributes to such publications as *The New York Times, The Wall Street Journal, The Huffington Post, Psychology Today*, and *Nature* and appears on broadcast and online media. In 2008–2009 he served as a member of President Barack Obama's transition team. *The American Journal of Bioethics* has called him "the most interesting bioethicist of our time."

Sabine Müller is an Assistant Professor of Neurophilosophy and Medical Ethics at the Charité–Universitätsklinik Berlin. Her research interests are ethical issues of psychiatric neurosurgery, particularly of deep brain stimulation, as well as ethical issues of psychiatry, neurology, and neurology.

Saskia K. Nagel is an Assistant Professor of Philosophy and Ethics of Technology at the University of Twente, Netherlands. Her research interests are at the intersection of ethics, philosophy, life sciences, and technologies. Her background is in cognitive science and philosophy. She works on approaches to individual and societal challenges in a technological culture, with a focus on the ethical, anthropological, and social consequences of neuroscientific progress. She is particularly interested in how technologies influence our self-understanding, and how they impact our understanding of autonomy and responsibility.

Brad Partridge is a Senior Research Fellow in the Research Development Unit at Caboolture Hospital, Queensland. His research interests are ethical issues in sport, and the use of human enhancement technologies.

Rémi Quirion is a Professor and Québec's first chief scientist since 2011. Until his appointment, Rémi Quirion was the vice-dean for science and strategic initiatives at McGill University. He was the scientific director of the Douglas Mental Health Institute and the executive director of the International Collaborative Research Strategy for Alzheimer's Disease of the CIHR. Professor Quirion was the first scientific director of the Institute of Neurosciences, Mental Health and Addiction (INMHA). He has received several awards and honours, including the Order of Canada (OC) and is a member of the Royal Society of Canada.

Eric Racine is Director of the Neuroethics Research Unit and Full Research Professor at the Institut de recherches cliniques de Montréal, Canada. Inspired by philosophical pragmatism, his research aims to understand and bring to the forefront the experience of ethically problematic situations by patients and stakeholders and then to resolve them collaboratively through deliberative and evidenced-informed processes.

Vasiliki Rahimzadeh is a PhD Candidate with the Centre of Genomics and Policy at McGill University and a 2016 Vanier Canada Graduate Scholar. Her doctoral thesis explores ethics governance of research with children in the data-intensive, biomedical sciences and its impact on national and international data sharing.

Khara M. Ramos is a Senior Science Policy Analyst in the Office of Scientific Liaison within the Office of the Director of the National Institute of Neurological Disorders and Stroke, a part of the National Institutes of Health (NIH). A neuroscientist and former AAAS Science and Technology Policy Fellow, she has a broad interest in neuroscience research, with a specific focus on how advances in neuroscience intersect with society. She serves as Executive Secretary of the Neuroethics Division of the NIH BRAIN Initiative˙ Multi-Council Working Group.

Alexander Rauscher is a Physicist and Assistant Professor in the Department of Pediatrics at the University of British Columbia. His research interests are the development and utilization of quantitative magnetic resonance imaging techniques for the characterization of tissue damage and repair in the central nervous system.

Peter B. Reiner is a Professor at the National Core for Neuroethics at the University of British Columbia. His research interests include cognitive and moral enhancement, autonomy, and technologies of the extended mind.

Urs Ribary is the BC LEEF Leadership Chair in Cognitive Neuroscience, Director of the Behavioral and Cognitive Neuroscience Institute (BCNI), and Professor of Psychology, Pediatrics and Psychiatry at SFU and UBC in Vancouver, Canada. After graduating from neuroscience and neuropharmacology at the Swiss Federal Institute of Technology in Zurich, Switzerland, he was Professor in Neuroscience and Director of the functional Brain Imaging Center at the NYU Medical Center in New York. He is known as a pioneer in human brain imaging with expertise in functional brain network connectivity and dynamics within the typical and pathological human brain.

Julie M. Robillard is Assistant Professor of Neurology at the University of British Columbia and Faculty of the Djavad Mowafaghian Centre for Brain Health and of the National Core for Neuroethics at the University of British Columbia. She brings her multidisciplinary background in neuroscience and biomedical ethics to the study of issues at the intersection of dementia, technology, and ethics. Dr. Robillard has developed innovative techniques for the analysis of brain health and new media and has unveiled areas of critical need in the field of neuroscience communication.

Karen S. Rommelfanger is an Assistant Professor in the Departments of Neurology and Psychiatry and Behavioral Sciences and the Neuroethics Program Director at Emory University's Center for Ethics, and Neuroscience Editor-in-Residence at the *American Journal of Bioethics Neuroscience*. She also serves on the National Institutes of Health BRAIN Initiative˙ Multi-Council Working Group Neuroethics Division. Her research explores how evolving neuroscience and neurotechnologies challenge societal definitions of disease and medicine. A key part of her work is fostering communication across multiple stakeholders and, as such, she edits the largest international online neuroethics discussion forum, *The Neuroethics Blog*.

Karine Sénécal is an Academic Associate at the Centre of Genomics and Policy at McGill University. Specializing in ethical, legal, and social issues arising in the domains of pediatric research and genetic research, Karine's work focuses on genetic testing and screening of minors, as well as pediatric biobanking.

Vesna Sossi is a Professor in the University of British Columbia Physics and Astronomy Department. Her main areas of expertise are development of positron emission tomography (PET) instrumentation, data quantification, and image analysis methods as well as application of PET to the study of neurodegeneration in rodent models of disease and human disease. She is interested in exploiting the synergy provided by multiparameter, multimodality imaging.

Dan J. Stein is a Professor of Psychiatry at the University of Cape Town. His research interests lie in anxiety and related disorders, and more recently in the intersection of global mental health, neuroscience, and neuroethics.

Cathie L.M. Sudlow is Professor of Neurology and Clinical Epidemiology at the University of Edinburgh, Chief Scientist of UK Biobank, and Honorary Consultant Neurologist in the National Health Service. Her research interests include prospective epidemiological studies, genetic and environmental determinants of stroke and neurodegenerative diseases, and deriving clinical phenotypes from routine healthcare datasets. As UK Biobank's Chief Scientist, she has been responsible for developing and evaluating procedures for feedback of incidental findings as part of UK Biobank's large multimodal imaging study. Professor Sudlow also leads the UK Biobank follow-up program, coordinating linkages to national death, cancer, hospital, primary care and other datasets, and expert outcomes adjudication subgroups.

Michael N. Tennison completed his MA in bioethics at Wake Forest University, NC, and his JD at the University of Maryland Carey School of Law. He has held positions at the Presidential Commission for the Study of Bioethical Issues, University of Pennsylvania Department of Medical Ethics & Health Policy, Wake Forest University's Center for Bioethics, Health, & Society, and University of Maryland's Thurgood Marshall Law Library. Mr. Tennison researches issues at the intersections of law, ethics, and neuroscience, including biomedical human enhancement and the role of neuroscience as an empirical basis for law and legal policy.

Jordan Tesluk is a Postdoctoral Fellow at the University of British Columbia, in the National Core for Neuroethics and the Department of Sociology. His research interests include cross-cultural translation of information related to the brain, relationships between environmental change and the human brain, and the protection of brain and mental health in occupational settings.

Christine M. Tipper is an Assistant Professor of Psychiatry at the University of British Columbia and Neuroimaging Scientist at the BC Children's Hospital Research Institute. Her research utilizes cognitive neuroscience, multimodal brain imaging, high-density EEG, and emerging network analysis techniques to map and quantify dynamic brain systems that support complex social and cognitive functions. She investigates how brain systems supporting how we respond to subtle social cues, understand actions, and manage attention to pursue goals are compromised across a range of neurodevelopmental, neuropsychiatric, and neurological conditions that are characterized by social and cognitive symptoms.

Anthony Traboulsee is an Associate Professor and Research Chair of the MS Society of Canada at the University of British Columbia in Vancouver. He is the Director of the MS/NMO Clinic and Clinical Trials Research Group at UBC Hospital. His research focus is on clinical trial design, the development of practice guidelines for the use of magnetic resonance imaging (MRI) in the management of multiple sclerosis, establishing advanced MRI imaging outcomes in multiple sclerosis, and the treatment of neuromyelitis optica.

Robert D. Truog is the Frances Glessner Lee Professor of Medical Ethics, Anesthesia and Pediatrics at Harvard Medical School, where he serves as Director of the Center for Bioethics. He is also a Senior Associate in Critical Care Medicine at Boston Children's Hospital, where he has practiced pediatric intensive care medicine for more than 30 years.

Lawrence M. Ward is a Full Professor of Psychology at the University of British Columbia. He investigates brain regional interactions underlying a variety of visual, auditory, and higher cognitive processes using EEG and MEG and advanced techniques such as independent component analysis, phase synchronization, and transfer entropy. He is currently focusing on the cognitive neuroscience of attention, memory, reading, and consciousness, the effects of challenges such as drugs and dying on brain network interactions, the effects of neural noise on cognition and perception, and mathematical and computer modeling of neuronal oscillations and synchronization.

Joanna M. Wardlaw is a Professor and Chair of Applied Neuroimaging and Director of Edinburgh Imaging, Neuroimaging Sciences and the Brain Research Imaging Centre at the University of Edinburgh, and Honorary Consultant Neuroradiologist in the National Health Service. Her research focuses on the pathophysiology, diagnosis, treatment, and prevention of stroke, most recently focusing specifically on cerebral small vessel disease. She led a UK-wide initiative to establish standards for management of incidental findings during research imaging.

Sarah S. Welsh is a Fellow in Pediatric Critical Care Medicine at The Children's Hospital of Philadelphia. Her research interests include drug monitoring and pharmacokinetics in pediatric status epilepticus, as well as the neuromonitoring and neuroprognostication of children with acute brain injury.

Emily Wight is Communications Manager at the Djavad Mowafaghian Centre for Brain Health at the University of British Columbia. Her focus is on plain-language science and academic communication, and her background is in nonfiction writing and publishing.

Todd S. Woodward is a Professor of Psychiatry at the University of British Columbia. His research interests are in cognitive neuropsychiatry and functional neuroimaging. The objectives of the cognitive neuropsychiatry research are to identify the cognitive operations underlying the primary symptoms of psychosis and schizophrenia. The objectives of the functional neuroimaging research are to gain an understanding of the cognitive systems involved in psychosis and schizophrenia, and to develop new multivariate methods for analyzing fMRI data, with applications to integrating information from fMRI, EEG, and MEG.

Part I

Neurotechnology: Today and tomorrow

When emerging biomedical technologies converge or collide

Debra J.H. Mathews

Introduction

Neuroethics, like bioethics more broadly, has grown up alongside of and has, in many ways, been propelled by emerging biomedical technologies. Over time, they have sorted themselves into broad areas largely along the lines of clinical practice, public health, and biomedical and research ethics. Each of these broad research areas can be further divided into more specialized areas of research and scholarship. Within the ethics of biomedical research, subdisciplines have emerged around genetics and genomics, regenerative medicine, synthetic biology, nanotechnology, and neuroscience. Neuroethics coalesced in the early 2000s to focus on ethical issues raised by the neurosciences and their application in medicine and society, as well as on issues related to the emerging neuroscience of ethics (Marcus 2002; Illes & Bird 2006). Over the past 10 or 15 years, neuroethics has founded an international professional society (http://www.neuroethicssociety.org/), created specialist journals, and supported the development of an active community of scholars. Neuroethics scholars have prompted and contributed to critical debates about disorders of consciousness, research with brain-interfacing devices, criminal justice, and moral decision-making, to name a few.

Neuroscience sometimes employs other emerging biomedical technologies, such as genetics, regenerative medicine, and synthetic biology, which generate their own ethics and policy conundrums. When emerging biomedical technologies converge or collide, the challenges inherent to each can be compounded. These challenges are of at least two types: First, emerging technologies are new by definition, fast moving, and have only nascent and rapidly evolving theoretical and empirical scholarship on the associated ethical issues. When two or more such technologies are combined, existing ethics, governance, and public engagement challenges are magnified. Second, communities of ethicists[1] tend to coalesce around specific technologies, and study the ethical, legal, and social implications (ELSI), and the governance landscape related to that particular technology or class of technologies; however, these communities often have little overlap. Neuroethics scholars are frequently, though certainly not always, different from ELSI scholars, who study the ethical, legal, and social implications of genetics and genomics, who are often different from ethicists who study regenerative medicine, synthetic biology, or nanotechnology.

The separation is particularly acute for the latter two, originating, as they have, largely outside biomedicine. This siloing limits information exchange and the benefit that can come from cross-pollination and collaboration not only between biomedical scientists and ethicists, but also across scholars working on different emerging technologies.

As technologies mature and become tools for other areas of science, emerging biomedical technologies will increasingly converge in ways that bring these challenges into sharp relief. In addition, the difficulty of staying up to date on bleeding-edge science, the siloing of the science itself, and personal academic interest all suggest that the challenges of siloing in the ethics community will persist. However, both sets of challenges might be attenuated by a more systematic approach to ethical analysis across subdisciplines, and at least some focus not on the details of the technologies but rather on the nature of the technologies. The development of frameworks for ethical analysis based on shared characteristics, including that the technologies are emerging, rapidly evolving, and ethically contentious, provide a process for mutual learning and growth across a variety of emerging biomedical technologies, easing the challenges of bringing together not just two, but many cultures (Snow 1998).

Multiple morally contested areas

The sorts of technologies with which neuroethicists and bioethicists engage are rarely straightforward, conventional, modest innovations. The emerging biomedical technologies at issue here are those associated with fundamental moral disagreements about their development and use, for example, deep brain stimulation, whole genome sequencing, neural imaging, embryonic stem cells, and genome editing. When a new technology emerges and gains traction, scientists flock to the new research area, the science is propelled forward, and the ethics community engages. The science evolves in parallel with the theoretical and empirical ethics scholarship, both often running ahead of public awareness, understanding, and engagement. Morally contested terrain takes time and effort to map, understand, explore, and develop norms and governance for use. Technologies can mature and diffuse rapidly, becoming tools in other areas of science before norms and governance structures are in place, leading to the convergence of two, or more, emerging and morally contested technologies in ways that can magnify existing ethics, governance, and public engagement challenges. Potential effects of such convergence, and sometimes collision, can be seen in the examples of whole genome sequencing and regenerative medicine that follow.

Whole genome sequencing

As the cost of whole genome sequencing has fallen from about $100 million USD per human genome sequenced at the end of the Human Genome Project, to about $1000 USD today, the technology has increasingly been integrated into research and clinical care. In the past, only targeted sequencing was feasible—the sequencing of the gene or genes of interest—such that the test results were narrowly targeted to the question being asked. Whole genome and whole exome (the part of the genome that codes for proteins) sequencing, however, produces a huge amount of data, much of which has unclear or unknown implications for the health of the individual. Some may have very serious, concrete, and medically actionable implications; the remainder of the data falls somewhere in between these two ends of the spectrum. One particularly vexing type of data includes genetic variants associated with serious adult-onset neurological conditions identified in young people and in particular, in children (e.g., Anderson et al. 2015). The genetics and ELSI communities have been working mightily to try to understand how best to manage these data—whether, how, and in what format any or all of it should be returned (e.g., Haga & Zhao 2013; Kleiderman et al. 2014; Kullo et al. 2014; Lewis & Goldenberg 2015). There remain many unsettled questions, and despite a strong and growing literature and multiple frameworks and recommendations for managing these data (e.g., Fabsitz et al. 2010; Wolf et al. 2012; Sénécal et al. 2015), a review suggests that changes in practice responding to these recommendations have been slow (Franca & Mathews, manuscript in preparation).

Regenerative medicine

In the field of regenerative medicine, meanwhile, ethicists had been exploring the issues raised by embryonic stem cell research, and developing relevant guidelines and governance structures (Institute of Medicine and National Research Council 2005), when Shinya Yamanaka announced at the annual meeting of the International Society for Stem Cell Research that he had developed a way to turn adult cells into pluripotent stem cells with properties similar to those of embryonic stem cells (Takahashi & Yamanaka 2006). Unlike embryonic stem cells, to which so many scholars and members of the public voiced moral objections due to their origins in the destruction of human embryos, these induced pluripotent stem cells (iPSCs) required only somatic or body cells from a person, and were therefore thought to be less problematic. In particular, somatic tissue such as that from the skin is much easier to obtain and is readily available, unlike human embryos that are usually obtained from couples who have completed in vitro fertilization treatment and have leftover embryos stored in a clinic freezer. Due to this lower barrier to entry, many investigators began studying and using iPSCs in their research, and the field accelerated rapidly. Foreskin from infant circumcisions is a frequent source. Another source is leftover pathology samples. Not only are there large institutional banks of such tissue, but if the tissue was obtained in the course of clinical care and has had identifiers stripped from the sample, its use in research does not require human subject research oversight in the United States, under current regulations (US Department of Health & Human Services 2008). However, current regulations are under revision, recent events related to the family of Henrietta Lacks (Skloot 2010; Callaway 2013), and existing data on patients' views of different uses of their archived tissue samples have led to new recommendations about the use of archived tissue in iPSC research (Dasgupta et al. 2014; Lomax et al. 2015). As such, institutions and stem cell research oversight committees are still trying to determine the appropriate path forward.

Convergence

Now, imagine the convergence of these two rapidly evolving areas of biomedicine. Each raises vexing ethics and policy issues in the absence of sufficient theoretical and empirical research for norms or policy-making, or both, that come together under one umbrella of research. Imagine, for example, a project that uses de-identified pathology samples to derive iPSC lines and then uses those to derive neurons. As part of this research, whole genome sequencing is completed for each cell line. Further, the data are posted to a publicly available database to maximize the utility of the (possibly publicly funded) resource that is this collection of lines. Sequence data reveal that several of the lines contain serious mutations that would result in an immediate medical response, if known by the person from whom the tissue was taken. Several contain mutations that guarantee that the person from whom the tissue was taken will develop a serious neurological condition later in life. How does the researcher respond?

Because the barrier to entry for iPSC research is so low and genome and exome sequencing are now so affordable, this situation is likely to become increasingly common. The genetics community may be well aware of the debate around the return of results and incidental findings (e.g., Wolf et al. 2012; Sénécal et al. 2015), in particular those associated with neurological disease (e.g., Illes et al. 2004), but the stem cell scientists, who are not technically working with human subjects, and perhaps never have, may not have any familiarity with these debates. Neither of these communities may be well acquainted with the debates within neuroethics about neurodegenerative disease, advance care planning, and loss of cognitive capacities and self (e.g., Mathews et al. 2009; Pierce 2009; Robillard et al. 2011).

This is but one example of how the convergence of emerging technologies can magnify existing ethics, governance, and public engagement challenges. While the above-mentioned example is hypothetical, real cases exist and are likely to become more prevalent, both across biomedical technologies, such as deep brain stimulation, fetal tissue research, novel treatments for psychiatric disease, genome editing, and including other emerging technologies, as well (e.g., Chen et al. 2014; Hutmacher et al. 2015; Meshi et al. 2015; Sproul 2015; Teoh et al. 2015).

Siloed research and analysis

A second challenge raised by converging technologies is the problem of siloing. Although pervasive in academia, siloing has particular implications for ethics and policy work related to emerging biomedical technologies. Neuroethics has grown and evolved in parallel with other subdisciplines of biomedical ethics, including those who work in ELSI, regenerative medicine, synthetic biology, and nanotechnology. Each of these subdisciplines has grown up around the technology at its center. ELSI is arguably the largest and most mature subdiscipline, having been established and supported by a large extramural funding program associated with the Human Genome Project (Meslin et al. 1997). Though the many subdisciplines have common features and often face analogous ethical issues, there is relatively little overlap in the researchers generating the related scholarship. The separation is particularly acute for synthetic biology and nanotechnology for example, originating as they have largely outside biomedicine and more closely associated with engineering. This sorting of bioethics into increasingly specialized subdisciplines often means that the scholars who read, write, and research ethics and policy issues related to neuroscience do not necessarily—and often do not—overlap with the scholars who focus on other emerging biomedical technologies.

It is not that there is no cross-talk or overlap across the subdisciplines—in the example in the prior section, the work on return of results in the ELSI literature was helpfully informed by parallel prior work in neuroimaging (e.g., Illes & Chin 2008; Illes et al. 2008). Indeed, there are a number of prominent cross-pollinators between neuroethics and ELSI, including several authors involved in the current volume (e.g., Cho and Knoppers). Likewise, there are cross-pollinators between ELSI and regenerative medicine, but these are emissaries, not robust conduits for data and knowledge exchange. For example, those in regenerative medicine and the associated ethics community will not necessarily be familiar with the theoretical and empirical research that has been done in neuroethics in response to a search for ways to diagnose dementia and other cognitive disorders earlier and earlier, in an effort to intervene before irreversible damage is done, and when a better chance at delaying onset or otherwise mitigating the damage wrought by disease might exist (e.g., Illes et al. 2007; Arias et al. 2015). Concerns have been raised and studied regarding the psychological implications of presymptomatic diagnosis, discrimination, stigma, and how preferences for individual care, life, and death might change as capacities are lost and the self changed.

Another example of potentially problematic siloing may be found in synthetic biology, for which many of the individuals working on the ethical implications of the science are members of fields outside of ethics. This is not to disparage those scholars or their work in synthetic biology in any way, only to note that synthetic biology, despite the explicit intentions of the field to develop health-related biomedical applications, has not thus far been a major focus of the biomedical ethics community. This disconnect exists despite continuity with prior work in genetics, recombinant DNA and the Asilomar conference of 1975, and existing governance structures in biomedicine, most notably the Recombinant DNA Advisory Committee. Instead, the ethics touchstones in synthetic biology are more likely to be genetically modified organisms, such as golden rice (Enserink 2008), or dual-use concerns, such as those raised by the introduction of new mutations that increased the transmissibility of the H1N1 virus (Enserink 2012; Herfst et al. 2012). While these are important and relevant cases, as synthetic biology moves more into biomedical science, there is critical history and literature tied to their origins and future in genetics, about which they must know. The story of Henrietta Lacks, her descendants, and the HeLa cell line, and the human genetic modification debates that have been going on in genetics for 30 years (President's Commission for the Study of Ethical Problems in Medicine Biomedical Behavioral Research 1982; Skloot 2010) are merely examples. Again, that is not to say that there is not any cross-pollination—the Hastings Center had a large project on synthetic biology (http://www.thehastingscenter.org/publications-resources/special-reports-2/synthetic-future-can-we-create-what-we-want-out-of-synthetic-biology/)—but rather that cross-pollination is the exception rather than the norm, and this comes with serious opportunity costs.

Though frustrating, siloing is understandable. Members of the Academy—whether in the sciences or humanities—cannot all be expert in everything. It takes a great deal of time to keep up both on the ethics and scientific literature related to work in one rapidly evolving area, let alone multiple sets of such literature pairs. Furthermore, there is clear value in specializing, given the importance of the technical and scientific details to forecasting, identifying, and analyzing related ethical issues. But for funding and time, one could spend many weeks attending even only to the major ethics and science conferences relevant to multiple subdisciplines. Furthermore, there are long-standing, entrenched reasons for such siloing, including institutional structures and constraints, and incentives and recognition in academia (Snow 1998; Kahn 2011). Understanding the origins of the problem does not mitigate its effects, however. Disciplinary segregation continues to limit information exchange and the benefits that can come from collaboration across the ethics communities associated with different technologies.

By relying on emissaries rather than principal conduits or stronger connections, scientists and ethicists risk losing relevant data, knowledge, and experience that can help an emerging field, for example, through understanding how early decisions (e.g., about shared, open resources) can shape downstream development and use. Further, direct connections can lead a new field to avoid reinventing a preexisting wheel (e.g., a useful and well-justified governance model, or an approach to incidental findings). Finally, the impact of ethics research and scholarship is very much linked to the scholars themselves (Mathews, unpublished data): if the scholars with valuable historical understanding, research experience, and networks are not at the table in other subdisciplines—at professional meetings, in governance committees, on advisory boards—if the right people are in parallel but non-overlapping fields, the potential impact is lost.

There have been attempts to counteract this siloing, most notable perhaps, James Crow's efforts at Arizona State University (https://newamericanuniversity.asu.edu/), where interdisciplinarity has been embraced and elevated, and academic divisions are being nucleated around problems to be solved, rather than traditional academic disciplines. Institutional issues can be mitigated by such innovative restructuring and culture change, but the change must extend beyond the borders of a single institution if the broader challenges faced by the many existing divisions of biomedical ethics writ large will be addressed.

A shared analytic framework

It is unlikely that there will ever be a grand, unifying framework for the analysis and structuring of all emerging biotechnologies: histories differ, the constellation of leading voices and stakeholders differ, technical details differ. Context matters deeply. However, new morally contentious biomedical fields, the ethicists who study them, and the public who funds and ultimately lives in the world that is shaped by the science, would all benefit from a more systematic approach to information exchange and ethical analysis. One possible solution is a shared method or analytic framework, akin to technology assessment but on a broader scale, to help emerging fields and the associated ethics community to ask the right questions, learn from analogous prior cases in other fields, identify shared challenges, and benefit from knowledge gained elsewhere, and to help both scientists and ethicists to gain a broader understanding of the relationship of their research to other areas of biomedical science. This analytic framework would focus not on the details of the technology but rather on its nature, based on shared characteristics, including that the technologies are emerging, rapidly evolving, and ethically contentious.

Late in its life in the US Congress, the Office of Technology Assessment (OTA) produced a report assessing the content and quality of its policy analysis (Office of Technology Assessment 1993). The OTA was a small shop that served the Congress internally, and was a model for many similar entities internationally. The non-partisan agency was charged with providing "thorough, objective information and analysis to help Members of Congress understand and plan for the short- and long-term consequences of the applications of technology, broadly defined." One of the findings of the 1993 report was that there was little consistency in how different programs within the OTA approached policy analysis. There were, however, commonalities. In developing the context for their reports to Congress, the OTA often evaluated the historical context and status quo, conducted a stakeholder analysis (including individuals, groups, and institutions), identified key assumptions around the technology, determined the international context and identified relevant international comparisons, and assessed the legal and regulatory landscape. While some of these components would be more easily addressed for a specific technology than for an entire field, these components are a solid starting point. It should be noted, though, that there are other models of, and exercises in, technology assessment and ethical analysis that can also serve as resources for the development of a shared analytic framework (e.g., Tairyan & Illes 2009; Sclove 2010; Chameau et al. 2014).

Based on my own work across subdisciplines in bioethics, policy analysis, and technology assessment, I would add the following: First, stakeholder analysis, in particular the role and views of the public, is critical. Scientists and ethicists need to know who cares about this area of science, why, and what their views are. The public will be heard at some point, so better earlier than later; besides, it is the right thing to do. Second, a discussion about intellectual property and precompetitive public resources, their support, and use should start early, though collective action may take time (Hinxton Group 2010). This is particularly true in morally contentious areas of science that work with a scarce or contested resource, or both, such as human embryos. Overall, from an ethics standpoint, researchers engaged with contentious emerging technologies, in particular when developing in pluralistic democracies, should be expected to identify key stakeholders, to articulate what is morally at stake, and to ask and answer questions about the social goals of the research, the anticipated benefits and harms and their distribution, the development of shared resources, and governance.

In addition to facilitating ethics and policy analysis of converging technologies and mitigating the negative effects of siloing, the broadest goal of this sort of systematic effort at assessment or mapping of subdisciplines should be the same as the overarching goal of ethics in biomedicine, including neuroethics: to identify, frame, and solve problems in medicine, public health, and the biosciences (Mathews et al. 2016). Ideally, the process of working through the model would also facilitate stakeholder engagement.

Table 1.1 combines the OTA components and the ethics-oriented components, and provides an outline for a shared analytic framework. The approach is designed to produce a series of reports on different emerging biotechnologies that would facilitate both information flow across subdisciplines and comparison of different emerging biotechnologies converging in a new project, application, or product, as well as the identification of novel issues raised by that convergence.

The development of a shared analytic framework is, of course, only the first step. The framework must also be put into practice, and there remain many practical questions; in particular, how would these analyses be conducted and by whom? Two possibilities are that this work could be centralized or distributed. The Human Fertilization and

Table 1.1 Potential components of a shared analytic framework for emerging biomedical technologies

Framework component	Key question(s)
Historical context	What are the key scientific antecedents and ethics touchstones?
Status quo	What are the key questions, research areas, and products/applications in the field today?
Stakeholder analysis	Which individuals, groups, and institutions have an interest or role in the emerging biomedical technology?
Key assumptions around the technology	What are the key assumptions of both the scientists and the other stakeholders that may impede communication and understanding or illuminate attitudes?
International context and relevant international comparisons	How is the technology and associated ethics and governance landscape evolving internationally? Are there useful international models?
Legal and regulatory landscape	What are the laws and policies that currently apply and what are the holes or challenges in current oversight?
Ethical and societal implications	What is morally at stake? What are the sources of ethical controversy in the field? What are the other societal concerns?
Social goals of the research	What are the goals of the field that are oriented toward improving the human condition? Are there other goals?

Embryology Authority in the United Kingdom currently does much of this analytic and governance work across technologies related to reproductive medicine and research. In addition, technology assessment of varying sorts occurs at national academies, national bioethics commissions, and similar organizations internationally. But again, this work is currently ad hoc, often in response to concerns about a particular application or product, rather than a proactive, systematic effort to understand and help shape an emerging subdiscipline. However, these centralized bodies do have significant authority and convening power, which may allow them to both conduct the analysis and disseminate the products effectively.

An alternative model would be to distribute the work to the subdisciplines themselves through their relevant professional societies. These societies are likely to include many of the relevant stakeholders, to encompass a broad understanding of the field, and to hold the respect of their professional community. However, this model may perpetuate the problem of siloing. To counteract this risk, a networked institution model might be applied (Sclove 2010) wherein the professional societies ally with one another in a structured collaboration to exchange information and even partner in the work of analysis. This virtual network of professional societies would also then be well positioned to disseminate the products not only to the individual subdisciplines, but also across them, and even to embody that work through, for example, the creation of cross-pollination sessions at academic meetings or similar sections of specialist journals that explicitly solicit cross-disciplinary views.

Regardless of the model for implementation of the shared analytic framework, given the nature of emerging biomedical technologies, all analyses must be reviewed and refined periodically with the natural ebb and flow of science and ethics.

Conclusion

When emerging biomedical technologies converge or collide, new challenges can arise. These challenges are of at least two types: (1) Emerging technologies are new by definition, fast moving, and have only nascent and rapidly evolving theoretical and empirical scholarship on the associated ethical issues. The combination of two or more technologies can magnify existing ethics, governance, and public engagement challenges. (2) Communities of ethicists tend to coalesce around specific technologies, to study the ethical, legal, and social implications, and the governance landscape related to that particular technology or class of technologies; however, these communities often have little overlap.

Just as the founders of a scientific field are responsible for laying the experimental groundwork for that field, so too do they have a responsibility for laying the ethics and governance groundwork. This responsibility applies to both scientists and ethicists. Each new, morally contested field in biomedicine has the opportunity to try again to get it right, in terms of evolution and shaping of the field, as well as in its governance. This will only work, however, if we learn from the experience of other disciplines at different developmental stages. Creation and implementation of a shared analytic framework will accelerate this learning and improve the chances of getting it right—or some approximation thereof—more quickly and with fewer mistakes. Such a framework will also ease the assessment of projects, applications, or products in which multiple emerging biomedical technologies converge or collide, and mitigate the effects of academic siloing. Customization will always be required, and systematizing will never completely eliminate siloing in academia. Neuroethics is a prime testing ground for a shared analytic framework, and for the learning and collaboration that can benefit both biomedical research and ethics overall.

Note

1. For the purposes of this chapter, I will use the term ethicist to mean the diverse array of scholars from areas such as philosophy, theology, medicine, biology, public health, history, ethnography, economics, sociology, political science, law, literature, and more, who focus on ethics and policy issues related to biomedicine. I appreciate that this is an imperfect shorthand, but it has been adopted here for simplicity and brevity.

References

Anderson, J.A., Hayeems, R.Z., Shuman, C., et al. (2015). Predictive genetic testing for adult-onset disorders in minors: a critical analysis of the arguments for and against the 2013 ACMG guidelines. *Clinical Genetics*, **87**(4), 301–310.

Arias, J.J., Cummings, J., Grant, A.R., and Ford, P.J. (2015). Stakeholders' perspectives on preclinical testing for Alzheimer's disease. *Journal of Clinical Ethics*, **26**(4), 297–305.

Callaway, E. (2013). Deal done over HeLa cell line. *Nature*, **500**(7461), 132–133.

Chameau, J.L., Ballhaus, W.F., and Lin, H.S. (Eds.) (2014). *Emerging and Readily Available Technologies and National Security: A Framework for Addressing Ethical, Legal, and Societal Issues*. Washington, DC: National Academies Press.

Chen, L., Qiu, R., and Li, L. (2014). The role of nanotechnology in induced pluripotent and embryonic stem cells research. *Journal of Biomedical Nanotechnology*, **10**(12), 3431–3461.

Dasgupta, I., Bollinger, J., Mathews, D.J.H., Neumann, N.M., Rattani, A., and Sugarman, J. (2014). Patients' attitudes toward the donation of biological materials for the derivation of induced pluripotent stem cells. *Cell Stem Cell*, **14**(1), 9–12.

Enserink, M. (2008). Tough lessons from golden rice. *Science*, **320**(5875), 468–471.

Enserink, M. (2012, June 2). Free to speak, Kawaoka reveals flu details while Fouchier stays mum. [News story.] *Science*. Available from: http://news.sciencemag.org/2012/04/free-speak-kawaoka-reveals-flu-details-while-fouchier-stays-mum [accessed July 12, 2015].

Fabsitz, R.R., McGuire, A., Sharp, R.R., et al. (2010). Ethical and practical guidelines for reporting genetic research results to study participants: updated guidelines from a National Heart, Lung, and Blood Institute working group. *Circulation: Cardiovascular Genetics*, **3**(6), 574–580.

Franca, A. and Mathews, D.J.H. *Review of return of results policy, as reflected in clinical trial informed consent documents*. Manuscript in preparation.

Haga, S.B. and Zhao, J.Q. (2013). Stakeholder views on returning research results. *Advances in Genetics*, **84**, 41–81.

Herfst, S., Schrauwen, E.J., Linster, M., et al. (2012). Airborne transmission of influenza A/H5N1 virus between ferrets. *Science*, **336**(6088), 1534–1541.

Hinxton Group (2010). *Statement on Policies and Practices Governing Data and Materials Sharing and Intellectual Property in Stem Cell Science.* Available from: http://hinxtongroup.org/Consensus_HG10_FINAL.pdf [accessed May 29, 2016].

Hutmacher, D.W., Holzapfel, B.M., De-Juan-Pardo, E.M., et al. (2015). Convergence of regenerative medicine and synthetic biology to develop standardized and validated models of human diseases with clinical relevance. *Current Opinion in Biotechnology,* **35**, 127–132.

Illes, J. and **Bird, S.J.** (2006). Neuroethics: a modern context for ethics in neuroscience. *Trends in Neurosciences,* **29**(9), 511–517.

Illes, J. and **Chin, V.N.** (2008). Bridging philosophical and practical implications of incidental findings in brain research. *The Journal of Law, Medicine & Ethics,* **36**(2), 298–304.

Illes, J., Rosen, A., Huang, L., et al. (2004). Ethical consideration of incidental findings on adult MRI in research. *Neurology,* **62**(6), 888–890.

Illes, J., Rosen, A., Greicius, M., and **Racine, E.** (2007). Ethics analysis of neuroimaging in Alzheimer's disease. *Annals of the New York Academy of Sciences,* **1097**, 278–295.

Illes, J., Kirschen, M.P., Edwards, E., et al. (2008). Practical approaches to incidental findings on brain imaging research. *Neurology,* **70**(5), 384–390.

Institute of Medicine and National Research Council (2005). *Guidelines for Human Embryonic Stem Cell Research.* Washington, DC: National Academies Press.

Kahn, J. (2011). The two (institutional) cultures: a consideration of structural barriers to interdisciplinarity. *Perspectives in Biology and Medicine,* **54**(3), 399–408.

Kleiderman, E., Knoppers, B.M., Fernandez, C.V., et al. (2014). Returning incidental findings from genetic research to children: views of parents of children affected by rare diseases. *Journal of Medical Ethics,* **40**(10), 691–696.

Kullo, I.J., Haddad, R., Prows, C.A., et al. (2014). Return of results in the genomic medicine projects of the eMERGE network. *Frontiers in Genetics,* **26**(5), 50.

Lewis, M.H. and **Goldenberg, A.J.** (2015). Return of results from research using newborn screening dried blood samples. *The Journal of Law, Medicine & Ethics,* **43**(3), 559–568.

Lomax, G.P., Hull, S.C., and **Isasi, R.** (2015). The DISCUSS Project: revised points to consider for the derivation of induced pluripotent stem cell lines from previously collected research specimens. *Stem Cells Translational Medicine,* **4**(2), 123–129.

Marcus, S. (Ed.) (2002). *Neuroethics: Mapping the Field, Conference Proceedings.* New York: Dana Press.

Mathews, D.J.H., Bok, H., and **Rabins, P.V.** (Eds.) (2009). *Personal Identity and Fractured Selves: Perspectives from Philosophy, Ethics, and Neuroscience.* Baltimore, MD: Johns Hopkins University Press.

Mathews, D.J.H., Hester, M., Kahn, J., et al. (2016). A conceptual model for the translation of bioethics research and scholarship. *Hastings Center Report* **46**(5), 34–39.

Meshi, D., Tamir, D.I., and **Heekeren, H.R.** (2015). The emerging neuroscience of social media. *Trends in Cognitive Sciences,* **19**(12), 771–782.

Meslin, E.M., Thomson, E.J., and **Boyer, J.T.** (1997). The Ethical, Legal, and Social Implications Research Program at the National Human Genome Research Institute. *Kennedy Institute of Ethics Journal,* **7**(3), 291–298.

Office of Technology Assessment (1993). *Policy Analysis at OTA: A Staff Assessment.* Available from: http://ota.fas.org/reports/PAatOTA.pdf [accessed August 14, 2015].

Pierce, R. (2009). A changing landscape for advance directives in dementia research. *Social Science & Medicine,* **70**(4), 623–630.

President's Commission for the Study of Ethical Problems in Medicine Biomedical Behavioral Research (1982). *Splicing Life: A Report on the Social and Ethical Issues of Genetic Engineering with*

Human Beings. Washington, DC: President's Commission for the Study of Ethical Problems in Medicine and Biomedical and Behavioral Research.

Robillard, J.M., Federico, C.A., Tairyan, K., Ivinson, A.J., and Illes, J. (2011). Untapped ethical resources for neurodegeneration research. *BMC Medical Ethics*, **12**. Available from: http://www.biomedcentral.com/1472-6939/12/9 [accessed May 27, 2016].

Sclove, R. (2010). *Reinventing Technology Assessment: A 21st Century Model*. Washington, DC: Science and Technology Innovation Program, Woodrow Wilson International Center for Scholars.

Sénécal, K., Rahimzadeh, V., Knoppers, B.M., Fernandez, C.V., Avard, D., and Sinnett, D. (2015). Statement of principles on the return of research results and incidental findings in paediatric research: a multi-site consultative process. *Genome*, **58**(12), 541–548.

Skloot, R. (2010). *The Immortal Life of Henrietta Lack*. New York: Crown Publishing Group.

Snow, C.P. (1998). *The Two Cultures*. New York: Cambridge University Press.

Sproul, A.A. (2015). Being human: the role of pluripotent stem cells in regenerative medicine and humanizing Alzheimer's disease models. *Molecular Aspects of Medicine*, 43–44, 54–65.

Tairyan, K. and Illes, J. (2009). Imaging genetics and the power of combined technologies: a perspective from neuroethics. *Neuroscience*, **164**(1), 7–15.

Takahashi, K. and Yamanaka, S. (2006). Induction of pluripotent stem cells from mouse embryonic and adult fibroblast cultures by defined factors. *Cell*, **126**(4), 663–676.

Teoh, G.Z., Klanrit, P., Kasimatis, M., and Seifalian, A.M. (2015). Role of nanotechnology in development of artificial organs. *Minerva Medica*, **106**(1), 17–33.

US Department of Health & Human Services (2008). Code of Federal Regulations Title 45, Section 46. In: *Federal Policy for the Protection of Human Subjects*. Washington, DC: US Government.

Wolf, S.M., Crock, B.N., Van Ness, B., et al. (2012). Managing incidental findings and research results in genomic research involving biobanks and archived data sets. *Genetics in Medicine*, **14**(4), 361–384.

Chapter 2

Emerging neuroimaging technologies: Toward future personalized diagnostics, prognosis, targeted intervention, and ethical challenges

Urs Ribary, Alex L. MacKay, Alexander Rauscher,
Christine M. Tipper, Deborah E. Giaschi,
Todd S. Woodward, Vesna Sossi, Sam M. Doesburg,
Lawrence M. Ward, Anthony Herdman,
Ghassan Hamarneh, Brian G. Booth,
and Alexander Moiseev

Introduction

In this chapter, we highlight a selection of advances, future challenges, and limitations of emerging multimodal neuroimaging technologies, data signal processing, and visualization strategies. We bring these examples to the forefront as experts in the field who are also attentive to the ethical challenges they pose for personalized diagnostics, prognosis, and targeted intervention into clinical practice.

Advances in structural magnetic resonance imaging: Toward the quantification of tissue properties

The basis of structural magnetic resonance imaging (MRI) is the three conventional magnetic resonance (MR) contrasts: proton density, T1 weighting, and T2 weighting. While the detailed mechanisms of T1 and T2 in brain tissue are not well understood, these processes provide qualitative contrast weightings that yield superb in vivo visualization of central nervous system tissue and have proved invaluable as a diagnostic and patient management tool. These days, MR is evolving into a quantitative tool due to incorporation of improvements in both hardware and software that allow for much faster scanning, such as parallel imaging (Pruessmann et al. 1999) and simultaneous multislice imaging techniques (Barth et al. 2016). These approaches facilitate the introduction of otherwise time-consuming MRI scans into clinical research and potentially into routine clinical practice. New MRI sequences have vastly expanded the types of information available by MRI and the latest developments blur the boundaries between structural and functional information. The new frontier in MRI is the quantification of changes in tissue due to development, aging, disease, and therapy. In this very brief overview we highlight a few of the most exciting developments.

A variety of imaging techniques measure brain perfusion (MacDonald & Frayne 2015); for example, arterial spin labeling can yield estimates of cerebral blood flow rates without requiring injection of a contrast agent. While most MR images arise from water, by using magnetization transfer techniques, one can access protons resident on other molecules; chemical exchange saturation transfer (CEST) (Ward K.M. 2000) promises to be able to produce images of the distribution of several biologically important molecules like glucose (Nasrallah et al. 2013) and amino acids residues in proteins. Inhomogeneous magnetization transfer produces images that are specific for membrane lipids in brain (Varma et al. 2015).

There is a large literature on measurement of water diffusion in brain (Jones 2016). Diffusion tensor imaging (DTI) enables tractography, which is the visualization of nerve tracts in brain. DTI provides mean diffusivity and fractional anisotropy, which are highly sensitive to subtle changes in microscopic brain structure. Despite limitations in terms of specificity (Beaulieu 2002; Jones et al. 2013), DTI has been used widely in neuroimaging. Advanced diffusion techniques, for example, diffusion-based size imaging (Wang Y. et al. 2011), enable more quantitative analyses including the proportion of axonal water and of inflammatory cells.

A specific measurement of myelin can be accomplished by myelin water imaging (MacKay et al. 1994; Prasloski et al. 2012). Assessment of myelin using this technique has been performed in multiple sclerosis (Laule et al. 2008; Vavasour et al. 2009; Manogaran et al. 2015), schizophrenia (Lang et al. 2014), stroke (Borich et al. 2013), and mild traumatic brain injury (Wright et al. 2016).

Probing the central nervous tissue using the phase information of MRI scans has become an active field of research. Phase provides information on the magnetic properties of tissue at high spatial resolution (Rauscher et al. 2005; Duyn et al. 2007). Considerable research efforts have been going into turning the phase images into maps of underlying magnetic tissue properties (quantitative susceptibility mapping (QSM) (Schweser et al. 2011, 2016; Bilgic et al. 2012; Liu C. et al. 2015)). Applications of phase-based contrast and QSM range from the investigation of the developing (Lodygensky et al. 2011) and aging brain (Li W. et al. 2014) to the assessment of changes in tissue structure and composition in neurodegenerative diseases (Wiggermann et al. 2013; Li X. et al. 2015; Guan et al. 2016).

Using MR spectroscopy (Oz et al. 2014), researchers can create images of the concentrations of important brain metabolites, for example, *N*-acetylaspartate (measure of neuronal integrity), the creatine/phosphocreatine pool (brain energetics), and glutamate (an important excitatory neurotransmitter).

Surgical procedures can become much less invasive and more effective when the surgeon has access to real-time MR images of the subject. For example, abnormal brain tissue may be ablated using high-intensity focused ultrasound while monitoring the health of adjacent organs of risk using MR images sensitive to temperature (Coluccia et al. 2014).

Imaging of nuclei other than hydrogen introduces novel new imaging contrasts: hyperpolarized MRI with carbon-13 nuclei enables detailed measurement of metabolic pathways (Ross et al. 2010) and sodium imaging (Shah et al. 2016) provides an indication of cellular and metabolic integrity.

Fueled by the imagination of MRI scientists, advancements in MRI hardware have been driving progress in MRI software (i.e., scans) and vice versa, transforming MRI from a tool for the visualization of structure into a science of quantitative mapping of central nervous system tissue properties.

Advances in functional magnetic resonance imaging

By measuring brain activity-related changes in blood oxygenation that are linked to specific sensory or cognitive events, functional MRI (fMRI) provides a remarkable window into the mind, and has revealed a great deal about patterns of functional brain anatomy underlying attention, memory, perception, social understanding, and affective responses. The first fMRI studies used simple blocked designs and echo-planar imaging to measure changes in blood oxygenation level-dependent (BOLD) contrast in response to robust visual (Kwong et al. 1992; Ogawa et al. 1992) or motor (Bandettini et al. 1992) stimulation. Advances in scanner technology, experimental design, and data analysis techniques have led to improved spatial and temporal resolution, making it possible to image more complex aspects of perception and cognition. These advances have also paved the way for the use of fMRI as a tool to establish the neural basis and determine the effects of treatment of neurodevelopmental disorders such as amblyopia, dyslexia, and autism spectrum disorder.

Typical event-related fMRI analyses rely on comparing brain activity engaged by experimental and control tasks, with the assumption that the only difference is the particular cognitive process of interest. While this approach has been fruitful in revealing brain activity associated with broad cognitive domains, such as attention and memory, advancing fMRI depends on developing nuanced techniques for quantifying complex interactions between concurrently engaged cognitive processes that support these broader cognitive domains (Decety & Cacioppo 2010; Poldrack 2010).

Embracing cognitive complexity

Significant advances in the past decade in the investigation of the brain basis of complex cognitive functions have opened up new domains of cognitive and clinical neuroscience research. There is growing momentum toward investigating the brain basis of complex, deeply human aspects of the mind that define experience in both health and illness, such as attitudes (McCall et al. 2012), morality (Kaplan et al. 2016), creativity (Beaty et al. 2015), and social understanding (Tipper et al. 2008; Immordino-Yang & Singh 2013; Saggar et al. 2016). A notable example is that of observing and understanding the actions of other people. Observing actions engages a distributed brain network of parietal, frontal, and temporal sensorimotor brain regions known as the action observation network (Buccino et al. 2001). An experimental design and analysis technique known as *repetition suppression/fMRI adaptation* (Grill-Spector et al. 2006) exploits the greater sensitivity of neural circuits to novel information than repeated information to reveal neural populations that decode distinct dimensions of a stimulus. During an action observation task, several dimensions of the action are processed concurrently (e.g., kinematics, goals, outcomes, intentions, expressive meaning). Using repetition suppression, brain processes that decode each of these dimensions have been localized to distinct, hierarchically organized functional modules nested within the larger action observation network (Grafton & Hamilton 2007; Tipper et al. 2015). By identifying more precise functional granularity within functional brain networks, the precision of the measuring sticks we use to assess brain pathology across of broad range of clinical conditions is greatly increased.

Advances in fMRI study design are made possible in part by an increasing willingness to utilize ecologically valid, complex, dynamic stimuli and tasks that employ video displays, interactive tasks, physical manipulanda, virtual reality, and wide-field three-dimensional (3D) immersion. As tasks and stimuli better match real-world conditions, the richness of available behavioral measures is opening a window onto the significance of individual variability. The result is a growing appreciation that clever new tools are needed to quantify this meaningful social, cognitive, and behavioral information to define nuanced *sociocognitive phenotypes*, and map them to distinct patterns of brain function. This new perspective is critical to developing new tools for improving diagnostics and individualized interventions for numerous neurodevelopmental, neurological, and neuropsychiatric conditions.

Clinical application: Localizing neuropathology in amblyopia

Amblyopia is a common developmental visual disorder that is characterized by reduced visual acuity in one eye that cannot be immediately corrected with lenses. Based on animal models, amblyopia is classically attributed to changes in the primary visual cortex (V1), such as shrinkage of ocular dominance columns driven by the amblyopic eye and reduction in the number of binocular neurons (Hubel et al. 1977; Crawford & Harwerth 2004). Very high-resolution fMRI has been used to confirm the shrinkage of ocular dominance columns in humans with amblyopia (Goodyear et al. 2002) and fMRI has also established the involvement of extrastriate visual areas outside of V1 in human amblyopia (Muckli et al. 2006; Lerner et al. 2006; Ho & Giaschi 2009; Thompson et al. 2012). These studies extend our understanding of the neural basis of amblyopia beyond that available through animal models (Kiorpes & McKee 1999), and confirm the need for a revised definition of amblyopia to include a broader range of behavioral deficits (Wong 2012). For example, we used an event-related, *parametric* fMRI design to study the neural correlates of a behavioral deficit in multiple-object tracking (the ability to use attention to track several randomly moving targets in a field of moving distractors) (Ho et al. 2006), measurable through both amblyopic and non-amblyopic eyes. Area MT in lateral occipitotemporal cortex, the anterior intraparietal sulcus, and the frontal eye fields, but not V1, were implicated in the visual deficit (Secen et al. 2011). Other studies used the subvoxel resolution provided by repetition suppression to confirm the involvement of V1 and extrastriate regions in amblyopia (Li X. et al. 2011a), and to demonstrate abnormal binocular interactions in these regions (Jurcoane et al. 2009).

Functional MRI is now being used to assess new treatment approaches to amblyopia aimed at augmenting visual cortex activity. A recent study reported an increase in BOLD response in visual cortex with amblyopic eye viewing after 30 days of perceptual learning treatment, but no structural change measurable with DTI (Zhai et al. 2013). In another study, anodal direct current stimulation was found to improve vision in the amblyopic eye in some adults with amblyopia; it also equalized the BOLD response of the visual cortex to inputs from each eye (Spiegel et al. 2013). In a more controversial study, oral administration of levodopa produced a small improvement in visual acuity in the amblyopic eye and a correspondingly small increase in V1 activation measured with fMRI (Yang et al. 2003).

From localization to organization

As novel imaging technologies, experimental paradigms, and analytical procedures enable the investigation of the finer granular substructure within functional brain networks, there is growing interest in not only mapping the *localization* of specific brain functions to specific brain anatomy, but also in quantifying patterns of functional *organization* of brain networks. For example, studies using *resting state* fMRI have reported disrupted spontaneous activity patterns and/or reduced *functional connectivity* between visual and visuomotor areas in amblyopia (Lin et al. 2012; Ding et al. 2013; Wang T. et al. 2014). One study used task-activated fMRI and found reduced *effective connectivity* involving thalamic, V1, and extrastriate visual areas for both feed-forward and feedback connections (Li X. et al. 2011b). Such analytical advancements in data analysis enable a more complete understanding of the neural basis of amblyopia, which is essential to improving treatment interventions. More generally, examining how functional brain networks interact under changing task demands may provide new insights into neuropathology linked to neurodevelopmental, neurological, and neuropsychiatric conditions.

A range of functional connectivity analysis techniques is now available to quantify brain organization, each taking a slightly different approach. Multivariate techniques such as constrained principal component analysis for fMRI (fMRI-CPCA) (Woodward et al. 2006) reveal distinct functional networks based on coherent patterns of task-related variation in whole-brain activity. Other task-based approaches include seed-voxel connectivity maps, which reveal functional brain networks based on correlations in activity under various task conditions, psychophysiological interaction maps, which reveal differences in inter-regional connectivity associated with the experimental manipulation, and Granger causality maps, which estimate the direction of information flow through various regions of a network (Rogers et al. 2007).

Another key advance in characterizing functional brain network organization is the emerging field of topological network analysis (Bullmore & Bassett 2010). This approach relies upon parcellating a brain network into a set of nodes, computing the functional relationships between each node-pair (e.g., activity correlations, coherence, mutual information, transfer entropy; Wang H.E. et al. 2014), and then quantifying the overall organizational structure, or architecture, of the network as a whole. A range of graph metrics is available to quantify properties of brain network organization, such as strength, efficiency, clustering, modularity, path length, centrality, and assortativity (Rubinov & Sporns 2010). An important advantage of these metrics is their ability to reveal organizational features that are unobservable with traditional fMRI analyses. As fMRI is used to characterize brain pathology linked to heterogeneous symptom profiles that occur with neurodevelopmental and neurological disorders, topological network analysis provides a promising analytical tool (Kaiser 2011).

Topological network metrics can be applied not only to functional brain organization, but also to anatomical brain organization based on tractography analyses of diffusion imaging data, which estimates probabilistic measures of white matter connectivity between brain regions (Basser et al. 2000). Functional and anatomical network metrics can then be compared to identify structure–function relationships related to task performance (Hermundstad et al. 2013) and individual variability in how underlying brain architecture is utilized to achieve particular cognitive states (Hermundstad et al. 2014). This approach is also being applied to move beyond identifying static patterns of functional organization to reveal dynamic brain changes that unfold over time (Bassett et al. 2011; Betzel et al. 2016). New techniques for characterizing interactions and transitions between identifiable coherent brain states are providing promising avenues for enabling distinctions between neurologically healthy and pathological brain function.

Task-based fMRI signal reliability and validity

Task-based fMRI data provides well-localized and compelling heat images depicting brain activity in response to a wide range of tasks. Typically, brain images depict the magnitude and reliability of differences of parameter estimates derived from individual conditions. However, task-based fMRI data provides richer information than brain activation (e.g., difference) images, such as task-based network information, and network-level hemodynamic response (HDR) shape estimates that may be informative in terms of understanding the cognitive operations driving their shape. The percentage of variance in raw BOLD signal that is predictable from task timing typically ranges from 5% to 15%, but this percentage depends on the task design, task timing, and the nature of the task. Even if the percentage of task-related signal in the BOLD data is as low as 5%, as long as that signal can be accurately extracted, it can still be reliable and valid, a concept identical in principal to the case of true-score correlations between two tests and unattenuated correlations (Spearman 1904).

In order to determine that reliable and valid task-related signal is being extracted from the BOLD signal, we suggest that task-based fMRI data should meet at least two and ideally three validity checks, and one reliability check, before being interpreted; namely, spatial, temporal, and possibly experimental validity, as well as reliability over subjects at the level of task-based networks. *Spatial validity* requires observation of network configurations that resemble those well known in the literature, first described as the task-positive and -negative networks (Fox et al. 2005). *Temporal validity* requires observation of a biologically plausible HDR shape for each network (Logothetis et al. 2001). *Experimental validity* requires observation of HDR shape and magnitude changes with important experimental manipulations or measured cognitive operations, or both (Friston et al. 1998). Finally, *reliability in HDR shape over subjects* at the level of whole-brain networks should be observed (Metzak et al. 2012; Lavigne et al. 2015b; Woodward et al. 2015). Functional MRI-CPCA efficiently meets all of these requirements as part of its standard analysis procedure that involves two essential steps: (1) constraining variance in BOLD signal to that predictable from task timing using multivariate multiple regression, and (2) extracting multiple task-related functional brain networks using principal component analysis. The percentage of variance in BOLD signal that can be predicted from task timing is part of the fMRI-CPCA output, as is this value for each brain network extracted. Component loadings are displayed on the brain images, and component scores are combined with the design matrix to estimate network-, subject-, and condition-specific HDR shape estimates, averaged over trials. For each task-related functional brain network, observed HDR shapes can be submitted to repeated-measures analysis of variance (ANOVA) to test temporal validity (i.e., the HDR shape is not a random shape that changes between subject) and experimental validity (i.e., differences in HDR shape between experimental conditions. This step can also be used to carry out group fMRI-CPCA, for testing differences between clinical samples in the HDR shape (Woodward et al. 2015, 2016).

Experimental manipulation/spatial and temporal replication

If the task-related BOLD signal displays spatial, temporal, and possibly experimental validity, as well as reliability in HDR shape over subjects at the level of task-based networks, a very powerful methodology is available through manipulation of functional brain networks with a variety of different tasks (and/or task conditions), and comparison of the nature of the BOLD signal associated with different tasks and/or task conditions. We quantify these principles as *spatial* and *temporal replication*, and in this section we outline how these concepts can be harnessed to identify the cognitive function of brain networks.

In order to assess spatial and temporal replication of network configurations, and take advantage of spatial replication combined with temporal differences to interpret function of brain networks, two tasks (or task conditions) can be compared though simultaneous analysis of two datasets (Lavigne et al. 2015a). When two (or more) experiment versions elicit the same underlying cognitive operation (e.g., visual attention), *spatial and temporal replication* would be observed if HDR shapes for the network in question were not distinguishable between the two experiment versions, and this should be the case if the timing elicited by that cognitive operation does not differ between experiments. In contrast, *spatial but not temporal replication* would be observed if HDR shapes were reliably different between the two experiment versions, and this should be the case if the timing of the cognitive operation differs between experiments. This case (viz., spatial but not temporal replication) provides an important scientific opportunity to use differences between experiments to help test the theorized timing of cognitive functions of brain networks. That is to say, if a theorized cognitive function can be manipulated using (for example) variation in task timing between experiments, it can be determined whether or not the BOLD signal pattern supports the account of the cognitive operation performed by that functional brain network, at least for the task in question. This is tested using the ANOVA method already mentioned above with experiment as a within- or between-subject factor. Finally, if a cognitive operation is elicited by only one version of the experiment but not the other, the version not eliciting this cognitive operation would show a flat HDR shape for that functional brain network, and therefore it could be concluded that *neither spatial nor temporal replication* has been observed. This methodology has been used in past work to test theoretical accounts of the function of brain networks across closely matched experiments (Lavigne et al. 2015a).

Future directions

Moving forward, the combination of more nuanced experimental paradigms to assess complex cognitive functions and characterize individual sociocognitive phenotypes and more sophisticated analyses of functional and anatomical brain network organization open up a new frontier for clinical neuroscience research. The potential to improve diagnostic and assessment capabilities and enhance individualized care across a range of neurodevelopmental, neurological, and neuropsychiatric conditions is unprecedented.

Advances in functional and biochemical positron emission tomography imaging

Many brain diseases are known to start well before clinical symptoms become manifest. The ability to correctly diagnose disease early is expected to enhance the probability of successful intervention. In this context, the network degeneration hypothesis (Palop et al. 2006), which suggests that initiation and progression of disease-specific pathological changes occur within specific brain networks and are mediated by protein misfolding, inflammation, and impaired cellular energetics coupled to abnormal neurotransmission, suggests that multiparameter information may lead to significantly enhanced understanding of brain function and thus disease. This approach is encouraging development of novel positron emission tomography (PET) imaging tracers that target abnormal protein aggregation, multimodality imaging, either sequential or simultaneous, such as that enabled by hybrid PET/MRI scanners, and application of analysis methods that are able to extract relevant features from multiparametric information.

Aggregations of the beta-amyloid protein, considered a hallmark of Alzheimer's disease, are fairly well studied with several established PET tracers (Villemagne 2016). Imaging of tau aggregation, implicated in Alzheimer's disease, chronic traumatic encephalopathy, progressive supranuclear palsy, and corticobasal syndrome is more recent; three promising novel tracers are currently used in human studies, ^{11}C-PPB3, ^{18}F-T807 or ^{18}F-AV-1451, and ^{18}F-THK5351 (Villemagne et al. 2015). The tau tracers are not equivalent in their ability to detect deposits of different tau isoforms that are selectively associated with different aspects of neurodegeneration: the precise relevance of each tracer to specific diseases is thus still under investigation. Development of PET tracers for other targets of interest such as alpha-synuclein, whose aggregations are implicated in the pathogenesis of Parkinson's disease, is a very active research area, but no successful candidates have been developed as yet (Eberling et al. 2013, Kikuchi et al. 2010).

The ability to identify abnormal protein aggregation or alterations in metabolic pathways observable with ^{18}F-FDG, does not always by itself lead to accurate differential diagnosis or to improved understanding of pathogenic mechanisms. Informed combination of data obtained from different imaging modalities has been shown to yield superior information; for example, a recent study performed on a hybrid PET/MRI demonstrated that the combination of regional metabolism, functional connectivity, and gray matter volume, which were derived from disease characteristic networks, was able to achieve high classification accuracies for separating patients with different neurodegenerative syndromes of 77.5% for Alzheimer's disease versus others, 82.5% for behavioral variant frontotemporal dementia versus others, 97.5% for semantic dementia versus others, and 87.5% for progressive nonfluent aphasia versus others—this multimodal classification was superior to unimodal approaches (Tahmasian et al. 2016). In addition to combing data from different modalities, novel image analysis approaches for PET images are also emerging and are proving complementary to the traditional ones based either on kinetic modeling or radioactivity concentration values. Such approaches include information about specific tracer distribution patterns within anatomically relevant regions of interest (Klyuzhin et al. 2016; Liu S. et al. 2016). Combinations of such patterns or features applied to the analysis of ^{18}F-FDG data coupled with MRI-derived information, resulting in the definition of multiparameter-based biomarkers, have been shown to be associated with different aspects of Alzheimer's disease progression (Liu et al. 2016). Such biomarkers will be of great relevance when stratifying patients and evaluating mechanisms of action and effectiveness of emerging treatments.

The hybrid PET/MRI provides access to novel measurements, such as direct investigation of the interactions between neurotransmitter activity and brain function either at rest or as stimulated by a pharmacological or behavioral intervention. This is a very exciting area of research relevant to many aspects of brain function that will see growth as the availability of the hybrid PET/MR scanners increases. For example, it will be possible to investigate correlations between responses to reward, pain, or treatment in terms of dopamine release and activation of cognitive networks and how such correlations differ amongst different subject populations. Ideally, it will be possible to identify imaging-based biomarkers that identify subjects at risk and thus promote development of disease prevention strategies. Optimization of such approaches will drive development of joint PET/MRI image processing methods, either in the image reconstruction step or image analysis step (Sander et al. 2013; Kang et al. 2015). In summary, over the last decade PET imaging has been enhanced by development of novel tracers, technical improvements that allow simultaneous PET/MRI data acquisition and novel data processing methods, thus catalyzing and reflecting the increased recognition of the interactions between neurochemistry and distributed brain networks.

Advances in functional and dynamic magnetoencephalography imaging

New horizons for mapping spontaneous neuromagnetic networks

Magnetoencephalography (MEG) offers a direct measure of neural activity, and enables mapping of neural activity with a combination of spatial and temporal resolution that is superior to any other noninvasive neuroimaging modality. This makes MEG ideal for mapping neural oscillations and their coordination among brain areas (Palva & Palva 2012), which has been proposed as a mechanism supporting dynamic functional interactions among neuronal populations to support cognition and perception (Fries 2015). The theoretical perspective of understanding brain function through neural oscillations and their inter-regional synchrony, together with the technical advantages for mapping such large-scale brain networks and dynamics provided by MEG, has opened new frontiers for understanding distributed human brain function, dysfunction, and development.

Like most human functional brain imaging approaches, much initial research in MEG focused on how the brain responds under various types of sensory stimulation, or cognitive or motor demands. It has been increasingly appreciated that most brain activity is comprised of spontaneous background fluctuations rather than task or stimulation-dependent activation, or both, and that much can be gleaned by mapping intrinsic functional connectivity at rest (Raichle 2015). It is also becoming increasingly clear that the dynamics of intrinsic connectivity are highly informative regarding the organization of brain function (Zalesky et al. 2014). This highlights the advantages of MEG, as the timing accuracy of MEG enables connectivity dynamics to be measured with millisecond precision and in fast physiological frequency ranges critically relevant for cognition, perception, and behavior.

Mapping intrinsic neurophysiological amplitude network correlations and dynamics

Correlations in slow fluctuations in neuromagnetic amplitudes, particularly in beta and alpha frequency bands, have been shown to recapitulate many of the resting state networks previously observed using other imaging modalities such as fMRI (Brookes et al. 2011). These functional networks, expressed at rest, are also modulated by cognitive activity (Brookes et al. 2012). Graph analysis of resting MEG amplitude correlations also conform to major anatomical divisions of the brain, with global hubs for different frequency ranges concentrated in different brain loci (Hipp et al. 2012). Inter-regional amplitude correlations have been shown to increase with age throughout childhood and adolescent development, particularly in beta and alpha frequency ranges (Schäfer et al. 2014). EEG studies suggest that such increasing spontaneous electrophysiological connectivity may be associated with the acceleration of the alpha peak (Miskovic et al. 2015), suggesting that acceleration of spontaneous oscillatory connectivity may lead to the predominant expression of intrinsic connectivity networks in the beta band in adulthood.

One important advantage MEG has over fMRI is temporal resolution. This allows mapping of the dynamics of intrinsic connectivity networks at faster time scales. For example, this approach has been useful for exploring the expression of connectivity in these networks in the time domain, and for elucidating sequential transitions between relative dominance of specific intrinsic connectivity networks (Baker et al. 2014). For example, it has been observed that it is much more likely for there to be a transition between a default mode network (DMN) dominant state and the predominance of another intrinsic connectivity network than it is to transition directly between two non-DMN resting state networks, and that these internetwork transitions primarily involve inter-regional amplitude correlations in the beta frequency range (de Pasquale et al. 2012). This could be interpreted as indicating that the DMN serves to shift the pattern of inter-regional communication of the brain between different functional states. For example, beta-amplitude correlations involving the DMN could mediate transitions between a motor-focused state supported by the motor network and a spatial attention-focused state supported by the dorsal attention network. Another advantage conferred by the millisecond timing accuracy of MEG network mapping is the ability to determine that certain networks, such as the motor network, may actually be comprised of a larger number of subnetworks that serially express periods of relative predominance (O'Neill et al. 2015). Notably, expression of local neuromagnetic spectral power has been shown to have clinical utility for detecting and characterizing conditions such as mild traumatic brain injury (Huang et al. 2014). Recent observations that inter-regional amplitude correlations are altered in mild traumatic brain injury and associated with cognitive and affective symptoms in this group (Dunkley et al. 2015) buttresses the hope that this type of connectivity measure may also hold future clinical applicability.

Spontaneous phase synchrony in neuromagnetic networks and its alteration in brain disorders

In addition to mapping of amplitude correlations in spontaneous MEG, another dominant approach has been to measure inter-regional phase locking, which is typically investigated at faster time scales. It has been proposed that the slower amplitude correlations are more closely related to underlying structural brain connectivity, whereas phase–phase relations encode more transient cognitive and/or computation specific information (Engel et al. 2013). Though much remains to be discovered regarding the relevant roles of amplitude and phase coherence, it is worth noting that spontaneous phase locking effects are often observed at high gamma frequencies that are much more spatially refined (Jerbi et al. 2009) and have been proposed to play a more computational role relative to oscillations at lower frequencies (Fries 2009). Phase and amplitude should not be viewed as independent in this context, however, as inter-regional correlations in cross-frequency phase amplitude coupling in spontaneous MEG has also been shown to correspond to resting-state networks first identified using fMRI (Florin and Baillet 2015).

Spontaneous neuromagnetic phase locking has been particularly successful for capturing atypical connectivity in neurological and neuropsychiatric populations, adding to the emerging network perspective on neurological disorders (Stam 2014), and adding credence to the notion that brain oscillations and their coherence may constitute a type of neural syntax for information processing and communication that is critical for understanding the brain basis for psychiatric disorders (Buzsáki and Watson 2012). For example, it has been shown that resting-state MEG synchrony is altered in post-traumatic stress disorder (Dunkley et al. 2014). Source mapping of such atypical neural synchrony has demonstrated that such altered neural synchrony prominently involves areas such as amygdala and hippocampus, and is associated with symptom severity as well as associated cognitive and affective sequelae (Dunkley et al. 2014). An exhaustive review of atypical resting MEG phase synchrony is too long for the present chapter, but it appears to be relevant for a broad array of conditions including neurodegenerative diseases such as Alzheimer's disease (de Haan et al. 2012), movement disorders such as Parkinson's disease (Berendse & Stam 2007), and epilepsy (Elshahabi et al. 2015), suggesting that such measures are tapping a fundamental aspect of neuronal communication and its disruptions in brain disorders.

Electroencephalographic research using a large sample has demonstrated reorganization of spontaneous network phase synchrony during child development (Boersma et al. 2011). Clinical child populations have been shown to express characteristic alterations in spontaneous MEG synchrony. Children with dyslexia have been shown to exhibit atypical neural synchrony, which was also shown to be associated with language difficulties (Dimitriadis et al. 2013). Frequency- and network-specific resting neuromagnetic phase synchrony have been reported in adolescents with autism (Ye et al. 2014), and altered MEG synchrony has also been reported autistic children (Kikuchi et al. 2015). Such alterations of inter-regional MEG synchrony have been related to autistic symptomatology (Kitzbichler et al. 2015). Spontaneous MEG synchrony has been shown to be reduced in school-age children born very preterm across multiple physiologically relevant frequency ranges (Ye et al. 2015), which may be related to well-characterized alterations in white matter development in this group (Pannek et al. 2014).

Bright future through big data? Toward quantitative translational MEG

Increasing emphasis in neuroimaging is being placed on the importance of large sample sizes to improve the accuracy and reproducibility of findings. This will be aided by large open source databases of resting state MEG recordings. For example, the Human Connectome Project will provide hundreds of such scans with accompanying high quality structural and functional MRI data (Larson-Prior et al. 2013). Open source data repositories more explicitly focused on MEG such as "OMEGA" will also help to provide large samples of MEG data collected across multiple centers (Niso et al. 2016). Hope is increasing that big data in neuroimaging, combined with techniques such as machine learning, will provide new translational opportunities for diagnosis and predicting treatment response with sufficient sensitivity and specificity to enable translation (Deco and Kringelbach 2014). Major future trends in resting MEG network mapping will likely include mapping normative networks, their development and relation to function, together with how such refined population data could be used for detection of neural network abnormalities and to guide treatment and rehabilitation for those suffering from disorders of the brain.

Advances in functional and dynamic electroencephalography imaging

From identification of neural sources of electrical activity to fast neural network dynamics

Functional imaging involves describing how the brain implements sensory, perceptual, cognitive, motor, and emotional processes at several scales of space and time. Functional MRI is efficient at doing this at several spatial scales but only at time scales from several seconds to several minutes; MEG can do a great job at shorter time scales of tens of milliseconds to seconds, and at spatial scales comparable to those of fMRI, but it is expensive and not portable. Arguably the greatest promise for bedside functional imaging is provided by current and near-future electroencephalography (EEG) technologies. It is relatively inexpensive, portable, and operates at time scales of tens of milliseconds. Its major drawback, spatial inaccuracy, is being addressed by new techniques of data analysis and recording technology. Here we describe recent advances and promise for this elderly technology that is being reborn.

Electroencephalography has been used in medicine for diagnosis of epilepsy foci, for example, for many years. These uses will continue, but EEG is now poised to contribute significantly to the growing knowledge of the fast dynamics of brain networks and their vicissitudes in challenged brains. First, new techniques of data analysis allow increasingly accurate identification of the neural sources of electrical activity recorded at the scalp by EEG. Independent component analysis (ICA) coupled with single dipole fitting (Delorme et al. 2012) has successfully recovered specific brain loci, associated with perceptual and cognitive processing, that have been previously identified by fMRI (e.g., Bedo et al. 2014). Similarly, beamformer analysis can perform this function well when applied to EEG data (e.g., Doesburg et al. 2009; Green et al. 2011). Moreover, some of the sources of inaccurate localization can be removed by using a wearable, dry, rigid electrode array rather than one that flexes or requires wet electrolytes, or both (e.g., Mullen et al. 2013). The latter paper also presents evidence that ICA can remove much of the movement-related artifact that plagues all neuroimaging techniques. Similarly, ICA also can be used to remove the artifacts generated by tiny shivers and so on of the muscles around the eyes, head, and neck (electromyography (EMG)) (Fitzgibbon et al. 2016). Thus, EEG is becoming a cleaner neuroimaging technique that can begin to approach the more expensive fMRI and MEG results, but more flexibly and at much less expense.

Second, the fast time scale at which EEG can be recorded allows the analysis of moment-to-moment dynamics in the neural networks that support cognition, emotion, and behavior. Two currently available techniques are especially useful in this regard. Various synchronization indices are available to measure phase synchronization of oscillatory neural activity. The first evidence of neural oscillations recorded by EEG was the alpha wave (around 10 Hz; Berger 1929). Since then oscillatory activity in several frequency bands has been closely associated with perceptual and cognitive processes (e.g., Varela, et al. 2001; Ward L.M. 2003). Phase synchronization between oscillations at the same or different frequencies is an indication of functional coupling between the brain areas generating the oscillations. Although the most direct evidence for such functional coupling comes from studies with implanted electrodes (e.g., Lewis et al. 2016), EEG data treated with advanced analysis techniques can also detect transient functional connectivity between neural sources at fast time scales of tens of milliseconds (e.g., Doesburg et al. 2009; Bedo et al. 2014) (Figure 2.1). In addition, new techniques of measuring directed information flow between oscillatory sources, such as transfer entropy (Schreiber 2000; Vicente et al. 2011; Wibral et al. 2011) and phase transfer entropy (Lobier et al. 2014), can actually allow us to make (weak, or Granger-type) causal inferences about interactions between brain areas, for example, in reading (Bedo et al. 2014). Finally, when there are competing network topologies or interaction models based on previous data, dynamic causal modeling can be applied to choose the most useful model (Friston et al. 2003).

There are still significant limitations to be overcome, however. One of the most important is the inability of EEG to visualize subcortical areas—MEG also is not good at this, but fMRI does it well. Thus, it will be important to develop linkages between fMRI and EEG/MEG results to allow inferences of fast dynamics from fMRI when subcortical brain regions are heavily involved. This is especially important for the thalamus, which is implicated in nearly every brain operation and network (Ward L.M. 2013). Another limitation is the relatively low signal to noise ratio enjoyed by EEG. Measures to increase this, such as ICA, have been successful, but still more is needed. Synchronization and transfer entropy analysis suffer particularly from this problem and new methods of filtering out the noise are needed. But even given what is now available it should be possible to use EEG both to characterize normal fast dynamics in neural networks, and to discover how aberrations in these networks can lead to dysfunction. Of course, it will then be necessary to decide what actions to take to ameliorate or rehabilitate these dysfunctions. The neuroimaging techniques will help point to the necessary treatment foci, and they can help assess whether the applied treatment has succeeded. But they will not answer the difficult ethical problems given rise to by the knowledge of the source or extent or cause of a brain dysfunction, especially if treatment is expensive, risky, or impossible. This is particularly true for EEG-facilitated discoveries of residual functions, and even indications

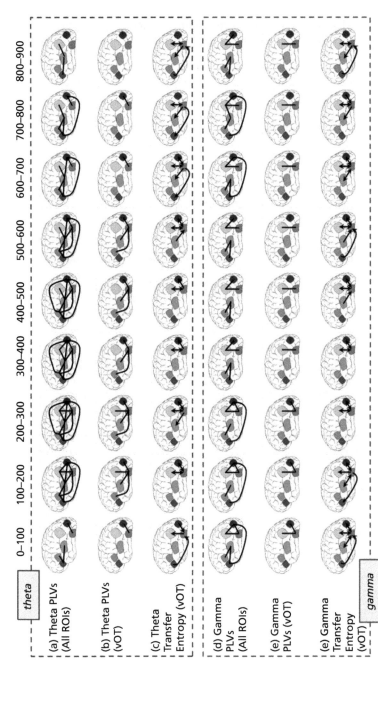

Figure 2.1 (See also color plate section.) Example of dynamic EEG synchronization and transfer entropy results from a word reading task. Synchronization was measured with phase locking value (PLV), and narrow band transfer entropy was measured using an adaptation of the Kullback entropy (for vOT to/from other areas only). Theta band is 3–7 Hz and gamma band is 35–45 Hz. Black lines (arrows) indicate significant interaction (directed information flow) between the connected reading-related brain areas during the time period (ms after stimulus presentation) indicated at the top of the figure. vOT indicates ventral occipital-temporal brain region, the putative word-form area.

Reproduced from Bedo N, Ribary U, and Ward LM. (2014) 'Fast dynamics of cortical functional and effective connectivity during word reading', *PLoS ONE*, Volume 9, Issue 2, e88940, doi:10.1371/journal.pone.0088940, Copyright © 2014 Bedo et al. Reproduced under the terms of the Creative Commons Attribution Version 4.0 International (CC BY 4.0), https://creativecommons.org/licenses/by/4.0/.

of conscious processes, in severely compromised brains, such as those of patients with unresponsive wakefulness or minimal consciousness (e.g., Cruse et al. 2011).

Applications and challenges in clinical neuroimaging: Localization and synchronization

As scientists, we are ethically responsible to conduct the necessary basic research to be able to take the next steps in conducting clinical research. With the growing interest in personalized healthcare, we need to provide tools that allow clinicians to make diagnosis and recommend management options at the individual patient level. Many neuroimaging tools exist (MRI, fMRI, PET, MEG, EEG, near infrared spectroscopy, etc.) and many are clinically used to screen, diagnose, and/or prognosticate many types of neurological diseases and disorders (e.g., epilepsy, schizophrenia, dementia, and dyslexia). Several of these neuroimaging methods have a long history of clinical use, while other methods have a more recent history. The literatures on these different neuroimaging methods are extensive and beyond the scope of this brief communication to be able to cover them in any great depth. Therefore, we very briefly highlight one neuroimaging method, the auditory event-related potential (ERP), that is used to assess sensorineural hearing loss and neurological disorders (e.g., vestibular schwannoma screening). We do this in order to stress the ethical responsibility of the scientific community to conduct basic and clinical research on neuroimaging tools that can be easily and effectively used clinically. We then go on to briefly discuss the importance of thoroughly investigating and validating neuroimaging methods prior to recommending their use clinically. We provide an example from our recent methodological work in beamformer source modeling and its role in functional connectivity imaging to emphasize this point.

Auditory ERPs directly reflect the electrical signal generated from neural responses in the auditory system and thus can be used as a neuroimaging tool. One example of an auditory ERP that is used clinically is the auditory brainstem response (ABR). Testing of ABR has been used for screening and diagnosing sensorineural hearing loss in infants for many years (see Picton 2011, pp.213–246, pp.449–492). This testing is based on decades of basic research that helped inform clinical researchers to investigate its utility in evaluating hearing function in humans across all ages. The ABR is commonly used to assess hearing thresholds in infants because infants younger than 8 months old are unable to provide reliable behavioral responses (i.e., raising a hand to a sound). The reason for using ABR to evaluate infant hearing thresholds is that research showed that detection of a hearing impairment in infants by 3 months of age and intervention by 6 months of age significantly improves the child's later language development (Yoshinaga-Itano et al. 1998, 2000). Beginning in the 1990s, another ERP method called the 80 Hz auditory steady-state response (ASSR) was suggested to improve hearing diagnostics because they could be a faster, more reliable, and more objective test of hearing thresholds in infants (see Picton 2011, pp.285–334). However, this excitement diminished because basic research helped inform the scientific community that ASSR testing of infant hearing was as reliable as the gold-standard ABR method but

ASSR testing had a noteworthy caveat; artifactual responses could be elicited at high sound intensities. This would lead to falsely indicating that individuals with profound sensorineural hearing loss had residual hearing. In other words, ASSRs would show residual hearing in individuals who were known to have no residual hearing. Although ASSR testing is not recommended currently for diagnostic hearing threshold testing in infants, enough basic and clinical research was conducted to provide evidence for its use as a tertiary-stage hearing screening tool. In addition, research is continuing to identify other potential clinical uses of ASSR in evaluating hearing functions (see Picton 2011, pp.320–325).

Neurological ABR testing is another auditory ERP method that is used clinically to screen for vestibular schwannomas (aka acoustic neuromas; Selters & Brackmann 1977; see also Picton 2011, pp.500–505). Although neurological ABR testing is not a spatial neuroimaging method for detecting tumors, it is a useful neuroimaging method for detecting delayed or disrupted neural communication along the auditory pathway (i.e., between the cochlear nerve and brainstem nuclei). There are clear guidelines and normative data collected in non-clinical and clinical populations (Stockard et al. 1978; Chiappa et al. 1979; Campbell et al. 1981). This evidence has informed researchers and clinicians about the validity of ABR to screen for tumors and brainstem lesions along the auditory pathway. Importantly, clinical research has clearly identified that the sensitivity and specificity for neurological ABR in detecting 1–2 cm tumors are 90–100% and 60–75%, respectively (see Picton 2011, p.503). Notably, this performance is based on large samples of patients. The tumor-detection performance of the ABR led early researchers to recommend that it was a good screening tool for discriminating between individuals with and without retrocochlear pathologies (e.g., vestibular schwannomas). As MRI became a more prevalent neuroimaging tool and its diagnostic performance was to be superior to ABR, MRI became the recommended tool for screening and diagnostics. However, the financial expense of MRI precluded much of its use as a screening tool (until recently) and when MRI is contraindicated in certain populations (e.g., patients with cardiac pacemakers). Thus, neurological ABR is still routinely used as a screening tool before using more expensive MRI diagnostic methods or when MRI is contraindicated. When MRI testing is available, MRI screening is recommended over ABR screening because of the added benefits of better sensitivity and better ability to identify the etiology of auditory pathology (Fortnum et al. 2009). However, a current challenge for using MRI as a screening tool is that MRI testing facilities typically reside only in major urban areas, which add barriers and costs to providing healthcare in rural communities. Neurological ABR testing, on the other hand, is easily performed in any clinical setting with significantly less expensive equipment as compared to a MRI scanner. Thus, screening in rural communities using neurological ABR and then referring patients for MRI testing in urban areas can be more cost-effective and easier for the patients and healthcare system. The history of basic and clinical research for using ABR and MRI to diagnose retrocochlear pathologies highlights the importance of initial and continued validation of using neuroimaging tools for clinical purposes (see Fortnum et al. 2009). This evidence inevitably reduces the uncertainty that clinicians often face when making decisions about diagnoses, prognoses, and management.

These brief historical accounts of using auditory ERPs showcase some of the basic principles that are needed for recommending clinical uses of specific neuroimaging tools. Some of the basic challenges to the tests are to determine (1) specific validity and reliability relating to technical algorithms used; (2) diagnostic performances relating to the applications on pathological subpopulations (e.g., sensitivity, specificity, predictive value); and (3) cost-effectiveness and cost-utility (Medina et al. 2013, pp.3–18). The histories of using auditory ERPs to diagnose hearing loss and screen for tumors are interesting with respect to how neuroimaging can be used as tools for screening and diagnosing sensory and neurological impairments. These histories are also a reminder of the diligence required to sufficiently evaluate and validate neuroimaging tools in order to recommend their use in healthcare at single-patient and population-based levels.

Validating the neuroimaging tools from a methodological point of view is essential before recommending their use for clinical purposes. One major goal of methodological research in neuroimaging is to improve our ability to confidently identify local and global brain function. For example, we are investigating the reliability of specific neuroimaging analyses to localize the neural generators underlying EEG and MEG and to identifying the neural communication within the human brain. However, some recent caveats have come to light that we are currently investigating and attempting to overcome. Here we highlight one of the major challenges in neural communication analyses of EEG and MEG data (aka functional connectivity analyses). We briefly present this research here to illustrate the point that basic methodological research is highly important before using tools for clinical purposes. A major challenge in functional connectivity analyses is making sure that the brain regions (aka, sources) under investigation do not contain activity from other brain regions because of the inherent physics of the neuroimaging measure or the mathematics underlying its analyses, or both. When using some neuroimaging tools and analyses, nearby (<20 mm) brain regions can contain a mixture of signals that are separately generated from different brain regions. This is known as source mixing and can lead to false conclusions regarding how these regions are functionally connected and how they communicate information among themselves (Sekihara & Nagarajan 2015). Traditional beamforming and minimum-norm source modeling methods (Van Veen & Buckley 1988; Hamalainen & Ilmoniemi 1994; Van Veen et al. 1997; see also Sekihara & Nagarajan 2015, figure 2) cannot avoid source mixing where neighboring sources will be modeled to contain contributions of each other's activity. Unfortunately, this source mixing can generate deceptively false oscillatory phase-coherence between sources; thereby causing misleadingly strong inter-source connectivities (i.e., type I errors). To overcome this source-mixing challenge, different connectivity measures of phase-coherence (e.g., phase-lag index or imaginary-phase coherence) (Stam et al. 2007; Ewald et al. 2012; Sekihara and Nagarajan 2015) have been suggested and implemented with some success. However, these compensatory algorithms become increasingly blinded to identifying true connectivities as inter-source phase differences approach zero (type II errors). Another method to overcome the possible type I errors of source mixing is to average across many participants. However, this often leads to type II errors or spatial smearing of the connectivity maps because individual participant's inter-source connection pairs usually do not exactly line up spatially across participants brains. Moreover, providing personalized healthcare would require neuroimaging tools to provide reliable connectivity maps at the individual patient level, not at the population level. Overall, source mixing leads to traditional beamformers being less sensitive and less specific to detecting true functional connections in local networks. Thus, to be able to confidently evaluate functional connectivity among locally connected brain regions in individual patients, we need better tools that minimize or negate source mixing (Figure 2.2).

Multiple constrained minimum-variance (MCMV) beamforming (Moiseev et al. 2011) is one such neuroimaging method that overcomes source mixing. The MCMV beamformer technology mathematically constrains the sources to be independent of each other; thereby, preventing source mixing (see Figure 2.2; for mathematical proofs and methods see Moiseev et al. 2011 and 2013). Figure 2.3 demonstrates the ability of MCMV-connectivity analyses to accurately resolve network connectivity in real EEG data by overcoming source mixing among nearby brain regions. Phase-locking connectivity maps (Lachaux e al. 1999) using traditional LCMV beamformers revealed significantly strong alpha-band (9–10 Hz) synchronization (red lines) among nearby sources in the posterior left visual network (Figure 2.3, left brain map). However, when analyzed using a MCMV beamformer, the connectivity map for these nearby sources had significantly larger alpha-band desynchronizations (blue lines; Figure 2.3, right brain map). The MCVM results in Figure 2.3, are likely the more valid result because the EEG data analyzed were from a visual discrimination task and from a post-stimulus interval (380 ms) where event-related desynchronization in the scalp recordings occurred. In addition, this interval is well known to contain event-related desynchronizations over occipital scalp regions. Thus, it is highly unlikely that the local visual network would be synchronized within this interval as revealed by the traditional LCMV beamformer. Therefore, the LCMV connectivity map on the left is likely showing *false* synchronizations among nearby posterior visual sources due to source mixing. Given the advantage of MCMV beamforming not having source mixing, we can now be more confident that our results of local functional connections represent the true underlying functional connectivity. By combining MCMV beamforming with functional connectivity measures (such as phase-locking values (Lachaux et al. 1999)), we can significantly enhance our ability to image local and global network connections. This will further improve the validity of our results obtained from network graph-theory analysis (Sporns 2003). Overall, thoroughly investigating and validating our methods will enhance our ability to accurately represent the local and global network dynamics underlying the EEG and MEG recorded from individual patients.

This brief overview of ABR testing provided examples of how basic research has provided suitable evidence for using neuroimaging tools for clinical purposes. In addition, continued research in validating neuroimaging analyses (such as source modeling methods) is essential so that we can obtain the most confident and accurate depiction of brain

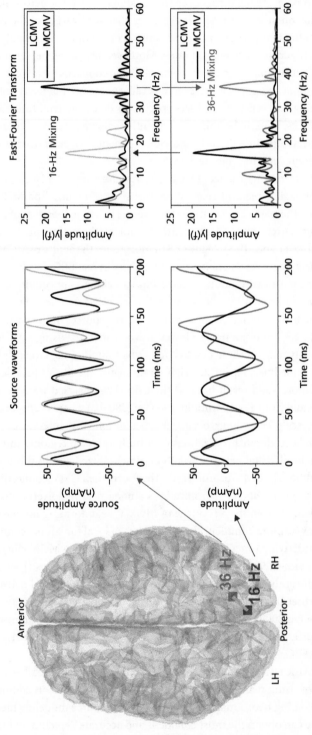

Figure 2.2 (See also color plate section.) EEG localization and connectivity accuracy. Left panel shows a simulated source location (blue square) that had only a 16-Hz oscillation and a neighboring (within 15 mm) source location (red square) that had only a 36-Hz oscillation. Source mixing of 16- and 36-Hz oscillations were clearly visible in the traditional beamformer (LCMV) reconstructed source waveforms (blue and red lines; middle panel). Amplitude spectra (right panels) showed that both red and blue LCMV sources contained energies at 16 and 36 Hz (i.e., source mixing). Because of this source mixing, phase-coherence between the sources existed and functional connectivity was falsely revealed (green line; left panel). Our MCMV beamformer (black lines), however, prevented source mixing. MCMV sources only had 16-Hz (lower panels' black lines) or 36-Hz oscillations (upper panels' black lines). Obviously, false-connections were not found for the MCMV beamformer.

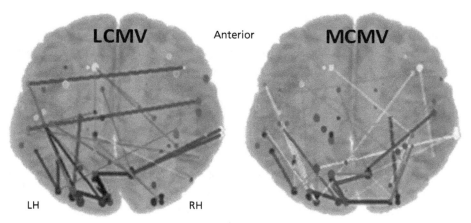

Figure 2.3 (See also color plate section.) Challenge in EEG connectivity maps. Traditional LCMV-beamformer connectivity map (left) for alpha-band (9–10 Hz) revealed highly synchronized (red lines) local couplings in visual cortices at 380 ms after letter onset. However, MCMV-beamformer connectivity map (right) revealed strongly desynchronized (blue lines) couplings. The MCMV-connectivity maps fit better with known event-related desynchronizations that occur after a visual stimulus onset for this band and latency. Thus, the MCMV map is likely the real map.

function. It is imperative that basic and clinical researchers work together to provide clinicians with strong evidence that validates new neuroimaging tools in order to improve patient care.

Advances in brain imaging signal processing

In past years, the main focus of analysis of brain signals was on studying responses of individual brain areas, associated with certain functions, to external stimuli. More recently, however, this focus shifted to investigating interactions between a large number of brain regions, often functionally connected with each other, in both task-related environments and in the case of resting state brain activity. As a consequence, modern processing methods are more oriented on analysing large sets of correlated and interacting brain sources, rather than on individual ones.

Such analyses in MEG/EEG brain imaging, which we broadly call *brain network* or *connectivity analyses*, typically involve the following steps. First, an inverse problem must be solved, which means reconstructing signals from specific locations within the brain, using data from physical MEG/EEG sensors, located outside the brain. Second, connectivity (or interaction) between different areas should be estimated. Third, this connectivity data should be statistically evaluated, to detect true connections and discard spurious ones. Let us briefly discuss each step in more detail.

Start with the source reconstruction. There are several methods to solve the inverse problem, which are popular in MEG/EEG community (Baillet et al. 2001). An important example that we will consider here is called minimum variance beamformer solution

(Van Veen et al.1997; Robinson & Vrba 1999; Sekihara & Nagarajan 2008), but similar considerations apply to other methods. Technically, the original beamformer solution was developed assuming that electromagnetic field registered by the sensor array is produced by a single brain source of interest, plus noise. This is called a single-source beamformer approach. Of course in reality even performing very simple tasks engages many brain areas, and consequently many sources are simultaneously activated. It turned out though that single-source beamformer, applied to each source individually, is still quite successful in locating and reconstructing multiple active sources, as long as they are not too close to each other and their correlations are negligible. However, if sources are close or they are strongly synchronized, or both, localization accuracy and spatial resolution of the beamformer degrades. Besides, reconstructed signals become contaminated with artifactual contributions from other brain areas. This phenomenon is called *source leakage*. To overcome these problems, more advanced multisource beamforming methods were developed (Dalal et al. 2006; Hui et al. 2010; Moiseev et al. 2011; Moiseev and Herdman 2013). In particular, these methods allow for strong interactions (correlations) between the sources, and they take care of signal leakage by rejecting any direct contributions from one reconstructed signal to another.

The next step is estimating connectivity between the brain signals. No matter which one of many known measures of synchronization between signals is applied (Lachaux et al. 1999; Palva & Palva 2012; Vinck et al. 2011; Schoffelen & Gross 2009), it is important to ensure that such synchronization is not artificial, occurring, for example, due to the common noise components in the signals, or due to the signal leakage. One possible way to achieve that is to use connectivity measures that are insensitive to correlations between signals, which have no time delay. Linear mixing due to leakage involves no time delay; therefore, measures in question are immune to those artifacts. The flip side of this approach is that these measures are almost insensitive to connections with time lags that are small compared to the time period of the signal frequency. Consider an alpha frequency, for example. Its period is around 100 ms, therefore in the alpha range, connections with physiologically feasible propagation delays of several or even 10–20 ms may not be detected. An alternative way to exclude spurious connectivity due to linear mixing is to eliminate the leakage at the very beginning—that is, at source reconstruction step. In this case, a broader set of connectivity measures, including those sensitive to zero-delay connections, can be applied. This is where the multisource beamformers become handy, because direct leakage from one source to another is rejected.

The third step in the analyses consists of recognizing distinct groups of interacting brain regions, or brain networks, based on the connectivity data. The difficulty here is that the number of possible pairwise connections between different regions is usually very large, while existence of each connection can only be established with certain probability (say, 95%)—in other words, involves statistical error. It follows then that some of the found connections will be not real, occurring simply because of random fluctuations. For example, suppose we consider locations corresponding to approximately 50 known Brodmann areas. The number of possible pairwise connections between those is around 1200. If our statistical error per connection is 5%, then around 5% of all detected connections—say, 50 or 60 connections—will be false, and there is no way to distinguish between the real and spurious connections. The situation we described is called a "multiple-comparisons problem" in statistics (Shafer 1995). There is no general way to overcome the multiple comparison problem. Depending on specific situation, different approaches can be used. Here we will just mention a few popular ones.

The most straightforward approach is to try to reduce the number of statistical tests (comparisons) using prior knowledge. For example, if one can limit the number of brain areas to look at from 50 to 5 based on neurophysiological considerations, the number of pairwise connections will go down from approximately 1200 to 10. Then by limiting statistical error for each pair to 0.5%, one can guarantee with 95% certainty that all detected connections are the true ones. When such reduction of the number of connections is not possible, another useful approach is to apply statistical tests pertaining to (sub)networks as a whole rather than to individual connections. For example, one can look at subsets, or "clusters" of locations with mutual connectivity values above a certain threshold. The probability of such a cluster to occur by chance can be estimated (Zalesky et al. 2010), and if it is small enough, one can argue that the locations involved constitute a brain network. This way, interconnected brain areas can be identified in spite of the curse of multiple comparisons, even though it might be not possible to make conclusions about individual pairwise connections within the network. Another method is to consider a set of areas strongly connected to a single location (a seed point) to form a separate network (Brookes et al. 2012). Yet another set of approaches involves representing a signal from each location as a linear combination of a certain number of uncorrelated or statistically independent components (so-called principal or independent component analysis). When any such component mostly consists of contributions from a well-defined, distinct subset of locations, one can argue that corresponding areas operate synchronously and constitute a single network (Luckhoo et al. 2012; Whitman et al. 2016).

In summary, we discussed how the focus on interactions and connectivity within the human brain affected MEG/EEG analyses of brain signals. We showed that new, more sophisticated signal processing techniques were developed to deal with the new challenges. However, this journey is far from over, and we need more novel, even more advanced techniques to emerge in this field in coming years to further improve the quality of brain connectivity analysis.

Advances in brain imaging processing and visualization: Interpretation of diffusion MRI for the brain

As advances have been made in brain imaging techniques, the quantity and complexity of the outputted images has also increased, leading to data visualization and interpretation challenges, challenges most prominently seen in the area of diffusion MRI (dMRI) (Booth & Hamarneh 2015). In dMRI, image sequences are used to measure, non-invasively and in vivo, the rates of molecular diffusion in the brain. This diffusion is impeded by cell structure in such a way that fibrous tissue (e.g., white matter) shows a maximal rate of diffusion along the direction of the fibers (Stejskal & Tanner 1965; Le Bihan & Breton 1985). Uncovering these directions of maximal diffusion—and therefore the fiber directions—requires the use of diffusion-sensitizing gradient pulses to sample diffusion along tens, or even hundreds, of directions (as in high angular resolution diffusion imaging, e.g., Q-ball imaging) (Tuch 2004). The result is a high-dimensional 3D image with a (fiber) orientation distribution function (Tournier et al. 2007) at each voxel—an image that is difficult to manually examine and interpret.

Initial attempts to visualize dMRI datasets revolved around reducing their dimensionality to a single scalar-valued image by computing statistics across the sampled diffusion directions. By computing the mean and variance in diffusion across the sample directions, we obtain the commonly used mean diffusivity (MD) and fractional anisotropy (FA) measures (Figure 2.4a, b). The MD is sensitive to the amount of cell structure present while the FA is sensitive to how fibrous the tissue is. It is also common to see the FA image color-coded to show the directions of maximal diffusion, which approximate the fiber directions (Figure 2.4c). Other scalar measures include radial, axial, and relative diffusivity; linear, planar, and sphericity measures; mode; and volume ratio (Westin et al. 2002). Finally, it has also become customary to reduce the dimensionality of a dMRI dataset by fitting models, at each voxel, to the sampled diffusion measurements. Through the use of tensors (i.e., DTI) and spherical harmonics (SPHARM) (Basser et al. 1994: Jian & Vemuri 2007), we can reduce the dimensionality of a dMRI dataset, from the number of diffusion samples to the much lower number of parameters of the model, thus facilitating visualization (Figure 2.4d).

More recent work has looked at producing dMRI visualizations that highlight different geometric properties. For example, recent work has led to color visualizations of dMRI datasets that match the perceptual differences in color to the real differences in measured diffusion (Figure 2.4e) (Hamarneh et al. 2011). Other works have looked at extracting regions of interest—usually axon bundles—from a dMRI scan through the use of specially-designed structure detectors (Figure 2.4f) (Nand et al. 2011) or segmentation techniques (Figure 2.4g) (Hamarneh & Hradsky 2006; Weldeselassie & Hamarneh 2007; Booth & Hamarneh 2013). Similarly, those regions of interest can also be detected through the use of image registration. By aligning two dMRI scans and examining the differences in the image data at the level of individual voxels, we can flag regions of interest where the scans differ (Figure 2.4h) (Smith et al. 2006; Booth et al. 2016). In all of these techniques, the high-dimensional dMRI scan undergoes multiple processing steps in order to generate the visualization. As a result, various mathematical models have had to be extended to account for the dMRI's high dimensionality and to ensure that its measured diffusion values remain positive (Booth & Hamarneh 2011a). Further, these processing and analysis techniques benefit from pre-processing algorithms that remove imaging noise (Hamarneh & Hradsky 2007; Nand et al. 2012) and ones that capture relations between neighboring voxels (Booth & Hamarneh 2011b).

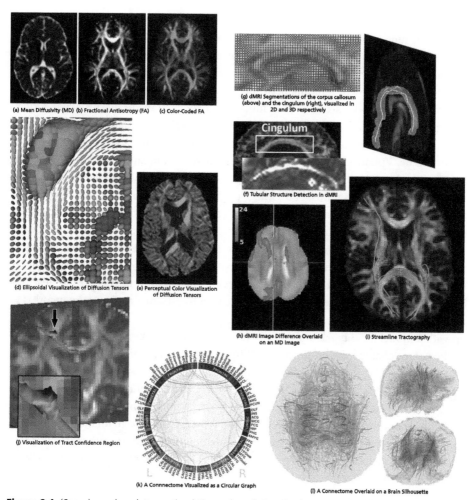

(a) Mean Diffusivity (MD) (b) Fractional Antisotropy (FA) (c) Color-Coded FA

(g) dMRI Segmentations of the corpus callosum (above) and the cingulum (right), visualized in 2D and 3D respectively

Cingulum

(f) Tubular Structure Detection in dMRI

(d) Ellipsoidal Visualization of Diffusion Tensors (e) Perceptual Color Visualization of Diffusion Tensors

(h) dMRI Image Difference Overlaid on an MD Image

(i) Streamline Tractography

(j) Visualization of Tract Confidence Region

(k) A Connnectome Visualized as a Circular Graph

(l) A Connectome Overlaid on a Brain Silhouette

Figure 2.4 (See also color plate section.) Examples of visualizations associated with diffusion MRI (dMRI) data. Some visualizations highlight the brain's diffusion properties (a–e), while others highlight its geometric properties (f–h), while still others highlight brain connectivity (i–l). See the text for a description of each visualization technique: (a) mean diffusivity (MD); (b) fractional anisotropy (FA); (c) color-coded FA; (d) ellipsoidal visualization of diffusion tensors; (e) perceptual color visualization of diffusion tensors; (f) tubular structure detection in dMRI; (g) dMRI segmentation of corpus callosum (above) and the cingulum (right), visualized in 2D and 3D respectively; (h) dMRI image difference overlaid on MD image; (i) streamline tractography; (j) visualization of tract confidence regions; (k) a connectome visualized as a circular graph; (l) a connectome overlaid on a brain silhouette.

The most appealing aspect of dMRI may be its unique ability to examine structural connectivity within the brain. As dMRI can identify the direction of tissue fibers, in a process called tractography (Mori & van Zijl 2002; Booth & Hamarneh 2012), we can trace out 3D curves called streamlines that estimate the location of white matter axonal pathways, thereby uncovering connections between different brain regions. Tractography techniques range from streamline-based techniques (Figure 2.4i) to probabilistic techniques that capture the uncertainty in the fiber tract location (Figure 2.4j) (Brown et al. 2013a, 2013b). Recently, groups have begun summarizing the structural connections tractography uncovers by building networks known as connectomes (Sporns et al. 2005). In a connectome, nodes are defined as functional regions in the brain (usually identified by registering a labeled atlas to the dMRI scan under evaluation) and weighted edges are placed between nodes based on how strongly connected the tractography results show the regions to be. Visualizing connectomes remains an area of open research, but two common approaches include the use of circular graphs (Figure 2.4k) and spatially overlaying the network's properties onto a brain image (Figure 2.l) (Brown et al. 2014).

While the visualization techniques presented here provide the ability for a human operator to examine and interpret a high-dimensional, complicated dMRI dataset, the value of these visualization techniques can be further strengthened by having complementary automated computational methods that identify which features or aspects would be most valuable to visualize. Recently, researchers have begun using machine learning algorithms (e.g., support vector machines, decision forests, or deep neural networks) to model relationships between the image data and clinical variables of interest. By coming up with a model for this relationship, we identify regions in the images, or edges in the connectomes, that are the most important to the model. We then visualize these image-based features using the above techniques. For example, Figure 2.4j shows the presence of abnormal geometry in a fiber bundle, while Figure 2.4h highlights voxels that are significantly different between a subject's scan and a healthy population, and Figure 2.4k shows edges in a connectome most predictive of motor outcome in preterm infants (Brown et al. 2015). By using machine-learning techniques, not only are we able to automatically identify relevant visualizations, but also seek imaging biomarkers of neurological disorders (Brown et al. 2015; Booth et al. 2016).

Emerging neurophysiological frameworks for unified thalamocortical processing

Sensory, motor, and cognitive functions are associated with oscillatory changes within task-relevant cortical brain regions, as well as alterations in functional and effective connectivity between those regions associated with neuronal synchronization (Llinás & Ribary 1993; Varela et al. 2001; Ward L.M. 2003; Uhlhaas et al. 2009; Wang X.J. 2010; Ribary et al. 2014). Recording methodologies with sufficient temporal resolution to measure fast above 1 Hz neural oscillations, such as EEG and MEG, have shown that brain dynamics within and between regions often differ markedly across frequencies, and that brain rhythms often interact across widely separated frequency ranges (Llinás et al. 1999; Sauseng et al. 2008; Doesburg et al. 2009, 2012; Holz et al. 2010; Palva et al. 2010; Palva and Palva 2012). It has further been reported that cross-frequency interactions may play an important role in local processing within the thalamus and neocortex, as well as information transfer between them (FitzGerald et al. 2013). Strong commonalities in rhythmic network properties have been observed across recording techniques and task demands, but strong neuroscientific theories to situate such observations within a unified framework that has direct relevance to the explanation of neuropathologies remain scarce.

Our recent comprehensive review into the animal and human literature indicates a further thinking beyond synchrony and connectivity and the readiness for more hypothesis-driven research and modeling toward unified principles of thalamocortical processing (Doesburg et al. 2015). We further introduced such a possible emerging framework: "the alpha-theta-gamma (ATG) switch." We probed and introduced this neurophysiological framework to explain how coordinated cross-frequency and interregional oscillatory cortical dynamics may underlie typical and atypical brain activation, and the formation of distributed functional ensembles supporting cortical networks underpinning sensation and cognition (Doesburg et al. 2015). We also discussed evidence that alpha-theta-gamma dynamics emerging from thalamocortical and cortico-cortical interactions may be implicated in cognition across diverse contexts, and that the disruption of such processes is implicated in numerous neurological and neuropsychiatric conditions. Moreover, this emerging framework further challenges the dominant view in current neuroscience on the classical activation of the brain per se and rather suggests an inhibitory induced dis-inhibition of intrinsic cortical activity during sensory and cognitive processing, a novel plausible mechanism to integrating local and large-scale brain networks and its related alterations in clinical pathologies (Doesburg et al. 2015).

Such neuroscientific advances, emerging neurophysiological frameworks, and more future hypothesis-driven basic and clinical research and modeling will pose further future ethical challenges for the implementation of quantitative diagnostic and prognostic strategies into clinical practice according to best practice and the highest ethical standards.

Conclusion

This comprehensive chapter, highlighting advances in multimodal brain imaging, signal processing, and visualization from related expert perspectives, points to the next necessary step in cognitive and clinical neuroscience brain imaging, namely to more basic and clinical research and the quantification of the *five-dimensional human brain* (across 3D space, frequency, and time) in health and disease. In this chapter, we urge quantification of the biochemical, structural, functional, and dynamic network connectivity and causality of the brain underlying typical/atypical sensory and cognitive processing, to establish objective diagnostic neuromarkers for the identification and better characterization of subtypes of neurological disabilities and pathologies. In addition, we believe that it is essential to indicate new innovative ways of how such subtypes can be selectively and optimally targeted with existing or additionally modified neuroscience-based interventional training programs that exploit brain plasticity with neuropharmacological or neurosurgical interventional strategies, and that thereby validate and monitor their effectiveness with the highest possible precision and accuracy across five dimensions. It is incumbent on us, as experts in neuroscience, and on experts in neuroethics and biomedicine, to proactively consider and advance the best possible diagnostic evaluation, utilization of criteria, and prediction of interventional outcome to researchers, educators, clinicians, and healthcare practitioners at large.

Acknowledgments

The co-authors are listed in chronological order of their contributions: MacKay and Rauscher (MRI), Tipper, Giaschi and Woodward (fMRI), Sossi (PET), Doesburg (MEG), Ward (EEG-1), Herdmann (EEG-2), Moiseev (Signal Processing), Hamarneh and Booth (Processing and Visualization), Ribary, Doesburg and Ward (Emerging Neurophysiological Frameworks).

References

Baillet, S., Mosher, J.C., and Leahy, R.M. (2001). Electromagnetic brain mapping. *IEEE Signal Process Magazine*, **18**, 14–30.

Baker, A.P., Brookes, M.J., Rezek, I.A., et al. (2014). Fast transient networks in spontaneous human brain activity. *Elife*, **3**, e01867.

Bandettini, P.A., Wong, E.C., Hinks, R.S., Tikofsky, R.S., and Hyde, J.S. (1992). Time course EPI of human brain function during task activation. *Magnetic Resonance in Medicine*, **25**, 390–397.

Barth, M., Breuer, F., Koopmans, P.J., Norris, D.G., and Poser, B.A. (2016). Simultaneous multislice (SMS) imaging techniques. *Magnetic Resonance in Medicine*, **75**, 63–81.

Basser, P.J., Mattiello, J., and LeBihan, D. (1994). MR diffusion tensor spectroscopy and imaging. *Biophysical Journal*, **66**, 259–267.

Basser, P.J., Pajevic, S., Pierpaoli, C., Duda, J., and Aldroubi, A. (2000). In vivo fiber tractography using DT-MRI data. *Magnetic Resonance in Medicine*, **44**, 625–632.

Bassett, D.S., Wymbs, N.F., Porter, M.A., Mucha, P.J., Carlson, J.M., and Grafton, S.T. (2011). Dynamic reconfiguration of human brain networks during learning. *Proceedings of the National Academy of Sciences of the United States of America*, **108**, 7641–7646.

Beaty, R.E., Benedek, M., Silvia, P.J., and Schacter, D.L. (2015). Creative cognition and brain network dynamics. *Trends in Cognitive Sciences*, **20**, 87–95.

Beaulieu, C. (2002). The basis of anisotropic water diffusion in the nervous system—a technical review. *NMR in Biomedicine*, **15**, 435–455.

Bedo, N., Ribary, U., and Ward, L.M. (2014). Fast dynamics of cortical functional and effective connectivity during word reading. *PLoS One*, **9**(2), e88940.

Berendse, H.W. and Stam, C.J. (2007). Stage-dependent patterns of disturbed neural synchrony in Parkinson's disease. *Parkinsonism & Related Disorders*, **13**(Suppl 3), S440–445.

Berger, H. (1929). Uber das elektrenkephalogramm des menschen. *Archiv für Psychiatrie und Nervenkrankheiten*, **87**, 527–570.

Betzel, R.F., Gu, S., Medaglia, J.D., Pasqualetti, F., and Bassett, D.S. (2016). Optimally controlling the human connectome: the role of network topology. arXiv:1603.05261 [q-bio.NC]. Available from: http://arxiv.org/abs/1603.05261 [accessed February 3, 2017].

Bilgic, B., Pfefferbaum, A., Rohlfing, T., Sullivan, E.V., and **Adalsteinsson, E.** (2011). MRI estimates of brain iron concentration in normal aging using quantitative susceptibility mapping. *NeuroImage*, 59(3), 2625–2635.

Boersma, M., Smit, D.J., de Bie, H.M., et al. (2011). Network analysis of resting state EEG in the developing young brain: structure comes with maturation. *Human Brain Mapping*, 32(3), 413–425.

Booth, B.G. and Hamarneh, G. (2011a, July). Consistent information content estimation for diffusion tensor MR images. In *Proceedings of IEEE International Conference on Healthcare Informatics, Imaging and Systems Biology, HISB*, 166–173.

Booth, B.G. and Hamarneh, G. (2011b, April). Exact integration of diffusion orientation distribution functions for graph-based diffusion MRI analysis. In: *Proceedings/IEEE International Symposium on Biomedical Imaging: From Nano to Macro. IEEE International Symposium on Biomedical Imaging*, 935–938.

Booth, B.G. and Hamarneh, G. (2012, January). Multi-region competitive tractography via graph-based random walks. In: *2012 IEEE Workshop on Mathematical Methods in Biomedical Image Analysis (MMBIA)*, 73–78.

Booth, B.G. and Hamarneh, G. (2013). A cross-sectional piecewise constant model for segmenting highly curved fiber tracts in diffusion MR images. In: Mori, K., Sakuma, I., Sato, Y., Barillot, C., and Navab, N. (Eds.) *Medical Image Computing and Computer-Assisted Intervention – MICCAI 2013* (Lecture Notes in Computer Science Vol. 8151), pp.469–476. New York: Springer.

Booth, B.G. and Hamarneh, G. (2015). Diffusion MRI for brain connectivity mapping and analysis. In: Majumdar, A. and Ward, R.K. (Eds.) *MRI: Physics, Image Reconstruction, and Analysis*, pp.137–171. Boca Raton, FL: CRC Press.

Booth, B.G., Miller, S.P., Brown, C.J., et al. (2016). STEAM—statistical template estimation for abnormality mapping: a personalized DTI analysis technique with applications to the screening of preterm infants. *NeuroImage*, 125, 705–723.

Borich, M.R., Mackay, A.L., Vavasour, I.M., Rauscher, A., and Boyd, L.A. (2013). Evaluation of white matter myelin water fraction in chronic stroke. *NeuroImage Clinical*, 2, 569–580.

Brookes, M.J., Woolrich, M., Luckhoo, H., et al. (2011). Investigating the electrophysiological basis of resting state networks using magnetoencephalography. *Proceedings of the National Academy of Sciences of the United States of America*, 108(40), 16783–16788.

Brookes, M.J., Liddle, E.B., Hale, J.R., et al. (2012a). Task induced modulation of neural oscillations in electrophysiological brain networks. *NeuroImage*, 63(4), 1918–1930.

Brookes, M.J., Woolrich, M.W., and Barnes, G.R. (2012b). Measuring functional connectivity in MEG: a multivariate approach insensitive to linear source leakage. *NeuroImage*, 63, 910–920.

Brown, C.J., Booth, B.G., and Hamarneh, G. (2013a, April). K-confidence: assessing uncertainty in tractography using k optimal paths. In Proceedings of *IEEE International Symposium on Biomedical Imaging: From Nano to Macro. IEEE International Symposium on Biomedical Imaging*, 250–253.

Brown, C.J., Booth, B.G., and Hamarneh, G. (2013b, January). Uncertainty in tractography via tract confidence regions. In: *Proceedings of Medical Image Computing and Computer-Assisted Intervention Workshop on Computational Diffusion MRI (MICCAI CDMRI)*, 13–22.

Brown, C.J., Miller, S., Booth, B.G., et al. (2014). Structural network analysis of brain development in young preterm neonates. *NeuroImage*, 101, 667–680.

Brown, C.J., Miller, S., Booth, B.G., et al. (2015, October). Prediction of motor function in very preterm infants using connectome features and local synthetic instances. In: *Proceedings of Medical Image Computing and Computer Assisted Intervention (MICCAI) 2015*, 9349, 69–76.

Buccino, G., Binkofski, F., Fink, G.R., et al. (2001). Action observation activates premotor and parietal areas in a somatotopic manner: an fMRI study. *European Journal of Neuroscience*, 13, 400–404.

Bullmore, E.T. and Bassett, D.S. (2010). Brain graphs: graphical models of the human brain connectome. *Annual Review of Clinical Psychology*, 7, 16–18.

Buzsáki, G. and Watson, B.O. (2012). Brain rhythms and neural syntax: implications for efficient coding of cognitive content and neuropsychiatric disease. *Dialogues in Clinical Neuroscience*, 14(4), 345–367.

Campbell, K.B., Picton, T.W., Wolfe, R.G., Maru, J., Baribeau-Braun, J., and Braun, C. (1981). Auditory potentials. *Sensus*, 1, 21–31.

Chiappa, K.H., Gladstone, K.J., and Young, R.R. (1979). Brainstem auditory evoked responses: studies of waveform variations in 50 normal human subjects. *Archives of Neurology*, 36, 81–87.

Coluccia, D., Fandino, J., Schwyzer, L., et al. (2014). First noninvasive thermal ablation of a brain tumor with MR-guided focused ultrasound. *Journal of Therapeutic Ultrasound*, 2, 17.

Crawford, M.L. and Harwerth, R.S. (2004). Ocular dominance column width and contrast sensitivity in monkeys reared with strabismus or anisometropia. *Investigative Ophthalmology & Visual Science*, 45, 3036–3042.

Cruse, D., Chennu, S., Chatelle, C., et al. (2011). Bedside detection of awareness in the vegetative state: a cohort study. *The Lancet*, 378: 2088–2094.

Dalal, S., Sekihara, K., and Nagarajan, S. (2006). Modified beamformers for coherent source region suppression. *IEEE Transactions on Biomedical Engineering*, 53, 1357–1363.

Decety, J. and Cacioppo, J. (2010). Frontiers in human neuroscience: the golden triangle and beyond. *Perspectives on Psychological Science*, 5, 767–771.

Deco, G. and Kringelbach, M.L. (2014). Great expectations: using whole-brain computational connectomics for understanding neuropsychiatric disorders. *Neuron*, 84(5), 892–905.

DeHaan, W., van der Flier, W.M., Wang, H., Van Mieghem, P.F., Scheltens, P., and Stam, C.J. (2012). Disruption of functional brain networks in Alzheimer's disease: what can we learn from graph spectral analysis of resting-state magnetoencephalography? *Brain Connectivity*, 2(2), 45–55.

Ding, K., Liu, Y., Yan, X., Lin, X., and Jiang, T. (2013). Altered functional connectivity of the primary visual cortex in subjects with amblyopia. *Neural Plasticity*, 2013, 1–8.

Delorme, A., Palmer, J., Onton, J., Oostenveld, R., and Makeig, S. (2012). Independent EEG sources are dipolar. *PLoS One*, 7, e30135.

DePasquale, F., Della Penna, S., Snyder, A.Z., et al. (2012). A cortical core for dynamic integration of functional networks in the resting human brain. *Neuron*, 74(4), 753–64.

Dimitriadis, S.I., Laskaris, N.A., Simos, P.G., et al. (2013). Altered temporal correlations in resting-state connectivity fluctuations in children with reading difficulties detected via MEG. *NeuroImage*, 83, 307–317.

Doesburg, S.M., Green, J.J., McDonald, J.J., and Ward, L.M. (2009). Rhythms of consciousness: binocular rivalry reveals large-scale oscillatory network dynamics mediating visual perception. *PLoS One*, 4(7), e6142.

Doesburg, S.M., Green, J.J., McDonald, J.J., and Ward, L.M. (2012). Theta modulation of inter-regional gamma synchronization during auditory attention control. *Brain Research*, 1431, 77–85.

Doesburg, S.M., Ward, L.M., and Ribary, U. (2015). The alpha-theta-gamma (ATG) switch: toward unified principles of cortical processing. *Current Trends in Neurology*, 9, 1–12.

Dunkley, B.T., Doesburg, S.M., Sedge, P.A., et al. (2014). Resting-state hippocampal connectivity correlates with symptom severity in post-traumatic stress disorder. *NeuroImage Clinical*, 5, 377–384.

Dunkley, B.T., Da Costa, L., Bethune, A., et al. (2015). Low-frequency connectivity is associated with mild traumatic brain injury. *NeuroImage Clinical*, 7, 611–621.

Duyn, J.H., Van Gelderen, B., Li, T.Q., de Zwart, J.A., Koretsky, A.P., and Fukunaga, M. (2007). High-field MRI of *Brain Cortical Substructure Based on Signal Phase. Proceedings of the National Academy of Sciences of the United States of America*, 104, 11796–801.

Eberling, J.L., Dave, K.D., and Frasier, M.A. (2013). Alpha-synuclein imaging: a critical need for Parkinson's disease research. *Journal of Parkinson's Disease*, 3, 565–567.

Elshahabi, A., Klamer, S., Sahib, A.K., Lerche, H., Braun, C., and Focke, N.K. (2015). Magneto-encephalography reveals a widespread increase in network connectivity in idiopathic/genetic generalized epilepsy. *PLoS One*, 10(9), e0138119.

Engel, A.K., Gerloff, C., Hilgetag, C.C., and Nolte, G. (2013). Intrinsic coupling modes: multiscale interactions in ongoing brain activity. *Neuron*, 80(4), 867–886.

Ewald, A., Marzetti, L., Zappasodi, F., Meinecke, F.C., and Nolte, G. (2012). Estimating true brain connectivity from EEG/MEG data invariant to linear and static transformations in sensor space. *NeuroImage*, 60(1), 476–488.

Fitzgerald, T.H.B., Valentin, A., Selway, R., and Richardson, M.P. (2013). Cross-frequency coupling within and between the human thalamus and neocortex. *Frontiers in Human Neuroscience*, 7, 1–13.

Fitzgibbon, S.P., DeLosAngeles, D., Lewis, T.W., et al. (2016). Automatic determination of EMG-contaminated components and validation of independent components analysis using EEG during pharmacologic paralysis. *Clinical Neurophysiology*, 127, 1781–1793.

Florin, E. and Baillet, S. (2015). The brain's resting-state activity is shaped by synchronized cross-frequency coupling of neural oscillations. *NeuroImage*, 111, 26–35.

Fortnum, H., O'Nei, C., Taylor, R., et al. (2009). The role of magnetic resonance imaging in the identification of suspected acoustic neuroma: a systematic review of clinical and cost effectiveness and natural history. *Health Technology Assessments*, 13, 1–154.

Fox, M.D., Snyder, A.Z., Vincent, J.L., Corbetta, M., Van Essen, D.C., and Raichle, M.E. (2005). The human brain is intrinsically organized into dynamic, anticorrelated functional networks. *Proceedings of the National Academy of Sciences of the United States of America*, 102, 9673–9678.

Fries, P. (2009). Neuronal gamma-band synchronization as a fundamental process in cortical computation. *Annual Review of Neuroscience*, 32, 209–224.

Fries, P. (2015). Rhythms for cognition: communication through coherence. *Neuron*, 88(1), 220–235.

Friston, K.J., Fletcher, P., Josephs, O., Holmes, A.P., Rugg, M.D., and Turner, R. (1998). Event-related fMRI: characterizing differential responses. *NeuroImage*, 7, 30–40.

Friston, K.J., Harrison, L., and Penny, W. (2003). Dynamic causal modelling. *NeuroImage*, 19, 1273e1302.

Goodyear, B.G., Nicolle, D.A., and Menon, R.S. (2002). High resolution fMRI of ocular dominance columns within the visual cortex of human amblyopes. *Strabismus*, 10, 129–136.

Grafton, S.T. and Hamilton, A.F. (2007). Evidence for a distributed hierarchy of action representation in the brain. *Human Movement Science*, 26, 590–616.

Green, J.J., Doesburg, S.M., Ward, L.M., and McDonald, J.J. (2011). Electrical neuroimaging of voluntary audio-spatial attention: evidence for a supramodal attention control network. *Journal of Neuroscience*, 31(10), 3560–3564.

Grill-Spector, K., Henson, R., and Martin, A. (2006). Repetition and the brain: neural models of stimulus-specific effects. *Trends in Cognitive Sciences*, 10, 14–23.

Guan, X., Xuan, M., Gu, Q., et al. (2016). Regionally progressive accumulation of iron in Parkinson's disease as measured by quantitative susceptibility mapping. *NMR in Biomedicine*. Advance online publication. doi:10.1002/nbm.3489.

Hamalainen, M.S. and Ilmoniemi, R.J. (1994). Interpreting magnetic fields of the brain: Minimum norm estimates. *Medical & Biological Engineering & Computing*, 32(1):35–42.

Hamarneh, G. and Hradsky, J. (2006, August). DTMRI segmentation using DT-snakes and DT-livewire. In: *Proceedings of IEEE International Symposium on Signal Processing and Information Technology (IEEE ISSPIT)*, 513–518.

Hamarneh, G. and Hradsky, J. (2007). Bilateral filtering of diffusion tensor magnetic resonance images. *IEEE Transactions on Image Processing*, 16, 2463–2475.

Hamarneh, G., McIntosh, C., and Drew, M. (2011). Perception-based visualization of manifold-valued medical images using distance-preserving dimensionality reduction. *IEEE Transactions on Image Processing*, 30, 1314–1327.

Hermundstad, A.M., Bassett, D.S., Brown, K.S., et al. (2013). Structural foundations of resting-state and task-based functional connectivity in the human brain. *Proceedings of the National Academy of Sciences of the United States of America*, 110, 6169–6174.

Hermundstad, A.M., Brown, K.S., Bassett, D.S., et al. (2014). Structurally-constrained relationships between cognitive states in the human brain. *PLoS Computational Biology*, 10, e1003591.

Hipp, J.F., Hawellek, D.J., Corbetta, M., Siegel, M., and Engel, A.K. (2012). Large-scale cortical correlation structure of spontaneous oscillatory activity. *Nature Neuroscience*, 15(6), 884–890.

Ho, C.S. and Giaschi, D.E. (2009). Low- and high-level motion perception deficits in anisometropic and strabismic amblyopia: evidence from fMRI. *Vision Research*, 49, 2891–2901.

Ho, C.S., Paul, P., Asirvatham, A., Cavanagh, P., Cline, R., and Giaschi, D. (2006). Abnormal spatial selection and tracking in children with amblyopia. *Vision Research*, 46, 3274–3283.

Holz, E.M., Glennon, M., Prendergast, K., andSauseng, P. (2010). Theta-gamma phase synchronization during memory matching in visual working memory. *NeuroImage*, 67:331–343.

Huang, M.X., Nichols, S., Baker, D.G., et al. (2014). Single-subject-based whole-brain MEG slow-wave imaging approach for detecting abnormality in patients with mild traumatic brain injury. *NeuroImage Clinical*, 5, 109–119.

Hubel, D.H., Wiesel, T.N., and LeVay, S. (1977). Plasticity of ocular dominance columns in monkey striate cortex. *Philosophical Transactions of the Royal Society B: Biological Sciences*, 278, 377–409.

Hui, H.B., Pantazis, D., Bressler, S.L. and Leahy, R.M. (2010). Identifying true cortical interactions in MEG using the nulling beamformer. *NeuroImage*, 49, 3161–3174.

Immordino-Yang, M.H. and Singh, V. (2013). Hippocampal contributions to the processing of social emotions. *Human Brain Mapping*, 34, 945–955.

Jerbi, K., Ossandón, T., Hamamé, C.M., et al. (2009). Task-related gamma-band dynamics from an intracerebral perspective: review and implications for surface EEG and MEG. *Human Brain Mapping*, 30(6), 1758–1771.

Jian, B. and Vemuri, B.C. (2007). A unified computational framework for deconvolution to reconstruct multiple fibers from diffusion weighted MRI. *IEEE Transactions on Medical Image Analysis*, 26, 1464–1471.

Jones, D.K. (2016). *Diffusion MRI: Theory, Methods, and Applications*. New York: Oxford University Press.

Jones, D.K., Knösche, T.R., and Turner, R. (2013). White matter integrity, fiber count, and other fallacies: the do's and don'ts of diffusion MRI. *NeuroImage*, 73, 239–254.

Jurcoane, A., Choubey, B., Mitsieva, D., Muckli, L., and Sireteanu, R. (2009). Interocular transfer of orientation-specific fMRI adaptation reveals amblyopia-related deficits in humans. *Vision Research*, 49, 1681–1692.

Kaiser, M. (2011). A tutorial in connectome analysis: Topological and spatial features of brain networks. *NeuroImage*, 57, 892–907.

Kang, J., Gao, Y., Shi, F., Lalush, D.S., Lin, W., and Shen, D. (2015). Prediction of standard-dose brain PET image by using MRI and low-dose brain [18F]FDG PET images. *Medical Physics*, 42, 5301–5309.

Kaplan, J.T., Gimbel, S.I., Dehghani, M., et al. (2016). Processing narratives concerning protected values: a cross-cultural investigation of neural correlates. *Cerebral Cortex*. Advance online publication. doi:10.1093/cercor/bhv325

Kikuchi, A., Takeda, A., Okamura, N., et al. (2010). In vivo visualization of alpha-synuclein deposition by carbon-11-labelled 2-[2-(2-dimethylaminothiazol-5-yl)ethenyl]-6-[2-(fluoro)ethoxy]benzoxazole positron emission tomography in multiple system atrophy. *Brain*, **133**, 1772–1778.

Kiorpes, L. and McKee, S.P. (1999). Neural mechanisms underlying amblyopia. *Current Opinion in Neurobiology*, **9**, 480–486.

Kitzbichler, M.G., Khan, S., Ganesan, S., et al. (2015). Altered development and multifaceted band-specific abnormalities of resting state networks in autism. *Biological Psychiatry*, **77**(9), 794–804.

Klyuzhin, I.S., Gonzalez, M., Shahinfard, E., Vafai, N., and Sossi, V. (2016). Exploring the use of shape and texture descriptors of positron emission tomography tracer distribution in imaging studies of neurodegenerative disease. *Journal of Cerebral Blood Flow & Metabolism*, **36**(6), 1122–1134.

Kwong, K.K., Belliveau, J.W., Chesler, D.A., et al. (1992). Dynamic magnetic resonance imaging of human brain activity during primary sensory stimulation. *Proceedings of the National Academy of Sciences of the United States of America*, **89**, 5675–5679.

Lachaux, J.P., Rodriguez, E., Martinerie, J., and Varela, F.J. (1999). Measuring phase synchrony in brain signals. *Human Brain Mapping*, **8**, 194–208.

Lang, D.J.M., Yip, E., MacKay, A.L., et al. (2014). 48 echo T2 myelin imaging of white matter in first-episode schizophrenia: evidence for aberrant myelination. *NeuroImage Clinical*, **6**, 408–414.

Larson-Prior, L.J., Oostenveld, R., Della Penna, S., et al. (2013). Adding dynamics to the Human Connectome Project with MEG. *NeuroImage*, **80**, 190–201.

Laule, C., Kozlowski, P., Leung, E., Li, D.K.B., MacKay, A.L., and Moore, G.R.W. (2008). Myelin water imaging of multiple sclerosis at 7 T: correlations with histopathology. *NeuroImage*, **40**, 1575–1580.

Lavigne, K.M., Metzak, P.D., and Woodward, T.S. (2015a). Functional brain networks underlying detection and integration of disconfirmatory evidence. *NeuroImage*, **112**, 138–151.

Lavigne, K.M., Rapin, L.A., Metzak, P.M., et al. (2015b). Left-dominant temporal-frontal hypercoupling in schizophrenia patients with hallucinations during speech perception. *Schizophrenia Bulletin*, **41**, 259–267.

LeBihan, D. and Breton, E. (1985). Imagerie de diffusion in vivo par résonance magnétique nucléaire. *Compte Rendus de l'Académie de Sciences Paris*, **301**, 1109–1112.

Lerner, Y., Hendler, T., Malach, R., et al. (2006). Selective fovea-related deprived activation in retinotopic and high-order visual cortex of human amblyopes. *NeuroImage*, **33**, 169–179.

Lewis, C.M., Bosman, C.A., Womelsdorf, T., and Fries, P. (2016). Stimulus-induced visual cortical networks are recapitulated by spontaneous local and interareal synchronization. *Proceedings of the National Academy of Sciences of the United States of America*, **113**, E606–E615.

Li, W., Wu, B., Batrachenko, A., et al. (2014). Differential developmental trajectories of magnetic susceptibility in human brain gray and white matter over the lifespan. *Human Brain Mapping*, **35**, 2698–2713.

Li, X., Coyle, D., Maguire, L., McGinnity, T.M., and Hess, R.F. (2011a). Long timescale fMRI neuronal adaptation effects in human amblyopic cortex. *PLoS ONE*, **6**, e26562.

Li, X., Mullen, K.T., Thompson, B., and Hess, R.F. (2011b). Effective connectivity anomalies in human amblyopia. *NeuroImage*, **54**, 505–516.

Li, X., Harrison, D.M., Liu, H., et al. (2015). Magnetic susceptibility contrast variations in multiple sclerosis lesions. *Journal of Magnetic Resonance Imaging*, **43**(2), 463–473.

Lin, X., Ding, K., Liu, Y., Yan, X., Song, S., and Jiang, T. (2012). Altered spontaneous activity in anisometropic amblyopia subjects: revealed by resting-state fMRI. *PLoS ONE*, **7**, e43373.

Liu, C., Wei, H., Gong, N.J., Cronin, M., Dibb, R., and Decker, K. (2015). Quantitative susceptibility mapping: contrast mechanisms and clinical applications. *Tomography: A Journal for Imaging Research*, **1**, 3–17.

Liu, S., Cai, W., Pujol, S., et al. (2016). Cross-view neuroimage pattern analysis in Alzheimer's disease staging. *Frontiers in Aging Neuroscience*, **8**, 23.

Llinás, R. and Ribary, U. (1993). Coherent 40-Hz oscillation characterizes dream state in humans. *Proceedings of the National Academy of Sciences of the United States of America*, **90**, 2078–2081.

Llinás, R., Ribary, U., Jeanmonod, D., Kronberg, E., and Mitra, P.P. (1999). Thalamocortical dysrhythmia: a neurological and neuropsychiatric syndrome characterized by magnetoencephalography. *Proceedings of the National Academy of Sciences of the United States of America*, **96**, 15222–15227.

Lobier, M., Siebenhühner, F., Palva, S., and Palva, J.M. (2014). Phase transfer entropy: A novel phase-based measure for directed connectivity in networks coupled by oscillatory interactions. *NeuroImage*, **85**, 853–872.

Lodygensky, G.A., Marques, J.P., Maddage, R., et al. (2011). In vivo assessment of myelination by phase imaging at high magnetic field. *NeuroImage*, **59**(3), 1979–1987.

Logothetis, N.K., Pauls, J., Augath, M., Trinath, T., and Oeltermann, A. (2001). Neurophysiological investigation of the basis of the fMRI signal. *Nature*, **412**, 150–157.

Luckhoo, H., Hale, J.R., Stokes, M.G., et al. (2012). Inferring task-related networks using independent component analysis in magnetoencephalography. *NeuroImage*, **62**, 530–541.

MacDonald, M.E. and Frayne, R. (2015). Cerebrovascular MRI: a review of state-of-the-art approaches, methods and techniques. *NMR in Biomedicine*, **28**, 767–791.

MacKay, A., Whittall, K., Adler, J., Li, D., Paty, D., and Graeb, D. (1994). In vivo visualization of myelin water in brain by magnetic resonance. *Magnetic Resonance in Medicine*, **31**, 673–677.

Manogaran, P., Vavasour, I., Borich, M., et al. (2015). *Corticospinal Tract Integrity Measured Using Transcranial Magnetic Stimulation and Magnetic Resonance Imaging in Neuromyelitis Optica and Multiple Sclerosis. Multiple Sclerosis.* Basingstoke, England: Houndmills.

McCall, C., Tipper, C.M., Blascovich, J., and Grafton, S.T. (2012). Attitudes trigger motor behavior through conditioned associations: neural and behavioral evidence. *Social Cognitive and Affective Neuroscience*, **7**, 841–849.

Medina, L.S., Balckmore, C.C., and Applegate, K.E. (2013). Evidence-based imaging: principles. In: Medina, L.S., Balckmore, C.C., and Applegate, K.E. (Eds.) *Evidence-Based Imaging: Improving Quality in Patient Care*. New York: Springer.

Metzak, P.D., Riley, J.D., Wang, L., Whitman, J.C., Ngan, E.T.C., and Woodward, T.S. (2012). Decreased efficiency of task-positive and task-negative networks during working memory in schizophrenia. *Schizophrenia Bulletin*, **38**, 803–813.

Miskovic, V., Ma, X., Chou, C.A., et al. (2015). Developmental changes in spontaneous electrocortical activity and network organization from early to late childhood. *NeuroImage*, **118**, 237–247.

Moiseev, A. and Herdman, A. (2013). Multi-core beamformers: derivation, limitations and improvements. *NeuroImage*, **71**, 135–146.

Moiseev, A., Gaspar, J., Schneider, J., and Herdman, A. (2011). Application of multi-source minimum variance beamformers for reconstruction of correlated neural activity. *NeuroImage*, **58**, 481–496.

Mori, S. and van Zijl, P.C.M. (2002). Fiber tracking: principles and strategies—a technical review. *NMR in Biomedicine*, **15**, 468–480.

Muckli, L., Kieb, S., Tonhausen, N., Singer, W., Goebel, R., and Sireteanu, R. (2006). Cerebral correlates of impaired grating perception in individual, psychophysically assessed human amblyopes. *Vision Research*, **46**, 506–526.

Mullen, T., Kothe, C., Chi, Y.M., et al. (2013, July). *Real-Time Modeling and 3D visualization of Source Dynamics and Connectivity Using Wearable EEG.* Presented at the 35th Annual International Conference of the IEEE EMBS Osaka, Japan.

Nand, K.K., Abugharbieh, R., Booth, B.G., and **Hamarneh, G.** (2011). Detecting structure in diffusion tensor MR images. *International Conference on Medical Image Computing and Computer-Assisted Intervention*, **14**(Pt 2), 90–97.

Nand, K., Hamarneh, G., and **Abugharbieh, R.** (2012). Diffusion tensor image processing using biquaternions. In: *Proceedings of 9th IEEE International Symposium on Biomedical Imaging (ISBI) 2012*, 538–541.

Nasrallah, F.A., Pagès, G., Kuchel, P.W., Golay, X., and **Chuang, K.H.** (2013). Imaging brain deoxyglucose uptake and Metabolism by glucoCEST MRI. *Journal of Cerebral Blood Flow and Metabolism*, **33**, 1270–78.

Niso, G., Rogers, C., Moreau, J.T., et al. (2016). OMEGA: The Open MEG Archive. *NeuroImage*, **124**(Pt B), 1182–1187.

Ogawa, S., Tank, D.W., Menon, R., et al. (1992). Intrinsic signal changes accompanying sensory stimulation: Functional brain mapping with magnetic resonance imaging. *Proceedings of the National Academy of Sciences of the United States of America*, **89**, 5951–5955.

O'Neill, G.C., Bauer, M., Woolrich, M.W., et al. (2015). Dynamic recruitment of resting state sub-networks. *NeuroImage*, **115**, 85–95.

Oz, G., Alger, J.R., Barker, P.B., et al. (2014). Clinical proton MR spectroscopy in central nervous system disorders. *Radiology*, **270**, 658–679.

Palop, J.J., Chin, J., and **Mucke, L.** (2006). A network dysfunction perspective on neurodegenerative diseases. *Nature*, **443**, 768–773.

Palva, J.M. and **Palva, S.** (2012). Discovering oscillatory interaction networks with M/EEG: challenges and breakthroughs. *Trends in Cognitive Science*, **16**(4), 219–230.

Palva, J.M., Monto, S., Kulashekhar, S., and **Palva, S.** (2010). Neuronal synchrony reveals working memory networks and predicts individual memory capacity. *Proceedings of the National Academy of Sciences of the United States of America*, **107**, 7580–7585.

Pannek, K., Scheck, S.M., Colditz, P.B., Boyd, R.N., and **Rose, S.E.** (2014). Magnetic resonance diffusion tractography of the preterm infant brain: a systematic review. *Developmental Medicine & Child Neurology*, **56**(2), 113–124.

Picton, T.W. (2011). *Human Auditory Evoked Potentials*. San Diego, CA: Plural Publishing Incorporated.

Poldrack, R.A. (2010). Mapping mental function to brain structure: how can cognitive neuroimaging succeed? *Perspectives on Psychological Science*, **5**, 753–761.

Prasloski, T., Rauscher, A., MacKay, A.L., et al. (2012). Rapid whole cerebrum myelin water imaging using a 3D GRASE sequence. *NeuroImage*, **63**, 533–539.

Pruessmann, K.P., Weiger, M., Scheidegger, M.B., and **Boesiger, P.** (1999). SENSE: sensitivity encoding for fast MRI. *Magnetic Resonance in Medicine*, **42**, 952–962.

Raichle, M.E. (2015). The restless brain: how intrinsic activity organizes brain function. *Philosophical Transactions of the Royal Society B: Biological Sciences*, **370**(1668), 20140172.

Rauscher, A., Sedlacik, J., Barth, M., Mentzel, H.J., and **Reichenbach, J.R.** (2005). Magnetic susceptibility-weighted MR phase imaging of the human brain. *AJNR. American Journal of Neuroradiology*, **26**, 736–742.

Ribary, U., Doesburg, S.M., and **Ward, L.M.** (2014). Thalamocortical network dynamics: a framework for typical/atypical cortical oscillations and connectivity. In: Supek, S. and Aine, C.J. (Eds.) *Magnetoencephalography—From Signals to Dynamic Cortical Networks*, pp.429–450. Heidelberg: Springer Verlag.

Robinson, S. and **Vrba, J.** (1999). Functional neuroimaging by synthetic aperture magnetometry (SAM). In: Yoshimoto, Y. (Ed.) *Recent Advances in Biomagnetism*, pp.302–305. Sendai, Japan: Tohoku University Press.

Rogers, B.P., Morgan, V.L., Newton, A.T., and Gore, J.C. (2007). Assessing functional connectivity in the human brain by fMRI. *Magnetic Resonance Imaging*, **25**, 1347–1357.

Ross, B.D., Bhattacharya, P., Wagner, S., Tran, T., and Sailasuta, N. (2010). Hyperpolarized MR imaging: neurologic applications of hyperpolarized metabolism. *American Journal of Neuroradiology*, **31**, 24–33.

Rubinov, M. and Sporns, O. (2010). Complex network measures of brain connectivity: uses and interpretations. *NeuroImage*, **52**, 1059–1069.

Saggar, M., Vrticka, P., and Reiss, A.L. (2016). Understanding the influence of personality on dynamic social gesture processing: an fMRI study. *Neuropsychologia*, **80**, 71–78.

Sander, C.Y., Hooker, J.M., Catana, C., et al. (2013). Neurovascular coupling to D2/D3 dopamine receptor occupancy using simultaneous PET/functional MRI. *Proceedings of the National Academy of Sciences of the United States of America*, **110**, 11169–11174.

Sauseng, P., Klimesch, W., Gruber, W.R., and Birbaumer, N. (2008). Cross-frequency phase synchronization: a brain mechanism of memory matching and attention. *NeuroImage*, **40**, 308–317.

Schäfer, C.B., Morgan, B.R., Ye, A.X., Taylor, M.J., and Doesburg, S.M. (2014). Oscillations, networks, and their development: MEG connectivity changes with age. *Human Brain Mapping*, **35**(10), 5249–5261.

Schoffelen, J.M. and Gross, J. (2009). Source connectivity analysis with MEG and EEG. *Human Brain Mapping*, **30**, 1857–1865.

Schreiber, T. (2000). Measuring information transfer. *Physical Review Letters*, **85**, 461e464.

Schweser, F., Deistung, A., Lehr, B.W., and Reichenbach, J.R. (2011). Quantitative imaging of intrinsic magnetic tissue properties using MRI signal phase: an approach to in vivo brain iron metabolism? *NeuroImage*, **54**, 2789–2807.

Schweser, F., Deistung, A., and Reichenbach, J.R. (2016). Foundations of MRI phase imaging and processing for quantitative susceptibility mapping (QSM). *Zeitschrift Für Medizinische Physik*, **26**, 6–34.

Secen, J., Culham, J., Ho, C., and Giaschi, D. (2011). Neural correlates of the multiple-object tracking deficit in amblyopia. *Vision Research*, **51**, 2517–2527.

Sekihara, K. and Nagarajan, S. (2008). *Adaptive Spatial filters For Electromagnetic Brain Imaging*. Berlin: Springer.

Sekihara, K. and Nagarajan, S.S. (2015). *Electromagnetic Brain Imaging: A Bayesian Perspective*. Switzerland: Springer International Publishing.

Selters, W.A. and Brackmann, D.E. (1977). Acoustic tumor detection with brainstem electric response audiometry. *Archives of Otolaryngology*, **103**, 181–187.

Shafer, J.P. (1995). Multiple hypothesis testing. *Annual Review of Psychology*, **46**, 561–584.

Shah, N.J., Worthoff, W.A., and Langen, K.J. (2016). Imaging of sodium in the brain: a brief review. *NMR in Biomedicine*, **29**, 162–174.

Smith, S.M., Jenkinson, M., Johansen-Berg, H., et al. (2006). Tract-based spatial statistics: voxelwise analysis of multi-subject diffusion data, *NeuroImage*, **31**, 1487–1505.

Spearman, C. (1904). The proof and measurement of association between two things. *American Journal of Psychology*, **15**, 72–101.

Spiegel, D.P., Byblow, W.D., Hess, R.F., and Thompson, B. (2013). Anodal transcranial direct current stimulation transiently improves contrast sensitivity and normalizes visual cortex activation in individuals with amblyopia. *Neurorehabilitation and Neural Repair*, **27**, 760–769.

Sporns, O. (2003). Graph theory methods for the analysis of neural connectivity patterns. In: Kötter, R. (Ed.) *Neuroscience Databases: A Practical Guide*, pp.171–185. New York: Springer US.

Sporns, O., Tononi, G., and Kötter, R. (2005). The human connectome: a structural description of the human brain. *PLoS Computational Biology*, **1**, e42.

Stam, C.J. (2014). Modern network science of neurological disorders. *Nature Reviews Neuroscience*, **15**(10), 683–695.

Stam, C.J., Nolte, G., and Daffertshofer, A. (2007). Phase lag index: assessment of functional connectivity from multi-channel EEG and MEG with diminished bias from common sources. *Human Brain Mapping*, **28**(11), 1178–1193.

Stejskal, E.O. and Tanner, J.E. (1965). Spin diffusion measurements: spin echoes in the presence of a time-dependent field gradient. *Journal of Chemical Physics*, **42**, 288–292.

Stockard, J.J., Stockard, J.E., and Sharbrough, F.W. (1978). Nonpathologic factors influencing brainstem auditory-evoked potentials. *American Journal of EEG Technology*, **18**, 177–193.

Tahmasian, M., Shao, J., Meng, C., et al. (2016). Based on the network degeneration hypothesis: separating individual patients with different neurodegenerative syndromes in a preliminary hybrid PET/MR Study. *Journal of Nuclear Medicine*, **57**, 410–415.

Thompson, B., Villeneuve, M.Y., Casanova, C., and Hess, R.F. (2012). Abnormal cortical processing of pattern motion in amblyopia: evidence from fMRI. *NeuroImage*, **60**, 1307–1315.

Tipper, C.M., Handy, T.C., Giesbrecht, B., and Kingstone, A. (2008). Brain responses to biological relevance. *Journal of Cognitive Neuroscience*, **20**, 879–891.

Tipper, C.M., Signorini, G., and Grafton, S.T. (2015). Body language in the brain : constructing meaning from expressive movement. *Frontiers in Human Neuroscience*, **9**, 450.

Tournier, J.D., Calamante, F., and Connelly, A. (2007). Robust determination of the fibre orientation distribution in diffusion MRI: non-negativity constrained super-resolved spherical deconvolution. *NeuroImage*, **35**, 1459–1472.

Tuch, D.S. (2004). Q-ball imaging. *Magnetic Resonance in Medicine*, **52**, 1358–1372.

Uhlhaas, P.J., Pipa, G., Lima, B., et al. (2009). Neural synchrony in cortical networks: history, concept and current status. *Frontiers in Integrative Neuroscience*, **3**(17), 1–19.

Van Veen, B.D. and Buckley, K.M. (1988). Beamforming: a versatile approach to spatial filtering. *IEEE Transactions on Signal Processing*, **5**, 4–24.

Van Veen, B.D., van Drongelen, W., Yuchtman, M., and Suzuki, A. (1997). Localization of brain electrical activity via linearly constrained minimum variance spatial filtering. *IEEE Transactions on Biomedical Engineering*, **44**, 867–880.

Varela, F., Lachaux, J.P., Rodriguez, E., and Martinerie, J. (2001). The brainweb: phase synchronization and large-scale integration. *Nature Reviews Neuroscience*, **2**(4), 229–239.

Varma, G., Duhamel, G., de Bazelaire, C., and Alsop, D.C. (2015). Magnetization transfer from inhomogeneously broadened lines: a potential marker for myelin. *Magnetic Resonance in Medicine*, **73**, 614–622.

Vavasour, I.M., Laule, C., Li, D.K.B., et al. (2009). Longitudinal changes in myelin water fraction in two MS patients with active disease. *Journal of the Neurological Sciences*, **276**, 49–53.

Vicente, R., Wibral, M., Lindner, M., and Pipa, G. (2011). Transfer entropy-a model-free measure of effective connectivity for the neurosciences. *Journal of Computational Neuroscience*, **30**, 45–67.

Villemagne, V.L. (2016). Amyloid imaging: past, present and future perspectives. *Ageing Research Reviews*, **30**, 95–106.

Villemagne, V.L., Fodero-Tavoletti, M.T., Masters, C.L., and Rowe, C.C. (2015). Tau imaging: early progress and future directions. *The Lancet Neurology*, **14**, 114–124.

Vinck, M., Oostenveld, R., van Wingerden, M., Battaglia, F., and Pennartz, C. (2011). An improved index of phase-synchronization for electrophysiological data in the presence of volume-conduction, noise and sample-size bias. *NeuroImage*, **55**, 1548–1565.

Wang, H.E., Bénar, C.G., Quilichini, P.P., Friston, K.J., Jirsa, V.K., and Bernard, C. (2014). A systematic framework for functional connectivity measures. *Frontiers in Neuroscience*, 8, 1–22.

Wang, T., Li, Q., Guo, M., et al. (2014). Abnormal functional connectivity density in children with anisometropic amblyopia at resting-state. *Brain Research*, 1563, 41–51.

Wang, X.J. (2010). Neurophysiological and computational principles of cortical rhythms in cognition. *Physiological Reviews*, 90, 1195–1268.

Wang, Y., Wang, Q., Haldar, J.P., et al. (2011). Quantification of increased cellularity during inflammatory demyelination. *Brain*, 134, 3590–3601.

Ward, K.M., Aletras, A.H., and Balaban, R.S. (2000). A new class of contrast agents for MRI based on proton chemical exchange dependent saturation transfer (CEST). *Journal of Magnetic Resonance*, 143, 79–87.

Ward, L.M. (2003). Synchronous neural oscillations and cognitive processes. *Trends in Cognitive Science*, 7, 553–559.

Ward, L.M. (2013). The thalamus: gateway to the mind. *Wiley Interdisciplinary Reviews Cognitive Science*, 4(6), 609–622

Weldeselassie, Y. and Hamarneh, G. (2007). DT-MRI segmentation using graph cuts. *SPIE Medical Imaging*, 6512–1K, 1–9.

Westin, C.F., Maier, S.E., Mamata, H., Nabavi, A., Jolesz, F.A., and Kikinis, R. (2002). Processing and visualization for diffusion tensor MRI. *Medical Image Analysis*, 6, 93–108.

Whitman, J.C., Takane, Y., Cheung, T.P.L., et al. (2016). Acceptance of evidence-supported hypotheses generates a stronger signal from an underlying functionally-connected network. *NeuroImage*, 127, 215–226.

Wibral, M., Rahm, B., Rieder, M., Lindner, M., Vicente, R., and Kaiser, J. (2011). Transfer entropy in magnetoencephalographic data: quantifying information flow in cortical and cerebellar networks. *Progress in Biophysics and Molecular Biology*, 105, 80–97.

Wiggermann, V., Torres, E.H., Vavasour, I.M., et al. (2013). Magnetic resonance frequency shifts during acute MS lesion formation. *Neurology*, 81, 211–218.

Wong, A.M. (2012). New concepts concerning the neural mechanisms of amblyopia and their clinical implications. *Canadian Journal of Ophthalmology*, 47, 399–409.

Woodward, T.S., Cairo, T.A., Ruff, C.C., Takane, Y., Hunter, M.A., and Ngan, E.T.C. (2006). Functional connectivity reveals load dependent neural systems underlying encoding and maintenance in verbal working memory. *Neuroscience*, 139, 317–325.

Woodward, T.S., Tipper, C., Leung, A., Lavigne, K.M., Sanford, N., and Metzak, P.D. (2015). Reduced functional connectivity during controlled semantic integration in schizophrenia: a multivariate approach. *Human Brain Mapping*, 36, 2949–2964.

Woodward, T.S., Leong, K., Sanford, N., Tipper, C.M., and Lavigne, K. M. (2016). Altered balance of functional brain networks in schizophrenia. *Psychiatry Research*, 248, 94–104.

Wright, A.D., Jarrett, M., Vavasour, I., et al. (2016). Myelin water fraction is transiently reduced after a single mild traumatic brain injury—a prospective cohort study in collegiate hockey players. *PloS One* 11(2), e0150215.

Yang, C.I., Uang, M.L., Huang, J.C., Wan, Y.L., Tsai, R.J.F., Wai, Y.Y., and Liu, H.L. (2003). Functional MRI of amblyopia before and after levodopa. *Neuroscience Letters*, 339, 49–52.

Ye, A.X., Leung, R.C., Schäfer, C.B., Taylor, M.J., and Doesburg, S.M. (2014). Atypical resting synchrony in autism spectrum disorder. *Human Brain Mapping*, 35(12), 6049–6066.

Ye, A.X., AuCoin-Power, M., Taylor, M.J., and Doesburg, S.M. (2015). Disconnected neuromagnetic networks in children born very preterm. *NeuroImage Clinical*, 9, 376–384.

Yoshinaga-Itano, C., Sedey, A.L., Coulter, D.K., and Mehl, A.L. (1998). Language of early- and later-identified children with hearing loss. *Pediatrics*, **102**, 1161–1171.

Yoshinaga-Itano, C., Coulter, D.K., and Thomson, V. (2000). The Colorado newborn hearing screening project: effects on speech and language development for children with hearing loss. *Journal of Perinatology*, **20**(8 Pt 2), S132–137.

Zalesky, A., Fornitoa, A., and Bullmorec, E.T. (2010). Network-based statistic: Identifying differences in brain networks. *NeuroImage*, **53**, 1197–1207.

Zalesky, A., Fornito, A., Cocchi, L., Gollo, L.L., and Breakspear, M. (2014). Time-resolved resting-state brain networks. *Proceedings of the National Academy of Sciences of the United States of America*, **111**(28), 10341–10346.

Zhai, J., Chen, M., Liu, L., Zhao, X., Zhang, H., Luo, X., and Gao, J. (2013). Perceptual learning treatment in patients with anisometropic amblyopia: a neuroimaging study. *British Journal of Ophthalmology*, **97**, 1420–1424.

Chapter 3

Incidental findings: Current ethical debates and future challenges in advanced neuroimaging

Lorna M. Gibson, Cathie L.M. Sudlow, and Joanna M. Wardlaw

Introduction

Incidental findings (IFs) on neuroimaging are the subject of widespread debates, and with the increase in use of imaging across research, clinical, and commercial sectors, there is a pressing need to address knowledge gaps to inform the development of appropriate strategies for their management.

Incidental findings may be defined as "observations of potential clinical significance unexpectedly discovered in healthy subjects or in patients recruited to any imaging research study, and unrelated to the purpose or variables of the study" (Illes et al. 2006). However, with the recent surge of interest in and experience of managing participants with IFs, IFs can no longer be deemed unexpected, and should be anticipated (Medical Research Council & Wellcome Trust 2014; Bunnik & Vernooij 2016). Instead, an IF may be defined as "a finding concerning an individual research participant that has potential health or reproductive importance and is discovered in the course of conducting research but is beyond the aims of the study" (Wolf et al. 2008). The increasing use of imaging as a research tool, a clinical diagnostic tool, and within the commercial sector increases the likelihood of detecting IFs, and necessitates the development of a robust evidence base to inform policies for detecting, classifying, and feeding back information on IFs to individuals, and evaluation of the feasibility and impact of those policies.

In this chapter, we will describe the scale of the problem of IFs; summarize current knowledge on the prevalence and determinants of prevalence of IFs, the impact of IFs on key stakeholders, and the expectations of the public; and highlight gaps in the evidence that will inform debates on the management of IFs. While we focus on IFs detected during research imaging, we also highlight issues related to the management of IFs detected during clinical and commercial imaging.

Scale of the problem of incidental findings

Our understanding of the scale of the problem of IFs is informed by reports of recent and ongoing large imaging research studies, and studies of the prevalence of IFs.

Large imaging research projects

Population-based research projects are underway that include imaging of large subsets of their participants. Two of the largest projects are the UK Biobank and the German National Cohort. The UK Biobank is a cohort of 500,000 British adults aged 40–69 years at the time of recruitment that began in 2006 and ended in 2010. The cohort participants have been extensively characterized using data from questionnaires, blood samples, physical measurements, and cognitive tests, and data on health-related outcomes will also be generated via linkage to health records (Sudlow et al. 2015). The UK Biobank data resource will image 100,000 of its participants using brain, cardiac, and abdominal magnetic resonance imaging (MRI), carotid ultrasound, and dual-energy X-ray absorptiometry (Matthews & Sudlow 2015), and will generate the world's largest multimodal imaging dataset. The German National Cohort is collecting questionnaire and medical examination data and biological samples from 200,000 population-based participants aged 20–69 years, and will perform whole-body MRI in a subset of 30,000 of these individuals (German National Cohort (GNC) Consortium 2014).

Whole-body MRI has already been completed in 2500 population-based Study of Health in Pomerania participants (Hegenscheid et al. 2013), and multiple large-scale, population-based, neuroimaging-only initiatives are also complete (Ikram et al. 2011; Wardlaw et al. 2011; Smith et al. 2015; Stephan et al. 2015). The population-based, longitudinal Rotterdam Study introduced neuroimaging for all participants in 2005. It performed imaging in random subsets in 1995 and in 1999, and had obtained 5886 brain MRI scans as of January 2011(Ikram et al. 2011). Brain MRI was performed in 803 out of 10,455 participants enrolled in the Canadian Prospective Urban Rural Epidemiological (PURE) study in order to investigate associations between small vessel disease and cognition (Smith et al. 2015). Brain MRI was also performed in 1923 participants from the Three City Study, a French longitudinal cohort study that recruited participants in 1999–2001 and was designed to investigate dementia and cognitive impairment and vascular risk factors (Stephan et al. 2015). The Lothian Birth Cohort 1936 includes 1091 individuals who reside in Edinburgh, were born in 1936, and underwent the Scottish Mental Survey general intelligence test in 1947 (Deary et al. 2007). The age 70 years cohort underwent medical interviews, physical examination, fitness testing, and cognitive testing, and answered questionnaires on personality, quality of life, and diet (Deary et al. 2007). Three years later, surviving members of the cohort were invited for retesting and brain MRI in order to provide imaging data that could be used to investigate associations between structural brain variables and cognitive function (Wardlaw et al. 2011).

Prevalence of incidental findings on neuroimaging

Current knowledge about the prevalence and types of IFs on brain imaging conducted in asymptomatic populations is informed by a recent systematic review. From a meta-analysis of 16 studies of 19,559 apparently asymptomatic people, the pooled prevalence of neoplastic IFs on brain MRI was 0.7% (95% confidence interval (CI) 0.5–1.0%), and in 15 studies of 15,559 people the pooled estimate of non-neoplastic IFs was 2.0% (95% CI 1.1–3.1%) (Morris et al. 2009). Additional data from 2000 participants from the Rotterdam Scan Study estimated that 94.6% of 45–59-year-olds and 98.9% of people aged 75 and over had white matter hyperintensities, and 7.2% of the total cohort had silent infarcts (Vernooij et al. 2007).

In contrast, our knowledge of the prevalence of IFs on neuroimaging in patient cohorts is relatively limited. Estimates of IFs on brain MRI performed in patients is informed by small (range n = 12–931, mean n = 274) single-center studies using either 1.5 or 3.0 Tesla scanners, with estimates ranging from 7.9% to 62.5% (Wahlund et al. 1989; Lubman et al. 2002; Papanikolaou et al. 2010; Powell & Choa 2010; Koppelmans et al. 2011; Parker et al. 2011; Jaremko et al. 2012; Khandanpour et al. 2013; Proctor et al. 2013; Winston et al. 2013; Trufyn et al. 2014). While MRI is the most commonly used modality for performing brain imaging in research, computed tomography (CT) remains the workhorse of clinical brain imaging. A retrospective study of 3000 CT brain examinations performed in trauma patients identified 30 cases with IFs, which included 11 with tumors (of which three were benign meningiomas and three lipomas, the remainder being malignant) (Eskandary et al. 2005). Of 2195 cases who underwent cranial CT angiography for neurological symptoms, trauma, or known intracranial lesions, 39 (1.8%) had incidental unruptured intracranial aneurysms, who went on either to treatment or annual surveillance (Agarwal et al. 2014).

Estimates of the prevalence of IFs on spinal MRI also vary widely, from 8.4% to 99.1% (Cheung et al. 2009; Kamath et al. 2009; Park et al. 2011; Quattrocchi et al. 2013; Cieszanowski et al. 2014). The prevalence of suspected malignant lesions on spinal MRI in a population-based sample of 666 participants was 0.4% (Cieszanowski et al. 2014). Lumbar spine MRI in 1043 population-based participants in China demonstrated lumbar disc degenerative changes in 42% of participants aged 18–30 years, rising to 88% in participants aged 50–55 years (Cheung et al. 2009). Such a high prevalence of IFs in this population-based sample raises the question of which IFs could be deemed normal.

The wide ranges in estimates of prevalence of IFs on brain and spine imaging may be partly explained by variation in investigators' case definitions and sources of data for IFs. Prevalence estimates may be influenced by different decisions to classify normal variants as IFs, such as mega cisterna magna (Khandanpour et al. 2013), or to exclude lesions that were not deemed clinically relevant (Koppelmans et al. 2011). Data on IFs may be sourced from IFs documented in available reports, or by dedicated review of images to identify IFs, and use of these different data sources likely also affects estimates of the prevalence of IFs. For example, the prevalence of extra-spinal IFs described in clinical reports of 3000 lumbar spine MRI examinations performed in patients aged 16–91 years (mean age 59.3,

1453 (48.4%) men) was 7.2% (217/3000), which rose almost by a factor of ten (68.7%, 2060/3000) on dedicated review of images to identify IFs. The lower prevalence in clinical reports may partly be explained by a lack of reporting of nonserious IFs in clinical practice, such as diverticulosis (330/351 unreported), renal cysts (702/732 unreported), or postsurgical appearances (13/16 unreported), but also lack of reporting of a smaller number of potentially clinically significant IFs that could be reasonably expected to be well characterized, including abdominal aortic aneurysms (10/11 unreported), and enlarged lymph nodes (36/38 unreported) (Quattrocchi et al. 2013).

Potentially clinically significant incidental findings

A meaningful and useful definition of IFs should include a statement about clinical significance, or potential health importance (Wolf et al. 2008). Data on the prevalence of potentially clinically significant IFs could be used to inform potential research participants' decisions about enrolling in neuroimaging research, and to inform the design of study policies for detecting, handling, and feeding back information about IFs to participants and their healthcare providers. Some findings, such as sinus opacification, are common and of little or no clinical significance (Katzman et al 1999; Alphs et al. 2006). It could be argued that IFs should only be fed back to individuals if they are clinically significant, or if feedback will provide a strong net benefit (Wolf et al. 2008). These might include IFs likely to be life-threatening, where adverse consequences could be avoided or treated, or that could inform reproductive decision-making (Wolf et al. 2008). Therefore, accurate data on the prevalence of clinically significant IFs is paramount.

However, the clinical significance of an IF cannot always be accurately determined. Research imaging protocols are designed to optimize the collection of data relevant to the study's particular question or questions. In contrast, clinical imaging is optimized to demonstrate a suspected diagnosis. Given the range of potential IF diagnoses, optimization to detect each of these would not be practical, affordable, or, due to the duration of scan time, tolerable to participants. There have been some calls for the addition of clinical-type imaging sequences in research imaging protocols as standard (Milstein 2008), which should reduce both the rate of false positives (Hegenscheid et al. 2013) (i.e., IFs that are finally diagnosed as nonserious conditions, artifacts, already known conditions, or are not confirmed on follow-up) and the necessity for public healthcare systems to meet the costs of further investigation and diagnosis. However, adding additional sequences to research protocols may yield significant extra cost to the study and be impractical to implement, for example, if scanning time within a center is simply not available, or by lengthening scan time beyond that which is tolerable to participants (Booth et al. 2010). Indeed, adding clinical standard imaging sequences to research protocols may be perceived as screening. This would generally be deemed inappropriate given the potential range of IFs that may be detected, with their differing clinical consequences, variation in the necessity for or availability of treatment options, and limited evidence that such treatments positively benefit morbidity or mortality (Zealley 2015).

The difficulty of handling IFs of potential clinical significance is that lack of a firm diagnosis is likely to prompt further investigations. There is limited reporting of systematic follow-up of participants with indeterminate IFs. At the very least, potential imaging research participants should be made aware of the possibility of demonstrating an IF and that further tests and referrals may be required to accurately characterize it (Illes et al. 2006, 2008). In the event that an IF is finally diagnosed as not clinically significant, in retrospect, any follow-up that occurred may be deemed unnecessary. Such unnecessary follow-up may have exposed the individual to harm, for example, through the use of ionizing radiation for further imaging, or an invasive procedure, without clinical benefit to the individual. Unnecessary follow-up may also impact resources of individuals and society, in terms of incurring costs of unnecessary healthcare, unnecessary use of healthcare services, and time away from employment.

The clinical significance of some IFs is currently uncertain due to lack of data on their natural history. A systematic review of the clinical significance of white matter hyperintensities found associations with risk of incident stroke in six studies of the general population (hazard ratio 3.1, 95% CI 2.3–4.1) and with risk of incident dementia in three studies of the general population (hazard ratio 2.9, 95% CI 1.3–6.3), leading the authors to conclude that the identification of white matter hyperintensities should prompt assessment for other risk factors for stroke and dementia in such individuals (Debette & Markus 2010). However, the methods of measuring white matter hyperintensities varied between studies and risk of incident stroke and dementia with varying volumes was not presented. Whether it is the presence versus absence, or a particular proportion of white matter affected by hyperintensities that is associated with increased risk of incident disease remains unclear. In addition, there is limited evidence of effective preventive therapies for stroke and dementia in people with white matter hyperintensities. With further research, judgment of the clinical significance of these and many other IFs is likely to evolve.

In contrast to lesions that can be detected by visual assessment of images, abnormalities of brain tissue volumes, which require segmentation and computation, tend not to be considered as IFs. Recent studies suggest an association between small hippocampal volumes with a first episode of major depressive disorder (Cole et al. 2011), and with genetic risk for Alzheimer disease (Lupton et al. 2016), but there is not enough evidence yet to support advising assessment for depressive symptoms or cognitive impairment in participants identified to have hippocampal volumes below a particular level. A small hippocampal volume may not be readily apparent on visual assessment of images, and measuring hippocampal volume seems to have a more "active" nature of seeking IFs compared to visual assessment alone. However, it is not clear how performing measurements on imaging to identify IFs differs from the act of, for example, calculating a cognitive test score, on which a low score on a clinically validated instrument may also be deemed an IF. Researchers should consider how to handle abnormal results generated by any data collection activity or measure that may have clinical implications.

Further studies of the natural history of different IFs and robust evidence for benefits of treatment are needed. In the meantime, for IFs in which the clinical significance and benefit of treatments is not known, the usefulness of disclosing information on these to participants is likely limited. Care must be taken with using current estimates of prevalence of IFs, as variability of case definitions will affect any overall prevalence, and instead, data on the prevalence of potentially clinically significant IFs should guide designs of policies to handle IFs, and inform potential participants of the likelihood of identifying IFs. Judgments of the clinical significance of IFs should be informed by empirical data generated from long-term follow-up and studies of the natural history of findings.

Determinants of prevalence

With better understanding of the subject, imaging, and image reader characteristics that affect prevalence of potentially serious IFs, better predictions will be possible about the likely prevalence of IFs generated by newly planned studies. Such estimates could then inform the design of appropriate consent materials and policies for detecting, classifying, and feeding back IFs in imaging studies, which are now mandated by major funding bodies including the Medical Research Council and the Wellcome Trust (Farrar & Savill 2014; Medical Research Council & Wellcome Trust 2014).

Subject characteristics, such as age and sex, may affect the prevalence of IFs. Such characteristics are selection criteria likely to be considered at study planning stage, therefore data on the prevalence of IFs in different groups will inform future studies' estimates of the likely prevalence of IFs they will generate, and help them to design appropriate consent materials and policies to handle IFs. Data from the Rotterdam Scan Study participants show that the prevalence of asymptomatic brain infarcts increases with age, from 30/750 (4.0%) in those aged 45–59, to 47/257 (18.3%) in those aged 75–97. Median volume of white matter lesions also increased, from 1.8 mL in those aged 45–59 years, 3.1 mL in those aged 60–74 years, to 7.7 mL in those aged 75–97 years (Vernooij et al. 2007). In contrast, the prevalence of intracranial aneurysms did not differ significantly between participants within these age groups (Vernooij et al. 2007).

The prevalence of IFs may vary between men and women. In the Lothian Birth Cohort, brain imaging of 700 adults aged 73 years (SD 1.5 years) found that the prevalence of IFs was higher in men (134/368, 36.4% of men versus 89/332, 26.8% of women, p = 0.007) (Sandeman et al. 2013). However, these prevalence estimates include IFs that may not be clinically significant, such as arachnoid cysts. Of 1113 consecutively imaged healthy adults aged 22–84 years, no significant difference in distribution of men and women was observed between groups with and those without IFs (Hoggard et al. 2009).

Less is known about the influence of imaging and imaging reader characteristics on the prevalence of IFs. Sequences and combinations of sequences may influence the perceptibility and characterization, and therefore the frequency and diagnostic certainty, of IFs. T2-weighted sequences are helpful in distinguishing between solid and liquid lesions, and the use of a contrast agent can make some lesions more conspicuous, or demonstrate the extent of a lesion more completely. The volume of tissue coverage and the resolution of images will also likely increase the prevalence of IFs

(Maxwell et al. 2015). In one study, there was no difference in the prevalence of IFs in healthy volunteers imaged on a 1.5 T (31/374) compared to a 3.0 T (15/151, p = 0.6) MRI scanner (Hoggard et al. 2009).

The effect of reader experience on the prevalence of IFs is not known. Reader experience is likely to affect prevalence for two reasons: first, perception of a demonstrable lesion may increase with experience; second, interpretation of the clinical significance of a demonstrated lesion may be more accurate by people with medical, or more specifically, radiological training. For these reasons, it is natural to assume that consultant radiologists would be the most appropriate readers of imaging for IFs, being doctors who are highly trained in perception and interpretation of images. However, there is a lack of evidence to inform this view. While radiologists may detect many IFs, the proportion of these that turn out to be serious is not known, and the sensitivity and specificity of radiologists for detecting potentially clinically serious IFs on nonoptimized, nondiagnostic-grade research imaging is not known. Logistically, given the shortage of staff in clinical practice (The Royal College of Radiologists 2011), it is not feasible for radiologists to report all research imaging, particularly now that much larger imaging studies are being conducted. One solution may be a network of local reporting radiologists although, where reported, there appears to have been limited uptake of such services by local researchers (Cramer et al. 2011). While inexperienced readers may be expected to fail to perceive or misinterpret some IFs, this is not always the case. For example, correct classification of a set of 14 CT brain scans, performed in patients with suspected stroke documented in the European Co-operative Acute Stroke Study trial book that details imaging findings as normal or abnormal did not differ significantly between observers classed as experienced (consultant neuroradiologists, consultant neurologists with an interest in stroke, consultant stroke physicians), and those classed as inexperienced (consultant neurologists, general practitioners, trainee physicians) (68% correctly classified compared to 63%, p = 0.3) (Wardlaw et al. 1999). To date, there have been no published head-to-head comparisons of detection of IFs, or those which are finally diagnosed as serious, between readers of differing experience, such as trainee radiologists, radiographers, or nonmedically trained scientists, to inform study policies on the detection of IFs.

Research is needed on participant, imaging, and reader-related determinants of prevalence, to inform estimates of the likely prevalence of potentially serious IFs when planning new neuroimaging studies. Such data could be generated from prespecified subgroup analyses of participants, head-to-head comparisons between different neuroimaging studies, or between different types of readers within the same study.

The impact of incidental findings

Incidental findings have the potential to impact participants, health services, researchers, and the wider scientific community. Data are needed on the extent of these impacts, and on how impact may vary between different methods of handling IFs, to inform the design of appropriate policies and participant consent materials.

Feedback of an IF could potentially impact an individual's emotional well-being and financial situation, by affecting their employment and ability to obtain and the cost of various forms of insurance. The Study of Health in Pomerania surveyed 471 participants who received notification of an IF, with data returned from 405 respondents. Of these, 46.9% (190/405) reported some distress while waiting for potential notification of an IF, with 9.9% reporting strong distress during this period (Schmidt et al. 2013). Nondisclosure of an IF to an insurer would likely render the insurance void (Apold & Downie 2011), with similar issues being faced by patients and participants undergoing genetic tests (Thomson 1998; Raeburn 2002). However, robust data on the actual impact of receiving feedback of an IF on insurance and employment are not available, or how these different types of impact may vary by age, gender, or employment status.

Neuroimaging IFs have the potential to impact the use of a variety of services and generate associated costs. These include clinic attendance at family doctor and specialist hospital clinics; additional investigations such as blood tests, further imaging, and invasive diagnostic procedures; medical and surgical interventions; and hospital admissions. Arranged follow-up may vary by availability across different health services and healthcare systems, and by the diagnostic certainty of the IFs generated from research imaging. For example, a study that includes sequences close to diagnostic standard may reasonably detect an uncomplicated arachnoid cyst and confidently advise that further imaging is not required. However, an intracranial mass is unlikely to be characterized to the extent required for clinical management without dedicated, optimized, post-contrast, clinical diagnostic imaging, clinical review of the participant, and discussion between doctors from multiple clinical specialties. Use of gadolinium-based MRI contrast agents is not without risk, with 0.2–3.3 per 1000 injections resulting in an immediate adverse reaction, and a risk of severe reaction, such as cardiac arrest, in approximately 1 in 40,000 injections (Prince et al. 2011). Although rare, the risk of such severe reactions to MRI contrast is offset by the benefits of diagnostic information in patient settings, but given the lack of benefit to research participants, MRI contrast is generally not used in research studies, particularly in healthy volunteers. Given the potential costs, feedback of IFs should perhaps only occur when there is potential benefit for the individual, to minimize unnecessary use of resources, particularly within publicly funded healthcare systems (Medical Research Council & Wellcome Trust 2014).

No published assessment of the service use and cost of follow-up of IFs takes into account all of the potential sources described previously. However, particular types of IFs may lend themselves to standardized clinical management pathways, the costs of which may be assessed. For example, at the University of Utah Hospital, Salt Lake City, United States, follow-up of patients with an incidentally detected nonfunctioning microadenoma includes blood tests, repeat MRI after 1 year with radiologist report, and consultation with a physician, at a total estimated cost of $6062—$6215 USD per patient (Randall et al. 2010). Decision analytical modeling has also been used to estimate costs of follow-up of IFs. In a decision analysis study of different strategies for handling incidentally detected intracranial aneurysms on functional MRI (fMRI), cost-effectiveness was found to vary with patient sex and family history, and experience of reporter (Sadatsafavi et al. 2010). For example, imaging review by a nonspecialist, such as a nonmedical postdoctoral fellow, with confirmation or refutation of detected abnormalities by a radiologist, was not deemed cost-effective (Sadatsafavi et al. 2010). However, decision analytical modeling is limited by the availability of data to inform the model, and in this case the sensitivity and specificity of detection of incidental aneurysms of a variety of sizes by a nonspecialist reader was informed by expert opinion rather than empirical data.

It is extremely difficult to estimate the cost of follow-up and any cost-benefit of detecting IFs given the variety of possible detected IFs, variation in clinical practice between centers, necessity to tailor management to the patient, and differences in costs incurred between countries and even between different institutions within those countries. The magnitude of increase of service use and cost by people who receive feedback of an IF, compared to people of similar age, sex, and comorbidities, might usefully inform our understanding of the cost and service use impact on healthcare systems, but to our knowledge, such comparisons have not been published.

The impact of IFs on a particular research study and on the wider scientific community, in terms of cost of administering an IFs handling policy, and the impact on public trust and involvement in research, is also unknown. Procedures for handling IFs vary widely among research imaging centers (The Royal College of Radiologists 2011). Recent guidance and funding bodies have forced researchers to develop policies to handle IFs that must be submitted during the application for funding (Medical Research Council & Wellcome Trust 2014). It is emphasized that such plans be appropriate, in terms of time and cost (Medical Research Council & Wellcome Trust 2014), but there is currently little guidance or training to assist researchers with the development and administration of IFs policies (Wardlaw et al. 2015). Methods and descriptions of the results of current policies for handling IFs have been published (Hegenscheid et al. 2013; Sandeman et al. 2013; Boutet et al. 2016), but head-to-head comparisons of different methods of handling IFs, the time and cost they take to administer, and their impacts, have not. Appropriate handling of IFs and clear communication with participants is particularly important for longitudinal cohort studies, where attrition of the cohort could seriously bias the results of research arising from the study. In the burgeoning era of big data and biobanking, research data, including biobanked images, are generated for and are being used by secondary researchers, that is, those not involved in the process of the data collection. Such "secondary research" on biobanked images may yield IFs, potentially some years after the initial data collection. Researchers must consider how to handle IFs detected during secondary research (Richardson & Cho 2012), and the implications for consent processes. Biobank participants may give broad consent at the time of data collection for their biobanked data to be used for as-yet-unspecified future projects, or alternatively, participants may be contacted to give repeat consent to use their biobanked data at the start of each new project, so-called dynamic consent (Steinsbekk et al. 2013). Either way, the implications for participants and the financial and reputational costs to studies of detecting an IF months or even years after imaging should be considered.

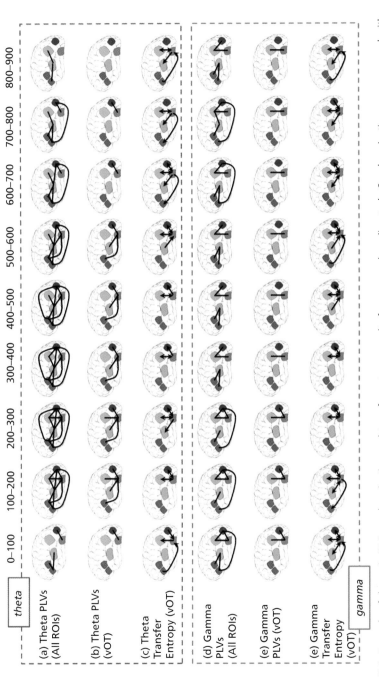

Figure 2.1 Example of dynamic EEG synchronization and transfer entropy results from a word reading task. Synchronization was measured with phase locking value (PLV), and narrow band transfer entropy was measured using an adaptation of the Kullback entropy (for vOT to/from other areas only). Theta band is 3–7 Hz and gamma band is 35–45 Hz. Black lines (arrows) indicate significant interaction (directed information flow) between the connected reading-related brain areas during the time period (ms after stimulus presentation) indicated at the top of the figure. vOT indicates ventral occipital-temporal brain region, the putative word-form area.

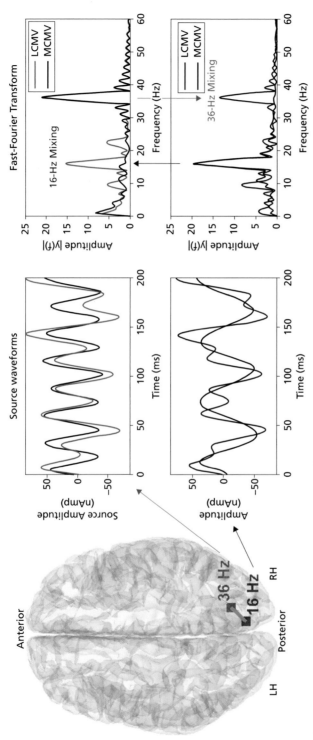

Figure 2.2 EEG localization and connectivity accuracy. Left panel shows a simulated source location (blue square) that had only a 16-Hz oscillation and a neighboring (within 15 mm) source location (red square) that had only a 36-Hz oscillation. Source mixing of 16- and 36-Hz oscillations were clearly visible in the traditional beamformer (LCMV) reconstructed source waveforms (blue and red lines; middle panel). Amplitude spectra (right panels) showed that both red and blue LCMV sources contained energies at 16 and 36 Hz (i.e., source mixing). Because of this source mixing, phase-coherence between the sources existed and functional connectivity was falsely revealed (green line; left panel). Our MCMV beamformer (black lines), however, prevented source mixing. MCMV sources only had 16-Hz (lower panels' black lines) or 36-Hz oscillations (upper panels' black lines). Obviously, false-connections were not found for the MCMV beamformer.

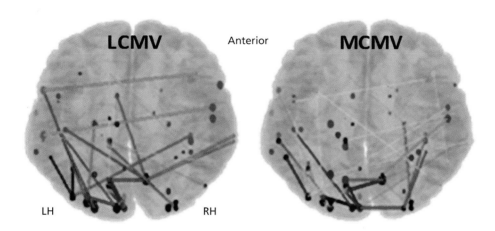

Figure 2.3 Challenge in EEG connectivity maps. Traditional LCMV-beamformer connectivity map (left) for alpha-band (9–10 Hz) revealed highly synchronized (red lines) local couplings in visual cortices at 380 ms after letter onset. However, MCMV-beamformer connectivity map (right) revealed strongly desynchronized (blue lines) couplings. The MCMV-connectivity maps fit better with known event-related desynchronizations that occur after a visual stimulus onset for this band and latency. Thus, the MCMV map is likely the real map.

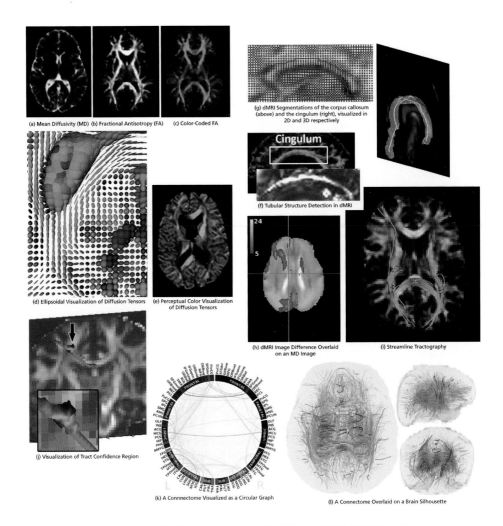

(a) Mean Diffusivity (MD) (b) Fractional Antisotropy (FA)   (c) Color-Coded FA

(g) dMRI Segmentations of the corpus callosum (above) and the cingulum (right), visualized in 2D and 3D respectively

Cingulum

(f) Tubular Structure Detection in dMRI

(d) Ellipsoidal Visualization of Diffusion Tensors

(e) Perceptual Color Visualization of Diffusion Tensors

(h) dMRI Image Difference Overlaid on an MD Image

(i) Streamline Tractography

(j) Visualization of Tract Confidence Region

(k) A Connnectome Visualized as a Circular Graph

(l) A Connectome Overlaid on a Brain Silhouette

Figure 2.4 Examples of visualizations associated with diffusion MRI (dMRI) data. Some visualizations highlight the brain's diffusion properties (a–e), while others highlight its geometric properties (f–h), while still others highlight brain connectivity (i–l). See the text for a description of each visualization technique: (a) mean diffusivity (MD); (b) fractional anisotropy (FA); (c) color-coded FA; (d) ellipsoidal visualization of diffusion tensors; (e) perceptual color visualization of diffusion tensors; (f) tubular structure detection in dMRI; (g) dMRI segmentation of corpus callosum (above) and the cingulum (right), visualized in 2D and 3D respectively; (h) dMRI image difference overlaid on MD image; (i) streamline tractography; (j) visualization of tract confidence regions; (k) a connectome visualized as a circular graph; (l) a connectome overlaid on a brain silhouette.

Future financial and reputational costs relating to IFs should be considered by any research imaging study, not just biobanks. Months or years after imaging, participants may be diagnosed with conditions and may return to study investigators wanting to know whether those conditions were in fact visible on their research scan. While there are no documented cases of this within the literature, the scenario is not unrealistic, and in such a case a participant may seek legal action or financial compensation, or both, for a clinically significant finding if it was deemed to have been missed on a research scan.

To inform the design of future neuroimaging studies, further data are needed on the impact of IFs, and how these may vary between different policies to handle IFs. Studies of neuroimaging could build in robust data collection on the impact of IFs on a participant's emotional well-being, employment and finances, clinical follow-up, and final diagnoses in order to generate evidence that will help inform future potential research participants' consent. Data on the impact on health service use and cost of different policies of handling IFs would inform future policymaking. It is in the interests of science that processes for handling IFs are sensible, appropriate, and clearly communicated to potential and actual research participants, and to their healthcare providers, in order to maintain trust in and cooperation with researchers and the wider scientific community.

Public expectations

Maintaining trust in research is paramount as the progress of clinical science depends on the participation of human volunteers. However, several studies show that participants' expectations of research may be unrealistic, with research imaging being associated with clinical diagnosis, and consequently raising expectations that such images will be assessed by a competent, clinical professional (The Royal College of Radiologists 2011).

The National Child Development Study is a longitudinal cohort study of approximately 17,000 people born during 1 week of 1958 in England, Scotland, and Wales. In 2010, 133 of the participants living in and around Cambridge were invited to complete further cognitive assessments, of whom assessments were conducted in 45 people, and of these, 7 had previously undergone MRI for clinical diagnosis, and none for research. A qualitative interviewer explained the idea that taking part in a fMRI study "could reveal something that suggests that there could be a more serious problem [...] in about one in 40 scans." The researcher explained people may have different views on the feedback that they would like or expect, and offered the 45 respondents five options: not to take part in a fMRI research study; take part regardless of the provision or not of feedback; take part only if all potential problems were fed back; take part only if serious, treatable problems were fed back; or take part only if the study provided no feedback. Of 45 potential imaging research participants, 40.9% reported that they would only participate in a study that provided feedback on all IFs, with a minority (11.4%) responding that they would only participate in a study that fed back only IFs that were considered serious and treatable (Brown & Knight 2010).

Findings were similar in participants who had already undergone neuroimaging for research. Participants were invited to complete a web-based questionnaire about IFs on neuroimaging if they had acted as healthy controls for fMRI research studies at one of two imaging facilities in the United States. Over 90% of 104 respondents (mean age 25 years, 76/104 (73.1%) were undergraduate or postgraduate students) who had undergone research brain imaging wished to be informed of any IFs, regardless of its clinical significance, and 94% wished to be informed by a member of the research team, and 54% expected any abnormality present to be detected although 84% did not expect images to be reviewed by a doctor (Kirschen et al. 2006). These expectations would be extremely costly to meet, particularly in large studies, both in terms of researchers' time and study costs. Furthermore, given the range of the definition of IFs by some investigators to include postsurgical appearances (Sandeman et al. 2013), and normal variants such as cavum septum pellucidum (Lubman et al. 2002), this approach is likely to be of limited benefit to participants. While many potential research participants might anticipate that their research imaging would be reviewed by a competent professional, the provision for review of images varies considerably throughout centers, ranging from full review of images by radiologists, to no review of images by radiologists at all (The Royal College of Radiologists 2011).

Members of the public perceive the benefits of receiving feedback on IFs to outweigh potential disadvantages. Detecting a condition before symptoms arise is seen as advantageous by potential research participants as it may allow early intervention, decision-making around family planning, or prompt screening for conditions in relatives (Wellcome Trust et al. 2012). This may be founded on the belief that to detect a disease before it becomes symptomatic must confer a survival benefit, a concept introduced to the public domain by widespread public health messages related to screening programs (McCaffery et al. 2016). As discussed previously, research imaging is not a substitute for clinical assessment and diagnostic imaging, therefore IFs demonstrated on research imaging may be incompletely characterized, and may require clinical follow-up and generate anxiety. However, despite considering that feedback of IFs may generate shock and worry, and that this may be unnecessary if an IF turned out to be a false positive (i.e., a nonserious condition, artifact, condition which turns out already to be known, or a finding which is not confirmed on follow-up investigations), there remains overwhelming support from the UK public to receive feedback on IFs (Wellcome Trust et al. 2012).

People who have health problems but are unable to afford healthcare within a private healthcare system might view research as an opportunity to have a free scan (Wardlaw et al. 2015). Not only would such participants be at risk of failing to understand consent, researchers may find themselves handling more potentially serious IFs than expected. A variation of "research tourism" may also exist within publicly funded healthcare systems, whereby participants may view research imaging either as a faster means to access health services in order to investigate specific health concerns, or as an opportunity to gain peace of mind, which would not warrant investigation within such publicly funded health services (Townsend & Cox 2013). Within developing countries, participation in research may be driven by either therapeutic misconception, defined as the mistaken belief by participants that they are receiving clinical care, or in order to access treatments and health services as ancillary care, whereby being a research participant enables them to access free treatments or diagnostic services for problems unrelated to the purposes of the study (Mfutso-Bengo et al. 2008). However, participants may also recognize the potential detrimental effects of receiving health-related information from research studies. For example, research participants in private healthcare settings in the United States have raised concerns that disclosure of results of genetic research may impact health insurance coverage (Murphy & Thompson 2009). To our knowledge, there are no data on the association of participants' motivation to take part in neuroimaging and prevalence of IFs, or direct comparisons of the prevalence of IFs in studies performed within different healthcare systems. Researchers should consider participants' existing access to healthcare and potential motivations for participating in research when preparing information materials and when obtaining consent and managing participants' expectations about handling of IFs.

When developing policies for handling IFs, researchers should consider balancing feasibility, in terms of cost and time of administration, with the potential impacts of IFs, and with participants' expectations. Such expectations may need to be managed during the consent process, and evaluation of understanding consent is needed to ensure this process is adequate.

Looking to the future

Despite the progress to date, difficult issues around handling IFs remain, due to the complex and variable nature of IFs and the lack of robust, empirical data to inform practice. The usefulness of current estimates of the prevalence of IFs on neuroimaging is limited by variability in case definitions, methods of determining clinical significance, and lack of information on natural history. Research on participant, imaging, and reader-related determinants of prevalence are also needed to inform our understanding of the likely prevalence of potentially serious IFs in any new neuroimaging study. Data are needed on the extent of the impact of IFs on individuals, health services, and the scientific community, and how impact may vary between different methods of handling IFs. Together, such data would build an evidence base to inform the design of feasible policies for handling and feeding back IFs that do not unduly encumber individual research studies or health services and that are acceptable to the multiple stakeholders. Such an evidence base would also facilitate informed consent of participants prior to participating in neuroimaging research studies while managing their expectations and preserving public trust in medical research.

References

Agarwal, N., Gala, N.B., Choudhry, O.J., et al. (2014). Prevalence of asymptomatic incidental aneurysms: a review of 2685 computed tomographic angiograms. *World Neurosurgery*, **82**(6), 1086–1090.

Alphs, H.H., Schwartz, B.S., Stewart, W.F., and Yousem, D.M. (2006). Findings on brain MRI from research studies of occupational exposure to known neurotoxicants. *American Journal of Roentgenology*, **187**(4), 1043–1047.

Apold, V.S. and Downie, J. (2011). Bad news about bad news: the disclosure of risks to insurability in research consent processes. *Accountability in Research*, **18**(1), 31–44.

Booth, T.C., Jackson, A., Wardlaw, J.M., Taylor, S.A., and Waldman, A.D. (2010). Incidental findings found in "healthy" volunteers during imaging performed for research: current legal and ethical implications. *British Journal of Radiology*, **83**(990), 456–465.

Boutet, C., Vassal, F., Celle, S., et al. (2016). Incidental findings on brain magnetic resonance imaging in the elderly: the PROOF study. *Brain Imaging and Behavior*. Advance online publication. doi:10.1007/s11682-016-9519-4

Brown, M.A. and Knight, H.M. (2010). *Attitudes Towards Participating in fMRI Studies Amongst Participants in a Birth Cohort Study*. CLS Working Paper 2010/8. Available from: http://www.cls.ioe.ac.uk/library-media/documents/CLS_WP_2010_8_.pdf [accessed January 30, 2017].

Bunnik, E.M. and Vernooij, M.W. (2016). Incidental findings in population imaging revisited. *European Journal of Epidemiology*, **31**(1), 1–4.

Cheung, K., Chan, D., Song, Y., et al. (2009). Prevalence and pattern of lumbar magnetic resonance imaging changes in a population study of one thousand forty-three individuals. *Spine*, **34**(9), 934–940.

Cieszanowski, A., Maj, E., Kulisiewicz, P., et al. (2014). Non-contrast-enhanced whole-body magnetic resonance imaging in the general population: the incidence of abnormal findings in patients 50 years old and younger compared to older subjects. *PLoS One*, **9**(9), e107840.

Cole, J., Costafreda, S.G., McGuffin, P., and Fu, C.H.Y. (2011). Hippocampal atrophy in first episode depression: a meta-analysis of magnetic resonance imaging studies. *Journal of Affective Disorders*, **134**(1–3), 483–487.

Cramer, S.C., Wu, J., Hanson, J.A., et al. (2011). A system for addressing incidental findings in neuroimaging research. *NeuroImage*, **55**(3), 1020–1023.

Deary, I.J., Gow, A.J., Taylor, M.D., et al. (2007). The Lothian Birth Cohort 1936: a study to examine influences on cognitive ageing from age 11 to age 70 and beyond. *BMC Geriatrics*, **7**, 28.

Debette, S. and Markus, H.S. (2010). The clinical importance of white matter hyperintensities on brain magnetic resonance imaging: systematic review and meta-analysis. *BMJ*, **341**, p. c3666.

Eskandary, H., Sabba, M., Khajehpour, F., and Eskandari, M. (2005). Incidental findings in brain computed tomography scans of 3000 head trauma patients. *Surgical Neurology*, **63**(6), 550–553.

Farrar, J. and Savill, J. (2014). UK funders' framework for health-related findings in research. *The Lancet*, **383**(9933), 1954–1955.

German National Cohort (GNC) Consortium (2014). The German National Cohort: aims, study design and organization. *European Journal of Epidemiology*, **29**(5), 371–382.

Hegenscheid, K., Seipel, R., Schmidt, C.O., et al. (2013). Potentially relevant incidental findings on research whole-body MRI in the general adult population: frequencies and management. *European Radiology*, **23**(3), 816–826.

Hoggard, N., Darwent, G., Capener, D., Wilkinson, I.D., and Griffiths, P.D. (2009). The high incidence and bioethics of findings on magnetic resonance brain imaging of normal volunteers for neuroscience research. *Journal of Medical Ethics*, **35**(3), 194–199.

Ikram, M.A., van der Lugt, A., Niessen, W.J., et al. (2011). The Rotterdam Scan Study: design and update up to 2012. *European Journal of Epidemiology*, **26**(10), 811–824.

Illes, J., Kirschen, M.P., Edwards, E., et al. (2006). Incidental findings in brain imaging research. *Science*, **311**(5762), 783–784.

Illes, J., Kirschen, M.P., Edwards, E., et al. (2008). Practical approaches to incidental findings in brain imaging research. *Neurology*, **70**(5), 384–390.

Jaremko, J.L., MacMahon, P.J., Torriani, M., et al. (2012). Whole-body MRI in neurofibromatosis: incidental findings and prevalence of scoliosis. *Skeletal Radiology*, 41(8), 917–923.

Kamath, S., Jain, N., Goyal, N., Mansour, R., and Mukherjee, K. (2009). Incidental findings on MRI of the spine. *Clinical Radiology*, 64(4), 353–361.

Katzman, G.L., Dagher, A.P., and Patronas, N.J. (1999). Incidental findings on brain magnetic resonance imaging from 1000 asymptomatic volunteers. *JAMA*, 282(1), 36–39.

Khandanpour, N., Hoggard, N., and Connolly, D.J. (2013). The role of MRI and CT of the brain in first episodes of psychosis. *Clinical Radiology*, 68(3), 245–250.

Kirschen, M.P., Jaworska, A., and Illes, J. (2006). Subjects' expectations in neuroimaging research. *Journal of Magnetic Resonance Imaging*, 23(2), 205–209.

Koppelmans, V., Schagen, S.B., Poels, M.M., et al. (2011). Incidental findings on brain magnetic resonance imaging in long-term survivors of breast cancer treated with adjuvant chemotherapy. *European Journal of Cancer*, 47(17), 2531–2536.

Lubman, D.I., Velakoulis, D., McGorry, P.D., et al. (2002). Incidental radiological findings on brain magnetic resonance imaging in first-episode psychosis and chronic schizophrenia. *Acta Psychiatrica Scandinavica*, 106(5), 331–336.

Lupton, M.K., Strike, L., Hansell, N.K., et al. (2016). The effect of increased genetic risk for Alzheimer's disease on hippocampal and amygdala volume. *Neurobiology of Aging*, 40), 68–77.

Matthews, P.M. and Sudlow, C. (2015). The UK Biobank. *Brain*, 138(12), 3463–3465.

Maxwell, A.W.P., Keating, D.P., and Nickerson, J.P. (2015). Incidental abdominopelvic findings on expanded field-of-view lumbar spinal MRI: frequency, clinical importance, and concordance in interpretation by neuroimaging and body imaging radiologists. *Clinical Radiology*, 70(2), 161–167.

McCaffery, K.J., Jansen, J., Scherer, L.D., et al. (2016). Walking the tightrope: communicating overdiagnosis in modern healthcare. *BMJ*, 352, p. i348.

Medical Research Council and Wellcome Trust (2014). *Framework on the Feedback of Health-Related Findings in Research*. Available from: https://www.mrc.ac.uk/documents/pdf/mrc-wellcome-trust-framework-on-the-feedback-of-health-related-findings-in-researchpdf/ [accessed January 30, 2017].

Mfutso-Bengo, J., Ndebele, P., Jumbe, V., et al. (2008). Why do individuals agree to enrol in clinical trials? A qualitative study of health research participation in Blantyre, Malawi. *Malawi Medical Journal*, 20(2), 37–41.

Milstein, A.C. (2008). Research malpractice and the issue of incidental findings. *Journal of Law, Medicine and Ethics*, 36(2), 356–360.

Morris, Z., Whiteley, W.N., Longstreth, W.T., et al. (2009). Incidental findings on brain magnetic resonance imaging: systematic review and meta-analysis. *BMJ*, 339, b3016.

Murphy, E. and Thompson, A. (2009). An exploration of attitudes among black Americans towards psychiatric genetic research. *Psychiatry*, 72(2), 177–194.

Papanikolaou, V., Khan, M.H., and Keogh, I.J. (2010). Incidental findings on MRI scans of patients presenting with audiovestibular symptoms. *BMC Ear, Nose and Throat Disorders*, 10, 6.

Park, H.-J., Jeon, Y.-H., Rho, M.-H., et al. (2011). Incidental findings of the lumbar spine at MRI during herniated intervertebral disk disease evaluation. *American Journal of Roentgenology*, 196(5), 1151–1155.

Parker, J., Wolansky, L.J., Khatry, D., Geba, G.P., and Molfino, N.A. (2011). Brain magnetic resonance imaging in adults with asthma. *Contemporary Clinical Trials*, 32(1), 86–89.

Powell, H.F. and Choa, D. (2010). Should all patients referred for magnetic resonance imaging scans of their internal auditory meatus be followed up in ENT clinics? *European Archives of Oto-Rhino-Laryngology*, 267(9), 1361–1366.

Prince, M.R., Zhang, H., Zou, Z., Staron, R.B., and Brill, P.W. (2011). Incidence of immediate gadolinium contrast media reactions. *American Journal of Roentgenology*, 196(2), W138–W143.

Proctor, R.D., Gawne-Cain, M.L., Eyles, J., Mitchell, T.E., and Batty, V.B. (2013). MRI during cochlear implant assessment: should we image the whole brain? *Cochlear Implants International*, 14(1), 2–6.

Quattrocchi, C., Giona, A., Di Martino, A., et al. (2013). Extra-spinal incidental findings at lumbar spine MRI in the general population: a large cohort study. *Insights into Imaging*, 4(3), 301–308.

Raeburn, S. (2002). Implications of genetic testing for the insurance industry: the UK example. *Community Genetics*, 5(2), 102–109.

Randall, B., Kraus, K., Simard, M., and Couldwell, W. (2010). Cost of evaluation of patients with pituitary incidentaloma. *Pituitary*, 13(4), 383–384.

Richardson, H.S. and Cho, M.K. (2012). Secondary researchers' duties to return incidental findings and individual research results: a partial-entrustment account. *Genetics in Medicine*, 14(4), 467–472.

Sadatsafavi, M., Marra, C., Li, D., and Illes, J. (2010). An ounce of prevention is worth a pound of cure: a cost-effectiveness analysis of incidentally detected aneurysms in functional MRI research. *Value in Health*, 13(6), 761–769.

Sandeman, E.M., Hernandez, M.D.C.V., Morris, Z., et al. (2013). Incidental findings on brain MR imaging in older community-dwelling subjects are common but serious medical consequences are rare: a cohort study. *PLoS One*, 8(8), e71467.

Schmidt, C.O., Hegenscheid, K., Erdmann, P., et al. (2013). Psychosocial consequences and severity of disclosed incidental findings from whole-body MRI in a general population study. *European Radiology*, 23(5), 1343–1351.

Smith, E.E., O' Donnell, M., Dagenais, G., et al. (2015). Early cerebral small vessel disease and brain volume, cognition, and gait. *Annals of Neurology*, 77(2), 251–261.

Steinsbekk, K.S., Kare Myskja, B., and Solberg, B. (2013). Broad consent versus dynamic consent in biobank research: is passive participation an ethical problem? *European Journal of Human Genetics*, 21(9), 897–902.

Stephan, B.C.M., Tzourio, C., Auriacombe, S., et al. (2015). Usefulness of data from magnetic resonance imaging to improve prediction of dementia: population based cohort study. *BMJ*, 350, p. h2863.

Sudlow, C., Gallacher, J., Allen, N., et al. (2015). UK Biobank: an open access resource for identifying the causes of a wide range of complex diseases of middle and old age. *PLoS Medicine*, 12(3), e1001779.

The Royal College of Radiologists (2011). *Management of Incidental Findings Detected During Research Imaging*. London: The Royal College of Radiologists.

Thomson, B. (1998). Time for reassessment of use of all medical information by UK insurers. *The Lancet*, 352(9135), 1216–1218.

Townsend, A. and Cox, S.M. (2013). Accessing health services through the back door: a qualitative interview study investigating reasons why people participate in health research in Canada. *BMC Medical Ethics*, 14, 40.

Trufyn, J., Hill, M.D., Scott, J.N., et al. (2014). The prevalence of incidental findings in multiple sclerosis patients. *Canadian Journal of Neurological Sciences*, 41(1), 49–52.

Vernooij, M.W., Ikram, M.A., Tanghe, H.L., et al. (2007). Incidental findings on brain MRI in the general population. *New England Journal of Medicine*, 357(18), 1821–1828.

Wahlund, L.O., Agartz, I., Saaf, J., Wetterberg, L., and Marions, O. (1989). Magnetic resonance tomography in psychiatry-clear benefits for health care services. *Lakartidningen*, 86(46), 3991–3994.

Wardlaw, J.M., Dorman, P.J., Lewis, S.C., and Sandercock, P.A. (1999). Can stroke physicians and neuroradiologists identify signs of early cerebral infarction on CT? *Journal of Neurology, Neurosurgery and Psychiatry*, **67**(5), 651–653.

Wardlaw, J.M., Bastin, M.E., Valdes Hernandez, M.C., et al. (2011). Brain aging, cognition in youth and old age and vascular disease in the Lothian Birth Cohort 1936: rationale, design and methodology of the imaging protocol. *International Journal of Stroke*, **6**(6), 547–559.

Wardlaw, J.M., Davies, H., Booth, T.C., et al. (2015). Acting on incidental findings in research imaging. *BMJ*, **351**, h5190.

Wellcome Trust, Medical Research Council, and Opinion Leader (2012). *Assessing Public Attitudes to Health Related Findings in Research*. London: Opinion Leader.

Winston, G.P., Micallef, C., Kendell, B.E., et al. (2013). The value of repeat neuroimaging for epilepsy at a tertiary referral centre: 16 years of experience. *Epilepsy Research*, **105**(3), 349–355.

Wolf, S.M., Lawrenz, F.P., Nelson, C.A., et al. (2008). Managing incidental findings in human subjects research: analysis and recommendations. *Journal of Law, Medicine and Ethics*, **36**(2), 219–248.

Zealley, I.A. (2015). Re: Incidental abdominopelvic findings on expanded field-of-view lumbar spinal MRI: frequency, clinical importance, and concordance in interpretation by neuroimaging and body imaging radiologists. *Clinical Radiology*, **70**(6), 679.

Chapter 4

Vulnerability, youth, and homelessness: Ethical considerations on the roles of technology in the lives of adolescents and young adults

Niranjan S. Karnik

Adolescence and digital culture

Case scenario: Justin[1] is a 14-year-old male with a prior history of attention deficit hyperactivity disorder. He has been managed on low-dose stimulants for some time, and is emotionally stable and doing well academically. He is social, joins sports, and enjoys spending time with his friends. To this point, his parents have limited his use of social media and Justin does not own a smartphone but is eager to have one like his friends. His parents are consulting his physician to ascertain whether giving him a smartphone is wise at this age, and they are debating whether to allow him to start using social media. They express fears about cyberbullying, online stalkers, and other elements that they have read and heard about in regards to social media and teens.

There is understandable anxiety about the nature of technology and its impacts on the lives of young people. Over 90% of adolescents now use social media with a significant number using it on a regular basis and a marked subset reporting nearly constant connection (Lenhart 2015). Prior to the development of social media and mobile technologies, television and radio were triggers for similar waves of anxiety. It may be true that every new societal technology change prompts concern for the impact on adolescents and young people. Much of this might be traced to a developmental truism—that young people seek to differentiate themselves from the prior generation. All children go through a phase of development wherein they seek to establish themselves as autonomous individuals. One requisite to establish an individual's identity is to differentiate and separate from other adults to whom an individual has been connected up to this point in life. For most children, this process of individuation involves creating separation from parents or guardians. Technology is often an easy way to achieve this goal. Young people often claim a cultural space in new technology, and this space can be difficult for parents and other adults to enter. Adults have less incentive to use these spaces for their everyday needs as they have already established patterns of work, communication, cultural meaning, and social interaction with their peers. Youth are not only early adopters of technology for the desire to be in novel spaces, but also precisely due to the fact that their parents and other adults are less likely to be on these media.

When I present talks on social media and adolescence, I often take an informal poll of the providers in the room to see which platforms they prefer. Facebook, Twitter, and Instagram often lead the current line-up. I then assure the audience that if they are on these platforms their cool factor is likely diminishing and their youngsters are likely now moving to Snapchat and other newer and more advanced media.

Nevertheless, there are certain heuristics that are useful in considering how adolescents use new social media. One notion that is important to dispel is the traditional distinction between "real" and "virtual." For people who grew up prior to the advent of social media, the separation between the real world—that space where they live, work, and have social interactions—feels very different from the virtual world built within social media platforms. For adolescents who have grown up in the period of an omnipresent social media landscape, these worlds are not necessarily separate: they are continuous—the physical world connecting with new social media in a rather seamless way. The communications that they have in the presence of other people feel very much parallel to those online. Hence, why it is easy to observe young people sending text messages and links to one another while physically being in the same space.

One of the issues that youth, generally, are apt to not consider is that in the virtual space people may not be who or what they seem. They see this space as a means to connect with their peers and friends. As youth emerge from childhood which is traditionally in contact with friends and family that are relatively narrowly circumscribed by their schools and families' social lives, they are then in a developmentally sensitive period of time where they trust others around them to be like their prior contacts. The virtual space of social media introduces the potential for people to represent themselves in an alternative way. Briefly, they may not be who they say they are, and can instead create a character.

This fluidity of expression and projection of self presents a risk for adolescents in that other adolescents or adults might pose as peers or age-appropriate friends online when in fact they may be different. The stories of these types of behaviors have been told many times in news stories across multiple platforms and need not be repeated here. The counter to this nefarious behavior is to give adolescents an opportunity to explore themselves. Adolescents can alter and change their hobbies, interests, gender identity, and even claim a different sexual orientation within the space of social media. Such exploration can happen under their own name or handle, or could be created with another pseudonym. While individuals interacting with a somewhat fictionalized character might justifiably claim that they are being misled, the youth projecting the character could be learning what their choices might engender in terms of relationships, people, and connections. They may be learning the types of people that share their interests and may be using the experience as a chance for personal growth.

Clinically, I often liken virtual spaces for adolescents and young people to extensions of their rooms. The computer or smartphone is a portal into their lives and their communication with others is often taking place in the privacy of their rooms or other unobserved venues. This leads to a situation where young people often believe they have privacy and in fact can interact in ways that expose them very publically. Social media creates a new form of public space or commons. Many people believe that when they post text, video, or photos that only their friends can see this. The degree to which these communiques are limited to a small group is directly related to the degree to which privacy settings have been enabled. For many adolescents, privacy is not a key concern, especially when weighed against peer acceptance and social popularity. Hence, young people are more likely to unknowingly take risks in the space and be unaware of the ways in which their private information might be used or misused.

Digital literacy—the education and teaching about the ways that digital media work and impact our lives—is an absent part of present educational practices. Society desperately needs a pedagogical approach to digital culture that starts at an early age because we can easily witness children as young as 2 or 3 years of age interacting with iPads and iPhones.

Mental health, adolescence, and social media

Case scenario: Debbie is a 16-year-old female who presents to her therapist with significant anxiety and depressive symptoms. She is presently entering her junior year in high school and admits that her academic performance has suffered over the past year. She has had increasing difficulties getting along with a group of popular girls in her class. Debbie is unsure as to why she has become a target but notes that they regularly post comments, taunts, and threats about her on Facebook and other social media. She is dysphoric and tearful as she describes the ongoing patterns, and notes that she has largely kept this information to herself and a few friends. She has not yet informed her parents or her teachers who rarely use these media sources.

The data on the use of online social media and development of mental health problems is a vexed area. While many popular commentators like the actor Wentworth Miller have argued for a link between depression, suicide, and other behavioral problems, and the online environment (Grant 2016), the academic research in this area is more subtle and less clear. One of the few reviews of this literature concluded that the methods and time lags for this field lead to an inability to draw conclusions on the direct impact of social media on mental health (Pantic 2014). While the majority of studies reviewed did find correlations between amount of time spent on social media and symptoms of depression, low self-esteem, and online addiction, the directionality of these relationships is unclear. A recent study by Lin and colleagues examined 1787 young adults and determined that there appeared to be a relationship between the amount of social media use and degree of depressive symptoms (Lin et al. 2016). Nevertheless, the authors were unable, using their methodology, to assess the directionality of this effect. They speculate that there is a dose relationship to this effect but a number of questions remain. For example, does use of social media lead to mental health and behavioral issues or is the mental health problem driving the youth toward social media use? Correlational studies are unable to assess the directionality due to their cross-sectional nature. A set of longitudinal studies is needed to better understand the nature of these relationships.

One issue that has emerged is cyberbullying. This form of bullying has become an issue of deep concern for child mental health professionals (Sivashanker 2013) because of the multiple impacts that it has on youth in terms of depression and suicidal ideation (Pham & Adesman 2015; Mitchell et al. 2016; Ranney et al. 2016). In contrast to the broad impacts of social media on mental health, the specific impact of cyberbullying has been well documented as being at least as potent as in-person bullying or even more toxic because of the extent and remoteness of the attacks. In other words, while social media mitigates the physical component of bullying, the psychological component may become more toxic because of the frequency of attacks and the often relative isolation of the victim.

Cyberbullying of teens and adolescents emerges largely from the social context of education and peer relationships. The study of social relationships in schools and the sociology of education have been significant parts of the social science canon. The introduction of social media has likely changed this milieu but it also offers a new opportunity to study and understand adolescent social relationships. These connections or social networks may be adaptive and promote self-esteem or resiliency, and they might also be maladaptive as in the case of cyberbullying.

Ethically, studying adolescent social relationships is hard. Traditionally, these youth cannot provide consent because they lack adult decision-making capabilities. Most scholars in this area are required to obtain informed consent from the parents or guardians of adolescents, and then the adolescent may be allowed to opt out by virtue of an assent process. The challenge arises in that the youth who are most at risk of, or being, victimized may be the least likely to share their true experiences. As in the case outlined at the beginning of this section, Debbie has yet to inform adults. Is there a scenario where she is going to be willing to work with researchers who, depending on jurisdiction, may be required to report the bullying to authorities? Such disclosure would require that her parents learn about the ongoing issues. While this disclosure may be necessary and even therapeutic, the researcher is caught in a dilemma of facing likely refusal from the teen and also failing to capture the realities of what is happening in this context. Based on these structural issues, studies of adolescence and social media likely overestimate positive outcomes and underestimate negative effects. This observation bias cannot be easily overcome under current research ethics standards. Even anonymous online surveys that are used may suffer from a degree of observation bias and they are likely at risk for other types of bias due to their low response rates.

Autism spectrum disorders and social media

Case scenario: James is a 13-year-old boy with high-functioning autism. He attends regular classes at the local school and has a special education plan that provides additional support. He has been a victim of bullying in the past due to his autism but seems to have found a good circle of peers that are protective of him. He generally spends several hours per day online and lately has been spending time connecting with other youth with autism using online social media and chat forums. He tells his counselor that he prefers to talk through the computer because it produces less anxiety than talking to people directly.

Youth with autism often face a variety of challenges. The core illness is characterized by difficulties in social communication, stereotypic or repetitive behaviors, lack of emotional reciprocity, and, in extreme cases, difficulties in language development. Autism is now seen as a spectrum illness with youth having significant variation from mild to severe presentations. Recent data from the US Centers for Disease Control and Prevention suggest that approximately 1 in 150 children are on the autism spectrum (Kuehn 2007). This estimate was a marked change from prior studies of autism and some degree of alarm was sounded about an epidemic of autism. Most clinicians in the field believe that this trend reflects better assessment methods and education about the disorder, and a near-universal screening that is now undertaken by primary care physicians.

As with most populations, technology and social media use for adolescents with autism is a two-sided experience. On the one side, it can be a source of empowerment, connection, and support from other youth and families similarly impacted. In contrast, it can also be a space of further isolation and reduction of interaction, and a space where social growth is not as essential as in other spaces with people present. Current research indicates that youth with autism spectrum disorders spend the majority of their time watching television or playing video games, while a minority (13.2%) spend time on social media (Mazurek et al. 2012). Similarly, a study by Kuo and colleagues found that nearly 80% of adolescents with autism spectrum disorders report that they watch television for an average of 2 hours per day (Kuo et al. 2014). In addition, the authors found that 98% of adolescents in their sample used computers for an average of 5 hours per day. Kuo and colleagues also noted that they most frequently watched cartoons, played computer or video games, or visited websites that are about video games. They found that adolescents who visited social networking websites or received emails from friends reported more positive friendships (Kuo et al. 2014). Based on this research, it appears that a minority of youth with autism spectrum disorders use social media to connect with friends while a majority of these youth spend a significant amount of time watching television and playing games on the Internet.

Studying the use of social media by youth with autism is clearly a challenging subfield and an area worthy of further investigation. There is some potential in this area to develop interventions keyed to the social network and Internet environments that could help young people on the spectrum to develop better social skills, and reduce the impact of autism in their lives. The challenge will be balancing the autonomy that should be accorded to adolescents who are mildly impacted by autism, and their ability to make decisions for themselves or with minimal support from their parents or guardians, against the need to protect more significantly impacted youth who may be more vulnerable to risk in social networks. Consent and assent processes for these youth need to follow the patterns for adolescents more generally, but further consideration may need to be given to the particularities of subgroups of youth with autism.

Lesbian, gay, bisexual, and transgender youth and social media

Case scenario: Amy is a 17-year-old, male-to-female (MTF) transgendered youth who began her transition about 3 years ago. She was supported by her mother and encountered significant resistance from her father who continues to refuse to acknowledge her new gender identity. She has been meet-ing young men to date via a social networking profile. She practices safer sex inconsistently and is often pressured by her partners to have unprotected sex. As her primary care physician, you are now giving some consideration to prescribing pre-exposure prophylaxis to better protect her against HIV transmission.

Lesbian, gay, bisexual, and transgender (LGBT) youth are a subgroup of adolescents and young adults who deserve special attention. LGBT youth face multiple challenges including rejection or violence in the coming out process, increased risk of mental health and substance use disorders, and, depending on the LGBT subgroup, increased risk of sexually transmitted infections (STIs).

Mustanski and colleagues have published a series of papers on the ethics of research with LGBT youth (Mustanski 2011; Fisher & Mustanski 2014; Ybarra et al. 2016). They generally argue that the health disparities facing these populations merit a change of approach to ethical standards for considering studies in this area. They note multiple challenges for investigators in this field including heightened perceptions of risk, cum-bersome requirements for parental consent, and a lack of expertise concerning this pop-ulation on IRBs.

Elements of Mustanski's critique are accurate especially in regard to the lack of experi-enced investigators in this subfield who are willing to serve on institutional review boards (IRBs). The lack of knowledgeable researchers who are willing to expend time and effort to serve on an IRB is likely a result of the small size of this field. In addition, Mustanski accurately outlines why adolescents under the age of 18 would potentially be at greater risk in the event of inadvertent disclosure to parents and others who may not be aware of the youth's sexual orientation or gender identity.

Mustanski's analysis of risk estimation for LGBT youth is more problematic. He and his colleagues argue in the absence of evidence that subjective estimates of suicide, sub-stance use, and HIV/STI risks are overestimated and that most research should be con-sidered minimal risk because of the techniques used by social and behavioral scientists. The difficulty with this risk analysis is that it ignores several lines of research that indicate higher risks of mental health issues, substance use, suicidality, and HIV/STI transmission. In order to mitigate these risks, researchers need to develop clear assessment plans for addressing mental health and substance use disorders. In regards to LGBT youth, it would be prudent to involve clinicians who routinely work with patients. Many researchers have clinical credentials but do not have active clinics or lead clinical teams. In the conduct of research with LGBT youth, attention needs to be paid to providing care to patients when the need is clear or the risks are high. This approach is more likely to create a safer pro-tocol for LGBT youth, and also maximize the potential for advancement of the research without having adverse events. Such an approach will likely also better address concerns that IRBs may raise and serve to convince reviewers who only see risk, that the research teams have protocols for working with these youth that acknowledge risks and address them directly.

Homeless youth

Case scenario: Gene is a 15-year-old male who was kicked out of his home when he was age 13 and announced to his family that he was gay. He ran away from his rural environment in a southern US state and hitchhiked his way to Chicago. He lived on the streets for several months, and survived by engaging in the sex trade and sex work. He would often sleep with people in exchange for a night in a bed, and a shower. Some of these people would sometimes give him money, and he often found himself using drugs with them, especially alcohol, marijuana, and methamphetamines. He comes to the homeless drop-in shelter at times to meet with a case manager as he is hoping to get clean and go to college.

Homeless youth are a major social problem with estimates varying in the range of 1.6 to 2 million youth annually experiencing homelessness (Edidin et al. 2011). The social forces that drive youth on to the streets include domestic violence, perceived LGBT identity, and economic instability or poverty. All of the groups described earlier in this chapter including youth with mental health issues, autistic youth, and LGBT youth are at disproportionately higher risk of becoming homeless. Most homeless youth in the US come out of the fractured foster care system where young people who turn age 18 "age-out" and are ineligible to remain in their foster homes or receive ongoing state support. This transition has been identified repeatedly as an ongoing source of homeless youth and small programs have been established to stem this trend but none of these have achieved the growth and coverage that are needed to reduce youth homelessness.

In this section I will review some of the homelessness research I have pursued with an interdisciplinary team across three institutions. In this manner, I hope to highlight practices we have developed or used to underpin our research from a strong ethically informed perspective. Our research team has been working with homeless youth for the past 7 years in Chicago and 2 years before that in San Francisco. We initially started by chronicling the mental health and substance use challenges faced by this population (Quimby et al. 2012; Castro et al. 2014). These studies demonstrated that this was a population in high need. Our research also began to show that many homeless youth use the Internet and social media at high rates (Goldberg et al. 2013). This finding was in line with our colleague Eric Rice and his team's findings in Los Angeles-based homeless youth (Rice et al. 2011).

We sought to develop an intervention based on cell phones that would enable us to better reach these youth and to intervene at an earlier time point. We discussed our project with staff members at several of the community-based organizations with which we have partnered with in the past, and also undertook some qualitative research to guide our work (Goldberg et al. 2013). As we assembled our research protocol we initially thought we would deliver a trauma-informed sleep intervention using cell phones that we would give homeless youth. We then launched a series of five focus groups with shelter-based youth (Kaiser et al. in press). The youth in the focus groups urged us to change the approach to our study. They agreed that sleep and trauma are major issues but they felt that their more pressing issues related to anxiety and depression. They advised us to direct our treatment toward these symptoms.

In the course of our protocol creation we heard questions from various individuals about three issues including (1) distribution of high-quality smartphones, (2) appropriate amount of compensation, and (3) monitoring youth using tracking software. First, we opted to give these youth the best smartphones we could afford to buy or have donated. This logic was challenged because many observers assume that homeless youth will use any technology that is given or donated to them. Our experience was different in that we discovered that homeless youth value the best and most current technology, and will go to great lengths to direct funds that they get toward their phone or obtaining cell/data time. The rationale is simple—when you lack a permanent physical address, a virtual one is even more essential and important. Homeless youth live through their phones and use them as means to search for resources, keep connected to friends and family, and to entertain themselves. Thus, from our perspective, giving a high-quality smartphone was an important part of our study.

Compensation for underserved or impoverished populations is always a point of concern for review committees. The balance point is to give compensation that is not coercive and yet acknowledges the time and effort that the youth put into working with our research team. While our distribution of a smartphone and 6 months of data plan are a major part of the compensation, we also knew that the measures we were asking youth to complete for follow-up require time and effort on their part. We opted to give gift cards to a local store where they could obtain food or clothing.

When considering compensation, there is also a justice issue that arises. It would be ethically inappropriate to give an impoverished youth less compensation than a wealthier youth who was participating in the same study or the same type of study. This issue arises even though the power to coerce may be greater for the impoverished youth. One of the issues that my team has debated repeatedly is how to value time in a research study that is spent completing measures or other questionnaires. There is no easy answer and hence why I have avoided specifying amounts—the answers to these questions will vary by the study, the topics under investigation, the time required, and locality. We work closely with our community-based partners to define appropriate compensation levels. We believe that it is important to repeatedly visit these issues, and consider them at regular intervals and in tandem with local human subject protection committees.

Our study makes use of novel software from Northwestern University's Center for Behavior Intervention Technologies (CBITs) called Purple Robot. This software is able to send comprehensive user data from the phone and tell us locations, contacts, texts, emails, and which apps are used. There are also algorithms that the CBITs teams have developed using smartphone sensor data to estimate factors like sleep and mobility. All of these capabilities give the appearance of a very invasive study. In the course of our focus groups, we reviewed the use of this software with the homeless youth. They asked good questions about what the data would be used for, and also requested that the software be removed at the end of the study period. The focus groups supported the use of this approach if the findings would help other homeless youth. We included specifics about the monitoring element of the study in our consent forms that all study participants sign.

Our study is presently in progress and to this point our team has not faced any major problems with the consent process. We are providing telephone-based brief psycho-therapy to homeless youth and also enabling them to contact their study therapist via a custom-designed text messaging interface. We hope that these interventions will increase the likelihood of homeless youth engaging in mental health services.

Conclusion

Vulnerable youth and social media is a new area of research. In our experience work-ing with homeless youth, there has been an exploration of appropriate research practices and approaches which we tried to define within existing standards and in tandem with community-based stakeholders. Further research is needed in developing appropriate ethics standards for research in this population. In particular, it is clear that adolescents and youth have a different view of privacy than prior generations. They are more apt to share deeply personal elements of themselves on social media and do so in a way that they see as empowering. Further research should examine the intersection of privacy and social media for young people in order to understand this space better and to develop appropriate standards for researchers to follow.

Finally, IRBs are advised to undertake to educate their members about the needs of these youth populations so as to be better positioned to judge protocols in this subfield. Knowledge about adolescents, LGBT youth, the impact of homelessness and trauma on brain development, and technology are all presently lacking on most IRBs. These com-mittees should also seek external consultation when protocols fall outside of their areas of expertise and draw from individuals who have worked in these subfields.

Note

1. All clinical examples in this chapter are created for educational purposes only and do not reflect any patients directly.

References

Castro, A.L., Gustafson, E.L., Ford, A.E., et al. (2014). Psychiatric disorders, high-risk behaviors, and chronicity of episodes among predominantly African American homeless Chicago youth. *Journal of Health Care for the Poor and Underserved*, 25(3), 1201–1216.

Edidin, J.P., Ganim, Z., Hunter, S.J., and Karnik, N.S. (2011). The mental and physical health of homeless youth: a literature review. *Child Psychiatry and Human Development*, 43(3), 354–375.

Fisher, C.B. and Mustanski, B. (2014). Reducing health disparities and enhancing the responsible conduct of research involving LGBT youth. *The Hastings Center Report*, 44(Suppl 4), S28–31.

Goldberg, S., Karnik, N.S., and Hunter, S.J. (2013, October). *Technology Usage and Social Support in a Population of Inner-City Homeless Youth*. Presented at the 60th Annual Meeting of the American Academy of Child & Adolescent Psychiatry, Orlando, Florida.

Grant, S. (2016, March 29). Wentworth Miller opens up about depression and suicidal thoughts in candid Facebook post. *MTV News*. Available from: http://www.mtv.com/news/2858392/wentworth-miller-facebook-weight-gain-memes-depression-suicide-the-lad-bible/

Kaiser, E., Zalta, A.K., Boley, R.A., Glover, A., Karnik, N.S., and Schueller, S.M. (in press). Exploring the potential of technology-based mental health services for homeless youth: a qualitative study. *Psychological Services*.

Kuehn, B.M. (2007). CDC: autism spectrum disorders common. *JAMA*, **297**(9), 940.

Kuo, M.H., Orsmond, G.I., Coster, W.J., and Cohn, E.S. (2014). Media use among adolescents with autism spectrum disorder. *Autism*, **18**(8), 914–923.

Lenhart, A. (2015). *Teens, Social Media & Technology Overview 2015*. Washington, DC: Pew Research Center.

Lin, L.Y., Sidani, J.E., Shensa, A., et al. (2016). Association between social media use and depression among U.S. young adults. *Depression and Anxiety*, **33**(4), 323–331.

Mazurek, M.O., Shattuck, P.T., Wagner, M., and Cooper, B.P. (2012). Prevalence and correlates of screen-based media use among youths with autism spectrum disorders. *Journal of Autism and Developmental Disorders*, **42**(8), 1757–1767.

Mitchell, S.M., Seegan, P.L., Roush, J.F., Brown, S.L., Sustaíta, M.A., and Cukrowicz, K.C. (2016). Retrospective cyberbullying and suicide ideation: the mediating roles of depressive symptoms, perceived burdensomeness, and thwarted belongingness. *Journal of Interpersonal Violence*. Advance online publication. doi:10.1177/0886260516628291

Mustanski, B. (2011). Ethical and regulatory issues with conducting sexuality research with LGBT adolescents: a call to action for a scientifically informed approach. *Archives of Sexual Behavior*, **40**(4), 673–686.

Pantic, I. (2014). Online social networking and mental health. *Cyberpsychology, Behavior and Social Networking*, **17**(10), 652–657.

Pham, T. and Adesman, A. (2015). Teen victimization: prevalence and consequences of traditional and cyberbullying. *Current Opinion in Pediatrics*, **27**(6), 748–756.

Quimby, E.G., Edidin, J.P., Ganim, Z., Gustafson, E., Hunter, S.J., and Karnik, N.S. (2012). Psychiatric disorders and substance use in homeless youth: a preliminary comparison of San Francisco and Chicago. *Behavioral Sciences*, **2**(3), 186–194.

Ranney, M.L., Patena, J.V., Nugent, N., et al. (2016). PTSD, cyberbullying and peer violence: prevalence and correlates among adolescent emergency department patients. *General Hospital Psychiatry*, **39**, 32–38.

Rice, E., Lee, A., and Taitt, S. (2011). Cell phone use among homeless youth: potential for new health interventions and research. *Journal of Urban Health*, **88**(6), 1175–1182.

Sivashanker, K. (2013). Cyberbullying and the digital self. *Journal of the American Academy of Child and Adolescent Psychiatry*, **52**(2), 113–115.

Ybarra, M.L., Prescott, T.L., Phillips, G.L. 2nd, Parsons, J.T., Bull, S.S., and Mustanski B. (2016). Ethical considerations in recruiting online and implementing a text messaging-based HIV prevention program with gay, bisexual, and queer adolescent males. *Journal of Adolescent Health*, **59**(1), 44–49.

Chapter 5

The neuroethical future of wearable and mobile health technology

Karola V. Kreitmair and Mildred K. Cho

Introduction

Imagine you come home to your apartment after a long day at work. It is a comfortable 72°F (22°C), just as you like it, because your smart phone has alerted your smart thermostat to your impending arrival and activated the heating in time for you to walk through the door. Your wrist-worn heart rate monitor, the size of a wrist watch, which also functions as an activity tracker, GPS device, and pulse monitor, detects a slightly elevated heart rate, and the sweat-sensing fabric of your smart shirt measures cortisol in your sweat. These data are sent wirelessly to your sound system, which chimes in with a soothing Chopin Nocturne. You just now remember that you had run out of almond milk this morning, but fortunately your smart fridge has ordered some online and it is now sitting in a little box in front of your door. Also waiting for you is your salmon dinner, delivered from the local Italian restaurant. Your blood nutrient wearable device, a noninvasive Band-Aid-like patch that continuously measures your glucose, iron, electrolytes, and proteins in your blood, has deduced that your high-density lipoprotein is low, but that your glucose is high. The computer in your smart phone matched this information with all nearby delivery-enabled restaurants, cross-referenced it with your taste preferences—which it learned via machine learning from accumulated data from your smart plates—then deprioritized recently consumed meals to ensure an element of novelty and made the optimal choice. After you finish your dinner the cortisol levels in your sweat decline and the smart music serenades you with a more vigorous Bach as you use your smart tooth brush to apply precisely the right amount of pressure during your 2-minute teeth cleaning. Tooth brushing movements, times, and patterns are also relayed to your designated smart phone app, where they are displayed in graphic format. Finally, you sink into your bed, which is fully enabled via your smart mattress pad to track your heart rate and sleep cycles, and offset disturbing light levels or detected noise pollution with soothing orange melatonin-activating light and lullabies of circadian rhythm-mimicking frequencies.

Of course, you need not imagine it, since (almost) all of this technology already exists.[1] Wearable and mobile technology, along with the Internet of Things (IoT), has firmly established its foothold in virtually every arena of twenty-first century Western life. In addition to

the near ubiquitous activity trackers (such as Fitbit°, Jawbone°, and Misfit°) that are estimated to constitute an over $5 billion USD industry by the year 2019 (Lamkin 2015), there
is a huge array of other wearable technology on both the consumer and industrial markets,
with new modalities seemingly appearing on a weekly basis. This technology includes smart
clothing, headphones, shoes, wristbands, ski goggles, and jewelry, as well as smart mouth
guards, ovulation trackers, personal trainers, and asthma monitors (Wearable Technologies
2016). Devices may be unobtrusive patches worn on the skin, sensors woven into clothing,
or even ingestible sensors swallowed inside pills (Belknap et al. 2013; Banerjee et al. 2015;
Lehmkuhle et al. 2015). Technologies can mine location, activity, breathing rate and volume,
sleep patterns, heart rate, sweat, blood chemistry, and even brain waves for biometric data
(Lupton 2013).

Furthermore, these wearable devices are inextricably connected with mobile technology, smartphones, smart watches, tablets, and personal computers in intelligent ways.
Through mobile apps and mobile platforms that allow the integration of data from multiple apps, biometric data is stored, processed, and represented in colorful visual formats.
The data can be shared online and on social platforms; one's performance can even be
used to compete in games or as part of an employer wellness incentive.

Healthcare emerges as a natural target for wearable and mobile technology. Indeed
thousands of wearable devices and mobile apps have been developed and are in use
for healthcare purposes (Zheng et al. 2014). Health-related platforms, such as Apple's
Health Kit (Apple Inc. 2016), allow for the integration of health and fitness data from
different mobile apps as well as wearable technologies. Furthermore, Apple's *Research
Kit* permits the creation and roll out of medical research apps, for example, Stanford's
My Heart Counts (Apple Inc. 2015; Stanford Medicine 2016). Proponents for the use of
this technology in healthcare see a future in which medically relevant data are wirelessly
relayed to one's physician or electronic health record on a continuous basis, obviating
the need for cumbersome in-office visits and patients' subjective logging (Lupton 2013).

Many of the same ethical issues that arise with wearable and mobile technology in general are exacerbated when the technology enters the domain of neuroscience, neurology,
and mental health. Technology here runs the gamut from transcranial magnetic stimulation (TMS) systems designed for the casual user, such as the neuro-stimulating wearable
Thync° (Zinko 2015), to seizure monitoring devices for epileptics, such as the EEG Patch™
(Lehmkuhle 2015), to mobile apps intended to provide nontraditional modes of mental
healthcare (see "A case study: Mental health apps").

In this chapter, we conduct an exploration into some of the neuroethical issues associated with wearable and mobile technology. In "Classifications," we classify neuro- and
mental health wearable and mobile technology according to a number of ethically relevant criteria, including the purpose, functionalities, and user profiles of the technology.
In "Ethical considerations," we consider the accuracy and reliability of these technologies,
and address the issue of privacy and confidentiality. We also provide a discussion of new
opportunities and issues with informed consent. We then broaden the angle to see how
this technology factors into the deprofessionalization trend in healthcare in general and

introduce a particular case of nontraditional technology-enhanced mental healthcare. We use the example of the company *7 Cups of Tea*, as it is for us, a case study for many of the ethical promises and concerns of this new technology. Finally, we conduct an analysis of the possible effect of wearable and mobile technology to our conception of self and our authentic experiencing of life.

Classifications

The concept of wearable devices eludes simple definition, so in this section we provide the parameters of what falls within this category. As already stated, any discussion of wearables must recognize that such technology is inextricably linked with the mobile technology, usually smartphones, that support it. Consequently, we cannot in any principled way separate wearable from mobile technology. In the arena of healthcare in particular, the mobile health aspects relevant to our discussion cut across mobile platforms and wearable devices. Indeed, in many ways the smartphone is itself a wearable device. It accompanies individuals almost everywhere, including during exercise and sleep, and can be used as an activity tracker, as well as a means to measure an array of other biometric and behavioral dimensions (Bonnington 2014). It is not feasible, therefore, to distinguish sharply between technology that is worn on the body, such as a wristband or a patch, and a smartphone. In this chapter we are thus looking at an inclusive class of technologies that covers mobile technology and devices traditionally thought of as wearables. In fact, some forms of wearables are not *worn* at all, but are ingested or inserted. For the purposes of brevity, we will refer to all members of this class as technology*. Roughly, we can think of the class of technology* as encompassing technology that is worn, adhered to, ingested, or inserted in the body, or that is contained on mobile computing platforms, that provides and/or manipulates data generated by the physiological, behavioral, and environmental features of an individual's quotidian activities. We delineate the parameters of this technology* in what follows.

It is useful to make categorizations along three dimensions: purpose, functionality, and usage profile. Purpose refers to the goal or aim of the particular technology in a given context. Functionality refers to the method by which that technology operates. Usage profile describes the kind of user using the technology.

Purposes

We have identified five general purposes of this technology*.[2] These are:

◆ education

◆ diagnosis

◆ treatment

◆ crisis management

◆ research.

A brief description of these five purposes is as follows:

Education: Technology* can be used to educate the user and to deliver customized information in easy, immediate fashion. For this purpose, technology* is not categorically different than traditional online sources of information, with the exception that the information is mobile and accessible in a way that PC-based access of online information is not.

Diagnosis: Technology* can be used as a diagnostic tool for neurological and mental health issues. Technology* can be employed to determine whether an individual is afflicted with a particular neurological condition or is suffering from a given mental health problem. The technology* may even be used to identify from which neurological or mental health problem the user is suffering.[3]

Treatment: Treatment can also be a purpose of this technology*. The notion of "treatment" is complicated, as it lies on a spectrum. In its strictest form, a "treatment" requires that it be applied to a "disease." However, here we mean "treatment" in a more inclusive sense. For instance, we can think of a glass of warm milk as a "treatment" for sleeplessness. This does not imply that sleeplessness is, in general, a condition that qualifies as a disease, or that a glass of warm milk is a proven method of alleviating this condition. Rather we use the idea here merely in the sense that the technology* considered to be a "treatment," is intended and/or used as a means of intervening with or addressing a state that is found to be undesirable. Thanks to the mobility of smart phones and the worn/embedded nature of wearables, devices can be used to administer anything from soothing words and sounds, to doses of nicotine or other drugs, to electronic stimulation, to virtual reality set-ups that are intended to reprogram one's thoughts (Luxton et al. 2011; Leonard 2014; Alba 2015b; Loria 2016). It should be noted that the notion of treatment, as we use it here, in no way implies that the means thus described are effective.

Crisis management: Though arguably a subset of treatment, the purpose of crisis management distinguishes itself by being a generally short-term use, intended for immediate effect, that the user employs in states of high mental distress or an acute neurological event (Samson 2015). Technology* used for crisis management tends to be either direct interventions of the technology, such as when the technology is used for a one-off calming effect or administering of medication, or a method for alerting relevant persons or authorities in the case of a crisis (Luxton et al. 2011).

Diagnosis: Finally, technology* may also be used to record data of a population without the intent of treatment. This constitutes the research purpose of this technology*. Research is also distinguished from clinical applications such as diagnosis and treatment, in that data are generally de-identified. Research, conducted either by medical researchers, or the technology provider, or even lay people, may often be occurring in parallel with one or more of the other purposes.

Functionalities

Having outlined the five purposes of this technology, let us now turn to the functionalities with which these purposes are to be achieved. Some of the most common functionalities of this technology are:

Tracking: First and foremost, this technology allows individuals to *track* a variety of bodily, mental, and environmental dimensions. Some of the most important ones are as follows:

Sleep—One of the most obvious ways in which technology* interacts with neurological and mental health issues is through the tracking of sleep. There is a wide array of sleep trackers available. Many are wrist-worn, such as Fitbit˚, Jawbone˚, and Misfit˚ (Stables 2016). These mainly use motion sensors to monitor sleep and sleep cycles, displaying a large array of data to one's smartphone or other mobile device. But there are also sensors that are placed in the bed and connect to a smartphone via Bluetooth˚, for example, Sense˚ and Beddit˚, or body-sized sensors that are placed under the bed sheets, track sleep patterns and are connected to lamps that activate to wake the sleeper during the lightest part of her sleep, for example, Withings Aura™ (Haslam 2015). Of course, smartphones also have sleep tracking apps, though purpose-built technologies appear to be more accurate (Winter 2014).

Activity and location—Besides their use for measuring fitness, which also has effects on mental health (Folkins & Sime 1981), activity trackers are used for detecting specific neurological or mental health conditions (Banerjee et al. 2015). There are apps that track activity for the purpose of monitoring an individual's bipolar disorder as reduced activity indicates a depressive phase, such as Mood Rhythm™ (Voida et al. 2013). There are sensors (worn on the wrist or elsewhere on the body) that are used to monitor the activity and whereabouts of individuals with dementia, such as 9Solutions products (9Solutions n.d.). Wearable clothing (e.g., LifeShirt™) is used to research behavior and activity in people with schizophrenia (Miniassian et al. 2010). Many such devices and technologies are also equipped with alarm systems that alert the user or a designated care giver when either the user's location or her activity is outside of pre-set values (Samson 2015).

Electroencephalogram (EEG)—Continuous EEG monitoring, such as EEG Patch™ (Lehmkuhle 2015), uses small, waterproof wearable patches that are placed on the scalp of a person suffering from epilepsy and enables continuous seizure tracking.

Electromyography (EMG)—Convulsive seizure detection and warning systems, such as Brain Sentinel˚ (Cavazos et al. 2015), are being developed. They are worn on the biceps and measure skeletal muscle electrical activity.

Electrocardiogram (ECG)—ECG tracking, for instance in wearable vests and shirts such as those produced by Hexoskin™ (Banerjee et al. 2015), allows for monitoring ECG which can be used in the diagnosis of dementia.

Breathing volume—Smart clothing, such as Hexoskin™, can also track breathing volume that may be correlated with mental and neurological events (Banerjee et al. 2015).

Heart rate (HR)—HR is also measured more simply by the smartphone itself using its camera, for example with ReachOut Breathe™ (iTunes App Store 2016b). This may be used to combat the physical symptoms of stress.

Sweat composition—The composition of an individual's sweat may be tracked via continuous sweat sensors. The user can measure glucose, lactate, sodium, potassium, and skin temperature using, for example, UC Berkeley's wearable sweat sensor (Geddes 2016). In future, sweat analysis of further chemicals is likely, giving insight into an individual's stress, hydration, and exertion levels.

Blood-alcohol levels—Alcohol levels can be tracked transdermally using, for example, the University of Washington's Vive (Jeffrey 2014). This may be useful in the monitoring of substance abuse.

Ingestion—Tiny ingestible sensors that are placed inside of pills transmit information about time and dosage of ingestion to a patch worn on the body, which transmits information via Bluetooth' to a mobile device, using, for example, Proteus Health (Belknap et al. 2013). This allows for the monitoring of adherence to medication schedules.

Logging: There are a plethora of mobile apps that allow an individual to log different aspects of their mental and physical ongoing. There are apps for logging daily mood, eating patterns, sleep patterns, and social interactions. These are in service of providing more insight into an individual's mental health problems such as anxiety, depression, post-traumatic stress disorder (PTSD), eating disorder, and substance abuse disorders. Logging differs from tracking in that it is requires active entering of information by the user, rather than the passive tracking that can occur without the user's ongoing active participation (Luxton et al. 2011).

Vocal analysis: By recording and analyzing short snippets of phone calls, mobile apps can detect speech patterns associated with bipolar disorder; Priori™ (Walker 2015) is one example.

Video and photography: Mobile and wearable technology allows users and healthcare provider to exchange video and photos (Luxton et al. 2011).

Communication: While this may seem like a fairly traditional use of this technology, the expansion of online messaging into the mental healthcare arena has allowed for more continuous access of a patient to her provider, for example, Breakthrough™ and Talkspace™ (Lamas 2014).

Information delivery: Many mobile mental health apps primarily provide the user with specialized information. These resource services are often combined with one or more of the other functionalities, such as PTSD Coach™ (Luxton et al. 2011).

Quizzes and questionnaires: Apps such as What's my M3™ (iTunes App Store 2016c) are designed both to diagnose and to provide ongoing progress reports on conditions such as mood disorders.

Activity tasks: Some mobile apps provide tasks that users have to complete for diagnostic purposes. These include tapping and breathing tasks, for example, ReachOut Breathe™ (iTunes App Store 2016b).

Games: Gamification is one way in which this technology* has made an impact on mental healthcare. Users can use values "achieved" on their technologies* to compete against others or themselves. SuperBetter™ (Roepke et al. 2015) is one example of this technology.

Medication delivery: Some technologies*, for example, Chrono Therapeutics (Alba 2015b), deliver treatment such as nicotine replacement therapy, and then tailor the coaching recommendations on the app to the amount of nicotine replacement consumed.

Augmented reality: While virtual reality headsets are not typically thought of as "wearable" technology*, developers are beginning to use this functionality as part of addiction treatment, by recreating trigger situations for the addict, and then providing cognitive skills for the patient to avoid succumbing to her addiction (e.g., University of Houston Virtual Reality Lab, Loria 2016). Moreover, augmented reality, such as Google Glass™, that overlays information onto reality and thereby reminds or nudges a user about relevant information, may be useful in addressing neurological and mental health issues.

Usage profiles

Technologies* can also be categorized based on the type of user. We have identified four relevant usage profiles:

Direct-to-consumer use: This category comprises the technology* that is commercially available, that consumers can purchase directly, and that is not mediated by a healthcare provider or other professional.

Provider-mediated clinical use: Technology* in this category may be either commercially available or not. However, it differs from the previous category in that it is administered and monitored by a healthcare professional, such as a neurologist or a mental healthcare provider.

Provider-mediated research use: Research may be conducted with this technology*, in which case it is monitored by a researcher. Research uses differ from the previous two use profiles in that they have no clinical (diagnostic or therapeutic) intent.

Third-party use: Although this practice is not yet common with neuro- and mental health wearable and mobile health technology, technology* may be used by employers as part of a wellness incentive, by universities and schools to encourage healthy behavior, by the military and hospitals to track the practices of soldiers/physicians, and by insurance companies to learn more about their subscribers.

We have thus identified the various purposes, functionalities, and usage profiles that are implicated in the generation of ethical issues of this technology*. As we will see, the ethical issues that arise, arise out of a combination of these three dimensions. In "A case study: Mental health apps," for example, we find that the purpose of *treatment* as pursued by the *direct-to-the-consumer user* emerges as particularly precarious. Having laid the groundwork we now move on to delineate the various ethical issues.

Ethical considerations

Accuracy and reliability

A primary consideration with this technology*, indeed all technologies involved in a person's well-being and healthcare, is accuracy and reliability. Here, the extent to which accuracy and reliability matters depends both on the usage profile as well as the purpose. Technology* that is used for clinical purposes and is provider mediated demands a high level of accuracy and reliability. In treating epilepsy, a clinician may base her treatment plan on the readings from an EEG patch—a small patch worn on the scalp that continually tracks seizures over the course of multiple days (Lehmkuhle 2015)—and in dealing with bipolar disorder, schizophrenia, or dementia, on the data of a wearable sleep and activity tracking sensors (Alba 2015a; Banerjee 2015). Any medical device that is used in clinical decision-making in the United States must conform with Food and Drug Administration (FDA) regulations for medical devices (FDA 2014). This ensures a certain degree of sensitivity and specificity of the device, which is obviously necessary to prevent harmful medical advice.

In general, the FDA's approach to regulation is risk-based, that is, dependent on a consideration of the potential risks of the technology not working as intended, as well as based on the specific health-related claims the developer makes about the product (FDA 2013). However, technology* may not always be used as specified or intended. Commercially available devices that are not regulated by the FDA can also be used by healthcare providers for clinical purposes (Sutner 2016). Here, accuracy and reliability are not regulated and devices have been found to be inaccurate (Geddes 2015; Cipriani 2016). This presents a problem when clinicians rely on such devices for the purposes of diagnosis or treatment.

Naturally, accuracy and reliability is also critical for the purpose of research. Often researchers have to rely on participants correctly employing the technology*, since it is used remotely and researcher have little to no oversight. A subject that is being tracked for dementia might forget to put on or activate her device and consequently is recorded as having been stationary all day. Moreover, the identity of the person who is wearing the device could also be at issue. This, however, becomes less of a concern when the technology* includes devices that are implanted, inserted, or even ingested.

An issue for research applications of this technology* may arise when it is intended for dual use. For example, some vocal analysis platforms are intended both for researching speech patterns associated with conditions such as depression or bipolar disorder and are simultaneously being used by insurance companies to evaluate their policy holders (Walker 2015).[4] Given this dual purpose of both research and diagnosis, incentives may not be aligned for the user, and data may be inaccurate as a result. For example, when an individual uses technology* to participate in a corporate wellness program, the researcher's goal of obtaining accurate data, may not be those of the user, who is trying to do well in a competitive, and possibly remunerated, setting. Moreover, given the remote nature of this technology*, which is of course touted as one of its benefits (e.g., performing EEG readings on-the-go rather than in-office), correct use is not always easy to ensure in research contexts.

Prima facie, technology* that is used directly by the consumer and is commercially available may seem to have the least stringent requirements for accuracy and reliability. Casual usage of an activity tracker, for example, even if the measurements are inaccurate, seems unlikely to have a major detrimental impact on a person's health and behavior. After all, if the idea is to motivate an individual to participate in healthier behavior, being off by a few percentage points should not matter too much. However, individuals are increasingly using this technology* in ways that amount to treatment, in the sense specified earlier of mental health problems, enabling individuals to address conditions such as depression, anxiety, obsessive–compulsive disorder, eating disorders, and PTSD (Luxton et al. 2011). Technology* used in this arena (mainly mobile or online services) are largely unregulated, and are therefore not legally required to meet minimum accuracy, reliability, safety, or efficacy standards.

With the presence of the internet in every facet of modern life and the spread of technology* advancing at a staggering pace, individuals are increasingly empowered with respect to their psychological health, taking treatment into their own hands. Traditional mental healthcare is often inaccessible particularly for young people (Melnyk 2007). Moreover, a do-it-yourself attitude toward healthcare, along with patient advocacy and citizen science groups, is gaining prominence (Sarasohn-Kahn 2008; Greene 2016). In the time of Uber and Airbnb, individuals are moving away from professional providers of services and toward the soliciting of services from technology* enthusiasts or financially motivated lay people. Individuals are replacing psychotherapy with online services in which anonymous "listeners" provide support. Accuracy and reliability matter to the extent that users are employing this technology* for diagnostic and treatment purposes, which indeed seems to be the case (Jacobson 2014). While some individuals may be helped in their mental healthcare by technology* that is accessible and cheap, or free, and provided by altruistic or well-meaning outlets, the lack of oversight of technology* used in this category, either through government regulation, legislation, or professional self-regulation, means that inaccurate and unreliable information is a serious risk.[5]

Privacy and confidentiality

Wearable and mobile health app technology used for neurological and mental health purposes also raises new concerns with respect to privacy and confidentiality of data. Much of the data involved in mental health is sensitive and it is not clear that this technology* has the requisite precautions in place to deal with data of this sort. In the direct-to-consumer category, for instance, the mobile, wearable nature of this technology* makes it easy for users to share sensitive information about themselves in a non-private, non-confidential way. Of course, voluntary sharing of identified sensitive medical information on social media, on platforms like *PatientsLikeMe*, or in games is always problematic. However, neurological and mental health information from wearable devices is arguably more sensitive to the extent that it can be coupled with GPS-tracked location and other information that gives unprecedented insight into an individual's behavior. This can make her vulnerable to criminal activity, social discrimination, and employer retaliation.

Users of this technology*, especially children, may not be aware just how accessible their data are, or what the potential effects of a loss of privacy may be. However, providers of the technology* arguably have a duty to apprise users of the non-confidentiality of their data. Given the lack of comprehension that is present when users click through usage agreements, it is not clear that this duty is being satisfied (Felt et al. 2012). In the United States, in the direct-to-consumer category, devices and apps that are not considered to be medical devices by the FDA are under no obligation to comply with Health Insurance Portability and Accountability Act (HIPAA). Individuals are likely to underestimate the accessibility of this information, which suggests greater efforts must be made to emphasize the non-private status of this information, as well as possible negative consequences.

A further concern is simply the security of data. Medically sensitive data is stored on personal devices, such as smart phones, that are not HIPAA compliant and security breaches both there and on the servers of the technology* developers who are receiving the data are unavoidable (McNeal 2015). Moreover, hackers may actually hijack the device itself and track the user or issue malicious guidance (Leavitt 2011), which would not only endanger the data, but could have dangerous consequences for the user's health and safety.

Users' privacy may also be at risk in the provider-mediated clinical category. When a user receives an alert or a signal from her technology*, this may be more intrusive than receiving an email or phone call. The likelihood of a bystander noticing a buzzing or flashing wearable device is high and might result in socially uncomfortable situations. This indicates both the promise of this technology*, since messages can be relayed to patients in a more time-sensitive manner than ever before, and the peril, since these messages may be less private than is desirable.

Third parties may also be employing this technology* in ways that are worrisome from a privacy standpoint. Insurance companies (e.g., Aetna and Humana) are using voice analysis during phone calls to detect whether a person may be exhibiting speech patterns that are indicative of depression (Hodson 2014; Walker 2015). The purpose is to monitor whether that person is engaged with "activities like physical therapy or taking the right kinds of medications" in an effort to ensure that individuals on short-term disability benefits return to work sooner (Sanicola 2015). Often, individuals are not even informed that their voices are being analyzed, because of the risk that that might cause the individual to modify his speech patterns.[6] Technology* that conducts this kind of voice analysis may be administered remotely or be installed directly on the user's smartphone. When voice data is being recorded for the purpose of monitoring smartphone users, privacy and confidentiality may not be regulated by HIPAA or other regulation, and there is a risk that the individual's privacy is not being adequately respected.

Finally, data security and privacy laws vary internationally, but technology* and smart phones move across borders. Europe, for instance, has strict, centrally mandated privacy laws that allows users to remove personal information about them from online searches as well as prevent the transfer of data between companies or across borders (European Union 1995), while such laws are nonexistent in the United States. Since European laws apply not only to hardware physically located in Europe, but similarly to all data accessed through search engines in Europe (European Commission 2014), this might mean that while an individual located in the United States has no recourse to prevent her employer from recording data pertaining to her practices, the same individual when travelling in Europe does, raising issues of justice and fairness.

Consent

Medical consent forms are notorious for being difficult to understand, and indeed misunderstood (Akkad et al. 2006). Given that consent with this technology*, in particular in the direct-to-consumer category, is usually indicated by a tap on a mobile screen, it is likely that users' comprehension of data usage, sharing, privacy, security, and other important issues may be similarly low (Felt et al. 2012). Users may believe they own or have control over the sharing of their data, while the app developer collecting the data is bound, in fact, only by the voluntary terms set forth in the poorly understood consent agreement (Rosner 2014). Users may be unaware that the data they generate may be sold for commercial gain (Kaye 2014). In addition, the context of a smart phone app may elicit less than rigorous consideration, due to an app being viewed not as a serious medical intervention. Moreover, should the development company go out of business—a commonplace occurrence in the fast-paced start-up world—the data may be bought by a different company with no obligation to honor the terms of the original consent.

Consent is a particular concern for technologies* that offer unregulated online or mobile mental health services. Users and participants as young as 15 or 13 years old are permitted to use these services and even here very little effort is made to verify the age or identity of the user (7 Cups of Tea 2016e). In fact, many technologies* stress the anonymous nature of their services and actively discourage users from revealing any identifying information, which raises concerns of a lack of accountability on the part of the provider. It is unlikely that such young individuals understand the complexities of consenting to releasing sensitive information about themselves and more protections probably ought to be in place.[7]

At the same time, the potential for two-way interaction between technology* users and researchers or clinicians presents an opportunity for novel approaches that may be more effective than existing, traditional means of obtaining consent. Early developers of some medical research apps, including *ResearchKit*, utilize multimedia and animation, and require research participants to go through a set of screens that explain the research prior to consent being obtained (Stanford Medicine 2016).

In the provider-mediated clinical use category, users also may not understand that while the primary goal of data generated will be clinical, physicians are permitted to share data for research purposes (US Department of Health and Human Services 2003). Electronic health records that have been de-identified, but linked to identifiable information have been consistently used in research settings without explicit informed consent, even though there is evidence that patients expect and prefer to be asked for such consent (Cho et al. 2015). Despite such research not being considered human subject research in the United States, it may nonetheless be a violation of patient autonomy and the principle of respect. This issue may be exacerbated when the correlated data include locations frequented, phone calls made, items purchased, or websites visited (Rothstein 2010).

In the instances where there is a lack of explicit informed consent with research conducted on technology*-generated data, the issue of incidental findings is also complicated (Illes et al. 2008). For instance, data gleaned from this technology*, intended to study something unrelated, might reveal that a particular user exhibits a pattern highly correlated with dementia—maybe going back in and out of the same room, or dialing a number wrong multiple times. Should the user be alerted to this? In the absence of an explicit directive, as may be the case with de-identified research data, the researcher cannot know whether the user wants to be informed of a finding that suggests she may have a potentially serious medical condition. Moreover, the participant will likely not have appropriate expectations for such findings, as her intentions may have been more casual; she may have worn the technology* or downloaded the app as light-hearted entertainment.

Democratization of healthcare

Technology* in this area lends itself to individuals taking both clinical decision-making and mental health research into their own hands. We look more closely at how this technology* is influencing traditional mental healthcare in the next section; here we consider some general features of how this technology* contributes to the democratization of healthcare.

Within the category of research, wearable and mobile technology in neurological and mental health area allows users to track large amounts of data about themselves. This copious amount of data has the potential to be mined either individually, or in conjunction with other users, outside of the umbrella of professional research teams or academic institutions. While there may not yet be a publicly accessible repository of wearable and mobile technology* data for neurological or mental conditions, models for such databases exist in Harvard's Personal Genome Project (Church 2005) and the American Gut Microbiome Project (Garbarino & Mason 2016). Moreover, individuals who are invested in the medical research for particular conditions are finding a voice on patient advocacy and data collection sites such as *PatientsLikeMe*, where they share data with each other, as well as with commercial entities (PatientsLikeMe n.d.).

Increasingly, the roles of patient, provider, subject, and researcher are becoming blurred. Big data projects where self-trackers contribute clinical data, which nonprofessional data scientists analyze, are becoming more widespread (Sarasohn-Kahn 2008; Kelty & Panofsky 2014). This raises issues regarding the roles and responsibilities of the actors engaged in this "citizen science." Since this research is typically not publicly funded and since individuals engaged in this research are generally not associated with academic or research institutions, the regulations and norms that usually apply to healthcare research may not apply. So far we have not seen massive citizen science endeavors in neurology or mental health arenas, as we have seen with genomic data, but it seems only a matter of time that data consolidated from this technology* become subject to big data analysis projects. It is thus important to determine whether in addition to general duties of non-maleficence to others, lay researchers possess any special duties, for example, duties of veracity, confidentiality, and data sharing, associated with scientific research.

Similarly, in the clinical category, nonprofessional individuals are taking on roles hitherto reserved for trained professionals (see "A case study: Mental health apps"). While this offers the promise of fillings gaps in clinical care, little is known so far about the kinds of responsibilities implicated in this kind of democratization of healthcare. Do nonprofessional treatment providers have the same duties of care as professionals? If not, is this a function of a difference in role, or a difference in training? Or is this difference in responsibilities due to a different set of intentions or aims inherent in the definition of a lay person healthcare provider? Does it matter what the expectations of the individuals seeking treatment are? And if the responsibilities of a professional and a lay person healthcare provider are similar, are these responsibilities being effectively discharged by such "citizen clinicians"? We address some of these questions in the case study we consider in the next section.

A case study: Mental health apps

In this chapter we have already pointed to some of the issues with accuracy, privacy, consent, and the democratization of healthcare involving this technology*. We now take a closer look at these issues with respect to a particular application of mental healthcare technology*. To this end, we look at 7 Cups of Tea as a case study for the issues we are considering. 7 Cups of Tea is an online and mobile service intended to help individuals cope with their emotional and mental health concerns. Launched as a start-up by psychologist Glen Moriarty in 2014, his vision is one of a democratized, nonprofessional mental healthcare counterpart of such sharing economy services as Uber and Airbnb. The service connects individuals seeking relief from mental health concerns, such as alcohol and drug abuse, bullying, anxiety, bipolar disorders, domestic violence, eating disorders, depression, and self-harm, with nonprofessional, volunteer listeners who provide emotional support and counselling (Google Play 2016; iTunes App Store 2016a).[8] Both users and listeners are anonymous. The simplest version of the service is free, while the cost for the premium version, offering a personalized progression path, ranges from $6.24 USD per month with a 2-year subscription to a lifetime subscription for $419.95 USD (7 Cups of Tea 2016f).

Given the recent advisory by the US Preventive Services Task Force (USPSTF) to screen the general adult population, including pregnant and postpartum women, for depression (Siu et al. 2016), and the fact that the USPSTF is poised to recommend similar screening in 12- to 18-year-old adolescents (Siu & US Preventive Services Task Force 2016) it stands to reason that a greater incidence of diagnosis and treatment will be in demand, which traditional mental health services, already overextended, will struggle to fulfill (Melnyk 2007). Moreover, it is likely that adolescents in particular will increasingly seek out cheap or free, nontraditional, readily-available treatment options, such as mobile apps and online services.

Let us thus take a closer look at such technologies*. There are indeed online therapy services that operate with licensed therapists and adhere to professional guidelines and regulations, utilizing the communication and information functionalities listed discussed previously. These services show considerable promise for patients dealing with affective disorders (Wagner et al. 2014) and include examples such as the US Veterans Affairs telemedicine service, but also private providers such as Talkspace™ and Breakthrough™. Additionally, other mobile and online technologies function as self-help guides and reference sources for mental wellness concerns. These technologies* utilize the information, quizzes, and questionnaires, and sometimes activity task functionalities. Such resources can provide useful information to individuals who may wish to enhance their well-being through approaches such as mindfulness or meditation, or learn about cognitive behavioral or dialectic behavioral therapy.

A third category of services, comprising technologies* such as 7 Cups of Tea, can reasonably be taken to offer something akin to therapy; they offer the advice or commentary of concerned but nonprofessional lay people, fellow sufferers, or both. This may look like a form of peer counseling, but it is problematic, as it intimates the guise of traditional therapy. For example, 7 Cups of Tea advertises that these listeners are specialized in issues "ranging from panic attacks and bullying to eating disorders, surviving a breakup, and more" (Google Play 2016a; iTunes App store 2016a).

Employing the categories of purposes, functionalities, and usage profiles we introduced earlier, we find that the purposes involved here are primarily education and treatment. Users seek out this technology* to learn more about the particular mental health issue they perceive themselves to be experiencing, and they aim to treat these issues by talking them through with and receiving guidance from nonprofessional listeners, as well as progressing along the games, quizzes, and activity tasks that constitute their progression paths. At the same time, research is occurring in the background with this technology* (Baumel 2015). Finally, while crisis management is explicitly a purpose 7 Cups of Tea rejects as being one of its applications, there are likely instances where this technology* is employed for the purpose of seeking out psychological help in times of critical mental health need.

The usage profile implicated in this technology* is direct-to-consumer. This technology* is not mediated by healthcare providers, and individuals mostly seek out these kinds of services at their own discretion. Of note is that with this technology* the providers that are implicated are themselves not professionals, but rather unpaid volunteers, who provide services and whose reputation is built on user reviews, or as the case may be 'cheers', 'badges', and 'compassion hearts' (7 Cups of Tea 2016c).

Ethical issues arise if services such as 7 Cups of Tea may reasonably be taken to be offering therapy or counseling. We ask, therefore, whether a reasonable individual could perceive 7 Cups of Tea and similar services as offering therapy. 7 Cups of Tea explicitly states that listeners are not psychiatrists or psychologists, and that it does not intend to establish a professional–patient relationship, or diagnose, test, treat, or recommend a course of treatment (7 Cups of Tea 2016b). However, there are at least three reasons why people may reasonably have the expectation that the company is offering a service akin to counseling.

First, the kinds of conditions that 7 Cups of Tea markets target fall within the purview of mental health therapy. Depression, anxiety, PTSD, eating disorders, bipolar disorder, and substance abuse are all conditions recognized by psychiatric associations around the world (World Health Organization 1992; American Psychiatric Association 2013). Such disorders have serious consequences for individuals and considerable time and effort have been expended to develop therapeutic and pharmaceutical treatments. Second, the kind of listener–user relationship 7 Cups of Tea is seeking to engender is one characterized by emotional support and (in some cases) a deeper ongoing relationship (Google Play 2016; iTunes App Store 2016a). These are similarly features of a healthy therapist–patient relationship, so to draw a comparison between the two is not unreasonable. Third, the motivations 7 Cups of Tea suggests for making use of its service are ones that seek to favorably compare this technology* with traditional therapy. Glen Moriarty, the founder of 7 Cups of Tea, states that travel to and from the therapist is cumbersome and that therapists tend to judge their patients—both undesirable aspects of traditional therapy that are absent from this technology* (Lufkin 2014). Moreover, traditional therapy is "expensive and carries an unfortunate stigma" (7 Cups of Tea 2016a). This technology* thus presents itself as pursuing similar aims as mental health therapy, while fulfilling desiderata that such traditional means cannot.

Given that the above-mentioned considerations provide sufficient cause to believe that a reasonable person might view this technology* to have aims similar to those of traditional mental health counseling or therapy, we ask whether the technology* meets those aims. If not, and if the failure of meeting these goals carries undesirable ethical implications with it, we may conclude that this technology* ought not to be permitted in its present form. The ethical issues that arise with this technology* are specific instances of issues we discuss in the aforementioned sections of this chapter.

Accuracy and reliability

One major ethical concern for this technology* is a lack of accuracy and reliability. Leaving aside the question of what the legal responsibilities of service providers like 7 Cups of Tea may be, the combination of minimal training of the nonprofessional listeners combined with the gravity of some of the conditions for which this service is advertised, increases the risk of harmful or misguided medical advice. We also worry that this technology* may be conveying a false sense of therapeutic efficacy. The problem here is not merely that these mobile or online services are not providing effective treatment, but that users are eschewing actual effective treatment for the sake of these services under the mistaken belief that these services provide effective interventions. Since users on a site like 7 Cups of Tea can be as young as 13 years—with parental consent, though this consent is not confirmed it is likely that users may believe that they are effectively treating their mental health issues, when in reality they are not.

In a recent study, Baumel (2015) finds that individuals who engaged both in sessions on 7 Cups of Tea and with a real-life psychotherapist, rated the support provided by online nonprofessional listeners as equally or more helpful than that of the professionals. This, Baumel argues, is because individuals find online listeners to be more genuine. But of course, whether an individual finds the therapist to be genuine is not the true measure of the efficacy of the intervention. The efficacy of nonprofessional listener services must be assessed according to actual medical outcomes rather than user satisfaction. Moreover, the perceived genuineness of listeners may actually exacerbate the false sense of therapeutic efficacy this technology* is providing.

While the efficacy of technology* in this third category is an important question that must be studied empirically, it is worth noting that nonprofessional listeners are not trained to handle individuals in situations of crisis with serious and potentially life-threatening conditions. As already noted, 7 Cups of Tea does state that its service is not intended for crisis situations (7 Cups of Tea 2016e). However, while such a disclaimer may be legally exculpating, we argue that 7 Cups of Tea nonetheless has an ethical obligation toward individuals in states of great mental distress. Particularly if users have strong expectations that listeners can provide assistance in critical situations, this technology* has some duty to offer precisely such assistance. It is a real concern to us that listeners may provide inaccurate and inadequate medical advice in such critical situations. Further exacerbating this concern is that there seems to be a large turnover amongst listeners, suggesting not only that there is little consistency or opportunity to build up expertise, but that accountability is also limited.

Privacy and confidentiality

As is the case with other mental health and neuro technology*, privacy and confidentiality considerations also arise with services such as 7 Cups of Tea. Since such technologies* do not technically constitute mental health counseling, official privacy regulations are not in place. This means that, for example in the United States, listeners and service providers such as 7 Cups of Tea are not bound by HIPAA privacy regulations that limit the extent to which psychiatrists and therapists may discuss medical information about their patients with third parties (US Department of Health and Human Services 2003).[9] Of course, technologies* like 7 Cups of Tea may have their own privacy agreements, but these are not legally mandated. As a private company, it is within its rights to pass on user data to third parties for the purposes of marketing or consumer research, which 7 Cups of Tea explicitly states it does (7 Cups of Tea 2016e). Listeners of course face similar privacy concerns as any information they reveal about themselves, either toward 7 Cups of Tea or a particular user, is also not legally protected. For both listener and user, this highlights precisely the concern we stressed above. Moreover, communications between listeners and users on 7 Cups of Tea are not encrypted and are thus under considerable threat of unauthorized interception and monitoring (7 Cups of Tea 2016b).

Democratization of healthcare

Having looked at issues with this technology* including accuracy and reliability, privacy and confidentiality, we now consider how this technology* illustrates the concerns discussed in "Democratization of healthcare." As we have already argued, technology* such as 7 Cups of Tea seeks to augment and possibly replace traditional mental healthcare with accessible, democratized, and nonprofessional avenues of counseling and therapy. We can now consider what sorts of responsibilities this might generate for the nonprofessional treatment providers—where treatment is intended in the sense discussed in "Purposes" in the "Classifications" section—who populate such sites.

Individuals who operate as listeners complete a multiple choice training guide (7 Cups of Tea 2016a). This training consists of clicking through a small number of screens featuring scenarios that the listener might encounter. Still, this does not equip them to respond to serious mental health issues that users might be experiencing; and, 7 Cups of Tea explicitly states that dealing with critical mental health situations is not the responsibility of the listener (7 Cups of Tea 2016g). For example, listeners are instructed that if a user appears to be in a state of crisis, they are immediately to interrupt the session, direct the user to an emergency resource, and terminate the interaction.

However, given that as we have already argued, a user can have a justifiable expectation of receiving mental healthcare, the question arises whether a given listener truly has no ethical responsibilities toward the struggling user with whom she is communicating. If we believe that responsibilities are generated by the kind of role that an individual occupies, and if this role shares some of the features of that of a professional mental healthcare provider, then the listener may well have at least some of the ethical responsibilities that a professional therapist or counselor has. For instance, it is worth considering whether listeners, as is the case with counselors and therapists, have a duty to report evidence of child, partner, or elder abuse (US Department of Health and Human Services 2009). Even though listeners may not have legal responsibilities in this arena, the role they assume may engender such ethical duties.

While professional mental healthcare providers are well-positioned to discharge such duties of care thanks to their training, it is not clear that nonprofessional listeners are similarly well-apportioned. Consider the instruction to cut off communication with a user who exhibits signs of severe mental anguish. As a listener, one may have been conversing with this user for many hours and have gotten a sense that one is providing vital support to a troubled person. In abruptly terminating the conversation with such an individual in what may be a moment of dire need, the user may feel abandoned. Moreover, due to the anonymous nature of the technology*, unless that user reaches out to the same listener again, the listener will have no way of knowing whether the user did seek professional services or, troublingly, acted on emotionally distressed impulses. In contrast, professional mental healthcare providers are both trained to deal with and support patients through crises, and in general are privy to a patient's mental health developments.

Similarly, many listeners are themselves individuals who are currently or have in the past struggled with mental health issues. Unlike their professional counterparts, who have been trained in coping with problematic countertransference, listeners may have negative emotions triggered and experience a worsening of their own condition as a result of their relationship with the user. 7 Cups of Tea does suggest that listeners steer away from subjects with which they themselves are struggling, but listeners cannot know in advance if or when such a triggering issue may arise in a given session.

A further worrisome feature of the anonymous nature of technologies* such as 7 Cups of Tea is that listeners have no way of ensuring that users are honest. Listeners may be trolled, that is to say, a seemingly legitimate user may guide conversations into inappropriate or hurtful areas. This can be emotionally taxing for a listener who may be torn between wanting to support the user and creating some self-protection against abuse by inappropriate questions or comments. We believe that this negotiation may be difficult and harmful particularly for younger listeners.[10] 7 Cups of Tea listener training does instruct listeners-in-training to terminate a chat when one suspects that a user is a troll, but there is a risk of cutting off a discussion with a person in actual need of help.[11] 7 Cups of Tea also releases itself of any legal responsibility for listeners' actions by stating that a listener is "solely and fully liable for all conduct, services, advice" and "solely [...] responsible and liable for any damages to any member to whom listener provides services and where that member suffers damages arising from or connected to such services" issued within the framework of the service (7 Cups of Tea 2016b). All this suggests that listeners, who are generally no more than concerned, nonprofessional volunteers and, with parental consent, can be as young as 15 years old (7 Cups 2016d) are expected to operate in the context of professional duties and responsibilities for which they are likely not equipped.

In conclusion, mobile and online mental health technology* has the potential to make care more accessible. For young people for whom mobile technology is already integrated into life, it may be particularly beneficial in enabling access to mental healthcare otherwise unavailable. The adoption of a nonprofessional technology* variants of mental healthcare, however, represents a troubling erosion of professional norms. Ethical duties and responsibilities are generated by technologies* such as 7 Cups of Tea. Regulations and oversight in this new area of democratized mental healthcare are urgently needed.

Authentic living

Wearable and mobile technology not only constitutes a new means for consumers and healthcare providers to acquire unprecedented continuous and detailed information about bodily and environmental goings-on, it also affects the very way an individual relates to herself. Thus, any neuroethical overview here would not be complete without a consideration of the psychological and existential effects this technology* precipitates.

Use of this technology* is touted, particularly in the direct-to-consumer category, as enabling individuals to take control of their quotidian habits and thus improve their over-all health and well-being. Increasingly, individuals use this technology* to continuously track their heart rates, glucose levels, stress levels, activity patterns, food intake, sleep activity, as well as numerous features of their environments (Boudway 2014). Adherents of the so-called quantified self movement, whose slogan may be taken to be "self-knowledge through numbers," aim to employ this wealth of self-centered data pertaining to their bodies, minds, and environments to subtly alter everyday behaviors and decisions in pur-suit of a more efficient, more productive, healthier, and happier life (Wolf 2009). In the final section of this chapter, we explore a number of concerns that arise as a result of this pursuit.

Approaching one's being-in-the-world, to use a Heideggerian notion, through quanti-fication and self-tracking presupposes a fundamentally reductionist model of the mind and mental well-being (Heidegger 1953).[12] Here we do not mean to imply that the mind is not identical to or supervenient on a physical entity. As do most philosophers these days, we embrace physicalism about the mind. This does not entail, however, that psycho-logical, behavioral, and social phenomena are reducible to quantities a wearable device or smart phone app can track. The difficulty is threefold. Firstly, it is not at all clear that aspects of daily living that matter to us correspond neatly to quantifiable dimensions. A feeling of elation after a love interest's text message may be accompanied by an increase in heart rate, but it is not therefore given that an increase in heart rate is indicative of a feeling of elation. Secondly, by quantifying aspects of mental or physiological processes, we necessarily strip away the complexity and entanglement inherent in the experienced present. In reality, the phenomenology of a particular instant is composed of a multitude of mental, environmental, physiological, and antecedent phenomenological experiences of the world. It is likely that by capturing a measurable data point, we are simply grasp-ing a partial and conceivably accidental feature of that experience. This means we may be missing the intended goal—assuming this goal is something like greater well-being or happiness and not merely the attainment of certain quantitative values—when we seek to adjust our behavior to attain particular numerical results. Lastly, it may not be the case that what amounts to a useful quantification of a facet of one individual's experience translates across individuals.

In general, the trend toward self-tracking and the quantified self demonstrates our changing relationship with health and sickness. As Schüll (2016, p.2) aptly notes, "once understood as a baseline state temporarily interrupted by anomalous moments of illness, health has been recast as a perpetually insecure state that depends on constant vigilance, assessment and intervention." This twenty-first century attitude toward our health and well-being brings with it the burden of constantly being in charge and aware of our bodily functioning. Traditionally harmless behaviors, for example, sleeping too long, eating irregularly, or skipping visits to the gym, become medicalized, and normal variation in everyday life is pathologized. In light of all the technology at our fingertips, it is really a shortcoming on the individual's part, we might think, if she fails to keep her physiological data points within levels that are hailed as desirable or at least condoned as normal. Motivational features of these technologies range from gentle reminders to, say, take the stairs instead of the elevator, to shaming-inducing social media posts of a person's undesirable behavior, to physical pain upon the measurement of an unacceptable tracked value, or the performance of an unsanctioned action (Leonard 2014). This trend may set up unrealistic, overly onerous expectations of the user, and engender obsession and anxiety.

There is an analogy to be drawn here with calorie counting. Counting calories was long thought to be a reliable way to reduce body weight. However, evidence is mounting that by tracking and logging calories as a long-term strategy, individuals become depressed, lonely, preoccupied with food, lose their intuitive abilities to listen to their inner signals, and are more prone to binge-eating and over-eating than if they were not counting calories (Smith et al. 1999). Similar adverse and counterproductive effects are likely when individuals attempt to control behavior by quantifying dimensions that previously resided in the realm of intuitive functioning.

In fact, it has been shown that externally rewarding, inherently pleasurable activities may negatively affect their enjoyableness. With respect to self-tracking specifically, Etkin (2016) provides preliminary experimental evidence that suggests that while measuring output can increase the amount of activity an individual achieves, it can simultaneously decrease how much the activity is enjoyed, to the point that it may reduce how happy and satisfied the individual feels overall. In one study, participants who knew they were tracking their progress on walks through nature experienced less joy than those who did not. It seems it is precisely the drawing of the attention toward output, rather than allowing the individual to enjoy the pleasurable activity in itself, that undermines motivation and overall happiness (Etkin 2016).

In a certain sense, we may think of this as an issue concerning the kind of epistemic access we have to our inner states. While in the past a person may have relied primarily on immediate access to inner states to know whether she was hungry or sated, tired or energized, nervous or calm, with the increasing quantification of the self, we are turning to external technology and thus, mediated access, to inform us of whether we are hungry or tired, how warm we would like our apartment to be, or the kind of music to which we would like to listen. This can be seen as an offloading of first-person capacities to third-person means.

Similarly, some philosophers have argued that this technological external cognitive scaffolding represents a development of the mind into something they call the *extended mind*, which consists not only within the biological confines of the skull, but extends into the world to the extent that information is readily available to the cognizer (Clark

& Chalmers 1998). As such, the mind, properly conceived, may encompass an individual's computing devices, smart phone, and arguably self-tracker. Rather than being merely a theoretical metaphysical stance, the extended mind hypothesis has moral and existential implications. First, a reliance on non-biological, external components of our minds makes us dependent on this technology. Thus, should we temporarily or permanently lose access to this technology, we may be rendered helpless in the face of cognitive challenges that were hitherto within our capacities. Second, it may erode evolutionarily established methods of determining bodily needs. Relying on GPS technologies for such tasks as navigation has been shown to have adverse effects on memory and hippocampal health (Edwards 2010). It is plausible that relying on technologies for other cognitive tasks may also have detrimental effects for our brain health. Finally, offloading the monitoring of our mental and physiological processes, and relying on third person means of acquiring self-knowledge, we risk becoming removed from our bodily selves, rather than experiencing the world in a present, in-the-moment fashion, characteristic of such phenomena as flow.

We may even think of the integration of self-tracking into everyday life as altering the very way individuals relate to and conceive of themselves. Swan (2013, p.95) calls this a "fourth-person perspective" on the self. An individual's being-in-the-world is no longer immediate, but is mediated by data, to constitute what Schüll (2016, p.8) calls the "self as database." This is a self not accessible via ordinary senses and thus is not phenomenologically parameterized. Rather the self is reconstituted as information, beyond the sensory domain. The body, rather than functioning as a sensing entity through which the world is experienced and "gains self-knowledge [is instead] [...] a data-generating device that must be coupled to sensor technology and analytic algorithms in order to be known" (Schüll 2016, p.10). Swan (2013, p.96) believes that the quantified self will be transformed in to an "extended exoself" as quantification and self-tracking make the body "a more knowable, calculable, and administrable object [...]. and individuals have an increasingly intimate relationship with data as it mediates the experience of reality." While this future vision constitutes an exciting scenario for some, the mediation of one's experience of reality through data risks alienating the individual from her biological, instinctual self. Much of the intelligence that drives our behavior is evolutionarily entrenched and below the threshold of conscious decision-making (Wilson 2004; Gigerenzer 2007). It is not clear that by inserting a quantified layer between behavior and processing, we will truly achieve greater mental and physical well-being.

Conclusion

Neurological and mental health applications of wearable and mobile health technology show great promise to benefit humans in their pursuit of health and knowledge. In particular, in the hands of healthcare providers and medical researchers, this technology* can aid in the diagnosis, treatment, and research of illness. However, for this technology to be most beneficial to individuals and healthcare in general, it is important to address the concerns raised in this chapter. These are (1) accuracy and reliability, the absence of which could have harmful consequences for users and researchers; (2) privacy and confidentiality, which are threatened due to the lack of regulation and the ease of data sharing; (3) consent, for which this kind of technology presents a particular challenge since there is a risk of a mismatch of expectations concerning the perceived and actual seriousness of the medical intervention; (4) duties and responsibilities that are generated by the democratization of healthcare, for which we particularly looked at the troubling case of direct-to-consumer mental health technology; and (5) authentic living, which may be under threat from the alienating potential of self-tracking. We believe that with the appropriate consideration of these issues, new technologies* can function as empowering tools for neurological and mental health. Not addressing these important issues as technologies* are developing would be a missed opportunity to bring about such beneficial effects.

Notes

1. Smart thermostats are available from Ecobee (2016), sweat sensing technology from UC Berkeley (Geddes 2016), and smart fridges from Samsung (Bohn 2016). Smart Plate makes smart plates (Pugh 2015). Smart toothbrushes are available from Kolibree (Kolibree 2016). The sleep system is the Withings Aura™ (iClarified 2014).

2. In general, the purpose of a technology* is determined by the developer. However, the developer-intended purpose may be subverted by the user and a new or extended purpose may be generated. The notion of "purpose" is meant to encompass all these meanings.

3. Examples here include activity and sleep tracking, in addition to mood-logging and games to test mental acuity, in an attempt to diagnose conditions such as bipolar disorder and schizophrenia.

4. This also raises privacy and consent issues, which we will look at in "Privacy and confidentiality" and "Consent."

5. We explore this issue further in "A case study: Mental health apps."

6. This also seems to be a violation of consent, even though individuals are usually informed that their conversations are being recorded for quality purposes.

7. We will take an in-depth look at these kinds of technologies in "A case study: Mental health apps."

8. A list of conditions addressed by this service is only accessible after signing on as a member, but anyone can join at https://www.7cups.com/member/.

9. It is possible that Europe, with its stronger privacy laws (European Union 1995), may actually prohibit the sharing of identifiable information of users. This illustrates the complexity of privacy considerations with a technology* that is not in any way geographically contained.

10. One of the authors experienced several presumed trolling experiences while acting as a listener.

11. Instructions on how to deal with trolling events are only accessible after one has signed into one's listener account. We therefore cannot provide a URL citation here. However, anyone can sign up as a listener (7 Cups of Tea 2016d).

12. Heidegger himself was concerned that technology may convert people into things that are measurable and can be manipulated, thus ultimately reducing *beings* to *not-beings*.

References

7 Cups of Teas (2016a). *FAQ*. Available from: http://www.7cups.com/about/faq.php [accessed April 4, 2016].

7 Cups of Tea (2016b). *Listener Agreement*. Available from: https://www.7cups.com/inc/listenerTOS.html [accessed April 12, 2016].

7 Cups of Tea (2016c). *Listener Reputation*. Available from: https://www.7cups.com/about/faqBadgesCheers.php [accessed April 5, 2016].

7 Cups of Tea (2016d). *Listener Signup*. Available from: https://www.7cups.com/listener/CreateAccount.php [accessed April 13, 2016].

7 Cups of Tea (2016e). *Privacy Policy*. Available from: http://www.7cups.com/memberAgreement.php [accessed April 5, 2016].

7 Cups of Tea (2016f). *Start Feeling Better Today! Enroll in the 7 Cups Self-Harm: Safer Solutions Path*. Available from: https://www.7cups.com/backus/upgrade7cups_19.php [accessed April 14, 2016].

7 Cups of Tea (2016g). *Support Desk*. Available from: http://www.7cups.com/supportDesk.php [accessed April 4, 2016].

9Solutions (n. d.). *Products*. Available from: http://www.9solutions.com/home [accessed February 24, 2016].

Akkad, A., Jackson, C., Kenyon, S., Dixon-Woods, M., Taub, N., and Habiba, M. (2006). Patients' perceptions of written consent: questionnaire study. *BMJ*, 333(7567), 528.

Alba, D. (2015a, November 20). How smartphone apps can treat bipolar disorder and schizophrenia. *Wired Magazine*. Available from: http://www.wired.com/2014/11/mental-health-apps/ [accessed April 3, 2015].

Alba, D. (2015b, January 15). This wearable helps you quit smoking by dosing you with nicotine. *Wired Magazine*. Available from: http://www.wired.com/2015/01/chrono-therapeutics/ [accessed April 14, 2016].

American Psychiatric Association (2013). *Diagnostic and Statistical Manual of Mental Disorders* (5th ed.). Arlington, VA: American Psychiatric Publishing.

Apple Inc. (2015). *Apple Introduces ResearchKit, Giving Medical Researchers the Tools to Revolutionize Medical Studies*. Available from: https://www.apple.com/pr/library/2015/03/09Apple-Introduces-ResearchKit-Giving-Medical-Researchers-the-Tools-to-Revolutionize-Medical-Studies.html [accessed January 22, 2016].

Apple Inc. (2016). *Health; An innovative way to use your health and fitness information*. Available from: http://www.apple.com/ios/health/ [accessed January 22, 2016].

Banerjee, T., Anantharam, P., Romine, W.L., and Lawhorne, L.W. (2015). *Evaluating a Potential Commercial Tool for Healthcare Application for People with Dementia*. Available from: https://drive.google.com/file/d/0By39GGXx-P9NR29MeUx3b19abDQ/view?pref=2&pli=1 [accessed April 13, 2016].

Baumel, A. (2015). Online emotional support delivered by trained volunteers: users' satisfaction and their perception of the service compared to psychotherapy. *Journal of Mental Health*, 24(5), 313–320.

Belknap, R., Weis, S., Brookens, A., et al. (2013). Feasibility of an ingestible sensor-based system for monitoring adherence to tuberculosis therapy" *PloS One*, 8(1), e53373.

Bohn, D. (2016, January 5). Samsung's new fridge can order Fresh Direct groceries from its humongous touchscreen. *The Verge*. Available from: http://www.theverge.com/2016/1/5/10708380/samsung-family-hub-fridge-mastercard-app-groceries-ces-2016 [accessed April 12, 2016].

Bonnington, C. (2014, January 28). Turn your smartphone into the ultimate fitness tracker. *Wired*. Available from: http://www.wired.com/2014/01/smartphone-fitness-tracker/ [accessed April 12, 2016].

Boudway, I. (2014, June 6). Is Chris Dancy the most quantified self in America? *Bloomberg News.* Available from: http://www.bloomberg.com/bw/articles/2014-06-05/is-chris-dancy-the-most-quantified-self-in-america [accessed April 11, 2016].

Breakthrough (2016). *We Cover Most Insurance Plans.* Available from: https://www.breakthrough.com/how-it-works/insurance [accessed April 22, 2016].

Cavazos, J., Girouard, M., and Whitmire, L. (2015). Novel ambulatory EMG-based GTC seizure detection device for home & hospital use (I6-4B). *Neurology,* **84,** 14 Supplement.

Cho, M.K., Magnus, D., Constantine, M., et al. (2015). Attitudes toward risk and informed consent for research on medical practices: a cross-sectional survey. *Annals of Internal Medicine,* **162**(10), 690–696.

Church, G.M. (2005). The personal genome project. *Molecular Systems Biology,* **1**(1).

Cipriani, J. (2016, January 6). Lawsuit says Fitbit fitness trackers are inaccurate. *Fortune.* Available from: http://fortune.com/2016/01/06/fitbit-heart-rate-accuracy-lawsuit/ [accessed March 15, 2016].

Clark, A. and Chalmers, D. (1998). The extended mind. *Analysis,* **58**(1), 7–19.

Ecobee (2016). *The Smarter Wifi Thermostat with Remote Sensors.* Available from: www.ecobee.com [accessed March 10. 2016].

Edwards, L. (2010). *Study Suggests Reliance on GPS May Reduce Hippocampus Function as We Age.* Available from: http://phys.org/news/2010-11-reliance-gps-hippocampus-function-age.html [accessed April 11, 2016].

Etkin, J. (2016). The hidden cost of personal quantification. *Journal of Consumer Research,* **42**(6), 967–984. Available from: https://jcr.oxfordjournals.org/content/early/2016/03/04/jcr.ucv095.full [accessed April 11, 2016].

European Commission (2014). *Factsheet EU-US Negotiations on Data Protection.* Available from: http://ec.europa.eu/justice/data-protection/files/factsheets/umbrella_factsheet_en.pdf [accessed April 11, 2016].

European Union (1995). *Directive 95/46/EC of the European Parliament and of the Council on the Protection of Individuals with Regard to the Processing of Personal Data and on the Free Movement of Such Data,* 24 October. Available from: http://www.refworld.org/docid/3ddcc1c74.html [accessed April 18, 2016].

FDA (2013). *Guidance for Industry Oversight of Clinical Investigations—A Risk-Based Approach to Monitoring.* Available from: http://www.fda.gov/downloads/Drugs/.../Guidances/UCM269919.pdf [accessed January 12, 2016].

FDA (2014). *Is the Product a Medical Device?* Available from: http://www.fda.gov/MedicalDevices/DeviceRegulationandGuidance/Overview/ClassifyYourDevice/ucm051512.htm [accessed March 22, 2016].

Felt, A.P., Ha, E., Egelman, S., Haney, A., Chin, E., and Wagner, D. (2012). Android permissions: user attention, comprehension, and behavior. In: *Proceedings of the Eighth Symposium on Usable Privacy and Security,* p.3, ACM. Washington, DC, July 11–13.

Folkins, C.H. and Sime, W.E. (1981). Physical fitness training and mental health. *American Psychologist,* **36**(4), 373–389.

Garbarino, J. and Mason, C.E. (2016). The power of engaging citizen scientists for scientific progress. *Journal of Microbiology & Biology Education,* **17**(1), 7–12.

Geddes, J. (2015, January 18). Fitness trackers up to 40% inaccurate: Fitbit, Jawbone, Nike, others tested in new study, which performed best? *Tech Times.* Available from: http://www.techtimes.com/articles/27248/20150118/fitness-trackers-up-to-40-inaccurate-fitbit-jawbone-nike-others-tested-in-new-study-which-performed-best.htm [accessed February 12, 2016].

Geddes, L. (2016, January 27). Wearable sweat sensor paves way for real-time analysis of body chemistry. *Nature News*. Available from: http://www.nature.com/news/wearable-sweat-sensor-paves-way-for-real-time-analysis-of-body-chemistry-1.19254 [accessed April 13, 2016].

Gigerenzer, G. (2007). *Gut Feelings: The Intelligence of the Unconscious*. New York: Penguin.

Google Play (2016). *7 Cups of Tea*. Available from: https://play.google.com/store/apps/details?id=com.sevencupsoftea.app&hl=en [accessed April 12, 2016].

Greene, J.A. (2016). Do-it-yourself medical devices—technology and empowerment in american health care. *New England Journal of Medicine*, **374**(4), 305–308.

Haslam, C. (2015). *Counting Sheep: The Best Sleep Trackers and Monitors*. Available from: http://www.wareable.com/withings/best-sleep-trackers-and-monitors [accessed April 14, 2016].

Heidegger, M. (1953). *An Introduction to Metaphysics*. [Manheim, R., trans, in 1959.] New Haven, CT: Yale University Press.

Hodson, H. (2014, May 7). Speech analyser monitors emotion for call centres. *New Scientist*. Available from: https://www.newscientist.com/article/mg22229683-800-speech-analyser-monitors-emotion-for-call-centres/ [accessed March 31, 2016].

iClarified (2014, January 6). *Withing Unveils the First Active Sleep System "Aura*. Available from: http://www.iclarified.com/37345/withings-unveils-the-first-active-smart-sleep-system-aura [accessed February 22, 2016].

Illes, J., Kirschen, M.P., Edwards, E., et al. (2008). Practical approaches to incidental findings in brain imaging research. *Neurology*, **70**(5), 384–390.

iTunes App Store (2016a). *7Cups of Tea*. Available from: https://itunes.apple.com/us/app/7-cups-anxiety-stress-depression/id921814681?mt=8 [accessed April 11, 2016].

iTunes App Store (2016b). *ReachOut Breathe*. Available from: https://itunes.apple.com/us/app/reachout-breathe/id985891649?mt=8 [accessed April 14, 2016].

iTunes App Store (2016c). *What's My M3*. Available from: https://itunes.apple.com/us/app/whatsmym3/id515945611?mt=8 [accessed April 13, 2016].

Jacobson, R. (2014, November 1). 10 mobile apps that deliver advice and therapy. *Scientific American*. Available from: http://www.scientificamerican.com/article/10-mobile-apps-that-deliver-advice-and-therapy/ [accessed December 10, 2015].

Jeffrey, C. (2014, July 18). Vive aims to help avoid alcohol fueled social dangers" *gizmag*. Available from: http://www.gizmag.com/vive-bracelet-concept/32993 [accessed April 13, 2106].

Kaye, K. (2014). *FTC: Fitness Apps Can Help You Shred Calories—and Privacy*. Available from: http://adage.com/article/privacy-and-regulation/ftc-signals-focus-health-fitness-data-privacy/293080/ [accessed April 3, 2016].

Kelty, C. and **Panofsky, A.** (2014). Disentangling public participation in science and biomedicine. *Genome Medicine*, **6**(1), 8.

Kolibree (2016). *Meet Kolibree: The Fun, Intelligent & Beautiful Tootbrush*. Available from: https://www.kolibree.com/en/ [accessed April 2, 2016].

Lamas, D. (2014, August 4). Virtual therapy expanding mental health care. *Boston Globe*, Available from: http://www.bostonglobe.com/lifestyle/health-wellness/2014/08/03/virtual-therapy-sessions-make-mental-health-care-more-widely-available/t5kWYFmQP609BohewW3HuK/story.html [accessed April 13, 2016].

Lamkin, P. (2015). *Fitness Tracker Market to Top $5bn by 2019*. Available from: http://www.wareable.com/fitness-trackers/fitness-tracker-market-to-top-dollar-5-billion-by-2019-995 [accessed January 12, 2016].

Leavitt, N. (2011). Mobile security: finally a serious problem? *Computer*, **44**(6), 11–14.

Lehmkuhle, M., Elwood, M., Wheeler, J., Fisher, J., and Dudek, F.E. (2015). *Development of Discrete, Wearable EEG Device for Counting Seizures*. Available from: https://www.aesnet.org/meetings_events/annual_meeting_abstracts/view/2327785 [accessed April 3, 2016].

Leonard, A. (2014). *Wearable technology that hurts you on purpose*. Available from: http://www.salon.com/2014/07/23/wearable_technology_that_hurts_you_on_purpose/ [accessed April 11, 2016].

Loria, K. (2016, March 2). Therapists have created a virtual reality 'heroin cave' in an attempt to help addicts. *Tech Insider*. Available from: http://www.techinsider.io/university-of-houston-heroin-cave-virtual-reality-2016-3 [accessed April 20, 2016].

Lufkin, B. (2014). A psychologist in Y combinator, and his controversial mission to revolutionize therapy. *Fast Company*. Available from: http://www.fastcompany.com/3026967/a-psychologist-in-y-combinator-and-his-controversial-mission-to-revolutionize-therapy [accessed February 6, 2016].

Lupton, D. (2013). Quantifying the body: monitoring and measuring health in the age of mHealth technologies. *Critical Public Health*, 23(4), 393–403.

Luxton, D.D., McCann, R.A., Bush, N.E., Mishkind, M.C., and Reger, G.M. (2011). mHealth for mental health: Integrating smartphone technology in behavioral healthcare. *Professional Psychology: Research and Practice*, 42(6), 505–512.

McNeal, G.S. (2015, February 4). Health insurer Anthem struck by massive data breach. *Forbes Online*. Available from: http://www.forbes.com/sites/gregorymcneal/2015/02/04/massive-data-breach-at-health-insurer-anthem-reveals-social-security-numbers-and-more/ [accessed April 3, 2016].

Melnyk, B.M. (2007, April/May). The child & adolescent mental health crisis: closing the service gap. *Imprint*. Available from: http://www.nsna.org/Portals/0/Skins/NSNA/pdf/Imprint_AprMay07_Feat_Child.pdf [accessed April 2, 2016].

Minassian, A., Henry, B.L., Geyer, M.A., Paulus, M.P., Young, J.W., and Perry, W. (2010). The quantitative assessment of motor activity in mania and schizophrenia. *Journal of Affective Disorders*, 120(1–3), 200–206.

PatientsLikeMe (n.d.). *Help Center*. Available from: https://support.patientslikeme.com/hc/en-us/articles/201245750-How-does-PatientsLikeMe-make-money [accessed March 5, 2016].

Pugh, C. (2015, June 16). A plate that can count your calories and more. *Huffington Post Blog* [Blog post]. Available from: http://www.huffingtonpost.com/caroline-pugh/a-plate-that-can-count-yo_b_7548948.html [accessed April 18, 2016].

Roepke, A.M., Jaffee, S.R., Riffle, O.M., McGonigal, J., Broome, R., and Maxwell, B. (2015). Randomized controlled trial of SuperBetter, a smartphone-based/Internet-based self-help tool to reduce depressive symptoms. *Games for Health Journal*, 4(3), 235–246.

Rosner, G. (2014). Who owns your data? In: *Proceedings of the 2014 ACM International Joint Conference on Pervasive and Ubiquitous Computing: Adjunct Publication*, pp. 623–628. ACM. New York, USA.

Rothstein, M.A. (2010). Is deidentification sufficient to protect health privacy in research? *American Journal of Bioethics*, 10(9), 3–11.

Samson, K. (2015). Wearing the detectives: wristbands, smartwatches, and other wearable devices allow for more real-time monitoring of seizures and other neurologic symptoms—and, possibly, more precise treatment. *Neurology Now*, 11(4), 34–36.

Sanicola, L. (2015). *The Fitbit for the Mind*. Available from: https://www.linkedin.com/pulse/fitbit-mind-lenny-sanicola?trk=prof-post&trkSplashRedir=true&forceNoSplash=true [accessed April 4, 2016].

Sarasohn-Kahn, J. (2008). *The Wisdom of Patients: Health are Meets Online Social Media*. Available from: http://www.chcf.org/publications/2008/04/the-wisdom-of-patients-health-care-meets-online-social-media [accessed January 5, 2016].

Schüll, N.D. (2016). Data for life: wearable technology and the design of self-care. *BioSocieties*. Available from: http://www.palgrave-journals.com/biosoc/journal/vaop/ncurrent/abs/biosoc201547a.html [accessed March 30, 2016].

Siu, A.L., Bibbins-Domingo, K., Grossman, D.C., et al. (2016). Screening for depression in adults: US Preventive Services Task Force recommendation statement. *JAMA*, **315**(4), 380–387.

Siu, A.L. and US Preventive Services Task Force (2016). Screening for depression in children and adolescents: US Preventive Services Task Force recommendation statement. *Pediatrics*, **137**(3), 1–8.

Smith, C.F., Williamson, D.A., Bray, G.A., and Ryan, D.H. (1999). Flexible vs. rigid dieting strategies: relationship with adverse behavioral outcomes. *Appetite*, **32**(3), 295–305.

Stables, J. (2016). *Best Fitness Trackers 2016: Jawbone, Misfit, Fitbit, Garmin and More*. Available from: http://www.wareable.com/fitness-trackers/the-best-fitness-tracker [accessed April 12, 2016].

Stanford Medicine (2016). *My Heart Counts iPhone Application*. Available from: https://med.stanford.edu/myheartcounts.html [accessed December 19, 2015].

Sutner, S. (2016). *Physicians, Patients Benefit From Wearable Activity Trackers*. Available from: http://searchhealthit.techtarget.com/feature/Physicians-patients-benefit-from-wearable-activity-trackers [accessed April 2, 2016].

Swan, M. (2013). The quantified self: fundamental disruption in big data science and biological discovery. *Big Data*, **1**(2), 85–99.

US Department of Health and Human Services (2003). *Protecting Personal Health Information in Research: Understanding the HIPAA Privacy Rule*. Available from: http://privacyruleandresearch.nih.gov/pdf/HIPAA_Booklet_4-14-2003.pdf [accessed December 6, 2015].

US Department of Health and Human Services (2009). *Mandatory Reporters of Child Abuse and Neglect*. Available from: http://www.childwelfare.gov/systemwide/laws_policies/statutes/manda.cfm [accessed April 7, 2016].

Voida, S., Matthews, M., Abdullah, S., et al. (2013, September). MoodRhythm: tracking and supporting daily rhythms. In: *Proceedings of the 2013 ACM Conference on Pervasive and Ubiquitous Computing Adjunct Publication*, pp.67–70. ACM, Zurich, Switzerland.

Wagner, B., Horn, A.B., and Maercker, A. (2014). Internet-based versus face-to-face cognitive-behavioral intervention for depression: a randomized controlled non-inferiority trial. *Journal of affective disorders*, **152**, 113–121.

Walker, J. (2015, January 5). Can a smartphone tell if you're depressed? *The Wall Street Journal*. Available from: http://www.wsj.com/articles/can-a-smartphone-tell-if-youre-depressed-1420499238?cb=logged0.14367316628029903 [accessed February 23, 2016].

Wearable Technologies (2016). *Innovation World Cup Series*. Available from: http://www.innovationworldcup.com/2016/01/the-wt-wearable-technologies-innovation-world-cup-crowned-the-six-most-innovative-wearable-companies/ [accessed March 24, 2016].

Wilson, T.D. (2004). *Strangers to Ourselves*. Cambridge, MA: Harvard University Press.

Winter, C. (2014, February 26). Personal sleep monitors: do they work? *Huffington Post Blog* [Blog post]. Available from: http://www.huffingtonpost.com/dr-christopher-winter/sleep-tips_b_4792760.html [accessed April 18, 2016].

Wolf, G. (2009, June 22). Know thyself: tracking every facet of life, from sleep to mood to pain, 24/7/365. *Wired Magazine*. Available from: http://www.wired.com/2009/06/lbnp-knowthyself/ [accessed December 1, 2015].

World Health Organization (1992). *The ICD-10 Classification of Mental and Behavioural Disorders: Clinical Descriptions and Diagnostic Guidelines*. Geneva: World Health Organization.

Zheng, Y.L., Ding, X.R., Poon, C.C.Y., et al. (2014). Unobtrusive sensing and wearable devices for health informatics. *IEEE Transactions on Biomedical Engineering*, **61**(5), 1538–1554.

Zinko, C. (2015, July 3). Seen: partygoers in Thync to alter moods. *SFGate*. Available from: http://www.sfgate.com/entertainment/article/Seen-Partygoers-in-Thync-to-alter-moods-6363869.php#photo-8248816 [accessed April 5, 2016].

Chapter 6

Technologies of the extended mind: Defining the issues

Peter B. Reiner and Saskia K. Nagel

Introduction

One of the primary concerns of the field of neuroethics has been the sanctity of the mind. Whether the worry is over others gaining access to a person's most private thoughts or manipulating memories, increasing cognitive abilities beyond species-typical functioning or the authenticity of modern life, the recurring ethical issues center around the question of whether people are masters of their own destinies. The progress that has been made in exploring this fascinating terrain has been substantial, as evidenced by the growing maturity of the field. Yet even while making great strides in defending the boundaries of what is neuroethically acceptable, many people seem to be yielding the sanctity of their minds to the convenience of modern devices. People do not rely only upon their 3-pound brains to navigate the world that surrounds them. Increasingly, a substantial portion of the population is blending their cognitive space with the algorithmic devices that are nearly always at hand. It is time that we recognize these devices as technologies of the extended mind (TEMs) (Fitz & Reiner 2016; Nagel et al. 2016).

The extended mind hypothesis

The intellectual forerunner to the concept of TEMs is the extended mind hypothesis (EMH) (Clark & Chalmers 1998), which suggests that cognition extends beyond the brain into the world at large. Importantly, the EMH specifies that for external cognitive processes to qualify as part of the mind, they must be active parts of the mind. Thus, the EMH goes beyond merely suggesting that human cognition relies on external structures for scaffolding and support. Rather, the EMH maintains that at least some of the physical vehicles that realize our cognitive processes lie outside of the bounds of the skull.

The classic example from the original paper by Andy Clark and David Chalmers—the case of Otto and Inga—illustrates the issue nicely (Clark & Chalmers 1998). Otto and Inga live in New York City. One day, Inga hears about an exhibition at a museum that she recalls is on 53rd Street and heads out the door, intent on seeing the artwork. Her neighbor Otto has been having trouble remembering things. In order to overcome this deficit, he has made a practice of storing important information in a small notebook that he carries with him in his breast pocket. When he hears about the exhibition, he consults his notebook, finds that the museum is on 53rd Street, and, just like Inga, sets off for the same destination. The similarity between the two situations should now be obvious: the cognitive function of storing information is mediated by neurons in one case and pen and paper in the other.

From this example, Clark and Chalmers develop the parity principle, asserting that if a process that unfolds in the external world would readily be classified as part of the cognitive toolkit when it goes on in the head, then it is, at least for that point in time, part of the cognitive process. Using the parity principle as a guide, Clark and Chalmers assert the equivalence of neuronal memory and paper memory as information storage strategies in the case of Otto and Inga.

The claims of the EMH are radical, and it remains a highly contentious theory in philosophy of mind (Adams & Aizawa 2001, 2010; Rupert 2004, 2013; Menary 2010). However, one need not fully accept the philosophical premise to appreciate that the concept resonates with a key feature of modern life: the growing sense that computers and smartphones (and soon "the Internet of Things") function as sophisticated extensions of the modern cognitive toolkit, even more so than Otto's dog-eared notebook. Moreover, rather than relying upon the parity principle to guide thinking in this regard, we will provide further specification on the features and uses of these algorithmic devices that serve to qualify them as TEMs.

Technologies of the extended mind

Before we consider the circumstances under which a device qualifies as a TEM, it is useful to explore what we mean when we use the word *mind*. A nuanced description of the term is beyond the scope of this chapter, and there exists an entire subdiscipline of philosophy that addresses the issue, but a few words of clarification are in order, in particular to place the notion of TEMs in context.

The *Oxford English Dictionary* defines mind as the element of a person that enables them to be aware of the world and their experiences, to think, and to feel: the faculty of consciousness and thought. Yet this definition is not satisfying, specifically because it dodges the elephant in the room: the mind–body problem. While we will use the term mind liberally in this chapter, it is not our intent to slip into some version of substance dualism in which there is brain-stuff and mind-stuff. But one specific distinction between brain and mind is in order: as we view it, brain is a *thing* while mind is a *concept*. Mixing those ontological levels is what often leads to confusion as to the relation between brain and mind.

One way of thinking about the issue is to say that the mind represents the full set of cognitive resources that we deploy in the service of thinking. Here we construe thinking to include what Keith Stanovich calls reflective, algorithmic, and autonomous thinking (Stanovich 2009). One can quickly see how this definition is friendly to the EMH, for once one uses the term full set of cognitive resources, one opens the door to things other than the brain contributing to mind. Such formulations represent direct challenges to the hard neuroessentialist perspective which, one of us has argued, suggests that there is no need to include anything other than the brain (Reiner 2011). Yet, as we shall see below, once one begins to give due consideration to things outside of the brain—in particular TEMs—it becomes difficult to conclude that they are not being deployed in the service of thinking. In this view, mental processes and mind cannot be fully reduced to cognitive processes, as they also refer to the plethora of affective, motivational, and social resources that can influence thinking and go beyond it in various ways. While many current TEMs aim at the extension of cognitive processes, this is by no means a principal constraint; to the contrary, design might explicitly focus on an extension of affective processes as well.

Having provided some conceptual clarity over what constitutes mind, we now address the question of what sorts of devices are TEMs. It is not the case that every algorithmic function carried out by devices external to the brain qualifies them as a TEM, but rather that there is a relatively seamless interaction between brain and algorithm such that a person *perceives* of the algorithm as being a bona fide extension of a person's mind. This raises the bar for inclusion into the category of algorithms that might be considered TEMs. It is also the case that algorithmic functions that do not qualify as TEMs today may do so at some future point in time and vice versa.

By way of illustration, consider the use of GPS in a smartphone, an example we have previously described in Nagel et al. (2016). Imagine a 35-year-old man John who has lived in Manhattan for the past decade. John has rarely driven a car since he moved to New York, as he now relies heavily on the subway or taxis to get about town. But John recently had a baby and needs a bit of extra income, so he asks his brother whether he thinks he could become a driver for Uber, a company that enables non-professionals to act as taxi drivers using their own vehicles. His brother, who has been making a fair bit of money doing just that over the past year, is quite encouraging and even offers John the use of his car in the evenings when it is idle. John signs up and within a week finds himself behind the wheel of his brother's car, answering calls for Uber drivers to come and ferry people all over New York City.

In the normal course of events, Uber drivers are highly reliant on GPS. On his very first day, John, an avid user of computer technology, is fascinated by how easy it is to enter addresses into his smartphone and have it show him the best route to his passengers' destinations. It is particularly helpful because otherwise John would often find himself lost, as many of his fares lead him to visit neighborhoods with which he is unfamiliar. Of course, John has heard stories that sometimes GPS can lead you to the wrong place, so he remains alert to his environment in order to be certain that he delivers his passengers to their destination without a hitch.

Is John's GPS functioning as a TEM? It is certainly carrying out computational work that is external to John's brain. But it is probably most appropriate to consider the GPS in John's smartphone as cognitive support, for neither the algorithmic calculations nor John's reliance upon them are seamlessly integrated with John's mind.

Now imagine that a week or two has passed. John has taken dozens of passengers to their destinations. Even though he now knows the city a bit better than before, he always uses the GPS in his smartphone, and it has not let him down even once. At this point, when he enters an address and the route flashes up on the screen, he doesn't give it a moment's notice before following it to the destination suggested by his smartphone. The GPS now functions very much as a TEM, for John has integrated its algorithmic output into the working of his mind.

Much as neuroessentialism challenges traditional views of how people view themselves (Reiner 2011), conceiving of the mind as a blend between brain and algorithm poses a challenge to a range of existing worldviews. At one end of the spectrum, if one is troubled attempting to reconcile concepts of ensoulment with mechanistic explanations of how the brain works (Bering 2006; Farah & Murphy 2009; Preston et al. 2013), the realization that an algorithm is now an extension of one's mind makes the issue even more problematic. At the other end of the spectrum, the very same mechanistic explanations of brain function that lead to the position that "we *are* our brains" (Greene & Cohen 2004; Roskies 2007; Reiner 2011) are somewhat destabilized by the emerging reality of TEMs. For these reasons, people may very well resist the disruptive concept of TEMs. However, the intrusion of algorithms upon daily life seems relentless, and in much the same way as John's GPS transitioned from external computational algorithm to bona fide TEM, we envision this concept becoming an emergent meme, with an ever-growing share of the population perceiving of their devices as TEMs, even if they do not explicitly conceptualize them as such.

The neuroethical issues

The neuroethical issues engendered by the EMH were first brought to light by Neil Levy in a prescient paper published in 2007 (Levy 2007). At the time, he allowed that the EMH was an obscure debate, but nonetheless suggested that it had substantial implications for the field. He points out that part of what makes neuroethics a discipline is the claim, sometimes explicit but nearly always implicit, that there is something different about intervening in the brain, that such interventions are different than traditional means of altering mental states. Levy suggests that if one accepts the EMH, the exceptionalism that is normally offered to worries about intrusions, manipulations, and surveillance of mental states in the brain naturally flow to the extended mind as well. He sums the situation up nicely in his concluding paragraph:

> These reflections on the *prima facie* parity between environmental manipulations and new technologies may seem to reduce the importance of neuroethics. On the contrary, I think that it dramatically increases it. It might seem that the extended mind thesis entails that neuroethics is less important inasmuch as, in its light, it becomes apparent that neurological interventions into the mind—that is, interventions that target neurons, neurotransmitters or brain structures—are not after all so special: they represent merely the latest means of doing something that is quite ubiquitous in human cultures. Although it is true that the extended mind thesis may dampen some of the hype surrounding these technologies, it should be seen as dramatically expanding the *scope* of

neuroethics, not detracting from its importance. Neuroethics focuses ethical thought on the physical substrate subserving cognition, but if we accept that this substrate includes not only brains, but also material culture, and even social structures, we see that neuroethical concern should extend far more widely than has previously been recognized. In light of the extended mind thesis, a great many questions that are not usually seen as falling within its purview—questions about social policy, about technology, about food and even about entertainment—can be seen to be neuroethical issues. (Levy 2007)

We agree with Levy that the EMH has profound implications for our thinking about neuroethics. Three issues are particularly worthy of further elaboration: the threat to autonomy posed by manipulation of our TEMs, the threat to privacy of thought by peering into TEMs, and the relevance of TEMs to questions regarding cognitive enhancement. Each has changed substantially in the decade since Levy's paper, a decade in which adoption of smartphones as ubiquitous computing has moved from concept to reality. This development is worth highlighting, as it seems to be more than just another technology, but rather has been a major influence that shapes our technological surroundings. Other influences are and will be relevant, but the prevalence of smartphones—nearly half the adult population on the planet has one (Pew Research Center 2016)—and our readiness to rely on them is certainly remarkable. These features make it a useful example with which to explore some of the neuroethical issues associated with TEMs.

Autonomy

The concept of autonomy is foundational to modern thinking about who we are as sentient beings. Particularly since the Enlightenment, Western societies have largely accepted the position that we—and only we—have the right to determine the course of our lives. Autonomy underpins many of the most celebrated political and social movements of modernity. In human rights, it is autonomy that beckons us to consider the needs and desires of the individual as having primacy. In politics, autonomy is the firmament upon which democratic governance rests. Autonomy shows no signs of losing its appeal, continually forcing us to modify our practices. This is best exemplified in the medical realm in which modern bioethical principles have invoked the concept of autonomy to produce a sea change in the way that physicians and their patients interact (Beauchamp & Childress 2012).

A fundamental feature of the concept of autonomy is that the autonomous individual should not be unduly influenced when making decisions (Frankfurt 1971; Dworkin 1981). One need look no further than the ideal of the rugged individualist who navigates their environment relying on nothing more than his or her wits to see how this works. In this common trope, the autonomy of decision-making by the individual is held as sacrosanct, and any infringement is considered a violation. It turns out that this picture of humans as self-sufficient rational actors who live their lives independently of others is at substantial variance with how people actually live in the real world. Rather, decisions are regularly influenced by the input of others, whether it is in the form of the books and newspapers that we read, the opinions of people that we listen to, or other features of the social and physical environment around us. Recognizing this, feminist and communitarian scholars have updated the concept, developing variants on what is commonly known as relational autonomy (Nedelsky 1989; Friedman 2003; Christman 2004; Meyers 2005; Mackenzie 2010). The relational account of autonomy suggests that when people make decisions, they often admit input from friends, family, colleagues, or professionals. That this occurs without demurral complicates the calculus for evaluating when an external influence might be considered due and undue. Indeed, "representing these two sorts of effects with roughly accurate proportionality is, however, a formidable project [since] matters of degree are notoriously difficult to specify philosophically" (Friedman 2003).

If it is a struggle to determine what influences are due and undue in the context of input from other people, the task is even more complicated when we consider the influence of TEMs upon these very same decisions. Before we do so, it is worth considering the general features of algorithms that might modify the degree to which an influence is perceived as violating autonomy. We have suggested that three factors are key: (1) the persuasiveness of the algorithm in the decision-making process, (2) the seriousness of the decision, and (3) the ability for the algorithm to learn about user preferences (Nagel et al. 2016).

Persuasiveness seems to be a central feature of technological autonomy violations. Decision-making can be influenced across a spectrum ranging from minimally to highly persuasive, and can even turn into being coercive (Fogg 2002; Verbeek 2006, 2009). If the ability to thoughtfully engage in the decision-making process and to reflect on the situation is ensured, the influence of the technology will not be perceived as a particularly problematic autonomy violation, as self-control does not seem to be harmed.

The seriousness of the decision also varies across a spectrum that differs depending upon the level of potential harm or benefit an individual may experience as a result of a particular choice. The lower the assumed potential harms or benefits, the lower the perceived seriousness of the decision. Imagine for a moment the grave real-world situation of an individual grappling with the decision of whether to undergo a course of chemotherapy that will briefly extend his or her life by some weeks but will compromise the quality of those weeks substantially. Many such patients consult—and trust—online resources to help them evaluate the relative risks and benefits. Under such circumstances, even a small misstep in influence can result in a substantial autonomy violation.

The ability to learn about user preferences is important, as it makes a great difference as to whether a system only follows a set of preprogrammed instructions or is able to monitor individual behaviors and preferences and learns from them. If we return to the situation of the individual confronting the decision of whether to embark upon a course of chemotherapy, we can easily imagine two versions of the online resource. In the static version where the information that is found online is offered by an unknown designer, the opportunity for the designers' preferences to be unwittingly smuggled into the algorithmic calculation is high, and the possibility that an autonomy infraction may occur not insubstantial. A more dynamic version would have the underlying algorithm learn something about the preferences of the user and then provide advice accordingly. Since the algorithm has learned at least something about the user's worldview, the advice that it offers might better be tailored to their second-order desires, and the likelihood of an autonomy violation diminished.

Having laid the groundwork for our thinking about the issue, we turn our attention to an example in which the external technology is a TEM. We will use a simple and relatively trivial example to illustrate the relevant issues, but the reader can readily imagine how modifying persuasiveness, seriousness, and learning user preferences can affect the degree of perceived autonomy violation.

To begin, consider an algorithm that is not a TEM, for example, the GPS app on John's smartphone in the example described in the previous section. Imagine that on the first evening that John used the GPS function (i.e., before it has transitioned into being a TEM), John passes a billboard advertising a bakery. Pushing the bounds of current practice (but only slightly), let's also imagine the smartphone calls attention to the billboard and alerts John to the fact that the bakery is just up ahead; sealing the deal, the smartphone "asks" if John wishes to drop in to get something sweet. John is tempted, and although he likes to manage his intake of sugar, he decides that there would be little harm. A moment later he stops at the bakery, purchases a donut, and soon is smiling as he tastes the sugary treat. In this scenario, the GPS program has clearly influenced John, sufficiently so that he altered his second-order desires about his food intake. Many people would call such an influence undue and, despite the relative triviality of the infraction, the situation represents an autonomy violation of sorts.

Now let us imagine that 2 weeks have passed, and the relationship between John and the GPS app has grown more intimate—John now integrates its algorithmic output into the working of his mind while driving his Uber cab. For all intents and purposes, the GPS algorithm is now functioning as a TEM. Of course, because John carries his smartphone everywhere, the device gathers a fair bit of information about his daily activities, and the abilities that this confers on the smartphone only serve to reinforce the feeling that the suite of apps on the phone are functioning as a more-or-less unified TEM. It requires little in the way of stretching credulity to imagine that the GPS might not only call attention to a billboard alongside the road, but having "consulted" a database that indicates that John frequently shops at an organic grocery store, instead of suggesting that John get something sweet, the smartphone now suggests that he may wish to get a piece of organic blueberry pie. Adding to the persuasiveness of the suggestion, the GPS program on his smartphone—which also monitors his activity levels during the day—reminds John that he went for a 5K run this morning and can probably afford the calories. A few minutes later, John is smiling as he enjoys a forkful of delicious pie. Once again, the GPS program has influenced John, sufficiently that he altered his second-order desires about his food intake. But given that the suggestions are aligned with John's overall desire to eat organic food, and that they include at least a rough calculation on calories burned earlier in the day, does the persuasive influence of the GPS app represent an autonomy violation?

The difference between the two scenarios described above depends to a substantial degree on how John perceives the algorithmic device that accompanies him everywhere. As his relationship to the smartphone becomes closer, as both he and device provide information that the other uses to help them navigate the world around them and John perceives more and more that he relies on the program's advice, the device increasingly becomes an integral part of his extended mind. Given the considerations from relational autonomy, the influence of the smartphone on John's behavior in the second scenario appears to be less "undue" than it was in the first instance. In this view, the more that devices become TEMs, the likelihood that their advice results in perceived autonomy violations diminishes.

However, there is something missing from this description, and that is that the very same algorithm that is an extension of John's mind is also an extension of the mind of an other—in this case the corporate entity that has designed the GPS app. It is not at all out of the bounds of modern economic activity that the corporate entity would be paid for directing John to the bakery. Thus, while one objective of that app is to align John's behavior with his second-order desires, a second objective of another entity is also smuggled into the relationship. In view of such potential conflicts of interest, it becomes harder to accept the premise that as our devices become TEMs, autonomy violations become less likely; if anything, they become more insidious. Serving two masters is certainly part of the issue, but it is also the case that the privacy implications are impossible to ignore.

Privacy of thought

One of the most controversial topics in neuroethics is the worry over *mind reading* (Wolpe et al. 2005; Illes 2006, 2007; Greely & Illes 2007; Parens & Johnston 2014). Frances Shen has summarized the sentiments of many by suggesting that using advances in neuroscience to access the most private of spaces—our minds—would have tremendous privacy implications (Shen 2013). Yet despite intriguing insights obtained with functional magnetic resonance imaging (Haynes & Rees 2006; Mitchell et al. 2008; Naselaris et al. 2009; Rissman et al. 2010), it remains difficult to obtain more than a rudimentary profile of the thoughts of individuals by measuring brain activity (Farah et al. 2009). In contrast, it seems that a great deal of what might be termed the inner life of the mind may be accessible via the technologies on our desktops, and even more so, in our pockets. Cognizant that this is a discourse that has encountered more than its fair share of hype, it seems prudent to consider whether access to our TEMs may represent intrusions upon privacy of thought.

Such considerations come in the context of the well-known fact that online privacy is generally under threat. Illegal breaches and over-sharing of digital information have grown from the occasional to everyday events. While any intrusion on privacy may be unwelcome, some transgressions are more problematic than others. When others gain access to information that reflects one's innermost thoughts, the incursion may go beyond privacy to intrude upon privacy of thought. Such intrusions have particular normative significance in neuroethics, and thus our analysis focuses upon this subset of privacy concerns.

The manifesto establishing the right to privacy was drafted in 1890 by Samuel Warren and Louis Brandeis (Warren & Brandeis 1890). As influential as it has been in legal circles, it is the historical sweep of the article that draws one's awe. As Warren and Brandeis incisively explain, at one time liberty meant freedom from physical restraint, but as society increasingly recognized the inner life of individuals, the right to life came to mean the right to enjoy life, and protection of corporeal property expanded to include protection of the products of the mind such as literature and art, trademarks, and copyrights. Warren and Brandeis noted that "recent inventions and business methods call attention to the next step which must be taken for the protection of the person." In 1890, their worry was the intrusiveness of photographers, but more generally their article suggested that as technology changes, societal norms might require updating. In the modern world, in which information all-too-readily moves online, we appear to be overdue for just such a review.

Legal arguments support such a view. Pointing out that under the third-party doctrine "an individual does not have a reasonable expectation of privacy with respect to information he voluntarily discloses to a third party, like a bank or a telecommunications carrier," Wittes and Chong suggest that the third-party doctrine is ill-suited to the modern situation in which devices generate data about their users (Wittes & Chong 2014). In a chilling demonstration of how much information is contained within metadata—the data about the numbers we call, or where our phone is at a particular moment, who we e-mail and the subject lines of those messages, our search queries and the websites we visit—Ton Siedsma, a researcher at the Dutch digital rights foundation Bits of Freedom, allowed an app on his cellphone to collect such metadata for 1 week (Tokmetzis 2014). The limited analysis of this information revealed not only Ton's circles of friends and the like, but also the types of information included in his search queries. That Ton might have an interest in bicycles is fairly innocuous, but what if he was searching for information about antidepressants or gender reassignment surgery, and these were topics that were in his head, but he had not—and more importantly did not wish to discuss with others?

Just such a situation is described by Wittes and Lui when alerting us to the fact that "the privacy that consumers value in practice is not always the privacy that activists devoted to privacy value on their behalf." To illustrate the matter, they point out that many people acquire medical information, salacious images, and more online because "they would rather be tracked online by commercial vendors than have to face parents, teachers, doctors, or the stern-faced old lady at a news stand" (Wittes & Liu 2015). In a similar vein, Frances Shen reminds us that what he terms the "privacy panic script" unfolds first with fear mongering regarding the power of technology, followed by the suggestion that institutions will use the technology in devious ways, and that current laws are insufficient to protect the average citizen (Shen 2013). These observations suggest that when advocating in favor of privacy, humility may be in order.

Given these varied interests, we suggest that explicitly recognizing some of our devices as TEMs may help us in arriving at a more reasonable state of affairs, especially if we arrive upon a clearer understanding of what we mean by privacy of thought. The philosopher Michael Lynch, writing as *Amicus Curiae* in support of the plaintiffs in *American Civil Liberties Union v. James Clapper*, the lawsuit that challenged the legality of the National

Security Agency's bulk phone metadata collection program traces the notion of privacy of thought back to Descartes and Locke, who both held that:

> what identifies your thoughts as your thoughts is that you have "privileged access" to them. This means at least two things. First, you access them in a way I cannot. Even if I could walk a mile in your shoes, I cannot know what you feel in the same way you can: you see it from the inside, so to speak. Second, you can, at least sometimes, control what I know about your thoughts. You can refrain from telling me the extent of your views and your feelings. (Lynch 2013)

In the context of TEMs, these criteria seem appropriate: if you have a private means of accessing information on your device that is not available to others, and you have the means of controlling that access, it seems as if you have demonstrated that, in your view, it is private. The problem arises, of course, when those very same devices are connected to others by virtue of the ever-present Internet.

To see how this may play out, it is worth recalling the infamous case of the discount retailer Target and the pregnant teen. By combining metadata that was packaged, bought, and sold (online, of course) with analysis of something as innocent as the shopping behavior of young women, Target was able to predict with high accuracy when women were in their second trimester of pregnancy (Hill 2012). The commercial objective was to target these potential customers with ads for diapers, baby clothes, and the like, but when Target did so, customers felt so invaded that they complained (in this case, it was a father who only discovered his daughter's pregnancy through this route). So Target modified their strategy; instead of sending ads just for diapers, etc., they mixed those ads in with ads for everyday items unrelated to pregnancy just to prevent the perception that they had invaded their customer's privacy.

Similar brinkmanship can be seen in the relationship between Facebook and its users. In a longitudinal study of 5076 Facebook users between 2005 and 2011, Facebook users increasingly exhibited privacy-seeking behavior, decreasing the amount of personal data that they shared publicly. At the same time, the amount of personal information they revealed privately to so-called connected profiles increased. The unintended result was that disclosures to "silent listeners" on the network—Facebook itself, third-party apps, and ultimately advertisers—grew. The authors conclude that these findings "highlight the tension between privacy choices as expressions of individual subjective preferences, and the role of the environment in shaping those choices" (Stutzman et al. 2012).

From these examples we can see that the networked nature of our TEMs makes them particularly vulnerable to intrusions upon privacy of thought. While people may take steps toward increasing the privacy of their online behavior, the corporate entities at the other end of the phone, as it were, are continually mining these very same actions in an effort to enhance their competitive position in the world of commerce. What is lost in the conversation is that along the way, people may *perceive* of their devices as extensions of their minds. We suggest that this implicit tagging of Internet-connected algorithmic devices as TEMs raises the bar for privacy, qualifying the information within the amalgam of our brains and our devices as private thoughts.

Cognitive enhancement

Discussion of the propriety of using advanced neurotechnologies to enhance cognitive abilities has been among the most protracted of debates in neuroethics (Parens 1998; Farah et al. 2004; Greely et al. 2008). Ethical arguments from different biopolitical positions (Reiner 2013) have been waged over a number of issues, but the cardinal concerns of safety, pressure, distributive justice, and authenticity have dominated (Fitz et al. 2014). In many ways, the debate is more pragmatic than philosophical, and so we will restrict our comments regarding the implications of TEMs to the cognitive enhancement debate to a short outline of the relevant issues.

The debate over cognitive enhancement is somewhat compromised by the observation that the agents that are available today—drugs and electrical stimulation devices—do not yield the sorts of enhancements that people seek: effect sizes tend to vary from minimal to none (Farah 2015; Ilieva et al. 2015). As we have recently suggested, this may be because attempts to bolster cognitive function from within run up against the hard limits of neurobiological reality (Fitz & Reiner 2016). Indeed, we suggest that a better strategy is to rely upon our TEMs to enhance cognitive function, not only because it is more effective but because we are already well down the road toward "commingling our cognitive space with technology."

The question that remains open, and one that merits further normative and empirical work in the field, is the degree to which considering TEMs as cognitive enhancers settles debates that have gone on for over a decade now. The answer is likely to be a mixed bag—the ubiquity of TEMs might help to dispel the distributive justice concern, but concerns over safety, pressure, and authenticity now morph into new territory. We predict that exploring these issues with a new technological lens may assist the field in clarifying how we should move forward, even if it involves playing catch-up to how we live today.

TEMs necessitate a new framework for neuroethics

The many important issues that have made neuroethics such a vibrant and fascinating field—ranging from questions on handling of incidental findings, brain reading to detect deception, brain intervention of offenders, to enhancement of healthy people by means of pharmacological agents or electrical brain stimulation—have largely been focused on the brain. Recognizing that TEMs qualify as extensions of the mind provides the impetus for "dramatically expanding the scope of neuroethics," as Neil Levy has argued (Levy 2007). This expansion is even more interesting the more one recognizes that the role of technologies is not just one of cognitive support but is fundamental to the modern way of being in our technological culture.

In philosophy, it has long been well accepted that humans cannot be understood and often do not conceive of themselves as isolated beings. Rather, people are *socially embedded*, existing as part of a collection of other beings who shape us and our image of ourselves. This idea of our deeply social nature can be found already in the writings of Hegel and Marx, and has been extensively developed by the critical psychologist Klaus

Holzkamp (Holzkamp 1985). Importantly, in this view social embeddedness goes beyond the direct private relations between and among individuals, reaching out to the various social institutions that surround and influence us.

Analogously, the modern world with its increasingly technological surroundings defines us as *technologically embedded* beings. Consider that today the trajectory of a life begins surrounded by reproductive technologies and often ends with a different array of medical technologies easing us into death. Just try to imagine for a moment the technologies that are around you right now—from the lightbulb above you, to the heat in the corner, to the smartphone in your pocket. Perhaps you took a drug this morning to help normalize some bodily function, or maybe you will use a car or bus or train to move from place to place, or even have a video-chat with a colleague on a different continent. Indeed, humans rely on a vast array of technologies, and one might argue that they depend upon them— some of us more, some less. Such is the modern predicament (Scialabba 2011) that it is hard to hear silence, that many live in fear of the content of their food, that it has become a common experience to feel estranged from the natural world. No matter how one *evaluates* being surrounded by and dependent upon technologies, it is fair to say that we are increasingly embedded by technologies.

It is no surprise then that some of those technologies engage with our cognitive processes. Even more: that they serve as TEMs by functioning as part of the mind that resides outside of the brain. And as the mind is perceived as perhaps the most intimate part of us, conceptualizing technologies as extensions of the mind deeply challenges the sense of who we are as humans. The questions Who am I? Am I my brain? Am I my mind? Am I my relationships? (Haslam 2004; Rose 2005; Feinberg 2006; Glannon 2009; Burwood 2009; Noë 2009; Brand 2010; Brenninkmeijer 2010; Pardo & Patterson 2010; Reiner 2011) are viewed from a different angle when a person recognizes that technologies serve as a functional part of the self. Ultimately, with this perspective, humans must find a new mode of being, and consequently a new mode of understanding ourselves and others. For if our cognitive processes are not restricted to our biological brains, we need to give serious consideration to their role in our being.

An important result is that the relationship that people have with those devices merits much closer attention than they have received to date. We might go as far as arguing that we need to consider TEMs as new actors on the neuroethical stage. Thus, we call for a new framework for neuroethics that is informed by philosophy of mind which needs to reconsider how humans conceive of who they are today: What is at stake when studying neuroethical questions? What is the "substrate" we are dealing with? If we take the nature of the intertwinement of modern humans with their surroundings seriously—and we argue that we should—we cannot any longer focus on the human brain as isolated or, at most, integrated into a body. The body and the brain are not only in constant interaction with the environment (Varela et al. 1991; Gallagher 2005; Fuchs 2011; Nagel 2015), they also constantly make use of their surroundings, of significant and non-significant others, and of technologies. Thus, in this new framework, neuroethics is not so much concerned with an organ but ultimately with what makes us human beings, and recognizing the impact of technology upon what makes us "us" today matters for the relevant questions in the field.

References

Adams, F. and Aizawa, K. (2001). The bounds of cognition. *Philosophical Psychology*, 14(1), 43–64.

Adams, F. and Aizawa, K. (2010). Defending the bounds of cognition. In: Menary, R. (Ed.) *The Extended Mind*, pp.67–80. Cambridge, MA: MIT Press.

Beauchamp, T.L. and Childress, J.F. (2012). *Principles of Biomedical Ethics* (7th edn.). New York: Oxford University Press.

Bering, J. (2006). The folk psychology of souls. *Behavioral and Brain Sciences*, 29(5), 453–462.

Brand, C. (2010). Am I still me? Personal identity in neuroethical debates. *Medicine Studies*, 1(4), 393–406.

Brenninkmeijer, J. (2010). Taking care of one's brain: how manipulating the brain changes people's selves. *History of the Human Sciences*, 23(1), 107–126.

Burwood, S. (2009). Are we our brains? *Philosophical Investigations*, 32, 113–133.

Christman, J. (2004). Relational autonomy, liberal individualism, and the social constitution of selves. *Philosophical Studies*, 117(1), 143–164.

Clark, A. and Chalmers, D. (1998). The extended mind. *Analysis*, 58(1), 7–19.

Dworkin, G. (1981). The concept of autonomy. *Grazer Philosophische Studien*, 12, 203–213.

Farah, M.J. (2015). The unknowns of cognitive enhancement. *Science (New York, NY)*, 350, 379–380.

Farah, M.J. and Murphy, N. (2009). Neuroscience and the soul. *Science (New York, NY)*, 323(5918), 1168.

Farah, M.J., Illes, J., Cook-Deegan, R., et al. (2004). Neurocognitive enhancement: what can we do and what should we do? *Nature Reviews Neuroscience*, 5(5), 421–425.

Farah, M.J., Smith, M.E., Gawuga, C., Lindsell, D., and Foster, D. (2009). Brain imaging and brain privacy: a realistic concern? *Journal of Cognitive Neuroscience*, 21(1), 119–127.

Feinberg, T.E. (2006). Our brains, our selves. *Dædalus*, 135(4), 72–80.

Fitz, N.S. and Reiner, P.B. (2016). Time to expand the mind. *Nature*, 531, S9.

Fitz, N.S., Nadler, R., Manogaran, P., Cong, E.W.J., and Reiner, P.B. (2014). Public attitudes toward cognitive enhancement. *Neuroethics*, 7(2), 173–188.

Fogg, B.J. (2002). *Persuasive Technology: Using Computers to Change What We Think and Do*. San Francisco, CA: Morgan Kaufmann.

Frankfurt, H.G. (1971). Freedom of the will and the concept of a person. *Journal of Philosophy*, 68(1), 5–20.

Friedman, M. (2003). Autonomy and social relationships: rethinking the feminist critique. In: *Autonomy, Gender, Politics*, pp.81–97. Oxford: Oxford University Press.

Fuchs, T. (2011). The brain—a mediating organ. *Journal of Consciousness Studies*, 18, 196–221.

Gallagher, S. (2005). *How the Body Shapes the Mind*. New York: Oxford University Press.

Glannon, W. (2009). Our brains are not us. *Bioethics*, 23(6), 321–329.

Greely, H.T. and Illes, J. (2007). Neuroscience-based lie detection: the urgent need for regulation. *American Journal of Law & Medicine*, 33(2–3), 377–431.

Greely, H.T., Sahakian, B., Harris, J., et al. (2008). Towards responsible use of cognitive-enhancing drugs by the healthy. *Nature*, 456(7223), 702–705.

Greene, J. and Cohen, J. (2004). For the law, neuroscience changes nothing and everything. *Philosophical Transactions of the Royal Society B: Biological Sciences*, 359(1451), 1775–1785.

Haslam, N. (2004). Essentialist beliefs about personality and their implications. *Personality and Social Psychology Bulletin*, **30**(12), 1661–1673.

Haynes, J.-D. and **Rees, G.** (2006). Decoding mental states from brain activity in humans. *Nature Reviews Neuroscience*, **7**(7), 523–534.

Hill, K. (2012). *How Target Figured Out A Teen Girl Was Pregnant Before Her Father Did*. Available from: http://www.forbes.com/sites/kashmirhill/2012/02/16/how-target-figured-out-a-teen-girl-was-pregnant-before-her-father-did/4/#583f593d34c6 [accessed March 27, 2016].

Holzkamp, K. (1985). *Grundlegung der Psychologie*. Frankfurt: Campus-Verlag.

Ilieva, I.P., Hook, C.J., and **Farah, M.J.** (2015). Prescription stimulants' effects on healthy inhibitory control, working memory, and episodic memory: a meta-analysis. *Journal of Cognitive Neuroscience*, **27**(6), 1069–1089.

Illes, J. (Ed.) (2006). *Neuroethics: Defining the Issues in Theory, Practice, and Policy*. Oxford: Oxford University Press.

Illes, J. (2007). Empirical neuroethics. Can brain imaging visualize human thought? Why is neuroethics interested in such a possibility? *EMBO Reports*, **8**, S57–S60.

Levy, N. (2007). Rethinking neuroethics in the light of the extended mind thesis. *American Journal of Bioethics*, **7**(9), 3–11.

Lynch, M. (2013). *Brief of Michael P. Lynch as Amicus Curiae in Support of the Plaintiffs, American Civil Liberties Union v. James R. Clapper*. Available from: https://www.academia.edu/4579430/_The_Privacy_Brief_Amicus_Curiae_Brief_in_Support_of_the_ACLU_vs._NSA.

Mackenzie, C. (2010). Imagining oneself otherwise. In: Mackenzie, C. and Stoljar, N. (Eds.) *Relational Autonomy: Feminist Perspectives on Autonomy, Agency, and the Social Self*, pp.124–150. New York: Oxford University Press.

Menary, R. (2010). *The Extended Mind*. Cambridge, MA: MIT Press.

Meyers, D. (2005). Decentralizing autonomy: five faces of selfhood. In: Christman, J. and Anderson, J. (Eds.) *Autonomy and the Challenges to Liberalism: New Essays*, pp.27–55. New York: Cambridge University Press.

Mitchell, T.M., Shinkareva, S.V., Carlson, A., et al. (2008). Predicting human brain activity associated with the meanings of nouns. *Science (New York, NY)*, **320**(5880), 1191–1195.

Nagel, S.K. (2015). Thickening descriptions with views from pragmatism and anthropology. In: Metzinger, T. and Windt, J.M. (Eds.) *Open MIND* (Vol. 1), pp.829–839. Cambridge, MA: MIT Press.

Nagel, S.K., Hrincu, V., and **Reiner, P.B.** (2016, May 13–14). *Algorithm Anxiety—Do Decision-Making Algorithms Pose a Threat to Autonomy?* Presented at 2016 IEEE International Symposium on Ethics in Engineering, Science and Technology, Vancouver, BC.

Naselaris, T., Prenger, R.J., Kay, K.N., Oliver, M., and **Gallant, J.L.** (2009). Bayesian reconstruction of natural images from human brain activity. *Neuron*, **63**(6), 902–915.

Nedelsky, J. (1989). Reconceiving autonomy: sources, thoughts and possibilities. *Yale Journal of Law & Feminism*, **1**(1), 5.

Noë, A. (2009). *Out of Our Heads: Why You Are Not Your Brain, and Other Lessons from the Biology of Consciousness*. New York: Hill and Wang.

Pardo, M.S. and **Patterson, D.** (2010). Philosophical foundations of law and neuroscience. *University of Illinois Law Review*, **5**, 1211–1250.

Parens, E. (1998). *Enhancing Human Traits: Ethical and Social Implications*. Washington, DC: Georgetown University Press.

Parens, E. and **Johnston, J.** (2014). Neuroimaging: beginning to appreciate its complexities. *Hastings Center Report*, **44**(s2), S2–S7.

Pew Research Center (2016). *Smartphone Ownership and Internet Usage Continues to Climb in Emerging Economies*. Available from: http://www.pewglobal.org/2016/02/22/smartphone-ownership-and-internet-usage-continues-to-climb-in-emerging-economies/ [accessed January 31, 2017].

Preston, J.L., Ritter, R.S., and **Hepler, J.** (2013). Neuroscience and the soul: competing explanations for the human experience. *Cognition*, **127**(1), 31–37.

Reiner, P.B. (2011). The rise of neuroessentialism. In Illes, J. and Sahakian, B.J. (Eds.) *Oxford Handbook of Neuroethics*, pp.161–175. Oxford: Oxford University Press.

Reiner, P.B. (2013). The biopolitics of cognitive enhancement. In: Hildt, E. and Franke, A. (Eds.) *Cognitive Enhancement: An Interdisciplinary Perspective*, pp.189–200. New York: Springer Science+Business Media.

Rissman, J., Greely, H.T., and **Wagner, A.D.** (2010). Detecting individual memories through the neural decoding of memory states and past experience. *Proceedings of the National Academy of Sciences of the United States of America*, **107**(21), 9849–9854.

Rose, S.P.R. (2005). Human agency in the neurocentric age. *EMBO Reports*, **6**(11), 1001–1005.

Roskies, A.L. (2007). The illusion of personhood. *American Journal of Bioethics*, **7**(1), 55–57.

Rupert, R.D. (2004). Challenges to the hypothesis of extended cognition. *Journal of Philosophy*, **101**(8), 389–428.

Rupert, R.D. (2013). Distributed cognition and extended mind theory. In: Kaldis, B. (Ed.) *Encyclopedia of Philosophy and the Social Sciences*, pp.209–213. Thousand Oaks, CA: Sage Publications Inc.

Scialabba, G. (2011). *The Modern Predicament*. Boston, MA: Pressed Wafer.

Shen, F.X. (2013). Mind, body, and the criminal law. *Minnesota Law Review*, **97**, 2036–2175.

Stanovich, K.E. (2009). Distinguishing the reflective, algorithmic, and autonomous minds: is it time for a tri-process theory. In Evans, J. and Frankish, K. (Eds.) *In Two Minds: Dual Processes and Beyond*, pp.55–88. Oxford: Oxford University Press.

Stutzman, F., Gross, R., and **Acquisti, A.** (2012). Silent listeners: the evolution of privacy and disclosure on Facebook. *Journal of Privacy and Confidentiality*, **2**, 7–41.

Tokmetzis, D. (2014). *How Your Innocent Smartphone Passes on Almost Your Entire Life to the Secret Service*. Available from: https://www.bof.nl/2014/07/30/how-your-innocent-smartphone-passes-on-almost-your-entire-life-to-the-secret-service/ [accessed March 27, 2016].

Varela, F.J., Thompson, E., and **Rosch, E.** (1991). *The Embodied Mind: Cognitive Science and Human Experience*. Cambridge, MA: MIT Press.

Verbeek, P.-P. (2006). Persuasive technology and moral responsibility toward an ethical framework for persuasive technologies. *Persuasive*, **6**, 1–15.

Verbeek, P.-P. (2009). Ambient intelligence and persuasive technology: the blurring boundaries between human and technology. *NanoEthics*, **3**(3), 231–242.

Warren, S.D. and **Brandeis, L.D.** (1890). The right to privacy. *Harvard Law Review*, **4**(5), 193–220.

Wittes, B. and **Chong, J.** (2014). *Our Cyborg Future: Law and Policy Implications*. Center for Technology Innovation at Brookings. Available from: https://www.brookings.edu/research/our-cyborg-future-law-and-policy-implications/ [accessed January 31, 2017].

Wittes, B. and **Liu, J.C.** (2015). *The Privacy Paradox: The Privacy Benefits of Privacy Threats*. Center for Technology Innovation at Brookings. Available from: https://www.brookings.edu/wp-content/uploads/2016/06/Wittes-and-Liu_Privacy-paradox_v10.pdf [accessed January 31, 2017].

Wolpe, P., Foster, K., and **Langleben, D.** (2005). Emerging neurotechnologies for lie-detection: promises and perils. *American Journal of Bioethics*, **5**(2), 39–49.

Chapter 7

Neuromodulation ethics: Preparing for brain–computer interface medicine

Eran Klein

Introduction

An expressed aim of brain–computer interface (BCI) or brain–machine interface (BMI) research is to develop devices that improve lives of people with disabilities (Hochberg and Anderson 2012; Shih et al. 2012; Murphy et al. 2016).[1] BCIs are integrated input–output devices that acquire and translate brain signals in order to change physiologic function or the environment through control of an output device. BCI research has experienced explosive growth in recent years (Mak & Wolpaw 2009) due to a confluence of advances, including computational power for modeling brain function, electrode design, neuro-imaging and electroencephalography (EEG) capabilities, and clinical experience with implanted neural devices such as deep brain stimulation (DBS). BCI technology is being envisioned as a way to address unmet or suboptimally met medical needs. BCIs can be wearable, such as EEG-based devices that detect brain waves through the skull, or implant-able, such as electrodes inserted into the brain to electrically stimulate or record neuronal activity. Though there are few current clinical applications of BCI and significant techni-cal challenges to wider translation remain, implantable and nonimplantable BCI-based devices promise to be used for a wide range of disorders of movement, affect, sensation, and cognition (Mak & Wolpaw 2009; Wolpaw & Wolpaw 2012; Soekedar et al. 2014).

Human BCI devices raise interesting ethical questions, some of which have been explored at the research and policy level:

* Privacy of thought (Clausen 2011)
* Security of brain data (Denning et al. 2009)
* Changes to identity (Goering 2014)
* Responsibility for action (Haselager 2013)
* Access to expensive technology and post-study obligations to subjects (Schneider et al. 2012)
* Informed consent for research participation (Klein 2016).

Though ethical issues raised by BCI research overlap with those of other neural interven-tions, such as DBS, neural cell transplantation, and neuropharmacology, the coupling of recorded brain data to output interventions (e.g., electrical, mechanical, molecular, and sensory-mediated) opens new possibilities for precise and timely modulation of brain function. A goal for BCI-based applications is to use algorithms to automatically adjust device function in a closed-loop fashion, substituting machine learning for manual interrogation and adjustment of devices (Hebb et al. 2014). The prospect of closed-loop medical devices with embedded artificial intelligence raises particularly challenging ethical questions for BCI research and policy such as determining agency and responsibility for behavior related to device function (Glannon 2016).

The transition of BCI devices from research into clinical practice raises clinical ethics challenges as well. A few of these challenges are already beginning to come into view. For instance, seizure detection and stimulation devices—an early kind of BCI—raise questions related to predictive, advisory, and automated functionalities (Gilbert 2015). How will new capabilities affect a patient's sense of autonomy? What would it mean to over-rely on such a device? How central to an individual's autonomy is feeling in control over a device (Goering 2015)? Such concerns provide a hint of the kinds of ethical questions that clinicians and patients are likely to face. The development of BCI technology and the prospect of its wider incorporation into the practice of medicine raise clinical ethics challenges that are worth trying to anticipate and preemptively address.

The purpose of this chapter is to sketch out some of the clinical ethics challenges of BCI medicine and to facilitate a discussion about how best to approach these. In the first section, I give an overview of the kinds of BCI devices that are likely to become a part of medicine in the near term—devices that target motor, communication, sensation, cognition, consciousness and affective impairments. This is not meant to be a comprehensive survey of BCI technology, but to give a glimpse of the current state and ambitions of the field and the prospect for clinically meaningful devices. In the second section, I explore some of the ethical issues around consent to BCI-based medical devices. In the third section, I discuss some of the ethical challenges of neuromodulation. In the final section, I argue that looking to an engineering model of medicine for ethical guidance in BCI medicine is problematic but that a rehabilitation model of medicine holds promise.

Near-term horizon of brain–computer interface medicine

BCI devices can be wearable or surgically implanted with different potential clinical uses and capabilities. Different kinds of brain signals can be obtained through implanted sensors versus nonimplantable technology, such as EEG, functional magnetic resonance imaging (fMRI), or near-infrared spectroscopy (NIRS), as well as different intended outputs of the BCI, for instance, implanted versus skull surface neurostimulation (Soekadar et al. 2014). For some BCI-based devices, brain signal information obtainable through nonimplanted means may be sufficient whereas for others the fine-grained, high signal-to-noise ratio neural data available from implanted electrodes may be necessary. Similarly for the output side of BCI, implanted electrodes or other implanted devices may be needed to appropriately target a specific brain region for some clinical indications, whereas this may be unnecessary in other kinds of BCIs, for instance, to apply skull surface stimulation or to control a robotic prosthetic limb or communication speller. The technological and functional tradeoffs of wearable versus implantable BCI take place against a backdrop of risks associated with surgical implantation and ongoing care of implanted devices.

The neurophysiologic brain signals used in BCI can be metabolic or magnetic, but currently the principal type is electrophysiologic (Mak & Wolpaw 2009). Four types of electrophysiologic brain signals are currently being investigated as a basis of wearable BCI technology (Soekadar & Birbaumer 2015): (1) slow cortical potentials, (2) sensorimotor or motor-related beta rhythms, (3) event-related potentials, and (4) steady-state or auditory-evoked potentials. Surgically implanted electrodes, on the other hand, whether strips, grids, or other arrays, can provide brain signal information at a different level, such as local field potentials, (2) single unit activity, or (3) multiunit activity. The brain signals can be generated actively, reactively, or passively, depending on whether the user needs to attend to or respond to a task or whether background cognitive monitoring is involved (Soekadar et al. 2014). Once the desired brain signal is acquired, the features of the signal encoding the intent of the user must be extracted, transformed by algorithm into a device command, and transmitted to a device for execution (Wolpaw et al. 2002).

Potential clinical applications of BCI span a wide range. BCI is currently being explored for use in controlling computer cursors, wheelchairs, robotic arms, or reanimated limbs (Mak & Wolpaw 2009). Depending on how widely BCI is defined and how the future of technological development and acceptance proceeds, BCI devices one day could be pervasive in medicine. Near-term clinical uses of BCI can be schematically grouped by functional impairment domains: motor, communication, sensation, cognition, consciousness, and affect.[2] Technology for BCI devices—wearable, implantable, or both—is being developed currently within each of these domains. In this section, I will briefly give examples of potential BCI devices being explored in each of these impairment domains.[3] It should be noted that many of these devices exist at an early stage of development, some merely at a proof-of-principle stage, and whether any of these particular devices will prove clinically beneficial or commercially viable is an open question.

Motor impairment

Examples of BCI for improving or substituting for impaired motor function include BCI-controlled exoskeletons, prosthetics, or robotics. Functional electrical stimulation represents an example of a BCI system for motor control in which EEG signals can be used to control electrical stimulation of impaired muscle systems, for instance, to facilitate the grasping motion of a paralyzed hand (Pfurtscheller et al. 2003) or assist in walking (King et al. 2014). The Braingate system is a well-known example of robotic control using brain-implanted electrodes. Hochberg and colleagues implanted dense electrode arrays in two individuals with quadriplegia that allowed them to use local field potentials to control computer cursors (Hochberg et al. 2006). Other subjects have been able to use devices with implanted electrodes to control prosthetic limbs (Collinger et al. 2013b).

A hope for BCI technology is that it may improve on neurostimulation devices already in clinical use, such as DBS for Parkinson's disease, essential tremor, dystonia, and Tourette's syndrome. An example of this is a BCI-based closed-loop DBS system for essential tremor. Essential tremor is a condition characterized by episodic tremor of the head or extremities made worse by initiation of movement, such as holding a pen or coffee cup. When medications are ineffective or not tolerated, open-loop DBS can be an effective treatment for essential tremor. Standard open-loop DBS systems for essential tremor involve always-on neurostimulation through electrodes placed into the ventral intermediate nucleus of the thalamus. Though effective, open-loop DBS systems provide a fixed pattern of electrical stimulation, even though essential tremor is only episodically symptomatic. BCI-based systems are being explored that could combine DBS with additional implanted electrodes capable of detecting the onset of tremor (Herron et al. 2015). BCI-based DBS applies stimulation only when needed, when characteristic neural patterns of tremor are detected. Potential advantages of a BCI-based system over current open-loop DBS include battery conservation and reduction in short- and long-terms side effects experienced by individuals with fixed neurostimulation settings.

Communication

BCI offers a way to restore or enhance communication in individuals with communicative impairment (Akcakaya et al. 2014). BCI-based devices use (1) slow cortical potentials, (2) sensorimotor rhythms, (3) event-related potentials, or (4) steady-state visual-evoked potentials to control communication devices that allow individuals to answer yes/no questions or select semantic elements, such as letters, words, or phrases (Soekadar & Birbaumer 2015). The first clinically useful BCI allowed two individuals with severe paralysis due to amyotrophic lateral sclerosis to communicate using slow cortical potentials to move a computer cursor to select letters (Birbaumer et al. 1999). Individuals who are incompletely locked-in due to conditions like amyotrophic lateral sclerosis or stroke can communicate through muscle twitches or eye movements. Such communication can be laborious and inefficient and individuals can lose these means of communication if their conditions deteriorate and they become completely locked-in and unable to communicate through volitional movement at all. BCI communication promises a way to maintain and potentially restore communication for this and other conditions affecting communication.

An example of BCI for communication is an EEG-based speller for individuals with locked-in state due to conditions like amyotrophic lateral sclerosis or brainstem stroke (Chaudhary & Birbaumer 2015). Event-related potentials, such as steady-state evoked potentials or P300, can allow individuals to communicate by selecting language elements. As this BCI paradigm is based on an individual focusing on and hence selecting from language elements from those presented on a screen, such as letters or words, measured by a P300 or other response, training and cognitive demands involved may be significantly less than other BCI communication approaches. Development of spellers that can be individualized to user neurophysiologic and language patterns and can incorporate lessons from natural language models may substantially improve accuracy and speed of BCI-based communication (Mainsah et al. 2015).

Sensation

BCI may be a way to bypass, modify, or substitute for impaired or aberrant sensory processes (Konrad & Shanks 2010). Sensory neural implants with the widest clinical use are cochlear implants for sensorineural hearing loss, where direct electrical stimulation is applied outside the central nervous system to the basilar membrane of the cochlea. Auditory brainstem implants are a form of BCI that have shown success at restoring some level of speech recognition in individuals with acoustic nerve damage (Colletti et al. 2009). Visual prosthetics have been developed with electrodes capable of stimulating the retina, visual cortex, or lateral geniculate body of the thalamus (Cohen 2007).

Stroke is the leading cause of disability worldwide and BCI is emerging as a promising area of research for stroke-related needs (Soekadar et al. 2015). BCI research after stroke has focused on (1) bypassing corticospinal pathways damaged by stroke to activate distal muscles or nerves directly or to control neuroprostheses or (2) facilitating neuroplasticity and motor learning for functional motor recovery (Mak & Wolpaw 2009). There is increasing recognition that sensory feedback is a critical element of stroke recovery and in control of stroke-related motor prostheses (Patil & Turner 2008).

Cognition and consciousness

BCI may be used to improve cognition or states of consciousness that support cognitive processes. Neurodegenerative disease, such as Alzheimer's disease, dementia with Lewy bodies, frontotemporal dementia, and nondegenerative conditions, like stroke or traumatic brain injury, are potential targets of both nonimplantable and implantable BCI devices. Implantable neurostimulation is an active area of research in disorders of consciousness (Giacino et al. 2014). Up to 30–40% of individuals diagnosed as being in a vegetative state retain conscious awareness (Seel et al. 2010). DBS has been used in one subject diagnosed as minimally conscious due to traumatic brain injury to improve arousal and restore some functional movements of upper extremities and the ability to self-feed (Schiff et al. 2007). Nonimplantable neurostimulation, such as transcranial magnetic stimulation (TMS), has been explored in disorders of consciousness as well (Giacino et al. 2014).

Treatment of dementia is a potential application of BCI (Freund et al. 2009; Mirzadeh et al. 2015). Individuals with Alzheimer's disease exhibit changes in neurophysiological measures, including delta, theta, and beta band frequencies (Soekedar et al. 2014). These differences relative to controls may be useful in diagnosis but could also be used in a BCI device to target brain regions for neurostimulation therapy. For instance, nonimplantable neurostimulation using TMS and transcranial direct current stimulation (Elder & Taylor 2014) and implantable neurostimulation using a vagus nerve stimulator (Sjögren et al. 2002) and DBS (Kuhn et al. 2015) have demonstrated modest cognitive benefits in subjects with dementia. Neurostimulation may activate neurotransmitter systems (e.g., acetylcholine) and stem loss of hippocampal volumes, as has been shown in some subjects with DBS (Sankar 2015; Sankar et al. 2015). Although the clinical significance of cognitive benefits in this population is far from clear (Nardone et al. 2015), the results suggest that neurostimulation could be useful. For instance, a BCI device might be used to promote neuroplasticity for cognitive rehabilitation in combination with future interventions that slow or stop neurodegenerative processes. Although admittedly speculative, BCI devices could provide a way to measure substrates of impaired cognition and provide targeted therapeutic neurostimulation.

Affect

Depression is the most common psychiatric illness, with a lifetime prevalence of nearly 20% (Bromet et al. 2011). Standard pharmacologic therapy and psychotherapy fail to alleviate symptoms in a substantial fraction of patients (Warden et al. 2007). Neurostimulation is an active area of research for treatment-resistant psychiatric disease, including depression. Electroconvulsive therapy is a well-recognized therapy for treatment-resistant depression. Newer methods of nonimplantable neurostimulation, such as TMS and transcranial direct current stimulation, have shown modest benefits (George & Aston-Jones 2010). Repetitive TMS received approval from the US Food and Drug Administration for mild treatment-resistant depression in 2008. DBS has been studied in major depressive disorder, with mixed results. Open-label studies of DBS showed success (Mayberg et al. 2005; Malone et al. 2009) while two formal clinical trials did not meet primary outcomes (Dougherty et al. 2015). Differences in psychiatric symptoms across individuals and within individuals over time have been postulated to explain this discrepancy (Widge et al. 2017).

BCI-based (closed-loop) DBS offers a way to address the heterogeneity of depression (Widge 2016). With embedded algorithms designed to automatically adjust stimulation to need, BCI-based DBS may be a significant improvement on open-loop DBS used in the failed clinical trials. Rather than employing a one-size-fits-all approach, which may be responsible for common side effects, such systems could detect real-time fluctuations in depressive symptoms and individualize stimulation to meet clinical need. The type, and even the location, of stimulation could be fine-tuned to meet clinical need or neurophysiological markers of happiness or quality of life. The potential to further incorporate machine learning into BCI systems offers a way to not only treat depression in increasingly effective ways but detect patterns of brain activity correlated with depression triggers and intervene before symptoms can take hold.

Clinical ethics and consent to brain–computer interface devices

The incorporation of BCI technology into medical devices, as with any new medical technology, raises ethical issues, not just at the level of research and policy, but at the level of patient and clinician interaction. Individual patients and their clinicians make decisions about whether or when to opt for medical devices. Consent to a medical device is not an isolated event, but a decisional process situated against a backdrop of relationships with family, friends, clinicians, and others—and is ethically freighted. What responsibilities and expectations come with accepting a device? Will adoption require a change to daily routine? Will the device come with direct or ancillary financial or other costs? How will adopting (or declining) the device change burdens on or relationships with caregivers? Will having the device engender unwanted expectations of family, clinicians, or even self? For implanted devices, what are the risks of surgery? What are the alternatives to or opportunity costs of adopting (or declining) a device? These kinds of ethical considerations are important to the consent process and fall under the provenance of what is commonly called clinical ethics.

Clinical ethics is a subdiscipline of bioethics concerned with ethical issues arising within the practice of medicine, such as decisions about diagnosis and therapy (Miller et al. 1996). Clinical ethics, insofar as it takes the clinical relationship as a focal point, provides a lens through which to view clinician responsibilities within the consent process. A principal responsibility of clinicians is to help patients understand risks associated with potential interventions. In the case of consent to BCI research, six risk domains have been identified (Klein 2016). Several of these risk domains—safety, cognitive and communicative impairment, inappropriate expectations, and privacy and security—are particularly relevant to *clinical* adoption of BCI devices as well. It is worth exploring these risk domains as they pertain to BCI medicine and what responsibilities clinicians have or will have in helping patients navigate the choice to adopt or decline a BCI.

The first challenge pertains to safety. Even after meeting regulatory standards for approval, important short- and long-term safety uncertainties will remain with respect to BCI devices. Just as certain side effects or adverse events are not known or fully appreciated until after a medication reaches wide clinical use (Klein & Bourdette 2013), so too with devices. For instance, biocompatibility and durability of implantable electrodes is an ongoing concern as electrodes can fail or lose fidelity over time due to breakdown of electrode components or to brain tissue reactive processes, such as gliotic encapsulation of electrodes (McGie 2013). Such longevity and long-term reliability uncertainties of BCI systems are important considerations in the consent process. This is particularly true when weighing potential benefits against risks of surgical implantation. Some individuals may be willing to undergo neurosurgery and tolerate uncertainties about how long a device will last, whereas others may have a lower risk tolerance and be willing to pass up potential benefits. There are safety risks of nonimplantable devices as well. For instance, transcranial direct current stimulation may cause electrical burns of the scalp. Clinicians have a responsibility to not only stay abreast of and effectively communicate the current state of known safety risks, but to explore how safety risks fit with patient values and future plans (Klein 2016).

A second challenge is that of progressive cognitive impairment. BCI may one day prove useful in stopping, slowing, or reversing pathological processes leading to cognitive impairment, such as Alzheimer's disease, or in treating noncognitive symptoms in patients with cognitive disorders, such as tremor in patients with Parkinson's disease dementia. Consent to BCI therapy in individuals with mild to moderate levels of cognitive impairment will require special attention to decision-making capacity. Do individuals demonstrate adequate understanding of the risks and benefits of a BCI device? Do they appreciate what impact adopting a BCI device will or will not likely have on their symptoms, level of functioning, and quality of life? Can they reason through BCI as a potential therapeutic option versus other available options? Can individuals make and express a choice about whether to go forward with or decline a BCI device? Though assessment of decisional capacity can be less formal and woven into ongoing discussions between patients, families, and clinicians, more formalized tools are available to assist with capacity assessments in individuals with known cognitive impairment, such as the MacArthur Competence Assessment Tool for Treatment (Grisso & Appelbaum 1998). In addition, multidisciplinary teams that incorporate individuals with relevant expertise, such as neuropsychologists, nurses, and social workers, can provide more comprehensive assessment of capacity to consent to BCI therapy. Models of multidisciplinary involvement in the consent process have been developed for DBS (Ford & Kubu 2006).

When decision-making capacity is absent or sufficiently diminished due to cognitive impairment, clinicians identify and work with appropriate surrogate decision-makers. While severe cognitive impairment may limit the range of applicable BCI devices, some BCI devices may still be of benefit. For instance, individuals with dementia can experience somatic pain that is undertreated in part due to difficulty communicating the experience of pain to caregivers (Malloy & Hadjistavropoulos 2004). A BCI that senses brain patterns characteristic of pain and in turn controls a neurostimulation device to treat the pain could address this problem. In this case, clinicians and surrogates will need to work together to determine whether BCI is in the overall best interest of patients. The matter will be more complicated when BCI is proposed as a way to *restore* cognitive capabilities in individuals who have lost them. This may require designing consent processes that make room for a revisiting of the decision to pursue a BCI or how it operates as elements of decision-capacity progressively improve. Something similar has been suggested with respect to DBS research in minimally conscious states (Giacino et al. 2012).

BCI may also have adverse effects on cognition. Some individuals with DBS for Parkinson's disease, for instance, experience more cognitive decline than would be expected as a result of the natural course of the disease (Combs et al. 2015). Concerns about potential cognitive side effects of BCI need to be approached with caution. Not enough is currently known about long-term cognitive effects of neurostimulation and the terms "cognition" and "BCI" are exceeding broad and often imprecisely used. It would not be surprising, however, if BCI devices led simultaneously to beneficial and detrimental effects on different areas of cognition depending on the type and target location of BCI and how cognition is measured.

Clinicians will have a responsibility to counsel patients on the range of cognitive effects of BCI-based therapies. Individuals with impaired communication but normal cognitive capabilities face barriers to meaningful informed consent (Fenton & Alpert 2008).

In individuals with partially locked-in syndrome, standard informed consent processes that involve demonstrating—through words or gestures—an understanding of disclosed information and a reasoning process about potential options, do not meet clinical reality. Extremely limited and intermittent communication is often incompatible with exigent demands of clinical decision-making. Even if such individuals are able to communicate by using a BCI, it is unclear whether yes/no or other linguistically constrained communication methods are sufficient for the nuanced and emotionally complex decisions about treatment in locked-in syndrome, such as declining life-sustaining treatment (Glannon 2016). Clinicians may lack the experience in how to engage in these discussions using a BCI device or how to assess the reliability and authenticity of patient responses. Consent to a BCI communication device by individuals with completely locked-in syndrome—where no speech or gestural communication is possible at all—presents a different issue. The ability to communicate is so central to the exercise of individual autonomy and to quality of life (Birbaumer et al. 1999) that it may be reasonable for surrogates to consent for interventions that aim to restore communicative ability, even if they involve significant risks.

A third challenge is that of managing expectations. Unrealistic expectations are a problem in BCI research (Haselager et al. 2009) and, if the transition to clinical use of DBS is a guide (Bell et al. 2010), this will be true of BCI as well. Most BCI devices are not implemented as off-the-shelf technology, but require significant training that involves physical and emotional investment on the part of the user (Mak & Wolpaw 2009; McGie et al. 2013). Clinicians have a responsibility to prepare patients and families for the work that is required of them and for the unfortunate possibility that a BCI device may not work as well as a patient or family hopes. Helping patients titrate expectations is a general responsibility of clinicians, but is particularly important and difficult with BCI devices when overly optimistic portrayals of BCIs in the media create headwinds for reasonable patient expectations. Conversely, inappropriately low expectations are a problem in BCI medicine as well. For instance, individuals with locked-in syndrome who require chronic artificial respiration and feeding demonstrate a higher quality of life than is typically presumed by significant others (Kübler et al. 2005). A poor understanding of quality of life and disability among clinicians and families could influence decisions about life sustaining therapy, advance directives, and BCI-based therapies, such as BCI-assisted communication.

A fourth challenge is privacy and security. BCI devices have the potential to record and store vast amounts of data on individual brain function. This raises data security concerns (Denning et al. 2009). Brain data can be collected, stored, and transmitted in more or less secure ways and hence have the potential to be stolen and put to nefarious uses. The hacking of implantable devices is a noted risk. For instance, former US Vice President Richard Cheney reportedly had the wireless capability of his implanted cardiac defibrillator disabled out of concern that it might be susceptible to hacking (Kolata 2013). Technological solutions to prevent hacking are worth pursuing but the presence of recordable brain data creates an ineliminable risk.

Privacy concerns exist beyond the security challenges. A BCI device may collect data over which an individual wants to exert control. For instance, an individual with a BCI device for communication may not want every conversation recorded or may want to dump brain data associated with conversations, perhaps to protect a family member's feelings or because of embarrassment. The desire to provide individuals with such a sphere of cognitive privacy may run up against competing interests, such as the need to establish competency for financial or healthcare decisions. Did the patient ever indicate a desire to change their last will and testament and give everything to charity? Were the preferences in their living will consistent (before falling into their current incapacitated state), or does a BCI record indicate ambivalent feelings? BCI data may become a trove of information to which parties seek access. The clinician has a role in preparing patients for privacy risks associated with BCI brain data, even if the scope and available protections of this privacy are not currently clear.

While it is clear that clinicians have ethical responsibilities to assist patients as they engage in the consent process for BCI devices, clinicians also have responsibilities *after* adoption of a BCI device. A critical feature of BCI technology is the need to monitor and adjust device function over time. This feature of BCI medicine—the need for neuromodulation—raises its own set of ethical issues (Ford & Henderson 2006; Fukushi 2012). How should the settings of a device be determined? Who determines whether a device is working well or needs to be adjusted? Who should have a say—patients, families, clinicians, others—in decisions about neuromodulation? In BCI systems that incorporate machine learning, when should clinicians or others be able to intervene, say, if patients are unaware or disagree that a device is malfunctioning?

Clinical ethics and neuromodulation

Neuromodulation can be understood as an iterative, goal-directed intervention in the nervous system. This description spans a wide range. Individuals can neuromodulate through various neural interventions, or what the Presidential Commission for the Study of Bioethics Issues calls neural modifiers (Presidential Commission 2015). This can range from reaching for a caffeinated beverage to counter a lag in concentration to implantation of DBS electrodes to treat Parkinson's disease. The term neuromodulation, however, is typically reserved for pharmacologic or mechanical interventions into the nervous system. Neuromodulation is "a technology that impacts upon neural interfaces and is the science of how electrical, chemical, and mechanical interventions can modulate or change central and peripheral nervous system functioning" (Krames et al. 2009, p.5). With the exception of one-off interventions, for example, a one-time dose of tissue plasminogen activator for acute stroke treatment, most neural modifiers are neuromodulatory insofar as they satisfy three criteria: (1) they are dynamic and ongoing (either continuous or intermittent), (2) their effects are mediated through neural networks, and (3) their clinical effect is continuously controllable by modifying device or intervention parameters (Holsheimer 2003).

Clinicians already play a central role in neuromodulation using implantable devices. Neural stimulators, microinfusion pumps, and other devices are currently used or being developed for diverse disorders, including disorders of cardiac pacing, gastric motility, urinary dysfunction, eyesight, hearing, epilepsy, movement, chronic pain, spasticity, and mental health (Kames et al. 2009). As a result, clinicians from across the medical spectrum—anesthesiologists, cardiologists, gastroenterologists, neurologists, neuro-surgeons, ophthalmologists, otolaryngologists, pain physicians, psychiatrists, physical medicine and rehabilitation specialists, and urologists—have developed skills at setting, monitoring, and adjusting parameters of implanted devices. A ready example of this is programming visits after implantation of DBS for movement disorders like Parkinson's disease. After surgical implantation, patients or research subjects with these devices are evaluated at set intervals and device settings—frequency, pulse width, and voltage—are adjusted to treat symptoms, such as a reduction in tremor, or otherwise improve quality of life. The need to develop such neuromodulatory skills was recognized soon after the introduction of DBS: "Movement disorders specialists becoming involved with this ther-apy need to acquire new skills to optimally adapt stimulation parameters and medication after implantation of a DBS system" (Volkmann et al. 2002, S181).

Decisions about how *best* to modulate brain function can be a source of ethical conflict between clinician and patient. Peter Kramer's *Listening to Prozac* describes the experi-ences of patients treated with antidepressants and the difficulties that arise in defining appropriate and successful treatment. Kramer notes that clinicians and patients can dis-agree on proper use of or indications for these medications. He describes the case of Sally, a shy, anxious, and socially isolated 41-year-old woman transformed by Prozac', becom-ing "brighter, calmer, self-assured, in firm control of herself" (Kramer 1993, p.147). The extent of this transformation led to conflict between the psychiatrist, Kramer, and Sally:

> I felt concern that Sally may have "overshot," that this new personality was too different from her old one. She demurred. She said Prozac had let her personality emerge at last—she had not been alive before taking an antidepressant. *Sally insisted I not stop the medication, but I tapered the dose slightly.* (Kramer 1993, pp.147–148, italics added)

Similar concerns about nonpharmacologic neuromodulation have been discussed in the context of DBS. Schermer (2013) describes the case of a Dutch patient treated with DBS for obsessive–compulsive disorder who failed to achieve relief from symptoms of obsessive–compulsive disorder, but who felt "better and happier" with the DBS. Her psy-chiatrist refused her request to continue DBS stimulation, and turned off the stimulator, noting that physicians are not in the business of "trading happiness." Similarly, Synofzik and Fins (2012) describe the case of a German man with obsessive–compulsive disorder and anxiety implanted with a DBS who experienced symptoms of pathologic euphoria at high voltage settings. Despite the patient's request for settings that would allow him to feel "a bit better," his clinician refused: "After extensive yet difficult deliberation about the benefits of an only intermediate stage of happiness and an intermediate voltage, the parameter settings were left unchanged" (Synofzik & Fins 2012, p.32).

This raises a number of questions about the role of the clinician in neuromodulation. Some of these questions relate to potential conflict between clinician and patient. What justifies a clinician abiding by a patient's neuromodulation request or refusing it? Should clinicians invoke notions of clinical integrity or the goals of medicine in refusing patient requests for neuromodulation (Woopen 2012; Schermer 2013)? Other questions relate to tradeoffs involved in neuromodulation. What is the clinician's role in advising about scientific uncertainties of BCI, for instance, if electrodes begin to fail and applying higher levels of current to overcome encapsulation resistance offers a potential remedy but at an unknown risk of longer-term tissue damage? Or what if uncertainty exists as to whether a device is failing or an individual instead is experiencing progression of their disease? Moreover, how should a clinician counsel about side effects of treatment if a patient and family disagree about the value of these side effects, for example, increased disinhibition due to BCI in a previously painfully shy individual like Kramer's Sally? How might these questions change if choice points for modulation are obviated by device algorithms that automatically decide how to adjust device settings? Does this change the obligations of clinicians to counsel and care for patients with BCI devices? These questions might be reframed at a more general level: what constitutes responsible neuromodulation?

Neuromodulation and the engineering model of medicine

One way to approach this question is to look at the clinical skills involved in neuromodulation. Just as clinicians have developed expertise in how to adjust antidepressants to treat symptoms of depression or DBS stimulation parameters to more effectively treat symptoms of tremor, clinicians will develop technical skills at modulating the function of medical BCI devices. Neuromodulation viewed as a set of technical skills makes sense from within a particular view of medicine. Robert Veatch coined this view of medicine the "engineering model" (Veatch 1972).[4] Veatch argues that clinicians within this view are applied scientists who possess a repository of technical skills. These skills range from manual dexterity of surgeons to deductive logic of medical diagnosticians, are developed and refined through medical education, and both help define the boundaries of medical practice and the identities of its practitioners.

Viewed from within an engineering model of medicine, the skills of neuromodulation are inherently non-normative. Such skills can be more or less developed or more or less effectively employed. Whether these technical skills in turn are used *responsibly* is a second-order question. Responsible neuromodulation, on the engineering view, depends on how technical skills are put to use. Do these skills advance the interest or well-being of a particular patient? Does the exercise of these skills in a particular instance serve the interest of the broader community? Are these skills developed and exercised as part of society's social contract with the medical profession? Put simply, the clinician's neuromodulatory skills are a "tool" that can be put to good or bad use.

Consider an analogy. The medical specialty that cares for people with multiple sclerosis (MS) is called clinical neuroimmunology (or just neuroimmunology).[5] Neuroimmunologists are clinicians specialized in understanding and modulating interactions between the nervous system and the immune system. Immune modulation therapies (or "disease-modifying therapies") for MS have expanded in recent years, starting with subcutaneous interferons in the early 1990s and expanding to a broad range of thera-pies (Ransohoff et al. 2015). Immunomodulatory therapies involve a tradeoff—in order to counter the body's autoimmune attack on the nervous system, a reduction in the ability of the immune system to perform its normal functions, such as fighting infections (Rizvi & Coyle 2011), can result. The introduction of one medication, natalizumab, typifies the immunomodulatory trade-off. When initially introduced, natalizumab demonstrated significantly better effectiveness at treating MS, but this increased effectiveness came with a cost: a 1:1000 risk of progressive multifocal leukoencephalopathy, a devastating, untreatable, and potential fatal brain infection (Clifford et al. 2010). For some patients, the potential benefits of natalizumab outweigh the rare chance of progressive multifocal leukoencephalopathy, for others, this is not the case. Patients and clinicians face similar, though perhaps less stark, tradeoffs with other immunomodulatory therapies.

Immune modulation decisions—whether to start, reduce, escalate, or abandon a therapy—in turn, raise ethical questions. What is a reasonable tradeoff? What level of consent is needed? Can a clinician refuse to start therapy, say, if adherence to follow-up surveillance is likely to be poor? The engineering model of medicine provides one way to approach these kinds of ethical questions. Clinicians bring technical expertise—the ability to assess symptoms, note physical signs, examine biomarkers of disease stability or progression, and effectively communicate information—and patients bring a moral framework—comprised of their preferences, belief systems, and modes of reasoning. Dan Brock describes the division of labor at the heart of the engineering model as:

> In this approach, both physicians and patients have essential roles in ideal treatment decision mak-ing. Physicians are to use their knowledge, training, and expertise to provide their patients with a diagnosis and a prognosis if no treatment is undertaken, together with information about alterna-tive treatments that might improve the prognosis, including the risks and benefits and attendant uncertainties of such treatments. Patients articulate their own aims, preferences, and values in order to evaluate which alternative is best for them. (Brock 1991, p.31)

This division of labor has implications for assignment of moral responsibility. On the engineering model, the primary obligation of the clinician is to develop technical exper-tise and to translate this expertise into information for patients. This has led Emanuel and Emanuel (1992) to describe the engineering model as primarily centered on delivery of information. This information consists of assessment of the patient's health, diagnostic and therapeutic options, and relevant probabilities. The information might even include the clinician's assessment of what she takes the patient's values to be (though without taking a stand on them). One sees this in discussions of shared decision-making in MS (Heesen et al. 2011). On this view, whatever subsidiary obligations clinicians may have can be viewed as deriving from a primary obligation: develop and exercise technical expertise.

On an engineering model, the role of the clinician in neuromodulation is to apply tech-nical skills to "fix" some dysfunctional aspect of the nervous system. Technical expertise is problem-oriented. Skills are put to use to solve or fix a particular problem or set of problems. This orientation centered on fixing is important because it helps structure how ethical issues are approached and how clinician responsibilities are understood. As an example, consider the individual shown to kick a soccer ball by using a BCI-based exoskeleton at the 2016 World Cup in Brazil (Smith 2014). A team of researchers identified a problem—the inability of an individual with lower extremity paralysis to stand independently and kick an object—and developed a plan to solve it. By framing this research in terms of a single fixable problem (or series of problems) ethical questions that might be raised about this research inherit this frame. What was the consent process like? Is this a just use of societal resources? How should the media be utilized to educate the public about an emerging technology? Questions about consent, justice, responsible conduct of research, and other matters have their meaning in reference to solving a particular technical problem: fixing the neural dysfunction that is preventing kicking a soccer ball. As such, an engineering model provides a simplifying framework for reflecting on the ethical responsibilities of a "neuromodulationist."

Limitations of the engineering model

The engineering model of medicine has been criticized within bioethics (Clouser 1983; Emanuel & Emanuel 1992; Brody 1997). An argument against the engineering model's commitment to fixing can be found outside bioethics in recent work in disability theory. Disability has long been understood along a medical model (Silvers 2009). Within a medical model, disability is a property of individual bodies. This contrasts with a social model of disability that locates disability outside individual biological bodies and locates it, at least in large part, in the relationship between the individual and the social and built environment. An individual in a wheelchair may have a disability if trying to board a bus that lacks a chair lift, but not have a disability if a lift is present or, more to the point, have something quite unlike a disability if the wheelchair allows one to speed down a long hill and catch a bus before departing. Disability is a way of understanding the mismatch between individual preferences and needs and environmental or social obstacles to their fulfilment. An outgrowth of the medical model of disability is to view loss of a function—or the body out of which it arises—as something to fix with medicine or science. Here the engineering model of medicine and the medical model of disability overlap. Both models share a subject matter: broken bodies or functions that are in need of a fix.

Applying the disability critique to BCI medicine helps shed light on several distorting features of the engineering model: a tendency to overlook the perspectives of end users, obscure the normative features of BCI problems, and misunderstand that desired outcomes of BCI device can be underdetermined.

The first way in which the fixing frame distorts is that it presumes that problems to be fixed are self-evident and easily identifiable from any standpoint, leading to such overly simplistic questions as: Who wouldn't want to restore walking ability with the use of a BCI exoskeleton? Who wouldn't want to restore the ability to write with the use of a BCI prosthetic device? Who wouldn't want to be fixed? Yet research in populations targeted by BCI devices challenge received assumptions about what individuals do and do not value. Anderson (2004) surveyed 681 individuals with spinal cord injury and found that regaining walking function was not a high priority among individuals with quadriplegia and paraplegia, whereas restoration of upper extremity and sexual function were. The value of end-user perspectives in BCI research is beginning to be recognized (Huggins et al. 2011; Collinger et al. 2013a; Kübler et al. 2015).

The second way in which the fixing frame can distort is in obscuring how problems addressed by BCI are normatively defined. Take the example of designing a BCI neuro-prosthetic arm that allows an individual to shake hands (Klein et al. 2015). The characteristics of handshakes are known to vary across cultures (Dibiase & Gunnoe 2004). What constitutes a handshake depends on norms and context. A grip that is strong and lingers may indicate confidence and competence in certain contexts, like a business meeting, but may be awkward or threatening in others, such as meeting a prospective son-in-law or daughter-in-law for the first time. That problems might not be definable without reference to social norms undermines the fixing frame. The promise of a purely technical fix only makes sense for problems that can be technically defined without remainder, and many of the functions that BCI devices target will be, at least in part, normatively defined.

The final limitation of the engineering model applied to BCI medicine derives from the iterative nature of neuromodulation. As discussed earlier, consent to obtain a BCI device is an important decision, and one that involves the patient and clinician in a discussion of goals, values, and expected outcomes. A process of shared decision-making is undertaken. One of the challenges of the engineering model is that while it may be of adequate fit for the initial decision to get a BCI device, it fits poorly decisions (or iterative consents) related to ongoing neuromodulation. The clinical benefit of a BCI device will turn to a great extent on a multitude of decisions about how to modulate the device function over time. The endpoint around which this modulation occurs is not fixed. An individual may start out wanting no tremor or no depression but over time develop a more nuanced set of preferences for BCI function, such as no tremor in church, or no depressive symptoms related to self-worth. What is important is that the goals of BCI therapy will be underdetermined at initiation of therapy. Patients, clinicians, and family will have to engage in an iterative practice of values exploration in order to modulate the device function effectively.

Brain–computer interfaces as enabling technology

An alternative framework for neuromodulation may be found in rehabilitation medicine. A central feature of rehabilitation is the use of enabling technology. Enabling technology includes traditional tools such as canes, walkers, and prosthetics, but increasingly includes tools with technological sophistication, such as BCI devices. What makes tools enabling technology (as opposed to enhancing or fixing technology) is that they alleviate the impact of disease or disability; such tools can be therapeutic, compensatory, assistive, or universal technology (Hansson 2007). The clinician's role in rehabilitation is not to fix patients with enabling technology but help them use technology to pursue goals relative to a given reference point. As a patient's abilities and possibilities change during the course of rehabilitation therapy, clinicians help patients reevaluate and reformulate goals over time in light of these changes. Most BCI devices under development are a form of assistive technology.

Consider a hypothetical case of Judith, a woman who receives a BCI-based DBS device for treatment-resistant depression. Judith's device has electrodes that record neural signals relevant to depression and uses these to adjust stimulation settings of DBS electrodes. At the outset of therapy, Judith and her clinician decide on the initial setting for her device. Whenever a certain pattern of activity is detected that correlates with her depressive symptoms, the DBS electrodes will apply stimulation to return her brain to a normal state. Over time, Judith and her clinician will reevaluate and adjust her device settings, that is, neuromodulate. What happens if Judith and her clinician disagree on new settings? Perhaps Judith comes to miss the fluctuations of her pre-device self or feels that a little dysthymia is helpful for her work as an artist. And maybe her clinician becomes wary, based on his prior relationship with her that a closed-loop system permissive of fluctuations may allow her to "spiral down" toward suicide. How should decisions about neuromodulation be made? What does consent mean in this context?

An advantage of looking to rehabilitation medicine is that it provides an alternative model of consent for neuromodulation. Purtilo (1988) argues that rehabilitation medicine provides a good example of why the event model of consent is inadequate. Decisions made in rehabilitation are most often iterative and incremental: as a patient's condition or functional capabilities change, therapies are adjusted, often in small ways. For example, the consent to a rehabilitation program to retrain one's upper extremity for self-feeding after a stroke is more accurately understood as a series of decisions that are made over time and represent small steps. "Today I will agree to work with the physical therapist." "Today I will try to move my fingers." "Today I will ignore spastic pain." "Today I will try to grasp an object." "Today I will try to lift an object." And so on. These decisions may not be linear or wholly predictable. They are a process. A process view of consent is a better fit for the iterative decisions that characterize BCI medicine.

The introduction of clinical BCI devices raises important ethical questions. Some of these questions will be faced by patients and clinicians as they decide on whether to pursue a BCI device. Other questions will arise in the practice of neuromodulation. How much control should individuals be given over their cognitive, affective, or motoric life? When should clinicians abide by or refuse patient requests for device settings? Though there are unlikely to be easy answers to these questions, it is worth anticipating now the kinds of ethical issues that patients and clinicians will face as BCI devices come into greater clinical use and exploring possible resources, such as those found in rehabilitation medicine, for addressing them.

Acknowledgments

This work was supported by a grant from the National Science Foundation (NSF Award #EEC-1028725). The views expressed are solely those of the author and do not represent the views of the NSF.

Notes

1. Although efforts have been made to establish different referents for BCI and BMI, the terms are used interchangeably in the literature. For instance, some prefer to reserve BMI for neural interfaces controlling robotic or prosthetic devices (Wolpaw & Wolpaw 2012). For the sake of simplicity, BCI will be used synonymously here.

2. This impairment-based schema has its limitations. For instance, if not put in context it may seem to reinforce certain questionable dichotomies in our thinking, such as between normal and abnormal function and able and disabled bodies. In addition, the domains are imperfect in that they are not comprehensive and in some cases overlap. For instance, does a seizure detection and stimulation BCI system belong in the motor, sensory, or cognition/consciousness domain? Nonetheless, this rough grouping maps onto both traditional medical and scientific categories and so is a useful preliminary framework.

3. The devices and technologies highlighted, both wearable and implantable, give a representative feel for work in each domain. This is helpful for exploring the ethical issues discussed in this section, even if the chosen examples are not the most representative technology or are not the closest to translation in a given domain. No position is taken on the warrant of pursuing wearable versus implantable devices in any given domain.

4. Veatch juxtaposes the engineering model (unfavorably) to the priestly, collegial, and contractual models.

5. Medical conditions falling within the scope of neuroimmunology typically include Guillain–Barré syndrome, myasthenia gravis, central nervous system vasculitis, dermatomyositis, and others, but multiple sclerosis, an inflammatory demyelinating and neurodegenerative disorder of the nervous system, is the paradigmatic example of a neuroimmunological disease.

References

Akcakaya, M., Peters, B., Moghadamfalahi, M., et al. (2014). Noninvasive brain–computer interfaces for augmentative and alternative communication. *Biomedical Engineering, IEEE Reviews*, 7, 31–49.

Anderson, K.D. (2004). Targeting recovery: priorities of the spinal cord-injured population. *Journal of Neurotrauma*, **21**(10), 1371–1383.

Bell, E., Maxwell, B., McAndrews, M.P., Sadikot, A., and Racine, E. (2010). Hope and patients' expectations in deep brain stimulation: healthcare providers' perspectives and approaches. *Journal of Clinical Ethics*, **21**, 112–124.

Birbaumer, N., Ghanayim, N., Hinterberger, T., et al. (1999). A spelling device for the paralysed. *Nature*, **398**(6725), 297–298.

Brock, D.W. (1991). The ideal of shared decision making between physicians and patients. *Kennedy Institute of Ethics Journal*, **1**(1), 28–47.

Brody, H. (1997). The physician-patient relationship. *Medical Ethics*, **2**, 75–101.

Bromet, E., Andrade, L.H., Hwang, I., et al. (2011). Cross-national epidemiology of DSM-IV major depressive episode. *BMC Medicine*, **9**(1), 1.

Chaudhary, U. and Birbaumer, N. (2015). Communication in locked-in state after brainstem stroke: a brain-computer-interface approach. *Annals of Translational Medicine*, **3**(Suppl. 1), S29.

Clausen, J. (2011). Conceptual and ethical issues with brain–hardware interfaces. *Current Opinion in Psychiatry*, **24**(6), 495–501.

Clifford, D.B., De Luca, A., Simpson, D.M., Arendt, G., Giovannoni, G., and Nath, A. (2010). Natalizumab-associated progressive multifocal leukoencephalopathy in patients with multiple sclerosis: lessons from 28 cases. *Lancet Neurology*, **9**(4), 438–46.

Clouser, K.D. (1983). Veatch, May, and models: a critical review and a new view. In: Shelp, E.E. (Ed.) *The Clinical Encounter*, pp.89–103. Netherlands: Springer.

Cohen, E.D. (2007). Prosthetic interfaces with the visual system: biological issues. *Journal of Neural Engineering*, 4(2), R14.

Colletti, V., Shannon, R.V., Carner, M., Veronese, S., and Colletti, L. (2009). Progress in restoration of hearing with the auditory brainstem implant. *Progress in Brain Research*, 175, 333–345.

Collinger, J.L., Boninger, M.L., Bruns, T.M., Curley, K., Wang, W., and Weber, D.J. (2013a). Functional priorities, assistive technology, and brain-computer interfaces after spinal cord injury. *Journal of Rehabilitation Research and Development*, 50(2), 145–160.

Collinger, J.L., Wodlinger, B., Downey, J.E., et al. (2013b). High-performance neuroprosthetic control by an individual with tetraplegia. *The Lancet*, 381(9866, pp.557–564.

Combs, H.L., Folley, B.S., Berry, D.T., et al. (2015). Cognition and depression following deep brain stimulation of the subthalamic nucleus and globus pallidus pars internus in Parkinson's disease: a meta-analysis. *Neuropsychology Review*, 25(4), 439–454.

Denning, T., Matsuoka, Y., and Kohno, T. (2009). Neurosecurity: security and privacy for neural devices. *Neurosurgical Focus*, 27(1), E7.

Dibiase, R. and Gunnoe, J. (2004). Gender and culture differences in touching behavior. *Journal of Social Psychology*, 144(1), 49–62.

Dougherty, D.D., Rezai, A.R., Carpenter, L.L., et al. (2015). A randomized sham-controlled trial of deep brain stimulation of the ventral capsule/ventral striatum for chronic treatment-resistant depression. *Biological Psychiatry*, 78(4), 240–248.

Elder, G.J. and Taylor, J.P. (2014). Transcranial magnetic stimulation and transcranial direct current stimulation: treatments for cognitive and neuropsychiatric symptoms in the neurodegenerative dementias? *Alzheimer's Research and Therapy*, 6(9), 74,

Emanuel, E.J. and Emanuel, L.L. (1992). Four models of the physician-patient relationship. *Journal of the American Medical Association*, 267(16), 2221–2226.

Fenton, A. and Alpert, S. (2008). Extending our view on using BCIs for locked-in syndrome. *Neuroethics*, 1(2), 119–132.

Ford, P.J. and Henderson, J.M. (2006). The clinical and research ethics of neuromodulation. *Neuromodulation: Technology at the Neural Interface*, 9(4), 250–252.

Ford, P.J. and Kubu, C.S. (2006). Stimulating debate: ethics in a multidisciplinary functional neurosurgery committee. *Journal of Medical Ethics*, 32(2), 106–109.

Freund, H.J, Kuhn, J., Lenartz, D., et al. (2009). Cognitive functions in a patient with Parkinson-dementia syndrome undergoing deep brain stimulation. *Archives of Neurology*, 66(6), 781–785.

Fukushi, T. (2012). Ethical practice in the era of advanced neuromodulation. *Asian Bioethics Review*, 4(4), 320–329.

George, M.S. and Aston-Jones, G. (2010). Noninvasive techniques for probing neurocircuitry and treating illness: vagus nerve stimulation (VNS), transcranial magnetic stimulation (TMS) and transcranial direct current stimulation (tDCS). *Neuropsychopharmacology*, 35(1), 301–316.

Giacino, J.T., Fins, J.J., Machado, A., and Schiff, N.D. (2012). Central thalamic deep brain stimulation to promote recovery from chronic posttraumatic minimally conscious state: challenges and opportunities. *Neuromodulation: Technology at the Neural Interface*, 15(4), 339–349.

Giacino, J.T., Fins, J.J., Laureys, S., and Schiff, N.D. (2014). Disorders of consciousness after acquired brain injury: the state of the science. *Nature Reviews Neurology*, 10(2), 99–114.

Gilbert, F. (2015). A threat to autonomy? The intrusion of predictive brain implants. *American Journal of Bioethics: Neuroscience*, 6(4), 4–11.

Glannon, W. (2016). Ethical issues in neuroprosthetics. *Journal of Neural Engineering*, 13(2), 021002.

Goering, S. (2014). Is it still me? DBS, agency, and the extended, relational me. *American Journal of Bioethics: Neuroscience*, **5**(4), 50–51.

Goering, S. (2015). Stimulating autonomy: DBS and the prospect of choosing to control ourselves through stimulation. *American Journal of Bioethics: Neuroscience*, **6**(4), 1–3.

Grisso, T. and Appelbaum, P.S. (1998). *MacArthur Competence Assessment Tool for Treatment (MacCAT-T)*. Sarasota, FL: Professional Resource Press.

Hansson, S.O. (2007). The ethics of enabling technology. *Cambridge Quarterly of Healthcare Ethics*, **16**(3), 257–267.

Haselager, P. (2013). Did I do that? Brain–computer interfacing and the sense of agency. *Minds and Machines*, **23**(3), 405–418.

Haselager, P., Vlek, R., Hill, J., and Nijboer, F. (2009). A note on ethical aspects of BCI. *Neural Networks*, **22**(9), 1352–1357.

Hebb, A.O., Zhang, J.J., Mahoor, M.H., et al. (2014). Creating the feedback loop: closed-loop neurostimulation. *Neurosurgery Clinics of North America*, **25**(1), 187–204.

Heesen, C., Solari, A., Giordano, A., Kasper, J., and Köpke, S. (2011). Decisions on multiple sclerosis immunotherapy: new treatment complexities urge patient engagement. *Journal of the Neurological Sciences*, **306**(1), 192–197.

Herron, J., Denison, T., and Chizeck, H.J. (2015). Closed-loop DBS with movement intention. In: *Proceedings of the 2015 7th International IEEE/EMBS Conference on Neural Engineering (NER)*, pp.844–847. Montpellier, France, April 2015

Hochberg, L.R., Serruya, M.D., Friehs, G.M., et al. (2006). Neuronal ensemble control of prosthetic devices by a human with tetraplegia. *Nature*, **442**(7099), 164–171.

Hochberg, L. and Anderson, K. (2012). BCI users and their needs. In: Wolpaw, J.R. and Wolpaw, E.W. (Eds.) *Brain-Computer Interfaces*, pp. 317–323. New York: Oxford University Press.

Holsheimer, J. (2003). Letters to the editor. *Neuromodulation*, **6**(4), 270–2.

Huggins, J.E., Wren, P.A., and Gruis, K.L. (2011). What would brain-computer interface users want? Opinions and priorities of potential users with amyotrophic lateral sclerosis. *Amyotrophic Lateral Sclerosis*, **12**(5), 318–324.

King, C., McCrimmon, C., Wang, P., Chou, C., Nenadic, Z., and Do, A.H. (2014). Brain-computer interface driven functional electrical stimulation system for overground walking: a case report. *Neurology*, **82**(Suppl. 10), P3–055.

Krames, E.S., Peckham, P.H., Rezai, A.R., and Aboelsaad, F. (2009). What is neuromodulation? In: Krames, E.S., Peckham, P.H., and Rezai, A.R. (Eds.) *Neuromodulation*, pp.3–8. San Diego, CA: Academic Press.

Klein, E. (2016). Informed consent in implantable BCI research: identifying risks and exploring meaning. *Science and Engineering Ethics*, **22**(5), 1299–1317.

Klein, E. and Bourdette, D. (2013). Postmarketing adverse drug reactions: a duty to report? *Neurology: Clinical Practice*, **3**(4), 288–294.

Klein, E., Brown, T., Sample, M., Truitt, A.R., and Goering, S. (2015). Engineering the brain: ethical issues and the introduction of neural devices. *Hastings Center Report*, **45**(6), 26–35.

Kolata, G. (2013, October 27). Of fact, fiction and Cheney's defibrillator. *The New York Times*. http://nytimes.com/2013/10/29/science/of-fact-fiction-and-defibrillators.html [accessed June 25, 2016].

Konrad, P. and Shanks, T. (2010). Implantable brain computer interface: challenges to neurotechnology translation. *Neurobiology of Disease*, **38**(3), 369–375.

Kübler, A., Winter, S., Ludolph, A.C., Hautzinger, M., and Birbaumer, N. (2005). Severity of depressive symptoms and quality of life in patients with amyotrophic lateral sclerosis. *Neurorehabilitation and Neural Repair*, **19**(3), 182–193.

Kübler, A., Holz, E., Kaufmann, T., and Zickler, C. (2013). A user centred approach for bringing BCI controlled applications to end-users. In: Fazel-Rezai, R. (Ed.) *Brain-Computer Interface Systems-Recent Progress and Future Prospects*. InTech, DOI: 10.5772/55802. Available from: http://www.intechopen.com/books/brain-computer-interface-systems-recent-progress-and-future-prospects/a-user-centred-approach-for-bringing-bci-controlled-applications-to-end-users

Kuhn, J., Hardenacke, K., Lenartz, D., et al. (2015). Deep brain stimulation of the nucleus basalis of Meynert in Alzheimer's dementia. *Molecular Psychiatry*, **20**(3), 353–360.

Mainsah, B.O., Collins, L.M., Colwell, K.A., et al. (2015). Increasing BCI communication rates with dynamic stopping towards more practical use: an ALS study. *Journal of Neural Engineering*, **12**(1), 016013.

Mak, J.N. and Wolpaw, J.R. (2009). Clinical applications of brain-computer interfaces: current state and future prospects. *IEEE Reviews in Biomedical Engineering*, **2**, 187–199.

Malloy, D.C. and Hadjistavropoulos, T. (2004). The problem of pain management among persons with dementia, personhood, and the ontology of relationships. *Nursing Philosophy*, **5**(2), 147–159.

Malone, D.A., Dougherty, D.D., Rezai, A.R., et al. (2009). Deep brain stimulation of the ventral capsule/ventral striatum for treatment-resistant depression. *Biological Psychiatry*, **65**(4), 267–275.

Mayberg, H.S., Lozano, A.M., Voon, V., et al. (2005). Deep brain stimulation for treatment-resistant depression. *Neuron*, **45**(5), 651–660.

McGie, S.C., Nagai, M.K., and Artinian-Shaheen, T. (2013). Clinical ethical concerns in the implantation of brain-machine interfaces: Part II: specific clinical and technical issues affecting ethical soundness. *Pulse, IEEE*, **4**(2), 32–37.

Miller, F.G., Fins, J.J., and Bacchetta, M.D. (1996). Clinical pragmatism: John Dewey and clinical ethics. *Journal of Contemporary Health Law & Policy*, **13**(27), 27–51.

Mirzadeh, Z., Bari, A., and Lozano, A.M. (2015). The rationale for deep brain stimulation in Alzheimer's disease. *Journal of Neural Transmission*, **123**(7), 775–783.

Murphy, M.D., Guggenmos, D.J., Bundy, D.T., and Nudo, R.J. (2016). Current challenges facing the translation of brain computer interfaces from preclinical trials to use in human patients. *Frontiers in Cellular Neuroscience*, **9**, 497.

Nardone, R., Höller, Y., Tezzon, F., et al. (2015). Neurostimulation in Alzheimer's disease: from basic research to clinical applications. *Neurological Sciences*, **36**(5), 689–700.

Patil, P.G. and Turner, D.A. (2008). The development of brain-machine interface neuroprosthetic devices. *Neurotherapeutics*, **5**(1), 137–146.

Pfurtscheller, G., Müller, G.R., Pfurtscheller, J., Gerner, H.J., and Rupp, R. (2003). Thought–control of functional electrical stimulation to restore hand grasp in a patient with tetraplegia. *Neuroscience Letters*, **35**(1), 33–36.

Presidential Commission for the Study of Bioethical Issues (2015). *Gray Matters: Topics at the Intersection of Neuroscience, Ethics, and Society*, 2. Washington, DC: United States Government.

Purtilo, R.B. (1988). Ethical issues in teamwork: the context of rehabilitation. *Archives of Physical Medicine and Rehabilitation*, **69**(5), 318–322.

Ransohoff, R.M., Hafler, D.A., and Lucchinetti, C.F. (2015). Multiple sclerosis—a quiet revolution. *Nature Reviews Neurology*, **11**(3), 134–142.

Rizvi, S.A. and Coyle, P.K. (2011). *Clinical Neuroimmunology: Multiple Sclerosis and Related Disorders*. New York: Springer.

Sankar, T. (2015). Alzheimer's disease: a novel application for deep-brain stimulation? *Future Neurology*, **10**(4), 297–300.

Sankar, T., Chakravarty, M.M., Bescos, A., et al. (2015). Deep brain stimulation influences brain structure in Alzheimer's disease. *Brain Stimulation*, **8**(3), 645–654.

Schermer, M. (2013). Health, happiness and human enhancement—dealing with unexpected effects of deep brain stimulation. *Neuroethics*, 6(3), 435–445.

Schiff, N.D., Giacino, J.T., Kalmar, K., et al. (2007). Behavioural improvements with thalamic stimulation after severe traumatic brain injury. *Nature*, 448(7153), 600–603.

Schneider, M.J., Fins, J.J., and Wolpaw, J.R. (2012). Ethical issues in BCI research. In: Wolpaw, J.R. and Wolpaw, E.W. (Eds.) *Brain-Computer Interfaces*, pp. 373–383. New York: Oxford University Press.

Seel, R.T., Sherer, M., Whyte, J., et al. (2010). Assessment scales for disorders of consciousness: evidence-based recommendations for clinical practice and research. *Archives of Physical Medicine and Rehabilitation*, 91(12), 1795–1813.

Shih, J.J., Krusienski, D.J., and Wolpaw, J.R. (2012). Brain-computer interfaces in medicine. *Mayo Clinic Proceedings*, 87(3), 268–279.

Silvers, A. (2009). An essay on modeling: the social model of disability. In: Ralston, D.C. and Ho, J.H. (Eds.) *Philosophical Reflections on Disability*, pp. 19–36. Netherlands: Springer.

Sjögren, M.J.C., Hellström, P.T.O., Jonsson, M.A.G., Runnerstam, M., Silander, H.C., and Ben-Menachem, E. (2002). Cognition-enhancing effect of vagus nerve stimulation in patients with Alzheimer's disease: a pilot study. *Journal of Clinical Psychiatry*, 63(11), 972–980.

Smith, S. (2014). Mind-controlled exoskeleton kicks off World Cup. *CNN*. Available from: http://www.cnn.com/2014/06/12/health/exoskeleton-world-cup-kickoff/ [accessed May 12, 2016].

Soekadar, S.R. and Birbaumer, N. (2015). Brain–machine interfaces for communication in complete paralysis: ethical implications and challenges. In: Illes, J. and Sahakian, B.J. (Eds.) *Handbook of Neuroethics*, pp.705–724. Netherlands: Springer.

Soekadar, S.R., Cohen, L.G., and Birbaumer, N. (2014). Clinical brain-machine interfaces. In: Tracy, J.I., Hampstead, B.M., and Sathian, K. (Eds.) *Cognitive Plasticity in Neurologic Disorders*, pp.347–362. New York: Oxford University Press.

Soekadar, S.R., Birbaumer, N., Slutzky, M.W., and Cohen, L.G. (2015). Brain–machine interfaces in neurorehabilitation of stroke. *Neurobiology of Disease*, 83, 172–179.

Synofzik, M., Schlaepfer, T.E., and Fins, J.J. (2012). How happy is too happy? Euphoria, neuroethics, and deep brain stimulation of the nucleus accumbens. *American Journal of Bioethics: Neuroscience*, 3(1), 30–36.

Veatch, R.M. (1972). Models for ethical medicine in a revolutionary age. *Hastings Center Report*, 2(3), 5–7.

Volkmann, J., Herzog, J., Kopper, F., and Deuschl, G. (2002). Introduction to the programming of deep brain stimulators. *Movement Disorders*, 17(S3), S181–S187.

Warden, D., Rush, A.J., Trivedi, M.H., Fava, M., and Wisniewski, S.R. (2007). The STAR* D Project results: a comprehensive review of findings. *Current Psychiatry Reports*, 9(6), 449–459.

Widge, A.S., Deckersbach, T., Eskandar, E.N., and Dougherty, D.D. (2016). Deep brain stimulation for treatment-resistant psychiatric illnesses: what has gone wrong and what should we do next? *Biological Psychiatry*, 79(4), e9–10.

Widge, A.S, Ellard, K.K, Paulk, A.C., et al. (2017). Treating refractory mental illness with closed-loop brain stimulation: progress towards a patient-specific transdiagnostic approach. *Experimental Neurology*, 287(Pt 4), 461–472.

Wolpaw, J.R. and Wolpaw, E.W. (2012). *Brain-Computer interfaces: Principles and Practice*. New York: Oxford University Press.

Wolpaw, J.R., Birbaumer, N., McFarland, D.J., Pfurtscheller, G., and Vaughan, T.M. (2002). Brain-computer interfaces for communication and control. *Clinical Neurophysiology*, 113(6), 767–791.

Woopen, C. (2012). Ethical aspects of neuromodulation. *International Review of Neurobiology*, 107, 315–332.

Chapter 8

Integrating ethics into neurotechnology research and development: The US National Institutes of Health BRAIN Initiative®

Khara M. Ramos and Walter J. Koroshetz

Introduction

The human brain contains approximately 84 billion neurons, collectively making trillions of connections. Fascination with this staggeringly complex organ has long drawn scientists to study the brain. Indeed, advances in neuroscience are emerging at an amazing rate, supported by investments in science such as the Brain Research through Advancing Innovative Neurotechnologies (BRAIN) Initiative® (http://braininitiative.nih.gov) that aims to accelerate the development and application of innovative neurotechnologies to revolutionize the understanding of the human brain.

The public in developed countries is becoming ever more interested in how the brain works and how scientists can bring meaningful solutions to those who suffer from disorders of the nervous system, which can be categorized in terms of neurological, mental health, and substance abuse (NMS) disorders. The disabling features of NMS disorders stem from dysfunction in neural circuits, but our current tools are unable to identify the abnormalities in signaling or normalize the circuit dysfunction. Around the world, disease burden continues to shift from communicable to noncommunicable diseases such as those in the NMS classification, and from early death to years spent living with a disability. Brain-based diseases and disorders comprise a significant share of this toll. In 2010, mental and behavioral disorders comprised 7.4% of the global burden of disease, stroke comprised 4.1%, and neurological disorders another 3% (Bloom et al. 2011; Murray et al. 2012). This rising burden causes immense human suffering and poses unprecedented challenges for healthcare. For example, Alzheimer's disease not only progressively destroys essential mental functions like memory but it is also an incredibly costly chronic disease. In 2015, the cost to American society of caring for Alzheimer's disease patients totaled an estimated $226 billion, with half of the costs borne by Medicare. If current trends continue, by 2050, that cost is estimated to rise to over $1.1 trillion (Alzheimer's Association 2015). These facts and figures underscore the need for action, and point to the importance and potential impact of neuroscience research for human health. The problems are difficult, and more powerful tools and greater fundamental knowledge of the nervous system are needed to solve them. Modern neuroscience is undergoing a revolution, one that is propelled by huge leaps in technological capabilities and focused on tool development. As Freeman Dyson famously noted (1997), new directions in science are launched by new tools much more often than by new concepts. That mindset, embodied in the BRAIN initiative*, will deliver powerful new tools and technologies that fall into two main classes: those that enable us to interrogate and monitor neural circuit activity, and those that enable us to modulate it. The ethical and societal issues attached to the medical and nonmedical use of these technologies are not subtle, nor are they entirely new.

The attempt to discern what a patient is feeling or experiencing is an inherent part of medicine. There are also nonmedical examples of trying to understand what is unfolding in someone's mind. For instance, scientists have long been interested in ferreting out the neurobiology of lying and deception. Early methods of lie detection rely on nonverbal cues, such as facial microexpressions and increased perspiration, but these are indirect reflections of the brain function that is at the core of processing and retelling information about the world. Modern neuroscience methods have enabled researchers to come closer to studying that brain function. Functional magnetic resonance imaging (fMRI) measures brain activity by detecting changes associated with blood flow, and Mohamed and colleagues (2006) found that fMRI could be used to discriminate patterns of brain activity associated with deception from those associated with telling the truth. They observed that deception triggered activity in brain regions associated with memory encoding and retrieval, response inhibition (i.e., suppressing the truth), and emotion. Though fMRI has taken us one step closer to accessing these types of higher-order cognitive functions, the BRAIN Initiative* holds promise of delivering tools to push this to another level, delivering both precision and accessibility (i.e., tools that are miniaturized compared to contemporary brain scanners, and widely available).

What might this look like in medical practice? As one example, psychiatrists might one day be equipped with lightweight brain-scanning helmets, which could rapidly, precisely, and noninvasively measure brain circuit function to aid diagnosis of autism, generalized anxiety disorder, post-traumatic stress disorder, manic versus depressive episodes in bipolar disorder, suicidality, or attention deficit hyperactivity disorder. New tools could give physicians and researchers ways to objectively measure pain, or assess the circuit abnormalities underlying epilepsy, cognitive impairment, dystonia, spasticity in cerebral palsy, and so on. From an ethics perspective, issues around the privacy and use of such data would be key, as the data will likely contain information linked to multiple core characteristics of the individual. Innovative neural circuit monitoring technologies may pose new challenges for ethicists and society, since they would far outstrip the current ability to record brain activity underlying thoughts, mood, and sensations, which is fairly primitive and not widely in use.

Of course, interrogating brain function is only half of the story. Dysregulated brain circuits are the most proximal cause of NMS disorders, and from a medical perspective, modulating brain circuits for health benefit is the basis of all treatments for NMS disorders. Currently, the most common symptomatic treatments are drugs that alter neural circuit function, but direct electrical modulation of brain circuits is also part of the therapeutic armamentarium. Following decades of study to identify the circuit abnormality underlying the motor symptoms of tremor and slowness of movement in Parkinson's disease, deep brain stimulation (DBS), in which a patient is implanted with a device to deliver electrical current focused on specific deep regions of the brain, is now a common neurosurgical treatment for Parkinson's disease (DeLong & Wichmann 2015). Major depressive disorder (often simply referred to as "depression"), a leading cause of disability around the world, has been treated to varying degrees of success with different types of brain stimulation (Murray et al. 2012; Deng et al. 2015). Electroconvulsive therapy, in which a generalized seizure is induced by delivering electricity directly to the brain, was originally developed in the 1930s and is still used for patients with severe depression, as it relatively rapidly elicits high rates of remission (Lisanby 2007). Attempts to understand the abnormalities in brain circuits that underlie depressed mood have already given researchers and clinicians brain targets for DBS as an option for treatment-resistant depression (Mayberg 2009). Beyond electroconvulsive therapy and DBS, researchers are also focused on noninvasive, subconvulsive methods of brain stimulation for depression, including transcranial direct current stimulation and transcranial magnetic stimulation.

What all of these techniques have in common is that they are aimed at modulating brain circuit function, and they vary in their precision in targeting brain circuitry. What is needed are the tools to both pinpoint the circuit dysfunction that underlies brain disorders and precisely modulate circuit activity to correct this dysfunction. Rendering these tools noninvasive would decrease risk and increase the number of people who could be treated, also expanding the pool of research subjects and thereby hastening the pace at which neuroscience unravels the mysteries of NMS disorders. Noninvasive technologies also open the door to nonmedical uses as illustrated by the current use of direct current brain stimulation for a variety of effects. The new circuit modulating technologies raise the same ethical and societal issues that surround drugs that have the potential to be mind or brain altering, but the magnitude and precision of the effects may be significantly greater. The questions that arise from medical use of the new technologies will be complicated, but they likely pale in comparison to the questions surrounding nonmedical uses of neurotechnologies that enable interrogating and modulating the neural circuits of individuals.

A brief history of the BRAIN Initiative®

The examples in the previous section illustrate how technology development can intersect with advances in scientific understanding of the brain. Indeed, over the past decade, technological advances in diverse fields such as physics and chemistry have yielded unprecedented opportunities for integration across scientific fields, and enabled powerful, entirely new ways of studying and modulating nervous system function. Recognizing this inflection point in neuroscience research, US President Barack Obama launched the

BRAIN Initiative˙ in April 2013. The BRAIN Initiative˙ aims to accelerate the development and application of new neurotechnologies, empowering researchers to produce a revolutionary, dynamic picture of the brain that, for the first time, shows how individual cells and complex neural circuits interact in time and space. A richer understanding of brain function will inform much-needed progress in diagnosing, treating, and even curing NMS disorders.

Given this bold goal, the President called for the BRAIN Initiative˙ to be an "all hands on deck" effort involving not only US government agencies (currently the National Institutes of Health (NIH), National Science Foundation (NSF), Intelligence Advanced Research Projects Activity (IARPA), Defense Advanced Research Projects Agency (DARPA), and the Food and Drug Administration (FDA)), but also companies, private research institutes, patient advocacy organizations, state governments, research universities, philanthropists, and more. Many groups signed on in an enthusiastic response to this call, even extending internationally. Currently, the Brain Canada Foundation, the Australian National Health and Medical Research Council, and the Lundbeck Foundation are NIH BRAIN Initiative˙ partners. The NIH, NSF, FDA, IARPA, the Kavli Foundation, Simons Foundation, and the Allen Institute for Brain Science coordinate communication activities through the BRAIN Initiative˙ Alliance with affiliates DARPA and the Howard Hughes Medical Institute at Janelia Research Campus.

To inform scientific planning for the BRAIN Initiative˙, NIH established a high-level working group of the Advisory Committee to the NIH Director (ACD) (http://acd.od.nih. gov/brain.htm). This working group sought broad input from the scientific community, patient advocates, and the general public. Their report, released in June 2014 and enthusiastically endorsed by the ACD, articulated the scientific goals of the NIH component of the BRAIN Initiative˙ and developed a multiyear scientific roadmap for achieving these goals, including timetables, milestones, and cost estimates (Advisory Committee to the National Institutes of Health Director (2014). *BRAIN 2025: A Scientific Vision* outlines a plan for accelerating technology development for neuroscience, with the intention of developing and using new tools to acquire fundamental new insights into how the nervous system functions in health and disease. The working group identified the analysis of neural circuits as being particularly suitable for potentially revolutionary scientific advances. Table 8.1 outlines how the NIH currently organizes scientific programming for the BRAIN Initiative˙.

Neurotechnology research and development

In thinking about neurotechnology research and development under the BRAIN Initiative˙, one can envision advances falling under three major categories: direct recording of neural circuit activity, direct stimulation of neurons and neural circuits, and anatomical techniques to define the wiring diagram of the brain. Layered onto these, the field will also need new conceptual foundations for understanding information flow and processing in the circuits that generate specific behaviors or sensations. By expanding the capability to record circuit activity on multiple scales, vast quantities of neural data will

Table 8.1 The scientific landscape of the US NIH BRAIN Initiative®

Scientific focus	Description
Cell census	Identify and provide experimental access to the different brain cell types to determine their roles in health and disease. Integrate information on molecular identity and gene expression with connectivity, morphology, location, and physiology
Circuit technologies	Develop tools and technologies for circuit mapping and functional, specifically targeted manipulation of circuit activity
Neural recording and modulation	Record and manipulate neural activity with cellular resolution, at multiple depths and locations across the brain
Human imaging and noninvasive neuromodulation	Develop next-generation imaging technologies, and advance understanding of the relationship between imaging signals and the underlying brain circuits and cellular activity. Promote the development, implementation, and integration of innovative new technologies for human neuroscience research
Understanding circuit function	Understand the circuits and patterns of neural activity that give rise to mental experience and behavior
Training and dissemination	Disseminate new neurotechnologies and tools to a wide scientific user base, along with the knowledge required to use them well
Neuroethics	Support neuroethics efforts that can both guide BRAIN Initiative® research and address ethical dilemmas raised through advancements in neuroscience

be generated and will challenge current methods of data analysis and intuitive interpretation of meaning. To date, 125 projects have been funded, exceeding an investment of $130 million USD. These include a systematic inventory of the different cell types of the brain; targeted genetic and nongenetic approaches for accessing specific cells and circuits; new and improved capabilities for recording from, and modulating activity of, rapidly firing collections of neurons; and development of new theories, models, and methods to analyze complex neural data. Some new tools have already sparked a significant shift in neuroscience research. For example, optogenetics and chemogenetics are two methods for cell type-specific modulation; the former uses light and the latter uses engineered receptor/ligand pairs to control the activity of cells, such as targeted neurons, in living tissue. In contrast to current tools such as DBS or available pharmaceuticals, these tools have unprecedented spatial and temporal control. They are currently used with much success in animal models to probe brain circuitry and modulate behaviors such as fear conditioning and susceptibility to stress (Haubensak et al. 2010; Perova et al. 2015). The use of these and other tools to interrogate and modulate circuits in animal studies may soon be used to study human circuits in ex vivo human brain tissue from surgical biopsies, autopsy tissue, or cerebral organoids derived from induced pluripotent stem cells. In the future, harnessing these tools to modulate brain circuit function in patients may provide therapeutic relief for NMS disorders (for discussion, see Whittle et al. 2014).

Of particular note when considering the potential ethical implications of NIH BRAIN Initiative* research are current efforts toward understanding human brain function and treating human brain disorders. To date, the NIH BRAIN Initiative* has funded 14 projects involving human subjects. These include studies to understand the signals underlying noninvasive imaging modalities, more sophisticated understanding of noninvasive neuromodulation techniques, and development of novel methods of noninvasive neuromodulation. Examples of the latter already studied in humans include focused ultrasound (Legon et al. 2014), and transcranial magnetic stimulation and direct current stimulation (Deng et al. 2015). In addition, BRAIN Initiative* projects are focused on developing genetic means to confer electromagnetic sensitivity (Stanley et al. 2016; Wheeler et al. 2016) or chemical sensitivity (Roth 2016) to precise brain circuits.

The NIH is also expanding its portfolio of research with implantable neuromodulation devices, including a new BRAIN Public–Private Partnership Program, which will connect academic researchers with manufacturers of next-generation invasive devices for recording and modulation in the human central nervous system. Further, to understand the unique properties and functions of human neural circuits, the NIH is supporting research opportunities for studies with neurosurgical patients. This pool of patients comprises three types: people who are receiving DBS, people implanted with brain–computer interfaces, and people with implanted devices that perform activity monitoring for epilepsy control.

One specific project proposes to use DBS to improve the level of consciousness in people suffering from severe traumatic brain injuries. Traumatic brain injury (TBI) afflicts hundreds of thousands of Americans each year, producing chronic cognitive disabilities that lack effective treatment. Preliminary studies with TBI patients and nonhuman primates suggest that these cognitive disabilities may be due to disrupted circuit function in the brain, specifically involving impaired connections between the thalamus and the frontal cortex (Kinomura et al. 1996; van der Werf et al. 2002; Schiff et al. 2007). Some investigators are using the latest device technology to deliver DBS to the thalamus, in hopes of developing a next-generation device for treating cognitive impairment associated with TBI.

Another project in the area of human imaging aims to develop a completely new, noninvasive method for measuring brain activity. This work is focused on the feasibility of sonoelectric tomography, in which ultrasound to tag specific locations in the brain is combined with conventional scalp electroencephalography to measure electrical activity. The information can then be used to construct a tomographic map of neuronal currents, leveraging the millisecond temporal resolution of electroencephalography and the millimeter spatial resolution of ultrasound. If successful, this would be a powerful next-generation imaging method, with the capacity not only to map the functions of brain circuits in healthy individuals, but also for understanding and potentially diagnosing complex neuropsychiatric disorders.

Neuroethical issues

The previously discussed example projects give us some insight into the potential ethical issues associated with neurotechnology research and development. Importantly, although we recognize that tools to understand and modulate brain function are and will continue to be used for a range of purposes, the NIH's perspective is focused solely on applications to improve health and reduce illness and disability—that is, medical applications. Many experts have given much thought to the various relevant ethical issues, and important scholarship is also stimulated by leadership of the International Neuroethics Society (a small sampling: Fins 2011; Illes & Bird 2006; Farah 2015; Sahakian et al. 2015). The NIH has hosted two meetings in Bethesda, MD, bringing together scientists and ethicists to discuss key issues in neuroethics (proceedings from these meetings, held in November 2014 and February 2016, can be found on the NIH BRAIN Initiative* website http://braininitiative.nih.gov/about/newg.htm). Some of these neuroethical issues are not new—that is, they can be thought of as classic bioethical issues that are not fundamentally different from those that ethicists and scientists grapple with in other areas of biomedical research.

For example, new methods for direct stimulation of neurons and neural circuits raise questions of safety and the acceptable degree of risk for human patients:

◆ What benchmarks should be adopted for testing new methods before they are put to use with humans, and who should make such decisions?

◆ What level of antecedent evidence is necessary to justify translating preclinical research into work with human subjects?

◆ When contemplating testing new methods in people who have brain disorders and diseases, what is the best way to balance decision-making between the patient and legal surrogate in deciding on an acceptable level of risk?

These important questions are not substantively different from those that would be posed when considering new methodologies for use in other areas of biomedical research, except that brain disorders can affect the circuits actually employed in decision-making and brain circuits uniquely define personhood.

A second classic bioethical issue to consider is the question of incidental findings. Current brain imaging techniques such as MRI are sufficiently sophisticated that, in the course of imaging undertaken for the purpose of research, an investigator might unintentionally find abnormalities or early indicators of disease. Indeed, incidental findings on brain scans with modern MRI technology are common, and prevalence increases with age (Morris et al. 2009). As neurotechnology research progresses and new methods for deciphering brain function develop, the domains of such incidental findings will undoubtedly expand. The question, of course, is what to do with this information, especially if the indicator reveals an uncertain degree of risk, or the disease or disorder is one for which there is no effective treatment. One field that has also spurred much thought on this question is genetics. Genes can be predictive, though often in complex patterns of association that modern science does not yet fully understand, making prognostication based on genetic testing nuanced

(Pasic et al. 2013). A further complicating factor, and one that may give a hint of what is to come for the field of neuroscience, is the growing usage of commercial direct-to-consumer genetic testing. In December 2013, the Presidential Commission for the Study of Bioethical Issues (Bioethics Commission) released its report *Anticipate and Communicate: Ethical Management of Incidental and Secondary Findings in the Clinical, Research, and Direct-to-Consumer Contexts*. The Commission argued that researchers, clinicians, and companies should expect incidental findings, plan ahead for how to address them, and communicate with patients or subjects about handling such findings in advance of conducting any tests. These recommendations echo those made by Illes and colleagues regarding incidental findings in neuroimaging (Illes et al. 2006). The authors argued that disclosure of suspicious neuroimaging-associated incidental findings to subjects is ethically desirable, though wide variability exists in such disclosure for the estimated tens of thousands of human subjects involved in neuroimaging research annually. Even with acknowledgment that disclosure is desirable, additional questions abound such as whether all neuroimaging studies should include a physician trained to interpret brain scans, and how to assist vulnerable populations (e.g., uninsured subjects) in navigating incidental findings.

Beyond such classic bioethical issues exist various ethical questions uniquely associated with neuroscience research that require careful thought. These questions arise because the brain is the organ of the mind, and our new neuroscience tools and technologies provide powerful, unprecedented ways of understanding and modulating brain function. Such scientific progress must be carefully considered, since it has the potential to fundamentally alter the understanding of our innate and autonomous sense of self, and could conflict with our deeply held beliefs about personal volition and responsibility. In recognition of these concerns, when launching the BRAIN Initiative* in 2013, President Obama charged his Bioethics Commission to identify proactively a set of core ethical standards to both guide neuroscience research and address ethical dilemmas raised by the application of neuroscience research findings. The Commission held public stakeholder meetings in 2013 and 2014, and issued a two-part report to respond to the rapidly emerging and evolving field of neuroscience (Presidential Commission for the Study of Bioethical Issues 2014, 2015). In the first volume of *Gray Matters*, the Commission focused on the importance of integrating ethics into neuroscience research across the life of a research endeavor. Preemptive integration of ethics into the scientific process can help guard against ethical setbacks that might undermine research and erode public support and trust. Thus their recommendation can be succinctly stated as: integration, not intervention. The European Union's Human Brain Project (HBP) offers one model of this approach in action: the HBP has a robust Ethics and Society Programme, which recognizes the social, ethical, and philosophical implications of HBP-related work, and aims to recognize concerns early and address them in an open and transparent manner. Additional recommendations from *Gray Matters* relate to the importance of integrating ethics into science education at all levels, the value of including ethicists on scientific advisory groups, and, from the second volume, specific analysis on the topics of cognitive enhancement, consent capacity, and neuroscience and the legal system.

Integrating ethics into research

At the NIH, we are engaged in efforts to integrate a rigorous program of ethics into the BRAIN Initiative*. The BRAIN Multi-Council Working Group (MCWG) provides ongoing oversight of the long-term scientific vision of the BRAIN Initiative* at NIH, in the context of the evolving neuroscience landscape (http://braininitiative.nih.gov/about/mcwg.htm). The BRAIN MCWG comprises distinguished neuroscientists, and additional at-large members appointed to supplement the group's expertise as appropriate, with ethics being a key area for inclusion. As of fall 2015, the MCWG now includes a Neuroethics Division. In broad terms, the group's purpose is to recommend overall approaches for how the NIH BRAIN Initiative* might handle issues and problems involving ethics. What this will look like in practice remains a work in progress. This work is informed not only by *Gray Matters*, but also by *BRAIN 2025*, which includes several recommendations regarding core ethical principles that will maximize the value of the BRAIN Initiative*. The authors of the latter publication noted that while there are well-established and agreed upon standards for human research, studies that involve recording from or stimulating the human brain require ongoing oversight. This need, they argue, is best served by a continual dialogue between researchers and ethical advisory groups that include members who are well informed in neuroscience research and aware of the historical context of brain manipulations and recordings.

In early 2016, as the Neuroethics Division has begun ramping up its efforts, our perspective has been for their deliberations and activity to be grounded in the science itself—that is, the existing portfolio of BRAIN research supported by the NIH, and the trajectory of the BRAIN Initiative* as described in *BRAIN 2025*. With that body of scientific work in mind, the group has discussed areas of BRAIN research that raise ethical concerns or questions of ethical risk, as well as different approaches to address these concerns. That fruitful conversation serves as the basis to shape our efforts in this space going forward, and is summarized here.

The group agreed on the complexity of these issues and the need to address them thoughtfully, with involvement of appropriate content experts as necessary. There are many relevant ethical questions, but four topics in particular emerged as having the highest priority. One area of focus is the ethics of research with invasive neurotechnologies:

◆ How might a clinician best determine a patient's competence for understanding the risks and benefits before deciding to participate in research?

◆ How can a patient consent to participate in research if the disease affects the very organ that grants consent?

◆ Once a patient is involved in a research study, what are best practices for managing unexpected physical or perceived nonphysical harm?

◆ Who should control devices that affect mood and the brain's reward systems? The patient? The doctor?

- If a device is placed in the brain as part of a research study, what is a researcher's responsibility for continued care of the patient and upkeep of the device?
- Who owns the data collected by such devices?

With widespread use of technologies such as fMRI, questions about the privacy of an individual's neural data and the need to protect those data are not new. As technology to record brain function and analyses to interpret those data become more sophisticated, the ethical, legal, and social challenges will undoubtedly intensify. Contemporary neuroscience may reveal information about individual traits that has social, medical, or even legal relevance. For example, a recent study suggested that patterns of resting state brain connectivity correlate with specific patterns of covariance between various behavioral and demographic indicators (Smith et al. 2015). The study received much press, since the behavioral and demographic indicators included data such as history of substance use, performance on tests of fluid intelligence, and level of education. The public may be uneasy with advances in neurotechnology if, beyond applications for reducing the burden of NMS disorders, these new techniques are perceived to be capable of parceling individuals into personality types or cognitive abilities. This is an area where the field of genetics may again provide a useful model. With rapid advances in genomic science at the turn of the millennium came widespread enthusiasm for the potential of genetic research to offer new insights into disease treatment and prevention. However, as researchers came to understand that each of us has some degree of genetic predisposition toward disease, many began to express concern about the potential for discrimination based on genetic testing. After years of discussion in the United States Congress, in 2008, President Bush signed the Genetic Information Nondiscrimination Act (GINA), which protects Americans against employment and health insurance discrimination based on genetic information. Many see GINA as a way to help citizens feel secure in taking part in modern genetic research, and taking advantage of the promise of genetic testing to inform personalized medicine (Slaughter 2008).[1] Some have called for a neuroscience analog of GINA, which could serve as a tool for grappling with the ethical, legal, and social implications of contemporary neuroscientific methods. A simple example underscores the current uneven landscape regarding discrimination associated with neurological disease prediction. Currently, if a person were found to be at high risk for Alzheimer's disease based on genetic testing, he or she would be protected against health insurance discrimination by GINA. If, however, high risk for developing Alzheimer's were revealed by a scan of brain metabolism, there is no legal protection against subsequent discrimination.

Additional questions are important points of focus. Novel neurotechnologies may present unprecedented ways of altering, enhancing, or manipulating the self and personal agency. How will society respond? There is a sense of fairness ingrained in the notion of working hard to ensure success; it remains to be seen how neural devices used to enhance memory, attention, or reasoning will fit with this mindset. From a clinical perspective, there could be great value in altering specific memories of traumatic experiences, for instance, to ease the suffering associated with post-traumatic stress disorder. In mice, scientists used optogenetics to prompt immediate recall of memory by activating a small population of neurons underlying a memory engram (Liu et al. 2012). Knowing this is possible, it seems only a matter of time before researchers adapt technologies for manipulating memories in humans, which raises concerns: Is it a violation of one's autonomous and continuous sense of self to alter specific memories? Does it impinge on the authenticity of personal identity?

In the longer term, one can readily imagine efforts to move neurotechnologies developed for the clinic or research contexts into the nonmedical commercial space. What are the necessary standards for validity and safety, or ensuring privacy? How should society collectively manage potential conflicts of interest as novel neurotechnologies are commercialized for nonmedical, public use? These are big questions for society to process. Philosophers, ethicists, legal scholars, journalists, and educators will be needed to navigate these issues, and to promote discussion and engagement with the general public, regulatory agencies, and policymakers.

Conclusion

We have described just some of the issues associated with neurotechnology research and development that a program of neuroethics can help navigate. The broad and varied neuroethical landscape will require different types of approaches. Some of the issues are particularly amenable to empirical or normative research. For example, it would be essential to analyze how participants, participants' families, and researchers differ in their perceptions of risk in clinical research with novel neurotechnologies, particularly in studies with children and cognitively impaired adults. These concrete data could then inform best practices for the consent process in these studies. Other areas, such as sharing human clinical research data or managing incidental findings, might benefit from the development of guidance documents. In some cases, additional scholarship may be needed, which could take the form of workshops with extensive discussion and published proceedings.

Going forward, the NIH BRAIN Initiative* will lean heavily on its Neuroethics Division to consider neuroethics issues and suggest ways to address them. Experts in philosophy, bioethics, psychology, and neuroscience will be consulted to inform future steps. Additional services may be important, such as providing ethics consultations to funded researchers working in this space. More broadly, we hope to foster dialogue between BRAIN-funded researchers and ethicists and their colleagues worldwide, empower scientists' and clinicians' consideration of the ethical issues, and urge them to be stewards of their own work. These efforts will help ensure that neurotechnology research and development rests on a solid ethical foundation.

Brain activity forms the physical basis of our humanity and neuroethics provides a framework for discussing the ethical use of technologies that monitor or modulate the fundamental features that make us unique as individuals. Society has engaged in a continuous dialogue over the centuries about accepting, rejecting, or regulating the tools that spring from science; that focus needs to sharpen on neuroscience as new tools open new windows into the brain.

Note

1. Editor's note: an act similar to GINA, the Genetic Non-Discrimination Act, was passed by the Parliament of Canada in March 2017.

References

Alzheimer's Association (2015). Alzheimer's disease facts and figures. *Alzheimer's & Dementia*, **11**(3), 332–384.

Advisory Committee to the National Institutes of Health Director (2014). *BRAIN 2025: A Scientific Vision*. United States Government. Available from: http://braininitiative.nih.gov [accessed March 2, 2016].

Bloom, D.E., Cafiero, E.T., Jané-Llopis, E., et al. (2011). *The Global Economic Burden of Noncommunicable Diseases*. World Economic Forum. Available from: http://www.weforum.org/EconomicsOfNCD [accessed March 2, 2016].

DeLong, M.R. and Wichmann, T. (2015). Basal ganglia circuits as targets for neuromodulation in Parkinson disease. *JAMA Neurology*, **72**(11), 1354–1360.

Deng, Z.D., McClintock, S.M., Oey, N.E., Luber, B., and Lisanby, S.H. (2015). Neuromodulation for mood and memory: from the engineering bench to the patient bedside. *Current Opinion in Neurobiology*, **30**, 38–43.

Dyson, F. (1997). *Imagined Worlds*. Cambridge, MA: Harvard University Press.

Farah, M.J. (2015). An ethics toolbox for neurotechnology. *Neuron*, **86**(1), 34–37.

Fins, J.J. (2011). Neuroethics, neuroimaging, and disorders of consciousness: promise or peril? *Transactions of the American Clinical and Climatological Association*, **122**, 336–346.

Haubensak, W., Kunwar, P.S., Cai, H., et al. (2010). Genetic dissection of an amygdala microcircuit that gates conditioned fear. *Nature*, **468**(7321), 270–276.

Illes, J. and Bird, S.J. (2006). Neuroethics: a modern context for ethics in neuroscience. *Trends in Neuroscience*, **29**(9), 511–517.

Illes, J., Kirschen, M.P., Edwards, E., et al. (2006). Working Group on Incidental Findings in Brain Imaging Research. Ethics. Incidental findings in brain imaging research. *Science*, **311**(5762), 783–784.

Lisanby, S.H. (2007). Electroconvulsive therapy for depression. *New England Journal of Medicine*, **357**(19), 1939–1945.

Liu, X., Ramirez, S., Pang, P.T., et al. (2012). Optogenetic stimulation of a hippocampal engram activates fear memory recall. *Nature*, **484**(7394), 381–385.

Kinomura, S., Larssen, J., Gulyas, B., and Roland, P.E. (1996). Activation by attention of the human reticular formation and thalamic intralaminar nuclei. *Science*, **271**, 512–515.

Legon, W., Sato, T.F., Opitz, A., et al. (2014). Transcranial focused ultrasound modulates the activity of primary somatosensory cortex in humans. *Nature Neuroscience*, **17**(2), 322–329.

Mayberg, H.S. (2009). Targeted electrode-based modulation of neural circuits for depression. *Journal of Clinical Investigation*, **119**(4), 717–725.

Mohamed, F.B., Faro, S.H., Gordon, N.J., Platek, S.M., Ahmad, H., and Williams, J.M. (2006). Brain mapping of deception and truth telling about an ecologically valid situation: functional MR imaging and polygraph investigation—initial experience. *Radiology*, **238**(2), 679–688.

Morris, Z., Whiteley, W.N., Longstreth, W.T., Jr., et al. (2009). Incidental findings on brain magnetic resonance imaging: systematic review and meta-analysis. *BMJ*, 339.

Murray, C.J., Vos, T., Lozano, R., et al. (2012). Disability-adjusted life years (DALYs) for 291 diseases and injuries in 21 regions, 1990–2010: a systematic analysis for the Global Burden of Disease Study 2010. *The Lancet*, **380**(9859), 2197–2223.

Pasic, M.D., Samaan, S., and Yousef, G.M. (2013). Genomic medicine: new frontiers and new challenges. *Clinical Chemistry*, **59**(1), 158–167.

Perova, Z., Delevich, K., and Li, B. (2015). Depression of excitatory synapses onto parvalbumin interneurons in the medial prefrontal cortex in susceptibility to stress. *Journal of Neuroscience*, **35**(7), 3201–3206.

Presidential Commission for the Study of Bioethical Issues (2013). *Anticipate and Communicate: Ethical Management of Incidental and Secondary Findings in the Clinical, Research, and Direct-to-Consumer Contexts*. United States Government. Available from: https://bioethicsarchive.georgetown.edu/pcsbi/node/3183.html [accessed February 12, 2017].

Presidential Commission for the Study of Bioethical Issues (2014). *Gray Matters: Integrative Approaches for Neuroscience, Ethics, and Society*. United States Government. Available from: https://bioethicsarchive.georgetown.edu/pcsbi/node/3543.html [accessed February 12, 2017].

Presidential Commission for the Study of Bioethical Issues (2015). *Gray Matters: Topics at the Intersection of Neuroscience, Ethics, and Society*. United States Government. Available from: https://bioethicsarchive.georgetown.edu/pcsbi/node/4704.html [accessed February 12, 2017].

Roth, B.L. (2016). DREADDs for neuroscientists. *Neuron*, **89**(4), 683–694.

Sahakian, B.J., Bruhl, A.B., Cook, J., et al. (2015). The impact of neuroscience on society: cognitive enhancement in neuropsychiatric disorders and in healthy people. *Philosophical Transactions of the Royal Society B: Biological Sciences*, **370**(1677), 20140214.

Schiff N.D., Giacino, J.T., Kalmar, K., et al. (2007). Behavioural improvements with thalamic stimulation after severe traumatic brain injury. *Nature*, **448**(7153), 600–603. [Erratum in: *Nature*, 2008, 452(7183), 120.]

Slaughter, L.M. (2008). The Genetic Information Nondiscrimination Act: why your personal genetics are still vulnerable to discrimination. *Surgical Clinics of North America*, **88**(4), 723–738.

Smith, S.M., Nichols, T.E., Vidaurre, D., et al. (2015). A positive-negative mode of population covariation links brain connectivity, demographics and behavior. *Nature Neuroscience*, **18**(11), 1565–1567.

Stanley, S.A., Kelly, L., Latcha, K.N., et al. (2016). Bidirectional electromagnetic control of the hypothalamus regulates feeding and metabolism. *Nature*, **531**(7596), 647–650.

van der Werf, Y.D., Witter, M.P., and Groenewegen, H.J. (2002). The intralaminar and midline nuclei of the thalamus. Anatomical and functional evidence for participation in processes of arousal and awareness. *Brain Research Reviews*, **39**, 107–140.

Wheeler, M.A., Smith, C.J., Ottolini, M., et al. (2016). Genetically targeted magnetic control of the nervous system. *Nature Neuroscience*, **19**(5), 756–761.

Whittle, A.J., Walsh, J., and de Lecea, L. (2014). Light and chemical control of neuronal circuits: possible applications in neurotherapy. *Expert Review of Neurotherapeutics*, **14**(9), 1007–1017.

Part II

Neuroethics at the frontline of healthcare

Chapter 9

What do new neuroscience discoveries in children mean for their open future?

Cheryl D. Lew

Introduction

At the time of publication of the first edition of *Neuroethics: Defining the Issues in Theory, Practice and Policy* (Illes 2006), there was focus on the implications of the new neurosciences on questions of self-determination, personal identity, and moral responsibility. There was also acknowledgment that both genetic and environmental factors were significant determinants in the neurocognitive capacity of children to achieve across domains of academic performance, social adaptations, and socioeconomic ends (Farah et al. 2006). Early concerns emerged about how this information should be and could be used, and what implications such information would have on how children should be viewed as individuals or as members of particular populations vulnerable to prejudicial labeling.

Observational studies in childhood of intellectual quotient, personality, and behavioral inventories date back to early in the twentieth century. Piaget's work began in the 1920s (Piaget 1926). In addition to characterizing patterns of infancy through childhood behavioral development, he also described preliminary findings of normal moral development (Piaget 1997) based on observations of games play. He was an important influence on subsequent generations of developmental investigators including Lawrence Kohlberg (Kohlberg 1981) who elaborated a highly controversial theory of moral development in children stratified according to gender. The expected stages of early infancy and childhood development learned by every pediatrician in the Anglo-European-American tradition of the twentieth century and the common screening inventories used in well-child pediatric care all derive from Piaget's insights. The first reported collaborative thinking about the intersections of developmental psychology and emerging developmental neurosciences occurred in 1989 during a National Institute of Child Health and Human Development/National Institute of Mental Health-sponsored meeting in which Adele Diamond brought together developmental psychologists and neuroscientists (Diamond 1990). Subsequent studies confirmed observations of variants in executive function in children under a variety of genetic and environmental influences (Farah et al. 2006). Over the last decade, there has been a flourishing of research centered on detailed neuroimaging correlates of these behavioral observations (Illes & Raffin 2002; Illes & Racine 2005). Further, there are emerging studies of the neurostructural changes associated with environmental and genetic variations. Since some of these changes appear following in utero environmental exposures, there are implications regarding the interactions between genetic endowment and potential adverse prenatal and postnatal environmental influences on the eventual capacity of children to have flourishing and productive lives.

The discovery of significant alterations in the normal structural development of the brain in children, from the fetal period into adolescence, raises great concerns about implications for behavior, cognitive functioning, and whether affected children are inevitably destined for diminished function. This question of inevitability or determinism is a key issue in the ethics of identity and personhood.

Because of the limitations of development and emotional immaturity on decisional capacity in children, there has been a tendency among bioethicists, in general, to discount the emerging expression of individual identity and self-determination in the pediatric age group. However, pediatric healthcare providers are acutely aware of the transforming effect of serious and/or chronic illness/disability on accelerated insight, self-awareness, and in many instances nuanced ability for self-determination (Campbell et al. 2012) even with respect to critical end-of-life decisions. Transformation is not limited to catastrophic illness and it has become apparent that children's behaviors and world views change in response to influences such as family networks, the nurturing of caring family members, and other positive experiences (Weithorn & Campbell 1982; Weir & Peters 1997). Therefore, the pediatric healthcare professional has been disposed to the perception that children remain malleable and that potential can be optimized despite or even in spite of genetic differences or neurostructural variances. However, the observations of neurostructural anomalies in the developing brain correlated with measurable deficiencies in cognitive performance call into question whether such individuals are owed more or less opportunity to express their self-determination. What is the meaning of self-determination, in fact, in the context of neurostructural developmental abnormalities? This question directly pertains to the concerns of pediatric bioethicists, who tend to focus on eventual potential outcomes and have interests in the imprecise and unknowable futures of children who may have experienced both genetic and environmentally determined compromise to neurocognitive function (Wilkinson 2013).

Specific questions of significance to pediatric bioethics emerge from recent work:

1. With the ample evidence that a variety of genetic determinants influence developing brain structure and behaviors, are these determinants immutable? That is, if we identify developing sociopathy or psychopathy, are there social and policy implications that may abridge individual liberty and the individual's open future?

2. The socioeconomic and environmental pollution data are stunning. But can the data be misconstrued or misused, or both, to support a case for genetic ancestry determinants of poorer neurocognitive functioning? What safeguards do we as health researchers

and healthcare providers have a moral obligation to engage in to assure that this extraordinary data is not used to oppress further already disadvantaged groups?

3. With the prospect of full understanding of both intrinsic and extrinsic determinants of the mechanics of brain development in childhood rearing, we presume that interventions may be devised to normalize brain development and hence psychological functioning. How do we define normalcy? Would fixing or tweaking neurocognitive function eliminate diversity in personality, temperament, cognitive precocity, or genius?

As a framework for examining these questions, I will begin with a description of the concept of the open future of the child (Feinberg 1980). Next will be a discussion of what is accepted as evolving cognitive competence of children and adolescents, particularly with respect to self-awareness, insight, and decision-making. The accelerated maturity of healthcare decision-making among adolescents experiencing lifelong chronic or life-limiting illnesses, or both, will also be described. Then I will summarize some of the recent work among neuroscience colleagues on genetic and environmental influences on brain development and behavior from the prenatal period through adolescence. An account of the pediatric bioethical theory pertinent to this work will follow, providing a context for discussion about how neuroscience findings might change how we look at the moral status and evolving neurocognitive competency of affected children. Finally, I will discuss the social justice implications and what social strategies ought to be put in place in order to assure the most open future for not only affected children but also all of our children.

The notion of a child's open future

In the context of a discussion of rights, the political and legal philosopher Joel Feinberg distinguished the possession of rights (i.e., to autonomy, self-determination) versus actual exercise of those rights (Feinberg 1980, 1992). These are rights that are inherent to the moral status of human beings who are in possession of full personhood. While children are born with rights shared with adults, or all human persons, there are rights of self-determination that cannot be exercised until children acquire sufficient developmental and cognitive maturity. This latter state usually means that they become of age through the social or legal conventions (or both) of their culture. Until that time, the actions of children are ordinarily constrained by their parents and caretakers. However, since children are not the chattels of their parents, there are essential entitlements of childhood that extend beyond the absolute authority of parents: basic rights to safety, well-being, and protection from adversity. The presumption is that parents/caregivers are the stewards of the best interests and welfare of the children in their charge. Accordingly, parents and caretakers are given substantial liberty around decision-making across a variety of domains in the rearing of their children, such as religion, lifestyle, and schooling. However, Feinberg asserts that there are some specific rights that parents, caretakers, and society must safeguard in trust for the anticipated future capacities of their children. That is, children must have their mature, adulthood capacities for self-determination preserved.

Feinberg proposed that societies have broad moral obligations to set into place laws and policies which would assure the future of children's potential self-determination or autonomy. Examples include legal prohibitions of physical and emotional abuse, and obligations to avoid risky behaviors that endanger the welfare of children. Thus there are minimum age standards for permitting cigarette smoking, imbibing of alcoholic beverages, and driving an automobile. Once the minimal legal age of permission is passed, individuals are free, more or less, to decide for themselves the level of risk they wish to undertake on their own recognizance. The variations in societal determination of these minimum ages for specific autonomous functions are an interesting reflection on a given society's beliefs about when children actually do come of age, and therefore may take on the benefits of full autonomy as well as the risks of potential adverse outcomes from those independent risk-taking behaviors.

In the context of healthcare, this is most frequently taken to mean that parents or other responsible parties (i.e., caregivers, guardians, and healthcare professionals) are obligated morally to make healthcare decisions on behalf of the children in their charge which are most likely to result in the greatest realization of their children's eventual potential for self-determination and flourishing. Therefore, parents and caregivers are usually required to participate in certain public health measures such as childhood immunizations, to provide adequate nutrition, and to seek healthcare for interval or chronic illnesses.

Nevertheless, particularly in a liberal society, there is wide discretion given to parents/guardians about the determination of the threshold of obligation toward the usual societal norms. For instance, in some polities, parents/guardians may forego routine childhood immunizations for their children on the basis of religious liberty, or often a misplaced anxiety about the very small risks of vaccines compared with the risks of adverse outcomes following previously widespread viral diseases such as measles which can be associated with risk for death and severe long-term morbidities such as pan-encephalitis.

On the other hand, the concerns of a liberal polity may result in misplaced criticism about specific treatments of severely cognitively impaired individuals that are intended to improve the quality of life of those individuals as well as facilitate the caregiving. One example would be elective sterilization to eliminate risk of pregnancy in cognitively impaired women, unable to consent to intercourse or to defend against rape. Another example is the infamous case of the "Pillow Angel," a profoundly encephalopathic and neurocognitively diminished child, where growth attenuation treatments were provided with the intent to facilitate continued home versus institutional care (Gunther & Diekema 2006). Critics asserted that the Pillow Angel possessed future reproductive interests which were now foreclosed with the growth attenuation treatments (Liao et al. 2007; Peace & Roy 2014). For many healthcare professionals, as well as their parents, the notion that a microcephalic child who had been non-ambulatory and in a state of profound static encephalopathy had any particular interests in the future, let alone reproductive interests, is difficult to understand. These examples highlight some of the ethical tensions within the practical application of the principles embedded in Feinberg's child's open future concept.

With this background, then, the new discoveries in pediatric neuroethics bring into sharp relief some of the problems of self-determination against a background of genetic and environmental modulation of brain structure and behavior. The new neuroscience can be interpreted to diminish substantially concepts of free will as well as call into question the meaning of personal identity and personal authenticity, and ultimate moral responsibility for actions. And the question is whether free will and autonomy can be preserved in the face of the new neuroscience.

Developmental issues in childhood identity and self-determination

The contemporary standard of pediatric practice is to engage in transparency of communication with all patients about their healthcare status and needs in language appropriate for the individual level of cognitive development and function. While it is clear that parents or other legal caregivers/responsible parties carry the burden of legal consent for healthcare management or treatments, there are discrete advantages to inclusion of patients in communication about their care. These advantages include improvements in adherence, cooperation, and decreased anxiety about treatments on the part of the patients. Furthermore, since maturation of insight and decision-making capacity changes with time and specific life experiences, pediatricians and social scientists have recognized that adolescents who have had longstanding experience with personal chronic illness may be particularly astute about their own self-determination and therapeutic goals (Grisso & Vierling 1978; Bishop 1980; Ladd & Forman 1995; Hester 2009). Ongoing debate persists regarding how much weight ought to be given to adolescent patients' judgments about their own healthcare (Ross 1997) particularly with respect to high-stake health decisions. For instance, one of the greatest controversy surrounds the appropriateness of allowing adolescent patients to forego potentially life-saving procedures regardless of the burdens or risks of those treatments. An example might be whether an adolescent, or even younger child, with a failed solid organ transplant should be allowed to refuse a second transplant. Those who advocate proceeding with such treatments even in the face of adolescent objection, argue that to allow a minor to refuse life-sustaining treatments would be to foreclose that individual's open future. In other words, minors who choose for likely early death will not have any opportunity to become fully empowered adults with presumed better grounds for making such decisions.

With the neuroscience evidence supporting impairment of insight, self-awareness, and judgment in individual children on the basis of genetic or environmental influences, or both, how do the questions of capacity for decision-making play out? Should individuals determined to be at increased risk for neurocognitive disability or psychopathy due to observable alterations in brain structural development be deemed less competent than their age-matched peers? Given the many sorts of intelligence (Gray & Thompson 2004) individuals may exhibit, which particular intelligences should be deemed of greater importance in making autonomous healthcare decisions, let alone all decisions pertinent to evolving independent living? Should fluid intelligence be privileged over crystalline intelligence (Gray & Thompson 2004)? Should there be some sort of consensus taxonomy or sliding scale that guides how much prerogative ought to be provided to children, adolescents, and youth who are important stake holders in their health futures? What would be the role of early educational intervention for individuals identified to be at risk for judgment difficulties? Certainly health literacy is a sparse commodity among those individuals deemed to be within the normal neurocognitive range of capacity (IOM 2015). Therefore, it would seem appropriate to consider strategies of education and policy formulation which can enhance the capabilities of pediatric individuals of all capacities to develop appropriate insight into their own needs and desires, thereby exercising appropriate judgment and choices as they mature. However, before attempting to deal with the bioethics of the new neuroscience for children, it is necessary to review some of the more recent findings of significance.

Recent work in pediatric neuroimaging in at-risk populations

Children are continuously changing, morphologically and physiologically. They demonstrate remarkable increases in neurocognitive capacity from birth until they achieve full maturity. Along the way, brain structures change and evolve with concomitant changes in behavior, personality, temperament, and different forms of intelligence. The shape of these changes is determined by both intrinsic genetic endowment and the experiences of living. In contrast to the neurocognitive state of adults, where growth and development have already been achieved, and where adverse events may degrade neurocognitive function significantly, the plasticity of the pediatric brain makes for less predictable outcomes in the face of adverse events when considered against age at time of insult and capacity for recovery/compensation. Therefore, in childhood, future neurocognitive function in the individual is open to varying degrees and may be optimized through considered interventions. However, in order to develop effective interventions, understanding must be achieved of both normal structural/functional development and of changes to the expected development in response to a variety of external influences.

The neuroscientists and neurogeneticists of the Institute of the Developing Mind (IDM) at Children's Hospital Los Angeles, for example, are actively engaged in research pertaining to factors that can alter neurocognitive potential in childhood. Much of the research correlates brain structural differences with neurocognitive functional differences. The domains of investigation include prenatal and postnatal influences, environmental factors, and genetic versus nurture factors. These projects are representative of the efforts in pediatric neurosciences over the last decade and the findings from these projects provoke considerable bioethical concerns about potential for maintaining the open futures of children.

Considerable work has been reported on the association between brain structure abnormalities and psychopathy in adult males. For example, Yang and collaborators studied a cohort of adolescents who are a part of the University of Southern California Risk Factors for Antisocial Behavior Twin Study (Baker et al. 2002, 2006, 2013; Yang et al. 2012). In an endeavor to separate genetic influences from environmental influences on developing brain structure, 14-year-old twin pairs were studied: monozygotic pairs were compared with dizygotic pairs, the latter of whom share only one-half of their genetic endowment. The investigators hypothesized that any environmentally derived differences in cortical thickness would appear in regions of low heritability (i.e., the motor and sensory cortices). As it turns out, 80% of interpersonal variations in cortical thickness could be explained through genetic effects and occurred in later developing areas associated with higher-order functions in the prefrontal cortex (i.e., fluid intelligence). On the other hand, there was evidence for environmental malleability in posterior parietal areas associated with development of spatial attention, topographic information, and language (i.e., capacities that are determined experientially).

Normal children and adolescents experience a rapid increase in cognitive processing of sensory, motor, cognitive, and emotional information. These behavioral changes are associated with changes in brain structure—including synaptic pruning (i.e., controlled reduction of the number of synapses) and myelination (i.e., increased axonal dimensions) (Sowell et al. 1999, 2003; Selemon 2013; Houston et al. 2014). In order to characterize variations from the normal pattern, the investigators (Yang et al. 2015) conducted structural magnetic resonance imaging (MRI) brain scanning of 14-year-olds and compared these findings with serial measures of psychopathy with the Childhood Psychopathy Scale made at ages 9–10, 11–13, 14–15, and 16–18 years. The investigators were interested in changes in the psychopathy scores over time. They found that higher psychopathy scores were associated with increased temporal cortical thicknesses, findings specific to adolescents. There were gender-related differences in regional variation in cortical thickness as well (i.e., female subjects also showed decreased frontal cortical thickness). Increased psychopathic tendencies over time in individuals, as measured with the Childhood Psychopathy Scale, were associated with increased temporal cortical thickness. These changes were more prominent in males versus females. Localization of these structural anomalies to temporal lobe regions involved in complex social cognition and emotion are consistent with the observed functional abnormalities in psychopathic tendencies. Further, these observations are consistent with a neurodevelopmental origin of psychopathy as opposed to or in addition to an experiential model of psychopathy. All the study subjects were deemed to be healthy, in contrast to well-studied adults who were already known to have overt psychopathic behaviors (substance abuse, criminality). Therefore, Yang and colleagues concluded that abnormal cortical development in the temporal lobes is a likely biomarker of psychopathy in adolescents and may represent delayed maturation, which is genetically driven, and results in social and moral dysfunction (Yang et al. 2015).

New and continuously improving treatment and neonatal intensive care techniques have resulted in marked improvement in survival of preterm infants born at lower gestational ages. Adverse neurological injury such as intracranial hemorrhage and ischemic injury (periventricular leukomalacia) are known to be associated with poorer neurological and neurocognitive outcomes. Accordingly, imaging techniques such as cranial ultrasound, computed tomography scanning, and structural MRI cranial scanning have become commonplace in the twenty-first century neonatal intensive care unit and those results are routinely used for outcomes prognostication (Wilkinson 2013). However, even in the absence of routine cortical neuroimaging abnormalities, preterm infants frequently exhibit abnormal cognitive, behavioral, and social function outcomes, such as problems with executive function, attentional deficit activity disorder (ADHD), and issues with empathy and social restraint as often exhibited in autism spectrum disorder. While there are clearly recognized genetic determinants of some of these behavioral disorders, the incidence in preterm infants appears to be disproportionate. Leporé and colleagues sought to determine if there were any deep brain structure anomalies in preterm infants which could be determinants of later neurocognitive function (Lao et al. 2016). They compared imaging of subcortical brain structures between apparently healthy surviving preterm infants as compared with age-matched control infants born at term. They looked specifically at subcortical structures—the thalamus and putamen— using surface-based morphometry analysis. They found reduction in volume in all the thalamic nuclei, more pronounced on the left than the right. In addition, connecting pathways between thalamus and putamen (i.e., the dorsolateral prefrontal–subcortical circuit and the lateral orbitofrontal–subcortical circuit) may be reduced in volume. It is not clear if these changes are due to imposed injury such as hypoxemia and other metabolic changes associated with preterm birth or due to a reduction of in utero stimulatory experience because of the shortened gestation. The relationship of such subcortical changes in brain morphological development and hypoxic-ischemic injury to white matter in preterm infants is not yet defined.

Although an ostensible motivation for these studies was to promote early identification of risks for cognitive and behavioral disabilities, it is not at all clear that "early intervention" from a behavioral perspective will prove to be effective in altering expected worsened outcomes.

Environmental pollutants and toxins are of great concern for the public's health. In order to elucidate the impact of such pollutants on developing fetal brains and subsequent neurocognitive function and behavior, Peterson and his colleagues embarked on carefully designed long-term studies of maternal-fetal pairs living in an urban setting beginning in 1997 (Peterson et al. 2015). Nonsmoking pregnant women, predominantly African American or Dominican, were recruited from prenatal clinics in New York City which provide services to low-income minority and recent immigrant communities. The authors acknowledged that outdoor pollution sources tend to be concentrated in such communities resulting in disproportionate exposure of these residents to outdoor pollution. In addition, the pollutants in question, polycyclic aromatic hydrocarbons (PAH) readily penetrate indoor residential environments; consequently, inhabitants of low-income urban neighborhoods cannot escape exposure. During the third trimester of pregnancy, information about PAH exposures was collected through questionnaires and from individual monitors carried by participants which sampled levels of the eight most

common PAH in each woman's environment for 48 hours. More than 600 women participated in this phase of the study.

Children exposed to higher levels of PAH in utero showed a variety of neurodevelopmental problems at ages 3, 5, and 7 years (Perera et al. 2006, 2008, 2009, 2012). The
investigators then hypothesized that they could identify specific abnormalities in brain
structure associated with prenatal PAH exposure which resulted in the neurodevelopmental abnormalities (information processing speed, ADHD) seen in a subset of these
children at ages 7–9 years (Peterson et al. 2015). They performed MRI on a subset of 40
children stratified into above- and below-median in utero PAH exposure. In addition
to the imaging, the Child Behavior Checklist and the Wechsler Scales of Intelligence for
Children (WISC-IV) were administered. The differences between children with low intrauterine exposure versus high intrauterine exposure were striking: white matter reductions
were associated with higher PAH exposure, specifically in the left hemisphere. Therefore,
the left hemisphere in utero appears to be more vulnerable to the neurotoxic effects of
PAH. The investigators go on to link these observed changes to what is now known about
the transcriptional pathways that control brain hemispheric structure by week 10 of gestation. Of interest, the regions of white matter reduction in this cohort of children exposed
to high in utero PAH and who also had ADHD syndrome are qualitatively different
from white matter abnormalities observed in adolescent youth with ADHD syndrome
(Peterson et al. 2015). The authors speculate that environmental neurotoxins produce a
distinct subtype of ADHD.

The investigators recognized some of the implications relating to their study population
of impoverished, principally immigrant peoples and peoples of color in the urban neighborhoods under investigation. However, through careful stratification of study subjects,
they were able to demonstrate independent environmental toxic effects on both brain
structure and neurocognitive function within this population. Therefore, any potential
arguments about intrinsic racial or ethnic differences could be dismissed.

That childhood experiences can modulate developmental functions, behavior and neurocognitive capacity has been amply demonstrated (McLoyd 1998; Rosenzweig 2003;
Noble et al. 2007; Duncan & Magnuson 2012; Kim et al. 2013; Luby et al. 2013). However,
brain structural correlates of the neurocognitive, neurodevelopmental associations with
socioeconomic variables have not previously been demonstrated in any large numbers of study subjects. The collaborators of the Pediatric Imaging, Neurocognition, and
Genetics (PING) Study have examined these relationships among more than 1000 individuals between ages 3 and 20 years (Noble et al. 2015). Previous studies have looked at
brain cortical volume. However, changes in the normal sequence of cortical surface area
change may be more reflective over individual experiences. Normally, in contrast with
cortical volume, cortical surface area increases from infancy through early adolescence,
after which there is progressive shrinkage through middle adulthood. Genetic programming of these changes is modulated by individual experience. Therefore, although normally cortical thickness decreases into adolescence, and thinness tends to be associated
with greater intelligence, cortical surface area increases with greater intelligence. Against

the above-mentioned background, the PING study group made the following observations: both parental education and family income were associated with increased cortical surface area. Genetic ancestry was not a relevant factor. Subcortical structures, for example, hippocampal size, are positively associated with parental educational achievement. Parental educational attainment was found to be linearly associated with cortical surface area. Specifically, any increment of parental education was associated with a commensurate increase in cortical surface area. The brain regions in question are important in development of language, executive functions, and memory. Because of the logarithmic associations, the lower parental educational attainment and income, the greater the impact on brain structure.

This study is particularly controversial and therefore it is not surprising that the results elicited a considerable amount of media attention. The very tight coupling of socioeconomic status with brain structure seems to have struck a nerve in mainstream media and within the greater North American public. Apparently, the specter of shrunken brains among large segments of our population (i.e., our socioeconomically disadvantaged citizens and their children) created significant unrest (King 2015; Mohan 2015). It is not yet clear whether that concern and unrest are sufficiently long-lasting to result in significant social policy changes and access to early educational and psychological interventions.

Common ethical issues emerge across all of these studies. What do the data say about personal identity? Is identity immutable if associated with structural changes in the brain? Do all such changes translate to clinically significant abnormalities in information processing, insight, creativity, and fluid intelligence? How do these variants in brain development play against the normal ontogeny of maturation of cognitive processes in children and adolescents?

Bioethical and neuroethical implications

Bioethics is about relationships: how we see each other, how we treat each other, and what we owe to each other in terms of respect and moral seriousness. Relationships in childhood are normally dynamic, changing continuously because infants and children are changing continuously. Therefore, in order to understand what ethical harms children who are already neurocognitively compromised may experience—the additional neuroethical piece—it is worthwhile reviewing the basic bioethics issues of childhood.

Core issues in bioethics for children are centered on potentials: moral status, personhood, authenticity (i.e., persistence of personality over time) and judgment, and, subsumed under rights or eventual rights—self-determination and decision-making. Of course, these are all key issues for bioethics in adults. But here, the concern is how to ensure that each child has a flourishing future that is unlimited except for innate capacity. These issues constitute our sense of self and self-esteem and determine whether others look at us as whole or damaged. The interactions between how we see ourselves and others see us can be reinforcing, constructive, or destructive. I will look first at the large category of rights for self-determination and decision-making.

A key component for individual rights is the understanding that free will or at least the possibility or potential for free will is an important determinant of individual person-hood and worthy moral status. Without the possibility of free will, the concerns within bioethics about self-determination and respect for autonomy would be irrelevant, mean-ingless. A discussion about the power of determinism (Rachels 2012) versus the actuality of free will is well beyond the scope of this chapter, but the dichotomy of determinism (or more properly, pre-determinism) versus potential for eventual full autonomy is one of the problems of the now demonstrated morphological aberrations in the brains of the young throughout childhood.

Returning to Feinberg's "open future of the child," while theoretically compelling, much of pediatric approaches to bioethics have involved pragmatic specifications about how to implement appropriate decision-making for minors in stewardship for the future mature individual. There have been at least two significant competing models with some pro-posed variants.

The most prevalent model in pediatrics is that articulated in the canonical work *Deciding for Others* (Buchanan & Brock 1990). For individuals apparently either without compe-tence or having lost their competence in the neurocognitive sense, surrogates can best serve their charges by making decisions on their behalf using a "best interests standard" (BIS) where the parents, or other legitimate responsible parties attempt to determine net value among alternative choices which balance benefits versus the risks or burdens with each choice. This standard requires an evaluation of quality of life considerations and the authors stress that they mean quality of life, as it is significant to the presumed interests of the individual in question as opposed to the interests of the family or the community at large that may sway the decision-making toward allocation of either more or less resources. In other words, decisions on behalf of those deemed unable to decide for themselves should be made solely on the basis of the worth or consequences of those decisions for those indi-viduals and not on the basis of the perceived social utility of those individuals.

Although Buchanan and Brock were frequently speaking of individuals severely neu-rocognitively disabled or those without developmental capacity for decision-making, depending on how we determine the threshold for neurocognitive incapacity, this surro-gate decision-making standard could be applied to some individuals now known to have structural brain anomalies associated with genetic, socioeconomic, or environmental adverse effects consisting of behavioral anomalies, processing difficulties, and sociopathy. Further, Buchanan and Brock wrote separately of the specific problem of older children who demonstrated evolving self-awareness, maturation of abstract reasoning skills, exer-cise of judgment and insight, and who would have increasing direct interests in their own healthcare decision-making.

Buchanan and Brock emphasized the priority of the individual's interests over all other parties. Therefore, they strictly privileged rights and interests of children/minors over all other interests including family and community interests.

On the other hand, others have weighed in on behalf of the priority of relationships and the continuity thereof on whose interests should be first served? Murray speaks eloquently of the importance of the minor child's place in the tapestry of relationships of the family and then the social community (Murray 1996). Friedman Ross introduced her concept of "constrained parental autonomy" (Ross 1998) as a means of bridging the divide between pure parental prerogative and pure self-interested primacy of their children's interests over and above the interests of the parents/family and larger community. Friedman Ross has contended that childhood autonomy cannot be appropriately expressed by children due to inadequately developed maturity, lack of insight into consequences of decisions, and perhaps inadequate foresight or imagination into the future. Therefore, she has stipulated that parents must make all decisions in trust, premised on the requirement that the constraints that parents must abide by relate to their obligations to provide a minimum of supportive and protective care which can optimize or at least present the prospect of the potential for maturation into full autonomous adulthood. Friedman Ross's interpretation of the child's open future is that parents must engage in constraints to their children's own choices so as to optimize putative quality of the future child's life. By doing so, parents may promote the ongoing health and survival of their children until they can develop sufficiently mature capacity to be empowered to make their own autonomous decisions.

Two extensive commentaries (Hester 2007; Salter 2012) on the value and utility of the BIS shed a great deal of light on the pragmatic problems of applying the BIS. Salter has produced a highly nuanced and detailed description of the BIS in its several versions. She lists a variety of self-regarding interests ascribed to children including both present and future interests. The most pertinent of interests of children against the background of neuroscience are those relating to cognitive-developmental function both in the present and in the future. Salter is also highly critical of the BIS because of the following problems: poorly defined terms, standards, and evaluation processes; there appears to be no uniformly accepted understanding of criteria for determining the best interests resulting in inconsistent and often too narrow application of the standard to specific cases; finally, the standard does not allow for adequate respect for the family. Both Salter and Hester reflect on cases where the specific contexts and circumstances are not amenable to solution through the BIS. Salter points out clearly that other interested parties, aside from the child, for example, the healthcare providers and the parents, may not agree or have clarity about what course of action(s) represent the child's best interests.

Hester writes of the difficulties of ascribing interests to very young or compromised infants, or both, because of the lack of any duration of lived experience. Therefore, he observes that the tendency is to restrict the discussion about the best interest choices for young infants to either avoiding net harm or to give all priority of decision-making to parents. He also asserts that parental interests may be connected to but not necessarily identical to the cultural and social interests of the community at large in the welfare of infants and children. Therefore, Hester believes that the sources of interests contributing to decision-making for very young children or children who may be neurocognitively compromised should be expanded to encompass the larger community. Hester invokes narrative as a means for ascertaining the relevant stories of the family, the child and the larger community, which have bearing on healthcare decision-making for the individual child. That is, only by identifying all of the relevant stakeholders can one determine what are the real best interests for the individual child.

It is clear from these arguments that, as a prevalent/dominant practice in deciding for children, the BIS can weigh more heavily in the construction of a decision-making future for children with evidence of neurocognitive compromise or difference. That is, there is more at stake in potential imposed constraints on the futures of compromised children.

The issues surrounding the emergence of personal identity and how others view the individual emerging personality continue to be complex and difficult. As a counter to the ambiguities of the BIS and the problems of recognizing developing individuation in children, Lindemann has proposed a model of how personal identity is recognized by others and how the perception of that identity as well as the maintenance of that perception or recognition held over time, enables authenticity of the individual personality (Lindemann 2014). She has described personhood as a composite of four elements: sufficient mental activity is present to constitute a personality; aspects of the personality are expressed bodily; others recognize those embodiments as expression of personality; and others respond to what they perceive. Lindemann then goes on to describe how we see ourselves and others as specific identities through "a web of stories depicting our most important acts, experiences, characteristics, roles, relationships, and commitments. This narrative tissue constitutes our personal identities, which play a crucial role in the practice of personhood" (Lindemann 2014, p.ix).

What Lindemann emphasizes is that our identities, and by connection, our moral status as human beings, is determined first by how we are perceived by others, from the time the gestating mother recognizes her emerging fetus, to the born child and emerging adult. How individuals are perceived shapes how they view themselves, develop personality, and how they engage in transactional activities or relationships with others: "Personal identities function as counters in our social transactions, in that they convey understandings of what those who bear them are expected to do"(Lindemann 2014, p.ix). That is, how individuals are perceived results in responses, or reciprocal behaviors that either reinforce the perceptions, edit those perceptions, eliciting further responses so that the development of identity and personality is seen in a relational/relationship context. Lindemann does not directly discuss temperament (perhaps because that is intrinsic). But she does speak about how the treatment of the other's identity can be positive—that is, supportive, reinforcing, stimulating behaviors that lead to flourishing. On the other hand, the treatment of the other can be negative, and result in the creation of what Lindemann refers to as a damaged identity (Nelson 2001).

Lindemann's model is quite potent in the context of identified neurostructural anomalies in at-risk children and youth for the following reasons: (1) at-risk subjects may embody abnormal behaviors, or psychopathies, associated with their brain anomalies; (2) how others perceive and respond to these behaviors can either reinforce negative behaviors or shape more positive, more functional behaviors within the limits of the individual's cognitive capacity; and (3) targeted education and training of both at-risk individuals and their families/community members could, over time, result in sustained alterations of behavior translating to authenticity in personality. Therefore, Lindemann would advocate development of relational social structures that are protective and supportive of individual personal identities.

Considerations of social justice

The other domain of bioethics, aside from the individual and personal, is that of the public square. This is where the interests of the individual may intersect with the interests of the greater polity. The recent findings of neuroimaging anomalies in the structure of developing brains are worrisome. However, the acknowledged resilience and plasticity of the immature central nervous system present opportunities, indeed, confer obligations to create a social and cultural milieu, which can favor optimal growth and development of all children, let alone those at significant risk for neurocognitive compromise. The new neuroscience, because of the plethora of knowledge emerging about mechanisms of neurocognitive compromise presents the prospects of new and more effective prevention and possible treatments for these adverse conditions (Shonkoff & Levitt 2010). How can such programs of early diagnosis and early intervention come about?

Models of social justice tend to be theoretical and abstract (Wenar 2013). However, two parallel and somewhat related theories have introduced more pragmatic approaches (Nussbaum 2003, 2004, 2006; Powers & Faden 2006). Part of the more practical nature of these theories is the use of narrative in telling specific stories, which provide concrete detail about individuals and subsets of individuals, which may inform justice policy strategies.

Nussbaum has based her approach on some of the early work of Amartya Sen (Nussbaum 2003) and a critique of John Rawls (Nussbaum 2006). She addresses, in the latter work, the weaknesses in social contract theories around disparities of access to social and health goods where those who are disabled either lack voice or whose voices are dismissed. Nussbaum proposes a noncontractarian view of basic entitlements owed to every individual within a society. Nussbaum's capabilities include the following:

1. Life (i.e., the ability to live)

2. Bodily health

3. Bodily integrity

4. Senses, imagination, thought (i.e., to engage in as fully cognitive function as possible)

5. Emotions (or valuation) or to form attachments, enjoy an affective life

6. Practical reason or to be able to engage in conceptual thought and critical reflection on one's life and future

7. Affiliation or to have relationships with others, other groups

8. Solidarity with other living creatures

9. Play

10. Control over one's environment in both a social and political sense.

Implicit is the notion that all individuals are deemed to have personal identity and moral worth. The capabilities 4 through 10 are particularly pertinent to the status of children with neurocognitive variant or disability. Nussbaum's prescription, then, is that identification of socially constructed barriers to expression or experience of these capabilities ought to allow communities to develop policy and practices which optimize/enhance capabilities. The Americans with Disabilities Act (ADA) would be an example of the sort of policy/practice approach endorsed by Nussbaum. She undoubtedly views the ADA as moving in the right direction, though not as comprehensive as she would like. Nussbaum's variant on Sen's capabilities approach is far more expansive and is not limited to considerations of physical or mental disability but also includes other forms of social and/or economic disparity. Recognizing the disproportionate burdens (time, money, labor) on the part of caretakers of child with varying levels of neuro-disability, Nussbaum extends her capabilities approach to caretakers and other affiliated personnel who may be recruited to help with the increased care needs of special needs children. By no means, should we assume that the increased burden of care involves only those children with the most severe disabilities. Children with educational and/or social/behavioral disabilities may also require more resources for education, supervision, and treatment. And Nussbaum would contend that a just society recognize such needs and accommodate in order that both at-risk children and their caretakers have flourishing lives:

> Capabilities belong first and foremost to individual persons, and only derivatively to groups. The approach espouses a principle of *each person as an end*. It stipulates that the goal is to produce capabilities for each and every person, and not to use some people as a means to the capabilities of others or of the whole. (Nussbaum 2011, p.35)

Well-being is the key goal in an alternative theory of social justice (Powers & Faden 2006). The premise is that a sound public health system can ensure a minimum sufficiency of goods and resources to all members of the society so that well-being can be achieved:

> Our theory starts with the assumption that inequalities beget inequalities, and existing inequalities—in the social determinants of well-being and ultimately in the essential dimensions of well-being themselves—can compound, sustain, and reproduce a multitude of deprivations in well-being, bringing some persons below the level of sufficiency for more than one dimension. (Powers & Faden 2006, p.7)

Powers and Faden assert that the moral imperative of public health practice is not just regulatory and prescriptive but is also to ensure social justice through allocation and distribution practices. They have defined well-being as a sufficiency of existence across a number of dimensions which is similar to but not identical to the capabilities list:

1. Health
2. Personal security
3. Reasoning
4. Respect
5. Attachment
6. Self-determination.

They assert that the dimensions interact and that serious insufficiency of even one dimension tends to influence negatively the other dimensions of well-being too. They do not equate well-being with absence of disease. They also identify, for some populations, such as children, specific periods during the life cycle where public allocation decisions are particularly important to ensure existing and developing well-being. They identify early infancy and early childhood as periods of great vulnerability in terms of suboptimal development of reasoning skills that can impair cognitive function lifelong. These are also periods during the life cycle of great opportunity for interventions that can optimize practical reasoning skills, above and beyond intrinsic capacity. Because of the priority assigned to the needs of children by Powers and Faden, they acknowledge that in a public health system designed to assure the well-being of all its citizens might result in conflicts of priorities of allocation among subpopulations, for example, children versus the elderly where there were constraints of resources.

There are always constraints on resources of one sort or another and Powers and Faden propose strategies using quality of life potential, cost-effectiveness analyses, and cost-utilization analyses to develop schemes for resource allocation which are fair and still give priority to children's needs. This is certainly a refreshing proposition for pediatric healthcare providers! However, Powers and Faden also stipulate that context is important in the final allocation decisions—that is, if specific situations place individuals in other age groups at greater risk to their well-being, then it would be wrong to privilege children always in terms of allocation of resources. Accounting of all dimensions of well-being needs to be analyzed closely in order to determine who should be privileged at particular times and under particular circumstances. On the other hand, children tend to live longer and therefore a suboptimal well-being state would represent a greater burden to the individual as well as the community if the needs of children were not addressed aggressively. Presumably this stance would apply even more so for children identified to have risk for poor neurocognitive outcomes.

With these marginally less abstract theories to consider, what we are left with are some common conclusions: all children are owed optimal services in the form of education, training, comprehensive care, suitable living conditions, and environments designed to promote flourishing. Because of the increased burdens associated with providing additional services to compensate for additional neurocognitive risks, services for support of families and caretakers should be allocated as well. In concrete terms, neither Nussbaum nor Powers and Faden provided real descriptions of comprehensive capabilities or social justice schemes.

Conclusion

Have we made any progress in answering the three questions posed at the beginning of this paper?

1. How might genetic determinants of brain structure influence social policy to abridge individual liberty and the individual's open future?

2. What are the moral obligations of health researchers and healthcare providers to ensure that socioeconomic and environmental pollution data not be misconstrued to oppress, further, already disadvantaged groups by ascribing poorer neurocognitive outcomes to genetic ancestry?

3. How should normalcy be defined and would correction of neurocognitive function eliminate diversity in personality, temperament, cognitive precocity, or genius?

For question 1, yes, there are risks of biased perceptions and creation, in Lindemann's words, of damaged identities that foreclose many possibilities for remediation through education, training, and improvement of the rearing environment in childhood. Social and policy measures which could foster damaged identities are intuitively obvious, such as scapegoating and restriction of access to appropriate educational and other support services to such children and their families. On the other hand, we are currently in an era of great libertarianism among the families of children who are neurocognitively compromised and families are often tremendous advocates on behalf of their special needs children. However, resources are not distributed at a federal level, but rather at individual states levels, in the United States, so that access to those resources are certainly not universally available and are highly variable according to a priori conditions determined by various insurance plans or state policy. Further, mental health services are notoriously poorly supported.

For question 2, there is no doubt that researchers and clinicians have obligations to ensure that the data are gathered and reported with sufficient controls so as to eliminate misconstrual of results and to minimize the possibility of adding social oppression to the plight of children and families who are less fortunate. Both Peterson and Sowell were able to implement sufficient controls to disabuse any notions that the neurocognitive impairments observed in their study populations could be ascribed to race, ethnicity or culture. These are practices that ought to be a part of the design of all such studies.

For question 3, there are no answers yet. This question and all the implications will have to wait for the next edition of this book.

To conclude, while much of the content of this chapter has been theoretical, there are some pragmatic interventions that ought to be considered at both individual and societal levels. They draw upon the different aspects of and strengths of neuroethical inquiry and methods, and what they can offer looking ahead:

1. Minimization of prenatal exposure to neurotoxins (Lanphear et al. 2005; Mezzacappa et al. 2011), and of all environmental sources of neurotoxins since adverse effects on neurofunction are not limited to fetuses or children (Illes et al. 2014; Cabrera et al. 2016).

2. Focus on maximal educational achievement across all socioeconomic classes for the purposes of enhancing the likelihood of economic achievement that may ameliorate the functional consequences of brain growth anomalies among children and adolescents (Diamond & Lee 2011; Schonert-Reichl et al. 2015).

3. Promotion of a societal focus on improving the capabilities of all members of the polity—whether using the model of Nussbaum or the public health model of Powers

and Faden—with the intent to ameliorate the effects of socioeconomic disparities: better education, better nutrition, and better quality of life could reduce the likelihood of neurocognitive compromise on basis of social disparities.

4. Establishment of strategies at the individual level to promote construction of positive identities for all individuals, and especially for those known to be at risk for neurocognitive problems. Even individuals who exhibit psychopathic behaviors may have insight (Horstkötter et al. 2012) and may be more or less open to supportive networks that may increase their ability to lead flourishing and successful lives.

References

Baker, L.A., Barton, M., and Raine, A. (2002). The Southern California Twin Register at the University of Southern California. *Twin Research*, **5**, 456–459.

Baker, L.A., Barton, M., Lozano, D.I., Raine, A., and Fowler, J.H. (2006). The Southern California Twin Register at the University of Southern California: II. *Twin Research and Human Genetics*, **9**, 933–940.

Baker, L.A., Tuvblad, C., Wang, P., et al. (2013). The Southern California Twin Register at the University of Southern California: III. *Twin Research and Human Genetics*, **16**, 336–343.

Bishop, S. (1980). Children, autonomy and the right to self-determination. In: Aiken, W. and Lafollette, H. (Eds.) *Whose Child? Children's Rights, Parental Authority, and State Power*, pp.154–176. Totowa, NJ: Littlefield, Adams & Co.

Buchanan, A.E. and Brock, D.W. (1990). *Deciding for Others: The Ethics of Surrogate Decision-Making*. New York: Cambridge University Press.

Cabrera, L.Y., Tesluk, J., Chakraborti, M., Matthews, R., and Illes, J. (2016). Brain matters: from environmental ethics to environmental neuroethics. *Environmental Health*, **15**, 1–5.

Campbell, A.T., Derrington, S.F., Hester, D.M., and Lew, C.D. (2012). Her own decision: impairment and authenticity in adolescence. *Journal of Clinical Ethics*, **23**, 47–55.

Diamond, A. (1990). Introduction: the development and neural bases of higher cognitive functions. In: *The Development and Neural Bases of Higher Cognitive Functions*, pp.xiii–lvi. New York: New York Academy of Sciences.

Diamond, A. and Lee, K. (2011). Interventions shown to aid executive function development in children 4 to 12 years old. *Science*, **333**, 959–964.

Duncan, G.J. and Magnuson, K. (2012). Socioeconomic status and cognitive functioning: moving from correlation to causation. *Wiley Interdisciplinary Reviews: Cognitive Science*, **3**, 377–386.

Farah, M.J., Noble, K.G., and Hurt, H. (2006). Poverty, privilege, and brain development: empirical findings and ethical implications. In: Illes, J. (ed.) *Neuroethics: Defining the Issues in Theory, Practice, and Policy*, pp.277–287. New York: Oxford University Press.

Feinberg, J. (1980). The child's right to an open future. In: Aiken, W. and Lafollette, H. (Eds.) *Whose Child? Children's Rights, Parental Authority, and State Power*, pp.124–153. Totowa, NJ: Littlefield, Adams & Co.

Feinberg, J. (1992). *Freedom & Fulfillment: Philosophical Essays*. Princeton, NJ: Princeton University Press.

Gray, J.R. and Thompson, P.M. (2004). Neurobiology of intelligence: science and ethics. *Nature Reviews Neuroscience*, **5**, 471–482.

Grisso, T. and Vierling, L. (1978). Minors' consent to treatment: a developmental perspective. *Professional Psychology*, **9**, 412–427.

Gunther, D.F. and Diekema, D.S. (2006). Attenuating growth in children with profound developmental disability: a new approach to an old dilemma. *Archives of Pediatrics & Adolescent Medicine*, **160**, 1013–1017.

Hester, D.M. (2007). Interests and neonates: there is more to the story than we explicitly acknowledge. *Theoretical Medicine and Bioethics*, **28**, 357–372.

Hester, D.M. (2009). Adolescent decisionmaking, part i: introduction. *Cambridge Quarterly of Healthcare Ethics*, **18**, 300–302.

Horstkötter, D., Berghmans, R., De Ruiter, C., Krumeich, A., and De Wert, G. (2012). "We are also normal humans, you know?" Views and attitudes of juvenile delinquents on antisocial behavior, neurobiology and prevention. *International Journal of Law and Psychiatry*, **35**, 289–297.

Houston, S.M., Herting, M.M., and Sowell, E.R. (2014). The neurobiology of childhood structural brain development: conception through adulthood. *Current Topics in Behavioral Neurosciences*, **16**, 3–17.

Illes, J. (Ed.) (2006). *Neuroethics: Defining the Issues in Theory, Practice, and Policy*. Oxford: Oxford University Press.

Illes, J. and Racine, E. (2005). Imaging or imagining? A neuroethics challenge informed by genetics. *American Journal of Bioethics*, **5**, 5–18.

Illes, J. and Raffin, T.A. (2002). Neuroethics: an emerging new discipline in the study of brain and cognition. *Brain and Cognition*, **50**, 341–344.

Illes, J., Davidson, J., and Matthews, R. (2014). Environmental neuroethics: changing the environment changing the brain. Recommendations submitted to the Presidential Commission for the Study of Bioethical Issues. *Journal of Law and the Biosciences*, **1**(2), 221–223.

IOM (2015). *Health Literacy: Past, Present, and Future: Workshop Summary*. Washington, DC: The National Academies Press.

Kim, P., Evans, G.W., Angstadt, M., et al. (2013). Effects of childhood poverty and chronic stress on emotion regulatory brain function in adulthood. *Proceedings of the National Academy of Sciences of the United States of America*, **110**, 18442–18447.

King, N. (Director) (2015). *Socioeconomic Status Could be Affecting Childrens' Brains*. Los Angeles, CA: American Public Media.

Kohlberg, L. (1981). *The Philosophy of Moral Development: Moral Stages and the Idea of Justice* (Essays on Moral Development, Vol. 1). San Francisco:, CA: Harper & Row.

Ladd, R.E. and Forman, E.N. (1995). Adolescent decision-making: giving weight to age-specific values. *Theoretical Medicine*, **16**, 333–345.

Lanphear, B.P., Vorhees, C.V., and Bellinger, D.C. (2005). Protecting children from environmental toxins. *PLoS Medicine*, **2**, e61.

Lao, Y., Wang, Y., Shi, J., Ceschin, R., Nelson, M., Panigrahy, A., and Leporé, N. (2016). Thalamic alterations in preterm neonates and their relation to ventral striatum disturbances revealed by a combined shape and pose analysis. *Brain Structure and Function*, **221**(1), 487–506.

Liao, S.M., Savulescu, J., and Sheehan, M. (2007). The Ashley treatment: best interests, convenience, and parental decision-making. *Hastings Center Report*, **37**, 16–20.

Lindemann, H. (2014). *Holding and Letting Go*. New York: Oxford University Press.

Luby, J., Belden, A., Botteron, K., et al. (2013). The effects of poverty on childhood brain development: the mediating effect of caregiving and stressful life events. *JAMA Pediatrics*, **167**, 1135–1142.

McLoyd, V.C. (1998). Socioeconomic disadvantage and child development. *American Psychologist*, **53**, 185–204.

Mezzacappa, E., Buckner, J.C., and Earls, F. (2011). Prenatal cigarette exposure and infant learning stimulation as predictors of cognitive control in childhood. *Developmental Science*, **14**, 881–891.

Mohan, G. (2015). Can money buy yours kids a bigger brain? *Los Angeles Times*, March 30.

Murray, T.H. (1996). *The Worth of a Child*. Berkeley, CA: University of California Press.

Nelson, H.L. (2001). *Damaged Identities: Narrative* Repair. New York: Cornell University Press.

Noble, K.G., McCandliss, B.D., and Farah, M.J. (2007). Socioeconomic gradients predict individual differences in neurocognitive abilities. *Developmental Science*, 10, 464–480.

Noble, K.G., Houston, S.M., Brito, N.H., et al. (2015). Family income, parental education and brain structure in children and adolescents. *Nature Neuroscience*, 18, 773–778.

Nussbaum, M. (2003). Capabilities as fundamental entitlements: sen and social justice. *Feminist Economics*, 9, 33–59.

Nussbaum, M.C. (2004). Beyond the social contract: capabilities and global justice. an Olaf Palme lecture, delivered in Oxford on 19 June 2003. *Oxford Development Studies*, 32, 3–18.

Nussbaum, M.C. (2006). *Frontiers of Justice: Disability, Nationality, Species Membership*. Cambridge, MA: Harvard University Press.

Nussbaum, M.C. (2011). *Creating Capabilities: The Human Development Approach*. Cambridge, MA: Harvard University Press.

Peace, W.J. and Roy, C. (2014). Scrutinizing Ashley X: presumed medical "solutions" vs. real social adaptation. *Journal of Philosophy, Science & Law*, 14, 33–52.

Perera, F.P., Rauh, V., Whyatt, R.M., et al. (2006). Effect of prenatal exposure to airborne polycyclic aromatic hydrocarbons on neurodevelopment in the first 3 years of life among inner-city children. *Environmental Health Perspectives*, 114, 1287–1292.

Perera, F.P., Li, Z., Whyatt, R., et al. (2008). Prenatal airborne polycyclic aromatic hydrocarbon exposure and child IQ at age 5 years. *Pediatrics*, 124, e195–202.

Perera, F.P., Li, Z., Whyatt, R., et al. (2009). Prenatal airborne polycyclic aromatic hydrocarbon exposure and child IQ at age 5 years. *Pediatrics*, 124, e195-e202.

Perera, F.P., Tang, D., Wang, S., et al. (2012). Prenatal polycyclic aromatic hydrocarbon (PAH) exposure and child behavior at age 6-7 years. *Environmental Health Perspectives*, 120, 921–926.

Peterson, B.S., Rauh, V.A., Bansal, R., et al. (2015). Effects of prenatal exposure to air pollutants (polycyclic aromatic hydrocarbons) on the development of brain white matter, cognition, and behavior in later childhood. *JAMA Psychiatry*, 72, 531–540.

Piaget, J. (1926). *The Language and Thought of the Child*. Paris: Kegan Paul, Trench, Trubner & Co.

Piaget, J. (1997). *The Moral Judgment of the Child*. New York: Simon & Schuster.

Powers, M. and Faden, R. (2006). *Social Justice: The Moral Foundations of Public Health and Health Policy*. New York: Oxford University Press.

Rachels, J. (2012). The case against free will. In: Rachels, J. and Rachels, S. (Eds.) *Problems from Philosophy* (3rd edn.). New York: McGraw Hill.

Rosenzweig, M.R. (2003). Effects of differential experience on the brain and behavior. *Developmental Neuropsychology*, 24, 523–540.

Ross, L.F. (1997). Health care decisionmaking by children is it in their best interest? *Hastings Center Report*, 27, 41–46.

Ross, L.F. (1998). *Children, Families, and Health Care Decision Making*. New York: Oxford University Press.

Salter, E.K. (2012). Deciding for a child: a comprehensive analysis of the best interest standard. *Theoretical Medicine and Bioethics*, 33, 179–198.

Schonert-Reichl, K.A., Oberle, E., Lawlor, M.S., et al. (2015). Enhancing cognitive and social-emotional development through a simple-to-administer mindfulness-based school program for elementary school children: a randomized controlled trial. *Developmental Psychology*, 51, 52–66.

Selemon, L.D. (2013). A role for synaptic plasticity in the adolescent development of executive function. *Translational Psychiatry*, 3, e238.

Shonkoff, J.P. and Levitt, P. (2010). Neuroscience and the future of early childhood policy: moving from why to what and how. *Neuron*, **67**, 689–691.

Sowell, E.R., Peterson, B.S., Thompson, P.M., Welcome, S.E., Henkenius, A.L., and Toga, A.W. (2003). Mapping cortical change across the human life span. *Nature Neuroscience*, **6**, 309–315.

Sowell, E.R., Thompson, P.M., Holmes, C.J., Batth, R., Jernigan, T.L., and Toga, A.W. (1999). Localizing age-related changes in brain structure between childhood and adolescence using statistical parametric mapping. *NeuroImage*, **9**, 587–597.

Weir, R.F. and Peters, C. (1997). Affirming the decisions adolescents make about life and death. *Hastings Center Report*, **27**, 29–40.

Weithorn, L.A. and Campbell, S.B. (1982). The competency of children and adolescents to make informed treatment decisions. *Child Development*, **53**, 1589–1598.

Wenar, L. (2013). John Rawls. In: Zalta, E.N. (Ed.) *Stanford Encyclopedia of Philosophy* (Winter 2013 edn.). Stanford, CA.

Wilkinson, D. (2013). *Death of Disability?: The "Carmentis Machine" and Decision-Making for Critically Ill Children*. Oxford: Oxford University Press.

Yang, Y., Joshi, A.A., Joshi, S.H., et al. (2012). Genetic and environmental influences on cortical thickness among 14-year-old twins. *Neuroreport*, **23**, 702–706.

Yang, Y., Wang, P., Baker, L.A., et al. (2015). Thicker temporal cortex associates with a developmental trajectory for psychopathic traits in adolescents. *PLoS ONE*, **10**, e0127025.

Chapter 10

Neuroprognostication after severe brain injury in children: Science fiction or plausible reality?

Sarah S. Welsh, Geneviève Du Pont-Thibodeau, and Matthew P. Kirschen

Introduction

Neuroprognostication with regard to functional and cognitive recovery after brain injury in children is highly complex, and occurs as a continuous, iterative process throughout the initial resuscitation, acute, and subacute and rehabilitative phases of injury and recovery. It requires constant and vigilant attention to translation of prognostication to meaningful action for parents and families making clinical decisions on behalf of their children. This translation is fraught with the implications for a child's cognitive and physical potential, any particular family's set of goals and acceptable outcomes for that individual child, and the context driven by personal experience and media exposure within which each family centers the medical team's prognostic framework (Racine & Bell 2008).

The spectrum of causes for brain injury in children is broad. For select etiologies, such as hypoxic-ischemic injury after cardiac arrest or traumatic brain injury, epidemiological and observational data exist which can guide initial prognostic discussions. For less common disorders such as infectious or autoimmune encephalitis, sepsis-associated encephalopathy, or status epilepticus, estimating a child's potential for neurologic recovery can be far more challenging as the outcomes are less predictable, and dependent on the response to targeted therapy. A major challenge to accurate and timely prediction of both global and specific neurologic potential after brain injury in children is the ability of the young, developing brain to form new neuronal connections and updated structure–function associations. This process is highly individualized and variable, with the same apparent injury producing widely different clinical outcomes. Physicians typically have confidence in their prognostic abilities at the deep ends of the spectrum of injury—they are able to diagnose brain death and what lies very close to it, and they are comfortable with diagnosing mild injury with normal or near-normal neurologic outcomes—but they struggle to elucidate the many variations of gray that lie between those two divergent ends.

In this chapter, we utilize a clinical case to outline the phases of injury and recovery through which a child progresses after sustaining a severe brain injury. Understanding these phases is crucial to elucidating how and where newer technologies may have the greatest influence on our prognostic capabilities. We describe the potential role of current technologies, including functional neuroimaging, in neuroprognostication after brain injury, and examine ethical issues that may arise as these instruments are transitioned from the research lab into clinical practice. Finally, we envision how current and future technologies could be applied to improve neuroprognostication accuracy and physician and family confidence in those assessments.

Phases of injury and recovery after severe brain injury

Resuscitation phase

Case scenario: *An 8-year-old previously healthy boy was found unresponsive and submersed in a pool. He was a strong swimmer and seen by his mother about 5 minutes prior playing with friends. Emergency medical services initiated cardiopulmonary resuscitation (CPR) and inserted an endotracheal tube. He was transported to a local emergency room with CPR ongoing. Return of spontaneous circulation was achieved after approximately 25 minutes of CPR and multiple doses of epinephrine. At the time of transfer to a pediatric intensive care unit (PICU), he had ongoing hemodynamic instability and was comatose with no movement to painful stimuli, absent cough and gag reflex, and enlarged pupils that were minimally responsive to light.*

During the resuscitation phase, which often traverses several hours, pediatric critical care physicians are methodically and systematically stabilizing critical organ systems including the brain. This may require invasive procedures to insert devices for close physiologic monitoring, CPR, or institution of therapies such as extracorporeal membrane oxygenation (ECMO) for maximal cardiopulmonary support. This is a stressful and anxious time for both physicians and families. Discussions between medical staff and families during this tenuous time are typically focused on orientation to the ICU environment, planned procedures and interventions, and presumed injury severity, although data regarding the extent of the brain injury is limited often to a clinical exam and an early head computerized tomography (CT) scan. While head CT performed early after injury may fail to show the full extent of brain injury, loss of gray–white matter differentiation, basilar cistern effacement, and sulcal effacement are in general associated with poor outcome after pediatric out-of-hospital cardiac arrest (Starling et al. 2015). Similarly, a physical exam performed during or immediately after resuscitation can be misleading with respect to the degree and permanence of brain injury. There is often limited time for in-depth discussions regarding neurologic outcome during an intense resuscitation, especially when survival is paramount.

Data exist to assist with survival prediction during the resuscitation phase depending on the etiology of brain injury, and can be further refined based on individual clinical characteristics (Kessler et al. 2011; Girotra et al. 2013). Additionally, the mortality trend for pediatric cardiac arrest has been improved over recent years with early recognition and CPR, emphasis on high-quality resuscitation, and post-resuscitation care (Topjian et al. 2013). Thus, especially in pediatrics, it is often unclear how long to continue CPR or a resuscitation. Technology that could be applied noninvasively at the bedside during a resuscitation that allowed for quantification of brain injury and an estimation of neurologic outcome could be invaluable in guiding medical teams regarding when to limit or withdraw medical interventions including CPR.

Acute phase

Case scenario, continued: In the PICU, he required mechanical ventilation with high pressures due to severe lung injury and high-dose vasopressor infusions for hemodynamic support secondary to cardiac dysfunction. He remained comatose with no response to stimuli and intermittently absent brainstem reflexes. A head CT scan obtained the following day showed subtle loss of gray–white differentiation with crowding of the basal cisterns, but no herniation or midline shift. A continuous electroencephalogram (EEG) was abnormal, with disorganization, but no epileptiform discharges, and some periods of preserved sleep architecture.

After surviving the initial resuscitation, patients progress to the acute phase, which typically lasts for several days. This period of multisystem instability requires close physiologic monitoring and observation, and the institution of therapies aimed at mitigating and preventing secondary or ongoing brain injury. A complete discussion of these neuroprotective and neurorestorative therapies (e.g., therapeutic hypothermia) is beyond the scope of this chapter, although the neuroprognostication process must account for the presence and timing of these treatments (Perman et al. 2012). The efficacy of these therapies depends on the etiology of brain injury, and is the subject of many completed and ongoing clinical trials in adults and pediatrics (Holzer 2002; Nielsen et al. 2013; Moler et al. 2015).

During this acute phase in the ICU, parents frequently ask probing questions regarding their child's potential functional and cognitive outcomes (Box 10.1). Will the child wake up? Will s/he speak, interact, smile, walk? The answers to these questions often directly guide decisions about care. Intensivists and neurologists partner together during this phase to provide information to families regarding the suspected degree of brain injury, discuss potential treatment options, and ascertain the family values for the child (Kirschen et al. 2014; Kirschen & Walter 2015). Communicating and managing prognostic uncertainty can be especially challenging during this acute phase (Marcin et al. 1999, 2004).

Information about a patient's neurologic status during this phase is typically based on a child's physical exam, possible head CT, and continuous EEG recordings. The neurologic exam and EEG recordings are often confounded by analgesic and sedative infusions, which are used routinely in the pediatric ICU environment. EEG background and presence of subclinical seizures have been associated with poor neurologic outcome and increased mortality after cardiac arrest (Abend et al. 2009; Topjian et al. 2016). More definitive neuroimaging such as magnetic resonance imaging (MRI) to characterize the extent of the brain injury is not often feasible during this phase due to the critical nature of the patient and the inability for safe transport to the scanner suite. Serum biomarkers which measure neuronal and astrocyte destruction (e.g., neuron-specific enolase and

Box 10.1 Key questions from parents of children with severe brain injury

- ◆ Will my child wake up?
- ◆ What will my child be able to do?
  - • Will s/he be able to dress, feed, and toilet him or herself? (Functional potential)
  - • Will s/he be able to stand, walk, or run? (Gross motor potential)
  - • Will s/he be able to talk or communicate in any way? (Language and emotional potential)
  - • Will s/he be able to attend school, graduate from college, hold a job? (Cognitive potential)
- ◆ Can my child currently feel pain?
- ◆ How aware is my child now? Can s/he hear our voices? Does s/he know that I am here?

S100B) have been studied in select pediatric populations, but are not validated or incorporated into prognostic algorithms as they are in adults (Topjian et al. 2009; Berger et al. 2010). Clinical characteristics about a child's cardiac arrest and physiologic data obtained during the resuscitation and acute phases have also been correlated with survival and neurologic outcome (Girotra et al. 2013; Topjian et al. 2013, 2014; van Zellem et al. 2015).

Various guidelines exist to assist with neurologic prognosis in patients who remain comatose after survival from cardiac arrest during this acute phase, although these have only been developed and validated in adults. For example, in 2006, the American Academy of Neurology released a sequential algorithm to predict poor neurologic outcome and thus guide decision-making that focused on the clinical exam, somatosensory evoked potentials, and biomarkers obtained within 72 hours of the arrest, but did not take into account the advent of targeted temperature management (i.e., therapeutic hypothermia), EEG assessment, or advanced imaging techniques (Wijdicks et al. 2006). Thus, the European Resuscitation Council and the European Society of Intensive Care Medicine issued a revised algorithm in 2014, incorporating the patient's clinical exam, EEG, biomarkers, somatosensory evoked potentials, and neuroimaging including CT and MRI. A poor outcome according to those guidelines is "very likely" if the patient has no pupillary or corneal reflexes, and bilaterally absent N20 somatosenory-evoked potentials (SSEPs). A poor outcome is "likely" with the presence of myoclonic status, high neuron-specific enolase levels, burst suppression or status epilepticus on EEG and/or diffuse anoxic injury on CT/MRI (even in the absence of abnormalities in pupillary exam or SSEPs). The algorithm also recommends only initiating prognostic testing after an appropriate period post confounders such as hypothermia and residual sedation has elapsed (Greer et al. 2014; Sandroni et al. 2014; Sandroni & Nolan 2015; Hindle et al. 2015).

No formal guidelines or algorithms have been developed in pediatrics to assist with prognosticating in comatose survivors of cardiac arrest or other types of severe brain injury. One study found that delaying prognostication for several days post injury, and basing prognosis on duration of CPR, components of the clinical exam, MRI findings, and SSEPs, had the highest positive predictive value for poor prognosis. Specifically, CPR greater than 10 minutes, Glasgow Coma Scale score less than 5, absent pupillary response, bilaterally absent N20 SSEPs, abnormal EEG, and watershed/basal ganglia/brainstem injuries on MRI were highly predictive of poor outcome (Abend & Licht 2008).

Unfortunately, current predictors of long-term neurologic function differentiate only against a "poor" neurologic outcome, usually defined as death, severe neurologic disability, or persistent vegetative state. Physicians can only provide families with a categorized prediction of their child's predicted neurologic outcome—whether it will be favorable or unfavorable—without being able to define the subtleties that can considerably affect families' decision of care. Families often carry the burden of having to make life-altering decisions for their child based on somewhat elusive predictive models and risk estimations.

The acute phase may also contain a critical period of physiological instability, during which time withholding or withdrawal of life-sustaining therapies has a higher likelihood of leading to death. This period is sometimes referred to as the window of opportunity, and is one reason why emphasis is often placed on the need for accurate and timely prognosis (Wilkinson 2009, 2011; Kirschen & Walter 2015). Prognostic certainty can often be improved by delaying clinical decisions; however, children may no longer require life-sustaining therapies at that time, and thus survive with profound neurologic injury. Additionally, newer therapies aimed at neuroprotection after brain injury may not have sufficient time to demonstrate effectiveness until after this window has passed. Technologies aimed at improving diagnostic and prognostic accuracy during this acute phase would be an invaluable asset to physicians, families, and patients. This is especially crucial since some data show that the neurologic capabilities of some patients with hypoxic-ischemic brain injury from cardiac arrest can change in both positive and negative directions over time (Michiels et al. 2016).

Subacute phase

Case scenario, continued: Magnetic resonance imaging (MRI) of the brain 72 hours post injury showed mild signal abnormality in the bilateral thalami with subtle restricted diffusion, indicating some ischemia in these brain regions. He was extubated 4 days after injury. On exam over the following weeks he was non-verbal, lacked purposeful movements, was unable to follow commands, and had frequent episodes of posturing. Electrocardiogram (ECG) showed a prolonged corrected QT (QTc) interval, and family history revealed long QT syndrome in a maternal aunt and death by drowning in a maternal cousin (a prolonged QT interval is a cardiac rhythm abnormality that may be associated with sudden death).

In the subacute phase of brain injury and recovery, patients typically demonstrate signs that they will breathe and maintain sufficient hemodynamic stability to survive independently of technological support, although some patients require more prolonged mechanical ventilation. This phase is often characterized by a disordered state of consciousness. Disorders of consciousness are a spectrum of clinical syndromes that result from brain injury that compromises wakefulness, or self- and environmental awareness, or both (Box 10.2). This alteration in consciousness can be acute and transient, or may vary in permutations up to and including irreversible and permanent. Most of the advanced neuroimaging techniques used to assess neurologic outcome potential have been employed in patients with a disorder of consciousness during this subacute phase, as patients are clinically stable enough to undergo the imaging procedure.

Patients in a vegetative state, recently renamed *unresponsive wakefulness syndrome*, may appear awake, but will not demonstrate any evidence of self- and environmental awareness and do not show any voluntary motor responsiveness (Calabro et al. 2016). In these patients, auditory or painful stimuli may activate primary somatosensory cortices, but not higher-order associative areas (Laureys et al. 2000, 2002). In contrast, minimally conscious patients will demonstrate evidence of self- or environmental awareness with a variable spectrum of voluntary behaviors such as visual tracking, localization of a painful stimulus, or may even follow simple commands (Calabro et al. 2016). The natural history of brain recovery is the progression from a vegetative state to that of a minimally conscious state, followed by a confused and amnesic phase, and finally to a post-confusional functional recovery phase (Katz et al. 2009). The degree of recovery and of residual functionality, if any, is highly variable and dependent on the underlying etiology of the brain injury. For example, patients who sustain traumatic brain injury will have improved trajectories for neurologic recovery compared to non-traumatic brain injured patients (Katz et al. 2009). Patients who have sustained a hypoxic-ischemic injury typically have a more diffuse and global brain injury, whereas patients with trauma have a more heterogeneous

Box 10.2 Disorders of consciousness

- *Delirium:* often temporary state of confusion, anxiety, decreased awareness of self and the environment, can be hyperreactive to stimuli
- *Coma:* an eye-closed pathologic state of unresponsiveness from which individuals cannot be aroused to wakefulness by stimuli
- *Unresponsive wakefulness syndrome* (or vegetative state; "persistent" after being present for more than one month): an eye-open state of wakefulness without awareness
- *Minimally conscious state:* a chronic state of poor responsiveness to stimuli, but evidence of awareness of themselves and environment is present
- *Locked-in syndrome:* the patient is conscious and aware of self and the environment, but is outwardly behaviorally unresponsive.

pattern of brain injury, with areas of unaffected neural tissue interspersed with the injured tissue (Fins 2011).

Making the correct clinical assessment of consciousness is essential, as it directly impacts decisions of care, rehabilitation, and prognostication (He et al. 2014; Bender et al. 2015). Knowing whether a patient is conscious or aware of their environment certainly plays a crucial role in families' decision-making process. Adult physicians commonly utilize the standardized Coma Recovery Scale-Revised clinical tool to categorize degrees of impaired consciousness, and in particular, to distinguish the vegetative from the minimally conscious patients (Giacino et al. 2004). Making these diagnoses is a very challenging task for the clinician, as discerning reflexive movements from intermittent and episodic voluntary movements can be subjective and dependent on the timing of exam. Furthermore, patients may suffer from cognitive (aphasia, apraxia) or sensory impairments (auditory or visual) limiting their ability to interact. The inability to elicit responsiveness does not necessarily mean that the patient is not conscious. Misdiagnosing patients in vegetative state when they are in fact minimally conscious has been reported to occur in up to 40% of cases (Childs et al. 1993; Andrews et al. 1996; Wilson et al. 2002; Schnakers et al. 2009; Calabro et al. 2016).

Appreciating the wide disparity in adult practitioners' reliability and accuracy in diagnosing these disorders makes conferring these diagnoses in pediatric patients that much more challenging. First, this difficulty in outcome prediction is magnified by any preexisting chronic neurologic or developmental condition (Kirschen & Feudtner 2012; Kirschen & Walter 2015). Second, the broad range of normal developmental stages in childhood further complicates neurologic status assignment, especially for young children. And third, clinical exams based on the age and developmental stage of the child can be incredibly challenging and produce highly variable results, particularly in a child's response to painful or uncomfortable stimuli.

In the prior section focusing on prognosticating during the acute phase of injury, we discussed how physicians often focus on the period of cardiopulmonary instability as an optimal time to discuss outcome prediction since withholding or withdrawal of life-sustaining therapies during this window will likely result in death. A large number of families are unable to make such an irreversible decision during the highly emotional and stressful initial days after a child's injury, especially since they are hoping for a better-than-expected outcome. However, this initial period of physiologic instability is not the only opportunity families have to limit or withhold aggressive care. Thus, technologies that could enhance prognostic accuracy and certainty during this subacute phase would continue to be of significant value. They could provide vital information for a family seeking treatment options aligned with their values for the child, which may include limiting or withholding care, or proceeding with aggressive brain-centered rehabilitation. Options for limiting or withholding care may include removal of an endotracheal or tracheostomy tube that some children still require in the subacute phase, and discussions regarding goals of care for future illnesses and hospitalizations. Additionally, for some children who permanently lack conscious awareness, withdrawal of artificial hydration and nutrition can be ethically permissible (Diekema et al. 2009).

Advanced neuroimaging techniques in outcome prediction

To aid in the neuroprognostication process, researchers have turned to advanced neuroimaging techniques to attempt to determine consciousness beyond the current clinical inspection for purposeful behavior. Several imaging techniques have demonstrated the ability to detect activity in high-level associative cortical centers in otherwise vegetative-appearing patients, which could be suggestive of awareness. These techniques include positron emission tomography (PET scans; measuring metabolic activity including glucose uptake and oxygen extraction), magnetoencephalography (MEG; assessing electrical activity in the brain able to generate a magnetic field), MR spectroscopy (magnetic resonance evaluation of cell turnover/energetic function), MR diffusion tensor imaging (measuring the density and integrity of white matter tracts), functional MRI (fMRI; quantifying blood oxygenation level-dependent (BOLD) images), and electroencephalography (Fins et al. 2008; Owen 2013; Northoff & Heiss 2015).

Functional MRI allows the visualization of areas in the brain with high functional connectivity, elevated cerebral blood flow, and high rate of glucose metabolism (Wu et al. 2015). It has been showed to elicit areas responsible for information integration, an essential function of consciousness (Tononi 2004; Tononi & Koch 2008). Multiple small studies have shown these areas to be disrupted in patients with acute brain injury (Achard et al. 2006; Liang et al. 2013; Tomasi et al. 2013). These islands of functional brain seem to exist in some persistently vegetative patients, suggesting that they may have some retained consciousness (Laureys et al. 2000, 2002; Wu et al. 2015). Minimally conscious patients may retain some cognition and may be able to process language despite their inability to interact and communicate appropriately (Schiff et al. 2005). In one study, spoken narratives by family members to minimally conscious patients resulted in large-scale network activation similarly to healthy controls, but this activation did not occur when narratives were presented by the same family member without linguistic content. These findings suggest that patients may have a preserved ability to process language without being consciously aware. In fact, studies in patients under anesthesia demonstrated that speech perception and word processing continued to occur during anesthesia and sleep despite the absence of conscious awareness (Owen et al. 2006).

Given the preservation of some language capabilities, fMRI has been further employed to attempt to interact with a patient in a vegetative state. For example, a 23-year-old vegetative woman from a traumatic brain injury was asked to perform two mental tasks: the first was to imagine herself playing tennis, and the other one was to imagine visiting every room in her house. These mental tasks demonstrated increased activation in supplementary motor area, and parahippocampal gyrus, the posterior parietal cortex, and the lateral premotor cortex, respectively. Those same areas were also observed in healthy volunteers performing the same task. This patient demonstrated via fMRI brain activation maps that she was able to understand and respond to commands (Owen et al. 2006). Monti and colleagues were able to elicit similar responses in 4 of 23 patients previously diagnosed as being vegetative (Monti et al. 2010). In a separate study, this team asked a patient that had been in a vegetative state for 5 years to imagine playing tennis to communicate one answer (e.g., yes) and to imagine moving around the rooms of his home to communicate the alternative answer (e.g., no). The patient then correctly answered five personalized questions (e.g., his father's name) confirming that he was conscious, and able to successfully recall historical details of his life (Monti et al. 2010; Owen 2013).

Functional MRI has also shown promising results in *predicting* functional recovery in patients clinically diagnosed as being in a vegetative state. The detection of activity in higher-level associative cortices predicts recovery with 93% specificity and 69% sensitivity (Di et al. 2008). The strength of functional connectivity of whole-brain networks also correlates with neurologic recovery. A recent study of nearly 100 patients with varying degrees of impaired consciousness and 34 healthy controls demonstrated a correlation between functional connectivity strength, consciousness, and recovery. Functional connectivity strength could predict recovery of consciousness with 81% accuracy (Wu et al. 2015). These studies have identified a population of patients that demonstrate some evidence of preserved cognitive function, consciousness, and awareness of self and the environment, but without the ability to communicate with the outside world (Wu et al. 2015). This syndrome, newly termed *functional locked-in syndrome*, raises the possibility that some vegetative patients are in fact conscious, aware of their environment, and able to understand speech.

Pediatric advanced neuroimaging studies

Several studies have sought to extend these functional neuroimaging protocols to detect language and higher-order cognitive processing to pediatric patients with a disorder of consciousness after severe brain injury (Ashwal et al. 2014; Roberts et al. 2014). Nicholas and colleagues performed a study in which a 7-year-old girl in a minimally conscious state from a traumatic brain injury was presented names (her own name and an unfamiliar name) using familiar (her mother) and unfamiliar voices. Activation was detected in the primary auditory cortex, a site commonly activated during auditory language processing tasks in healthy controls. Activation maps to the patient's own name versus an unfamiliar name, and using familiar versus unfamiliar voices, were inconsistent with adult studies using the same paradigm. She demonstrated activity in cortical midline structures and dorsolateral prefrontal cortex to her own name, which are thought to be activated in response to self-related stimuli, and when making evaluations of self (Nicholas et al. 2014; Wang et al. 2015). Further studies in children of varying ages and developmental stages are needed to build on these results and move this field forward.

Ethical considerations

The ramifications of these recent findings are fraught with ethical issues. Such neuroimaging studies suggest that a patient who has clinically been deemed to be in an unresponsive vegetative state may in fact demonstrate consciousness and be capable of interacting with their environment. This possibility forces us to question the autonomy and decision-making ability of patients with disorders of consciousness. It also raises questions about perceived quality of life, and whether evidence for covert consciousness argues for ongoing aggressive medical care or compassionate withdrawal of life-sustaining treatments. While only a small subset of these patients may retain sufficient cognitive neural networks to perform the types of tasks required in these studies, there is still a significant leap between imagining motor tasks and answering questions about a person's past, and demonstrating adequate understanding to provide informed consent or express a person's wishes for complex clinical decision making. If technology were to advance through this gap, could it be used to assess pain, emotions such as anxiety, fear, or depression, and whether a patient feels hunger or thirst? If so, perhaps medical teams would more effectively be able to titrate care to actual patient wishes rather than by proxy.

Prior to transitioning this research-based technology into routine clinical practice, we as researchers and clinicians must first feel confident that the behavioral stimuli presented map to known anatomical locations. We must be mindful during this process that normal structure–function relationships have been developed in the adult population and may not be directly transferable to the developing brain, or a brain that has begun to functionally reorganize as it recovers from a brain injury. Second, we must be sure that any particular patient or group of patients will respond to a similar stimulus in a consistent, replicable manner. Lastly, we currently have no way of understanding whether or not a meaningful, replicable, and consistent response indicating awareness is predictive of future functional recovery for any given patient (Wilkinson et al. 2009; Weijer et al. 2014).

Jox and colleagues formatted a deft explanation of this ethical and clinical conundrum based on three possible outcomes of functional neuroimaging: (1) neuroimaging shows less evidence of consciousness or awareness than found on clinical exam, which may allow families to feel more comfortable with limitations or withholding of life-sustaining technology, but may strain the therapeutic relationship with the clinician if she is more optimistic about the patient's potential outcome; (2) neuroimaging shows more evidence of awareness than found on clinical exam, which may on one hand encourage more aggressive treatment, but on the other hand may create false hope for a more complete recovery or prompt a family to withhold technologic support if there is perceived awareness of pain or suffering; or (3) neuroimaging shows the same evidence of awareness as clinical exam, which may also strain the physician-family therapeutic relationship with a perception of "useless" and expensive testing (Jox et al. 2012). Regardless, as we currently hold neither a complete understanding of functional neuroanatomy, nor a meaningful ethical construct of interpreting covert consciousness, we are faced with data that we are unable to reliably act upon (Lutkenhoff et al. 2015; Kowalski et al. 2015; Wintermark et al. 2015a, 2015b).

The field of functional neuroimaging holds great promise to assist with the assessment and potential clinical management of patients with disordered consciousness after severe brain injury. It may contribute objective data to aid with diagnostic classification of patients and provide families with a more concrete understanding of a patient's capabilities. It may have implications for assisting patients with pain management, communicating wants and needs, and end-of-life decisions of care. Additionally, when repeated serially after injury, functional imaging may provide crucial information about neuronal recovery and reorganization over time from brain injury, and could potentially be used to monitor the response of the brain to medications such as amantadine and zolpidem that have been shown to produce and accelerating neurologic recovery (Giacino et al. 2014).

Envisioning future directions for neuroprognostication

Current frameworks for estimating neurologic outcome after severe brain injury are limited to a dichotomous poor or not likely to be poor metric. They are influenced by both patient and family factors, and physician-related factors including age, clinical experience, and medical specialty (Racine et al. 2009; Finley Caulfield et al. 2010; Perman et al. 2012). The development or application of technology that contributes meaningful objective data to the neuroprognostication process that can be applied clinically at the bedside would be an invaluable resource to physicians, families, and patients. In this section, we will review theoretical applications of current technology that have the potential to improve outcome prediction and aid families in the medical decision-making process. We envision these novel applications of current technologies to be of greatest utility in the acute and subacute phases of injury and recovery.

An initial step in advancing current techniques could be the creation of a multimodality database that incorporates signals from several neuromonitoring and neuroimaging devices, and follows trends from soon after injury through recovery. These signals could be compared to a similar database generated from typically developing children over time. By using EEG and fMRI data, a functional map of each individually injured brain could be generated, and changes in signals and activation patterns over time could be correlated with improvements in consciousness, awareness, behavior, and cognition. Once enough permutations of injuries with varying location and extent have been computationally analyzed, it is conceivable that data from a child with a new brain injury could be entered into the system and modeled for how that injury pattern will evolve and recover. This model would account for both normal brain development over time as well as the variable neuronal plasticity that occurs during recovery from brain injury. The model may be able to estimate, with some degree of certainty, the expected functional recovery of the patient. This information would be valuable for parents struggling with clinical decisions and determining goals of care that are in line with the values they have for their child.

These brain-derived signals could also provide information regarding a child's current perception of their environment and their degree of discomfort, pain, or hunger. This information could be transmitted to the medical team and family to improve clinical care and shared decision-making. A variety of platforms could be used to convey this information. Dramatic improvements in virtual reality technology have led to the ability to simulate a 360-degree representation of a space using a head-mounted display and a mobile phone. These systems have already been applied in medicine and the neurosciences to help physicians and patients enhance their understanding of disease processes and therapeutic options. For diseases such as multiple sclerosis and migraine headache, three-dimensional virtual reality systems allow entry into the central nervous system to visualize how the immune system has become dysregulated and neuronal transmission impaired. Procedural fields have used these platforms to demonstrate how catheter-based systems bore through blood clots or deploy stents. One could envision a scenario in which information gleaned from neuromonitoring and functional neuroimaging techniques could be uploaded onto a virtual reality platform to model a child's perceptions and awareness, and allow parents the opportunity to experience this for themselves. How does the child perceive him- or herself in the ICU? Is there pain? What brings pleasure or comfort? While this simulated environment may generate intense emotional responses from parents and family members, it may help clarify how they choose to apply the values they have for their child when making medical decisions.

Further information about brain function may be available through depth electrodes that can be surgically implanted in patients after severe acquired brain injury. These electrodes have primarily been used for seizure focus localization in patients being evaluated for epilepsy surgery, but more recently have been placed in patients after traumatic brain injury or subarachnoid hemorrhage to screen for subclinical seizures (Claassen et al. 2013; Vespa et al. 2016). Just as signals from these electrodes are more sensitive than routine scalp EEG to detect some types of seizures, these signals may carry additional information about awareness or cognition that could be incorporated into functional models and prediction algorithms.

By amalgamating these brain-derived signals, one could also create a meaningful interface that calculates not only the likelihood of overall survival, but answers crucial questions that could influence clinical decisions during the acute and subacute phases of injury. Will my child walk again after this injury? Will he or she regain consciousness? Recognize me? Smile and interact with me? Speak and understand language? For some parents, answers to these questions, even if delivered with some degree of uncertainty, would help them make these unfathomable decisions. Finally, the groundbreaking work with neural control of movement in animals and humans with powered prosthetics may provide an opportunity to create translational communication devices for children with disorders of consciousness. Such devices could truly facilitate communication of wants and desires in patients with covert awareness from brain injury.

These platforms to convert neurally derived signals into tangible reality for physicians and families are accompanied by considerable ethical considerations. While many of these technologies are currently in the domain of science fiction, the rapid growth of the neuroimaging and neuromonitoring fields may make these concepts plausible reality in the near future. They would require extensive investigation and vetting in the research domain prior to translation into the clinical arena. Additionally, these techniques will not be without inherent limitations and caveats, and the generated data and user interface will be susceptible to a wide spectrum of interpretation. Given the power of images and their influence in medical decision-making, it will be essential that families have a clear understanding of exactly what they are and are not viewing (Racine et al. 2005, 2006). Experiencing a virtual tour of their child's severely injured brain with lack of neural connections, abnormal neurotransmitter levels, and impaired blood flow, may be considered emotionally manipulative and unethical if used with the intent to encourage withdrawal of life-sustaining therapies, in the case of, for example, a family with deeply religious views valuing the preservation of life under any circumstances.

Conclusion

The neuroprognostication process is uncomfortable and anxiety provoking for many physicians. Conferring an overly optimistic prognosis may result in the prolonged survival of a neurologically devastated child, while an overly pessimistic prediction may result in a family withdrawing life-sustaining therapies in a child with a potentially more favorable neurologic outcome. The challenges associated with outcome prediction are increased with young children, especially when a child has an underlying neurologic or neurodevelopmental condition prior to brain injury. Predictions of neurologic outcome should be evidence-based whenever possible; however, to date, no single clinical exam feature, laboratory value, biomarker, or neuroimaging finding is highly reliable. Many intensivists and neurologists have memorable patients for whom initial outcome prediction was inaccurate, reaffirming both the challenging nature of the process and the incredible adaptability of the pediatric brain after injury.

The patient described in the case scenario survived his drowning and cardiac arrest. Several experienced physicians counseled the family about his poor prognosis for neurologic recovery given the ischemic injury to his thalami bilaterally, and his failure to make clinical improvements in the weeks after his injury. After months of intense inpatient rehabilitation, he has returned to school and is performing near grade level with some scholastic accommodations. He requires ongoing speech, physical, and occupational therapies, and has a full-time aid to help with his activities of daily living. His follow-up MRI showed sequelae of hypoxic-ischemic injury, with abnormal signal in the periventricular white matter and thalami bilaterally and generalized volume loss.

His example is not uncommon. While these types of patients illustrate the difficulty physicians have in accurately predicting functional outcome within the gray areas between a truly devastating brain injury and mild injury with minimal sequelae, they are a constant reminder that improved technology is essential to aid the neuroprognostic process. Our ability to provide an accurate and timely estimation of potential neurologic recovery directly impacts decision-making during all phases of injury. Functional neuroimaging provides a platform that may be able to detect covert awareness, predict neurologic recovery, and allow seemingly unresponsive patients to communicate. This technique, however, is often not feasible during the acute phase of injury due to the patient's clinical instability. Future directions and technologies for neuroprognostication may require synthesizing brain-derived signals from multiple sources and relaying that data to digital platforms easily accessible to physicians and families. These emerging techniques are on the verge of transforming our perception and understanding of human consciousness and cognition after severe brain injury. The associated ethical considerations are immense and should be discussed prospectively with stakeholders from both the medical community and the public. Clinical, scientific, and ethical caution is needed as these evolving technologies transition from science fiction to plausible reality.

References

Abend, N.S. and Licht, D.J. (2008). Predicting outcome in children with hypoxic ischemic encephalopathy. *Pediatric Critical Care Medicine*, 9(1), 32–39.

Abend, N.S., Topjian, A., Ichord, R., et al. (2009). Electroencephalographic monitoring during hypothermia after pediatric cardiac arrest. *Neurology*, 72(22), 1931–1940.

Achard, S., Salvador, R., Whitcher, B., Suckling, J., and Bullmore, E. (2006). A resilient, low-frequency, small-world human brain functional network with highly connected association cortical hubs. *Journal of Neuroscience*, 26(1), 63–72.

Andrews, K., Murphy, L., Munday, R., and Littlewood, C. (1996). Misdiagnosis of the vegetative state: retrospective study in a rehabilitation unit. *BMJ*, 313(7048), 13–16.

Ashwal, S., Tong, K.A., Ghosh, N., Bartnik-Olson, B., and Holshouser, B.A. (2014). Application of advanced neuroimaging modalities in pediatric traumatic brain injury. *Journal of Child Neurology*, 29(12), 1704–1717.

Bender, A., Jox, R.J., Grill, E., Straube, A., and Lule, D. (2015). Persistent vegetative state and minimally conscious state: a systematic review and meta-analysis of diagnostic procedures. *Deutsches Ärzteblatt International*, 112(14), 235–242.

Berger, R.P., Bazaco, M.C., Wagner, A.K., Kochanek, P.M., and Fabio, A. (2010). Trajectory analysis of serum biomarker concentrations facilitates outcome prediction after pediatric traumatic and hypoxemic brain injury. *Developmental Neuroscience*, 32(5-6), 396–405.

Calabro, R.S., Milardi, D., Cacciola, A., et al. (2016). Moving into the wide clinical spectrum of consciousness disorders: pearls, perils and pitfalls. *Medicina (Kaunas)*, 52(1), 11–18.

Childs, N.L., Mercer, W.N., and Childs, H.W. (1993). Accuracy of diagnosis of persistent vegetative state. *Neurology*, 43(8), 1465–1467.

Claassen, J., Perotte, A., Albers, D., et al. (2013). Nonconvulsive seizures after subarachnoid hemorrhage: multimodal detection and outcomes. *Annals of Neurology*, 74(1), 53–64.

Di, H., Boly, M., Weng, X., Ledoux, D., and Laureys, S. (2008). Neuroimaging activation studies in the vegetative state: predictors of recovery? *Clinical Medicine (London)*, 8(5), 502–507.

Diekema, D.S., Botkin, J.R., and Committee on Bioethics (2009). Clinical report: forgoing medically provided nutrition and hydration in children. *Pediatrics*, 124(2), 813–822.

Finley Caulfield, A., Gabler, L., Lansberg, M.G., et al. (2010). Outcome prediction in mechanically ventilated neurologic patients by junior neurointensivists. *Neurology*, 74(14), 1096–1101.

Fins, J.J. (2011). Neuroethics, neuroimaging, and disorders of consciousness: promise or peril? *Transactions of the American Clinical and Climatological Association*, 122), 336–346.

Fins, J.J., Illes, J., Bernat, J.L., Hirsch, J., Laureys, S., and Murphy, E. (2008). Neuroimaging and disorders of consciousness: envisioning an ethical research agenda. *American Journal of Bioethics*, 8(9), 3–12.

Giacino, J.T., Kalmar, K., and Whyte, J. (2004). The JFK Coma Recovery Scale-Revised: measurement characteristics and diagnostic utility. *Archives of Physical Medicine and Rehabilitation*, 85(12), 2020–2029.

Giacino, J.T., Fins, J.J., Laureys, S., and Schiff, N.D. (2014). Disorders of consciousness after acquired brain injury: the state of the science. *Nature Reviews Neurology*, 10(2), 99–114.

Girotra, S., Spertus, J.A., Li, Y., et al. (2013). Survival trends in pediatric in-hospital cardiac arrests: an analysis from Get With the Guidelines-Resuscitation. *Circulation Cardiovascular Quality and Outcomes*, 6(1), 42–49.

Greer, D.M., Rosenthal, E.S., and Wu, O. (2014). Neuroprognostication of hypoxic-ischaemic coma in the therapeutic hypothermia era. *Nature Reviews Neurology*, 10(4), 190–203.

He, J.H., Yang, Y., Zhang, Y., et al. (2014). Hyperactive external awareness against hypoactive internal awareness in disorders of consciousness using resting-state functional MRI: highlighting the involvement of visuo-motor modulation. *NMR in Biomedicine*, 27(8), 880–886.

Hindle, E.M., Dunn, M., Gillies, M., and Clegg, G. (2015). Neuroprognostication following out of hospital cardiac arrest – a retrospective study of departmental practice. *Resuscitation*, 94, e5–6.

Holzer, M. (2002). Mild therapeutic hypothermia to improve the neurologic outcome after cardiac arrest (Hypothermia after Cardiac Arrest Study Group). *New England Journal of Medicine*, 346(8), 549–556.

Jox, R.J., Bernat, J.L., Laureys, S., and Racine, E. (2012). Disorders of consciousness: responding to requests for novel diagnostic and therapeutic interventions. *Lancet Neurology*, 11(8), 732–738.

Katz, D.I., Polyak, M., Coughlan, D., Nichols, M., and Roche, A. (2009). Natural history of recovery from brain injury after prolonged disorders of consciousness: outcome of patients admitted to inpatient rehabilitation with 1-4 year follow-up. *Progress in Brain Research*, 177, 73–88.

Kessler, S.K., Topjian, A.A., Gutierrez-Colina, A.M., et al. (2011). Short-term outcome prediction by electroencephalographic features in children treated with therapeutic hypothermia after cardiac arrest. *Neurocritical Care*, 14(1), 37–43.

Kirschen, M.P. and Feudtner, C. (2012). Ethical issues. In: Abend, N. and Helfaer, M. (Eds.) *Pediatric Neurocritical Care*, pp.485–493. New York: Demos Medical Publishing.

Kirschen, M.P. and Walter, J.K. (2015). Ethical issues in neuroprognostication after severe pediatric brain injury. *Seminars in Pediatric Neurology*, 22(3), 187–195.

Kirschen, M.P., Topjian, A.A., Hammond, R., Illes, J., and Abend, N.S. (2014). Neuroprognostication after pediatric cardiac arrest. *Pediatric Neurology*, 51, 663–668.

Kowalski, R.G., Buitrago, M.M., Duckworth, J., et al. (2015). Neuroanatomical predictors of awakening in acutely comatose patients. *Annals of Neurology*, 77(5), 804–816.

Laureys, S., Faymonville, M.E., Degueldre, C., et al. (2000). Auditory processing in the vegetative state. *Brain*, 123 (Pt 8), 1589–1601.

Laureys, S., Faymonville, M.E., Peigneux, P., et al. (2002). Cortical processing of noxious somatosensory stimuli in the persistent vegetative state. *NeuroImage*, 17(2), 732–741.

Liang, X., Zou, Q., He, Y., and Yang, Y. (2013). Coupling of functional connectivity and regional cerebral blood flow reveals a physiological basis for network hubs of the human brain. *Proceedings of the National Academy of Sciences of the United States of America*, 110(5), 1929–1934.

Lutkenhoff, E.S., Chiang, J., Tshibanda, L., et al. (2015). Thalamic and extrathalamic mechanisms of consciousness after severe brain injury. *Annals of Neurology*, 78(1), 68–76.

Marcin, J.P., Pollack, M.M., Patel, K.M., Sprague, B.M., and Ruttimann, U.E. (1999). Prognostication and certainty in the pediatric intensive care unit. *Pediatrics*, 104(4 Pt 1), 868–873.

Marcin, J.P., Pretzlaff, R.K., Pollack, M.M., Patel, K.M., and Ruttimann, U.E. (2004). Certainty and mortality prediction in critically ill children. *Journal of Medical Ethics*, **30**(3), 304–307.

Michiels, E., Quan, L., Dumas, F., and Rea, T. (2016). Long-term neurologic outcomes following paediatric out-of-hospital cardiac arrest. *Resuscitation*, **102**, 122–126.

Moler, F.W., Silverstein, F.S., Holubkov, R., et al. (2015). Therapeutic hypothermia after out-of-hospital cardiac arrest in children. *New England Journal of Medicine*, **372**(20), 1898–1908.

Monti, M.M., Vanhaudenhuyse, A., Coleman, M.R., et al. (2010). Willful modulation of brain activity in disorders of consciousness. *New England Journal of Medicine*, **362**(7), 579–589.

Nicholas, C.R., Mclaren, D.G., Gawrysiak, M.J., Rogers, B.P., Dougherty, J.H., and Nash, M.R. (2014). Functional neuroimaging of personally-relevant stimuli in a paediatric case of impaired awareness. *Brain Injury*, **28**(8), 1135–1138.

Nielsen, N., Wetterslev, J., Cronberg, T., et al. (2013). Targeted temperature management at 33 degrees C versus 36 degrees C after cardiac arrest. *New England Journal of Medicine*, **369**(23), 2197–2206.

Northoff, G. and Heiss, W.D. (2015). Why is the distinction between neural predispositions, prerequisites, and correlates of the level of consciousness clinically relevant?: Functional brain imaging in coma and vegetative state. *Stroke*, **46**(4), 1147–1151.

Owen, A.M. (2013). Detecting consciousness: a unique role for neuroimaging. *Annual Review of Psychology*, **64**, 109–133.

Owen, A.M., Coleman, M.R., Boly, M., Davis, M.H., Laureys, S., and Pickard, J.D. (2006). Detecting awareness in the vegetative state. *Science*, **313**(5792), 1402.

Perman, S.M., Kirkpatrick, J.N., Reitsma, A.M., et al. (2012). Timing of neuroprognostication in postcardiac arrest therapeutic hypothermia. *Critical Care Medicine*, **40**(3), 719–724.

Racine, E. and Bell, E. (2008). Clinical and public translation of neuroimaging research in disorders of consciousness challenges current diagnostic and public understanding paradigms. *American Journal of Bioethics*, **8**(9), 13–15.

Racine, E., Bar-Ilan, O., and Illes, J. (2005). fMRI in the public eye. *Nature Reviews Neuroscience*, **6**(2), 159–164.

Racine, E., Bar-Ilan, O., and Illes, J. (2006). Brain imaging: a decade of coverage in the print media. *Science Communication*, **28**(1), 122–142.

Racine, E., Dion, M.J., Wijman, C.A., Illes, J., and Lansberg, M.G. (2009). Profiles of neurological outcome prediction among intensivists. *Neurocritical Care*, **11**(3), 345–352.

Roberts, R.M., Mathias, J.L., and Rose, S.E. (2014). Diffusion tensor imaging (DTI) findings following pediatric non-penetrating TBI: a meta-analysis. *Developmental Neuropsychology*, **39**(8), 600–637.

Sandroni, C. and Nolan, J.P. (2015). Neuroprognostication after cardiac arrest in Europe: new timings and standards. *Resuscitation*, **90**, A4–5.

Sandroni, C., Cariou, A., Cavallaro, F., et al. (2014). Prognostication in comatose survivors of cardiac arrest: an advisory statement from the European Resuscitation Council and the European Society of Intensive Care Medicine. *Resuscitation*, **85**(12), 1779–1789.

Schiff, N.D., Rodriguez-Moreno, D., Kamal, A., et al. (2005). fMRI reveals large-scale network activation in minimally conscious patients. *Neurology*, **64**(3), 514–523.

Schnakers, C., Vanhaudenhuyse, A., Giacino, J., et al. (2009). Diagnostic accuracy of the vegetative and minimally conscious state: clinical consensus versus standardized neurobehavioral assessment. *BMC Neurology*, **9**, 35.

Starling, R.M., Shekdar, K., Licht, D., Nadkarni, V.M., Berg, R.A., and Topjian, A.A. (2015). Early head CT findings are associated with outcomes after pediatric out-of-hospital cardiac arrest. *Pediatric Critical Care Medicine*, **16**(6), 542–548.

Tomasi, D., Wang, G.J., and Volkow, N.D. (2013). Energetic cost of brain functional connectivity. *Proceedings of the National Academy of Sciences of the United States of America*, 110(33), 13642–13647.

Tononi, G. (2004). An information integration theory of consciousness. *BMC Neuroscience*, 5, 42.

Tononi, G. and Koch, C. (2008). The neural correlates of consciousness: an update. *Annals of the New York Academy of Sciences*, 1124, 239–261.

Topjian, A.A., Lin, R., Morris, M.C., et al. (2009). Neuron-specific enolase and S-100B are associated with neurologic outcome after pediatric cardiac arrest. *Pediatric Critical Care Medicine*, 10(4), 479–490.

Topjian, A.A., Berg, R.A., and Nadkarni, V.M. (2013). Advances in recognition, resuscitation, and stabilization of the critically ill child. *Pediatric Clinics of North America*, 60(3), 605–620.

Topjian, A.A., French, B., Sutton, R.M., et al. (2014). Early postresuscitation hypotension is associated with increased mortality following pediatric cardiac arrest. *Critical Care Medicine*, 42(6), 1518–1523.

Topjian, A.A., Sanchez, S.M., Shults, J., Berg, R.A., Dlugos, D.J., and Abend, N.S. (2016). Early electroencephalographic background features predict outcomes in children resuscitated from cardiac arrest. *Pediatric Critical Care Medicine*, 17(6), 547–557.

Van Zellem, L., Utens, E.M., Legerstee, J.S., et al. (2015). Cardiac arrest in children: long-term health status and health-related quality of life. *Pediatric Critical Care Medicine*, 16(8), 693–702.

Vespa, P., Tubi, M., Claassen, J., et al. (2016). Metabolic crisis occurs with seizures and periodic discharges after brain trauma. *Annals of Neurology*, 79(4), 579–590.

Wang, F., Di, H., Hu, X., et al. (2015). Cerebral response to subject's own name showed high prognostic value in traumatic vegetative state. *BMC Medicine*, 13, 83.

Weijer, C., Peterson, A., Webster, F., et al. (2014). Ethics of neuroimaging after serious brain injury. *BMC Medical Ethics*, 15, 41.

Wijdicks, E.F., Hijdra, A., Young, G.B., Bassetti, C.L., and Wiebe, S. (2006). Practice parameter: prediction of outcome in comatose survivors after cardiopulmonary resuscitation (an evidence-based review): report of the Quality Standards Subcommittee of the American Academy of Neurology. *Neurology*, 67(2), 203–210.

Wilkinson, D. (2009). The window of opportunity: decision theory and the timing of prognostic tests for newborn infants. *Bioethics*, 23(9), 503–514.

Wilkinson, D. (2011). The window of opportunity for treatment withdrawal. *Archives of Pediatrics and Adolescent Medicine*, 165(3), 211–215.

Wilkinson, D.J., Kahane, G., Horne, M., and Savulescu, J. (2009). Functional neuroimaging and withdrawal of life-sustaining treatment from vegetative patients. *Journal of Medical Ethics*, 35(8), 508–511.

Wilson, F.C., Harpur, J., Watson, T., and Morrow, J.I. (2002). Vegetative state and minimally responsive patients – regional survey, long-term case outcomes and service recommendations. *NeuroRehabilitation*, 17(3), 231–236.

Wintermark, M., Coombs, L., Druzgal, T.J., et al. (2015a). Traumatic brain injury imaging research roadmap. *AJNR American Journal of Neuroradiology*, 36(3), E12–23.

Wintermark, M., Sanelli, P.C., Anzai, Y., et al. (2015b). Imaging evidence and recommendations for traumatic brain injury: conventional neuroimaging techniques. *Journal of the American College of Radiology*, 12(2), e1–14.

Wu, X., Zou, Q., Hu, J., et al. (2015). Intrinsic functional connectivity patterns predict consciousness level and recovery outcome in acquired brain injury. *Journal of Neuroscience*, 35(37), 12932–12946.

Chapter 11

No pain no gain: A neuroethical place for hypnosis in invasive intervention

Elvira V. Lang

A true story

A magnetic resonance imaging (MRI) technologist, let's call her AX, who had been trained in nonpharmacologic sedation techniques to help her claustrophobic patients through their tests, needed treatment of her varicose veins. Her vascular specialist was to use a minimally invasive procedure during which a laser or radiofrequency probe is inserted in the vein to heat its inner lining and damage the vessel wall so that it closes off. To ease the sensation, local anesthetic is applied sequentially along the entire course of the vein from the groin level down to below the knee. AX wanted to be comfortable and brought with her a colleague who had been trained in the same nonpharmacologic sedation techniques with which she was familiar. The doctor clearly did not want AX's colleague in the procedure room. He kept warning AX that the local anesthetic would be stinging and burning—repeating the expectations of hurt even after AX asked him to not use such wording since it was not helpful, and that she was prepared for the procedure. The doctor couldn't help himself and kept insisting that things would hurt until AX looked directly at him and told him, "You got this all wrong. I will just experience a delicious sense of tingling." This had the desired effect and the doctor became and remained silent. The procedure continued uneventfully and AX was proud of having been able to help herself.

This short interaction harbors ethics considerations which may even be at odds with each other depending on whether one analyzes the scenario from the point of morality ethics, concerning right or wrong behaviors, general societal belief systems, bioethics used by medical review boards in terms of beneficence/maleficence, risk/liability of various approaches, and professional standard of care based on prevailing values and practices. Further dilemmas arise when beliefs and expectations of patient and healthcare professional clash with each other and with scientific evidence, societal norms, or observable behaviors. This chapter examines how such interrelationships can shape the experience of acute pain versus comfort during medical procedures. Can there be gain without pain? Should there be?

Talking about pain

"Will it hurt?" How this question, whether asked or implied, is handled from the point of view of the recipient and the purveyor of potentially painful stimuli is deeply rooted in belief systems and cultures of practice. Introducing upcoming stimuli with wording such as, "This shouldn't hurt that much," or "It's just like a bee sting," "This is going to be the worst part," "A stick and a burn," or sympathizing, "That wasn't that bad?" are rampant in the medical environment and produce the opposite effect—more pain (Blankfield 1991; Lang et al. 2005; Cyna & Lang 2011). Research has shown that using the word "pain" or other verbiage with negative emotional content such as "sting" or" burn" will significantly increase pain and anxiety, even when preceded by "not much" or "little" (Lang et al. 2005). The mind is only able to call upon the image of something, not of nothing, just as one can imagine bananas but not no-bananas. Even if I were to tell you not to think of bananas (or pink elephants), their image and perhaps some appetite for bananas may have come to mind even though I suggested not to think of such.

Implicitly, the use of such negative suggestions before and during potentially uncomfortable procedures is sanctioned by the lack of official guidelines by professional governance bodies to replace it with more positive suggestions (Blease 2012). One should think that by inflicting more pain through one action as compared to another less hurtful one violates the medical ethics principle of nonmaleficence: to do no harm. In clinical practice, the maleficence potential is to be balanced against the beneficence principle: is the risk of greater harm justified by a greater good? In the case of negative suggestions this is not the case, at least not for the patient. So why is the use of negative suggestions so common?

There are various reasons that words and phrases with negative emotional content have become embedded in the vocabulary of many frontline medical caregivers. For some, it may be a desire to be perfectly honest with the patient along a general ethical principle of full disclosure. One procedure nurse once told me she likes to describe the upcoming procedure as more painful since she knows it is not the case and the patient will be relieved and feel good about the experience (an aspect of beneficence) as well as thank her for great care (a more questionable motive). Practitioners may truly believe, from training or experience, that the patient will feel pain and it is only fair to give forewarning. Feeling it then as a duty to warn the patient of the upcoming hurt and finding confirmation in the resulting report (more pain than if nothing had been said) will tend to perpetuate the belief in the need for warning in the future. It may even cause the practitioner to increase the negatively worded warning thus perpetuating and driving up the spiral of healthcare provider-induced pain.

Familiarity with routine ways of eliciting a response from patients about their experience of pain may be another powerful reason that practices do not change. Nurses and other medical personnel are required to query patients about their pain levels (Joint Commission on Accreditation of Healthcare Organizations 2016). Typically 0–10 pain scales are used. To make sure patients know about these instruments they may contain bright red pointers, a bold header such as "Pain Assessment Tool," and may be prominently posted in the treatment areas. The intent of better pain assessments in the hospital and ambulatory environment was a good one—to reduce suffering; the choice of advertisement of these instruments in clinical practice, however, may lead to the opposite effect.

There are simple solutions to the beneficence/nonmaleficence equation for this aspect. One study that assessed whether asking patients instead for their comfort levels after cesarean section found that this simple change in wording resulted in the women also reporting their overall postoperative experience as less bothersome or unpleasant compared to those being asked to rate their pain (Chooi et al. 2013). In the clinical trials I was involved with, we would introduce the customary hospital 0–10 pain scale with the question, "How is your comfort level on a scale of 0–10 with 0=no pain at all and 10= worst possible." At least this avoided setting an expectation that there should be pain. AX's story also highlights what may happen when expectations of pain are at a variance between healthcare/stimulus provider and recipient thereof. Do patients have a right to their own experience? Do they, as has happened in the case, have to be convinced that they are "wrong" and that it will hurt even when they present with a declared set of anodyne skills in which the provider does not believe or is unfamiliar with? Is it ok to trust the patient's perceptions even if they run counter to the provider's beliefs?

Expecting and experiencing pain

Feeling pain is not a static or predictable experience. The same type of stimulus or medical condition does not provoke the same sensation of hurt in different people (Burgstaller et al. 2016). Thus the same level of tissue injury or physical severity of stimulus may lead to a response of 8 out of 10 in one person and a response of 2 out of 10 in another. Functional brain imaging has shown the patient's self-report of pain in terms of suffering or being affected by the sensation correlates well with activation of pain-related processing in their brain (Peyron et al. 2000; Hayashi et al. 2016). Patients' self-reports of pain thus should be trusted.

The pain experience depends on a variety of factors. Beside the stimulus intensity, setting, meaning, and expectations play a considerable role and their effect also can be documented with neuroimaging (Atlas & Wager 2012). Words can be powerful in shaping a negative experience as pointed out earlier. Words can also be powerful in shaping positive expectations. A recent meta-analysis showed moderate to large effect sizes on reduction of experienced acute pain after verbal suggestions that promoted expectations of pain relief (Peerdeman et al. 2016). Effects were particularly prominent when associated with an active treatment such as placing an intravenous (IV) line or needles. Such verbal suggestions surrounding pain expectations have been shown just by themselves to produce associated changes in pain-related brain activity and connectivity that then subsequently diminish the negative effects of painful stimuli (Hashmi et al. 2014).

Societal norms or individual perceptions can create beliefs of "no pain, no gain." The term is particularly prized in the fitness movement where sufferance and feeling the burn are valued on the road to the ideal body sculpt and physical/mental toughness. Furthermore, aspects of the Puritan legacy may lead to the consideration of pain as deserved, and attempts to alleviate suffering as moral weakness (Kilwein 1989). While individuals are free to choose their actions, the situation becomes ethically more challenging when the standards of the community or healthcare professionals are forced on individual patients, and when hurtful approaches are rationalized on moral-ethical grounds.

When an individual who is engaged in researching and learning skills in how to adapt expectations positively encounters a healthcare professional with a different perception of what is appropriate (such as AX and the doctor in our story) both may be behaving morally and true to their convictions even though in this case the patient is subjected to maleficence based on research evidence. Conversely, when a patient's beliefs embrace a high-risk approach, inappropriate in the eye of conventional medicine—such as refusing blood transfusions or antibiotics for fear of spiritual/religious doom—how much physical sufferance is acceptable for the higher spiritual good? Lawsuits and ethics boards struggle over these questions and often there are no good answers. On occasion, it may even be that the aversion of a professional caregiver to inflict pain at all is in the gray area of what is ethically justifiable and what is not.

Inflicting pain

Is it ethical to inflict pain? Having been an interventional radiologist all my life, performing image-guided surgery on mostly awake patients, it is clear that it is sometimes necessary to violate tissue boundaries to be able to proceed with treatment. Even just applying lidocaine can be interpreted as painful by the patient. Again, belief systems will dictate whether avoiding potential pain altogether is justifiable, thereby making oneself feel good, while letting a treatable condition go untreated and endangering a patient's health or life. The ethical solution could be to not choose a profession where this may be a dilemma but there is also an ethical consideration to having only the disinhibited provide care. Certainly history has sufficient examples of atrocious behavior of physicians for the sake of science.

What of the dentist who is a soft-hearted individual who will only proceed if the local anesthetic produces complete numbness? I had the unfortunate experience to observe this when my husband, who in the beginning cloud of Alzheimer's, wasn't able to clearly distinguish between pressure and hurt when tested with a clanking instrument on a tooth fragment to be extracted. It was very clear that all would have been fine, based on a prior visit he had to the same dentist, and with me being there with him. However, the dentist was so scared that it would hurt that he would not do the extraction and let an underlying abscess smolder for another day until a colleague at a different practice was able to achieve the goal.

The pitfalls of empathy

Modern medical training highly stresses empathy; to feel for a patient is presumed to be the sign of a good provider. As a shared sense of suffering, empathy can even be found in laboratory animals and is present to varying degrees in humans (Singer et al. 2004; Goubert et al. 2005; Mogil 2012; Chen et al. 2015). Higher levels of empathy, however, do not necessarily produce better medical outcomes. In a study with nurses in cardiovascular units, levels of empathy among the staff did not correlate with patients' pain perception or the amount of medication administered (Watt-Watson et al. 2000). Misbeliefs and associated behavior were independent of the nurses' levels of empathy.

Seeing another person suffer may result in either withdrawal from the situation to avert one's own suffering, or behaviors that may be considered helpful in the given cultural environment but in themselves can be deleterious. We once had to halt a clinical trial because patients in the empathic attention control group had significantly more adverse events than those patients either left alone or supported with guidance in the self-hypnotic relaxation group (Lang et al. 2008). Review of taped interactions between providers and patients suggested that increased efforts of being nice likely provided more distraction from the patient's own internal coping mechanisms or may have signaled that things were more risky than perhaps portrayed, keeping adrenergic stimulation up. Thus just feeling for someone, as desirable, ethical, and humanly noble as this might be, not knowing how to express it may actually be hurtful.

In the context of this study I recall a taped interaction in which one of the nurses approached the patient on the procedure table and kindly stroked his hair. The psychologist on the team pointed out that such an approach should be avoided since it placed the patient at a child-like level and, in addition, risked potentially evoking past abuse. After all, we worked in a setting ideally suited to elicit such a memory: a darkened interventional radiology lab with the patient naked under surgical drapes, immobilized, and with little control while the procedure determined health and well-being. I also recall how difficult it was to convince the nursing staff that this was not great care and that unsolicited touching or touching without obtaining permission was inappropriate. It was a case of reducing the risk of the patient feeling uncomfortable or reliving old trauma versus the beneficence of good intentions in the hope of improving the experience through touch.

Reducing the pain of others

When going to the dentist, having surgery, or preparing for any other medical procedure, have you ever wondered who determines whether and how much drug you receive for managing pain and anxiety? You might think that your needs or your doctor's are the critical factors but that is not necessarily the case. A study in which the same physicians performed the same procedure for evaluation of the blood vessels with the same protocols at three different hospitals was able to illustrate that the culture of an institution is the overriding factor (Lang et al. 1998). The nursing staff did not rotate among the hospitals and supposedly was to provide medications under the umbrella of physician's orders. The amount of intravenous drugs given was relatively constant within any given hospital but significantly different among hospitals varying by a factor of as much as 3 with regard to pain medication. There was no correlation with patients' age, sex, medical condition, or physician identity, or with patients' levels of pain and anxiety. One might conclude that local hospital culture or habit was the driving force. This approach bears the risk of over-medicating individuals whose metabolism and constitution is sensitive to small dosages of sedatives and analgesics, with possibly fatal or serious adverse effects on breathing and blood pressure, and undertreating individuals who either experience greater pain and who need more medication for the same amount of pain relief.

The wide variety in the approach to drug usage for pain management during medical procedures is not limited to healthcare providers (McDermott et al. 1993) but is also reflected in patient expectations. Patients risk not to be given enough credence when, based on their own biases, healthcare professionals expect a different response in terms of pain severity as the one the patient indicated. Higher than expected pain ratings from the patient may be interpreted as drug-seeking behavior and withholding of needed medication, or possibly even as personal offense when the patient indicates pain but refuses drugs.

The medically ethical approach for a caregiver to minimize sufferance would be to be responsive to the patients' needs rather than one's own assumptions. Ethical behavior, though, relies on rational decisions. Encounters with nursing staff having to perform repeated painful dressing changes on author Dan Ariely in his youth led him to research human decision-making in this setting and become a major contributor to the field of behavioral economics (Ariely 2010). He concluded that humans behave predictably irrationally—a feature well exploited by the advertising industry. He describes his experience that the nurses tended to rather quickly rip the bandages off his heavily burnt skin rather than removing them slowly from the edges. They believed that a short spike in excruciating pain was better for the patient than a lengthier but less severe approach. There was no scientific evidence behind the nurses' behavior who were in general kind and generous individuals. Even when years later he presented scientific evidence based on his own experiments that people feel less pain when treatments are carried out with lower intensity and longer duration, no change in practice ensued. When he asked one of his more trusted nurses for a reason, she explained that pulling the bandages fast might be more understandable if it were indeed the nurses' way of shortening their own torment. Even after the discussion produced agreement that practice should be changed on the burn unit, the change never happened. The nurses, with all their experience and true compassion for their patients, kept misunderstanding the consequences of their behaviors and repeatedly made the wrong irrational decisions for their patients.

Pain management goes to the core of the self of healthcare professionals who often choose their vocations for wanting to help others and the biases emerge along the way through personal experience, customary behaviors, and social norms in their respective environments. However, convictions about specific approaches run deep. I had the opportunity to observe the interpersonal clashes after a hospital merger when staff from two interventional radiology units had to work together. One team had been used to injecting relatively large amounts of sedatives and narcotics for procedures, the other relatively smaller amounts. Nursing staff of the lower-use site would accuse their colleagues of being drug pushers and the nursing staff of the higher-use site would return the compliment by accusations of lack of caring. The fact that, as division chief, I introduced nonpharmacologic adjuncts to pain management only helped insofar as forming an alliance of high- and low-users against this new threat to entrained practice patterns. It took about half a year and weekly meetings supported by the hospitals' skilled organizational development facilitators to arrive at a new common and patient-centered practice.

Societal influence

In years past, most medical procedures and treatments were ordered by a physician and carried out in an appropriate hospital or office setting with very little consultation with the patient. What was offered was accepted as necessary, including any accompanying pain or discomfort. The attitude that the doctor knows best has eroded and the belief that the healthcare system will provide what is best for the patient has been largely shattered. Social norms, based on mutual trust between healthcare providers and recipients, are increasingly being replaced by material norms, expectations of quality outcomes in return of value for healthcare expenditures.

With healthcare reform in the United States and focus on National Quality Strategies, patients are being considered more as consumers, clients, and customers (Van Fleet & Peterson 2016), a concept that is clearly uncomfortable for some healthcare professionals (Leng 2015). However, the switch to a consumer's market has pushed the element of patient satisfaction into the forefront of various tools that evaluate medical facilities, including those used by the Centers for Medicare and Medicaid Services (CMS) when weighing reimbursement to hospitals for patient care (Centers for Medicare and Medicaid Services 2009, 2011). In the United States, patient satisfaction rankings contribute 30% to the overall quality scores. Hospitals ranking below the 50th national percentile are fined 1–2% of all payments through CMS, which is significant considering hospital margins commonly range between 1% and 5%.

Pain management and communication play prominently in the US patient satisfaction surveys (Lang et al. 2013). Questions include "How often was your pain well controlled?" "How often did the hospital staff do everything they could to help you with your pain?" The emphasis on "everything" acknowledges that drugs are not the only way opening the way to include nonpharmacologic approaches. Effective January 1, 2015, the Joint Commission on Accreditation of Healthcare Organizations clarified its Standard PC.01.02.07 on managing patient's pain in that both pharmacologic and nonpharmacologic strategies have a role in the management of pain.

The new emphasis on nonpharmacologic adjuncts is driven in part by the opioid epidemic. In 2014, opioids killed more than 28,000 people in the United States alone, with more than half involving prescription drugs (Centers for Disease Control and Prevention 2016). With greater pressure on medical professionals to see more patients in less time, it is often easier to prescribe large amounts of narcotics rather than to sit and discuss what else may work to improve comfort after dental or other surgery and risk a possible subsequent unreimbursed call or repeat visit of the same patient. Recent laws that attempt to prohibit prescription of narcotics beyond a certain number of days can be helpful in preventing overdoses but at the same time also risk leaving patients in need of them, such as terminal cancer patients stranded in how to get their prescriptions weekly from the pharmacy. Education in pain and opioid management is now mandatory in many states, an example of how legal pressure can lead practitioners to engage in socially and morally responsible behavior. At least this requires them to understand about the problem and learn new skills to fix it.

Changing the dynamics of pain

During medical procedures and tests, pain increases linearly over time relatively independent of stimulus severity and the amount of drugs given (Lang et al. 2014). The increase in pain over time may be explained by several mechanisms. In a setting of ambiguity, human nature tends to choose the worst possible interpretation born out of a protective mechanism of the subconscious (Ewin & Eimer 2006). Once a painful stimulus has occurred, one also tends to be more attentive to external cues that could indicate pain even when there is none (Bayer et al. 1998). Expectation then further shapes pain-intensity processing in the brain (Atlas & Wager 2012). If expectation is changed based on cues of a prior pain experience, neurobiological changes in the brain take place in the pain-related circuitry, such that stimuli that come with suggestions of hurt will be experienced more painfully than when preceded by nonpainful stimuli or absence of cues of pain (Keltner et al. 2006; Atlas et al. 2010). At the same time, brain regions are activated that are not only associated with pain perception but also affected by processes such as conflict, negative affect, and response inhibition (Atlas & Wager 2012).

Thus by changing expectations or other mechanisms of pain processing, a more comfortable experience may be created. As research with three large-scale prospective randomized trials demonstrated, it is possible to reframe the mind's approach to processing events during medical procedures through relatively simple means at the onset of the interaction (Lang et al. 2000, 2006, 2008). Using advanced rapport and a 1–3-minute script at the time patients were brought into their procedure suite prevented the processing that led to an increase of pain over time, remaining effective even for hours of procedure time to come. The effectiveness of a combination of positive-type suggestions and guidance in hypnotic or imagery-focused techniques has also been shown by other investigators (Kroger & DeLee 1957; Weinstein & Au 1991; Everett et al. 1993; Faymonville et al. 1995; Meurisse et al. 1999; Montgomery et al. 2002; Liossi & Hatira 2003; Kupers et al. 2005; Elkins et al. 2006; Liossi et al. 2006; Montgomery et al. 2007). One may thus wonder why these techniques are not used more broadly?

Shaping the experience

The proverbial term of the white coat trance, with the associated higher blood pressure and heart rate is an indication that the medical encounter itself produces a state of heightened suggestibility. One can consider the patient to be already in a state of self-hypnosis and being highly susceptible to everything that is said or not being said or the suggestive power of external cues. Thus, just as any negatively voiced statement or negative suggestion can make the experience worse in the sense of a nocebo, mere avoidance of such negative suggestions improves the experience, and helpful suggestions can improve the experience just as the expectations coming with placebo treatment.

Of particular benefit is the fact, as shown in our studies, that shaping the patient's experience *at the onset* of the provider interaction is the critical step and may be all that is needed, even for procedures that last for hours (Lang et al. 2000, 2006, 2008). The elements of this process are rapport, relaxation, and reframing in conjunction with awareness of what is said, injection of hypnoidal language, and, depending on the need of the situation, reading of a formal self-hypnotic relaxation script or comparative unscripted induction of a relaxed state.

One truly does not know how a patient will experience a stimulus or treatment, and there is the tendency to project one's own expectations onto the other person in concordance with one's own belief system. In going further to assess how the patient's pain experience can be improved, we found that the very first step is to establish rapport. Once in rapport, it makes it much easier to communicate and find the right words spontaneously and without awkwardness (Marchant 2016). For this reason we place heavy emphasis on advanced rapport skills when training medical teams in nonpharmacologic calmative and analgesic techniques (Lang 2012; Norbash et al. 2016).

Advanced rapport skills entail adapting to the patient's body position, for example, bending over toward sitting patients, but may also mean adapting to their emotional state, such as speaking louder or in an agitated manner if they do so, or softer if they are more subdued and then lead by modeling to a more resourceful state (Lang & Laser 2009). The same methods are found in some of the sales literature (Boothman 2010) and the question is often raised during workshops we give to medical professionals whether this would not be manipulative and therefore unethical.

Often the individuals that object the most to intentional matching with another person are the ones who do so very naturally on their own. Is it unethical to do something intentionally that one does by natural instinct? The answer will most likely be in the intent of the outcome, whether to deceive or to help. However, an interesting phenomenon happens in that regard. People who match each other, whether intentionally or not, are in rapport and are more open to each other's suggestions and leads. So even if one may start with less than best intentions, being so-to-speak willing to walk in the other person's shoes may produce a better mutual understanding than intentional lack of matching.

Feelings and external expression thereof are interrelated. For example, the application of botulinum toxin (Botox®) into the forehead muscles above the nose, mainly done for cosmetic reasons, has been found to have as a side effect the reduction in depression of the recipient, giving credence to a facial feedback mechanism on emotion (Wollmer et al. 2014). Furthermore, the facial expression of one person will affect the conversation partner in the sense of an emotional contagion (Hatfield & Cacioppo 1994). What may start as mimicry or matching ends as feeling in similar ways. If done intentionally, thus one could make another person feel worse, which could ethically be considered as objectionable, or make the other person feel better. However, when this requires an intentional effort, as we find in our training, it tends to be interpreted as more manipulative when done to make a patient feel more resourceful, as compared to when executed without noticing by radiating off the stress from work and making other people feel equally stressed.

Language

In the interest of full disclosure, consenting and listing of side effects of treatments has to be honest to enable a person to determine which risk ratio to accept with a treatment as compared to no treatment or alternative treatment. If done in a collaborative approach, discussing what will be done to manage potential complications and what omission of proceeding may evoke with the same conditions, patients can arrive at a decision that feels right for them (Lang 2014b). After the medical consent is given, full focus on avoidance of nocebo effects and use of more positive language is in order. How can this be achieved in practical terms?

Avoiding upfront use of verbiage with negative emotional content is a first step. Use of neutral statements such as, "I will give you the local anesthetic," is a further step up the comfort ladder, and mentioning the likelihood of a reasonably expectable desirable outcome can help even further, such as, "I will give you the anesthetic and the area should become numb shortly" (Lang et al. 2005; Dutt-Gupta et al. 2007; Varelmann et al. 2010). Consider the difference between such a statement and "I will stick you with a needle" (Lang 2014a).

At the other end of the suggestive spectrum are placebo statements that are known to the user to be untrue and may constitute a moral dilemma. One of the most impressive demonstrations of placebo, where need and belief coincide in a powerful combination, occurred at the warfront in World War II when Henry Beecher ran out of morphine while treating wounded American soldiers. Desperate to be supportive, nurses would inject saline, describing it as a potent pain killer. It worked in about 40% of the cases.

With regard to the use of placebo, the American Medical Association urges practitioners to be "extremely thorough in obtaining consent from patients" (American Medical Association 1997) echoed by the ethics code of the UK General Medical Council equally insisting on complete openness. Blease argues that medical treatment always occurs in the framework of interpersonal interactions between staff and patients and is influenced by the beliefs of those involved in the power of the treatment thus making the term "positive care effects" more appropriate than "placebo effect" (Blease 2012). "Positive care effects" may also take away the stigma that some perceive to be associated with a "sugar pill" or an element of deception. Although the literature suggests that even open labeling of placebo, for example revealing the placebo nature of a treatment doesn't interfere with its success in pain management and can still result in a 50% success rate (Kam-Hansen et al. 2014).

One can see how the element of known deception in some placebo approaches could interfere with the patient–provider relationship. After all, one really doesn't know what the patient will experience and thus it is wise to avoid suggesting that something will not or does not hurt or feel in a particular way. One way around this dilemma is the use of permissive language leaving options for an outcome, indications that something, for example, "might happen," "could happen," or "would be interesting to notice." This approach is often associated with Milton Erickson a psychiatrist and psychologist who greatly shaped modern hypnosis. The permissive approach avoids the possibility that a prediction fails and thereby undermines the trust a patient may have in the healthcare professional. It also adheres to a dictum in my own practice, to leave to the patients the right to their own experience. On the spectrum between purely neutral descriptors to the possible integration of stimuli in a deeper hypnotic experience is a language approach I like to call hypnoidal language. It is still fairly conversational while moving suggestions up to the next level of desired outcomes. Giving a choice, "Would you prefer the IV on the right arm or the left arm?" gives the patient a sense of control that otherwise may be lacking. Double binds target a combination of desired results but leave the perception of a choice without being authoritarian: "Some people first close their eyes, then they relax; others first relax then close their eyes." A sentence using presupposition of outcome in response to a patient's complaint may be, "I don't know when it will stop hurting, in a minute, in two minutes, in three minutes, or … NOW [emphasized]."

Offering choices for an experience can serve as a distraction by occupying the mind about what the sensation might be, thereby leading away from the more catastrophizing expectations, "You might feel some warmth or coolness, and then it should become numb where it needs to be." Including confusional elements can further help to take the mind off spinning around negative recurring thoughts. As such we use in our script or when announcing injections something like, "You might feel some warmth, or coolness. And some people even describe a delicious sense of tingling." The issue is that nobody really knows how a delicious sense of tingling feels, and the statement is sufficiently surprising that it helps to end unhelpful interactions, whether when voiced by the healthcare professional or by the patient, as in the example with AX above. Since AX's report I have had to use the "delicious tingling" reference myself once with a surgeon treating me in a similar scenario and I can attest that it does shut off unwelcome negative suggestions by others. Should we have asked our doctors' consent to such confusional induction to reset their own fears? Having requested them to stop in their unhelpful predictions of discomfort should have sufficed, and by the time a procedure is started one can assume all participants are already in a trance-like state.

Procedure hypnosis

Going one step further when guiding patients in nonpharmacologic pain and anxiety management is to have the patient associate with a pleasant scenario and integrate all stimuli. This tends to be mostly acceptable as long as it is called imagery, visualization, or meditation, but beware of using the word that actually would describe it best, "hypnosis," a state of focused concentration where awareness of outside stimuli become muted. To distinguish such use of language from hypnotherapy, where psychotherapy is bundled with hypnosis, I prefer the term "procedure hypnosis" as a practice limited to health-care professionals who routinely engage in "sedation without medication" in their daily practice.

There is considerable uproar in the American Society of Clinical Hypnosis (ASCH) about the use of "lay" hypnosis. The idea is that nobody except a mental healthcare professional, social worker, doctor, or master's degree nurse should be trained even in basic skills that could help their patients through everyday procedures. Thus, the front-line providers of medical care such as nurses, medical technologists (such as AX whom I trained), and first responders who deal on a daily basis with highly distressed patients not only should not be allowed to be trained, but anyone who would train them would be charged with unethical behavior and expelled from the ASCH. The Society of Clinical and Experimental Hypnosis and International Society of Clinical Hypnosis have been more open and accepted the reality that frontline providers such as bachelor-level nurses are ideal candidates to learn and be taught such techniques. In this ongoing debate, an argument of ethics is used to prevent approaches that would provide better outcomes to patients in an effort to preserve trade superiority.

The premise is again that one does not need permission to conduct negative hypno-sis in terms of powerful negative suggestions with their associated worse outcomes as compared to being allowed to use or being trained in evidence-based methods that have been tested in the environment. As our research has shown, the use of targeted hypnoi-dal language and advanced rapport skills can be applied quite effectively and safely by licensed medical professionals who are the frontline but do not have advanced medical degrees. Training of just a few such individuals can change the experience for the better of tens of thousands of patients over the course of a year as we have shown in a recent study after training personnel in three MRI facilities (Norbash et al. 2016). Studies show positive outcomes after training: a better patient experience, shorter procedures, and greater safety—fewer adverse events.

Looking at the use of nonpharmacologic pain and anxiety management techniques, there is much at stake by wanting to limit practice to mental healthcare professionals and to exclude others who are at the frontline of medical, surgical, and dental healthcare delivery. Not only are there not enough mental healthcare professionals to cover routine medical encounters, their time may not be well spent as extras in the critical first 90 sec-onds to a few minutes of patient encounters to enable the upcoming medical treatments or tests as compared to their regular patient approach of spending 45–60-minute sessions on the couch for deeper exploration of lives and behaviors. This is not necessary for pro-cedure hypnosis.

Not allowing frontline medical professionals to be trained overlooks tremendous societal cost. Outside hospitals the opioid crisis is raging. Extrapolating the risk of IV conscious sedation as found in a large collaborative study with greater than 21,000 endoscopies (Arrowsmith et al. 1991) to 8.7 million invasive medical procedures, it is predicted that 47,000 patients will suffer serious cardiorespiratory complications and 2600 patients will die each year in the United States alone. This number does not take into account the fatalities in dental offices using IV drug sedation (Yagiela 2001; Bennett et al. 2015). In addition, the cost of patient anxiety poses a considerable burden. The inability of patients to complete their MRI exams, leaving capacity unused, wastes about $310 million USD annually by conservative estimates (Norbash et al. 2016). These losses could be largely avoided by team training of medical personnel in nonpharmacologic calmative and analgesic techniques (Lang & Rosen 2002). Here the question of societal ethics clearly comes into play when considerable sums of healthcare expenditures are wasted because of preconceived notions. The situation is further aggravated by the fact that rationing of pain medication now occurs even in prestigious hospitals, mostly without the patient's knowledge. A recent article in *The New York Times* described that all sorts of drugs including anesthetics, pain killers, antibiotics, and cancer drugs that may be produced only by few companies because of poor profit margin, may suddenly become unavailable during production shortages, and that the practice of doctors of triaging and rationing critical medications have become the new normal (Fink 2016).

Conclusion

There is ample scientific evidence about the power of words and how small changes in behavior and wording can make huge differences in terms of medical outcomes and patient satisfaction. Any change, however small, will likely encounter resistance from engrained practice patterns often under the guise of the proposal violating prevailing ethics or not being authentic. It can be a dilemma when scientific findings counteract one's deepest held beliefs about how another person's discomfort should be managed or when one's own coping mechanisms clash with different preferences of friends or colleagues. The resulting disagreement becomes even more critical in a patient–provider relationship, which is inherently unequal. While giving a patient control can be therapeutic in itself for the patient, giving up control without giving up responsibility can be a challenge for the healthcare professionals accustomed to a more patriarchal model.

With patients' empowerment toward expectation of greater patient-centered care comes the recognition that patient satisfaction is not just a "customer" gimmick. Rather, it is intertwined with quality in any kind of healthcare system and can enhance patients' willingness to follow treatment recommendations and save healthcare expenditures.

The hope is that continued documentation of the efficacy of nonpharmacologic calmative and analgesic techniques in well-conducted clinical trials will overcome these hurdles and further stepwise integration in medical practice will facilitate the needed culture change of Western medical practice. The aim is to make gains for all concerned without unnecessary pains.

References

American Medical Association (1997). *Opinion 2.075—The Use of Placebo Controls in Clinical Trials.* Available from: http://www.ama-assn.org/ama/pub/physician-resources/medical-ethics/code-medical-ethics/opinion2075.page? [accessed March 8, 2016].

Ariely, D. (2010). *Predictably Irrational, Revised and Expanded Edition: The Hidden Forces that Shape our Decisions.* New York: Harper Perennial.

Arrowsmith, J.B., Gerstman, B.B., Fleischer, D.E., and Benjamin, S.B. (1991). Results from the American Society for Gastrointestinal Endoscopy/U.S. Food and Drug Administration collaborative study on complication rates and drug use during gastrointestinal endoscopy. *Gastrointestinal Endoscopy*, 37, 421–427.

Atlas, L.Y. and Wager, T.D. (2012). How expectations shape pain. *Neuroscience Letters*, 520, 140–148.

Atlas, L.Y., Bolger, N., Lindquist, M.A., and Wager, T. D. (2010). Brain mediators of predictive cue effects on perceived pain. *Journal of Neuroscience*, 30, 12964–12977.

Bayer, T.L., Coverdale, J.H., Chiang, E., and Bangs, M. (1998). The role of prior pain experience and expectancy in psychologically and physically induced pain. *Pain*, 74, 327–331.

Bennett, J.D., Kramer, K.J., and Bosack, R.C. (2015). How safe is deep sedation or general anesthesia while providing dental care? *Journal of the American Dental Association*, 146, 705–708.

Blankfield, R.P. (1991). Suggestion, relaxation, and hypnosis as adjuncts in the care of surgery patients: a review of the literature. *American Journal of Clinical Hypnosis*, 33, 172–186.

Blease, C. (2012). The principle of parity: the "placebo effect" and physician communication. *Journal of Medical Ethics*, 38, 199–203.

Boothman, N. (2010). *Convince Them in 90 Seconds or Less: Make Instant Connections that Pay Off in Business and Life.* Chapel Hill, NC: Workman Publishing Company.

Burgstaller, J.M., Schuffler, P.J., Buhmann, J.M., et al. (2016). Is there an association between pain and magnetic resonance imaging parameters in patients with lumbar spinal stenosis? *Spine*, 41(17), E1053–1062.

Centers for Disease Control and Prevention (2016). *Injury Prevention & Control: Opioid Overdose.* Available from: http://www.cdc.gov/drugoverdose/ [accessed April 16, 2016].

Centers for Medicare and Medicaid Services (2009). *Roadmap for Implementing Value Driven Healthcare in the Traditional Medicare Fee-for-Service Program.* Available from: http://www.cms.hhs.gov/QualityInitiativesGenInfo/downloads/VBPRoadmap_OEA_1-16_508.pdf [accessed December 5, 2012].

Centers for Medicare and Medicaid Services (2011). *Hospital Value-Based Purchasing Program (VBP).* Washington, DC: Medicare Learning Network. Available from: http://www.cms.gov/Hospital-Value-Based-Purchasing [accessed January 20, 2013].

Chen, J., Li, Z., Lv, Y.F., et al. (2015). [Empathy for pain: a novel bio-psychosocial-behavioral laboratory animal model]. *Sheng Li Xue Bao*, 67, 561–570.

Chooi, C.S., White, A.M., Tan, S.G., Dowling, K., and Cyna, A.M. (2013). Pain vs comfort scores after Caesarean section: a randomized trial. *British Journal of Anaesthesia*, 110, 780–787.

Cyna, M.A. and Lang, E.V. (2011). How words hurt. In: Cyna, M., Andrew, M.I., Tan, S.G.M., and Smith, F. (eds.) *Handbook of Communication in Anaesthesia and Critical Care*, pp.30–37. Oxford: Oxford University Press.

Dutt-Gupta, J., Bown, T., and Cyna, A.M. (2007). Effect of communication on pain during intravenous cannulation: a randomized controlled trial. *British Journal of Anaesthesia*, 99, 871–875.

Elkins, G., White, J., Patel, P., Marcus, J., Perfect, M.M., and Montgomery, G.H. (2006). Hypnosis to manage anxiety and pain associated with colonoscopy for colorectal cancer screening: Case studies and possible benefits. *International Journal of Clinical and Experimental Hypnosis*, 54, 416–431.

Everett, J.J., Patterson, D.R., Burns, G.L., Montgomery, B., and Heimbach, D. (1993). Adjunctive interventions for burn pain control: comparison of hypnosis and ativan: the 1993 Clinical Research Award. *Journal of Burn Care & Research*, **14**, 676–683.

Ewin, D.M. and Eimer, B.N. (2006). *Ideomotor Signals for Rapid Hypnoanalysis*. Springfield, IL: Charles C. Thomas Publishers.

Faymonville, M.E., Fissette, J., Mambourg, P.H., Roediger, L., Joris, J., and Lamy, M. (1995). Hypnosis as adjunct therapy in conscious sedation for plastic surgery. *Regional Anesthesia*, **20**, 145–151.

Fink, S. (2016). Drug shortages forcing hard decisions on rationing treatments. *New York Times*, January 29.

Goubert, L., Craig, K.D., Vervoort, T., et al. (2005). Facing others in pain: the effects of empathy. *Pain*, **118**, 285–288.

Hashmi, J.A., Kong, J., Spaeth, R., Khan, S., Kaptchuk, T.J., and Gollub, R.L. (2014). Functional network architecture predicts psychologically mediated analgesia related to treatment in chronic knee pain patients. *Journal of Neuroscience*, **34**, 3924–3936.

Hatfield, E. and Cacioppo, J. T. 1994. *Emotional Contagion (Studies on Emotion and Social Interaction)*. New York: Cambridge University Press.

Hayashi, K., Ikemoto, T., Ueno, T., et al. (2016). Higher pain rating results in lower variability of somatosensory cortex activation by painful mechanical stimuli: an fMRI study. *Clinical Neurophysiology*, **127**, 1923–1928.

Joint Commission On Accreditation Of Healthcare Organizations (2016). *Joint Commission Statement on Pain Management*. Available from: https://www.jointcommission.org/joint_commission_statement_on_pain_management/ [accessed February 8, 2016].

Kam-Hansen, S., Jakubowski, M., Kelley, J.M., et al. (2014). Altered placebo and drug labeling changes the outcome of episodic migraine attacks. *Science Translational Medicine*, **6**, 218ra5.

Keltner, J.R., Furst, A., Fan, C., Redfern, R., Inglis, B., and Fields, H.L. (2006). Isolating the modulatory effect of expectation on pain transmission: a functional magnetic resonance imaging study. *Journal of Neuroscience*, **26**, 4437–4443.

Kilwein, J.H. (1989). No pain, no gain: a Puritan legacy. *Health Education Quarterly*, **16**, 9–12.

Kroger, W.S. and Delee, S.T. (1957). Hypnoanesthesia for cesarean section and hysterectomy. *JAMA*, **163**, 442–444.

Kupers, R., Faymonville, M.E., and Laureys, S. (2005). The cognitive modulation of pain: hypnosis- and placebo-induced analgesia. *Progress in Brain Research*, **150**, 251–269.

Lang, E.V. (2012). A better patient experience through better communication. *Journal of Radiology Nursing*, **31**, 114–119.

Lang, E.V. (2014a). *How (Not) to Place an IV*. Comfort Talk d/b/a/ Hypnalgesics, LLC. Available from: http://www.youtube.com/watch?v=526N-3Xqkwg [accessed February 8, 2016].

Lang, E.V. (2014b). *Managing Your Medical Experience: The Information You Need Plus Self-Hypnosis for Finding Comfort With Your Tests and Treatments*. Middletown, DE: Create Space.

Lang, E.V. and Laser, E. (2009). *Patient Sedation Without Medication. Rapid Rapport and Quick Hypnotic Techniques. A Resource Guide for Doctors, Nurses, and Technolgists*. Raleigh, NC: Lulu.

Lang, E.V. and Rosen, M.P. (2002). Cost analysis of adjunct hypnosis with sedation during outpatient interventional radiologic procedures. *Radiology*, **222**, 375–382.

Lang, E.V., Chen, F., Fick, L.J., and Berbaum, K.S. (1998). Determinants of intravenous conscious sedation for arteriography. *Journal of Vascular and Interventional Radiology*, **9**, 407–412.

Lang, E.V., Benotsch, E.G., Fick, L.J., et al. (2000). Adjunctive non-pharmacological analgesia for invasive medical procedures: a randomised trial. *Lancet*, **355**, 1486–1490.

Lang, E.V., Hatsiopoulou, O., Koch, T., et al. (2005). Can words hurt? Patient-provider interactions during invasive procedures. *Pain*, **114**, 303–309.

Lang, E.V., Berbaum, K.S., Faintuch, S., et al. (2006). Adjunctive self-hypnotic relaxation for outpatient medical procedures: a prospective randomized trial with women undergoing large core breast biopsy. *Pain*, **126**, 155–164.

Lang, E.V., Berbaum, K.S., Pauker, S.G., et al. (2008). Beneficial effects of hypnosis and adverse effects of empathic attention during percutaneous tumor treatment: when being nice does not suffice. *Journal of Vascular and Interventional Radiology*, **19**, 897–905.

Lang, E.V., Yuh, W.T., Ajam, A., et al. (2013). Understanding patient satisfaction ratings for radiology services. *AJR American Journal of Roentgenology*, **201**, 1190–1195.

Lang, E.V., Tan, G., Amihai, I., and Jensen, M.P. (2014). Analyzing acute procedural pain in clinical trials. *Pain*, **155**, 1365–1373.

Leng, S (2015, February). Patients are not customers. Here are 6 reasons why. *KevinMD.com*. Available from: http://www.kevinmd.com/blog/2015/02/patients-not-customers-6-reasons.html [accessed February 8, 2016].

Liossi, C. and Hatira, P. (2003). Clinical hypnosis in the alleviation of procedure-related pain in pediatric oncology patients. *Journal of Clinical and Experimental Hypnosis*, **51**, 4–28.

Liossi, C., White, P., and Hatira, P. (2006). Randomized clinical trial of local anesthetic versus a combination of local anesthetic with self-hypnosis in the management of pediatric procedure-related pain. *Health Psychology*, **25**, 307–315.

Marchant, J. (2016). *The Cure: A Journey into the Science of Mind Over Body*. New York: Crown.

McDermott, V.G.M., Chapman, M.E., and Gillespie, I. (1993). Sedation and patient monitoring in vascular and interventional radiology. *British Journal and Radiology*, **66**, 667–671.

Meurisse, M., Hamoir, E., Defechereux, T., et al. (1999). Bilateral neck exploration under hypnosedation. A new standard of care in primary hyperparathyroidism? *Annals of Surgery*, **229**, 401–408.

Mogil, J.S. (2012). The surprising empathic abilities of rodents. *Trends in Cognitive Sciences*, **16**, 143–144.

Montgomery, G.H., David, D., Winkel, G., Silverstein, J.H., and Bovberg, D.H. (2002). The effectiveness of adjunctive hypnosis with surgical patients: a meta-analysis. *Anesthesia & Analgesia*, **94**, 1639–1645.

Montgomery, G.H., Bovbjerg, D.H., Schnur, J.B., et al. (2007). A randomized clinical trial of a brief hypnosis intervention to control side effects in breast surgery patients. *Journal of the National Cancer Institute*, **99**, 1304–1312.

Norbash, A., Yucel, K., Yuh, W., et al. (2016). Effect of team training on improving MRI study completion rates and no show rates. *Journal of Magnetic Resonance Imaging*, **44**(4), 1040–107.

Peerdeman, K.J., Van Laarhoven, A.I., Keij, S.M., et al. (2016). Relieving patients' pain with expectation interventions: a meta-analysis. *Pain*, **157**, 1179–1191.

Peyron, R., Laurent, B., and Garcia-Larrea, L. (2000). Functional imaging of brain responses to pain. A review and meta-analysis. *Clinical Neurophysiology*, **30**, 263–288.

Singer, T., Seymour, B., O'Doherty, J., Kaube, H., Dolan, R.J., and Frith, C.D. (2004). Empathy for pain involves the affective but not sensory components of pain. *Science*, **303**, 1157–1162.

Van Fleet, D.D. and Peterson, T.O. (2016). Improving healthcare practice behaviors. *International Journal of Health Care Quality Assurance*, **29**, 141–161.

Varelmann, D., Pancaro, C., Cappiello, E.C., and Camann, W.R. (2010). Nocebo-induced hyperalgesia during local anesthetic injection. *Anesthesia & Analgesia*, **110**, 868–870.

Watt-Watson, J., Garfinkel, P., Gallop, R., Stevens, B., and Streiner, D. (2000). The impact of nurses' empathic responses on patients' pain management in acute care. *Nursing Research*, **49**, 191–200.

Weinstein, E.J. and Au, P.K. (1991). Use of hypnosis before and during angioplasty. *American Journal of Clinical Hypnosis*, **34**, 29–37.

Wollmer, M.A., Kalak, N., Jung, S., et al. (2014). Agitation predicts response of depression to botulinum toxin treatment in a randomized controlled trial. *Frontiers in Psychiatry*, **5**, 36.

Yagiela, J.A. (2001). Making patients safe and comfortable for a lifetime of dentistry: frontiers in office-based sedation. *Journal of Dental Education*, **65**, 1348–1356.

Chapter 12

Placebo beyond controls: The neuroscience and ethics of navigating a new understanding of placebo therapy

Karen S. Rommelfanger

Why discuss placebos in neuroethics?

Neuroethics as a field exists because the brain enjoys a privileged position unlike any other organ. For better or worse, the predominant belief is that the brain is synonymous with personality and the essence of who people are; the brain is not an object, but a subject and agent (Roskies 2007; Racine et al. 2010). Adina Roskie's appraisal of the impact of neuroscience in society holds that "the rhetorical force of the neuroscientific understanding, rather than what it can actually reveal … causes the potential ethical issues to arise" (Roskies 2007, p.S54). Kathinka Evers advocates not for a total reduction of the human experience to the material brain, but that a valuable integration of neuroscience into ethical discussions might be toward an "informed materialism" that could be used to "oppose both dualism and naïve reductionism" (Evers 2007, p.S48). The role that individuals feel their brains play in their lives, from free will to personhood, is culturally bound and informs fundamental personal and societal values and impacts ethical decision-making from when a person comes into being and to when a person is pronounced to be dead (Adam et al. 2015; Yang & Miller 2015). In this chapter, I discuss placebo as a case study of neuroethical inquiry. I explore how neuroscience evidence informs and advances existing ethical debates, and how these debates have real-world impacts on daily well-being.

Introduction

The powers of placebo are long known. In the popular imagination, placebos have become synonymous with snake oil, deception and lying, and other nefarious immoral endeavors (the primary exception being a mother's kiss to the bumps and bruises of childhood). The scientific and medical community has assumed a stance that placebos are by definition inert and their use should be reserved to that of an experimental control; therapeutically, placebo use is condemned as a transgressive practice and punishable based on the assumed deception necessary with placebo administration—a breach of ethical conduct.

This stance is now being challenged by the emerging scientific evidence demonstrating that placebos are far from inert as well as by studies revealing widespread, although largely clandestine, therapeutic use (Sherman & Hickner 2008; Tilburt et al. 2008; Kermen et al. 2010; Rommelfanger 2013a). Perhaps even more surprising is the frequently reported patient approval of therapeutic use of placebos even under the context of deception (Lynoe et al. 1993; Fassler et al. 2011; Hull et al. 2013; Pugh et al. 2016). For those who unequivocally believe deception is immoral, emerging data suggest that placebo effects can be maintained even without deception (Kaptchuk et al. 2010; Schafer et al. 2015). Collectively, these new data on placebo require reappraisal.

What are placebos?

History

In the medical context, placebo is typically described as an inert treatment of no remedial value that is given to reinforce a patient's expectation to get well (Diederich & Goetz 2008; Finniss et al. 2010). It seems that simply the act of taking an intervention meant for healing—with a suggestion that medicine might work—can improve patient outcomes. With such clinical effects, why has the research and development of placebo therapies not been the subject of more open conversation in clinical care?

The powerful therapeutic benefits of placebo have made it a long-standing subject of bewilderment and ethical scrutiny. As far back as 1621, in the *Anatomy of Melancholy* Robert Burton wrote that "a silly chirurgeon, doth more strange cures than a rational physician ... because the patient puts his confidence in him" (Burton 1621). In 1807, Thomas Jefferson described therapeutic placebo as a pious fraud, yet, "one of the most successful physicians ... used more of bread pills, drops of colored water, and powders of hickory ashes, than all other medicines put together" (Ford 1898). Arguments about the ethics of therapeutic placebos during the nineteenth and twentieth centuries have remained largely faithful to the following problem: placebos seem to work, but not in the right way—an argument that has pervaded medicine for more than 200 years (Brody 1982). Patients may report being or feeling better after placebo intervention, yet placebos are by definition inert. The restorative effect requires deceptive manipulation of a vulnerable patient's mind by implanting false beliefs about an inert remedy to overcome the inert properties of placebos.

In popular imagination, neither the placebo nor its effect is considered real. Placebos are conceptualized as inert and their powerful effects are caused by the imagination of the individual under their influence. Indeed, the effects are fundamentally a product of the mind. In this conceptual framework, the affected person is influenced by expectation and false belief, an insult to otherwise rational intelligence (a greater affront is when a person is duped by a trusted authority and violated in a vulnerable state). In coming to recognize the power differential historically held in the practice of medicine, there has been a significant shift away from paternalism in medical practice and today a key ethical principle in medicine is to ensure the patient has autonomy and is self-determining (Beauchamp & Childress 2009). This is achieved largely through transparency in and shared decision-making between the healthcare provider, patients, and their families, seemingly the antithesis of placebo therapy.

This evolution of anti-paternalism and patient empowerment has not been without its challenges. A well-noted and natural tension in placebo use lies between the principles of beneficence and respect of autonomy (Beauchamp & Childress 2009). If placebo use requires deception and efficacy in the patient's best interest, is it not the physician's obligation to deceive? Adding to the complexity of the placebo story, patients report supporting deception, permissible particularly in therapeutic encounters with unresolved chronic conditions (Lynoe et al. 1993; Fassler et al. 2011; Hull et al. 2013; Pugh 2015). Perhaps in the context of placebo therapy, shared decision-making between patient, families, and providers might provide some level of resolution (Brody et al. 2012). What information needs to be shared and under what contexts are topics discussed later in this chapter.

Placebos in practice

Placebo therapy use is a global phenomenon, and according to some reports, up to 97% of physicians administer placebo therapeutically (Fassler et al. 2010; Fent et al. 2011; Meissner et al. 2012; Howick et al. 2013). Fifty percent of US physicians have used placebo in clinical practice (Sherman & Hickner 2008; Tilburt et al. 2008) for an array of disorders from gastrointestinal and immune disorders to cancer and neurological disorders (Sherman & Hickner 2008). Placebo treatments include saline, sugar pills, vitamins, over-the-counter analgesics, antibiotics, and sedatives (Tilburt et al. 2008). Physicians often describe placebo therapies as "a substance that may help and will not hurt" or "it is medication" or "it is medicine with no specific effect," "a medicine not typically used for your condition but might benefit you," or might say, "This may help you, but I am not sure how it works" (Sherman & Hickner 2008; Tilburt et al. 2008). These statements demonstrate a mode of communicating information that falls somewhere on a spectrum between acceptable professional norms of oversimplifying treatment descriptions (i.e., physician states that a pill is akin to an antibiotic or antihistamine) to deliberate attempts at verbal misdirection.

Professional values

While placebo therapy has been widely employed across the globe (Hrobjartsson & Norup 2003; Tilburt et al. 2008; Fassler et al. 2010; Fent et al. 2011; Meissner et al. 2012; Howick et al. 2013), this practice is quite contrary to the values self-professed to be held by professional medical associations such as the American Medical Association (AMA)—although the AMA only in 2006 categorically prohibited deceptive placebo use (Bostick et al. 2008; Blease 2012). Interestingly, no pronouncement has been made by similar global organizations and the AMA's stance is by far the most explicit and stringent.

According to the Report of the AMA Council on Ethical and Judicial Affairs, placebo is defined as "a substance that the physician believes had no known specific pharmacological activity against the condition being treated" (Bostick et al. 2008, p.58). The Council states that placebo use with deception "directly conflicts with contemporary notions of patient autonomy and the practice of shared decision making (Bostick et al. 2008, p.59). Aside from the harm of violating the patient–physician alliance, the Council argue that patients may also encounter nocebo effects, adverse side effects rather than the positive effects associated with placebo use. These statements reveal the inherent professional and conceptual confusion about placebo in two ways. First, the authors warn against placebos as treatments with no specific pharmacological activity that may be accompanied by legitimate adverse effects, implying that while benefits are somehow fake, the harms are somehow real. Second, the placebos are defined by the physician's belief that the placebo has no known specific pharmacological activity, not characterized or defined by the scientific evidence that might explain mechanisms of placebo. However, in the concluding statements of the report, the Council states that placebos "may be used in clinical practice to determine a diagnosis or appropriate treatment in the face of clinical uncertainty," but that physicians should avoid deception. The report also recommends that physicians should avoid placebos, but are nonetheless supported in their attempt to elicit a "placebo-like effect through the skillful use of reassurance and encouragement … In this way, the physician builds respect and trust, promotes the patient-physician relationship, and improves health outcomes" (Bostick et al. 2008, p.60). One may well wonder, which is it? Do placebos violate the patient–physician relationship and cause harm or do placebos build respect and trust and improve health outcomes? Perhaps it is no wonder in this linguistic equivocation that clinicians find themselves in practice clandestinely using placebo.

The AMA has also created explicit guidelines in the form of an Opinion under their Code of Medical Ethics on the clinical use of placebos. In their Opinion statement there is one loophole for deception under the condition that the patient is informed *at some point* that placebo will be used but that the patient will be blind to the precise timing of its implementation (AMA 2007). In this case the Opinion appears to sanction a kind of deception that requires patient ignorance of some details such as timing with the patient's approval. Again such an Opinion also highlights lingering ambiguity about the role both of deception in routine practice and for deceptive placebo use. The AMA Council report states that "physicians may utilize placebos within their clinical practice without relying on the act of deception" (Bostick et al. 2008). This statement in the Council Report in combination with the Opinion seems to suggest that there is some form of acceptable deception, or appears to suggest that authorized deception is not really deception (Blease et al. 2016), just like placebos are not really placebos if the physician does not believe they are, as with the case of active placebos where active agents such as vitamins or off-label prescriptions are given as placebo.

How and why does the prevalent use of (deceptive) therapeutic placebo prevail even in the absence of solidarity from professional medical societies? Have physicians betrayed their oaths to care for patients and to try to make them well or improve their quality of lives? Is there an ethical framework that would make such a behavior morally permissible? A consequentialist approach where the ends justify the means seems to explain popular physician use of placebo therapy. A deontological approach championing the duty not to lie about an inert treatment seems to drive concerns for deception. The result is a relatively widespread use of placebo in a clandestine manner or a compromised use of active placebos that utilize off-label or reduced doses of real drugs (Sherman & Hickner 2008; Tilburt et al. 2008; Rommelfanger 2013a). While these arguments around placebo are not new, the definition and what we know about placebos must be updated with the accumulating body of evidence showing specific neurobiological effects from placebo (Benedetti 2014) as well as the necessity—or lack thereof (Blease et al. 2016)—for deception with placebo administration. These data will be discussed in "Explanations for placebo effects." Just as remarkably absent as the neuroscience data in these ethical evaluative schemas, is the patient's voice and her reflections about the use of placebo therapy. These data too must be incorporated into our future reappraisals of placebo therapy and in determinations of what is best for individual patients.

Explanations for placebo effects

Prejudicial terminology and placebo

Much of ethical demise of placebo is owed to its conceptual entanglements.

In 2011, Miller and Brody outlined common conceptual distinctions of placebo that produce misunderstanding about the placebo; such awkward terminology undermines a critical understanding of placebo in medicine and almost universally work to devalue placebo (Miller & Brody 2011). These conceptual framings serve as policing agents for an outdated and outmoded Cartesian model that attempts to separate the mind from the body, a view that is so prevalent in the practice of Western medicine. These terms are summarized as follows.

The first is *verum* (what is true) versus placebo. If *verum* is truth, placebo is necessarily positioned as false. Another confusing set of terms, often used in clinical trials, is active versus inactive wherein the inactive agent is considered placebo. Such a distinction is misguided because, as discussed, placebo controls are used *because* of their activity and known–unknown arbiters to any drug effect. Another framing to describe clinical effects is to distinguish specific from nonspecific effects where placebo effects are designated as nonspecific contamination of desired clinical effects. In this framework, it follows that a disease is not really a legitimate diagnosable disease until the specific biochemical pathways are understood. Reporting of symptoms is not enough. The need for specificity seems to coincide with redefining of diseases with "specificity and specificity of mechanism" (Rosenberg 2002, p.243) and by extension, a specific therapy (Rosenberg 2002). Rosenberg calls this the "Tyranny of Diagnosis."

By extension, utilizing nonspecific therapies—such as off-label use of drugs or even something like acupuncture—whose mechanism is not understood is to practice medicine that was not scientifically grounded. However, the nature of practicing medicine is full of ambiguity. The goal of research with scientific studies is to identify generalizable findings. This scientific strategy will not necessarily map cleanly onto real-world individual experience and illness. In fact, those who subscribe to such a model are guilty of reducing the human experience in a way that devalues the individual patient's subjective experience while over-privileging objective measures. The long-suffering road of pain researchers who have struggled to find objective measures that can replace subjective reports of pain demonstrate the downfall of such a strategy.

Another way of conceptualizing placebo is to suggest that placebos are largely effects of context (Miller & Kaptchuk 2008). So much so, perhaps, that Di Blasi and colleagues suggest that placebo effects would more appropriately be referred to as "context effects" (Di Blasi et al. 2001). It would seem that the ritual around the intervention, rather than the intervention itself, is what determines the placebo effect. Anthropologist Daniel Moerman advocates for a reconceptualization of placebo effects as meaningful responses, and in doing so one can avoid the ethical quandaries of both placebo itself and the embedded connotations of deception (Moerman 2002; Moerman & Jonas 2002). With Western biomedicine, technological interventions are privileged and the placebo is understood as the non-technological context, somehow a nonspecific variable and background noise. Placebo effects, in other words, are psychological contamination (Colloca & Benedetti 2005). A discussion of how context impacts learning and expectation associated with placebo is discussed further in "Mechanisms of placebo."

Regardless, placebo, conceptualized as nonspecific and subjective phenomena, has demonstrably specific and objectively measurable effects. Given these demonstrable effects, should placebos be given more credibility as a viable intervention? Can we overcome the historical linguistic trappings with placebo and its counterparts? The term placebo is in grave need of updating and perhaps at best only serves as an anachronistic placeholder, like *thingamajig*, a relic of a troubled legacy of misunderstanding.

Mechanisms of placebo

Placebo effects can be thought of in terms of mechanisms of expectation, learning, and conditioning (Colloca & Miller 2011b). In Colloca and Miller's model, placebo is a vehicle for therapy not because of its internal properties, but because they are a collection of salient signals or symbols. An intervention, whether considered a placebo or not, cannot be separated from the context in which it is given. Colloca and Miller point out that, "The patient does not come to the clinical encounter as a blank slate but with a history of experiences and memories evoked by prior responses to signals related to the milieu of therapy" (Colloca & Miller 2011b, p.1860). In this sense, placebo effects are not unique to placebo interventions per se and can happen with any clinical intervention—with or without physical placebo.

According to a classical conditioning framework, repeated exposure to contextual factors that coincide with ingestion of a medication—such as visual, tactile, and gustatory stimuli (unconditioned stimuli)—become associated conditioned stimuli (Pavlov 1927; Colloca & Miller 2011b). Placebos, as a consequence, are in turn able to elicit conditioned responses in the presence of their conditioned stimuli. For example, were one to visit a physician's office and receive pain relief from an injection of an anti-inflammatory drug, a syringe can become the conditioned stimulus—it becomes the symbol for pain relief. Then future observation of the syringe (or injection from the syringe regardless of its content) at a doctor's office may then elicit pain relief for a patient so conditioned (Montgomery & Kirsch 1997; Colloca & Miller 2011b).

Learning associations for placebo, like any learning, allows for environmental cues or elements to be connected with social cues. There appear to be several different cues or conditions that can elicit a placebo response. These include verbal cues, prior experience, observation, and social learning, each of which lead to expectation and conditioned learning (Colloca & Miller 2011b). Early studies explored the effect of verbal language on bronchoconstriction in patients with asthma (Luparello et al. 1968, 1970). When patients were instructed to inhale a solution (of nebulized saline) and told that the solution was an allergen, nearly half of the patients demonstrated increased airway resistance. When the same patients were instructed to inhale the same solution (again of nebulized saline) and told that the solution was a medicine for asthma, airway bronchoconstriction was reversed. In an additional study, carbachol, a bronchoconstrictor, was given to patients who were told that it was a bronchoconstrictor. These patients had greater effects than when patients were told carbachol was indeed a bronchodilator. In another series of studies for postoperative pain relief, patients (who were told that a placebo was a potent pain reliever) were able to reduce their opioid intake more than patients who were told nothing about the analgesic effect of the placebo or patients who were told that they *might* receive a placebo or pain reliever (Pollo et al. 2001). Verbal suggestion also has been shown to have significant placebo (and nocebo, or negative, unpleasant effects) effects in an experimental model of itch (Van Laarhoven et al. 2011).

One example of conditioning by prior experience was demonstrated by Benedetti and colleagues. Participants given placebo for migraine relief were able to experience subjective pain relief accompanied by the same profile of significant biochemical effects (albeit not as dramatic) of increased growth hormone release and decreased cortisol release (Benedetti et al. 2003). The placebo was only able to elicit this response after prior administration of the migraine medication, that is, no effect was seen if placebo was administered before administration of migraine medication. This effect was also independent of verbal instructions about the placebo/migraine medication that would have primed expectations. Similar conditioned placebo responses have been seen with immunosuppressors (Goebel et al. 2002, 2005), dopamine agonists (De La Fuente-Fernandez et al. 2001; Benedetti et al. 2004; Lidstone et al. 2010), and benzodiazepines (Petrovic et al. 2005). Conditioned placebo responses have also been modeled in nonhuman animal studies in mice (Guo et al. 2010) and rats (Zhang et al. 2013; Lee et al. 2015). Interestingly, in one follow-up study, Guo and colleagues suggest that placebo effects can be transferred from one domain to another, from pain to emotion in their rodent model (Guo et al. 2011).

In a recent placebo analgesia study involving rodents, Lee and colleagues attempted to parse out learning neurophysiology from analgesia physiology (Lee et al. 2015). Rodents were trained in a conditioned place preference paradigm to associate a place/room with low or high pain induced by heat. Rats then formed an association between the room cue (one room had a rough floor with a black and white wall and the other room had black wall with a soft floor) and the pain stimulus following exposure to the room. Given the choice post training, the rats came to prefer the room that had preceded lower pain exposure during training. The pain response was measured with hind paw withdrawal to a stimulus. Longer latencies are associated with lower pain. Administration of the dopamine antagonist haloperidol blocked cue learning preference for the room preceding low pain exposure. Both dopamine and opioid antagonist naloxone blocked the expression of placebo analgesia as measured by biomarkers in the ventral tegmental area and nucleus accumbens. From these data, researchers concluded that dopamine pathways were critical in the acquisition of association of cues with the placebo response. Corroborating human study data (Wager et al. 2007), Zhang and colleagues further concluded through direct microinjections of various subtype opioid antagonists into the rat anterior cingulate cortex, an area associated with registered physical pain, that mu-opioid receptors mediate placebo analgesia (Zhang et al. 2013).

Prior experience with active drugs or even placebos can influence the magnitude of subsequent placebo responses (Colloca & Benedetti 2006). In one study of placebo analgesia, two groups of participants were exposed to painful stimuli. In one group, individuals were told they were given an analgesic agent while the researcher simulated a placebo effect by surreptitiously reducing the intensity of the painful stimuli. In a second group the intensity of the painful stimuli was not reduced. When both groups were given the placebo again, but without reduction of painful stimuli, the participants who initially were primed to believe the placebo was an effective pain reliever by surreptitiously reducing intensity of the painful stimuli in the initial exposure, experienced a greater placebo response. These placebo responses and the anticipation of analgesia were correlated to activity in the dorsolateral prefrontal cortex suggesting that anticipation of placebo involves higher-order cognitive activities such as executive functioning and perception (Lui et al. 2010).

In another study by Lidstone and colleagues, participants with Parkinson's disease, told a numerical probability of receiving dopamine therapy while always receiving placebo, demonstrated significant dopamine release in both the nigrostriatal system that is characteristically impacted in Parkinson's disease, as well as the mesoaccumbens, a node in the reward pathway, when the probability was high (75%) (Lidstone et al. 2010). The strength of expectation by giving participants a numerical expectation value impacted the degree of dopamine release in both the dorsal striatum associated with therapeutic dopamine release, and ventral striatum, associated with reward/expectation mediated dopamine release (Lidstone et al. 2010). Cortical excitability has also been associated with expectation of treatment (Lou et al. 2013).

Reward learning is the subject of significant study for drug addiction, attention deficit hyperactivity disorder, and basic learning mechanisms. Critical to reward learning is the recruitment of the ventral striatum (Schultz 2006), a region also thought to underlie placebo effects (De La Fuente-Fernandez et al. 2004) by moderating salience and expectation. Dopamine release in the ventral striatum has been identified with placebo analgesia

(Scott et al. 2008) as well as in response to the probability of receiving active drug as mentioned earlier (Lidstone et al. 2010).

While some scholars have attempted to draw clear distinctions between expectancy and conditioning (Kirsch 2004; Stewart-Williams & Podd 2004)—that is, conditioning requires unconscious participation while expectancy may require conscious processing—it is also recognized by critics (Kirsch 2004) that some form of conditioning likely underlies critical aspects of the generation of expectancy. Classical conditioning is also likely to involve a higher order of cognition in order to form logical and perceptual relationships between events, stimuli, and representations (Shanks 2010).

Amanzio and Benedetti attempted to dissect components of expectation-based analgesia and conditioning-based analgesia, eliciting recruitment of specific neurophysiological changes related to pain analgesia (Amanzio & Benedetti 1999). In this human study, expectations of receiving pain relievers that work by two different mechanisms induced opioid- and nonsteroidal anti-inflammatory drug (NSAID)-mediated placebo analgesia respectively. However, conditioned responses recruited specific neurophysiological responses dependent on the type of analgesia; morphine-induced opioid-mediated pain relief was blocked by the opioid antagonist naloxone and NSAID-mediated analgesia was not blocked by naloxone. Importantly, the researchers found that the two components were intertwined and that reduced expectancy. By telling participants they would receive an NSAID rather than morphine while they were being given the opioid antagonist that blocks opioid-mediated pain relief, the conditioned placebo analgesia response was reduced but not eliminated (Amanzio & Benedetti 1999). In this case, the expectation that the drug was a pain reliever still resulted in placebo analgesia likely though a specific NSAID-mediated mechanism (as a conditioned response to the NSAID) that could not be blocked by an antagonist for another pain reliever (morphine).

Finally, Colloca and Miller note models of social learning for placebo (Colloca & Miller 2011a), rather than direct first-hand experience as described in the studies previously described. The magnitude of the placebo response for those who were conditioned through observing pain relief in another was similar to conditioned placebo responses via first-hand experience (Colloca & Benedetti 2009). In addition, those participants with higher empathy scores had greater placebo responses upon observing placebo pain relief.

Another form of social learning with placebo can occur through patient–provider interaction. Placebo effects have been shown to be augmented when participants had more supportive practitioner engagement as enacted though active listening (e.g. repeating patients' words and asking for clarification) and attempts to behave in a warm, friendly, and empathetic (e.g., by saying, "I understand how this could be difficult for you") way as well as setting positive expectations (e.g., saying, "I believe this will have a positive impact for you") and 20 seconds of thoughtful silence after interventions (Kaptchuk et al. 2008). It seems the ritual of therapeutic encounters provides a learning mechanism wherein patients can learn to produce placebo responses through having an interaction with a healer/clinician (Miller et al. 2009). In this model of "interpersonal healing," the doctor's beliefs about the efficacy of treatment can be extremely influential on the patient's expectation and placebo response to an intervention.

Societal influence and impact of branding can influence placebo effects. In a recent study on placebo analgesia for migraine, by Kam-Hansen and colleagues, investigators noted that placebo bottles labeled as "Maxalt" reduced headache severity as well as Maxalt labeled as placebo (Kam-Hansen et al. 2014). Further placebo, labeled as placebo, was more effective in reducing headaches than no treatment at all. A recent study exploring the price attached to a medication for Parkinson's showed that patients told they were given expensive medication manufactured at $1500 USD/dose demonstrated a greater reduction in motor symptoms than patients told they were given cheaper medication manufactured at $100 USD/dose, both of which were placebos (Espay et al. 2015). Clearly these placebo effects are modulated both by direct and indirect experience and embedded meanings in those experiences. In a global study of placebo effects in clinical trials for pain medications, researchers identified an increasing effect in the United States and only United States (Tuttle et al. 2015). These effects are stronger in longer and larger trials that tended to occur in the United States, suggesting the strong influence of cultural values and meaning of pharmacology on eliciting placebo responses.

Neurobiological insights from disease with established biochemical pathways

A lingering mindset among some skeptics, even with the information above-mentioned, is that placebos may alter subjective illness but would never have an impact on the disease or the biological processes of that disease (Spiro 1986). This has all been controverted by a growing and impressive set of data outlining changes in human and nonhuman neurophysiology underlying placebo effects that are related to neurological disorders whose biochemical pathways are well-characterized (Benedetti 2014). These data also build on, but should not replace, the body of research exploring the psychosocial dimensions of placebo therapy.

Placebo responses have been identified in 16–55% of Parkinson's disease patients (Goetz et al. 2008), 30% of migraine patients (Bendtsen et al. 2003), and 50% of depressed patients (Dworkin et al. 2005). Studies in Parkinson's disease have demonstrated enhanced endogenous release of neurotransmitters such as dopamine (De La Fuente-Fernandez et al. 2001; Mercado et al. 2006; Lidstone et al. 2010) as well therapeutically associated changes in neuronal activity upon placebo administration (Benedetti et al. 2004). In one of the most compelling studies on placebo neurophysiology, Benedetti and colleagues showed that aberrant neuronal activity in the subthalamic nucleus associated with Parkinson's disease recorded in single neurons was normalized after placebo treatment (Benedetti et al. 2004). The physician's examination corroborated the patient's reported improvements (Benedetti et al. 2004).

Opiate release (Zubieta et al. 2005; Wager et al. 2007) has also been documented upon placebo administration, and pharmacological antagonists, such as naloxone (Amanzio & Benedetti 1999) can inhibit placebo-induced analgesia in both humans (Amanzio &

Benedetti 1999) and animal models (Nolan et al. 2012) indicating that placebos have specificity with regard to endogenous neurochemistry.

As noted earlier, studies in human patients suggest that placebo effects utilize reward pathways in the ventral striatum and that these reward pathways are modulated by expectation. De la Fuente-Fernandez and colleagues demonstrated in several occasions through positron emission tomography that placebo can activate the specific nigrostriatal dopamine system whose degeneration and disruption characterizes Parkinson's disease, and greater placebo-induced dopamine release associated with patient-reported clinical improvement (De La Fuente-Fernandez et al. 2001, 2004). Regardless of whether patients reported clinical improvement, all participants demonstrated placebo-induced increases in dopamine in the ventral striatum, an area typically associated with reward functioning and expectation of rewards (De La Fuente-Fernandez et al. 2004; Schultz 2006). These studies suggest that while expectation plays a key role in facilitating downstream placebo effects, the degree of recruitment of the specific relevant biochemical pathway can impact the perceived therapeutic benefit. Critically, intact top-down cognitive processing may mediate these placebo effects, as some are lost in patients with Alzheimer's disease (Benedetti 2006; Enck et al. 2008).

While many placebo studies have been conducted in pain studies, Parkinson's disease allows unique attributes for exploring the biological correlates to the placebo pathway. For example, with Parkinson's disease versus pain, treatment responses can be evaluated objectively by the experimenter evaluations of motor performance (Frisaldi et al. 2014). The second unique advantage is that neurophysiological correlates of placebo response can be measured at the level of the individual brain cell in awake patients by using electrodes implanted for deep brain stimulation. In a study building on that by Benedetti and colleagues (Benedetti et al. 2004), Fisaldi and colleagues examined the placebo response in Parkinson's disease through single-unit recordings throughout the basal ganglia (Frisaldi et al. 2014). These studies were critical to connect previous studies that separately indicated changes in downstream neurophysiology in the subthalamic nucleus (Benedetti et al. 2004) and the facilitation of this effect mediated by changes in expectation and reward pathways (De La Fuente-Fernandez et al. 2001). They were able to demonstrate that critical areas affected in Parkinson's disease, subthalamic nucleus, and the substantia nigra pars reticulata as well as the ventral anterior thalamus were involved in the placebo response. Frisaldi and colleagues were also able to compare patterns of activity between placebo responders and nonresponders (Frisaldi et al. 2014).

In sum, we see the development of an evolving model of placebo (Diederich & Goetz 2008) that suggests that high-order brain regions (such as the prefrontal and cingulate cortices) modulate pathways involved in expectation and reward processing (such as the ventral striatum). In turn, changes in these pathways affect disease-specific or condition-specific pathways—the dorsal striatum in Parkinson disease, the midbrain in modulation of opioid release for pain relief, and the amygdala and limbic areas to mimic serotonin reuptake inhibitors in depression (Mayberg et al. 2002). Enck and colleagues postulate that placebo is a kind of evolved "endogenous healthcare system" (Enck et al. 2008).

More recent studies have explored possible biomarkers and genetic polymorphisms such as enzymes that metabolize catecholamines like catechol-O-methyl transferase, that would make one more susceptible to placebo responses, the so-called placebome (Hall et al. 2012, 2015). These studies, along with the notion of placebo effects as a product of social learning or classical conditioning (even in nonhuman animals), suggest that placebo responses are biological and evolutionarily conserved (Colloca & Miller 2011a). Evolutionary psychologist Nicholas Humphrey proposes that typical bodily response to stressors such as pain or anxiety are counteracted by placebo responses using hope to counteract pain and anxiety and engaging internal healing mechanisms instead of shutting down (Humphrey 2002). While there is no shortage of future empirical research needed to further elucidate such mechanisms, this research has moved past its infancy and warrants more sophisticated integration into practical applications of placebo therapy.

Deception

A seemingly universal tenet of medical ethics is the belief that good practice requires honesty and forbids deceit. The clearest ethical violation of placebo therapy is not necessarily the use of the placebo itself, but the secreted nature of this use. The use is hidden not just to the patient, but also from other physicians in the healthcare team.

Deception and physician belief

The AMA's recommendation against placebos defines placebo as "a substance provided to a patient that the physician believes has no specific pharmacological effect upon the condition being treated." Therefore the troubling act of placebo administration is that physicians are knowingly using substances that they believe are inert to treat patients. Given the body of mechanistic evidence of placebo and its effect outlined earlier, one might argue that to assert that placebos are inert is either willful ignorance or an active choice to not practice evidence-based medicine, its own brand of ethical violation. If physicians believe placebos are not inert, is it still a violation of AMA guidelines? According to some studies, 96% of physicians who utilize placebo believe that the placebo offers physiological benefit to patients (Sherman & Hickner 2008; Kermen et al. 2010). This is also an important counterargument to placebo critics who voice concerns related to the physicians mal-intent, that is, physicians are trying to punish patients, relieve their own frustrations, or trick patients into revealing that their symptoms are not real when using placebo (Purtilo 1993). Studies suggesting physicians are using placebos because they do not believe placebos are inert does not eliminate the possibility of mal-intent. Were placebo to be adopted openly into routine use, mal-intent would need to be monitored, but mal-intent is likely not to be the primary driving force behind placebo therapy.

Deception and patient autonomy

Protecting patient autonomy is a prominent ethical concern around placebo practice. To truly empower patients, healthcare practitioners must be sure of what patients want;

the answer to this question must be empirically informed. A handful of studies report actual collected data on patient preferences about placebo treatment. A study of patients from Sweden revealed that patients were, perhaps surprisingly, stronger (25% of patients) advocates of deceptive placebo than physicians (7% of physicians) (Lynoe et al. 1993). In a more recent study, US respondents reported significant acceptance of placebos with some respondents supportive of deceptive placebo (Hull et al. 2013). It remains unclear precisely how patients are conceptualizing placebos. In a study by Bishop and colleagues, while many patients reported supporting placebo usage, particularly if there were perceived beneficial outcomes, those who had negative views of placebo interpreted the term "placebo" as synonymous with "ineffective" and requiring deception (Bishop et al. 2014). Similar studies will need to be conducted before generalizing results, but these data should give us pause before categorically prohibiting placebo usage, deceptive or otherwise, in the name of protecting patient autonomy.

In a paper advocating for "Paternalism and partial autonomy," O'Neill argued that deception associated with placebo does not infringe upon patient autonomy because it does not infringe upon fundamental aspects of care (O'Neill 1984), and Barnehill has argued that the "fundamental aspects" of care are the purpose of the treatment and that the mechanism of action is ancillary (Barnhill 2011). Not unlike Barnehill, I have argued, the deciding factor of what is ancillary information or not should be based on empirical investigation of patient's attitudes about placebo administration (Rommelfanger 2013b), and at the very least determinations of what counts as fundamental should be strongly informed by such data. A way forward is for medicine to strive for a scenario in which patients are informed about the nature of placebo and its potential uses rather than purely relying on patient intuitions (Blease et al. 2016). However, making decisions, that is, joint care decisions, about whether deceptive or open-label placebo is the appropriate way forward should be informed by patient values.

Deception and disclosure

As discussed in "Deception and physician belief," the AMA Code of Medical Ethics prohibited the deceptive use of placebo as of 2006. However, what if deception could be eliminated from the equation and was not required to gain benefit from placebo? In a study from 2010, researchers found that placebo effects were maintained even when patients were aware that they are being given placebo (Kaptchuk et al. 2010). Patients with irritable bowel syndrome experienced relief from their symptoms after being told they would be given "placebo (inert) pills, which were like sugar pills which [sic] had been shown to have self-healing properties." While not as well known, an earlier study from 1965 demonstrated similar results in patients with somatic symptoms in a non-blind placebo trial of a week's worth of sugar pills (Park & Covi 1965).

Sandler and colleagues used placebo in a conditioned placebo dose reduction paradigm in 99 children between the ages of 6 and 12 with attention deficit hyperactivity disorder (Sandler et al. 2010). In this study, children who began on their full optimal stimulant dose were then given 50% of this dose for 1 month either with or without a placebo. Both children and their parents were told that they would receive placebo as a dose extender. Importantly, they were told that the placebo contained no active pharmaceutical ingredient. The reduced dose plus placebo was not only more effective than the 50% dose alone, but also was just as effective as the full optimal dose, once again demonstrating that that placebo can be used without deception to elicit positive therapeutic results. These findings also suggest placebo need not replace existing treatments, but can also be used to modify existing treatments—in this case to reduce the dose of medication that is potentially addictive to children.

A common intuition is that once placebo is revealed, it can no longer have its effects. Schafer and colleagues addressed this topic by first giving participants deceptive topical cream placebo to elicit the placebo analgesia. Then participants were told explicitly that the cream was simply petroleum jelly with blue food coloring and participants were even shown how the placebo cream was prepared (Schafer et al. 2015). Participants continued to experience placebo-induced analgesia even after being told about and visually seeing the inert components of the cream, demonstrating that revealing that an intervention is a placebo, after experiencing placebo effects during blinded administration, does not necessarily impact the placebo's future ability to elicit positive effects. Taken together, we see several models of effective non-deceptive placebo therapy applications.

A new way forward?

Given the emerging mechanistic information on placebo and that deception is not necessary to elicit a placebo response, at least in some paradigms (Kaptchuk et al. 2010; Schafer et al. 2015), how might we conceptualize an ethical way forward? How might we refine our definitions of deception, autonomy, and informed consent to ultimately advance patient care?

Placebo studies may be used to investigate how to improve general clinical practices. For example, in one experimental model of pain, participants were divided into three groups and given a novel or established analgesic—both placebos: choice, wherein participant could choose the analgesic; no-choice, wherein participants were instructed which analgesic to use; and a control group who was not given a choice but also not given expectations of the analgesic's efficacy (Rose et al. 2012). The choice group had the highest placebo effect suggesting that enhanced patient involvement elicited greater placebo responses and perhaps might even be used to inform patient-centered medicine (Rose et al. 2012; Fleurence et al. 2013). Varying role and levels of choice a patient is given might help to improve placebo responses and perhaps even standard treatment outcomes.

It is also known that open administration of analgesics delivered by a computer-controlled infusion pump can result in greater pain relief than concealed administration. While hidden administration of the analgesic resulted in pain relief and associated decreased activity in regions of the brain associated with pain processing, pain relief of open administration of the drug doubled this activation with the recruitment of the pregenual anterior cingulate and dorsolateral prefrontal cortex associated with expectation

(Benedetti et al. 2011; Bingel et al. 2011). Conversely, describing potential adverse side effects of interventions can elicit nocebo effects or a higher likelihood of experiencing those negative side effects (Amanzio et al. 2009). Nocebo has become such a prominent concern that researchers self-identified as the "placebo competence team" have argued for a professional imperative to decrease negative expectations of interventions (Bingel & Placebo Competence Team 2014). This suggests not just open, but conspicuous administration of standard drugs may be helpful in augmenting their effects and possibly reduce adverse effects of drugs where long-term use is detrimental to patient health. These results also suggest a need for changes in the way informed consent is conceptualized and perhaps even a need to inform of potential nocebo effects as a consequence of over-informing. Placebos may become an attractive candidate for personalized therapy and based on previous studies about expectancy and conditioning. However, it would be imperative to acknowledge that each patient may have a completely different history and experience with medication, therapeutic encounters, and series of symbols for understanding their therapeutic encounters.

To be clear, administration of placebos is not harmless. Deceptive placebos can be harmful in the case of active or impure substances that are subthreshold doses of medication. Foddy argues that "placebos are always safe," and "are sometimes the best treatment" (Foddy 2009). If placebos do have specific physiological effects that could possibly impact disease pathways then not only are they not inert, but also they could possibly be contraindicated (Rommelfanger 2013b). If placebo is surreptitiously administered, it may hamper the rest of the healthcare team's efforts to address adverse events related to placebo. Further, as others and I (Freedman 1990; La Vaque and Rossiter 2001; Foddy 2009; Rommelfanger 2013b) have argued, it would be unethical to prescribe placebos in the place of established successful standards of care or if a better treatment became available. Placebos need not replace existing successful therapies; they may be able to augment or supplement existing therapies (Ader et al. 2010; Sandler et al. 2010). Future placebo research could be directed toward reducing side effects or mitigating addictive potential of select therapies.

Perhaps one of the greatest potential gains in devoting research to placebo and its mechanism is to break down silos and move toward perceiving health more holistically. Current secretive placebo practices can largely be attributed to an entrenched belief in our medical care system that the mind is somehow separate from the body. Bodily disorders enjoy the privilege of being legitimate problems and in turn escape the stigma, blameworthiness, and illness invalidation that accompany the stigma of a disorder that is considered to be psychological (Kendell 2001; Miresco & Kirmayer 2006). Placebo effects have been similarly dismissed as a phenomenon of the mind, and therefore non-physiological or nonspecific in origin and illegitimate (Lichtenberg et al. 2004). The emerging neuroscience data suggest the placebo story is more multifaceted. Therefore, a renewed exploration of placebo with these data in mind offers an opportunity to disrupt this division of mind versus body and provides an opportunity to redress the resulting confusion.

It is also critical to note that the rise of biomedicine has been critiqued as creating an impractical dichotomy between disease seen as objectively measurable, and illness seen as the patient's subjective experience (Kleinman 1988), and there certainly is a danger of using neuroscience of placebo (Miller & Brody 2011) to reinforce this dichotomy. Contemporary Western medicine is besotted with biomarkers and biological mechanisms and some have argued to the detriment of the patient (Kleinman 1988); the patient's subjective experience of their improvement is confirmation of the measured biological outcome and not converse.

However, until recently, the perceived improvement or placebo effects reported by patients were interpreted in the popular imagination as purely subjective and without a biomarker of their effects, were limited to subjective improvement. The emerging neurophysiological correlates to placebo should not be interpreted as neuroscience usurping the import of social and psychological science, anthropology, or humanities in the conversation, but should be a reminder that conversations about human health must continue to be conducted with humility with an openness toward cross-disciplinary discussion and design. These data invite a broader conversation about the healthcare context, what we mean by evidence-based medicine, and how we, as a society, balance the value of holistic patient care.

In some ways, to keep abreast of the emerging data on placebo is extremely threatening in a healthcare model where physicians are pressured to increase the number of patients and feel pressures to shorten visits—devaluing the intricacies of the context of the therapeutic encounter—to meet productivity quotas in an assembly line style of healthcare (Konrad et al. 2010). While these financial concerns are legitimate and pragmatic, there may be a missed opportunity to utilize placebo therapy to minimize unnecessary expenditures. The clear way forward is that placebos become a therapeutic target and a subject of empirical research—rather than psychological contamination—and a fertile area of investigation in its own right.

References

Adam, H., Obodaru, O., and Galinsky, A.D. (2015). Who you are is where you are: antecedents and consequences of locating the self in the brain or the heart. *Organizational Behavior and Human Decision Processes*, **128**, 74–83.

Ader, R., Mercurio, M.G., Walton, J., et al. (2010). Conditioned pharmacotherapeutic effects: a preliminary study. *Psychosomatic Medicine*, **72**, 192–197.

Amanzio, M. and Benedetti, F. (1999). Neuropharmacological dissection of placebo analgesia: expectation-activated opioid systems versus conditioning-activated specific subsystems. *Journal of Neuroscience*, **19**, 484–494.

Amanzio, M., Corazzini, L.L., Vase, L., and Benedetti, F. (2009). A systematic review of adverse events in placebo groups of anti-migraine clinical trials. *Pain*, **146**, 261–269.

American Medical Association (2007). *Opinion 8.083: Placebo Use in Clinical Practice*. American Medical Association. Available from: http://www.ama-assn.org/ama/pub/physician-resources/medical-ethics/code-medical-ethics/opinion8083.page? [accessed April 25, 2016].

Barnhill, A. (2011). What it takes to defend deceptive placebo use. *Kennedy Institute of Ethics Journal*, 21, 219–250.

Beauchamp, T.L. and Childress, J.F. (2009). *Principles of Biomedical Ethics*. New York: Oxford University Press.

Bendtsen, L., Mattsson, P., Zwart, J.A., and Lipton, R.B. (2003). Placebo response in clinical randomized trials of analgesics in migraine. *Cephalalgia*, 23, 487–490.

Benedetti, F. (2006). Placebo analgesia. *Neurological Sciences*, 27(Suppl. 2), S100–S102.

Benedetti, F. (2014). Placebo effects: from the neurobiological paradigm to translational implications. *Neuron*, 84, 623–637.

Benedetti, F., Pollo, A., Lopiano, L., et al. (2003). Conscious expectation and unconscious conditioning in analgesic, motor, and hormonal placebo/nocebo responses. *Journal of Neuroscience*, 23, 4315–423.

Benedetti, F., Colloca, L., Torre, E., et al. (2004). Placebo-responsive Parkinson patients show decreased activity in single neurons of subthalamic nucleus. *Nature Neuroscience*, 7, 587–588.

Benedetti, F., Carlino, E., and Pollo, A. (2011). Hidden administration of drugs. *Clinical Pharmacology & Therapeutics*, 90, 651–661.

Bingel, U. and Placebo Competence Team (2014). Avoiding nocebo effects to optimize treatment outcome. *JAMA*, 312, 693–694.

Bingel, U., Wanigasekera, V., Wiech, K., et al. (2011). The effect of treatment expectation on drug efficacy: imaging the analgesic benefit of the opioid remifentanil. *Science Translational Medicine*, 3, 70ra14.

Bishop, F.L., Aizlewood, L., and Adams, A.E. (2014). When and why placebo-prescribing is acceptable and unacceptable: a focus group study of patients' views. *PLoS One*, 9, e101822.

Blease, C. (2012). The principle of parity: the "placebo effect" and physician communication. *Journal of Medical Ethics*, 38, 199–203.

Blease, C., Colloca, L., and Kaptchuk, T.J. (2016). Are open-label placebos ethical? informed consent and ethical equivocations. *Bioethics*, 30(6), 407–414.

Bostick, N.A., Sade, R., Levine, M.A., and Stewart, D.M., Jr. (2008). Placebo use in clinical practice: report of the American Medical Association Council on Ethical and Judicial Affairs. *Journal of Clinical Ethics*, 19, 58–61.

Brody, H. (1982). The lie that heals: the ethics of giving placebos. *Annals of Internal Medicine*, 97, 112–118.

Brody, H., Colloca, L., and Miller, F.G. (2012). The placebo phenomenon: implications for the ethics of shared decision-making. *Journal of General Internal Medicine*, 27, 739–742.

Burton, R. (1621). *The Anatomy of Melancholy*. [Reprinted in 2001 by The New York Review of Books, New York.]

Colloca, L. and Benedetti, F. (2005). Placebos and painkillers: is mind as real as matter? *Nature Reviews Neuroscience*, 6, 545–552.

Colloca, L. and Benedetti, F. (2006). How prior experience shapes placebo analgesia. *Pain*, 124, 126–133.

Colloca, L. and Benedetti, F. (2009). Placebo analgesia induced by social observational learning. *Pain*, 144, 28–34.

Colloca, L. and Miller, F.G. (2011a). Harnessing the placebo effect: the need for translational research. *Philosophical Transactions of the Royal Society B: Biological Sciences*, 366, 1922–1930.

Colloca, L. and Miller, F.G. (2011b). How placebo responses are formed: a learning perspective. *Philosophical Transactions of the Royal Society B: Biological Sciences*, 366, 1859–1869.

De La Fuente-Fernandez, R., Ruth, T.J., Sossi, V., et al. (2001). Expectation and dopamine release: mechanism of the placebo effect in Parkinson's disease. *Science*, 293, 1164–1166.

De La Fuente-Fernandez, R., Schulzer, M., and Stoessl, A.J. (2004). Placebo mechanisms and reward circuitry: clues from Parkinson's disease. *Biological Psychiatry*, **56**, 67–71.

Di Blasi, Z., Harkness, E., Ernst, E., Georgiou, A., and Kleijnen, J. (2001). Influence of context effects on health outcomes: a systematic review. *The Lancet*, **357**, 757–762.

Diederich, N.J. and Goetz, C.G. (2008). The placebo treatments in neurosciences: new insights from clinical and neuroimaging studies. *Neurology*, **71**, 677–684.

Dworkin, R.H., Katz, J., and Gitlin, M.J. (2005). Placebo response in clinical trials of depression and its implications for research on chronic neuropathic pain. *Neurology*, **65**, S7–19.

Enck, P., Benedetti, F., and Schedlowski, M. (2008). New insights into the placebo and nocebo responses. *Neuron*, **59**, 195–206.

Espay, A.J., Norris, M.M., Eliassen, J.C., et al. (2015). Placebo effect of medication cost in Parkinson disease: a randomized double-blind study. *Neurology*, **84**, 794–802.

Evers, K. (2007). Towards a philosophy for neuroethics. An informed materialist view of the brain might help to develop theoretical frameworks for applied neuroethics. *EMBO Reports*, **8**(Spec No.), S48–51.

Fassler, M., Gnadinger, M., Rosemann, T., and Biller-Andorno, N. (2011). Placebo interventions in practice: a questionnaire survey on the attitudes of patients and physicians. *British Journal of General Practice*, **61**, 101–107.

Fassler, M., Meissner, K., Schneider, A., and Linde, K. (2010). Frequency and circumstances of placebo use in clinical practice – a systematic review of empirical studies. *BMC Medicine*, **8**, 15.

Fent, R., Rosemann, T., Fassler, M., Senn, O., and Huber, C.A. (2011). The use of pure and impure placebo interventions in primary care—a qualitative approach. *BMC Family Practice*, **12**, 11.

Finniss, D.G., Kaptchuk, T.J., Miller, F., and Benedetti, F. (2010). Biological, clinical, and ethical advances of placebo effects. *The Lancet*, **375**, 686–695.

Fleurence, R., Selby, J.V., Odom-Walker, K., et al. (2013). How the Patient-Centered Outcomes Research Institute is engaging patients and others in shaping its research agenda. *Health Affairs (Millwood)*, **32**, 393–400.

Foddy, B. (2009). A duty to deceive: placebos in clinical practice. *American Journal of Bioethics*, **9**, 4–12.

Ford, P.L. (1898). *The Writings of Thomas Jefferson*. New York: Putnam.

Freedman, B. (1990). Placebo-controlled trials and the logic of clinical purpose. *IRB*, **12**, 1–6.

Frisaldi, E., Carlino, E., Lanotte, M., Lopiano, L., and Benedetti, F. (2014). Characterization of the thalamic-subthalamic circuit involved in the placebo response through single-neuron recording in Parkinson patients. *Cortex*, **60**, 3–9.

Goebel, M.U., Trebst, A. E., Steiner, J., et al. (2002). Behavioral conditioning of immunosuppression is possible in humans. *FASEB Journal*, **16**, 1869–1873.

Goebel, M.U., Hubell, D., Kou, W., et al. (2005). Behavioral conditioning with interferon beta-1a in humans. *Physiology & Behavior*, **84**, 807–814.

Goetz, C.G., Wuu, J., McDermott, M.P., et al. (2008). Placebo response in Parkinson's disease: comparisons among 11 trials covering medical and surgical interventions. *Movement Disorders*, **23**, 690–699.

Guo, J.Y., Wang, J.Y., and Luo, F. (2010). Dissection of placebo analgesia in mice: the conditions for activation of opioid and non-opioid systems. *Journal of Psychopharmacology*, **24**, 1561–1567.

Guo, J.Y., Yuan, X.Y., Sui, F., et al. (2011). Placebo analgesia affects the behavioral despair tests and hormonal secretions in mice. *Psychopharmacology (Berlin)*, **217**, 83–90.

Hall, K.T., Lembo, A.J., Kirsch, I., et al. (2012). Catechol-O-methyltransferase val158met polymorphism predicts placebo effect in irritable bowel syndrome. *PLoS One*, **7**, e48135.

Hall, K.T., Loscalzo, J., and Kaptchuk, T.J. (2015). Genetics and the placebo effect: the placebome. *Trends in Molecular Medicine*, **21**, 285–294.

Howick, J., Bishop, F.L., Heneghan, C., et al. (2013). Placebo use in the United Kingdom: results from a national survey of primary care practitioners. *PLoS One*, **8**, e58247.

Hrobjartsson, A. and Norup, M. (2003). The use of placebo interventions in medical practice – a national questionnaire survey of Danish clinicians. *Evaluation & the Health Professions*, **26**, 153–165.

Hull, S.C., Colloca, L., Avins, A., et al. (2013). Patients' attitudes about the use of placebo treatments: telephone survey. *BMJ*, **347**, f3757.

Humphrey, N. (2002). *Great Expectations: The Evolutionary Psychology of Faith Healing and Placebo Effect*. Oxford: Oxford University Press.

Kam-Hansen, S., Jakubowski, M., Kelley, J.M., et al. (2014). Altered placebo and drug labeling changes the outcome of episodic migraine attacks. *Science Translational Medicine*, **6**, 218ra5.

Kaptchuk, T.J., Kelley, J.M., Conboy, L.A., et al. (2008). Components of placebo effect: randomised controlled trial in patients with irritable bowel syndrome. *BMJ*, **336**, 999–1003.

Kaptchuk, T.J., Friedlander, E., Kelley, J.M., et al. (2010). Placebos without deception: a randomized controlled trial in irritable bowel syndrome. *PLoS One*, **5**, e15591.

Kendell, R.E. (2001). The distinction between mental and physical illness. *British Journal of Psychiatry*, **178**, 490–493.

Kermen, R., Hickner, J., Brody, H., and Hasham, I. (2010). Family physicians believe the placebo effect is therapeutic but often use real drugs as placebos. *Family Medicine*, **42**, 636–642.

Kirsch, I. (2004). Conditioning, expectancy, and the placebo effect: comment on Stewart-Williams and Podd (2004). *Psychological Bulletin*, **130**, 341–343.

Kleinman, A. (1988). *The Illness Narratives: Suffering, healing and the Human Condition*. New York: Basic Books.

Konrad, T.R., Link, C.L., Shackelton, R.J., et al. (2010). It's about time: physicians' perceptions of time constraints in primary care medical practice in three national healthcare systems. *Medical Care*, **48**, 95–100.

La Vaque, T.J. and Rossiter, T. (2001). The ethical use of placebo controls in clinical research: the Declaration of Helsinki. *Applied Psychophysiology and Biofeedback*, **26**, 23–37.

Lee, I.S., Lee, B., Park, H.J., et al. (2015). A new animal model of placebo analgesia: involvement of the dopaminergic system in reward learning. *Scientific Reports*, **5**, 17140.

Lichtenberg, P., Heresco-Levy, U., and Nitzan, U. (2004). The ethics of the placebo in clinical practice. *Journal of Medical Ethics*, **30**, 551–554.

Lidstone, S.C., Schulzer, M., Dinelle, K., et al. (2010). Effects of expectation on placebo-induced dopamine release in Parkinson disease. *Archives of General Psychiatry*, **67**, 857–865.

Lou, J.S., Dimitrova, D.M., Hammerschlag, R., et al. (2013). Effect of expectancy and personality on cortical excitability in Parkinson's disease. *Movement Disorders*, **28**, 1257–1262.

Lui, F., Colloca, L., Duzzi, D., et al. (2010). Neural bases of conditioned placebo analgesia. *Pain*, **151**, 816–824.

Luparello, T., Lyons, H.A., Bleecker, E.R., and Mcfadden, E.R., Jr. (1968). Influences of suggestion on airway reactivity in asthmatic subjects. *Psychosomatic Medicine*, **30**, 819–825.

Luparello, T.J., Leist, N., Lourie, C.H., and Sweet, P. (1970). The interaction of psychologic stimuli and pharmacologic agents on airway reactivity in asthmatic subjects. *Psychosomatic Medicine*, **32**, 509–513.

Lynoe, N., Mattsson, B., and Sandlund, M. (1993). The attitudes of patients and physicians towards placebo treatment – a comparative study. *Social Science & Medicine*, **36**, 767–774.

Mayberg, H.S., Silva, J.A., Brannan, S.K., et al. (2002). The functional neuroanatomy of the placebo effect. *American Journal of Psychiatry*, 159, 728–737.

Meissner, K., Hofner, L., Fassler, M., and Linde, K. (2012). Widespread use of pure and impure placebo interventions by GPs in Germany. *Family Practice*, 29, 79–85.

Mercado, R., Constantoyannis, C., Mandat, T., et al. (2006). Expectation and the placebo effect in Parkinson's disease patients with subthalamic nucleus deep brain stimulation. *Movement Disorders*, 21, 1457–1461.

Miller, F.G. and Brody, H. (2011). Understanding and harnessing placebo effects: clearing away the underbrush. *Journal of Medicine and Philosophy*, 36, 69–78.

Miller, F.G. and Kaptchuk, T.J. (2008). The power of context: reconceptualizing the placebo effect. *Journal of the Royal Society of Medicine*, 101, 222–225.

Miller, F.G., Colloca, L., and Kaptchuk, T.J. (2009). The placebo effect: illness and interpersonal healing. *Perspectives in Biology and Medicine*, 52, 518–539.

Miresco, M.J. and Kirmayer, L.J. (2006). The persistence of mind-brain dualism in psychiatric reasoning about clinical scenarios. *American Journal of Psychiatry*, 163, 913–918.

Moerman, D.E. (2002). The meaning response and the ethics of avoiding placebos. *Evaluation & the Health Professions*, 25, 399–409.

Moerman, D.E. and Jonas, W.B. (2002). Deconstructing the placebo effect and finding the meaning response. *Annals of Internal Medicine*, 136, 471–476.

Montgomery, G.H. and Kirsch, I. (1997). Classical conditioning and the placebo effect. *Pain*, 72, 107–113.

Nolan, T.A., Price, D.D., Caudle, R.M., Murphy, N.P., and Neubert, J.K. (2012). Placebo-induced analgesia in an operant pain model in rats. *Pain*, 153, 2009–2016.

O'Neill, O. (1984). Paternalism and partial autonomy. *Journal of Medical Ethics*, 10, 173–178.

Park, L.C. and Covi, L. (1965). Nonblind placebo trial: an exploration of neurotic patients' responses to placebo when its inert content is disclosed. *Archives of General Psychiatry*, 12, 36–45.

Pavlov, I.P. (1927). *Conditioned Reflexes: An Investigation of the Physiological Activity of the Cerebral Cortex*. London: Oxford University Press.

Petrovic, P., Dietrich, T., Fransson, P., et al. (2005). Placebo in emotional processing – induced expectations of anxiety relief activate a generalized modulatory network. *Neuron*, 46, 957–969.

Pollo, A., Amanzio, M., Arslanian, A., et al. (2001). Response expectancies in placebo analgesia and their clinical relevance. *Pain*, 93, 77–84.

Pugh, J. (2015). Ravines and sugar pills: defending deceptive placebo use. *Journal of Medicine and Philosophy*, 40, 83–101.

Pugh, J., Kahane, G., Maslen, H., and Savulescu, J. (2016). Lay attitudes toward deception in medicine: theoretical considerations and empirical evidence. *AJOB Empirical Bioethics*, 7, 31–38.

Purtilo, R. (1993). *Ethical Dimensions in the Health Care Professions*. Philadelphia, PA: WB Saunders.

Racine, E., Waldman, S., Rosenberg, J., and Illes, J. (2010). Contemporary neuroscience in the media. *Social Science & Medicine*, 71, 725–733.

Rommelfanger, K.S. (2013a). Attitudes on mind over matter: physician views on the role of placebo in psychogenic disorders. *American Journal of Bioethics Neuroscience*, 4, 9–15.

Rommelfanger, K.S. (2013b). Opinion: a role for placebo therapy in psychogenic movement disorders. *Nature Reviews Neurology*, 9, 351–356.

Rose, J.P., Geers, A.L., Rasinski, H.M., and Fowler, S.L. (2012). Choice and placebo expectation effects in the context of pain analgesia. *Journal of Behavioral Medicine*, 35, 462–470.

Rosenberg, C.E. (2002). The tyranny of diagnosis: specific entities and individual experience. *Milbank Quarterly*, 80, 237–260.

Roskies, A.L. (2007). Neuroethics beyond genethics. Despite the overlap between the ethics of neuroscience and genetics, there are important areas where the two diverge. *EMBO Reports*, **8**(Spec No.), S52–56.

Sandler, A.D., Glesne, C.E., and Bodfish, J.W. (2010). Conditioned placebo dose reduction: a new treatment in attention-deficit hyperactivity disorder? *Journal of Developmental and Behavioral Pediatrics*, **31**, 369–375.

Schafer, S.M., Colloca, L., and Wager, T.D. (2015). Conditioned placebo analgesia persists when subjects know they are receiving a placebo. *Journal of Pain*, **16**, 412–420.

Schultz, W. (2006). Behavioral theories and the neurophysiology of reward. *Annual Review of Psychology*, **57**, 87–115.

Scott, D.J., Stohler, C.S., Egnatuk, C.M., et al. (2008). Placebo and nocebo effects are defined by opposite opioid and dopaminergic responses. *Archives of General Psychiatry*, **65**, 220–231.

Shanks, D.R. (2010). Learning: from association to cognition. *Annual Review of Psychology*, **61**, 273–301.

Sherman, R. and Hickner, J. (2008). Academic physicians use placebos in clinical practice and believe in the mind-body connection. *Journal of General Internal Medicine*, **23**, 7–10.

Spiro, H.M. (1986). *Doctors, Patients, and Placebos*. London: Yale University Press.

Stewart-Williams, S. and Podd, J. (2004). The placebo effect: dissolving the expectancy versus conditioning debate. *Psychological Bulletin*, **130**, 324–340.

Tilburt, J.C., Emanuel, E.J., Kaptchuk, T.J., Curlin, F.A., and Miller, F.G. (2008). Prescribing "placebo treatments": results of national survey of US internists and rheumatologists. *BMJ*, **337**, a1938.

Tuttle, A.H., Tohyama, S., Ramsay, T., et al. (2015). Increasing placebo responses over time in U.S. clinical trials of neuropathic pain. *Pain*, **156**, 2616–2626.

Van Laarhoven, A.I., Vogelaar, M.L., Wilder-Smith, O.H., et al. (2011). Induction of nocebo and placebo effects on itch and pain by verbal suggestions. *Pain*, **152**, 1486–1494.

Wager, T.D., Scott, D.J., and Zubieta, J.K. (2007). Placebo effects on human mu-opioid activity during pain. *Proceedings of the National Academy of Sciences of the United States of America*, **104**, 11056–11061.

Yang, Q. and Miller, G. (2015). East-West differences in perception of brain death. Review of history, current understandings, and directions for future research. *Journal of Bioethical Inquiry*, **12**, 211–225.

Zhang, R.R., Zhang, W.C., Wang, J.Y., and Guo, J.Y. (2013). The opioid placebo analgesia is mediated exclusively through mu-opioid receptor in rat. *International Journal of Neuropsychopharmacology*, **16**, 849–856.

Zubieta, J.K., Bueller, J.A., Jackson, L.R., et al. (2005). Placebo effects mediated by endogenous opioid activity on mu-opioid receptors. *Journal of Neuroscience*, **25**, 7754–7762.

Chapter 13

Ethical challenges of modern psychiatric neurosurgery

Sabine Müller

Introduction

Until the 1970s, rather crude forms of psychosurgery had been used in hundreds of thousands of mentally ill patients (Valenstein 1986; Diering & Bell 1991; Chodakiewitz et al. 2015). The abuse of psychosurgery was brought to the public consciousness by Ken Kesey's famous novel *One Flew Over the Cuckoo's Nest* (Kesey 1964). Due to public criticism and, coincidentally, the development of antipsychotic drugs, psychosurgery was nearly completely abandoned and forbidden in many countries. Since 1999, after a nearly 30-year hiatus, there is a renaissance of psychiatric neurosurgery in a much more refined and safer form. Today it is developing quickly as an experimental therapy for medically refractory psychiatric disorders (Lévêque et al. 2013; Luigjes et al. 2013; Lévêque 2014; Sun & De Salles 2015) that are not caused directly by identified brain anatomical or functional pathologies, such as brain tumors or epileptogenic tissue, and for which biological underpinnings are unknown. Nonetheless, psychiatric neurosurgery is based on the assumption that certain dysfunctional brain areas or structures play a crucial role in psychiatric disorders, and that lesioning or deactivating them can alleviate psychiatric symptoms. For example, the cortico-striato-thalamo-cortical loop is strongly implicated in the pathogenesis of obsessive–compulsive disorder (OCD). Decreased frontal-striatal control of limbic structures such as the amygdala might account for the inadequate fear response seen in OCD patients with fear of contamination (Na et al. 2015). Thus, knowledge of the interconnected neural circuits underlies hypothesis-driven rationales for choosing particular targets of neurosurgical intervention (Na et al. 2015).

Many researchers and clinicians expect that modern psychiatric neurosurgery has the potential to become a safe and effective treatment option for severe, therapy-refractory psychiatric disorders. According to a recent survey, about 90% of functional neurosurgeons feel optimistic about the future of psychiatric neurosurgery (Lipsman et al. 2011; Mendelsohn et al. 2013).

Currently, such psychiatric neurosurgical procedures include deep brain stimulation (DBS) and ablative neurosurgical procedures. Because DBS is an established therapy for treatment-refractory Parkinson's disease, essential tremor and dystonia, and because it is considered reversible, its experimental use for treating psychiatric disorders was seen as justifiable by most commentators. Indeed, high-frequency DBS creates temporary lesions by deactivating the targeted brain areas through chronic electrical current. Thus, areas that are assumed to be hyperactive in people with psychiatric disorders can be inhibited as long as the stimulation is activated. Since its effect can be adjusted through fine-tuning the stimulation parameters, DBS is considered also to be a neuromodulatory technique. In the wake of psychiatric DBS, ablative neurosurgical procedures are seeing a revival. The ablative neurosurgical procedures currently in use comprise thermocoagulation or radiofrequency ablation procedures, and Gamma Knife® radiosurgery. The latter is a non-invasive and very precise method for creating confined brain lesions. Moreover, a new technique has recently been introduced into the field: magnetic resonance-guided focused ultrasound (MRgFUS), which has been tested in the first four psychiatric patients (Na et al. 2015). Most recent developments aim at controlling the brain situation-specifically. Such developments comprise closed-loop DBS (Tass 2003; Rosin et al. 2011), DBS combined with optogenetics (Deisseroth 2012; Walter and Müller S. 2013), as well as implantable devices that measure continuously the brain activity and initiate counter-measures (e.g., warning signals or painful stimuli) as soon as suspicious arousal is detected (Cook et al. 2013; Gilbert 2015). Since modern psychiatric neurosurgery procedures are much safer and more effective than their historical predecessors, it would not be justified to condemn them generally. Rather, a careful re-evaluation of these techniques is necessary. Characterizing modern ablative procedures as successors of historical psychosurgery while considering DBS as something quite different is unwarranted. In reality, both psychiatric DBS as well as modern ablative psychiatric neurosurgery are significantly improved successors of the historical psychosurgery.

From a medical perspective, each of these types of procedures has different profiles of advantages and disadvantages such that none can be considered absolutely superior over another. Specifically, DBS advantages include adjustability and a high degree of reversibility, whereas microsurgical ablative procedures have a rapid onset of action, and radiosurgery and MRgFUS are noninvasive and have low rates of adverse effects.

The use of all methods of psychiatric neurosurgery raises difficult ethical and legal issues. A fundamental issue is whether it is justified to intervene directly in the brain of mentally ill patients with the risk or even with the aim to change personality and behavior. Even if this question is answered in the affirmative, many ethical issues remain:

- What is the risk:benefit ratio of the different methods?
- Who shall decide how the patient's mood and personality are to be changed and according to which criteria?
- Can the autonomy of patients be respected, in particular if their capacity for autonomous decision-making is affected by their psychiatric disorder or as a consequence of the intervention?
- For which kind of disorders is psychiatric neurosurgery justified?
- Can these procedures be justified in cases of self-inflicted disorders such as drug addiction or anorexia nervosa?

- How should researchers and physicians deal with conflicts of interest, for example, conflicts between research interests and benefits to patients?
- Should psychiatric neurosurgery be applied in forensic contexts, for example, to reduce the risk of re-offense of pedophilic child offenders or violent psychopaths?

Whereas medical ethicists have discussed such questions intensively for DBS, they have neglected ablative psychiatric neurosurgery. This blind spot is astonishing. Rather than ignoring the widespread practice of ablative psychiatric neurosurgery, a comprehensive and differentiated ethical analysis of the pros and cons of the distinct approaches is necessary, which is based on rational evidence, not on fear or hopes or outdated prejudices.

In response, in the rest of this chapter, I provide a continuing overview of the different techniques of psychiatric neurosurgery and explain their various underlying paradigms and follow with an overview and comparison of the efficacy and the adverse effects of the different techniques. Next, I summarize the ethical debate about DBS, and discuss the neglect in ethics with regard to ablative psychiatric neurosurgery, and conclude with some preliminary recommendations for the future.

Overview of the different techniques of psychiatric neurosurgery

Modern psychiatric neurosurgery differs with regards to the paradigm used. Deep brain stimulation is based on the adjustability paradigm, because it allows for optimizing the treatment outcome individually. Microsurgical ablative procedures are based on a quick-fix paradigm aimed at a very rapid onset of action. In comparison, Gamma Knife® radiosurgery and MRgFUS are grounded in the paradigm of minimal invasiveness as they do not require a craniotomy, and their side effects are very limited (Müller S. et al. 2015a).

Deep brain stimulation

Deep brain stimulation has been investigated in several clinical studies for treating various severe psychiatric disorders, including major depression (Morishita et al. 2014), OCD (Kohl et al. 2014), anorexia nervosa (Müller S. et al. 2015b), schizophrenia (Corripio et al. 2016; Salgado-López et al. 2016), aggressive disorder (Franzini et al. 2013; Torres et al. 2013), drug addiction (Müller U.J. et al. 2013), Tourette syndrome (Andrade & Visser-Vandewalle 2016), dementia (Mirzadeh et al. 2016), and severe obesity (Dupré et al. 2015). The procedure requires drilling small burr holes in the skull to insert usually two electrodes deep into the brain. These electrodes are connected with a stimulator that continuously delivers electrical current to the targeted brain area in order to activate or deactivate these areas. The result is also known as a reversible lesion. Indeed, high-frequency DBS has a similar effect to lesions in that the activity of brain areas that are believed to be hyperactive in psychiatric disorders is inhibited. However, the precise mechanism of action is still the subject of ongoing debate. Several hypotheses are discussed to explain the blocking effect of stimulation (Lévèque 2014).

When subcortical motor areas of the basal ganglia and the thalamus are targeted for treating movement disorders, the chief benefit of DBS over ablative procedures is that DBS has a lower risk of side effects on speech, swallowing, cognition, and balance (Pepper et al. 2015). However, the rationale for DBS is less compelling when non-motor areas are targeted such as for treating psychiatric disorders (Pepper et al. 2015). There is not much more in favor of DBS "other than it is perceived as nonablative and assumed to be reversible and more forgiving than lesions, and therefore more acceptable" (Pepper et al. 2015, p.1029). However, many experts doubt that DBS is always reversible and, thus, the case for DBS is attenuated. In an expert survey conducted by Markus Christen and me, results showed that about 40% of the participating DBS experts (strongly) disagree with the statement that DBS is a completely reversible procedure, whereas 43% (strongly) agree (Christen et al. 2014). The reversibility is questionable because, first, the insertion of the electrodes might cause irreversible lesions, even with the consequence of death or of permanent neurological damage; and second, because the brain will adapt to the stimulation.

Therefore, arguably, the main advantage of DBS is not its reversibility after all, but its adjustability, that is, that the stimulation parameters like voltage, pulse width, and frequency can be fine-tuned individually to optimize the effect on the target symptoms with respect to the disease progression over time. Consequently, DBS requires extensive follow-up for programming and adjusting stimulation parameters, which can only be done by specialists. Furthermore, the implantable pulse generator (including the battery) has to be replaced every 3–5 years requiring regular surgeries with full anesthesia. The cost of DBS device implantation is estimated at $50,000–$120,000 USD. Each battery replacement costs between $10,000 and $25,000 USD (Chodakiewitz et al. 2015). Therefore, the treatment is both time-consuming and, also in the long-run, very expensive.

Ablative microsurgical procedures

Ablative microsurgical procedures require a craniotomy. The tiny ablations are performed after neuroimaging (computer tomography, magnetic resonance imaging (MRI), and diffusion tensor imaging) and can be performed by either thermocoagulation or radiofrequency heating.

The purpose of ablative microsurgical procedures is to enhance brain function and reduce psychiatric symptoms by disconnecting limbic system circuits related to different psychiatric disorders (Martínez-Álvarez 2015). The effects occur rapidly, often within days. If the effect is not sufficient, a second intervention can be performed in order to enlarge the lesion or create an additional lesion in another target. However, if the lesions are too large or misplaced, it is not possible to correct them.

Although today the field has advanced substantially, these procedures have a long tradition. In 1949, Talairach introduced anterior capsulotomy in France. In parallel, anterior cingulotomy was introduced in the United States (Pepper et al. 2015). The main targets for treating OCD are the anterior cingulate gyrus (cingulotomy) and the anterior limb of the internal capsule (anterior capsulotomy) (Na et al. 2015). Further indications for contemporary ablative microsurgical procedures comprise anxiety disorder, major depression, anorexia nervosa, drug addiction, hyperaggressivity, and schizophrenia.

Gamma Knife® radiosurgery

In 1968, Leksell and Larsson developed the first Gamma Knife˙ unit in Sweden (De Salles & Gorgulho 2015). Gamma Knife˙ radiosurgery is a noninvasive procedure which does not actually involve surgery with a knife; instead, external photon beam radiation is used to concentrate a radiation dose to a pathological area while peripheral structures are spared. The Gamma Knife˙ focuses about 200 beams of gamma radiation emitted by cobalt-60 sources in the target to be treated. The radiation dose of each single beam is low, but focusing beams with high precision for several hours results in doses of 120 Gy or more and produces a lesion at the target, while the adjacent radiation exposure is low (Martínez-Álvarez 2015). Radiosurgery can also be performed with a CyberKnife˙ or a linear accelerator (e.g., LINAC˙), but only the Gamma Knife˙ and the CyberKnife˙ have been used so far for treating psychiatric disorders.

The main indications of Gamma Knife˙ radiosurgery are tumors and arteriovenous malformations of the brain. However, the Gamma Knife˙ is also used for treating functional disorders of the brain, particularly Parkinson's disease, essential tremor, trigeminal neuralgia, intractable tumor pain, some forms of epilepsy, and psychiatric illness (Friehs et al. 2007). In 1988, the Karolinska Institute in Stockholm started to treat OCD patients with bilateral radiosurgical anterior capsulotomy (Friehs et al. 2007). Today, capsulotomies and cingulotomies are performed with the Gamma Knife˙ in specialized centers in the United States, Spain, Brazil, and Asia.

The main advantage of radiosurgery compared to DBS or ablative microsurgery is its minimal invasiveness. Gamma Knife˙ radiosurgery does not require a craniotomy. Normally, it is conducted as an ambulant treatment. Hair is not shaved. Patients stay fully awake during the whole treatment and can listen to music during the procedure. It is suitable even for patients with advanced age, with medical conditions that preclude surgery, or who receive anticoagulation therapy (Friehs et al. 2007).

Alongside these benefits, limitations are known: neurophysiological confirmation of the target area is not possible and targeting relies completely on neuroimaging; lesion sizes may vary; and, it might be difficult or impossible to shield adjacent structures against the radiation (Friehs et al. 2007).

Radiosurgery is generally categorized as an ablative treatment. MRI studies have verified formation of small lesions after radiosurgical treatments, with necrosis volume of 100 mm^3 or less when doses of 120–180 Gy were used (Friehs et al. 2007). Recent neurophysiological, radiological, and histological research, however, calls into question whether all Gamma Knife˙ radiosurgery procedures should indeed be considered ablative (Régis 2013). Whether it is ablative or not appears to depend on the dose used. Radiosurgical protocols for functional disorders are assumed to have differential effects on various neuronal populations and the glial environment, some of which may play a modulatory role on neuronal function while preserving basic processing. Thus, depending on the context, the paradigm used in modern functional radiosurgery might be more appropriately categorized as neuromodulatory rather than ablative (Régis 2013).

Magnetic resonance-guided focused ultrasound

Magnetic resonance-guided focused ultrasound (MRgFUS) is a novel thermal ablation method that is performed without craniotomy. It is a non-ionized MR-guided procedure with real-time MRI and intraoperative feedback on the temperature at the target (Na et al. 2015). The patients remain fully awake during the treatment. The first tests of MRgFUS for a psychiatric indication have been conducted in South Korea (Na et al. 2015).

A beam of ultrasound is used in MRgFUS that is concentrated with its focal point on the target area. The MRgFUS system combines a focused ultrasound delivery system with a 1.5 or 3 Tesla MRI scanner. This system provides a real-time therapy planning algorithm, thermal dosimetry, and closed-loop therapy control. During the treatment, a specific MR scan, which can be processed to identify changes in tissue temperature, provides a thermal map of the treatment volume. With this map, the treatment in progress is monitored in order to confirm that the ablation is proceeding according to plan (Na et al. 2015).

The method promises several advantages: like Gamma Knife[*] radiosurgery, MRgFUS is noninvasive and does not require scalp incision, holes, and electrodes penetrating the brain, thus reducing the risk of hemorrhagic complications and of infectious complications (Na et al. 2015). Unlike Gamma Knife[*] radiosurgery, MRgFUS does not use ionizing radiation, which precludes the risk of radiation-induced tumorigenesis. However, MRgFUS is a novel procedure, with very limited data available regarding its efficacy and side effects.

Efficacy

Reliable estimation and comparison of the efficacy of the different psychiatric neurosurgery procedures remains elusive due to a number of reasons such as publication bias, methodological heterogeneity, and use of weak methodology of the studies (Müller S. et al. 2015a).

Publication bias

Due to a strong publication bias in the DBS literature (Schläpfer & Fins 2010), the efficacy of psychiatric neurosurgery procedures is certainly overestimated. Most reviews consider only cases published in Anglophone medical journals, and cases with an unfavorable outcome are often not reported (Schläpfer & Fins 2010). The presumed efficacy of psychiatric neurosurgery procedures is thus very likely an overestimation. The publication bias might be the explanation for another astonishing phenomenon. A recent review, comprising 22 papers with data from 188 patients and 6 stimulation targets reported responder rates ranging from 29% to 92%, thus suggesting that DBS is efficacious in ameliorating treatment-refractory major depression (Morishita et al. 2014). In contrast, however, these positive results could not be verified in the first two studies that fulfilled scientific quality criteria such as randomization, double blinding, and placebo control. Indeed, two randomized, controlled, prospective multicenter studies were discontinued because of inefficacy. These studies had two different targets: (1) the ventral capsule/ventral striatum (Dougherty et al. 2015) and (2) the subgenual cingulate cortex. Only a short report exists about DBS in the subgenual cingulate cortex (Cavuoto 2013).

As concerns ablative procedures, there is probably an even stronger publication bias. Most data on ablative neurosurgery are not published at all, since the procedure is predominantly applied in clinical practice and not in research. Public websites of private clinics in Asia reveal that ablative neurosurgery for psychiatric disorders is offered there as a part of clinical routine rather than clinical trials. Recently, Sun and De Salles published a book with more than 100 previously unpublished ablative psychiatric neurosurgery case reports from around the world (Sun & De Salles 2015). However, most data in this book do not contain sufficient detail for consideration in systematic reviews.

Publication bias in psychiatric neurosurgery literature is a fundamental problem that compromises the systematic evaluation and comparison of the different procedures (Müller S. et al. 2015a). It also undermines the ethical evaluation, which critically depends on adequate and objective information of evidence-based risk:benefit ratios.

Methodological and study heterogeneity

Deep brain stimulation targets for major depression include a variety of brain regions, including the VC/VS, the nucleus accumbens (NAcc), the Brodmann area 25 (Cg25) in the subgenual cingulate cortex (SCC), the superolateral medial forebrain bundle (slMFB), the lateral habenula, the inferior thalamic peduncle (ITP), and the anterior limb of the internal capsule (ALIC) (Morishita et al. 2014; Bergfeld et al. 2016). Similarly, brain regions targeted for OCD include the ALIC, the NAcc, VC/VS, ITP, and the nucleus subthalamicus (STN) (Kohl et al. 2014). Several of these regions are also targets for treating other psychiatric disorders such as drug addiction or anorexia nervosa. Thus, whereas many targets are tested for the same disorder, the same targets are tested for different disorders. Additional methodological and study heterogeneity results from continuous development and refinement of psychiatric neurosurgical methods. In radiosurgery, for example, radiation doses have been reduced significantly over the years. Diversity in targeted brain regions for various psychological disorders as well as rapid methodological progress both contribute to pragmatic challenges in comparing efficacy of different psychiatric neurosurgery approaches.

Methodological weaknesses

Studies published thus far in English-language journals have very small patient numbers, most with fewer than ten patients, and so lack of statistical power is thus a major concern in these studies. Furthermore, most studies are neither placebo-controlled nor double-blinded, and notably, none of the psychiatric neurosurgery approaches is strictly evidence-based (Pepper et al. 2015). Observer bias in reporting results also presents a methodological concern as the evaluation of treatment outcomes have not yet been conducted by independent parties who were not involved in patient selection, surgery, or follow-up (Pepper et al. 2015).

Despite the positive results of DBS studies in open-label trials, the above-mentioned failure of two high-quality studies may indicate the typical overestimation of efficacy that is associated with open-label trials with missing placebo control, and biases due to lack of blinding and randomization (Morishita et al. 2014). The failure of these studies, however, does not prove that DBS is in general ineffective for treatment-refractory major depression. The study of Dougherty and colleagues (Dougherty et al. 2015), for example, had severe deficits such as a short treatment period (4 months) and unsatisfactory adjustment of stimulation parameters (Schläpfer 2015). A current study is investigating the slMFB (Schläpfer et al. 2013) as a DBS target for major depression in a randomized, controlled, prospective study; the outcome of this study may shed light on the efficacy of DBS for major depression. For Gamma Knife' radiosurgery, several very small double-blind, randomized controlled trials have been performed thus far (Lévèque et al. 2013).

Rigorous evidence-based comparison of the efficacy of the different psychiatric neurosurgery approaches is not yet possible due to methodological hurdles and publication bias. Despite these limitations, however, attempts to review efficacy of procedures have utilized pertinent data, reviews, and a book (Sun & De Salles 2015).

Efficacy of the different techniques

Deep brain stimulation

Two reviews have been recently published on the efficacy of DBS for OCD. Kohl and colleagues analyzed 25 papers comprising 109 DBS patients and 5 targets (NAcc, VC/VS, ITP, STN, and ALIC) and found response rates ranging from 45.5% to 100% (Kohl et al. 2014). Pepper and colleagues compared DBS and ablative neurosurgery that included more or less homogeneous anatomical areas to ensure a fair comparison (Pepper et al. 2015). Thus, they included DBS studies with the targets of VC/VS and NAcc, and only anterior capsulotomy (but not cingulotomy, subcaudate tractotomy, or limbic leucotomy). Their analysis included 10 studies with a total of 108 patients, who were treated with anterior capsulotomy, and 10 studies with a total of 62 patients, who underwent DBS. The response rate of DBS patients was 52%, and of the anterior cingulotomy patients 62% (response = improvement of Yale–Brown Obsessive Compulsive Scale (Y-BOCS) score ≥ 35%) (Pepper et al. 2015).

For treatment-refractory major depression, Morishita and colleagues reviewed data from 22 papers comprising 188 DBS patients and 6 targets (NAcc, VC/VS, SCC, lateral habenula, ITP, and slMFB) (Morishita et al. 2014). Very recently, an additional target, namely the ALIC, has been tested in 25 patients (Bergfeld et al. 2016). The reported response rates ranged from 29% to 92%. However, as mentioned earlier, the failure of two multicenter, randomized, controlled, prospective studies evaluating the efficacy of VC/VS DBS and SCC DBS (Cavuoto 2013; Dougherty et al. 2015) raises questions about the efficacy of DBS for depression.

For anorexia nervosa, we have reviewed 6 papers comprising 18 patients and three targets (NAcc, SCC, and VC/VS) (Müller S. et al. 2015b). Remission in terms of normalized body mass index occurred in 61% of patients, and psychiatric comorbidities improved in 88.9% of the patients as well. However, Sun and colleagues have recently published less favorable results in which only 20% (3/15) of their patients treated with NAcc DBS showed improvements in symptoms. The other 80% underwent anterior capsulotomy, which improved eating behavior and psychiatric symptoms in all patients (Sun et al. 2015).

Microsurgical ablative procedures

A review by Greenberg and colleagues reported that 45–65% of patients with intractable OCD benefitted from ablative procedures (Greenberg et al. 2003). According to Martínez-Álvarez (2015), response rates between 36% and 89% have been published for ablative neurosurgery for OCD. Martínez-Álvarez (2015) reports data of 100 of their own OCD patients of whom 71% responded. According to the review of Pepper et al. (2015), 62% of the patients treated with anterior capsulotomy (micro- or radiosurgery) showed a clinically significant improvement (Y-BOCS score ≥ 35%).

For treatment-refractory depression, 40–60% of patients responded to bilateral capsulotomy or cingulotomy (Eljamel 2015). Response was defined as at least 50% improvement on depression scoring system such as Hamilton Rating Scale for Depression, or the Montgomery-Asberg Depression Rating Scale.

For anorexia nervosa, three papers with a total of nine patients reported a remission rate of 100% with regard to both weight normalization and psychiatric comorbidities. Different targets were used (dorsomedial thalamus, anterior capsule, and NAcc) (Müller S. et al. 2015b). Sun and colleagues report 150 patients treated with capsulotomy, of which 85% experienced an improvement in symptoms (Sun et al. 2015).

Gamma Knife® radiosurgery

Martínez-Álvarez (2015) reported a response rate of 100% in five OCD patients treated with Gamma Knife® radiosurgery. Friehs and his coauthors reported long-term and significant reduction of OCD symptoms in roughly two-thirds of their patients who were treated in a currently ongoing open-label study (Friehs et al. 2015). A review by Lévèque et al. (2013) found that at least 55% of OCD patients responded to Gamma Knife® capsulotomy. The review by Pepper and colleagues found that 62% of patients responded (Y-BOCS score ≥ 35%) when treated with anterior capsulotomy using either microsurgery or radiosurgery for OCD (Pepper et al. 2015).

Magnetic resonance-guided focused ultrasound

Presently, data from only four OCD patients who have been treated with MRgFUS is available (Na et al. 2015), and in all patients, the Y-BOCS score improved. The mean improvement in Y-BOCS was 33%, whereby the range was 24–47%. Two of the four patients achieved a full response (Y-BOCS ≥ 35%) during the 6 months of follow-up (Na et al. 2015). The Y-BOCS of all patients was still falling at the end of the study. Perhaps, it takes more than 6 months until the full treatment effect can be seen (Na et al. 2015).

Another interesting result is the strong reduction of anxiety and depression after MRgFUS. The mean Hamilton Rating Scale for Depression score improved about 61% at 6 months after treatment, and the mean Hamilton Rating Scale for Anxiety score about 69%. Anxiety and depression improved significantly within 1 week after MRgFUS, and this improvement was maintained throughout the 6 months (Na et al. 2015).

Comparative efficacy

Pepper and colleagues found greater response rates for OCD patients who underwent anterior capsulotomy (micro- or radiosurgery) than for patients treated with DBS (62% versus 52% (response = improvement on Y-BOCS score ≥ 35%) (Pepper et al. 2015). Additionally, patients who underwent anterior capsulotomy were more likely to go into remission than DBS patients. This difference in efficacy cannot be explained by group differences in patient age, symptom severity, or disease duration (Pepper et al. 2015). A possible explanation is the differential surgical experience. Surgeons have greater experience with ablation and anterior capsulotomy is more standardized, whereas there is still a learning curve in the practice of DBS for OCD (Pepper et al. 2015).

Adverse effects

Deep brain stimulation

Adverse effects of DBS are usually differentiated into surgery-related, device-related, and stimulation-related effects. Because the implantable pulse generator has to be replaced every 3–5 years, risks of surgery and infection are not limited to the implantation of the DBS system. Furthermore, infections near the implanted device can always occur. Furthermore, DBS devices are sensitive to high-energy electrical fields, which can switch them off or even cause a reset of the device (Na et al. 2015). The adverse effects depend less on the indication for the intervention; therefore, I recapitulate them in the following collectively.

Serious adverse events during or shortly after surgery occurred in a few reported cases. These include intracerebral hemorrhages, which in one case, resulted in a temporary hemiparesis (Kohl et al. 2014; Morishita et al. 2014; Pepper et al. 2015); intraoperative seizure; intraoperative panic attack; and cardiac air embolus (Lipsman et al. 2013a). In several cases, wound infections or inflammation occurred (Kohl et al. 2014; Pepper et al. 2015).

Several device-related adverse effects have been reported, namely breaks of electrodes, stimulating leads or extension wires requiring replacement (Kohl et al. 2014; Pepper et al. 2015). Further device-related side effects are dysesthesia in the subclavicular region, feelings of the leads or stimulators (Kohl et al. 2014), or allergy to the pulse generator (Pepper et al. 2015).

Some patients suffered temporarily from vertigo, olfactory hallucinations, insomnia, headache, micturition problems, weight loss or gain, long-lasting fatigue, and visual disturbances (Kohl et al. 2014; Pepper et al. 2015). Patients suffering from anorexia nervosa had a particularly high rate of severe complications, namely an epileptic seizure during electrode programming, further weight loss, pancreatitis, hypophosphataemia, hypo-kalaemia, a refeeding delirium, cardiological disturbances, and worsening of mood (Lipsman et al. 2013a).

Several patients had cognitive problems following DBS, particularly forgetfulness and word-finding difficulties (Pepper et al. 2015). Many patients suffered from stimulation-induced adverse effects, particularly from depression, anxiety, worsening of OCD, sui-cidality, panic attacks, fatigue, hypomania, increased libido, and problems at home. In some cases, these adverse effects were caused either by a change of stimulation parameters or by battery depletion, and were reversible by respective adjustments (Kohl et al. 2014; Morishita et al. 2014). Some DBS patients reported feelings of self-estrangement (Gilbert 2013a). A great problem after DBS is the high number of suicides and suicide attempts (Kohl et al. 2014; Morishita et al. 2014; Pepper et al. 2015).

Microsurgical ablative procedures

Ablative procedures bear the risk of surgical complications such as intracerebral hemor-rhage, epilepsy, and hydrocephalus (Na et al. 2015). Adverse side effects of microsurgi-cal ablative surgery for major depression comprised epilepsy (up to 10%), weight gain, transient confusion, transient mania, and transient incontinence. Further side effects reported by only one or two studies are personality change, lethargy, hemiplegia, and suicide (Eljamel 2015). Following microsurgical ablative surgery for treating OCD, the following transient adverse effects have been observed: mood disturbances, somnolence, lethargy, frontal syndrome, headaches, hallucinations, urinary incontinence, pneumonia, impaired cognitive function, confusion, and epileptic seizures (Martínez-Álvarez 2015; Pepper et al. 2015). The following adverse long-term effects have been reported: weight gain, incontinence, long-term hemiplegia, cognitive deficits, sexual disinhibition, apathy, memory problems, seizures, and personality or behavior change (Pepper et al. 2015). In the case of anorexia nervosa, the published journal papers reported only transient adverse effects: bradycardia, mild disorientation, moderate somnolence, loss of concentration, apathy, emotional emptiness and mild loss of decorum, headaches, and centric fever (Müller S. et al. 2015b). However, Sun and colleagues reported intracranial hematomas in 1.9% of their patients (4/216); one patient died thereof (0.5%) (Sun et al. 2015).

Gamma Knife® radiosurgery

According to Lévèque (2014), adverse effects such as fatigue, weight gain, or apathy did not occur in newer studies with radiation doses of not more than 180 Gy. The dose effect is also supported by the review of Pepper and colleagues (Pepper et al. 2015). In stud-ies in which radiation doses of maximum 160 Gy were applied, no relevant side effects occurred, whereas in studies with 180–200 Gy, a few patients suffered from adverse long-term effects such as worsening scores on tests of executive function (without negative impact on life), weight gain, tinnitus, facial paresthesia, sexual disinhibition (Rück et al. 2008), or urinary incontinence.

Magnetic resonance-guided focused ultrasound

There is a potential risk that the scalp, the dura, the arachnoid, and brain tissues adjacent to the target can become heated to the point where tissue damage or a burn might occur. Furthermore, there is a risk that the blood–brain barrier is disrupted, and that edema, swelling, and hemorrhage outside and remote from the targeted area occur (Na et al. 2015).

There is a long list of contraindications of MRgFUS: standard contraindications for MRI , dialysis treatment, unstable cardiac status, severe hypertension, a history of abnormal bleeding or coagulopathy, risk factors for intra- or postoperative bleeding, a cerebrovascular disease, a history of intracranial bleeding, increased intracranial pressure, neurodegenerative diseases, brain tumors, history of aneurysms, an acute or chronic uncontrolled infection or known life-threatening systemic disease, history of immunocompromise (including HIV-positive status), seizures within the past year, remarkable atrophy and poor healing capacity of the scalp, ethanol or drug abuse, pregnancy or lactation, prior treatment with DBS, or stereotactic ablation of the anterior cingulated gyrus (Na et al. 2015).

A frequent adverse effect is vertigo that is induced by high-field (≥2 Tesla) MRI scanners. Symptoms include nausea, vomiting, and dizziness. About half of the patients (5 of 11 patients) who have been treated with MRgFUS for essential tremor suffered from MR-induced vertigo (Na et al. 2015).

In the four patients treated so far with MRgFUS for OCD, no significant and/or permanent complications including physical, neurological, and psychological changes occurred (Na et al. 2015).

Comparison of the different techniques of psychiatric neurosurgery

Table 13.1 summarizes the results of the comparison between DBS, ablative neurosurgery, and radiosurgery (Müller S. et al. 2015a; slightly modified and column "MRgFUS" added).

What Pepper and colleagues have stated with regard to OCD (Pepper et al. 2015) can be generalized to several further psychiatric disorders such as depression and anorexia nervosa. The superiority of DBS cannot be the reason for contemporary enthusiasm for DBS, neither in terms of efficacy nor with regard to side effects and complications. Rather, the enthusiasm for DBS might be grounded in the opportunity to study brain circuitry and individual optimization of outcomes. An absolute superiority of one of the different methods cannot be established, therefore. Each method has its own profile of advantages and disadvantages. The significant differences between neurosurgical approaches clearly suggest that they should be investigated differentially from an ethical, legal, and societal perspective. In particular, arguments developed with regard to DBS cannot just be transferred to ablative neurosurgery and radiosurgery.

Table 13.1 Comparison of the different approaches of psychiatric neurosurgery

	DBS	Microsurgery	Radiosurgery	MRgFUS
Paradigm	Adjustability	Quick fix	Minimal invasiveness	Minimal invasiveness
Adjustability	Very high	Low (second intervention to create another lesion or enlarge the lesion)	Low (second intervention to create another lesion) to medium (through a step-by-step approach)	Low (second intervention to create another lesion) to medium (through a step-by-step approach)
Multiple targets in a single session	No	Yes	Yes	Not yet done
Reversibility	High (exception: permanent adverse effects due to lesions, infections, bleeding)	No	No	No
Invasive craniotomy	Yes	Yes	No	No
Onset of action	Hours to 12 months	Days or weeks	6–12 months	1–6 months
Appropriateness for patients with special needs	No	Patients who would not comply with long-term follow-up	Patients: ◆ who would not comply with long-term follow-up ◆ for whom open surgery is precluded ◆ with higher risks of anesthesia ◆ with higher infection risks ◆ who must receive anti-coagulation therapy	Patients: ◆ who would not comply with long-term follow-up ◆ for whom open surgery is precluded ◆ with higher risks of anesthesia
Time and effort of the procedure	High (single surgery; several days in hospital; parameter adaption)	Medium (single surgery; several days in hospital)	Low (ambulatory treatment, single session)	Medium (single surgery; several days in hospital)
Long-term treatment	Frequent consultation of specialists required (parameter adjustment, regular device exchange)	Not necessary	Not necessary	Not necessary

(continued)

Table 13.1

	DBS	Microsurgery	Radiosurgery	MRgFUS
Costs	Very high direct and life-long costs (first implantation: $50,000–$120,000 USD, each battery exchange: $10,000–$25,000 USD)	Medium	Low (in Germany: about €4500)	Probably medium
Mortality risk	Yes	Yes	No	Probably no
Short-term risks	◆ Anesthesia ◆ Infection ◆ Intracerebral hemorrhage ◆ Hardware complications	◆ Anesthesia ◆ Infection ◆ Hemorrhage	◆ Development of cysts ◆ Edemas	◆ Damage to the scalp, the dura, the arachnoid, or brain tissue adjacent to the target ◆ MR-induced vertigo
Possible adverse effects	◆ Suicidality ◆ Neurological disorders, e.g., hemiparesis ◆ Mood disturbance ◆ Anxiety ◆ Panic attacks ◆ Hypomania ◆ Increased libido ◆ Weight loss or gain ◆ Long-lasting fatigue ◆ Headaches ◆ Visual disturbances ◆ Cognitive problems	◆ Suicidality ◆ Neurological disorders, e.g., hemiplegia ◆ Headaches ◆ Seizures ◆ Drowsiness ◆ Urinary incontinence ◆ Cognitive impairment ◆ Personality change ◆ Weight gain ◆ Sexual disinhibition	◆ Transient cognitive impairment ◆ Transient apathy ◆ Radiation dose >180 Gy: fatigue, weight gain, apathy, tinnitus, facial paresthesia, sexual disinhibition, urinary incontinence	Not known
Risks of long-term treatment	Yes: ◆ Infection risks (due to biofilms and battery exchange every few years) ◆ Hardware complications	No	No	No

(continued)

Table 13.1 Continued

	DBS	Microsurgery	Radiosurgery	MRgFUS
Disadvantages in daily life	Yes: ◆ device-related problems in daily life (e.g., at airport controls)	No	No	No
Disadvantages for further medical treatment	Yes: ◆ Special MRI required	No	No	No
Possible problems of psychosocial adaptation	Possible; self-estrangement, feeling of being manipulated; burden of normality syndrome	Possible; burden of normality syndrome	Improbable	Improbable

Ethical discussion

Ethical, legal, and social context

The clinical, ethical, legal, and societal issues of psychiatric neurosurgery are strongly intermingled. The desperate situation of many people with mental disorders, who suffer not only from their disabling condition, but also from stigma, social exclusion, compulsory treatment, and frequently from incarceration, needs to be taken into account. Thus, development and use of psychiatric neurosurgery has to be considered in a broader social context, which will undoubtedly impact the development and acceptance of these new treatment options. Moreover, the context for evaluating neurosurgical developments also comprises the societal burden of chronic mentally ill patients, the trend of dehospitalization, as well as financial and research interests of device producers, hospitals, and physicians.

Framed within the legal context, national legislation and international law limit the field of psychiatric neurosurgery. In some countries, for example, ablative neurosurgery is forbidden. Important legal issues include the admissibility of offering ablative psychiatric neurosurgery in general; the admissibility of offering psychiatric neurosurgery to criminal offenders in jail or forensic psychiatry, particularly, if combined with the offer of reducing the duration of imprisonment; and the admissibility of treating minors with psychiatric neurosurgery, particularly, if performed in children with cognitive disabilities or against the will of a minor (e.g., in cases of anorexia nervosa).

Media reports play a crucial role for the public acceptance or refusal of psychiatric neurosurgery. The media also influences legal theorists and politicians who are tasked with deciding the admissibility of these procedures. The media in turn are influenced by success stories from researchers and clinicians, and by diverging statements from bioethicists. Social media disseminate information and opinions rapidly to the traditional media in addition to the general public. For example, a critical article about psychosurgery in China in *The Wall Street Journal* was published in November 2007. It reported that thousands of patients had been treated with ablative brain surgery, and in many cases, in an unethical way. The article reports that this was not done with political intention to suppress dissenters, but mainly because of profit-interests of physicians. In April 2008, the Ministry of Health of China issued a strict regulation on neurosurgery for psychiatric disorders that forbids neurosurgical treatments of schizophrenia (Wu et al. 2012). By contrast, DBS seems to be portrayed over-optimistically in the media, and ethical issues are largely ignored (Racine et al. 2007; Gilbert & Ovadia 2011). Many DBS researchers are very aware of the importance of the image of DBS in the media where negative reports on psychiatric neurosurgery pose a genuine threat to this field.

Bioethics principles in discussion of deep brain stimulation

Many ethics papers on (psychiatric) DBS are based on the bioethics principles (Beauchamp & Childress 2013) and generally focus on issues of autonomy, beneficence, nonmaleficence, and justice.

An important issue of the autonomy-related debate is informed consent, particularly in patients with psychiatric diseases (Skuban et al. 2011). The usual methods to determine the capacity for informed consent are insufficient for psychiatric patients, who often do not have cognitive, but emotional and value-related difficulties of decision-making (Tan et al. 2006; Breden & Vollmann 2006). Physicians cannot communicate risks and expectations to patients in an objective and comprehensive manner due to the severe publication bias (Müller S. et al. 2015b). Another factor that confounds acquisition of informed consent in DBS studies is therapeutic misconception (Fisher et al. 2012), that is, that the subjects of a clinical study fail to recognize adequately the key differences between treatment and clinical research (Lidz et al. 2004). Furthermore, many patients are influenced by overoptimistic media reports about DBS which makes counseling very difficult, ultimately challenging the validity of informed consent in these cases.

From the very beginning, ethicists have commented on the development of DBS for psychiatric disorders. Many ethicists are either directly involved in DBS research projects or participate in committees that publish guidelines and recommendations. Some ethicists have demanded an investigation of the publication bias in the DBS literature (Schläpfer & Fins 2010; Gilbert & Dodds 2013) as well as better regulations for the disclosure of conflicts of interests (Schermer 2011). In addition, the enrollment criteria (Bell et al. 2009) as well as the adequate selection of outcome endpoints (Woopen et al. 2012) for clinical trials in psychiatric DBS have been critically discussed. Furthermore, several authors have discussed the need to assess equitable distribution of treatment options (Goldberg 2012), and the influence of economic interests that drive the development of DBS (Christen & Müller S. 2012; Erickson-Davis 2012; Christen et al. 2014).

Personality change and legal responsibility following deep brain stimulation

The standard interpretation of Beauchamp and Childress' ethics principles does not suffice to deal with specific ethical problems of interventions that can alter the personality or the capacity for autonomy (or both of these), or make the patients more dangerous for third persons. Beauchamp and Childress' concept of autonomy is too narrow for an adequate ethical evaluation of such issues, because they do not consider sufficiently the dependence of the capacity for autonomy from neurological prerequisites (Müller S. & Walter 2010; Müller S. 2014, pp.91–106). Furthermore, the ethical evaluation of personalities and personality changes outreaches any ethics of mid-level moral principles; but such an evaluation is required for ethically evaluating personality-changing interventions into the brain (Müller S. 2014, pp.129–137). Difficult questions of autonomy are posed with regard to the question of who should determine the stimulation parameters (Mackenzie 2011a), in particular concerning patients with addiction risk (Glannon 2014a), or patients whom DBS could make more dangerous to third persons (Müller S. et al. 2014).

Further ethical questions arise from technological advancements such as optogenetics and closed-loop DBS, which aim at continuously measuring brain activity in order to provide patients with feedback about their brain states or to alter these brain states directly. Such automatic regulation prevents the patients from becoming aware of the automatic processes controlling their emotions, and thus influencing their behavior. This raises new dimensions to questions regarding patient autonomy given that their emotions, mood, decision-making, and behavior may be influenced automatically according to parameters set by physicians.

Moreover, these issues lead to questions about the moral and legal responsibility of people acting under DBS (Klaming and Haselager 2013) or of people who have been treated with ablative psychiatric neurosurgery. Although factual litigable cases are very rare, they occur. For example, a patient, who had undergone capsulotomy, became severely sexually disinhibited immediately after surgery, and was convicted of rape 5 months postoperatively (Rück et al. 2008). Perhaps one of the most controversial issues, both from an ethical and a legal perspective, is psychiatric neurosurgery for the purposes of law enforcement and social control. Is it ethically justified to offer psychiatric neurosurgery to forensic patients or offenders with pedophilia, psychopathy, or impulsive aggressiveness, particularly with the offer of earlier release? Voluntariness, capacity for informed consent, and responsibility for reoffending in spite of the device implantation or ablative surgery are critical issues to consider in this context (Merkel et al. 2007; Gilbert et al. 2013b; Giordano et al. 2014; Glannon 2014b; Vincent 2014).

Furthermore, much of the bioethical debate on DBS concerned with personality changes are differently conceptualized, for example, in terms of threats to the personal identity, self-alienation, or inauthenticity (Northoff 2004; Merkel et al. 2007; Focquaert & DeRidder 2009; Schechtman 2009; Mackenzie 2011b; Baylis 2013; Kraemer 2013; Lipsman & Glannon 2013b; Witt et al. 2013). I think that the transfer of the concept of personal identity from analytic philosophy to neurosurgical patients is problematic for two reasons: first, because this debate does not adequately capture the clinical reality and, second, because it might have negative effects on the patients. If legal theorists take metaphysical theories about personal identity changes through DBS seriously, then psychiatric advance directives of DBS patients would become invalid. I raised a plea for using more realistic descriptions in order to capture the different kinds of mental or behavioral changes following neurosurgery instead of speculating about the metaphysical concept of personal identity (Müller S. 2014, pp.106–129).

Neglect of modern ablative psychiatric neurosurgery

Although ethicists have extensively investigated DBS, there is a neglect in ethics with regard to ablative psychiatric neurosurgery comprising thermocoagulation or radiofrequency ablation procedures, and Gamma Knife radiosurgery. Whereas ethical issues of DBS are well analyzed, there is nearly no ethical work on modern psychiatric neurosurgery other than DBS. Our recent literature research on ethical papers on psychiatric neurosurgery delivered more than 200 papers on ethical issues of DBS, but not a single paper from ethicists about modern ablative neurosurgery and radiosurgery (apart from our own papers: Müller S. et al. 2015a, 2015b).

I assume that this neglect in ethics is due to most ethicists having adopted the prevailing position of DBS researchers that all other methods of psychiatric neurosurgery are forms of psychosurgery, and therefore outdated and ethically obsolete, whereas DBS is significantly different. Another explanation would be that bioethicists have failed to see the crucial differences between different psychiatric neurosurgery approaches and that these differences matter ethically. With regard to the principles of beneficence and nonmaleficence, for example, evidence-based risk:benefit ratios are decisive and, as has been pointed out earlier in this chapter, these ratios may differ dramatically for the various distinct neurosurgical procedures. A last explanation might be that most bioethicists wrongly presuppose that ablative neurosurgery is not in use anymore in Western countries. This last opinion, however, is challenged both empirically and normatively in a number of ways:

1. The fraction of ablative procedures in psychiatric neurosurgery is large. In North America, 50% of psychiatric neurosurgeons use lesioning exclusively or combined with DBS (Lipsman et al. 2011), and outside of North America the number is even higher at 54.9% (Mendelsohn et al. 2013). For OCD, the number of patients treated with Gamma Knife is more than twice the number of published DBS cases (250 Gamma Knife[1] patients versus 109 DBS patients (Kohl et al. 2014)).

2. The claim that DBS is clearly ethically superior is questionable. Although DBS has the advantages of adjustability and high reversibility, it is inferior with regard to adverse effects, contraindications, and the required time and effort. Accordingly, two expert panels have affirmed stereotactic ablative procedures as important alternatives for appropriately selected patients (for Parkinson's disease: Bronstein et al. 2011; for psychiatric disorders: Nuttin et al. 2014). A clear superiority of any procedure in all

relevant aspects cannot be established (Müller S. et al. 2015a). The authors of a recent comparative review of DBS and anterior capsulotomy conclude "that the current popularity of DBS over ablative surgery for OCD is not due to clinical superiority over AC [anterior capsulotomy], but rather that clinicians and patients find DBS to be more acceptable" (Pepper et al. 2015, p.1035).

3. The much higher long-term costs of DBS, particularly for long-term treatment, exclude this option for the majority of patients worldwide.

4. Ablative neurosurgery and Gamma Knife® radiosurgery might avoid several difficult ethical issues of DBS, which are raised because DBS allows for fine-tuning the mood and certain personality aspects of patients. If such a fine-tuning is not possible, severe ethical issues referring to the authority over the stimulation parameters, and to conflicts between the patients' will and the physicians' attitudes with regard to parameters that might cause hypomania or addiction are not raised. In particular, Gamma Knife® radiosurgery does not raise the challenging ethical problem of "inauthenticity" and the so-called burden of normality syndrome (Wilson et al. 2001; Gilbert 2012), just because the onset of its effect is delayed for several months (Lindquist et al. 1991; De Salles and Gorgulho 2015).

Advantages of the different methods

In the following section, I will analyze the main advantages for DBS as well as for the ablative neurosurgical procedures from an ethics point of view.

Reversibility

The reversibility that has been proclaimed for DBS is presumably the most important argument that has prevented it from general moral condemnation. Only a few commentators consider psychiatric DBS as a continuation of the discredited historical psychosurgery, whereas the majority highlight the important differences between previous psychosurgery and DBS with regard to reversibility, invasiveness, adjustability, and ethical orientation (Müller S. 2014, p.5). The reversibility is the main argument in many statements. For example, in Germany, the working team of ethical review committees has written in its recommendation for the appraisal of clinical studies that DBS research projects are only ethically justifiable if the reversibility of the intervention is guaranteed (Raspe et al. 2012). The German Association for Psychiatry, Psychotherapy and Psychosomatics (DGPPN) states in its recently published guidelines for the treatment of OCD that neurolesional or ablative methods have a very controversial history, and should be considered very critically because of their irreversibility. The authors recommend not using capsulotomy by thermocoagulation or gamma radiation because the risk:benefit ratios were too negative in comparison with DBS, and because of the sometimes severe, irreversible side effects (Hohagen et al. 2015). However, this recommendation is based on only one follow-up study of 25 patients treated with capsulotomy between 1988 and 2000, in which two patients became sexually disinhibited resulting in one case in a conviction of rape and in the other case in job loss (Rück et al. 2008). Furthermore, the technique has been significantly refined since then, and the high radiation doses used by Rück and colleagues (e.g., three bilateral doses of 200 Gy) are not used anymore.

In spite of the broad acceptance of the reversibility argument in favor of DBS, it can be questioned whether the reversibility of psychiatric neurosurgery is ethically demanded at all, since generally, a permanent effect of a medical treatment is strongly preferred over a reversible effect, which requires continuous treatment to maintain the effect. Even in comparable areas of stereotactic neurosurgery (e.g., for epilepsy or Parkinson's disease), reversibility is never demanded, although these interventions often change the patients' personalities significantly. For these indications, the aim is a permanent cure or relief and the reversibility of possible adverse effects and personality changes has never been demanded by ethicists.

I suppose that there are at least three reasons why there is such a strong demand for reversibility in the case of psychiatric neurosurgery: first, the experimental character of all procedures and the insecurity about the optimal targets; second, the fear of perma-nent undesirable personality changes; and third, that some stakeholders consider severe, therapy-refractory psychiatric disorders as part of the personality or personal identity instead of considering them as diseases that compromise the personality.

Adjustability

The other main argument that is brought forward for the ethical superiority of DBS over ablative psychiatric neurosurgery is its adjustability. However, adjustability, in fact, leads to one of the most challenging ethical problems of DBS, namely that it allows for fine-tuning the mood and personality properties of patients. This raises difficult ques-tions, such as the following: Who shall decide about the patients' mood and personality traits and according to which criteria? How can the patients' autonomy be respected optimally: by giving the patients full power about their stimulation parameters, or by limiting their power in order to prevent addiction to the stimulation? These difficult problems, which can cause severe conflicts between patients and physicians, do not occur after ablative neurosurgery and radiosurgery, just because they do not allow for fine-tuning.

Quick onset of effect

Compared to Gamma Knife˙ radiosurgery, ablative microsurgery and DBS (at least DBS in the slMFB) have a quick onset of effect. This is advantageous for patients, who need a rapid symptom reduction, for example, for a mother with OCD, who is forced by her obsession to disinfect her baby before she can touch him. However, for other patients a slow, gradual onset of action is preferable, for example, for a patient, who has integrated his Tourette syndrome into his self-concept and is afraid of radical changes in his per-sonality. Thus, for a subgroup of patients, a gradual development of effects might in fact be advantageous, because it alleviates the psychological adjustment to the new situation (Lindquist et al. 1991; De Salles and Gorgulho 2015). The delayed onset of action might protect against feelings of being manipulated, self-estrangement, and the burden of normality syndrome (Müller S. et al. 2015a).

The conclusion must therefore be that an absolute superiority of one of the different methods cannot be established. Each method has a different profile of advantages and disadvantages. Furthermore, what counts as an advantage or disadvantage may differ from patient to patient, and depends on his or her individual situation and condition (Müller S. et al. 2015a).

Minimal invasiveness

Minimal invasiveness is proclaimed particularly for Gamma Knife˚ radiosurgery and for MRgFUS; although even these methods bear some risks for damaging the brain, these risks are much less than for DBS and microsurgery. With regard to the nonmaleficence principle, the minimal invasiveness speaks in favor of these two methods.

Since the claimed advantages are not advantages in each respect, it would be unjustified to identify any one of the different methods as superior. Therefore, individual factors should be crucial in decision-making. Which approach is optimal, depends significantly on individual medical and nonmedical factors, particularly on the patient's general health status, social situation, individual preferences, and individual attitudes (Müller S. et al. 2015a).

Recommendations

Further research in psychiatric neurosurgery is warranted given the need for effective therapies for treatment-refractory psychiatric disorders and because of the preliminary evidence for its efficacy. However, neither the high mortality rates of severe psychiatric disorders nor their socioeconomic burden can justify therapeutic adventurism. Since psychiatric neurosurgery has both the goal and the potential to change core features of the patients' personalities, these interventions have to fulfill the highest ethical and scientific standards. The current research practice in this area does not fulfill these standards in all cases. Therefore, I suggest several protective measures to ensure that psychiatric neurosurgery research proceeds in accord with the ethical demands (Müller S. 2014; Müller S. et al. 2015a, 2015b).

1. *Hypothesis-driven interventions*

Testable hypotheses on dysfunctional circuits in the different psychiatric disorders should be rigorously developed and tested (Müller S. et al. 2015b).

2. *Cooperation and competition of different approaches*

In spite of the competition of the different approaches of psychiatric neurosurgery, they do also enrich each other mutually. Most of the DBS targets are known from previous ablative procedures, whereas successful DBS targets could in turn be used to inform ablative procedures. In the long-run, DBS could become the preferable method for exploring new targets, just because it is largely reversible, whereas well-established targets could become candidates for ablative microsurgery, Gamma Knife° radiosurgery, or MRgFUS (Müller S. et al. 2015a).

3. *Case registries*

Independent case registries should become obligatory. These should contain all clinical studies and all individual treatment attempts, in order to avoid a publication bias and the negative consequences thereof, namely faulty evaluations of therapies, flawed therapy recommendations, unpromising treatment attempts, and unneeded clinical studies (Morishita et al. 2014; Müller S. 2014, p.132).

4. *No individual treatment attempts*

Individual treatment attempts should generally not be performed. Rather, all new applications should be investigated in clinical trials of the appropriate size and statistical power and approved by an ethics committee (Müller S. 2014, p.132). In this context, Fins and coauthors have rightly criticized the FDA's humanitarian device exemption for DBS for OCD (Fins et al. 2011). A humanitarian device exemption allows a manufacturer to market a device under certain conditions without subjecting it to a clinical trial. This designation is available only for devices intended to diagnose or treat conditions that annually affect 4000 or fewer people in the United States. In case of DBS for OCD, the humanitarian device exemption is misused for bypassing the rigors of clinical trials, since OCD is not an orphan, but a prevalent condition (Fins et al. 2011). The designation allows an institution to conduct an appropriately powered, hypothesis-driven clinical trial to assess the device's safety, efficiency, and mechanism of action (Fins et al. 2011). According to Fins and his coauthors, the current market-driven regulatory strategy undermines patient safety, scientific discovery, and research integrity (Fins et al. 2011).

5. *Evidence-based comparison of different therapeutic approaches*

For a valid evaluation and comparison of the efficacy and adverse effects of the different psychiatric neurosurgery approaches, head-to-head comparative studies with sufficient statistical power, placebo control, and as far as possible, double-blinded design are necessary. The scientific standard of randomizing patients to different treatment groups cannot be met for practical and ethical reasons. However, this limitation is less problematic, since head-to-head comparative studies that match patients who undergo different treatments can provide a valid efficacy comparison, too. Double-blind studies can be easily performed with radiosurgery. For DBS, sham-controlled trials can be performed, in which patients are randomized to active versus sham DBS treatment in a blinded fashion for several months followed by an open-label continuation phase (Dougherty et al. 2015).

6. *Inclusion and exclusion criteria*

Future studies should use similar inclusion and exclusion criteria and similar standardized assessment instruments in order to make the outcome straightforwardly comparable (De Zwaan & Schlaepfer 2013). Beyond the accepted medical criteria, it is a large challenge to establish criteria for selecting patients for a given intervention, since the criteria have to be both responsible and just, that is, they have to protect vulnerable patients, but should not exclude patients who could profit from a given intervention. Although the criteria should be formulated as general rules, they should also allow for individual exceptions (Müller S. & Christen 2011).

7. *No compulsory psychiatric neurosurgery*

Psychiatric neurosurgery should not be used as a compulsory treatment for adults or adolescents (even if demanded by their legal guardians). This prohibition is in accordance with the recommendations on psychosurgery of the National Commission for the Protection of Human Subjects of Biomedical and Behavioral Research (1977) and the consensus paper on stereotactic neurosurgery for psychiatric disorders (Nuttin et al. 2014; Müller S. et al. 2015b).

8. *Comprehensive investigation of adverse effects, particularly of sociopsychological sequelae*

Since the risk:benefit profile of a therapeutic approach is decisive for recommending it to patients, research about adverse effects has an important role. Information about psychosocial and economic consequences of interventions in the brain must be gathered (Müller S. 2014, p.134).

9. *Development of instruments for measuring subtle mental alterations*

Since personality changes are a main ethical concern and a central factor of patients' satisfaction with an intervention, instruments to evaluate even subtle changes should be developed further (Müller S. 2014, pp.134–135).

10. *Long-term, prospective follow-up*

Long-term, optimally prospective, follow-up studies are necessary for ensuring study quality. Nuttin and colleagues recommend that research and clinical protocols should include support for long-term safety and efficacy studies on psychiatric neurosurgery for at least 5–10 years of follow-up (Nuttin et al. 2014).

11. *Investigation of single cases*

Cross-sectional group research does not reveal the different individual trajectories and provides only limited clues about which factors are most relevant in effecting positive change for an individual. It is important to study individual outcomes, particularly by identifying subgroup patterns that can become lost in whole-group analyses (Wilson et al. 2005; Baxendale et al. 2012). Both positive and poor outcomes should be reported separately. Particularly cases with unfavorable or unexpected outcome should be investigated, since they offer extraordinary chances for scientific discovery and improving the techniques used (Christen & Müller S. 2011; Fins et al. 2011; Müller S. 2014, p.133).

12. *Comprehensive information for patients*

Patients need independent, evidence-based information about risks, benefits, and chances of different therapy options for decision-making. As we have proposed for DBS (Müller S. & Christen 2011), a living database should be developed and continuously updated, ideally for all neurosurgical therapies, with open access for physicians and patients. It should contain comparative data of single centers about the morbidity, the incidence of complications and adverse effects, the neuropsychological outcome, and quality of life following neurosurgery (Müller S. 2014, p.135).

13. *Patient-centered, multidisciplinary counselling and shared decision-making*

Patients should be informed comprehensively about different treatment options and their respective evidence-based risk:benefit profiles. Individual factors have to be taken into account because they play a crucial role for decision-making. These individual factors comprise the patients' social situation, individual preferences, and individual attitudes, for example, whether they can tolerate implanted devices, and whether they are more afraid of the irreversibility of an ablative procedure or of the medical risks of a craniotomy (Müller S. et al. 2015a). The counselling should also include therapy options, which are not affiliated with the institution that performs the procedure, even if the resulting consequence is that the patient will be treated in another institution or lost as a research participant. Optimally, a multidisciplinary team assists the patients in the decision-making process; if such a team does not exist locally, the patients should be referred to specialists of all relevant therapies. Counseling should also consider the individual situation of the patients, especially their professional activity and social integration (Müller S. 2014, p.136).

14. *Advance directives or Ulysses contracts*

Advance directives should be offered to the patients, particularly for those at risk of personality changes or risk of hampered autonomy. Advance directives and the determination of attorneys are helpful tools for preparing for the worst case. Ulysses contracts are particularly important for dealing with transient states of psychosis or mania following neurosurgery, which make consent to any medically indicated change of stimulation parameters or device explanation problematic (Müller S. 2014, p.137).

15. *Individual instead of institution-specific target selection*

Since multiple circuits seem to be involved in psychiatric disorders, targets of DBS or ablative procedures should be selected specifically with regard to the prominent symptoms instead of using the institution-specific target for all patients with certain diagnoses (Müller S. et al. 2015a).

16. *Temporary DBS*

Whether the DBS application should be chronic or whether it is possible to limit its application to a confined time-period is an open question. If a psychiatric disorder is a personality trait-related condition of a given patient, then it may be necessary to stimulate him or her throughout life. In this case, an ablative procedure might be preferable. But if the psychiatric disorder is a development-related disorder that occurs only under given sociocultural circumstances in biologically vulnerable individuals (e.g., anorexia nervosa or drug addiction), then the stimulation might be stopped when the patient is fully remitted for a certain period of time. If no relapse occurs after a longer period without stimulation, the whole DBS system might be removed, which would also be advantageous for medical, psychological, and financial reasons (Müller S. et al. 2015b).

Conclusion

Modern psychiatric neurosurgery has the potential to become a further method for treating otherwise treatment-refractory severe psychiatric disorders. However, to be broadly accepted in both the medical community and the society, it has to conform to rigorous scientific and ethical standards.

Acknowledgments

The research of Sabine Müller is funded by the German Federal Ministry of Education and Research (01 GP 1621A).

Note

1. Written communication of Catherine Gilmore-Lawless, Vice President of Elekta (the only producer of Gamma Knife®).

References

Andrade, P. and Visser-Vandewalle, V. (2016). DBS in Tourette syndrome: where are we standing now? *Journal of Neural Transmission (Vienna)*, **123**(7), 791–796.

Baxendale, S., Thompson, P.J., and Duncan, J.S. (2012). Neuropsychological function in patients who have had epilepsy surgery: a long-term follow-up. *Epilepsy & Behavior*, **23**, 24–29.

Baylis, F. (2013). "I Am Who I Am": on the perceived threats to personal identity from DBS. *Neuroethics*, **6**(3), 513–526.

Beauchamp, T.L. and Childress, J.F. (2013). *The Principles of Biomedical Ethics* (7th edn.). Oxford: Oxford University Press.

Bell, E., Mathieu, G., and Racine, E. (2009). Preparing the ethical future of deep brain stimulation. *Surgical Neurology*, **72**(6), 577–586.

Bergfeld, I.O., Mantione, M., Hoogendoorn, M.L.C., et al. (2016). Deep brain stimulation of the ventral anterior limb of the internal capsule for treatment-resistant depression. A randomized trial. *JAMA Psychiatry*, **73**(5), 456–464.

Breden, T.M. and Vollmann, J. (2006). The cognitive based approach of capacity assessment in psychiatry: a philosophical critique of the MacCAT-T. *Health Care Analysis*, **12**(4), 273–283.

Bronstein, J.M., Tagliati, M., Alterman, R.L., et al. (2011). Deep brain stimulation for Parkinson disease: an expert consensus and review of key issues. *Archives of Neurology*, **68**(2), 165–171.

Cavuoto, J. (2013, December). Depressing innovation. *Neurotech Business Report*. Available from: http://www.neurotechreports.com/pages/publishersletterDec13.html [accessed May 9, 2016].

Chodakiewitz, Y., Williams, J., Chodakiewitz, J., and Cosgrove, G.R. (2015). Ablative surgery for neuropsychiatric disorders: past, present, future. In: Sun, B. and De Salles, A. (Eds.) *Neurosurgical Treatments for Psychiatric Disorders*, pp.51–66. Dordrecht: Springer.

Christen, M., Ineichen, C., Bittlinger, M., Bothe, H.W., and Müller, S. (2014). Ethical focal points in the international practice of deep brain stimulation. *AJOB Neuroscience*, **5**(4), 65–80.

Christen, M. and Müller, S. (2011). Single cases promote knowledge transfer in the field of DBS. *Frontiers in Integrative Neurosciences*, **5**, 13.

Christen, M. and Müller, S. (2012). Current status and future challenges of deep brain stimulation in Switzerland. *Swiss Medical Weekly*, **142**, w13570.

Cook, M., O'Brien, T.J., Berkovic, S.F., et al. (2013). Prediction of seizure likelihood with a long-term, implanted seizure advisory system in patients with drug-resistant epilepsy. *Lancet Neurology*, **12**, 563–571.

Corripio, I., Sarró, S., McKenna, P.J., et al. (2016). Clinical improvement in a treatment-resistant patient with schizophrenia treated with deep brain stimulation. *Biological Psychiatry*, **80**(8), e69–70.

De Salles, A. and Gorgulho, A.A. (2015). Radiosurgery for psychiatric disorders. In: Sun, B. and De Salles A (Eds.) *Neurosurgical Treatments for Psychiatric Disorders*, pp.217–225. Dordrecht: Springer.

De Zwaan, M. and Schlaepfer, T.E. (2013). Not too much reason for excitement: deep brain stimulation for anorexia nervosa. *European Eating Disorders Review*, **21**, 509–5011.

Deisseroth, K. (2012). Optogenetics and psychiatry: applications, challenges, and opportunities. *Biological Psychiatry*, **71**, 1030–2.

Diering, S.L. and Bell, W.O. (1991). Functional neurosurgery for psychiatric disorders: a historical perspective. *Stereotactic and Functional Neurosurgery*, **57**, 175–194.

Dougherty, D.D., Rezai, A.R., Carpenter, L.L., et al. (2015). A randomized sham-controlled trial of deep brain stimulation of the ventral capsule/ventral striatum for chronic treatment-resistant depression. *Biological Psychiatry*, **78**, 240–248.

Dupré, D.A., Tomycz, N., Oh, M.Y., and Whiting, D. (2015). Deep brain stimulation for obesity: past, present, and future targets. *Neurosurgical Focus*, **38**(6), E7.

Eljamel, S. (2015). Ablative surgery for depression. In: Sun, B. and De Salles, A. (Eds.) *Neurosurgical Treatments for Psychiatric Disorders*, pp.87–94. Dordrecht: Springer.

Erickson-Davis, C. (2012). Ethical concerns regarding commercialization of DBS for OCD. *Bioethics*, **26**(8), 440–446.

Fins, J.J., Mayberg, H.S., Nuttin, B., et al. (2011). Misuse of the humanitarian device exemption in stimulation for obsessive-compulsive disorder. *Health Affairs*, **30**, 2302–2311.

Fisher, C.E., Dunn, L.B., Christopher, P.P., et al. (2012). The ethics of research on DBS for depression. *Annals of the New York Academy of Sciences*, **1265**(1), 69–79.

Focquaert, F. and DeRidder, D. (2009). Direct intervention in the brain: ethical issues concerning personal identity. *Journal of Ethics in Mental Health*, **4**(2), 1–6.

Franzini, A., Broggi, G., Cordella, R., Dones, I., and Messina, G. (2013). Deep-brain stimulation for aggressive and disruptive behavior. *World Neurosurgery*, **80**(3–4), S29.e11–14.

Friehs, G.M., Park, M.C., Goldman, M.A., Zerris, V.A., Norén, G., and Sampath, P. (2007). Stereotactic radiosurgery for functional disorders. *Neurosurgery Focus*, **23**(6), E3.

Gilbert, F. (2012). The burden of normality. *Journal of Medical Ethics*, **38**(7), 408–412.

Gilbert, F. (2013a). Deep brain stimulation for treatment resistant depression: Postoperative feeling of self-estrangement, suicide attempt and impulsive-aggressive behaviours. *Neuroethics*, **6**(3), 473–481.

Gilbert, F. (2015). A threat to autonomy? The intrusion of predictive brain implants. *AJOB Neuroscience*, **6**(4), 4–11.

Gilbert, F. and Dodds, S. (2013). How to turn ethical neglect into ethical approval. *AJOB Neuroscience*, **4**(2), 59–60.

Gilbert, F. and Ovadia, D. (2011). Deep brain stimulation in the media: over-optimistic portrayals call for a new strategy involving journalists and scientists in ethical debates. *Frontiers in Integrative Neuroscience*, **5**, 16.

Gilbert, F., Vranic, A., and Hurst, S. (2013b). Involuntary & voluntary invasive brain surgery. *Neuroethics*, **6**(1), 115–128.

Giordano, J., Kulkarni, A., Farwell, J. (2014). Deliver us from evil? *Theoretical Medicine and Bioethics*, **35**, 73–89.

Glannon, W. (2014a). Neuromodulation, agency and autonomy. *Brain Topography*, **27**(1), 46–54.

Glannon, W. (2014b). Intervening in the psychopath's brain. *Theoretical Medicine and Bioethics*, 35, 43–57.

Goldberg, D.S. (2012). Justice, population health, and deep brain stimulation. *AJOB Neuroscience*, 3(1), 16–20.

Greenberg, B.D., Price, L.H., Rauch, S.L., et al. (2003). Neurosurgery for intractable obsessive-compulsive disorder and depression: critical issues. *Neurosurgery Clinics of North America*, 14, 199–212.

Hohagen, F., Wahl-Kordon, A., Lotz-Rambaldi, W., and Muche-Borowski, C. (Eds.) (2015). *S3-Leitlinie Zwangsstörungen*. Berlin: Springer-Verlag.

Kesey, K. (1964). *One Flew Over the Cuckoo's Nest*. New York: The Viking Press, Inc.

Klaming, L. and Haselager, P. (2013). Did my brain implant make me do it? *Neuroethics*, 6(3), 527–539.

Kohl, S., Schönherr, D.M., Juigjes, J., et al. (2014). Deep brain stimulation for treatment-refractory obsessive compulsive disorder: a systematic review. *BMC Psychiatry*, 14, 214.

Kraemer, F. (2013). Me, myself and my brain implant. *Neuroethics*, 6(3), 483–497.

Lévèque, M. (2014). *Psychosurgery: New Techniques for Brain Disorders*. Dordrecht: Springer.

Lévèque, M., Carron, R., and Régis, J. (2013). Radiosurgery for the treatment of psychiatric disorders: a review. *World Neurosurgery*, 80(3/4), S32.e1–e9.

Lidz, C.W., Appelbaum, P.S., Grisso, T., and Renaud, M. (2004). Therapeutic misconception and the appreciation of risks in clinical trials. *Social Science & Medicine*, 58, 1689–1697.

Lindquist, C., Kihlström, L., and Hellstrand, E. (1991). Functional neurosurgery—a future for the Gamma Knife? *Stereotactic and Functional Neurosurgery*, 57, 72–81.

Lipsman, N., Mendelsohn, D., Taira, T., and Bernstein, M. (2011). The contemporary practice of psychiatric surgery: results from a global survey of North American functional neurosurgeons. *Stereotactic and Functional Neurosurgery*, 89, 103–110.

Lipsman, N., Woodside, D.B., Giacobbe, P., et al. (2013a). Subcallosal cingulate deep brain stimulation for treatment-refractory anorexia nervosa: a phase 1 pilot trial. *The Lancet*, 381 (9875), 1361–1370.

Lipsman, N. and Glannon, W. (2013b). Brain, mind and machine: what are the implications of deep brain stimulation for perceptions of personal identity, agency and free will? *Bioethics*, 27(9), 465–470.

Luigjes, J., de Kwaasteniet, B.P., de Koning, P.P., et al. (2013). Surgery for psychiatric disorders. *World Neurosurgery*, 80(3/4), S31.e17–e28.

Mackenzie, R. (2011a). Who should hold the remote for the new me? *AJOB Neuroscience*, 2(1), 18–20.

Mackenzie, R. (2011b). Must family/carers look after strangers? Post-DBS identity changes and related conflicts of interest. *Frontiers in Integrative Neuroscience*, 5, 12.

Martínez-Álvarez, R. (2015). Ablative surgery for obsessive-compulsive disorders. In: Sun, B. and De Salles, A. (Eds.) *Neurosurgical Treatments for Psychiatric Disorders*, pp.105–112. Dordrecht: Springer.

Mendelsohn, D., Lipsman, N., Lozano, A.M., Taira, T., and Bernstein, M. (2013). The contemporary practice of psychiatric surgery: results from a global survey of functional neurosurgeons. *Stereotactic and Functional Neurosurgery*, 91, 306–313.

Merkel, R., Boer, G., Fegert, J., et al. (2007). *Intervening in the Brain: Changing Psyche and Society*. Berlin: Springer Verlag.

Mirzadeh, Z., Bari, A., Lozano, A.M. (2016). The rationale for deep brain stimulation in Alzheimer's disease. *Journal of Neural Transmission (Vienna)*, 123(7), 775–783.

Morishita, T., Fayad, S.M., Higuchi, M., Nestor, K.A., and Foote, K.D. (2014). Deep brain stimulation for treatment-resistant depression: systematic review of clinical outcomes. *Neurotherapeutics*, 11, 475–484.

Müller, S. (2014). *Personality and Autonomy in Light of Neuroscience [Cumulative habilitation]*. Available from: http://www.diss.fu-berlin.de/diss/receive/FUDISS_thesis_000000097489 [accessed January 31, 2017].

Müller, S. and Christen, M. (2011). Deep brain stimulation in Parkinsonian patients—ethical evaluation of cognitive, affective and behavioral sequelae. *American Journal of Bioethics*, 2(1), 3–13.

Müller, S. and Walter, H. (2010). Reviewing autonomy. Implications of the neurosciences and the free will debate for the principle of respect for the patient's autonomy. *Cambridge Quarterly of Health Ethics*, (2), 205–217.

Müller, S., Walter, H., and Christen, M. (2014). When benefitting a patient increases the risk for harm for third persons—the case of treating pedophilic Parkinsonian patients with DBS. *International Journal of Law and Psychiatry*, 37(3), 295–303.

Müller, S., Riedmüller, R., and van Oosterhout, A. (2015a). Rivaling paradigms in psychiatric neurosurgery. *Frontiers in Integrative Neuroscience*, 9, 27.

Müller, S., Riedmüller, R., Walter, H., and Christen, M. (2015b). An ethical evaluation of stereotactic neurosurgery for anorexia nervosa. *AJOB Neuroscience*, 6(4), 50–65.

Müller, U.J., Voges, J., Steiner, J., et al. (2013). Deep brain stimulation of the nucleus accumbens for the treatment of addiction. *Annals of the New York Academy of Sciences*, 1282, 119–128.

Na, Y.C., Jung, H.H., and Chang, J.W. (2015). Focused ultrasound for the treatment of obsessive-compulsive disorder. In: Sun, B. and De Salles, A. (Eds.) *Neurosurgical Treatments for Psychiatric Disorders*, pp.125–141. Dordrecht: Springer.

Northoff, G. (2004). The influence of brain implants on personal identity and personality—a combined theoretical and empirical investigation in 'Neuroethics'. In: Schramme, T. and Thome, J. (Eds.) *Philosophy and Psychiatry*, pp.311–325. Berlin: De Gruyter.

Nuttin, B., Wu, H., Mayberg, H., et al. (2014). Consensus on guidelines for stereotactic neurosurgery for psychiatric disorders. *Journal of Neurology, Neurosurgery, and Psychiatry*, 85(9), 1003–1008.

Pepper, J., Hariz, M., and Zrinzo, L. (2015). Deep brain stimulation versus anterior capsulotomy for obsessive-compulsive disorder: a review of the literature. *Journal of Neurosurgery*, 122, 1028–1037.

Racine, E., Waldman, S., Palmour, N., Risse, D., and Illes, J. (2007). "Currents of hope": neurostimulation techniques in U.S. and U.K. print media. *Cambridge Quarterly of Health Ethics*, 16, 312–316.

Raspe, H., Hüppe, A., Strech, D., and Taupitz, J. (2012). *Arbeitskreis Medizinischer Ethikkommissionen: Empfehlungen zur Begutachtung klinischer Studien*. Köln: Deutscher Ärzte-Verlag.

Régis, J. (2013). Radiosurgery as a modulation therapy! *Acta Neurosurgery Supplement*, 116, 121–126.

Rosin, B., Slovik, M., Mitelman, R., et al. (2011). Closed-loop deep brain stimulation is superior in ameliorating Parkinsonism. *Neuron*, 72, 370–384.

Rück, C., Karlsson, A., Steele, D., et al. (2008). Capsulotomy for obsessive-compulsive disorder. Long-term follow-up of 25 patients. *Archives of General Psychiatry*, 65(8), 914–922.

Salgado-López, L., Pomarol-Clotet, E., Roldán, A., et al. (2016). Letter to the Editor: Deep brain stimulation for schizophrenia. *Journal of Neurosurgery*, 125(1), 229–230.

Schechtman, M. (2009). Getting our stories straight: self-narrative and personal identity. In: Mathews, S.J.H., Bok, H., and Rabins, P.V. (Eds.) *Personal Identity and Fractured Selves*, pp.66–92. Baltimore, MD: Johns Hopkins University Press.

Schermer, M. (2011). Ethical issues in deep brain stimulation. *Frontiers in Integrative Neuroscience*, 5, 17.

Schläpfer, T.E. (2015). Deep brain stimulation for major depression—steps on a long and winding road. Commentary. *Biological Psychiatry*, **78**, 218–219.

Schläpfer, T.E., Bewernick, B.H., Kayser, S., Madler, B., and Coenen, V.A. (2013). Rapid effects of deep brain stimulation for treatment-resistant major depression. *Biological Psychiatry*, **73**, 1204–1212.

Schläpfer, T.E. and Fins, J.J. (2010). Deep brain stimulation and the neuroethics of responsible publishing: when one is not enough. *JAMA*, **303**(8), 775–776.

Skuban, T., Hardenacke, K., Woopen, C., and Kuhn, J. (2011). Informed consent in deep brain stimulation. *Frontiers in Integrative Neuroscience*, **5**, 7.

Sun, B. and De Salles, A. (Eds.) (2015). *Neurosurgical Treatments for Psychiatric Disorders*. Dordrecht: Springer.

Sun, B., Li, D., Liu, W., Zhan, S., Pan, Y., and Lin, G. (2015). Surgical treatments for anorexia nervosa. In: Sun, B. and De Salles, A. (Eds.) *Neurosurgical Treatments for Psychiatric Disorders*, pp.175–187. Dordrecht: Springer.

Tan, D.J., Hope, P.T., Stewart, D.A., and Fitzpatrick, P.R. (2006). Competence to make treatment decisions in anorexia nervosa: thinking processes and values. *Philosophy, Psychiatry, & Psychology*, **13**(4), 267–282.

Tass, P.A. (2003). A model of desynchronizing deep brain stimulation with a demand-controlled coordinated reset of neural subpopulations. *Biological Cybernetics*, **89**, 81–88.

Torres, C.V., Sola, R.G., Pastor, J., et al. (2013). Long-term results of posteromedial hypothalamic deep brain stimulation for patients with resistant aggressiveness. *Journal of Neurosurgery*, **119**, S. 277–287.

Valenstein, E.S. (1986). *Great and Desperate Cures: The Rise and Decline of Psychosurgery and Other Radical Treatments for Mental Illness*. New York: Basic Books, Inc. Publishers.

Vincent, N.A. (2014). Restoring responsibility. *Criminal Law and Philosophy*, **8**(1), 21–42.

Walter, H. and Müller, S. (2013). Optogenetics as a new therapeutic tool in medicine? A view from the principles of biomedical ethics. In: Hegemann, P. and Sigrist, S. (Eds.) *Optogenetics*, pp.201–211. Berlin: De Gruyter.

Wilson, S.J., Bladin, P.F., and Saling, M.M. (2001). The "burden of normality": concepts of adjustment after surgery for seizures. *Journal of Neurology, Neurosurgery, and Psychiatry*, **70**, 649–656.

Wilson, S.J., Bladin, P.F., Saling, M.M., and Pattison, P.E. (2005). Characterizing psychosocial outcome trajectories following seizure surgery. *Epilepsy & Behavior*, **6**, 570–580.

Witt, K., Kuhn, J., Timmermann, L., Zurowski, M., and Woopen, C. (2013). Deep brain stimulation and the search for identity. *Neuroethics*, **6**(3), 499–511.

Woopen, C., Timmermann, L., and Kuhn, J. (2012). An ethical framework for outcome assessment in psychiatric DBS. *AJOB Neuroscience*, **3**(1), 50–55.

Wu, H., Gabriels, L., and Nuttin, B. (2012). Neurosurgery for psychiatric disorders in the People's Republic of China. *AJOB Neuroscience*, **3**(1), 56–59.

Chapter 14

At the crossroads of civic engagement and evidence-based medicine: Lessons learned from the chronic cerebrospinal venous insufficiency experience

Shelly Benjaminy and Anthony Traboulsee

Introduction

Contemporary formulations of science are increasingly shifting from a top-down model of expertise toward a more pluralistic approach that encourages the engagement of publics in the co-creation and governance of science. This conceptualization, in which non-scientific actors are regarded as agents capable of meaningfully contributing to scientific debates, evinces a systematic shift from a model of public understanding of science to a new paradigm marked by public participation in science.

A participatory and egalitarian orientation is central in the field of neuroethics, which at its core aims to align neurotechnological development with societal values. The brain is widely regarded as the seat of the mind, and is inextricably linked with abstractions of individuality and personal identity (Leshner 2005). It is therefore not surprising that new brain technologies that reside at the intimate frontier of personhood are compelling widespread public interest and engagement (Illes et al. 2005).

It is the civic duty of neuroscientists and neuroethicists to engage with publics in deliberation about the development of new research trajectories (Illes et al. 2005). Not only is this logical, but it is also socially conscientious to ensure that the publics who bear the burdens and risks of neurotechnological development are included in dialogue about its formulation and application. As such, public engagement has been a focal area in the field of neuroethics since its contemporary inception (Marcus 2002). Ongoing neuroethical debates about controversial issues including the trade-offs of cognitive enhancement, implications of brain–computer interfaces on autonomy, and the management of incidental findings in neuroimaging require the input of both scientists and lay citizens to inform the field in a sustainable and socially responsible capacity.

Successful civic engagement in science, however, is not without its risks and challenges. Public engagement may unveil divergence between expert opinion and public beliefs. Differing world views—which may be shaped by disciplinary divides or competing stakes—can be difficult to reconcile. Moreover, maintaining public trust in the face of conflicting priorities may pose threats to the sustainability of scientific enterprises.

Here, we explore the opportunities and challenges of community engagement in the neurosciences. We discuss prospects for public participation in science through media outlets, research initiatives, and advocacy. We draw on the contentious case study of the chronic cerebrospinal venous insufficiency (CCSVI) research trajectory that generated both hope and skepticism, galvanized substantial international attention, and was heavily criticized for privileging scientific inquiry driven more by public pressure than by empirical evidence. We conclude with lessons learned from the cautionary CCSVI tale, and suggest opportunities for reciprocal and impactful engagement that the field of neuroethics may foster as novel neurotechnologies are developed.

The media as a platform for public participation in science

Public participation in science and science policy is often supported by a complex infrastructure of media communications. Being the most accessible source of information about science in the public sphere, the media have long served as a prominent means for civic engagement in scientific advances. The media operate at the interface between scientific communities and lay publics, and thus serve as key gatekeepers of information between science and the society. The media may exert a substantial influence over public perceptions and opinions of controversial biotechnologies through deliberate reporting techniques. Two interrelated methods are prominently used in science reporting: agenda setting and framing. Agenda setting theory posits that the media can lend salience to certain topics through selective coverage. Media outlets cover certain issues with greater prominence (e.g., front-page news) or frequency to direct public attention to certain issues while omitting others (McCombs & Shaw 1972). In a similar capacity, frames, or simplified interpretive packages, are used to direct public opinions about important issues. Frames help audiences organize and process complex information by drawing attention to some considerations surrounding controversial topics while downplaying others (Scheufele 1999). In this manner, the media not only tell people what to think about (agenda setting), but also shape how people think about scientific controversies (framing). While the media engage in deliberate processes to direct public attention and opinions, there is a reciprocal exchange between popular media portrayals and public priorities. Indeed, public priorities and concerns also direct media attention just as media coverage shapes public discourse.

Citizens' voices have been receiving increasing exposure with the advent of social media. New forms of media, such as Web 2.0 platforms that allow publics to actively create content rather than passively absorb information (e.g., blogs, YouTube, Twitter, Facebook, and wikis) are now reforming traditional models of top-down science communication and introducing additional avenues for publics to directly contribute to debates about science and policy. While social media platforms have only been widely utilized for less than two decades, their accessibility and global reach have made them widely impactful in health-information sharing and decision-making. Citizen-generated content on online platforms is a powerful information tool for health researchers and policymakers. Indeed, public use of social media platforms has also been the subject of a growing body of empirical inquiry, with innovative approaches such as web ethnographic and infodemiological methodologies that aim to utilize public knowledge to inform healthcare delivery and science policy (Eysenbach 2009). For example, real-time infodemiological analyses of Twitter content have been used to track public perspectives about immunization during the H1N1 epidemic, and respond to public concerns using risk communication approaches to promote health (Chew & Eysenbach 2010). Publics also increasingly utilize social media platforms in advocacy efforts. In March of 2014, Josh Hardy, a 7-year-old boy who suffered a lifethreatening adenoviral infection, was denied compassionate access to brincidofovir, an antiviral drug that was under investigation in adult clinical trials. Brincidofovir's manufacturer, Chimerix, explained that the company did not have the resources to support a compassionate use program of the antiviral drug. In response, Hardy's parents launched a wide-reaching Facebook and Twitter campaign to pressure both the US Food and Drug Administration and Chimerix to provide access. Catalyzed by public pressure, Chimerix and the Food and Drug Administration designed an open-label clinical trial that provided Hardy and 19 other pediatric patients access to brincidofovir (Goodman 2014).

Success stories for community engagement in science and policy

While public participation in science is often promoted through traditional and social media outlets, initiatives that promote reciprocal and porous dialogue at the intersection of science and society are also bolstered by a rich variety of empirical approaches. These include community-based participatory action research, deliberative democracies, and consensus conferences, among others. Areas of such inquiry are equally diverse, ranging in focus from community-based research on immunization decision-making (Kowal et al. 2015) to deliberative engagements about sequencing of salmon genomes (O'Doherty et al. 2010). Such research trajectories are based on a celebration of diverse civic ways of knowing (Stilgoe et al. 2014).

Citizen engagement in science has also taken numerous community-initiated forms. For instance, proactive publics have had significant impact on science policy and healthcare delivery in the neuro-sphere through advocacy. In this capacity, engaged advocates have largely articulated a niche for InSite, a supervised injection facility for individuals who use illicit drugs with an evidence-based harm reduction mandate in Vancouver, despite numerous political and social hurdles (Boyd et al. 2009; Jozaghi 2014). Recent years have also seen patient engagement in science through processes that have blurred the distinction between patient and researcher. In 2008, a phase II clinical trial originating in Italy concluded that lithium carbonate delays the rate of progression of amyotrophic lateral sclerosis (ALS), an untreatable and invariably fatal neurodegenerative disease. This small trial, which consisted of a sample of 16 patients who received the experimental intervention and 24 controls followed over a period of 15 months, demonstrated that in contrast to a 29% incidence of morbidity in the control group, no patient deaths occurred in the experimental arm (Fornai et al. 2008). Given a scarcity of clinical interventions for ALS and a limited therapeutic window of opportunity for clinically meaningful intervention, neurologists began to prescribe lithium off-label. In response to demands for an ongoing evidentiary basis for the use of lithium, advocates in the ALS community in collaboration with PatientsLikeMe—an online patient network that aims to connect patients, improve outcomes, and promote research—initiated a platform for patients who had obtained lithium off-label to submit data about their usage and outcomes. Data probes included dosage, weight, and outcome measures on the ALS functional rating scale. Data from 149 treated ALS patients were matched with historical controls in an observational study that did not replicate the efficacy data suggested by Fornai and colleagues (Wicks et al. 2011). This patient-led research endeavor contributed to the early termination of clinical trials due to futility considerations, and spared the ALS community significant divestment of funds from promising areas of inquiry and opportunity costs for patients.

While numerous academic platforms are dedicated to public participation in science, and successes in community advocacy are many, public controversies have also revealed challenges in community engagement. We now direct focus to the case study of the CCSVI research trajectory, as an example of the challenges of community engagement.

The CCSVI story

Multiple sclerosis (MS) is a chronic and progressive neurological disease of the brain and the spinal cord that affects more than two million individuals worldwide (Multiple Sclerosis International Federation 2015). In people with MS, the immune system attacks myelinated axons in the central nervous system. This causes communication problems between the brain and the rest of the body that lead to a range of symptoms such as vision loss, fatigue, pain, sensory loss, spasticity, impaired mobility, and cognitive deficits. Initially these symptoms may fluctuate. However, over time these problems accumulate and are often irreversible. Since the 1990s, disease-modifying drugs have decreased the frequency of developing new symptoms and delayed the onset of progressive decline. Unfortunately, these treatments only appear to benefit those at the earliest stages of the disease, and have little impact on improving or reversing chronic symptoms that contribute to a decrease in quality of life. The unknown etiology of the disease, especially the progressive forms, presents challenges to finding a cure, and leaves many severely disabled.

Dr. Paulo Zamboni, an Italian vascular surgeon, began to tackle the underlying cause of MS when his wife was diagnosed with the disease in 1999. His primary area of inquiry, involving investigation into heavy metal aggregate damage to blood vessels in the leg, led him to suggest that iron deposits that are caused by narrowed or blocked veins in the neck (cerebrospinal venous insufficiency, or CCSVI) are at the root of the disease. Zamboni believed that such blockage hinders the efficient removal of blood from the brain and spinal cord, and causes a build-up of iron that induces inflammation and myelin sheath attack. Zamboni postulated two hypotheses. First, he proposed that all individuals with MS have narrowed or blocked veins, and second, that relieving such venous anomalies using a procedure similar to angioplasty, which he termed *liberation therapy*, would cure MS.

To test his first hypothesis, Zamboni performed a cross-sectional study where he looked to diagnose CCSVI in 65 MS patients and 235 controls using ultrasound and catheter venography. Indeed, his hypothesis was supported by results that established a strong association between MS and CCSVI with vein anomalies in virtually the entire MS cohort and none of the controls (Zamboni et al. 2009a). To test his second hypothesis, Zamboni conducted an open-label trial of the liberation procedure in 65 patients with MS, who showed symptomatic improvement, reduced clinical relapses and reduced formation of new lesions on brain magnetic resonance imaging (MRI) following the procedure (Zamboni et al. 2009b).

Since 2009, several studies have brought Zamboni's evidence into question. Scholars expressed skepticism about Zamboni's studies, pointing to several limitations in the methods employed. Indeed, Zamboni's liberation procedure was not conducted through a randomized control trial and did not account for a potential placebo effect. Additionally, patients who experienced the most improvement in Zamboni's study had the relapsing–remitting form of MS, where spontaneous remissions in disease activity commonly occur.

Moreover, although a meta-analysis of ultrasound diagnoses of CCSVI suggests an association between vein abnormality and MS (Laupacis et al. 2011), other studies demonstrate that CCSVI occurs rarely and with similar prevalence among both patients with MS and in healthy individuals (Doepp et al. 2010; Mayer et al. 2011; Traboulsee et al. 2014).

Commentators believe that the substantial evidence undermining Zamboni's hypotheses should have been the end of the CCSVI story (Reekers 2012). The cautionary word of the medical community, however, was seemingly dwarfed by optimistic anecdotes that amassed publicly accessible domains. Indeed, despite concerns about the validity of the data that bolstered the liberation procedure in the neurology community, news of Zamboni's study catalyzed a media frenzy and showcased a potent account of evidence: the anecdote.

The procedure was widely covered in international newspapers and was also commonly featured in television outlets (Favaro & Philip 2009). However, it was publicity through social media platforms that fueled widespread activism within the MS community. Indeed, CCSVI gained significant momentum in the blogosphere: within 2 years of the publication about the liberation procedure, thousands of people engaged with more than 500 Facebook groups, pages, and events about the intervention, and over 4000 YouTube videos were dedicated to CCSVI, many documenting the positive experiences of patients who underwent the procedure (Chafe et al. 2011). These powerful anecdotes showcased patient accounts before and after the procedure, which not only described the amelioration of their symptoms and improved quality of life, but also demonstrated gains in function (Mazanderani et al. 2013).

Anecdotal evidence motivated thousands of patients to seek access to CCSVI proce-dures, and when the intervention was unavailable through their local healthcare provid-ers, many chose to travel abroad at great personal cost to countries like Costa Rica and Poland that offered the CCSVI procedure (Snyder et al. 2011). Even in the face of several severe adverse events linked to CCSVI procedures, including deaths (Alphonso 2010; Samson 2010), patients persisted in advocacy efforts for access to the invasive procedure. Advocates continued to publicize personal anecdotes of the therapeutic benefits of the procedure through social media outlets, organize demonstrations, attack the credibility of parties that advocated for caution, and question potential conflicts of interest of both the MS societies and MS neurologists. Steadfast advocacy generated enormous pressures on granting agencies such as MS societies and government funding bodies in Europe, Australia, and North America to mobilize additional research on CCSVI. The controversy became particularly pronounced in Canada, where MS is very prevalent (Pullman et al. 2013). In 2010, the Canadian Institutes of Health Research, the primary government fund-ing agency for health research in Canada, formed an expert panel in collaboration with the Canadian MS Society about the CCSVI research agenda. The panel initially advised that only studies that examine the venous anomaly hypothesis receive funds, a stance that was also adopted by the United States. But despite expert caution, media pressures contin-ued to hail the potential of the liberation treatment and reputable media platforms began to exert political pressure to mobilize funding for CCSVI research (Pullman et al. 2013). For example, *The Globe and Mail*, one of Canada's most circulated newspapers, asserted: "Canada should fund medical trials of revolutionary treatment for multiple sclerosis, and not act as if rejection of those trials by an expert panel must be obeyed. This is a political question, not a purely scientific one" (Globe Editorial 2010).

Critics argue that CCSVI research was mobilized by intense social and political pres-sure, more so than by sound science (Chafe et al. 2011). Commentators believe that the CCSVI research trajectory has resulted in the unnecessary expenditure of scarce resources, divestment of funds from more promising areas of inquiry for MS, and caused signifi-cant therapeutic opportunity costs, adverse events, and preventable fatalities (Rasminsky 2013). Additionally, the CCSVI experience has contributed to unrealized hopes in the MS community, the impact of which is yet to be determined. Given these costs, it is essential that we examine the lessons learned from the CCSVI research trajectory to inform future public engagement in science and science policy.

What have we learned? Drawing lessons from community engagement in the neurosciences

The CCSVI experience illustrates a divergence between stakeholders' perspectives on what constitutes sound evidence. While the CCSVI hypotheses were largely met with skepticism in the neurology community, the evidence encouraged patient hope that mobilized intense public and political pressure to provide access to liberation interventions and invest in their evaluation. Educational initiatives may be needed to address public evaluation of diverse forms of evidence, particularly in light of the powerful and readily accessible anecdotes on online fora. The CCSVI experience also emphasizes the power of the media, particularly newer forms of social media, in influencing science policy and priority setting in research. Scholarly inquiry should continue to focus on harnessing the power of social media platforms to foster sustainable support for evidence-based science. Education about evidence should be a bidirectional effort. Scientists may reciprocally consider that the contemporary focus on evidence-based medicine sometimes overlooks a wide body of knowledge that has been gleaned from medicine-based evidence approaches. Indeed, fields such as surgery and oncology have largely seen innovation based on clinical experience. An educational initiative alone, however, would be a reductionist approach. Indeed, educational approaches do not adequately account for affective factors, such as hope, that extend beyond a deficit of understanding and exert a large influence on the decision-making process.

The CCSVI experience brings the delicate nature of hope to the surface—a force that ubiquitously surrounds biomedical research, and can serve to both mobilize and sometimes hinder successful scientific innovation. Hope largely motivated activism in the CCSVI context, as it often serves to promote developing biotechnologies. Hope is a natural and necessary component of technological development, and can also serve as an adaptive mechanism for managing daily living with serious illness (Groopman 2005; Kimmelman 2009). Successful community engagement about developing neurotechnologies will therefore need to address the complexities of hope with key end-users, such as patients and their families. Clinical conversations about hope may be particularly useful. Neurologists may best serve the interests of their patients by approaching conversations about hope through a posture of epistemic (knowledge-based) humility: a stance that acknowledges the uncertainty associated with decision-making in medical care, and one that also appreciates the lived experience that lends expertise to patients (Ho 2009; Schwab 2012). Honoring patient hope for therapeutic development while grounding practical advice in current clinical realities may strengthen collaboration between patients and their clinicians and support shared decision-making through an emphasis on informed hope (Reimer et al. 2010; Benjaminy et al. 2015). Conversations through a lens of epistemic humility that privilege openness and understanding over hierarchical assertion of expertise could also serve to strengthen relationships of trust between patients and physicians. Indeed, commentators argue that such relationships may have been strained in the CCSVI context (Pullman et al. 2013; Snyder et al. 2014). Additional research is necessary to explore the impact of the CCSVI research trajectory on community support in analogous non-pharmaceutical areas of inquiry that present the possibility for therapeutic development (e.g., stem cell interventions). Such knowledge may be helpful in understanding the response of communities that have been subject to disappointment following cycles of hope—a phenomenon that occurs all too often in biomedical research (Petersen 2009).

Finally, it is key that community engagement in science and policy be reciprocally complemented by responsible scientific citizenship. While the focus of this chapter is on community engagement, we should not underestimate the importance of engaged scholarship. Neuroscientists and neuroethicists have a social obligation to share their knowledge and expertise with the public, particularly as academic pursuits are often resourced through the public purse. As such, scientists should be encouraged to interface with citizens about the uncertainties, caveats, as well as the promise of their work. Scholars have suggested that the CCSVI experience illustrates the need to include the voices of scientists in social media platforms through which many public advocacy campaigns that mobilize science policy are launched, instead of disregarding such fora as untrustworthy or not evidence-based (Mazanderani et al. 2013). However, engaged scholarship will not be sustainable unless it is incentivized by the academic infrastructure (Boyer 1996; Illes et al. 2010). It is therefore important that a culture that values engaged scientific citizenship be embedded into the core of academic training in the long term, and into academic evaluation in the short term.

Conclusion

Public engagement in science is a central value in the field of neuroethics, and reflects a commitment to social responsibility and accountability in the process of democratizing science and policy. A commitment to citizen engagement recognizes the possibility that public controversies may arise as neuroethical issues are deliberated in the public sphere. The CCSVI research trajectory illustrates some of the challenges that may arise when public and scientific agendas diverge. Forging successful partnerships between neuroscientists, neuroethicists, and the broader community will require both trust and hope as well as reciprocity between mutually engaged publics and scholars.

Acknowledgments

The quotation attributed to Globe Editorial was reproduced by kind permission of *The Globe and Mail*, Copyright © 2016 The Globe and Mail Inc., http://www.theglobeandmail.com/opinion/editorials/funding-ms-trials-is-a-decision-that-goes-beyond-expertise/article1379043/.

References

Alphonso, C. (2010, November 19). Death of MS patient fuels debate over new treatment. *The Globe and Mail*. Available from: http://www.theglobeandmail.com/news/national/death-of-ms-patient-fuels-debate-over-new-treatment/article1314786/ [accessed January 22, 2016].

Benjaminy, S., Kowal, S.P., MacDonald, I.M., and Bubela, T. (2015). Communicating the promise for ocular gene therapies: challenges and recommendations. *American Journal of Ophthalmology*, 160(3), 408–415.

Boyd, S.C., MacPherson, D., and Osborn, B. (2009). *Raise Shit! Social Action Saving Lives*. Black Point, NS: Fernwood.

Boyer, E.L. (1996). The scholarship of engagement. *Bulletin of the American Academy of Arts and Sciences*, **49**(7), 18–33.

Chafe, R., Born, K.B., Slutsky, A.S., and Laupacis, A. (2011). The rise of people power. *Nature*, **472**(7344), 410–411.

Chew, C. and Eysenbach, G. (2010). Pandemics in the age of Twitter: content analysis of Tweets during the 2009 H1N1 outbreak. *PloS One*, **5**(11), e14118.

Doepp, F., Paul, F., Valdueza, J.M., Schmierer, K., and Schreiber, S.J. (2010). No cerebrocervical venous congestion in patients with multiple sclerosis. *Annals of Neurology*, **68**(2), 173–183.

Eysenbach, G. (2009). Infodemiology and infoveillance: framework for an emerging set of public health informatics methods to analyze search, communication and publication behavior on the Internet. *Journal of Medical Internet Research*, **11**(1), e11.

Favaro, A. and St Philip, E. (2009). The liberation treatment: a whole new approach to MS. *CTV News*. Available from: http://www.ctvnews.ca/the-liberation-treatment-a-whole-new-approach-to-ms-1.456617 [accessed February 12, 2016].

Fornai, F., Longone, P., Cafaro, L., et al. (2008). Lithium delays progression of amyotrophic lateral sclerosis. *Proceedings of the National Academy of Sciences of the United States of America*, **105**(6), 2052–2057.

Globe Editorial. (2010, September 2). Funding MS trials is a decision that goes beyond expertise. *The Globe and Mail*. Available from: http://www.theglobeandmail.com/opinion/editorials/funding-ms-trials-is-a-decision-that-goes-beyond-expertise/article1379043/ [accessed January 22, 2016].

Goodman, M. (2014). Twitter storm forces Chimerix's hand in compassionate use request. *Nature Biotechnology*, **32**(6), 503–504.

Groopman, J. (2005). *The Anatomy of Hope: How People Prevail in the Face of Illness*. New York: Random House.

Ho, A. (2009). 'They just don't get it!' When family disagrees with expert opinion. *Journal of Medical Ethics*, **35**(8), 497–501.

Illes, J., Blakemore, C., Hansson, M.G., et al. (2005). International perspectives on engaging the public in neuroethics. *Nature Reviews Neuroscience*, **6**(12), 977–982.

Illes, J., Moser, M.A., McCormick, J.B., et al. (2010). Neurotalk: improving the communication of neuroscience research. *Nature Reviews Neuroscience*, **11**(6), 61–69.

Jozaghi, E. (2014). The role of drug users' advocacy group in changing the dynamics of life in the Downtown Eastside of Vancouver, Canada. *Journal of Substance Use*, **19**(1–2), 213–218.

Kimmelman, J. (2009). *Gene Transfer and the Ethics of First-in-human Research: Lost in Translation*. Cambridge: Cambridge University Press.

Kowal, S.P., Jardine, C.G., and Bubela, T.M. (2015). 'If they tell me to get it, I'll get it. If they don't...': immunization decision-making processes of immigrant mothers. *Canadian Journal of Public Health*, **106**(4), E230.

Laupacis, A., Lillie, E., Dueck, A., et al. (2011). Association between chronic cerebrospinal venous insufficiency and multiple sclerosis: a meta-analysis. *Canadian Medical Association Journal*, **183**(16), E1203–1212.

Leshner, A.I. (2005). It's time to go public with neuroethics. *American Journal of Bioethics*, **5**(2), 1–2.

Marcus, S.J. (Ed) (2002). *Neuroethics: Mapping the Field, Conference Proceedings*. Chicago, IL: University of Chicago Press.

Mayer, C.A., Pfeilschifter, W., Lorenz, M.W., et al. (2011). The perfect crime? CCSVI not leaving a trace in MS. *Journal of Neurology, Neurosurgery & Psychiatry*, **82**(4), 436–440.

Mazanderani, F., O'Neill, B., and Powell, J. (2013). "People power" or "pester power"? YouTube as a forum for the generation of evidence and patient advocacy. *Patient Education and Counseling*, **93**(3), 420–425.

McCombs, M.E. and Shaw D.L. (1972). The agenda-setting function of mass media. *Public Opinion Quarterly*, **36**(2), 176–187.

Multiple Sclerosis International Federation. (2015). *What is MS?* Available from: http://www.msif.org/about-ms/what-is-ms/ [accessed February 10, 2016].

O'Doherty, K., Burgess, M., and Secko, D.M. (2010). Sequencing the salmon genome: a deliberative public engagement. *Life Sciences Society and Policy*, **6**(1), 15.

Petersen, A. (2009). The ethics of expectations: biobanks and the promise of personalized medicine. *Monash Bioethics Review*, **28**(1), 22–33.

Pullman, D., Zarzeczny, A., and Picard, A. (2013). Media, politics and science policy: MS and evidence from the CCSVI trenches. *BMC Medical Ethics*, **14**(1), 6.

Rasminsky, M. (2013). Goodbye to all that: a short history of CCSVI. *Multiple Sclerosis Journal*, **19**(11), 1425–1427.

Reekers, J.A. (2012). CCSVI and MS: a never-ending story. *European Journal of Vascular and Endovascular Surgery*, **43**(1), 127–128.

Reimer, J., Borgelt, E., and Illes, J. (2010). In pursuit of "informed hope" in the stem cell discourse. *American Journal of Bioethics*, **10**(5), 31–32.

Samson, K. (2010). Experimental multiple sclerosis vascular shunting procedure halted at Stanford. *Annals of Neurology*, **67**(1), A13–15.

Scheufele, D.A. (1999). Framing as a theory of media effects. *Journal of Communication*, **49**(1), 103–122.

Schwab, A. (2012). Epistemic humility and medical practice: translating epistemic categories into ethical obligations. *Journal of Medicine and Philosophy*, **37**(1), 28–48.

Snyder, J., Crooks, V.A., Adams, K., Kingsbury, P., and Johnston, R. (2011). The 'patient's physician one-step removed': the evolving roles of medical tourism facilitators. *Journal of Medical Ethics*, **37**(9), 530–534.

Snyder, J., Adams, K., Crooks, V.A., Whitehurst, D., and Vallee, J. (2014). "I knew what was going to happen if I did nothing and so I was going to do something": faith, hope, and trust in the decisions of Canadians with multiple sclerosis to seek unproven interventions abroad. *BMC Health Services Research*, **14**(1), 445.

Stilgoe, J., Lock, S.J., and Wilsdon, J. (2014). Why should we promote public engagement with science? *Public Understanding of Science*, **23**(1), 4–15.

Traboulsee, A.L., Knox, K.B., Machan, L., et al. (2014). Prevalence of extracranial venous narrowing on catheter venography in people with multiple sclerosis, their siblings, and unrelated healthy controls: a blinded, case-control study. *The Lancet*, **383**(9912), 138–145.

Wicks, P., Vaughan, T.E., Massagli, M.P., and Heywood, J. (2011). Accelerated clinical discovery using self-reported patient data collected online and a patient-matching algorithm. *Nature Biotechnology*, **29**(5), 411–414.

Zamboni, P., Galeotti, R., Menegatti, E., et al. (2009a). Chronic cerebrospinal venous insufficiency in patients with multiple sclerosis. *Journal of Neurology, Neurosurgery & Psychiatry*, **80**(4), 392–399.

Zamboni, P., Galeotti, R., Menegatti, E., et al. (2009b). A prospective open-label study of endovascular treatment of chronic cerebrospinal venous insufficiency. *Journal of Vascular Surgery*, **50**(6), 1348–1358.

Chapter 15

Ethical dilemmas in neurodegenerative disease: Respecting patients at the twilight of agency

Agnieszka Jaworska

> [A] man does not consist of memory alone. He has feeling, will, sensibilities, moral being ... And it is here ... that you may find ways to touch him.
>
> *(A.R. Luria, cited in Sacks 1985, p.32)*

Introduction

Neurodegenerative disease can significantly affect the requirements for an ethically appropriate treatment of an individual. The deterioration of the brain profoundly alters psychological functioning and capacities, and so profoundly alters the kind of being others are called upon to relate to and respond to from the point of view of ethics. Such transformations are easiest to see in cases in which the current preferences and interests of a patient in a degenerated state come into conflict with the values and choices the person professed before the neurodegeneration began. Which set of preferences should the caregiver follow? The answer will be different depending on the ethically relevant metaphysical and mental properties that the patient still possesses. Thus to resolve such dilemmas we must get the ethically relevant conceptual distinctions right, but we must also be true to empirical facts. To the extent that neuroscience has developed a detailed understanding of how various brain disorders undermine psychological functioning and capacity, it has increasingly become more pertinent to the untangling of such ethical puzzles. The following case study illustrates the interplay between ethical conceptual analysis and neuroscientific findings in the resolution of moral dilemmas that arise in Alzheimer's disease.

I defend the philosophical view that the immediate interests of an individual cannot be overridden as long as the individual possesses the capacity to value. In the context of each particular neurodegenerative disease, this recommendation must be guided by a scientifically informed assessment of when in the course of the disease the capacity to value could possibly be lost, and when it is likely to be retained. In the case of Alzheimer's disease, neuroscientific evidence indicates that the capacity to value is slowly and gradually weakened, and in some cases may not be completely lost until relatively far along in the disease's progression. Similar neuroethical analyses must be carried out for other diseases and disorders, and will probably yield different results.

Respecting the margins of agency: Alzheimer's patients and the capacity to value

Case scenario: Mrs. Rogoff was always an independent woman. Raised in an immigrant family, she was used to working hard for what she wanted. Most of her life she ran a successful business selling liquor. She also developed local fame as an outstanding cook and hostess. She was an introvert, liked living alone, and always carefully guarded the way she presented herself to others.

In her early eighties, Mrs. Rogoff developed severe motor impairments, which could only be corrected by a risky neurosurgery. She decided to undergo the procedure, insisting that she would rather die than be immobile. She prepared a living will, requesting not to have her life prolonged if she became a burden to her family or if she could no longer enjoy her current quality of life.

The surgery was successful, but shortly thereafter Mrs. Rogoff developed early signs of dementia: memory and word-finding difficulties. As she became more and more disoriented, her daughter hired a live-in housekeeper, Fran. Fran takes care of Mrs. Rogoff in the way one would take care of a child. Mrs. Rogoff enjoys the long hours she spends with Fran, and with her grandchildren when they visit, telling them somewhat disjointed stories about her earlier ventures. She watches television a lot, and her stories often incorporate the more exciting episodes from television as if they pertained to her own life. In her more lucid moments, Mrs. Rogoff tells her grandchildren that she is scared to die, that "she doesn't want to go anywhere."

Fran has to make day-to-day decisions for Mrs. Rogoff: Should Mrs. Rogoff get dressed if her family is coming to visit and she insists on wearing pajamas? Should she take a bath every day even if she is afraid of water? In general, should the current decisions reflect the care Mrs. Rogoff used to take in how she presented herself to others? Mrs. Rogoff's daughter faces the more weighty decisions: Should she use up Mrs. Rogoff's savings to pay Fran's salary, allowing Mrs. Rogoff to keep enjoying her companion, or should she place Mrs. Rogoff in a nursing home, increasing the likelihood that, when the time comes, there will be some money left to execute Mrs. Rogoff's will? What treatments should she authorize if Mrs. Rogoff develops a dangerous but treatable infection?[1]

People who care for Alzheimer's patients—family members, nursing home providers, physicians, medical researchers—face such dilemmas routinely, and these dilemmas are becoming more and more familiar as baby-boomers approach the age of high risk for Alzheimer's disease. The particulars of each dilemma may seem unique, but they typically have the same underlying structure. There is a conflict between the attitudes and values the patients espoused when they were still healthy and their later interests as people afflicted with dementia. The quandary, in a nutshell, is this: should we, in our efforts to best respect a patient with dementia, give priority to the preferences and attitudes this person held before becoming demented, or should we follow the person's present interests?

Competing theoretical perspectives

There are two dominant theoretical perspectives on how such dilemmas ought to be resolved, expressed most prominently by Rebecca Dresser and Ronald Dworkin. According to Dresser, decisions affecting a demented person at a given time must speak to the person's point of view as current at that time. Heeding values and wishes that the patient no longer espouses and that cannot be said to represent his present needs and interests can do no good for the patient (Dresser 1986).

Dworkin directly challenges this line of reasoning, adducing compelling reasons to adhere to the demented patients' earlier wishes and values (Dworkin 1993). In Dworkin's view, we fail to take seriously both the autonomy and the well-being of a demented patient unless we adhere strictly to the patient's earlier wishes, wishes that originated when he was still capable of acting autonomously and still able to judge what was required for his overall well-being.

In this chapter, I develop an alternative to both Dresser's and Dworkin's analyses. Like Dresser, I want to take seriously the current interests of demented patients, but for very different reasons: I believe that many of these patients may still be capable of autonomy to a significant degree and that they may still have authority concerning their well-being. Yet I emphasize very different aspects of both autonomy and well-being than Dworkin, who predicates autonomy on decision-making capacity and for whom well-being depends centrally on promoting one's own design for one's life as a whole. I associate potential for autonomy primarily with the capacity to value, and well-being with living in accordance with one's values (Jaworska 1997). Thus the central question for a caregiver attempting to best respect an Alzheimer's patient becomes not, "Can this patient reason thoroughly and come to a rational decision?" or "Does he grasp what's best for his life as a whole?" but "Does this patient still value?" I will argue that the capacity to value is not completely lost in dementia, and, to the extent that it is not, respect for the immediate interests of a demented person is contrary neither to his well-being nor to the respect for his autonomy.

Dworkin's arguments are a fruitful point of departure because their kernel is very plausible. After someone loses his status as an agent—as a creature capable of guiding his actions—his earlier autonomously chosen values should continue to govern what happens to him, despite his current inability to appreciate these values. This accurately locates the central issue of our dilemmas: at what point in the course of dementia are the attributes essential to agency lost? While I consider most of the ideas in Dworkin's arguments well founded, I challenge his two crucial premises. In the argument focused on the patient's well-being, I dispute the claim that demented patients are no longer capable of generating what Dworkin calls "critical interests." In the argument concerning autonomy, I question the premise that demented patients no longer possess the "capacity for autonomy." In each case, I will trace how the problematic premise arises within Dworkin's argument and then develop an alternative account of the relevant capacity.

Reconceiving well-being

Experiential interests versus critical interests

When we take enhancement of the demented patient's well-being as the caregiver's goal, we need to distinguish two types of prudential interest that the patient may have. Dworkin labels these two types of interest "experiential" and "critical." Experiential interests concern the quality of the person's experience, his state of mind. We have an interest in experiencing pleasure, satisfaction, enjoyment, contentment, lack of pain, and so forth; what states of mind count here and how they can be brought about is determined fully by how these experiences feel to the person in question. But most of us also have critical interests—interests in doing or having in our lives the things we consider good and in avoiding the things we consider bad, no matter what sorts of experiences result from fulfilling these interests. For example, a person's critical interests may include being a successful soldier or securing contentment and harmony for his family, however much stress or anguish the pursuit of these goals may engender.

Experiential interests are inherently time-specific—to satisfy them at a given time, the person must still espouse them at that time. For instance, it only makes sense to want to satisfy your experiential interest in having a good time with your guests if at the time they arrive you will still be interested in enjoying their company. Not so with critical interests; it may make sense to have your critical interest satisfied even if you are unaware of its being satisfied, even if you are dead at the time, or unconscious, or too demented to grasp what your critical interest has been all along. Fulfillment of a critical interest bears only on the object of that interest. It involves bringing about whatever state of affairs the person in question judged good; the fate of the person himself is not relevant to this, provided that the object of the interest is external to the person. Thus fulfilling a father's critical interest in the well-being of his family does not require his awareness of how well they are doing. This critical interest could be advanced, for example, by obeying a deathbed promise to take care of his wife and children after he passes away.

Dworkin readily grants that Alzheimer's patients, even at late stages of their illness, can experience pleasures as well as frustrations, and thus have the basis for contemporaneous experiential interests. He would interpret the dilemmas under discussion in this chapter as cases of conflict between these experiential interests and critical interests that the person professed when he was still healthy (e.g. a conflict between Mrs. Rogoff's experiential interest in continuing the enjoyable storytelling sessions and her critical interest in not being dependent on her family). And here Dworkin assumes that, at least in the types of cases he wants to address, the demented patient is not capable of generating contemporaneous critical interests. From this point, there follows a very plausible analysis. The fact that the demented patient no longer affirms critical interests in no way implies that he does not have critical interests. Since such interests are not inherently time-specific, the prudential importance of satisfying them may survive the person's unawareness of their satisfaction, whether due to unconsciousness, dementia, or even death. Thus a demented person who cannot generate contemporaneous critical interests may still have some of the same critical interests he professed when he was healthy. And this means that the conflict occurring in our dilemmas is best described as a conflict between the patient's *ongoing* experiential interests and his *ongoing* critical interests.

This description helps to clarify how the conflict ought to be resolved. In the case of an ordinary competent person, when his critical interests (his judgments and values) come into conflict with his experiential interests (what would lead to the optimal state of mind for him), we do not hesitate to give precedence to his well-considered values and judgments, and we concede that, overall, this is best for him. For example, we would accept that it is in the best interest of a devoted father to sacrifice his experiential interest in his current comfort for the sake of the future of his children, or that it is in the best interest of a patriotic soldier to forgo his experiential interest in a carefree life and sign up for demanding military training. The case of our demented person turns out to be no different: in *his* conflict between ongoing experiential and critical interests, it is also best to privilege the latter. We serve Mrs. Rogoff best by satisfying her critical interest in not being a burden to her family, at the expense of her experiential interest in enjoying television and storytelling.

This analysis stands or falls with the assumption that demented patients no longer originate critical interests. For if they do—if the conflict in our dilemmas is between the patient's contemporarily professed critical interests and the critical interests he professed before becoming demented—Dworkin's framework would readily allow that the contemporarily professed critical interests ought to take precedence. In this case, the demented person would be viewed as any other person whose values and commitments change over time and whose *currently* professed values are taken to have bearing on what is best for him.

The idea that a demented person can originate critical interests need not imply that the person generates brand new critical interests. What matters is whether the person still has an ongoing commitment to his critical interests. After all, the most likely scenario of a conflict between a demented person's current critical interests and his critical interests from the pre-dementia period is not one in which a person's demented mind generates completely new critical interests, but rather one in which dementia causes a person to lose some of his earlier more complex interests, so that in the new simpler configuration the remaining interests gain import.

Dworkin defends the claim that the demented cannot generate critical interests as follows:

> [B]y the time the dementia has become advanced, Alzheimer's victims have lost the capacity to think about how to make their lives more successful on the whole. They are ignorant of self—not as an amnesiac is, not simply because they cannot identify their pasts—but more fundamentally, because they have no sense of a whole life, a past joined to a future, that could be the object of any evaluation or concern as a whole. They cannot have projects or plans of the kind that leading a critical life requires. They therefore have no contemporary opinion about their own critical interests. (Dworkin 1993, p.230)

In contending that demented persons cannot have opinions about their critical interests, Dworkin presupposes that one needs to have a sense of one's life as a whole in order to originate critical interests, a sense that a person may begin to lose relatively early in the progression of dementia. Dworkin thinks of critical interests as stemming from "convictions about what helps to make a life good on the whole" (Dworkin 1993, pp.201–202). But do critical interests have to reflect the person's comprehensive design for the progression of his life? An alternative view, tacitly embedded in Dworkin, is more plausible.

Critical interests may well be understood to issue from something less grand—simply from convictions about what is good to have, which do not require the ability to grasp or review one's whole life. Dworkin himself describes "opinions about my critical interests" as "opinions about what is good for me" (Dworkin 1993, p.202), indicating that these are opinions about values, and that the ability to generate critical interests goes hand in hand with the ability to value. And it does seem possible for a person to value something at a given time, without referring this value to his conception of his life as a whole. This possibility is evident in patients with severe loss of memory and linguistic ability who are still aware of their decline and deeply regret it. I recently observed a patient who, in response to a simple question about what he did that day, had great difficulty keeping track of the sequence of his thoughts and of the sequence of words in the sentence he was composing, and after several starts and long pauses, said slowly, his voice trembling: "Here you can see Alzheimer's at work." There was no doubt that this man, who had little grip on "his life as a whole," nevertheless valued the abilities he could no longer command and was expressing profound regret.

Understanding values

Intuitively, it is easy to recognize when someone expresses a value—and not merely a simpler attitude such as a craving, a desire, or a wish. But to exhibit more clearly that valuing need not involve a grasp of one's life as a whole, let us characterize more systematically what valuing is and distinguish it from these simpler attitudes, which even Dworkin readily attributes to Alzheimer's patients.

The main difference between *mere* desiring and valuing is this: one way to deal with one's non-value-laden desires is to try to eliminate them—to try to bring it about that one does not feel them—but this would not be a satisfactory response to a valuing. A person could contemplate being free of a mere desire with a sense of relief, but one would always view the possibility of not valuing something one currently values as an impoverishment, loss, or mistake. We can all recognize clear cases when a strong desire is definitely not a value—think of a priest eager to rid himself of his sexual desires, or a temporarily depressed person seeking relief from his persistent wish to hurt himself. Also, even if one's attitude toward a desire is more neutral—perhaps one views it only as a nuisance—as long as one would not mind lacking it, it is still a mere desire. Cravings for specific food items are paradigmatic cases here. In contrast, when one values something, say, a particular friendship or a local community, one cannot be indifferent to whether one happens to value these things or not—a state in which one lacked one's feelings for one's friend or one's need for a sense of belonging would call for regret.[2] We can see this in our patient mourning the losses caused by Alzheimer's disease—he would view with horror the projected future in which he will no longer care about these losses.

Thus values have special attributes that do not apply to mere desires: we think that it would be a mistake to lose our current values—we hold our values to be correct, or at least correct for us. This means that we can typically give a rationale for why we consider something valuable or good, usually by situating this value in a larger normative framework. Also, since values are the sorts of attitudes that we allow could be correct or incorrect, they are open to criticism and revision. At minimum, there are consistency requirements on what one can consider good—if one values something and also values its opposite, one will be under rational pressure to resolve the conflict. For example, if you truly value a committed relationship, you cannot just as easily value the freedom of a lack of commitment; you may well see the merits of both, but you cannot be fully committed to your spouse unless you cease to care as deeply about your freedom. In contrast, as a matter of sheer desire, you may surely remain ambivalent without any rational impetus to settle the conflict—you may simply keep craving just as strongly the kind of intimacy possible only in an ongoing partnership as well as the excitement of always being able to walk out and entice a new partner.

Another mark of valuing as opposed to mere desiring is that a person's values are usually entangled with his sense of self-worth: a person values *himself* in terms of how well he lives up to his values. Some people pay little attention to their own value, so what I am now describing is not a necessary condition of having values. However, it is a sufficient condition: anyone who has a conception of himself, a set of ideals that he wants to live up to and in virtue of which he assesses his own value, is no doubt a valuer.

I have isolated two features essential to, or strongly indicative of, valuing: the person thinks he is correct in wanting what he wants, and achieving what he wants is tied up with his sense of self-worth. Nothing here suggests that valuing would require a grasp of the narrative of one's whole life.

Furthermore, for the purposes of the argument I outlined earlier, Dworkin does not need to interpret the capacity to generate critical interests as anything more than the so-specified capacity to value. As we have seen, the backbone of Dworkin's justification for disregarding current wishes of patients who can no longer originate critical interests is the perception that, ordinarily, critical interests take precedence over experiential interests in determining what is best for a person. But, presumably, critical interests are of such overriding importance because they stem from the person's values—because they reflect the person's opinion of what is correct for him. And this standing of critical interests is independent of whether they encompass the person's design for his life as a whole. For instance, a devoted father's critical interest in the well-being of his children overrides his interest in having optimal experiences, no matter whether he came to value his children by reflecting on the narrative of his whole life. Thus, to endorse Dworkin's compelling argument that deference to current wishes of a demented patient ought to depend on whether the patient can still originate critical interests, we have no need to understand critical interests in terms of the person's grasp of what is good for his life as a whole; we can just trace critical interests to the person's convictions about what would be good and correct for him—to the person's values as understood in the above-mentioned specifications.

Values do not require assessments of one's life as a whole

Of the three claims I have made—that critical interests are values, that conceptually such values may be understood as quite independent of the agent's grasp of his life as a whole, and that this is the interpretation relevant to Dworkin's argument—the second is the most contentious. However, it is confirmed by many real-life Alzheimer's cases in which valuing becomes uncoupled from the person's grasp of the narrative of his whole life.

Consider, for example, Mrs. D, a patient interviewed in a study by Sabat (1998), diagnosed with probable Alzheimer's for 5 years. As for her level of impairment:

> [S]he was moderately to severely afflicted (stage 4, Global Deterioration Scale, score of 7 on the Mini-Mental State Test). She could not name the day of the week, the date, the month, the season, year, the city and county she was in … She underestimated her age … and had difficulty finding her way to the bathroom at the day care center she attended two days each week. (Sabat 1998, p.46)

Mrs. D's memory deficiency was rather acute. Since she could not keep track of the passing time or of her own age, and had severe difficulties forming new memories, Dworkin could safely assume that she had lost grasp of the narrative of her whole life, that she lacked a sense of "a past joined to a future." However, Mrs. D still conducted herself as a valuer. She often volunteered as a research subject for tests and experiments at the National Institutes of Health. Although she did not choose to do so through systematic reflection on the whole trajectory of her life, she clearly felt that this was, for her, the right choice: "That was the nicety of it, cause I could have said, 'no,' but believe me, if I can help me and my [fellow] man, I would do it" (Sabat 1998, p.46). Her conviction that it would have been a mistake to say "no" comes across rather starkly here. And she had no need to review her life as a whole to affirm this conviction. What mattered for her was that this felt right to her, then and there. One has the sense that Mrs. D was simply enacting a basic part of her personality, one that had remained relatively intact despite her other impairments.

For a less altruistic example, consider another of Sabat's interviewees, Dr. B, an Alzheimer's patient who scored even lower than Mrs. D on cognitive tests. Like Mrs. D, he "could not recall the day of the week, the month, or the year" (Sabat 1998, p.41). His ability to evaluate his life as a whole could not have been better than that of Mrs. D. Yet he too proved capable of valuing. He became very interested in Sabat's research project. Although his grasp of its design was rather rudimentary, he thought of the project as his "salvation," as a way to engage, despite his impairments, in something other than "filler" (Sabat 1998, p.41), in something giving him a mark of distinction. He told Sabat more or less explicitly that he considered the project right and appropriate: "And you know I feel a way is that, I feel that this is a *real good*, big project, and I'm sure you do too. This project is a sort of scientific thing" (Sabat 1998, pp.41–42, italics mine). This assessment of the project went hand in hand with a boost to Dr. B's sense of pride and self-worth that ensued from his participation. The impact on his self-esteem was most evident whenever he compared the project to various "filler" group activities at the day care center: "If I'm working with you, I can—look, I can work in here for 30 times and all that, but in this group, *I'm nothing*" (Sabat 1998, p.41, italics mine). That his role in the project could so alter his self-image demonstrates most poignantly that he valued the project.

Mrs. Rogoff's case also demonstrates that the ability to value may outlast the patient's grasp of her life as a whole. Her confusion between a television-generated world and events of her own life easily rules out her having an adequate grasp of her life's narrative. However, she does remain a valuer, most clearly when her former reputation as a great cook is at stake. She invariably becomes upset and agitated seeing Fran usurp the mastery of the kitchen. One day, after Fran made a particularly delicious chicken leg roast, Mrs. Rogoff insisted that she would cook dinner herself, and asked her granddaughter, in secret, to buy "a hen with five legs," clearly in the spirit of one-upmanship with Fran. At such times, Fran arranges small make-work kitchen tasks that appease Mrs. Rogoff. Here, as before, the clearest indication of retained values comes from visible effects on the person's self-esteem: Mrs. Rogoff's self-image suffers whenever she realizes that Fran has taken over as a culinary expert, and these effects can be mitigated, at least temporarily, by a semblance of participation in culinary affairs.

Insights from neuroscience

My observations that valuing may be quite independent of grasping the narrative of one's life, and that this separation often occurs in Alzheimer's patients, are also supported by current findings in neurophysiology and in the neuropathology of Alzheimer's disease. The neuronal loss characteristic of Alzheimer's disease is not distributed evenly in the cerebral cortex. The disease affects most severely an area of the brain indispensable for maintaining the sense of one's life as a whole, but not particularly important for the ability to value.

In the early stages of Alzheimer's disease, the neuronal damage affects primarily the hippocampus. As the damage spreads, the hippocampus continues to be affected much more severely than other regions of the brain (Geula 1998; Laakso et al. 2000). The hippocampus is of crucial importance in the acquisition and processing of long-term explicit memory for facts and events. Although it is not involved in short-term memory or in the eventual storage of long-term memories, the hippocampus nonetheless plays an essential role in transforming a fresh short-term memory into a lasting long-term memory (Squire & Zola-Morgan 1991; Riedel & Micheau 2001). Accordingly, while damage to the hippocampus affects neither a person's processing of his immediate experience nor his memories of events that occurred long before the damage, it causes him to lose track of ongoing events soon after they happen, so that he typically has no recollection of the previous day (Squire & Zola-Morgan 1988). Such damage impairs a person's ability to come back to a recent thought or memory after a shift of attention to something new (Squire & Zola-Morgan 1991). These very impairments are, of course, the typical first clinical indications of Alzheimer's disease. They are also central to Dworkin's assessment that Alzheimer's disease destroys one's sense of one's life as a whole. Damage to the hippocampus alone leaves the person unable to update his autobiographical narrative. As he continually forgets his immediate past, he loses the sense of "a past joined to a future," which Dworkin deems necessary for the ability to formulate critical interests.

However, there is no reason to think that impairment of the hippocampus would obliterate one's ability to espouse critical interests when this is understood, following my recommendation, as the ability to value. While removal of the hippocampal formations leads to the memory defects described previously, it does not otherwise compromise the patient's mental functions (Young & Young 1997). Moreover, there is neurophysiological evidence that other regions of the brain are primarily responsible for interactions of reasoning and decision-making processes, especially those concerning personal and social matters, with feelings and emotions (Damasio 1994). It is damage to these regions that is most likely to directly compromise a person's ability to value.

Thus consider Elliot, a patient with brain damage localized in the ventromedial prefrontal cortices. He performed normally or even superiorly on a full battery of psychological tests (including intelligence, knowledge base, memory, language, attention, and basic reasoning), and yet was a very poor decision-maker in everyday life (Damasio 1994). He showed no abnormalities in means-ends reasoning and problem-solving; he was perfectly able to come up with a full array of options for action in a particular situation as well as to work out the consequences of each option. As it turned out, his impairment concerned the ability to choose among the options he could reason through so well. After a full analysis of all the options he would comment, "I still wouldn't know what to do!" (Damasio 1994, p.49). His emotional responses and feelings were severely blunted and this "prevented him from assigning different values to different options, and made his decision-making landscape hopelessly flat" (Damasio 1994, p.51). He lacked the very starting points of the ability to value: he was no longer sufficiently invested in anything; he ceased to care.

The ability that Elliot lacked is the indispensable core of the capacity to value. When you value something—be it a particular friendship or a local community—your commitment to these things is first and foremost a form of emotional engagement. You would not call it "valuing" or "being committed" unless there is some confluence between thinking about and acting upon what you say you value and your emotional life. True enough, since the conviction that it is right for you to care about these things and that you would be mistaken if you did not care is open to criticism, sophisticated and varied cognitive abilities are required to develop a robust set of values, values most immune from such criticisms. But having such convictions in the first place amounts to attaching emotional significance to the object of value; it involves having the corresponding emotional attitudes and reactions, so that some things simply "feel" important to you (Jaworska 1997). Elliot was unable to value because of the numbing of his affective response.

The neuronal destruction of Alzheimer's disease eventually reaches the regions of the brain most responsible for "giving emotional quality to occurrences which renders them important to the person concerned" (Souren & Franssen 1994 p.52). However, the destruction and the isolation of the hippocampus are always several steps ahead of the pathologies in the areas most likely to affect the capacity to value (Braak & Braak 1995).

Therefore, on the basis of neuropathological findings, one would expect Alzheimer's patients to lose their sense of life as a whole even as the essence of their ability to value remains relatively robust.

On the neuropathological picture of full-blown Alzheimer's disease, the loss of the sense of one's life as a whole is typically acute, while the destruction of the areas implicated in valuing remains more selective. Thus we would expect a patient in the moderate stage to have lost some, but not all, of his former values. With his ability to form new long-term memories compromised, he is unlikely to develop any new values. But his selective loss of former values may well result in changes of values pertinent to our dilemmas: as I previously observed, once some of the earlier values drop out, the exact content and the importance of the remaining values are typically reconfigured.

In this section I have chiefly argued that the ability to value is independent of the ability to understand the narrative of one's whole life, and that demented people may well retain the former ability long after they lose the latter. We also saw that, at least when the well-being of the demented is the focus, Dworkin's recommendation to disregard the patient's current wishes derives from the loss of the former capacity, the capacity to value. Thus, for a Dworkinian, the threshold capacity level necessary to lend prudential authority to a person's current wishes should not be set at the ability to grasp one's life as a whole, but rather at the ability to value. As long as the demented person still espouses values, we have seen no reason to override these in the name of values he professed earlier—Dworkin's recommendations do not apply.

Rethinking autonomy

Let us now turn to respect for the patient's autonomy as the primary goal of those caring for demented patients. How should we now approach our dilemmas? According to Dworkin, we need to consider whether the demented patient, in his current condition, still possesses the capacity for autonomy. The rationale here is that respecting people's autonomy is morally important only because a human being's very ability to act autonomously is morally important. If a person is not even capable of making autonomous decisions, allowing him to carry out his current wishes would do nothing to promote his autonomy. As Dworkin sees it, the only way to respect the autonomy of such patients is to respect their earlier ability to act autonomously; if their autonomous choices from that earlier time can still be satisfied now, these should be the focus of respect for autonomy. Of course, choices associated with critical interests are often still satisfiable, since, as we saw earlier, critical interests can be meaningfully fulfilled at a time when the person no longer espouses these interests. Thus, for Dworkin, the only way to respect the autonomy of patients who lost their capacity for autonomy is to adhere to the critical interests that they professed before this loss. He readily concludes that the types of demented patients he is interested in lack the capacity for autonomy, and hence that in order to respect their autonomy one must adhere to their earlier wishes, wishes that expressed this capacity.

The claim that demented patients no longer possess the capacity for autonomy is clearly pivotal to this part of Dworkin's analysis. But how plausible is the underlying interpretation of the capacity for autonomy?

Fundamentals of autonomy

Dworkin describes the capacity for autonomy as "the capacity to express one's own character—values, commitments, convictions, and critical as well as experiential interests—in the life one leads" (Dworkin 1993, p.224). So understood, this is the capacity to be fully in charge of one's life—to enact one's own values and convictions in the life one leads. Demented people may easily lose this capacity, because as they lose the understanding of the world around them and become increasingly disoriented, they no longer know how to translate their values and convictions into appropriate activity in the world. But suppose that a demented person who still espoused values and convictions received some help in enacting those values in his environment. Imagine, for instance, a demented man who values his independence above all else, but who is confused about what he is still able to do on his own. Were he left to make his own decisions, his choices would not ultimately enhance his independence, and perhaps would even lead to his harm. But imagine further that his family makes living arrangements for him that allow a maximum degree of independence feasible in his predicament. There is an important sense in which this man is still capable of exercising his capacity for autonomy, of living according to his convictions and values, albeit with some help in translating ends into means. Thus a possibility opens up that the capacity for autonomy ought not to be thought of as the capacity to carry one's convictions into action without external help, a capacity that requires reasoning through complex sets of circumstances to reach the most appropriate autonomous decisions; rather, that the capacity for autonomy is first and foremost the capacity to espouse values and convictions whose translation into action may not always be fully within the agent's mastery.

In his own elaboration of why the demented lack the capacity for autonomy, Dworkin emphasizes the claim that they have "no discernable even short-term aims" (Dworkin 1993, p.225). Presumably, Dworkin thinks that Alzheimer's patients cannot have even short-term aims because as soon as they embark on a course of action they forget what it was they were doing and are forced to start anew. But he should distinguish between an inability to form and then remember a plan for fulfilling one's purposes, and a lack of a stable set of purposes and preferences. For we can imagine an Alzheimer's patient who always wants the same set of things—say, he wants to feel useful and appreciated—and yet is unable to set up and carry through a plan for achieving any of them, partly because he cannot figure out the means to his ends, and partly because he cannot keep track of the steps he is taking. These latter deficiencies seem to be at stake in Dworkin's claim that the demented lack the capacity for autonomy, despite his explicit focus on their lack of consistent purposes.

For Dworkin, Alzheimer's patients cannot be autonomous because, left to their own devices, they cannot lead their lives by their own lights. This is largely because they have lost the ability to reason from their preferences to the appropriate decisions and actions—they have lost the adeptness for means-ends reasoning and planning.

However, there is no good reason to restrict the right to autonomy only to people who possess these abilities. After all, as the case of Elliot and other patients with prefrontal brain damage powerfully brings home, the very idea of governing one's life autonomously is a complete nonstarter unless the person knows how he wants his life to be governed—unless he has his own substantive principles or directives for running his life. These principles constitute the foundation of autonomy; means-ends reasoning and planning are mere tools for implementing the principles. Moreover, while having one's own directives is indispensable for exercising autonomy, we can well imagine that the tools of means-ends reasoning and planning could be supplied for the autonomous person from the outside. Accordingly, the essence of the capacity for autonomy consists in the ability to lay down the principles that will govern one's actions, and not in the ability to devise and carry out the means and plans for following these principles.[3]

Dworkin's analysis, then, focuses on peripheral rather than essential aspects of the capacity for autonomy. However, to offer a convincing alternative, we must specify more precisely what the essence of the capacity for autonomy amounts to, and in particular what counts as one's own principle or guideline for running one's life. Presumably this cannot be just any run-of-the-mill wish or desire, because a person may distance himself from a mere desire and not recognize it as truly his own.[4] So, at the very least, a principle suitable as a basis for self-governance must be viewed by the person as correct for him.

Values as starting points of autonomy

To explore this proposal, let us consider whether autonomy would be possible even for a hypothetical creature unable to judge the correctness or appropriateness of his desires.[5] Suppose that this creature can control whether or not he will act on a desire[6], and yet he experiences his desires simply as events in his body. There are two possibilities here. In one, the creature is ultimately indifferent to whether he will act on his desires—he finds himself inclined toward certain things, but upon reflection he never sees anything worthwhile in what he is inclined to do. In the other variant, although the creature finds himself inclined toward certain choices, he is altogether incapable of reflecting on the merits of these inclinations. In both cases, the desires, precisely because they lack the approval of the one who feels them, are too passive to be regarded as authentic directives for self-governance. Thus, indeed, to qualify as such an authentic directive, a principle must be viewed by the person as correct for him.

Following my earlier observations, this means that a principle qualifying as such an authentic directive must have the status of a value. As I explained previously, a person's values specify his own sense of what is correct and appropriate for him; they are his guidelines for behavior. Values are also directly connected to a person's sense of self, since the person measures his own worthiness according to how well he lives up to his values. This connection further confirms that values are genuinely the person's own principles of behavior, that they are the apt foundation for self-governance.

We can now restate in a more familiar form our earlier finding that the mere laying down of principles for one's conduct makes one capable of autonomy. Since such principles are values, the very ability to value, even if more instrumental abilities are absent, supplies the starting points for the exercise of autonomy, and thereby renders the person capable of autonomy.

Of course, possessing the capacity to value does not guarantee that the person can exercise autonomy to a full degree. Full-blown autonomy involves not only acting on one's own principles and convictions, but also the ability to scrutinize these principles and to revise them in light of critical evaluation, so that they are well articulated and robust. The capacity to value merely makes possible the most minimal and basic level of autonomy; other capacities are necessary to further develop and perfect autonomy. All the same, the capacity to value by itself does, in a very important sense, render a person capable of autonomy.

Autonomy of Alzheimer's patients

Alzheimer's patients usually retain their ability to value long after other capacities necessary for making their own decisions and fully directing their lives are gone. For example, Mrs. D's conviction that she ought to help her fellow man in any way she could certainly comes across in the interview as a truly self-given authentic principle of conduct. She talked of this conviction just as any other person would ordinarily talk of her commitments to her principles. Yet, since Mrs. D struggled even to find her way to the bathroom in a relatively familiar place, she clearly would have had trouble implementing her altruistic principles on her own; she would not have been able to figure out, for instance, how to enroll in a research study to help her "fellow man." However, with the assistance of others she was able to continue to lead a life informed by her valued self-selected principles, and thereby to continue to exercise a measure of self-government.

The paramount symptom of Alzheimer's disease, as we saw earlier, is the inability to form new long-term memories. This does not significantly affect a person's ability to value, but it does directly and severely hamper his efforts to implement his values, even more than it affects his grasp of his life as a whole. Even a modest impairment of long-term memory limits the person's access to information about the world around him and distorts his assessment of his own competence, compromising his ability to select the course of action befitting his values.

In her account of her husband's battle with Alzheimer's disease, Ann Davidson provides beautiful testimony that simultaneously illustrates Julian Davidson's relatively intact capacity to value and his slipping ability to implement his values.

Julian insisted that he had to compose a "Thank You" speech to be delivered at a banquet honoring his scientific contributions. On his own he was only able to produce phrases such as:

> [I]t will be a pleasure and joy to come back with and me tu omar see and and attend to the evening of June and its and day … Although I have not in worked in the day most loved and I will be a persual … strangely I was finished re this important and pleasure. (Davidson 1997, p.210)

But when Ann patiently interviewed him about what he wanted to say, he "spoke haltingly, but passionately, about leaving his career. In garbled phrases he described what he felt about science and why he was quitting. He needed his colleagues to know why he had left. He wanted them to know, too, that people with Alzheimer's are still 'regular folks'" (Davidson 1997, p.210).

Julian communicated his conviction that it was right for him to care deeply about science, and likewise that it was appropriate to give his colleagues an explanation. He was definitely conveying *values* here, his authentic ideals. At the same time, he needed Ann's help to translate these values into appropriate actions—he could not have figured out on his own that Ann needed to rewrite his speech and that it would be best if someone else delivered the speech on his behalf.

There are abundant variations of cases of Alzheimer's patients who remain valuers despite severe impairments in their decision-making capacity. On my analysis, these patients are still capable of fundamentals of autonomy. Accordingly, a caregiver committed to respect for autonomy must respect these patients' *contemporaneous* autonomy. This is a perfectly coherent goal, albeit respecting the autonomy of such patients usually requires much more active participation of the caregivers than what is required for ordinary competent patients. To properly respect the autonomy of many an Alzheimer's patient, one must do quite a bit to *enhance* his autonomy. One must help the person no longer able to do so on his own to lead his life according to his remaining values, to the extent that this is still feasible. This involves figuring out how his values would be best upheld in a reality he no longer fully understands, as well as helping him implement these solutions in practice. We saw this approach employed by Julian Davidson's wife, by Mrs. Rogoff's caregiver Fran, and by the researcher working with Dr. B. Sometimes enhancing a person's autonomy in this way may even involve going against his explicit choices. Ann Davidson did not simply allow Julian to try to deliver his jumbled speech, and Fran would not simply give in if Mrs. Rogoff insisted on cooking a potholder for dinner. The caregiver must learn to pay attention to the person's values rather than to her concrete, yet perhaps ill-informed, selection of options. All the same, as long as the patient is still able to value, respect for autonomy does not license the caregiver to disregard the patient's immediate interests altogether.

In sum, contrary to Dworkin's assumptions, in the context of dementia, the capacity for autonomy is best understood not as the ability to lead one's life by one's lights when one is left to one's own devices, and not as a full capacity to make a decision from the beginning to end, but as the capacity to value—to originate the appropriate bases for one's decisions that can then be, if the need arises, partly taken over by others. An Alzheimer's patient may be too disoriented to form a life plan or to choose specific treatment preferences, but as long as he still holds values, he is, in the most basic sense, capable of self-governance, and this fact about him commands utmost respect. Autonomy-based recommendations to disregard his contemporaneous interests have no purchase.

Conclusion

No matter whether we emphasize well-being or autonomy of Alzheimer's patients, we should approach our dilemmas mindful of Julian Davidson's insight that many people with Alzheimer's disease are still "regular folks." In some morally very important respects, many Alzheimer's patients remain, at least for a time, surprisingly similar to ourselves. As neuroscience allows us to corroborate, patients moderately afflicted with Alzheimer's disease may well remain valuers, and thus retain the essential characteristics by virtue of which persons and their interests enjoy an exalted moral status.

2016 Postscript

Since I formulated the ideas relayed in this chapter, exciting neuroscience research has emerged that not only corroborates my key claim that the capacity to care and value is relatively preserved at least up to the middle stages of Alzheimer's disease, but also, rather surprisingly, suggests that the core aspects of this capacity may even be enhanced by the degeneration typical of Alzheimer's disease.

This new evidence is tied to a broader finding about the functioning of the brain, namely, that distinct and often anatomically distant structures in the normal human brain tend to activate and deactivate together, forming integrated functional units or networks called intrinsically connected networks (Seeley et al. 2007). The intrinsically connected network most relevant to Alzheimer's disease is the so-called default mode network. Neuronal degeneration characteristic of Alzheimer's disease (the degeneration due to pathological accumulation of tau protein, to be exact) develops early on in one key node of this network—the hippocampus, as we saw earlier—and then spreads most prolifically along the pathways of this network (de Calignon et al. 2012; Raj et al. 2012). In other words, the pattern of the most acute degeneration in a typical Alzheimer's brain matches one of the patterns of co-activation found in a normal human brain (Seeley et al. 2009).

The default mode network supports not only the hippocampal function of transforming short-term memory into long term memory as emphasized in the original chapter, but also a broader set of abilities that may be loosely described as "time travel": the ability to sequence one's life as a temporally extended agent—to remember one's past and to imaginatively extend one's life timeline into the future, for example, through planning—as well as the broader ability to imaginatively project oneself into an alternative situation or perspective, be it the viewpoint of another person or even an alternative spatial perspective as in some forms of navigation (Buckner & Carroll 2007; Østby et al. 2012). These abilities are relevant downstream to valuing and caring, insofar as, for example, once you value or care about another person you are moved to understand their point of view and to plan for or with them, as their needs dictate. But these abilities are orthogonal to the ability to value and care about somebody or something in the first place; they merely affect how well you are able to live up to what you already value and care about. So the finding that

Alzheimer's disease spreads primarily in the default mode network is consistent with my claim that this trajectory of degeneration spares, at least for a time, the capacity to value and care.

However, once we look at the brain from the point of view of intrinsically connected networks, there is more to this story. It turns out that the default mode network is coupled in an interesting way with another intrinsically connected network, known as the salience network: in a normal human brain, states of high level of activity in the default network are associated with decreased activity in the salience network and vice versa (Raichle et al. 2001; Fox et al. 2005). In parallel with this finding, it also turns out that as the default mode network becomes compromised in Alzheimer's disease, this corresponds to increased connectivity and activity in the salience network (Zhou et al. 2010). The salience network plays a central role in social and emotional processing. More specifically, as the name "salience" suggests, its key function is thought to be isolating what the cognitive system will pay attention to and allocate more resources to, out of the vast amount of internal and external information that the brain is bombarded with and processing at any given time. The salience network swiftly identifies what is most relevant and responds accordingly by generating appropriate emotions that recruit other resources and guide our spontaneous behavior (Seeley et al. 2007). A paradigmatic example of a well-functioning salience network is a quick pivoting to information relevant to safety or security when one is absorbed in an unrelated task (e.g., running out of class upon hearing a fire alarm). However, the network has been conceptualized to have the broader function of focusing attention and cognitive resources not only on what impacts security or safety but, more generally, on what is personally relevant, and this maps naturally onto what one values and cares about. Thus, one would expect that the increased connectivity and activity in the salience network associated with the progression of Alzheimer's disease would result in a heightened emotional attunement to what is important to the person, in a manner characteristic of valuing and caring. While research in this area is only just beginning to emerge, Sturm and colleagues have already documented a suggestive result that emotional sensitivity to the emotions of others (so-called emotional contagion) is enhanced in mild Alzheimer's disease compared to normal controls and, perhaps even more surprisingly, it is further enhanced in moderate Alzheimer's disease as compared to mild (Sturm et al. 2013). The increase in emotional contagion was correlated with smaller volume in the default mode network. These results begin to corroborate the observations of clinicians and caregivers that some Alzheimer's patients become more sweet, loving, and attentive to them as their memory fades.

So far, I have interpreted the finding that connectivity in the salience network increases as a result of the progression of Alzheimer's disease as suggestive of an enhancement of the capacity to care and value, and thus as consistent with my claim that patients with moderate Alzheimer's disease possess intact the capacities on the basis of which we deem persons and their interests worthy of utmost moral respect. Nonetheless, construed differently, this finding may be thought to raise a worry for my account. Might the increased connectivity in the salience network lead to new cares and values that would be inauthentic in virtue of being mere artifacts of Alzheimer's disease? If so, the strengthening of the salience network might be seen not as enhancing the capacity to care and value but rather as disrupting it in a wayward way. It is one thing if Alzheimer's disease leads to a loss of some former values and to a reprioritization of the values that remain (in the manner I discussed earlier), or to a strengthening of some former values, but it is something quite different if the disease causes values to spring up de novo.

Addressing this issue adequately will need to await a more detailed understanding of the functioning of the salience network and of the specific ways in which the increased connectivity associated with Alzheimer's disease affects the operation of this network. For example, if the main effect of the increased connectivity is greater attunement to the emotions of others, this can plausibly be interpreted as an enhancement rather than disruption of the capacity to value. The attunement to the emotions of others is not, by itself, a new care or value. Rather, such attunement makes possible a better, more perceptive and appreciative understanding of other persons—and also of various ideals involving other persons such as tolerance, justice, etc.—as potential objects of care and value. Thus, greater attunement to the emotions of others can lead to new cares or values but through a path that, arguably, is not deviant, since it proceeds via greater appreciation of what the object of care or value is all about. However, if it turned out that increased connectivity in the salience network can lead in a more direct way to new cares or values, without the mediation of any new experience or reasoning, there would be much greater reason to worry that the capacity to care and value is disrupted.

Acknowledgments

A longer version of this chapter appeared in *Philosophy and Public Affairs*, Volume 28, Issue 2, pp.105–38, 1999 (see that version for acknowledgments).

The chapter sans Postscript appeared in *Neuroethics: Defining the Issues in Theory, Practice, and Policy*, edited by Judy Illes, pp.87–101, 2005 (https://global.oup.com/academic/product/neuroethics-9780198567202).

The epigraph at the start of this chapter was reprinted with the permission of Pan Macmillan via PLSclear, and with the permission of Touchstone, a division of Simon & Schuster, Inc. from *The Man Who Mistook His Wife for a Hat and Other Clinical Tales* by Oliver Sacks. Copyright © 1970, 1981, 1983, 1984, 1985 Oliver Sacks. All rights reserved.

The opinions expressed are those of the author and do not necessarily reflect the policies or positions of the National Institutes of Health, the Public Health Service, or the US Department of Health and Human Services. Special thanks to Winston Chiong, Katherine Rankin, and William Seeley for educating me about the science cited in the Postscript, and to Winston Chiong for comments that partly shaped the Postscript's content.

Notes

1. This case was reported to me by a family member. To protect the privacy of the patient, I use a fictitious name. I also do not list the name of my source, but I would like to thank her for her generous help.

2. In the case of some (often less central) values, one may anticipate a gradual transformation and loss of a value without thinking that it would be a mistake. This is not a counterexample to my characterization of valuing, because my claim concerns our attitudes to our present espousal of a value: It is part of valuing something that one would view the possibility of not caring about it *here and now* as an impoverishment, loss, or mistake. Imagining a future without the alleged value is a good test for this in many cases, but it does not work in every case because some values are dependent on context and are important only at a specific time in one's life.

3. None of what I say here should be taken to imply that, in a normal case, a person's own exercise of means-ends reasoning and planning could be ignored by those aiming to respect his autonomy. My point is that a person largely incapable of such reasoning may still be able to exercise autonomy, and not that such reasoning can be taken lightly in an ordinary person who exercises his autonomy through the use of such reasoning.

4. Some readers may think that this requirement is too stringent, and that the hypothetical individuals I go on to describe are still capable of autonomy. If so, it is only easier for me to claim that many Alzheimer's patients retain the capacity for autonomy.

5. I speak of a "creature" rather than a "person" here to allow for the view that the ability that my hypothetical creature lacks is essential to personhood.

6. Why this stipulation? If we imagined a creature who was also unable to control his response to a desire, it would have been easier to see him as lacking the capacity for autonomy. However, I want to show that our intuition that my imagined creature lacks the capacity for autonomy depends on something else; even a creature in control of his desires seems to lack the capacity for autonomy if he can never approve or disapprove of his desires.

References

Braak, E. and Braak, H. (1995). Staging of Alzheimer's disease-related neurofibrillary changes. *Neurobiology of Aging*, 16(3), 271–278.

Buckner, R.L. and Carroll, D.C. (2007). Self-projection and the brain. *Trends in Cognitive Sciences*, 11(2), 49–57.

Damasio, A.R. (1994). *Descartes' Error: Emotion, Reason, and the Human Brain*. New York: Putnam.

Davidson, A. (1997). *Alzheimer's, a Love Story: One Year in My Husband's Journey*. Secaucus, NJ: Carol.

de Calignon, A., Polydoro, M., Suárez-Calvet, M., William, C., Adamowicz, D.H., and Kopeikina, K.J. (2012). Propagation of tau pathology in a model of early Alzheimer's disease. *Neuron*, 73(4), 685–697.

Dresser, R. (1986). Life, death, and incompetent patients: conceptual infirmities and hidden values in the law. *Arizona Law Review*, 28(3), 373–405.

Dworkin, R. (1993). *Life's Dominion: An Argument about Abortion, Euthanasia, and Individual Freedom*. New York: Knopf.

Fox, M.D., Snyder, A.Z., Vincent, J.L., Corbetta, M., Van Essen, D.C., and Raichle, M.E. (2005). The human brain is intrinsically organized into dynamic, anticorrelated functional networks. *Proceedings of the National Academy of Sciences of the United States of America*, 102(27), 9673–9678.

Geula, C. (1998). Abnormalities of neural circuitry in Alzheimer's disease: hippocampus and cortical cholinergic innervation. *Neurology*, 51(Suppl. 1), S18–S29.

Jaworska, A. (1997). *Rescuing Oblomov: A Search for Convincing Justifications of Value* (Ph.D. thesis). Harvard University, Cambridge, MA.

Laakso, M.P., Hallikainen, M., Hänninen, T., Partanen, K., and Soininen, H. (2000). Diagnosis of Alzheimer's disease: MRI of the hippocampus vs delayed recall. *Neuropsychologia*, **38**(5), 579–584.

Østby, Y., Walhovd, K.B., Tamnes, C.K., Grydeland, H., Westlye, L.T., and Fjell, A.M. (2012). Mental time travel and default-mode network functional connectivity in the developing brain. *Proceedings of the National Academy of Sciences of the United States of America*, **109**(42), 16800–16804.

Raichle, M.E., MacLeod, A.M., Snyder, A.Z., Powers, W.J., Gusnard, D.A., and Shulman, G.L. (2001). A default mode of brain function. *Proceedings of the National Academy of Sciences of the United States of America*, **98**(2), 676–682.

Raj, A., Kuceyeski, A., and Weiner, M. (2012). A network diffusion model of disease progression in dementia. *Neuron*, **73**(6), 1204–1215.

Riedel, G. and Micheau, J. (2001). Function of the hippocampus in memory formation: desperately seeking resolution. *Progress in Neuro-Psychopharmacological and Biological Psychiatry*, **25**(4), 835–853.

Sabat, S.R. (1998). Voices of Alzheimer's disease sufferers: a call for treatment based on personhood. *Journal of Clinical Ethics*, **9**(1), 38–51.

Sacks, O. (1995). *The Man Who Mistook His Wife for a Hat and Other Clinical Tales*. New York: Summit Books.

Seeley, W.W., Crawford, R.K., Zhou, J., Miller, B.L., and Greicius, M.D. (2009). Neurodegenerative diseases target large-scale human brain networks. *Neuron*, **62**(1), 42–52.

Seeley, W.W., Menon, V., Schatzberg, A.F., et al. (2007). Dissociable intrinsic connectivity networks for salience processing and executive control. *Journal of Neuroscience*, **27**(9), 2349–2356.

Souren, L. and Franssen, E. (1994). *Broken Connections: Alzheimer's Disease, Part I* (van der Wilden-Fall, R.M.J., trans.). Lisse, the Netherlands: Swets and Zeitlinger.

Squire, L.R. and Zola-Morgan, S. (1988). Memory: brain systems and behavior. *Trends in Neurosciences*, **11**(4), 170–175.

Squire, L.R. and Zola-Morgan, S. (1991). The medial temporal lobe memory system. *Science*, **253**(5026), 1380–1386.

Sturm, V.E., Yokoyama, J.S., Seeley, W.W., Kramer, J.H., Miller, B.L., and Rankin, K.P. (2013). Heightened emotional contagion in mild cognitive impairment and Alzheimer's disease is associated with temporal lobe degeneration. *Proceedings of the National Academy of Sciences of the United States of America*, **110**(24), 9944–9949.

Young, P.A. and Young, P.H. (1997). *Basic Clinical Neuroanatomy*. Baltimore, MD: Williams and Wilkins.

Zhou, J., Greicius, M.D., Gennatas, E.D., et al. (2010). Divergent network connectivity changes in behavioural variant frontotemporal dementia and Alzheimer's disease. *Brain*, **133**(5), 1352–1367.

Chapter 16

Anticipating a therapeutically elusive neurodegenerative condition: Ethical considerations for the preclinical detection of Alzheimer's disease

Hervé Chneiweiss

A case for anticipation

Recent progress in predictive medicine opens new avenues for the concept of health. The extensive and idealistic definition given by the World Health Organization in 1946 that: "Health is a state of complete physical, social and mental well-being, and not merely the absence of disease or infirmity" is no longer pertinent in a world of anticipation where a biomarker of a disease exists in the absence of any symptom. This is a revolution in many fields but even more so for neurodegenerative disorders that happen late in a lifetime, making a person's entire life a period of anticipation.

Among neurodegenerative disorders, Alzheimer's disease (AD) has gained a special position during the last 40 years, due to the burden the condition imposes on more than 40 million affected people worldwide. Today's estimations might only be the beginning, in fact, since most of the patients are older than 65 years (prevalence of dementia before the age of 50 years is less than 1 in 4000 with only 30% of cases being attributed to AD), and projected increases in the prevalence of dementia are higher in developing countries with young populations (Winblad et al. 2016). Now that we, as a science community, know that the disease starts two or three decades before any symptoms occur, predictive medicine may bring forward possibilities for diagnosis or testing. Advances in screening for early detection may allow for early staging, intervention, and secondary prevention.

Asymptomatic individuals provide the research pool for preventive approaches. However, is there an interest in diagnosis in the absence of cure? Does no cure mean no care? Interestingly, results from several European cohort studies have provided evidence that the age-specific incidence of dementia has decreased in the past 20 years, thereby raising expectations for preventive interventions (Scheltens et al. 2016). Thus, the related ethical debate is not about the clinical stages of AD or the timing of interventions where early appears to be best, but instead, is on the multiple meanings and the impact of pre-clinical AD diagnosis before the onset of symptoms.

Alzheimer's disease still baffles scientists and tests the boundaries of therapeutic knowledge. Despite undeniable progress, many uncertainties remain, including the cause of the pathogenic process where the respective roles of amyloid deposition, tau protein, and other processes of neurodegeneration are not yet understood. The diagnostic criteria of the International Working Group (IWG) (Dubois et al. 2014) and the National Institute of Aging and the Alzheimer's Association (NIA-AA) (Albert et al. 2011) still rely on a clinical phenotype combined with biomarker information, mainly driven by amyloid and tau. The preclinical AD concept was initially defined from cognitively unimpaired individuals who displayed AD brain lesions on postmortem examination. With the development of AD biomarkers, the concept evolved and preclinical AD (NIA-AA) or at-risk for AD (IWG) are now considered when these markers are present in a cognitively normal individual. In addition, individuals known to carry an autosomal dominant monogenic mutation that will result in the development of full clinical AD are classified as having presymptomatic AD. These opportunities of anticipation of clinical AD open many questions that range from basic scientific and technical validity of the biomarkers and methods, toward ethical issues such as the impact on the individual, family, and society. Similarly challenging are the ethical issues associated with the set-up of clinical trials with asymptomatic individuals who will have to consent to a hard-to-bear treatment of unknown efficacy and unknown potential side effects for a long period, all while living with the status of preclinical/at-risk AD. Furthermore, while amyloidosis is necessary for an AD diagnosis, it is not sufficient to reliably predict further progression to a symptomatic stage of disease given postmortem evidence that demonstrates that a significant proportion of individuals with typical brain neuropathological AD lesions do not present symptoms of cognitive impairment (Price et al. 2009). Thus, it is important to understand progression drivers to determine the proportion of at-risk individuals who will progress to the clinical stage of AD. Considering the many uncertainties at hand, how then can protocolization for a possible early detection and the need for adaptation of taking care of individual peculiarities be reconciled? What are the implications for the family and the workplace? What reconfiguration of the care relationship and care course do such developments carry?

The short-term therapeutic horizon is further worrisome: effectiveness is not yet proven. Still, scientists, physicians, and patient organizations agree on the need to facilitate testing that is as reliable as possible when the first symptoms manifest themselves. The precocity of care indeed involves several indisputable advantages. First, the tested treatments have been proven effective only on patients at the early stage of AD. Recent clinical trials of three different monoclonal antibodies directed against the 1-42 beta-amyloid peptide (Aß42) (aducanumab, crenezumab, and solanezumab) have suggested slowing of cognitive decline of early/mild but not moderate AD patients (Selkoe & Hardy 2016). Second, early care seems to afford economies of scale both for individuals and the collective. Added to this is the argument that early detection of the disease could offer patients and their families time to anticipate and look to the future to consider financial and legal issues, and assert control over the course of their lives that might not have been otherwise considered or desired. To examine the values and choices that underlie aspirations for anticipation in more detail, I turn to lessons from other fields of medicine, such as cancer and the identification of high-risk individuals bearing pathogenic genetic mutations that predispose them to disease. To conduct this examination, I draw upon fruitful interactions from my work with a multidisciplinary network of scientists, neurologists, psychologists, philosophers, representatives of insurance companies, and carers. The effort was carried out under the auspices of the French national program "Distalz" headed by Professor Philippe Amouyel, and with the methods established by Professor Emmanuel Hirsch in the "Espace éthique régional Ile de France" (Forum for Ethics of Care Ile de France), that became for the occasion "Espace éthique maladies neurodégénératives" (Forum for Ethics of Neurodegenerative Diseases).

Biomarkers: The building blocks of the ethical debate

Reliable biomarkers allow us to diagnose the AD pathophysiological process in the absence of clinical symptoms. They also give rise to some ethical questions resulting in the opportunity for increasingly personalized medicine, such as:

♦ confirmation of diagnosis may avert useless treatment and mitigate confusing results

♦ identification of populations at high risk of developing AD can allow for the deployment of reinforced testing and preventive treatment based on the predisposition marker

♦ evaluation of prognostic variables can enable the estimation of speed of disease progression based on the prognostic marker

♦ prediction of therapeutic response can yield precision of intervention based on a prediction marker.

Diagnostic biomarkers

Alzheimer's disease develops clinically in several stages: a presymptomatic period when the biological process leading to the disease is active but the patient does not suffer any disorder, a prodromal phase where the patient suffers from mild cognitive impairment (MCI), and finally the gradual onset of increasingly severe dementia. In the past, the diagnosis of AD was based on postmortem neuropathological changes of the brain that included both positive and negative features. Positive lesions consist of abundant amyloid plaques and neurofibrillary tangles, and dystrophic neurites containing hyperphosphorylated tau protein. Amyloid plaques are found throughout the cortex whereas tangles are primarily located in limbic and association areas. Early lesions are found in the entorhinal cortex. Later in the progression of the disease, lesions diffuse to the hippocampus, then association cortex, and finally the primary neocortex. Astrogliosis and activation of microglia complete this scheme. Negative lesions include losses of neurons and synaptic elements. Today, biomarkers signal these lesions while the patient is still alive. They are used to validate the clinical diagnosis and to quantify the level of advancement of the disease through the concentration of the ß-amyloid protein and of the tau protein (total and under its hyperphosphorylated form) in the cerebrospinal fluid (CSF) and by brain imaging.

The pathophysiological process of AD has been modeled as a cascade of progressive neuropathological events initiated by ß-amyloid deposition that is mirrored by a low level of the CSF Aß42, followed by accumulation of tau (levels increased in CSF) preceding structural brain damage, decreased cerebral glucose metabolism, brain atrophy, and subsequent functional and cognitive impairments (Jack et al. 2016). Structural magnetic resonance imaging (MRI) allows brain medial temporal cortex atrophy to be identified, and rules out intracranial causes of dementia such as meningioma or subdural hematoma. Additionally, MRI remains the modality of choice for the assessment of vascular brain changes, such as white matter hyperintensities, lacunes, and microbleeds that have gained increasing attention because they are frequent side effects in antiamyloid trials (Winblad et al. 2016). New techniques of cerebral imaging help to visualize hypometabolism, amyloid plaques, neurofibrillary tangles, and the microglial activation. [18F] Fluorodeoxyglucose positron emission tomography (FDG PET) reveals focal cerebral hypometabolism in a very sensible and validated way. A normal FDG PET scan virtually excludes a diagnosis of neurodegenerative disease. Consequently, FDG PET is increasingly used to follow the disease progression and evaluate the effectiveness of interventions to modify the natural course of the disease. Amyloid PET, namely PET with ligands for Aβ such as florbetapir, florbetaben, and flutemetamol, which have been approved for clinical use, has a high negative predictive value since a patient with cognitive impairment and negative amyloid PET is unlikely to have AD (Jack et al. 2016).

Recent longitudinal studies on subjects without cognitive symptoms at the beginning of the study have shown that the coexistence of events in the amyloid biological cascade predict brain structural and functional declines (Bateman et al. 2012). Analysis of natural history in familial AD cases with mutations in the gene coding for the amyloid protein (APP), or in genes coding for presenilin (PSEN1 or PSEN2) that are the catalytic subunits to the γ secretases enzyme which cleave APP, show that Aß42 levels in CSF begin to decline 25 years before expected symptom onset (Selkoe & Hardy 2016). This is followed by the appearance of fibrillar amyloid deposits in the brain, increased levels of hyperphosphorylated tau in CSF, and progressive brain atrophy (particularly that of hippocampus) that occur 15 years before the onset of symptoms. Neuronal hypometabolism and short episodes of memory impairments begin 10 years before symptoms become clinically self-evident. According to current models, mutations in *APP, PSEN1*, and *PSEN2* genes in familial AD accelerate the generation of fibrillar amyloid deposits, whereas in sporadic AD, decreased clearance of Aβ associated with other pro-aggregating conditions might have roles in the accumulation of toxic Aβ species. But most authors consider that the time course of the disease should be similar in familial and sporadic AD (Villemagne et al. 2013). This is based on the conceptual framework that β-amyloid and tau pathologies synergistically potentiate neurodegeneration (Winblad et al. 2016). This synergistic framework challenges previous AD pathophysiological theories that focused on either ß-amyloid or tau pathologies as the main driving forces of disease progression, the so-called Baptists and Tauist perspectives respectively. Furthermore, data from animal models support the synergistic molecular interactions between ß-amyloid and P-tau peptides, which lead to a downstream toxicity. In a recent survey, Dubois and colleagues consider that: "AD exists and can be recognized before the onset of cognitive symptoms when there is little doubt about progression to clinical disease over a short period. This is the case when both tauopathy (tau PET ligand uptake spread to neocortex or CSF) and amyloidopathy (PET evidence of AD pattern or CSF measured) are present" (Dubois et al. 2016, p.297). A key lesson that emerges from such studies is that therapeutic interventions must come soon in the course of the disease and certainly before neuronal lesions are irreversible.

Interestingly, some authors still consider the amyloid hypothesis as controversial. Among their arguments are that many of the supporting animal studies used nonphysiologically high concentrations of Aß or transgenic mice significantly overexpress *APP* or any other gene. As such, they infer that the amyloid cascade hypothesis provides no explanation for the silent incubation period of AD and pinpoint the many discrepancies between neuropathological observations and clinical outcomes. In a recent review, De Strooper and Kerran (2016) suggest a model with three phases, a biochemical phase, a cellular phase, and the final clinical phase. In the biochemical phase, Aß and tau are still at the beginning of the pathological process. In the cellular phase, compensatory mechanisms allow a fragile but long-lasting equilibrium that involves neural networks, microglia, astroglia, and oligodendrocytes. Finally, during the clinical phase, the system is overwhelmed and collapses. Thus, they state that "AD is indeed not a biochemical or molecular problem but a physiological one of disrupted cellular connectivity. The disease can therefore only be fully understood in the context of the complex cellular interactions that maintain homeostasis in the brain—i.e., at the level of the neurovascular and glioneuronal units" (De Strooper & Karran 2016, p.612). Consequently, the therapeutic approach should move from targeting Aß toward a support of the many mechanisms involved in the cellular phase and allowing compensation: another good reason to make a diagnosis of AD before the onset of symptoms.

Predisposition, prognostic, and predictive biomarkers

To be of real interest for anticipation, AD biomarkers need not only to be reliable for diagnosis but also to help determine when the disease will turn from asymptomatic to symptomatic. Will everyone who presents an Aß decrease in the CSF develop a dementia syndrome if they live for long enough? If so, which factors determine when a patient will develop symptoms? We need measures that allow evaluation of a delay to clinical onset. Even if all PET Aß-positive individuals were to ultimately develop dementia, there is a difference between a person who is 80 years old with a predicted onset of dementia in 20 years, and a 50-year-old with the onset of cognitive decline imminent within the next 2 years. More importantly, what are actionable protective factors? Several recent findings should help to define individuals at higher risk of developing AD and the speed of progression of the disease (Lacour et al. 2017). Such data might help to set up a rationale framework for early interventions in the context of anticipation (Peters et al. 2013; Dubois et al. 2016).

More than 20 genetic loci associated with an increased risk of AD (Karch & Goate 2015) have been identified so far; a few others could have protective effects. The proteins encoded by these genes belong to pathways involved in the immune system and inflammatory responses, for example, the microglia surface receptor TREM2, cholesterol and lipid metabolism, and endosomal-vesicle recycling. Some factors could be age related. Identification of these genes is important to understand the pathophysiology of AD since they probably interact with the core mechanisms of the disease. However, they do not improve our evaluation of the risk for a given person since most of the reported polymorphisms are frequent in the population—more than 30% of the population bear the allele—and contribute little to the individual risk for AD (Lacour et al. 2017). That said, APOE4 is different. APOE proteins are involved in cholesterol metabolism and in the clearance of Aß. The E4 form has a reduced activity in comparison to E2 and E3, and APOE4 is considered the major genetic risk factor for AD. APOE4 protein expression has been associated with degeneration of brain blood vessels, leakage of the blood–brain barrier, and neurodegeneration independent of Aβ. Lifetime risk for AD is more than 50% for APOE4 homozygotes and 20–30% for APOE3 and APOE4 heterozygotes, compared with 11% for men and 14% for women overall irrespective of APOE genotype. Not only does the frequency matter, but so does age at clinical onset: 68 years in E4 homozygotes, 76 years in E4 heterozygotes, and 84 years in E4 noncarriers (Karch & Goate, 2015). Consequently, APOE analysis is of importance today to define a population with a clearly higher risk of AD.

Lifestyle-related factors also influence the age of onset and the delay to progression, with much evidence concerned with interventions for cardiovascular risk that improve cognitive health at the population level. In addition, the prevention or treatment of diabetes, obesity, depression, smoking, increasing physical exercise and prevention of mental inactivity, increasing level of education, and the so-called cognitive reserve, and improving diet, might result in up to a 30% reduction in dementia incidence (Scheltens et al. 2016). By contrast, the additional presence of a biomarker of tau pathology is associated with a more rapid progression to clinical AD (Vos et al. 2013).

No marker has been described that indicates a rapid onset of symptoms. However, subjective cognitive decline, defined as the personal sensation of cognitive decline despite results within healthy limits on cognitive tests, is associated with an increased risk of progression to dementia, especially when the patient is worried about the decline (Scheltens et al. 2016). Compared with healthy controls, patients with subjective cognitive decline have more brain abnormalities, including hippocampal volume loss and hypometabolism.

Some data suggest that an evaluation of the progression speed might be possible in the future, but the results remain at a population level and are not valid for a given individual (Vos & Fagan 2016). Most of the available data result from the observation of the progression of the disease in symptomatic patients, either with MCI or pro-dromal or moderate AD, which might not be informative for asymptomatic patients. Furthermore, these results need more patients to be really confirmative. One set of evidence correlates patient age with localization of the first lesions: elderly patients with a predominant limbic form of the disease generally have a slower progression compared to the neocortical form with a hippocampal-sparing type that can be observed in younger patients. Brain imaging studies have demonstrated that MCI patients with hippocampal atrophy or temporoparietal hypometabolism as seen on FDG PET, and CSF Aβ42, invariably developed dementia, whereas those with negative results for all three parameters almost never did (Vos et al. 2015). Among new biomarkers measurable in the CSF are neurogranin and SNAP25 (De Vos et al. 2015). Neurogranin is a dendritic protein involved in acquisition and consolidation of memory. High CSF concentrations of neurogranin predict progression to AD in patients with MCI and correlate with rapid cognitive deterioration. SNAP25 is a presynaptic protein and its concentrations increase during the prodromal stage. Their value in anticipating the clinical onset delay remains to be determined, and would require repeated CSF punctures.

Values and choices

Considering the recent knowledge accumulated on the biological processes underlying the progress toward clinical AD and the possibilities of early detection of some biomarkers, one may wonder if it is ethical to keep waiting for symptoms to make a diagnosis of AD or if it would be better to screen populations in search of at-risk individuals. Would it be ethical not to try to prevent AD for these at-risk individuals? On the other hand, one must consider the risk of destroying the quality of an asymptomatic life while doing little to efficiently prevent the disease, or worse, causing adverse effects associated with preventive treatments.

In oncology, the reflection on the expected useful screening was intense. Learning from the natural history of the disease led to some rationales for the ability to control the disease. Usually, this ability is stronger upstream of the pathological process and early symptoms. However, simplistic assertions such as "an early detection of a cancer makes it still curable" or "if a cancer is discovered too late it is always fatal" must be prevented to avoid false promises. Therapeutic efficacy may not always be correlated with early diagnosis. Observe that if, at some point, a therapeutic innovation brings about a cure even at an extremely late diagnosis, then screening no longer makes sense. If the patient can be cured at any stage, early diagnosis loses much of its interest.

In the case of AD, there is a consensus that antiamyloid drugs will be most effective early in the disease process. Supporting this hypothesis are the results of trials of aducanumab, crenezumab, and solanezumab, human monoclonal antibodies selective for aggregated and soluble forms of Aβ, with a dose-dependent decrease in amyloid charge and a slower decline in cognition (Selkoe & Hardy, 2016). However, results of trials of the antiamyloid monoclonal antibody gantenerumab and of the γ-secretase inhibitor avagacestat in patients with prodromal AD do not show the same results or slow the cognitive decline.

The negative effects of screening also need consideration. In oncology, overdiagnosis presents a major challenge. When lesions or anomalies are detected at a very early stage of tumor development, they may have no evolutionary properties. Sometimes, they are spontaneously regressive, whereas in other cases, they never evolve. The risk of overdiagnosis also entails the risk of overtreatment. Take the example of prostate cancer: if a physician treats a person with an early prostate cancer based on an increase in blood-level prostate-specific antigen, a life may be saved. The quality of life is often dramatically affected by treatment. However, there is a one in two chance that the treatment is useless because the cancer is not evolutive. In this perspective, it is clear that early diagnosis can also result in harm for some individuals.

What might be the most at stake for AD? Certainly, the question of disclosure of the AD status is central. Arguments against disclosure of biomarker status are numerous. For a given individual, knowing biomarker status is potentially harmful knowledge as anxiety and depression can develop. Consider also the impact on the close family whose emotions will be affected, the friendship networks that may shift when peers distance themselves, and broader consequences such as the right to drive a car or the insurance coverage.

In contrast, arguments in favor of a preclinical characterization and disclosure can be analyzed from the point of view of at-risk AD subjects, their families and relatives, the work place, clinicians, and society. Biomarker disclosure may serve as an argument of respect for individual autonomy since there are moral and legal rights to receive a specific diagnosis. For the at-risk individual, this should allow beneficial risk-lowering behaviors and the formulation of advance directives. For clinicians, this might facilitate the recruitment to preclinical trials for AD. In one study of public perception in France at the end of 2014 with more than 1000 people, 90% said that they would like to know their status. When asked why they would like to know, most answered that this should help them to choose a better life and alleviate their fears of the disease.

Other studies seem to confirm these rather surprising numbers although while facing a real at-risk situation, patients behave differently from those responding to a hypothetical case. In one study reported at a recent (April 28, 2016) meeting of the French research network Distalz, recruitment and consent in a preventive clinical trial were at only 25% of at-risk individuals. In this case, "at risk" was derived from people who complained about their memory but performed in the normal range for their age and education level on the memory tests. What were the determining factors for participation in this clinical trial using a drug-prevention protocol? The risk:benefit analysis by the research potential participants was not at all decisive: age, gender, and education were not discriminant, whereas sociological and cultural factors play a role. For example, it is significantly more difficult to recruit families in which one or more members are physicians. On the contrary, if someone had a close relative affected by the disease, that person was more likely to participate.

One concern about selecting at-risk individuals for preventive clinical trials is that this could involve coercive disclosure of unwanted information (Kim et al. 2015). A normative analysis applying four key ethical criteria—favorable risk:benefit ratio, informed consent, fair subject selection, and scientific validity—investigated the meaning of blinded enrollment where participants are tested but not told of their risk marker status (Kim et al. 2015). The authors concluded that none of these ethical criteria supports blinded enrollment over transparent enrollment. Furthermore, blinded enrollment protocols have been criticized ethically for non-respect of the fundamental rights of a person participating in biomedical research to have access to information. Among arguments to refuse access to the personal data is the context of research that does not reach the clinical grade of exams performed in a healthcare program.

Other studies reported that subjects with increased genetic risk for AD (i.e., carriers of the *APOE4* allele who knew their status), more frequently desired to reduce that risk than did counterparts who learned they were noncarriers. Moreover, the duration of treatment that is to be undertaken is crucial. A period of 1 year instead of 5 years can change the game in terms of acceptability. Inclusion in a clinical trial disrupts daily life. For example, often a person must be available for sampling in a 10-day window. Why would someone accept such constraints on daily life if not affected by the disease? The rationale that doctors and scientists need information about the long natural history and ongoing process appears insufficiently justified to many individuals. Furthermore, the needs of biomedical science and hopes of the prospective subjects who have their own life prospects may not always align.

Such considerations will be even more important in the context of drug trials with completely asymptomatic patients. The anticipation of consent is in the air. In North America, records of individuals who volunteer for clinical trials have been established for solicitation purposes, not necessarily on genetic bases. In a sense, pre-consent is in play. Registered individuals decide to volunteer depending on the protocol that is offered. This type of prearranged consent may be revoked or amended. A conceptual framework for prevention trials for AD has been proposed (Peters et al. 2013). A first dimension of this framework focuses on stratification of patients according to their genetic risk for AD and detection of AD biomarkers. The framework then proposes two gradients that refer to the risk of developing AD and to a therapeutic risk:benefit ratio. Interestingly, this framework also considers the critical need to revisit the risk:benefit ratio iteratively during the trial. The risks that individuals enrolled in an AD clinical trial will face are only beginning to be known. For example, repeated MRI scans sensitive to microbleeds are mandatory in clinical trials of antiamyloid drugs to monitor amyloid-related imaging abnormalities (ARIAs) (Arrighi et al. 2016). These include signal hyperintensities on FLAIR (fluid attenuation inversion recovery) sequences, which are thought to represent edema, and signal hypointensities on gradient-recalled echo/T2, which are thought to represent haemosiderin deposits, including microhemorrhage and superficial siderosis. ARIAs develop after removal of amyloid from cerebral microvessels, which leads to endothelial damage, increased vascular permeability, and leakage of water or blood in the brain parenchyma.

Are the risks involved in preventing AD justifiable? This has been the current practice since the middle of the twentieth century. The relationship between myocardial infarction and high cholesterol was demonstrated around 1955, resulting in prevention strategies to lower cholesterol. Several tested molecules such as thyroxin and estrogens performed worse than hypercholesterolemia. Many clinical trials conducted in the 1960s and 1970s would end up in court today. What risks are justifiable? In the late 1960s, side effects were tolerated a risk of 1 in 100. Today no company would want to take a risk of 1 in 10,000. The sensitivity of the public has changed considerably, calling into question which types of clinical research in AD prevention and early diagnosis should be accepted and which ones are unethical. Taking into consideration the conceptual framework mentioned previously (Peters et al. 2013), it seems highly justified to take some risk when uncertainties are low (i.e., familial AD or two copies of the *APOE4* allele) or when the toxicity is low (i.e., diet or exercise). Regarding AD, everyone should be aware of the risks inherent in the development of new treatments, including lawyers and journalists. Biomedical research in the field of neurodegenerative disorders may necessarily face devastating unexpected effects such as cases of encephalitis when testing new molecules. Omitting risk is no longer conceivable. We need to inform. Justifying the search every day is to be able to justify the risks we take. There are several ways to present a risk (absolute risk, relative, over a period) to a contact, and we must use appropriate words. Dedicated research on such communication should be developed and investigators will need to conceptualize the process of consent as being dynamic rather than static (Peters et al. 2013).

Finally, are we studying the right cases? I cannot stress enough the fact that the average age of onset of dementia is 80 years. In most cases, there is a combination of a degenerative neurological factor and a vascular factor. However, it is clear that clinical trials, at present, involve almost only subjects age 70 years or less with few comorbidities. A question of representativeness of the real disease in clinical experimentation is indispensable.

Conclusion

Considering the natural history of AD, there are more and more arguments to anticipate and try to prevent the onset of the disease. We may consider similarly a better diet, exercise, and increase of the cognitive reserve. Apart from the rather soft general control induced on our lifestyle, one might hardly see ethical issues in such early interventions. Much more complex remains the question of primary prevention and enrollment of at-risk individuals in clinical trials targeted against amyloid. Multidisciplinary research, including ethical, legal, and social aspects, is needed to investigate how best to inform and anticipate reactions of patients, and their families, to the knowledge of their status, and how to support individuals undergoing long periods of treatments with high constraints, unknown risks, and unknown benefits. They should be valued as ordinary heroes fighting against ignorance that is the main enemy. A new partnership needs to be set up between at-risk individuals, clinicians, scientists, and society to defeat AD for once and for all.

References

Albert, M.S., Dekosky, S.T., Dickson, D., et al. (2011). The diagnosis of mild cognitive impairment due to Alzheimer's disease: recommendations from the National Institute on Aging-Alzheimer's Association workgroups on diagnostic guidelines for Alzheimer's disease. *Alzheimer's & Dementia*, 7, 270–279.

Arrighi, H.M., Barakos, J., Barkhof, F., et al. (2016). Amyloid-related imaging abnormalities-haemosiderin (ARIA-H) in patients with Alzheimer's disease treated with bapineuzumab: a historical, prospective secondary analysis. *Journal of Neurology, Neurosurgery, and Psychiatry*, 87, 106–112.

Bateman, R.J., Xiong, C., Benzinger, T.L., et al. (2012). Clinical and biomarker changes in dominantly inherited Alzheimer's disease. *New England Journal of Medicine*, 367, 795–804.

De Strooper, B. and Karran, E. (2016). The cellular phase of Alzheimer's disease. *Cell*, 164, 603–615.

De Vos, A., Jacobs, D., Struyfs, H., et al. (2015). C-terminal neurogranin is increased in cerebrospinal fluid but unchanged in plasma in Alzheimer's disease. *Alzheimer's & Dementia*, 11, 1461–1469.

Dubois, B., Feldman, H.H., Jacova, C., et al. (2014). Advancing research diagnostic criteria for Alzheimer's disease: the IWG-2 criteria. *Lancet Neurology*, 13, 614–629.

Dubois, B., Hampel, H., Feldman, H.H., et al. (2016). Preclinical Alzheimer's disease: definition, natural history, and diagnostic criteria. *Alzheimer's & Dementia*, 12, 292–323.

Jack, C.R., Jr., Bennett, D.A., Blennow, K., et al. (2016). A/T/N: an unbiased descriptive classification scheme for Alzheimer disease biomarkers. *Neurology*, 87(5), 539–547.

Karch, C.M. and Goate, A.M. (2015). Alzheimer's disease risk genes and mechanisms of disease pathogenesis. *Biological Psychiatry*, 77, 43–51.

Kim, S.Y., Karlawish, J., and Berkman, B.E. (2015). Ethics of genetic and biomarker test disclosures in neurodegenerative disease prevention trials. *Neurology*, 84, 1488–1494.

Lacour, A., Espinosa, A., Louwersheimer, E., et al. (2017). Genome-wide significant risk factors for Alzheimer's disease: role in progression to dementia due to Alzheimer's disease among subjects with mild cognitive impairment. *Molecular Psychiatry*, 22(1), 153–160.

Peters, K.R., Lynn Beattie, B., Feldman, H.H., and Illes, J. (2013). A conceptual framework and ethics analysis for prevention trials of Alzheimer disease. *Progress in Neurobiology*, 110, 114–123.

Price, J.L., McKeel, D.W., Jr., Buckles, V.D., et al. (2009). Neuropathology of nondemented aging: presumptive evidence for preclinical Alzheimer disease. *Neurobiology of Aging*, 30, 1026–1036.

Scheltens, P., Blennow, K., Breteler, M.M., et al. (2016). Alzheimer's disease. *The Lancet*, 388, 505–517.

Selkoe, D.J. and Hardy, J. (2016). The amyloid hypothesis of Alzheimer's disease at 25 years. *EMBO Molecular Medicine*, 8, 595–608.

Villemagne, V.L., Burnham, S., Bourgeat, P., et al. (2013). Amyloid beta deposition, neurodegeneration, and cognitive decline in sporadic Alzheimer's disease: a prospective cohort study. *Lancet Neurology*, 12, 357–367.

Vos, S.J. and Fagan, A.M. (2016). Alzheimer's disease biomarker states. *Lancet Neurology*, 15, 25–26.

Vos, S.J., Xiong, C., Visser, P.J., et al. (2013). Preclinical Alzheimer's disease and its outcome: a longitudinal cohort study. *Lancet Neurology*, 12, 957–965.

Vos, S.J., Verhey, F., Frolich, L., et al. (2015). Prevalence and prognosis of Alzheimer's disease at the mild cognitive impairment stage. *Brain*, 138, 1327–1338.

Winblad, B., Amouyel, P., Andrieu, S., et al. (2016). Defeating Alzheimer's disease and other dementias: a priority for European science and society. *Lancet Neurology*, 15, 455–532.

Chapter 17

When bright lines blur: Deconstructing distinctions between disorders of consciousness

David B. Fischer and Robert D. Truog

Introduction

Of the many disorders of the brain, disorders of consciousness are among the most devastating. They have profound implications for whether an individual is considered capable of experiencing the world, or even considered alive. However, despite the critical implications, there remains substantial uncertainty about how they are defined and diagnosed, leading to uncertainty in how these patients are understood and managed. What makes a person conscious, and how can one person know if another is conscious? What makes a person dead, and how can one know if another is dead? These questions have challenged philosophers for centuries. Meanwhile, however, the ability to ethically manage patients who suffer from disorders of consciousness critically depends upon the answers to these questions.

Distinguishing between disorders of consciousness has been vital to guiding important clinical decisions. However, we argue that disorders of consciousness are not sufficiently distinct to dictate these decisions. After defining each disorder of consciousness and discussing the decisions that hinge upon the diagnostic distinctions, we will outline the factors that, upon closer inspection, obscure these distinctions. We argue that the non-distinct diagnostic boundaries reflect an inherent continuity between disorders of consciousness. Then we consider a new way of thinking about disorders of consciousness, permitting these diagnoses to more effectively guide decisions. We believe that these considerations bring clarity to disorders of consciousness, and can improve the ethical management of patients suffering from them.

Disorders of consciousness: Conceptual definitions and diagnostic criteria

Disorders of consciousness can be divided into four main categories. Each category can be understood in two distinct ways: through a conceptual definition and a set of diagnostic criteria. The conceptual definition of a condition describes what that condition is thought to fundamentally represent. In contrast, the diagnostic criteria of a condition are a set of empirical characteristics by which that condition is recognized clinically. The conceptual definitions and diagnostic criteria can sometimes be conflated, but explicitly distinguishing between the two is important. Discrepancies between them can lead to problems in the diagnosis and understanding of the disorders, as we will later explore.

Brain death

We will describe each disorder in order of severity, beginning with the most severe. Thus, the first we will consider is brain death. (Table 17.1) Though not classically included among other disorders of consciousness—as it is often considered a more severe form of brain injury with more grave implications—brain death represents a state of disordered consciousness following brain injury. We include it among the other disorders of consciousness here.

Conceptually, brain death is considered a form of death. Death has been a challenge to define precisely, and its medical definition has evolved continuously since the 1700s (Powner et al. 1996). In 1981, neurologist James Bernat built from these concepts to propose death as the "permanent cessation of functioning of the organism as a whole" (Bernat et al. 1981, p.390). Bernat argues that the "whole" of an organism consists in the "highly complex interaction of its organ subsystems," and "the spontaneous and innate activities carried out by the integration of all or most subsystems … and at least limited response to the environment" (Bernat et al. 1981, p.390). Wholeness of an organism can also be understood through thermodynamic principles: an organism perhaps can be considered living so long as it employs energy-consuming processes to oppose entropic forces and maintain homeostasis, thereby preserving itself as an entity distinct from the surrounding world (Truog 2015).

How does one recognize when an organism has permanently ceased to function as a whole? In the same year that Bernat offered his conceptual definition of death, the 1981 President's Commission proposed a legal framework by which this definition could be recognized, which led to the Uniform Determination of Death Act (Medical Consultants on the Diagnosis of Death 1981). Under this definition, any individual is considered dead if he or she has (1) irreversible cessation of circulatory and respiratory functions, or (2) irreversible cessation of all functions of the entire brain, including the brainstem. The latter criterion comprises the diagnostic criterion for brain death. That is, a person is thought to meet the conceptual definition of death (the permanent cessation of functioning as a whole) if he/she exhibits the irreversible cessation of all brain function. Bernat argues that the irreversible cessation of all brain functions is sufficient for the permanent cessation of functioning of the organism of the whole, "because the brain is necessary for the functioning of the organism as a whole. It integrates, generates, interrelates, and controls complex bodily activities" (Bernat et al. 1981, p.391).

The diagnostic criterion of brain death as the irreversible cessation of all brain functions was subsequently fleshed out with more specific criteria. In 1995, the American Academy of Neurology published a set of such criteria (Report of the Quality Standards

Table 17.1 Conceptual definitions and diagnostic criteria of disorders of consciousness

	Conceptual definition	**Diagnostic criteria**

Brain death	A form of death, where death is defined as the permanent cessation of functioning of the organism as a whole	◆ The irreversible cessation of all functions of the entire brain, including the brainstem
		• Prerequisites:
		  • Clinical or neuroimaging evidence of acute central nervous system injury
		  • Exclusion of confounding medical conditions
		  • No drug intoxication or poisoning
		  • Core temperature ≥32°C (≥90°F)
		• Coma
		• Absence of brainstem reflexes
		• Apnea

Coma	Absence of both wakefulness and awareness	◆ No eye opening spontaneously or to vigorous sensory stimulation ◆ Low cortical arousal on EEG (e.g., polymorphic delta, burst-suppression) ◆ Only reflexive and postural motor function ◆ No response to auditory or visual stimuli

Vegetative state/ unresponsive wakefulness syndrome	Wakefulness without awareness of self or environment	◆ Eye opening and closing that appear to follow sleep–wake cycles
		◆ Incapable of interaction with others
		◆ No language comprehension or expression
		◆ Reflexive behavior only, without any purposeful behaviors. Behaviors can include:
		  • Spontaneous breathing
		  • Move trunk or limbs non-purposefully, arch back, posturing
		  • Blink, roving eye movements, nystagmus, and brief, unsustained visual pursuit
		  • Move head or eyes briefly toward sound or movement
		  • Auditory startle and startle myoclonus
		  • Grimace to pain
		  • Yawn, chewing movements, swallowing saliva
		  • Reflexive crying, smiling without provoking stimuli
		  • Vocalizations without words
		  • Withdrawal from noxious stimuli

Minimally conscious state	Wakefulness with partial awareness of self or environment	◆ Eye opening and closing that appear to follow sleep–wake cycles
		◆ One or more of the following behaviors on a reproducible or sustained basis:
		  • Following simple commands
		  • Gestural or verbal yes/no responses (regardless of accuracy)

(*continued*)

Table 17.1 Continued

Conceptual definition	Diagnostic criteria

	• Intelligible verbalization
	• Purposeful behavior, including movements or affective behaviors that occur in contingent relation to relevant environmental stimuli and are not due to reflexive activity. Such behaviors include:
	  • Appropriate smiling or crying in response to emotional stimuli
	  • Vocalizations or gestures that occur in direct response to linguistic content of questions
	  • Reaching for objects that demonstrates a clear relationship between object location and direction of reach
	  • Touching or holding objects in a manner that accommodates the size and shape of the object
	  • Pursuit eye movement or sustained fixation that occurs in direct response to moving or salient stimuli

Source: data from Bernat et al. 1981; Medical Consultants on the Diagnosis of Death to the President's Commission for the Study of Ethical Problems in Medical and Biomedical and Behavioral Research 1981; Report of the Quality Standards Subcommittee of the American Academy of Neurology 1995; Wijdicks et al. 2010 (Brain death); Giacino et al. 2002; Laureys et al. 2004; Bernat 2006; Posner et al. 2007 (Coma); The Multi-Society Task Force on PVS 1994a; Giacino 2004; Bernat 2006 (Vegetative state/unresponsive wakefulness syndrome); Giacino et al. 2002 (Minimally conscious state).

Subcommittee of the American Academy of Neurology 1995; Wijdicks et al. 2010): in addition to coma (described in more detail later), patients must have an identifiable catastrophic brain injury, no confounding conditions (e.g., medical instability, drug toxicity, hypothermia), absent brainstem reflexes (e.g., pupillary, ocular, facial sensation/motor, pharyngeal and tracheal reflexes), and apnea (i.e., an absent respiratory drive). Optional ancillary assessments, which are not required for the diagnosis of brain death, include a repeat evaluation 6 hours after initial assessment (required by some US states), conventional angiography, magnetic resonance angiography, electroencephalography (EEG), transcranial Doppler ultrasonography, cerebral scintigraphy, and other technological assessments. Together, these clinical criteria were proposed as a means of determining that all functions of the entire brain, including the brainstem, have been irreversibly lost.

Coma

Coma, a component of the diagnostic criteria for brain death, represents the next most severe disorder of consciousness. To understand the conceptual definition of coma, it is first important to consider what comprises consciousness. Like death, consciousness is difficult to define, and so its precise definition has been a matter of debate among philosophers and scientists for centuries. However, there is general agreement that consciousness consists of two primary components: arousal (i.e., wakefulness) and awareness (i.e., the content of experience) (Posner et al. 2007; Laureys et al. 2009). Awareness requires arousal, in that one must be awake in order to be aware (Posner et al. 2007; Laureys et al. 2009).[1] However, one can be awake without being aware, as in other disorders of consciousness described later in this chapter.

Within this framework, coma is conceptually defined as the total absence of consciousness, where wakefulness, and therefore also awareness, are both lost (Giacino et al. 2002; Laureys et al. 2004; Posner et al. 2007). Diagnostically, the loss of wakefulness is identified by the absence of eye opening, either spontaneously or to vigorous sensory stimulation, and low cortical arousal (e.g., polymorphic delta, burst-suppression) on EEG (Giacino et al. 2002; Laureys et al. 2004; Bernat 2006; Posner et al. 2007). Behaviorally, comatose patients exhibit only reflexive and postural responses, without responding to auditory or visual stimuli (Giacino et al. 2002; Bernat 2006; Posner et al. 2007). These diagnostic features must persist for more than an hour, to ensure that they are not attributable to a transient state of unconsciousness, such as syncope or concussion (Laureys et al. 2004).

It should also be noted that, in comparison to the other disorders of consciousness described here, coma may be a more temporary state, rarely lasting longer than a month, with most patients dying, regaining consciousness, or progressing to another disorder of consciousness (Bernat 2006).

Vegetative state/unresponsive wakefulness syndrome

The term vegetative state was first proposed in 1972 by neurosurgeon Bryan Jennett and neurologist Fred Plum, based on the *Oxford English Dictionary* definition of "vegetative" as "an organic body capable of growth and development but devoid of sensation and thought" (Jennett & Plum 1972, p.736). Jennett and Plum used this term to describe patients who, conceptually, are awake but lack awareness of the self or environment.

In the 1990s, the Multi-Society Task Force on the Persistent Vegetative State established diagnostic criteria for the vegetative state, intended to outline objective measures of preserved wakefulness and absent awareness. Diagnostically, wakefulness is evidenced by spontaneous eye opening and eye closing, which appear to follow sleep–wake cycles (The Multi-Society Task Force on PVS 1994a). Patients are determined to lack awareness if they do not produce any purposeful or voluntary behavior in response to auditory, visual, tactile, or noxious stimulation; cannot interact with others; and show no evidence of language comprehension or expression (The Multi-Society Task Force on PVS 1994a). Importantly, such patients without awareness are not necessarily immobile, but their movements must be purely non-purposeful, that is, must be purely reflexive: for example, they may move their trunk or limbs (or both) in non-purposeful ways; display roving eye movements and brief, unsustained visual pursuit; yawn and make chewing movements; move head or eyes briefly toward sound or movement; startle to auditory stimuli; and smile, cry, make meaningless vocalizations, or a combination of these, in the absence of any provoking stimuli (The Multi-Society Task Force on PVS 1994a; Giacino 2004; Bernat

2006). Bowel/bladder incontinence and preserved autonomic/hypothalamic function are also thought to be diagnostically characteristic of the vegetative state, although they do not necessarily reflect the state's conceptual definition.

Variations of the vegetative state have been proposed. The term persistent vegetative state is used to describe a vegetative state that persists for greater than 1 month, and is not intended to portend prognosis (The Multi-Society Task Force on PVS 1994a, 1994b). The permanent vegetative state refers conceptually to an irreversible vegetative state (hence carrying an implication of prognosis), and is identified based on a vegetative state persisting over 3 months following non-traumatic injury, or over 12 months following traumatic injury (The Multi-Society Task Force on PVS 1994a, 1994b). As these subcategories of the vegetative state are frequently confused and intermix diagnosis and prognosis, they are contentious and inconsistently accepted, and thus will not be considered further here (Bernat 2006).

As the vegetative state, to much of the lay public and media, carries the pejorative connotation that affected patients are "vegetable-like," a new name for the condition was proposed in 2010: the unresponsive wakefulness syndrome (Laureys et al. 2010). This title better captures the conceptual definition of the state as a condition in which awareness is lost but wakefulness is preserved. Thus, we will hereafter refer to this state as the vegetative state/unresponsive wakefulness syndrome (VS/UWS).

Minimally conscious state

Following establishment of the VS/UWS, clinicians recognized that some awake patients were neither fully unaware (i.e., in a VS/UWS) nor fully aware, exhibiting signs of partial awareness. In 2002, neuropsychologist Joseph Giacino and colleagues proposed a new diagnostic category to describe such patients, labeled the minimally conscious state (MCS) (Giacino et al. 2002). The MCS is conceptually defined as a state of preserved wakefulness with only partial awareness of the self or environment.

In terms of diagnostic criteria, like VS/UWS patients, MCS patients exhibit sleep–wake cycles through spontaneous eye opening and eye closing. Unlike VS/UWS patients, MCS patients exhibit awareness of themselves or the environment, albeit only partially and intermittently. Giacino and colleagues proposed that empirical evidence of such awareness is the reproducible and sustained demonstration of one or more of the following behaviors: simple command following, answering of yes/no questions, intelligible verbalizations, or any other purposeful, non-reflexive behaviors, including appropriate smiling/crying in response to emotional stimuli, reaching for objects, or eye-tracking a target (Giacino et al. 2002).

Significance of diagnostic categories and their distinction

The disorders of consciousness described in the previous section have significant implications for how patients are understood and managed. Thus, the ability to distinguish between these various conditions is critical to ensuring that patients are treated ethically. Two distinctions between these disorders are especially crucial in dictating how patients are managed: the distinction between brain death and coma, and between the VS/UWS and MCS.

Coma versus brain death: The boundary between life and death

As described, brain death and coma are diagnostically similar. Indeed, the diagnostic criteria for brain death consist of coma, plus absent brainstem reflexes, apnea, and a few prerequisite conditions. However, conceptually, the distinction is stark: patients with brain death are considered dead, while patients in a coma are considered alive. This conceptual distinction carries tremendous implications for clinical management, not to mention for families and loved ones: patients who are dead no longer justify medical treatment, and can have their organs legally procured for transplantation.

Once a patient is considered dead, as in the case of brain death, there is understood to be the permanent cessation of functioning of the organism as a whole. Brain death therefore fulfills the most stringent interpretation of futility, as a situation where it is conceptually impossible for an intervention to work. When patients are declared brain dead, clinicians are therefore relieved of any obligation to treat, and so may remove life-sustaining treatment (LST) without the family's or surrogate's permission. In contrast, because patients in a coma are considered alive, clinicians caring for such patients are under a greater obligation to honor requests from families and surrogates for LST and intensive care.

The diagnosis of brain death also justifies organ procurement for transplantation. While patients with this diagnosis are claimed to no longer function as a whole, individual organs may remain intact and offer benefit to other ill individuals. On the assumption that these patients are dead, organ procurement is considered ethically appropriate. Though this intuition was implicitly accepted as an ethical and legal requirement for organ procurement since the beginning of transplantation in the 1960s (Truog 2015), it was explicitly articulated in 1999 as the "dead donor rule" (Robertson 1999). The dead donor rule states that vital organs for transplantation may only be procured from patients who are dead, or alternatively, that physicians may not cause death when procuring vital organs for transplantation (Robertson 1999). Though the dead donor rule is not, strictly speaking, a law, it is understood to follow logically from existing laws and ethical standards related to homicide. Thus, because brain death is considered death, diagnosing a patient as brain dead allows the procurement of their organs for transplantation, assuming it is consistent with that patient's wishes. In contrast, vital organs cannot be legally procured from comatose patients, who are considered alive.

MCS versus VS/UWS: The boundary between conscious and unconscious

As described, there are at least two aspects of consciousness: wakefulness and awareness. However, the colloquial understanding of consciousness aligns much more closely with the notion of awareness; an individual is typically considered conscious only if he or she is aware. Experts in disorders of consciousness will also often adhere to this conventional conception, defining consciousness as awareness of the self or environment. Furthermore, many more social and ethical implications hinge on one's capacity to experience the world than on whether one maintains sleep/wake cycles, as will be explored further later. Awareness of the self or environment is lost in the VS/UWS, but maintained (albeit only partially) in the MCS. Thus, to reflect the conventional conception of consciousness and for the purposes of these ethical implications, the conceptual distinction between these states (i.e., between unawareness and awareness of the self or environment) will be considered the distinction between unconsciousness and consciousness in the discussion that follows.

One major implication of this distinction pertains to LST. Decisions to withdraw LST often depend upon judgments about the quality of a patient's life. As quality of life depends upon conscious experience, VS/UWS patients, by the conceptual definition of the state, necessarily lack any quality of life. Thus, many may feel compelled to withdraw LST if the VS/UWS diagnosis is made. In 1989, the American Academy of Neurology issued a statement claiming that LST "provides no benefit to patients in a persistent vegetative state" (Executive Board of the American Academy of Neurology 1989, p.126), and other guidelines have claimed that the maintenance of consciousness should be the minimum objective of LST (Rubin & Bernat 2011). Accordingly, studies have shown that decisions to withdraw LST are frequently made soon after patients are found to have a disruption of consciousness and lose decision-making capacity (Fins et al. 1999; Becker et al. 2001). Ethicist and physician Joseph J. Fins summarizes:

> [T]he loss of consciousness in routine medical care takes on so much prognostic and ethical meaning. In the context of routine end-of-life care in the general hospital, most decisions about withholding or withdrawing life-sustaining therapy hinge on the presence or absence of consciousness. When consciousness (and resulting decision-making capacity) is lost, surrogate decision makers take it as an important prognostic sign and use this loss as a prompt to make end-of-life decisions … in the acute care setting DNR orders get written and ventilators are removed. (Fins 2015, p.184)

Thus, for many, determining whether a patient has lost consciousness (as in the VS/UWS) or retains consciousness (as in the MCS) plays an important role in determining whether LST is continued.

Second, some have suggested that the presence of consciousness has important implications for civil rights. Fins, in his book *Rights Come to Mind: Brain Injury, Ethics, and the Struggle for Consciousness*, writes that "to my mind, there is nothing more important than knowing that a patient may be conscious" (Fins 2015, p.201). Fins argues that being conscious means being entitled to civil rights. Because MCS patients are conscious (albeit minimally so) he argues that they are therefore entitled to the same civil rights as anyone else, and that to treat MCS patients otherwise is an act of segregation. Though Fins centers his arguments on MCS, the implication is that, by virtue of having lost conscious awareness, VS/UWS patients are no longer entitled to that same range of civil rights. If indeed the presence of consciousness is what dictates a person's entitlement to civil rights, the distinction between MCS and VS/UWS has major implications for whether patients are granted or denied these rights.

A third more mundane but still significant implication of this diagnostic distinction is a humanistic one. To family and loved ones, it matters if a patient is "in there" and aware of their presence and attempts at communication, which may influence how they interact with the patient (Fins 2015). For clinicians and healthcare providers, knowing that a patient is conscious, with a quality of life, may encourage them to talk to the patients, to reposition them, and to provide comfort in other ways. These everyday differences in how patients are perceived and treated critically depend upon whether they are felt to be conscious or unconscious, and therefore hinge on the distinction between the MCS and VS/UWS.

The importance of discrete distinctions in disorders of consciousness

These implications have profound social, legal, and ethical significance. To decide whether a patient is brain dead or comatose is to decide whether a patient is legally alive or dead. Similarly, to decide whether a patient is in a MCS or VS/UWS is to decide whether a patient is aware of the world, with a quality of life, or not. These determinations, in turn, have tremendous consequences, influencing whether LST is withdrawn, organs are procured, civil rights are granted, and even whether basic forms of respect and comfort are offered on an everyday basis.

Importantly, many of the critical decisions that hinge upon these diagnoses are discrete, binary, and without a middle ground. Withdrawal of LST and procurement of vital organs, for example, are serious and permanent decisions without gradation. Thus, these decisions require that the diagnostic distinctions that dictate them are similarly discrete. For example, the distinction between brain death and coma must be discrete enough to determine whether a patient is dead or alive, which in turn determines whether vital organ procurement is appropriate or forbidden. Otherwise, if it were possible for a patient to be 50% brain dead and 50% comatose, or even 99% brain dead and 1% comatose, then the patient would be neither clearly alive nor dead, and would therefore confound decisions about organ procurement. With regard to deciding whether or not to continue treatment, Fins echoes the necessity of discrete diagnostic distinctions, emphasizing that "those who were minimally conscious [merit] … a bright-line distinction with patients who were unconscious" (Fins 2015, p.128). Thus, to properly determine how significant, discrete decisions should be made, the distinction between brain death and coma, and between MCS and VS/UWS, must be clearly delineated.

Problems with discrete diagnostic boundaries

But are these diagnostic distinctions as discrete as they are presented or required to be? Closer inspection reveals that the distinctions between brain death and coma, and between VS/UWS and MCS, are both problematic, though in different ways.

Problems distinguishing between brain death and coma

In terms of their diagnostic criteria (see Table 17.1), brain death and coma can be distinguished relatively cleanly. To be considered brain dead a patient must exhibit, in addition to coma, absent brainstem reflexes and apnea, and must meet certain prerequisite conditions (e.g., minimum core temperature and absence of confounding medical conditions). Such criteria represent largely objective and discrete outcome measures, as they are either observably absent or present. While some of these findings may be subtle and require clinical expertise (e.g., to determine whether a brainstem reflex is diminished or totally absent), in principle, brain death and coma can be discretely distinguished diagnostically.

The problem, however, is that the diagnostic criteria do not accurately reflect the conceptual definition of brain death, and it is this conceptual definition that justifies major decisions such as organ procurement. There are at least two levels on which this discrepancy occurs. Recall that the diagnostic criteria for brain death (e.g., coma, apnea, and absent brainstem reflexes) are intended to represent an overarching diagnostic (and legal) criterion: the irreversible cessation of all functions of the entire brain, including the brainstem. This overarching diagnostic criterion, in turn, is thought to reflect the conceptual definition of death, as the permanent cessation of functioning of the organism as whole. Both connections are necessary in order for the diagnostic criteria for brain death to reflect its conceptual definition, but closer inspection reveals that both are flawed.

First, the diagnostic criteria for brain death do not necessarily represent the irreversible cessation of all functions of the entire brain, including the brainstem (Halevy & Brody 1993; Truog 1997, 2007, 2015). Many patients who fulfill the diagnostic criteria for brain death continue to show evidence of preserved brain function: many are capable of regulating body temperature, indicative of a functioning hypothalamus (indeed, ironically, the brain death criterion mandating a core temperature of at least 32°C often signifies intact temperature regulation and thereby brain function) (Truog 1997); many continue to control fluid and salt homeostasis through the neurologically mediated secretion of pituitary hormones (Schrader et al. 1980; Gramm et al. 1992; Halevy & Brody 1993; Truog 1997); many respond to surgical incisions with elevations in heart rate and blood pressure, reflecting a potential brainstem response to noxious stimuli (Pennefather et al. 1993; Truog 1997); and some (20% in one series) continue to exhibit electrical activity on EEG, which may represent continued function of some brain regions, though this remains controversial (Pallis 1983; Rodin et al. 1985; Grigg et al. 1987; Truog 1997). Based on these findings, there is now general agreement that the diagnostic criteria for brain death do not necessarily correspond to the overarching diagnostic criterion, and legal definition, which is the irreversible cessation of all brain function (Truog 1997, 2015).

Second, even if we assume that the diagnostic criteria for brain death do reflect the irreversible cessation of brain function, this overarching diagnostic criterion still does not adequately correspond to the conceptual definition of death, as the permanent cessation of functioning of the organism as a whole (Truog 2015). Bernat justified his proposal—that brain death meets the conceptual definition of death as the permanent cessation of functioning of the organism as a whole—by his observation that "[d]estruction of the brain produces apnea and generalized vasodilation; in all cases, despite the most aggressive support, the adult heart stops within 1 week, and that of the child within 2 weeks" (Bernat et al. 1981, p.392). While his observation was probably true at the time—patients invariably suffered cardiac arrest soon after the diagnosis of brain death—advances in intensive medical care now allow some patients to maintain cardiac and pulmonary function for years after the diagnosis (Shewmon 1998). In one case, a child who fulfilled the diagnostic criteria for brain death at 4 years old was kept on a ventilator and gastric tube for about 20 years, during which he grew and developed, maintained cardiopulmonary function, digested and excreted food, and mounted immune responses to infections; at autopsy, his brain had become a calcified mass, void of any remaining brain cells (Repertinger et al. 2006). There have also been numerous cases of women who suffered brain death during pregnancy, and were subsequently maintained on a ventilator and feeds throughout the gestation and delivery of their children (Lane et al. 2004; Ecker 2014). These cases illustrate that even without the brain, the body can maintain a level of integrated functioning and homeostasis, enough to support the gestation of a fetus. Thus, the overarching diagnostic criterion of brain death, as the loss of brain function, may not accurately represent the cessation of all functioning of the organism as a whole.

The President's Council on Bioethics confronted the discrepancy between the diagnostic criteria and conceptual definition of brain death in a 2008 report (The President's Council on Bioethics 2008). They acknowledged that:

> [i]f being alive as a biological organism requires being a whole that is more than the mere sum of its parts, then it would be difficult to deny that the body of a patient with [brain death] can still be alive, at least in some cases. (p.57)

They also asserted that:

> [t]he reason that these somatically integrative activities continue ... is that the brain is not the integrator of the body's many and varied functions... . [N]o single structure in the body plays the role of an indispensable integrator. Integration, rather, is an emergent property of the whole organism. (p.40)

Having acknowledged the inconsistency between the diagnostic criteria and conceptual definition of brain death, they attempted to solve the problem by proposing a revised conceptual definition of death. They suggested that death represents the inability to perform the "vital work" of the organism as a whole, where "vital work" consists of three components:

1. [R]eceptivity to stimuli and signals from the surrounding environment.
2. The ability to act upon the world to obtain selectively what it needs.
3. The basic felt need that drives the organism to act as it must, to obtain what it needs. (p.61)

However, this conceptual definition is also problematic (Miller & Truog 2009; Truog 2015). If "vital work" requires conscious awareness (as implied by an "ability to act upon the world" and a "felt need"), then VS/UWS patients and perhaps even MCS patients would be considered dead, a position the Council specifically rejected. If, alternatively, "vital work" only refers to unconscious bodily functions that respond to the environment (e.g., mounting an immune response to infection) and interact with the environment in accordance with the body's needs (e.g., absorbing oxygen and digesting nutrients), then the same diagnostic–conceptual discrepancy remains; as described, many patients who meet the diagnostic criteria for brain death are capable of several such bodily functions. Thus, the Council's proposition does not solve the incongruity between the diagnostic criteria and conceptual definition of brain death.

It is also worth noting that the irreversibility/permanence required of both the overarching diagnostic criterion of brain death (the *irreversible* cessation of brain functions) and conceptual definition (the *permanent* cessation of functioning of the organism as a whole) imposes an additional discrepancy between the diagnostic criteria and conceptual definition of brain death. Irreversibility and permanence are both statements of prognosis, or predictions for the future. While diagnoses may be *associated* with certain prognoses, even with high probability, it is problematic to *require* a prognosis in the very conceptual definition of a diagnosis, as doing so theoretically requires a certain prediction of the future. Even if, ignoring the earlier arguments, the diagnostic criteria for brain death actually reflected the cessation of all brain function or the cessation of all integrated functioning of the organism, and even if all patients who met the diagnostic criteria for brain death to date had never recovered those functions, it would still not be absolutely certain that those diagnostic criteria would predict permanent/irreversible cessation of those functions in 100% of future cases. Thus, while the diagnostic criteria for brain death may be *associated* with an overwhelmingly poor prognosis, the *requirement* of the prognosis in brain death's very definition impossibly demands a prediction of the future, thereby ensuring that the diagnostic criteria could, in principle, never match the overarching diagnostic criterion or conceptual definition with certainty.

To summarize, in order to effectively guide discrete decision-making, the distinction between brain death and coma must be similarly discrete. Although the diagnostic criteria for brain death and coma are in principle sufficiently discrete, the brain death criteria do not accurately reflect the conceptual definition of brain death. As it is the conceptual definition of brain death, as death, that justifies major decisions such as organ procurement, this diagnostic/conceptual discrepancy indicates that the current diagnostic criteria cannot adequately guide these decisions. That is, while one can definitively determine that a patient fulfills the diagnostic criteria for brain death, as distinct from coma, one cannot be confident that those diagnostic criteria meaningfully signify death in a way that justifies organ procurement. We should also note that while we currently focus on this problem as it pertains to the scientific and medical definition of death, this problem also extends to alternative religious and cultural definitions of death (Posner et al. 2007). Many, particularly those of orthodox religious groups and those with cultural roots in Asia and the Middle East, feel that the diagnostic criteria for brain death do not reflect their personal conceptions of death; in Japan, for example, two types of "death" are commonly recognized—one suitable for patients who wish to be organ donors, and another for everyone else (Lock 2002). Given the heterogeneity of views on brain death, US states such as New Jersey and New York have chosen to accommodate religious and moral objections to brain death testing (Kimura 1991; Posner et al. 2007). Thus, overall, the diagnostic distinction between brain death and coma does not accurately reflect their conceptual distinction as death and life, respectively, thus undermining the utility of these diagnostic distinctions in guiding decisions.

Problems distinguishing between VS/UWS and MCS

As described, the ability to discretely distinguish VS/UWS patients from MCS patients—that is, unconscious from conscious patients—is important in guiding discrete decisions that hinge upon whether consciousness has been lost or retained. However, closer inspection of the distinction between VS/UWS and MCS reveals two problems: first, the diagnostic distinction is not, in principle, discrete, and second, the diagnostic distinction does not wholly correspond to the conceptual distinction.

The diagnostic criteria distinguishing VS/UWS from MCS do not discretely delineate the two conditions. The crux of the diagnostic distinction is that while VS/UWS patients exhibit only reflexive behaviors, MCS patients exhibit both reflexive and purposeful behaviors, such as through intelligible verbalizations, answering yes/no questions, and reaching for objects in a goal-directed manner. However, it is unclear how purposeful behavior can be recognized and distinguished from reflexive behavior (Fischer & Truog 2015). It is not merely that reflexive behaviors are triggered by an external stimulus, while purposeful behaviors are internally motivated, as the "reflexive crying or smiling" (Table 17.1) that can be observed in the VS/UWS occurs without a triggering stimulus. Nor can they be distinguished by consistency, since in some instances purposeful behaviors are more consistent than reflexive behaviors (e.g., deliberate eye tracking versus transient pursuit), while in other instances the reverse is true (e.g., a patellar deep tendon reflex versus a volitional kick). Nor can they be distinguished by complexity, since, while purposeful behaviors can be complex, brain stimulation and seizures can also produce complex but non-volitional behaviors (Fischer & Truog 2015). Indeed, while some behaviors are clearly purposeful (e.g., holding a conversation) and others clearly reflexive (e.g., the pupillary light reflex), there is a wide range of behaviors between the two which cannot be clearly classified. Because the diagnostic distinction between the VS/UWS and MCS relies upon the distinction between reflexive and purposeful behaviors, the uncertainty in distinguishing these two types of behaviors creates uncertainty in clinically distinguishing the VS/UWS from the MCS.

This diagnostic ambiguity is compounded by an additional problem: as with brain death and coma, there is incongruity between the diagnostic criteria and conceptual definitions of VS/UWS and MCS. That is, even if purposeful behaviors and reflexive behaviors could theoretically be discretely distinguished, this diagnostic distinction does not clearly signify the presence or absence of consciousness, respectively. Again, the crux of the diagnostic distinction is that while VS/UWS patients exhibit only reflexive behaviors, MCS patients exhibit both reflexive and purposeful behaviors. The logic of these diagnostic criteria is that purposeful behaviors reflect volition, which can exist only in those who are conscious. While this is true, the same logic does not follow in the absence of these behaviors. Patients are given a VS/UWS if they exhibit only reflexive behaviors, that is, if they are not observed exhibiting any purposeful behaviors. However, first, just because purposeful behaviors are not observed or recognized does not necessarily mean that purposeful behaviors are absent. Second, the absence of purposeful behaviors does not necessarily imply the absence of volition. And third, the absence of volition does not necessarily imply the absence of awareness. We can imagine many scenarios—patients who are blind, deaf, aphasic, confused, apathetic, paralyzed, or even intermittently conscious, for example—where conscious patients are not recognized as such on the basis of observable purposeful behaviors. Thus, in clinically distinguishing VS/UWS from MCS, the absence of evidence is not necessarily the evidence of absence; the diagnostic criteria for these disorders imperfectly reflect their conceptual definitions.

Neuroimaging technology has poignantly revealed the danger of assuming that the absence of purposeful behaviors represents the absence of consciousness. In 2006, neuroscientist Adrian Owen and colleagues published a study on a 23-year-old woman who sustained traumatic brain injury after a road traffic accident (Owen et al. 2006). She continued to demonstrate sleep–wake cycles, but was behaviorally unresponsive, and so was given the diagnosis of VS/UWS. However, using functional magnetic resonance imaging (fMRI), Owen and colleagues found that she could reliably respond to commands through fMRI signals. When asked to imagine playing tennis, she exhibited activity in the supplementary motor area, a region associated with planning movement. When asked to imagine walking through the rooms of her house, she exhibited activity in regions associated with spatial navigation, such as the parahippocampal gyrus. These activity patterns were indistinguishable from those observed in healthy individuals. Moreover, these activations did not appear to be reflexive reactions to the words "tennis" or "house," but rather intentionally generated; the patient's responses lasted up to 30 seconds (longer than would be expected from a transient, automatic reaction) and only until the experimenters asked her to stop (Hopkin 2006). Another study found additional patients who, despite being behaviorally unresponsive, were similarly able to complete this fMRI task (Monti et al. 2010). One of these patients, who was behaviorally noncommunicative, could correctly answer yes-or-no questions with this task, imagining one scenario to signal "yes" and the other scenario to signal "no." These neuroimaging studies demonstrated that a subset of patients who are behaviorally unresponsive and meet the diagnostic criteria for VS/UWS in fact do have conscious awareness, thereby meeting the conceptual definition of MCS, and perhaps even retaining higher levels of consciousness (Fischer & Truog 2013). These instances further illustrate a significant discrepancy between the conditions' diagnostic criteria and conceptual definitions.

To summarize, in order to effectively guide discrete decision-making, the distinction between the conscious and the unconscious, that is, between MCS and VS/UWS, must be similarly discrete. However, the VS/UWS cannot be discretely distinguished from the MCS for at least two reasons. First, uncertainty in distinguishing between purposeful and reflexive behaviors obscures the diagnostic distinction between the VS/UWS and MCS. Second, the diagnostic criteria of these conditions do not wholly correspond to their conceptual definitions: their diagnostic criteria do not account for conscious individuals incapable of purposeful behaviors, a danger confirmed with neuroimaging studies. These sources of ambiguity undermine the utility of the current diagnostic criteria for VS/UWS and MCS in guiding decisions.

Non-discrete diagnostic boundaries reflect a continuum between conditions

The distinction between brain death and coma, as well as between VS/UWS and MCS, are important for guiding discrete decisions: the former represents a distinction between the dead and the living, and the latter between the unconscious and the conscious. However, neither distinction between conditions is, in reality, discrete. While the diagnostic distinction between brain death and coma is in principle discrete, this diagnostic distinction does not accurately correspond to the conceptual distinction between death and life. The distinction between VS/UWS and MCS is even more problematic: not only does the diagnostic distinction between VS/UWS and MCS inadequately reflect their conceptual distinction, but the diagnostic distinction itself is inherently ambiguous. Why are the distinctions between these conditions, which are often presented as discrete and relied upon to dictate discrete decisions, so problematic? It is likely that these difficulties stem from a continuity between the conditions, hampering strict delineation. This continuity can be considered in at least two ways: as an ontological continuity, or an epistemological continuity.

Ontological continuity between disorders of consciousness

By ontological continuity, we refer to the possibility that the conceptual distinctions between life and death, and between consciousness and unconsciousness, are inherently non-discrete. First we will consider how this applies to the life/death distinction. Death has been defined as the permanent cessation of the functioning of the organism as a whole, so life can be understood as the functioning of the organism as a whole. The "wholeness" of an organism, however, is challenging to define. Bernat defined wholeness as consisting in the "highly complex interaction of… [an organism's] organ subsystems" and "the spontaneous and innate activities carried out by the integration of all or most subsystems" (Bernat et al. 1981, p.390). This description reflects the philosophical view of wholism (also known as holism or organicism), which posits that:

> complex wholes are inherently greater than the sum of their parts in the sense that the properties of each part are dependent upon the context of the part within the whole in which they operate.

> Thus, when we try to explain how the whole system behaves, we have to talk about the context of the whole and cannot get away talking only about the parts. (Gilbert & Sarkar 2000, p.1)

Thus, the definition of life is an inherently wholistic one: life emerges not through independently functioning tissues and organ systems, but through the interactions and integration of those organ systems to produce a whole.[2] Importantly, because the whole must emerge from the interactions of its parts, wholism depends upon emergentism, the notion that:

> properties at one level of complexity (for instance, tissues) cannot be ascribed directly to their component parts but arise only because of the interactions among the parts. Such properties that are not those of any part but that arise through the interactions of parts are called *emergent properties*. (Gilbert & Sarkar 2000, p.2)

For example, wetness and temperature can be considered emergent properties of molecules; while an individual water molecule cannot be considered wet or to have a temperature, it is the larger-scale interaction of many water molecules through which wetness or temperature emerges. Similarly, life can be considered an emergent property of the interactions between tissues and organ systems, a view endorsed by The President's Council on Bioethics, which has written that the integrated functioning of life "is an emergent property of the whole organism" (The President's Council on Bioethics 2008, p.40).

Life, so considered, has important implications for death. Given that emergent properties typically arise gradually out of increasingly complex interactions between constituent parts, so too do emergent properties disappear gradually from decreasingly complex interactions. For instance, just as there is no defined number of water molecules above which wetness appears—wetness emerges gradually out of an increasing number of water molecules—there is also no defined number of water molecules below which wetness is suddenly lost. Similarly, just as life emerges gradually from the increasingly complex interactions of tissues and organ systems (as has been argued in the case of fetal development (Gilbert & Sarkar 2000)), so too may death occur continuously, not discretely, from the loss of those interactions. It should be noted that certain organ systems play a particularly critical role in sustaining the interactions between organ systems; if the heart stops beating, the loss of organ interactions, the loss of the organism as a whole, and hence the death of the organism, occurs more rapidly than if the kidneys stopped functioning. However, even in the case of cardiac arrest, death can still be understood as the continuous, albeit rapid, cessation of the organ interactions that sustain the emergent property of life.

Thermodynamic principles have been used to justify the discreteness of the life/death distinction. An organism perhaps can be considered whole so long as it employs energy-consuming processes to oppose entropic forces from the surrounding world. Thus, perhaps at the moment when the entropic forces of the surrounding world exceed the self-sustaining homeostatic forces of the organism, death can be said to discretely occur. However, an organism may continue to exist after entropic forces exceed homeostatic forces, just as beaches persist, while shrinking, after erosive forces exceed the accretive forces. Thus, the moment at which the entropic forces exceed the homeostatic forces of an organism may merely signify the beginning of dying, with death itself occurring more continuously as entropic forces erode at the wholeness of the organism. Moreover, because the entropic forces of the universe can vary in strength—natural disasters and extreme temperatures may suddenly give way to calm weather—one can imagine scenarios in which, at one moment the entropic forces exceed an organism's homeostatic forces, while in the next moment the entropic forces drop below that organism's homeostatic forces. It would seem implausible that that organism has died and returned to life, simply due to variability in environmental forces. Thus, though in many cases it occurs rapidly, death appears likely to be a fundamentally continuous occurrence.

If death is indeed a continuous occurrence, then no discrete diagnostic criteria can perfectly segregate all live patients from all dead patients. Selecting discrete criteria for death would be akin to selecting a number of water molecules below which wetness is lost: no matter how they are defined, they will oversimplify an inherently continuous range of conditions as binary categories. Depending on where the criteria are drawn, some patients will be incorrectly considered dead (when they continue to function partially as a whole), and/or some patients will be incorrectly considered alive (when their wholeness has been partially lost).

The distinction between consciousness and unconsciousness is likely also a continuous one. Like life, many consider consciousness to be an emergent property of the brain (Edelman 1999). Philosopher John Searle writes that:

> Just as one cannot reach into a glass of water and pick out a molecule and say "This one is wet," so, one cannot point to a single synapse or neuron in the brain and say "This one is thinking about my grandmother." (Searle 1992, cited in Gilbert & Sarkar 2000, p.2)

Conscious awareness, which involves the higher-order integration of different sensory modalities and cognitive faculties, is commonly considered to emerge from the dynamic interplay of neurons and networks, just as wetness emerges from the interactions of water molecules (Gilbert & Sarkar 2000). Thus, just as consciousness emerges continuously from the increasingly complex interactions between neurons, so too may consciousness be lost continuously through decreasingly complex interactions. Of note, just as certain organ systems (e.g., the heart) are particularly vital to maintaining the organ interactions necessary for life, so too are certain parts of the brain particularly vital to maintaining the interactions between neurons necessary for consciousness; injury to the brainstem and the ascending activating system, for instance, will more quickly diminish neuronal interactions, and thus consciousness, than injury to a region of cortex.

Moreover, even apart from considerations of its biological underpinnings, consciousness appears to exist along a continuum. Consider that the conceptual distinction between the MCS and VS/UWS is the presence or absence of consciousness, where the presence of consciousness means awareness of the self or environment. What constitutes awareness of the self or environment? Though never explicitly defined, a certain sophistication of awareness was likely implied by this conceptual definition when it was proposed.

Because the diagnostic criteria for MCS includes following simple commands, answering yes/no questions, smiling/crying in response to emotional content, or reaching for objects in a goal-directed manner, MCS patients were likely imagined to hear language and comprehend its meaning, to see objects and perceive their desirability/purpose, and/or to see people and recognize emotional valence. Awareness of the self also implies a relatively sophisticated integration of sensation and cognition into a coherent sense of self. However, one can imagine much more rudimentary forms of awareness: awareness, for instance, only of color, sounds, or sensations, without awareness of objects or even of a coherent environment per se, let alone of people, language, or a self. Thus, though the MCS was created as an intermediary between the VS/UWS and normal consciousness, the discrete distinction between VS/UWS and MCS may still be too crude, separated by a continuum of consciousness between complete unawareness and awareness of the self or environment. As in the distinction of life and death, if the distinction between consciousness and unconsciousness is indeed a continuous one, then no discrete diagnostic criteria for MCS or VS/UWS will perfectly distinguish the conscious from the unconscious.

Epistemological continuity between disorders of consciousness

One might reject the possibility that the distinction between life and death, or between consciousness and unconsciousness, is continuous, insisting upon a conceptual bright line between these states. However, even if such discrete distinctions exist, there remains an additional problem: we frequently do not know, and perhaps cannot know, when patients have crossed these lines. We do not have empirical characteristics that perfectly mark the transitions between these states, and thus while certain empirical characteristics clearly signify life versus death (e.g., someone holding a conversation versus someone with clinical findings of rigor mortis, respectively) or consciousness versus unconsciousness (e.g., someone holding a conversation versus someone whose brain has been removed, respectively), there remains a spectrum of empirical characteristics between these extremes which ambiguously signify whether a patient is alive versus dead, or conscious versus unconscious. We refer to this range of indeterminate characteristics (i.e., characteristics whose significance we do not know), and the resultant *appearance* of continuity between these states, as an epistemological continuity.

We have no metrics to determine when functioning of the organism as a whole has been permanently lost, and thus cannot distinguish life from death with perfect accuracy. Again, many characteristics can clearly signify life or clearly signify death, but of the range of characteristics in between, it is unclear which reliably mark the transition from life to death. Patients with brain death exhibit characteristics that fall within this spectrum. For the reasons discussed earlier, apnea and the loss of brainstem reflexes, while in principle measurable, do not clearly reflect the loss of functioning of the organism as a whole, and thus do not adequately represent death. Moreover, it is possible that, in principle, no characteristics within this spectrum could ever mark the life/death distinction with perfect accuracy; that is, the line between life and death in a person may be inherently unknowable, for at least three reasons. First, there is no clear "output" of a person functioning as a whole. For example, the heart has the function of pumping blood; thus, the functioning of the heart as a whole (i.e., the extent to which its component tissues are operating and interacting properly) can be assessed by its cardiac output, or its ability to effectively pump blood. In contrast, a person has no clear function, and thus we have no functional output from which to gauge a person's wholeness. Second, even if, in theory, there exists some empirical metric—a certain behavior, or a particular type of physiologic activity—that perfectly corresponds to the functioning of the organism as a whole, we have no way of identifying or validating it. That is, we could not know that this metric corresponds to life, because for us to know, we would need to observe a correspondence between this metric and life, and we have no way of knowing whether life is present. For us to know that life is present, we would need a metric that perfectly corresponds to the functioning of the organism as a whole, which traps us in circularity. Third, the required *permanence* of the cessation of functioning of the organism as a whole also makes death inherently unknowable, by demanding knowledge of prognosis (a prediction of the future), as described earlier. Thus, the distinction between life and death is empirically unknown, and may in principle be unknowable.

The detection of consciousness similarly suffers from an epistemological continuity, as we have no set of empirical criteria that will perfectly distinguish those aware of the self or environment from those who are not. Moreover, the line dividing consciousness from unconsciousness may also be inherently unknowable. The adage that "the only consciousness that we can truly know is our own" (Fins 2015, p.197) reflects the inherent subjectivity of consciousness. Thus, as in the distinction between life and death, the notion that an empirical metric could perfectly distinguish consciousness from unconsciousness generates circularity, as defining such a metric would require us to distinguish consciousness from unconsciousness in the first place. Thus, even if the distinction between consciousness and unconsciousness is in principle discrete, we do not, and perhaps cannot, have an empirical way of identifying it with precision.

How continuity problematizes diagnostic boundaries

Due to ontological continuity, epistemological continuity, or a combination of the two, we cannot discretely distinguish between life and death, or between consciousness and unconsciousness. The transitions between these states are inherently non-discrete, and/or we do not (and perhaps cannot) know when these transitions have occurred. Even Fins, who endorses a bright line distinction between the conscious and unconscious, echoes this sentiment, writing that neuroimaging data demonstrating consciousness in those who are behaviorally unresponsive:

> will add nuance and complexity to decision making and remove the haven of the simplicity afforded by dichotomous outcomes. It will not be as simple as we had all been led to believe: outcomes are not purely binary. Brain injuries are not just either catastrophic or miraculous. Like all things biological, they exist on a continuum with grey zones in between the extremes. (Fins 2015, p.112–113)

Given the continuity between these conditions, it is no wonder that the current diagnostic distinctions between brain death and coma, or between VS/UWS and MCS, do not successfully draw discrete delineations. One can imagine, however, why they were intended to. As described earlier, many significant, discrete decisions, such as whether to withdraw LST or to procure vital organs for transplantation, hinge upon the conceptual distinctions between life and death, or consciousness and unconsciousness. The desire to make these discrete decisions likely drove the diagnostic criteria to become similarly discrete. This is perhaps most clearly exemplified in the case of organ procurement in brain death. The very conception of brain death can be traced back to the first heart transplant, conducted in 1967 in South Africa (Barnard 1967; Truog 2015). The donor for the transplant was selected on the basis of devastating neurological injury, though she was maintained on a ventilator (Truog 2015). Recognizing that the criteria for devastating neurological injury would need to be formalized if organ transplantation was to ethically continue, anesthesiologist Henry Beecher organized and chaired a committee to propose the first definition of brain death (Ad Hoc Committee of the Harvard Medical School to Examine the Definition of Brain Death 1968; Truog 2015). Given that the very notion of brain death was born out of a need to definitively determine when organ procurement was ethical, it is not surprising that the diagnostic criteria for brain death were made to be discrete.

The drive to establish discrete diagnostic distinctions for these continuous disorders resulted in different problems in each case. For brain death and coma, the diagnostic distinction succeeded in being discrete, but failed to reflect their conceptual distinction. For VS/UWS and MCS, the diagnostic distinction was not discrete, nor did it reflect the conceptual distinction.

Rethinking decisions in disorders of consciousness

We would like to make decisions—such as whether to withdraw LST or procure organs for transplantation—based on whether patients are alive or dead, or conscious or unconscious. However, our current diagnostic criteria do not permit us to discretely distinguish between these states; perhaps, in principle, no empirical criteria ever could. If we are unable to distinguish between these fundamental, conceptual states, how are we to make these difficult decisions? We must rely upon what we have left: namely, the characteristics we can actually observe about patients.

Using diagnostic criteria as a legal fiction

One way of using empirical characteristics to guide decisions is by employing a "legal fiction." We may decide that despite the discrepancies we know to exist between diagnostic criteria and conceptual definitions, those who fulfill the diagnostic criteria for brain death should be considered dead, and those who fulfill the diagnostic criteria for VS/UWS should be considered unconscious. In other words, while we know that, technically speaking, the diagnostic criteria do not perfectly correspond to the conceptual definitions, we could treat them as if they do in order to guide decisions. In many ways, this approach best describes the mentality currently adopted, explicitly or implicitly, by the medical community.

While this type of pretending may seem outlandish, there is indeed legal precedent for it. Deliberately accepting a discrepancy between an entity's legal status and its biological reality for the purpose of the public good—called a legal fiction—can be traced back as far as the 1893 case of *Nix v. Hedden*, when the US Supreme Court had to decide whether a tomato should be considered a fruit or a vegetable (US Supreme Court 1893; Shah & Miller 2010; Shah et al. 2011; Truog 2015). At the time, a tax was imposed on the importation of vegetables, and tomato importers argued that they should be exempt from the tax because tomatoes are technically fruit. The Supreme Court stated that, "Botanically speaking, tomatoes are the fruit of the vine ... But in the common language of the people... these are vegetables" (US Supreme Court 1893, p.307), and so determined that tomatoes should legally be considered vegetables.

There are different types of legal fictions that could apply in the distinctions between brain death and coma, or between VS/UWS and MCS. The first is a "bright-line fiction," which refers to creating legally discrete categories by drawing a bright line at a point along a continuum. An example of a bright-line fiction already established in medicine and law is that of legal blindness (Shah & Miller 2010). While visual acuity falls along a spectrum, once an individual's visual acuity falls below 20/200 he or she is considered legally blind, despite technically retaining limited sight. This legal classification carries implications for rights (e.g., eligibility for disability benefits) as well as for restrictions (e.g., driving prohibitions). Thus, while the discrete, legal definition of blindness does not perfectly correspond to the biological definition of blindness, this bright line fiction serves an important function in dictating social and legal decisions.

A second type of legal fiction potentially relevant to disorders of consciousness is the status fiction, in which A is treated as if it were B (Shah et al. 2011). One example of a well-established status fiction is that corporations are assigned the same legal status as people (i.e., corporations are treated "as if they were" people), so that the body of law developed for people can be applied to corporations. While a corporation is not technically a person, this legal fiction is helpful in guiding many practical decisions, such as when a corporation can be sued, and how a corporation can contribute money to a political campaign (Shah et al. 2011).

Applying these legal fictions to disorders of consciousness would mean either considering the diagnostic criteria for brain death or VS/UWS to signify death or unconsciousness, respectively (in the case of a bright line fiction), or treating those who fulfill the diagnostic criteria for brain death or VS/UWS *as if* they were dead or unconscious, respectively (in the case of a status fiction). As described, these forms of pretending can serve useful social and legal functions. In the case of disorders of consciousness, the fiction that discrete diagnostic criteria reflect discrete conceptual categories (i.e., of life and death, or of consciousness and unconsciousness), could be useful in guiding discrete decisions that rely upon these conceptual distinctions.

But the problem with this approach is: who would we be pretending to? The legal fictions currently established are generally transparent to society and those involved. It is understood that a corporation is not literally a person, and those who are legally blind but retain vision understand that they are not literally blind. But in the case of disorders of consciousness, people could be unfairly misled. If we tell families and surrogates, or even medical students and physicians in training, that patients are dead if they meet the criteria for brain death, or are unconscious if they meet the criteria for VS/UWS—as we often currently do—these groups may not understand the fiction and actual uncertainty that underlie these statements. They may not understand that such patients are not *literally* dead or unconscious. Thus, such fictions could be more deceptive than clarifying, resulting in incompletely informed decisions. The alternative is to disclose these fictions to families and surrogates, to tell them that although we consider these patients to be dead or unconscious (or treat them as such) for legal purposes, we cannot actually be sure that they are dead or unconscious. In this case, these fictions would cease to effectively serve as fictions at all, and families and surrogates would need to decide if the diagnostic criteria themselves are sufficient to justify their decisions, bringing us to a second possible approach.

Using empirical characteristics that reflect values

Many of our decisions currently hinge upon whether a patient is alive or dead, or conscious or unconscious. However, these conceptual states cannot be distinguished for many patients. Though we could pretend that the diagnostic criteria signify these conceptual states, as through a legal fiction, the deception inherent to this approach raises significant ethical concerns. Thus, in order to continue making these decisions, we are forced to consider an alternative approach: to abandon the hope of definitively distinguishing between these conceptual states, and instead to rely upon empirical characteristics to guide decisions. That is, instead of trying, perhaps futilely, to determine whether any given patient is alive or dead, or conscious and unconscious, we must instead focus on what we can observe about a patient in order to make decisions. If we are to rely on empirical characteristics to justify decisions, we should be thoughtful about how those characteristics are selected. If they cannot reliably distinguish between conceptual states, then they must instead account for our *values*. What empirical characteristics *matter* to us when making decisions about a patient?

First, we will consider how this approach applies to decisions that typically hinge upon the distinction between brain death and coma—perhaps the most notable being whether or not to procure vital organs. What empirical characteristics should dictate whether organ procurement is acceptable? Or put another way, what empirical characteristics, if exhibited by a patient, would convince us that procuring his/her organs is ethically justified? Some may endorse the dead donor rule because of the value that patients should be dead before their organs are procured. This group may therefore favor an empirical criterion for organ procurement that signifies death as certainly as possible—perhaps something akin to clinical signs of rigor mortis. However, selecting such a criterion would significantly limit the value of the procured organs, since waiting for rigor mortis would cause substantial organ hypoxia and injury. Moreover, others might disagree that the certainty of death is the most important value in deciding whether organs can be procured. Perhaps instead, the values most important to our current system of organ procurement are that of autonomy (that one should be able to choose if his or her organs are donated) and nonmaleficence (that one should not be harmed by the donation of his or her organs). The empirical characteristics we currently use as the diagnostic criteria for brain death may already capture these values, since it is widely accepted that those who fulfill these criteria are not conscious (there is no documented case of such patients regaining consciousness) and therefore arguably cannot experience harm. There is indeed evidence that for most, these diagnostic criteria for brain death justify organ procurement, independently of what they signify about death. A survey of over 1000 American adults presented participants with a case of a man who, after an automobile accident, fulfilled the diagnostic criteria for brain death (including irreversible unconsciousness and apnea); however, the term "brain death" was not used, and the man was explicitly described as alive (Nair-Collins et al. 2015). The survey determined that 71% of the participants thought it should be legal for the patient to donate his organs, 70% thought that doctors should be able to procure organs from similar patients (assuming consent), and 67% reported wanting to donate their own organs if in the patient's circumstances. This study suggested that the empirical criteria for brain death and what they represent about an individual's neurological function, and not necessarily the conceptual state of death per se, for many justify the procurement of organs.

Second, how might this approach apply to decisions that typically hinge upon the distinction between VS/UWS and MCS? Perhaps the most critical decision that is often influenced by this distinction is whether or not to withdraw LST. There are numerous reasons why a surrogate might choose to withdraw LST for a patient, but as discussed, for many, a major determinant is whether a patient retains consciousness, as an unconscious patient necessarily lacks any quality of life. Such decisions are discrete, yet as we have shown, the current criteria distinguishing VS/UWS from MCS rely upon the inherently uncertain distinction between reflexive and purposeful behaviors, and as such do not form a discrete boundary on which such decisions can be based (Fischer & Truog 2015). Moreover, the current diagnostic criteria for the VS/UWS and MCS pertain only to behavior, and thus neglect patients who might only show signs of consciousness through emerging technologies such as fMRI. So, if we cannot rely upon the current diagnostic criteria for guiding these discrete decisions, what empirical characteristics can we rely upon that are both discrete and consistent with our values? Or in other words, what characteristics matter most when making decisions that hinge upon whether a patient is conscious or unconscious, when this conceptual distinction remains inherently unclear?

We have previously proposed one such criterion for guiding these decisions: interactive capacity, or the ability to receive communicated information and generate a coherent response (Fischer & Truog 2013, 2015). Under this proposal, a patient should be considered conscious only if they demonstrate interactive capacity. We have also stipulated that the patient's response must be potentially intentional in order to signify interactive capacity; that is, responses that are inherently non-volitional, even if performed by a conscious individual (e.g., automatic brain activity in response to voice or pain), cannot be considered interactive (Fischer & Truog 2015).

Interactive capacity circumvents the problematic distinction between reflexive and purposeful behaviors by falling safely among the latter, that is, by intuitively signifying consciousness. Fins relays the thoughts of a woman whose mother suffered from a disorder of consciousness:

> [S]he told us why restoring her mother's ability to communicate was most important to her. "I think," she said, "it's because speech and language are a clear sign that she's in there." (Fins 2015, p.199)

Fins elaborates:

> Ultimately, communication ... helps families know if their loved one is conscious ... Although the only consciousness that we can truly know is our own, in practice we know the consciousness of others through communication, which is a proxy for the cognitive experience within. Without communication, others cannot know a patient's level of awareness ... It limits knowledge of the patient's current state and their future capabilities. (Fins 2015, p.197)

Along the spectrum of reflexive and purposeful behaviors, interactive behaviors can be assumed to be purposeful, and thus trusted to signify consciousness. Perhaps this is because, while spontaneous behaviors without context can be ambiguously interpreted, behaviors in coherent response to external cues are more likely purposeful. The Turing test offers an analogous application of the same intuition (French 2000; Stins 2009). In this test, an examiner enters questions into a computer, and must determine whether the responses are generated by a human respondent or an automated algorithm. The Turing test thus assesses an algorithm's ability to simulate human consciousness based on its interactivity. Assessing the consciousness of a patient is conceptually similar: one communicates to a patient and evaluates whether the responses are sufficiently coherent to signify consciousness.

In intuitively falling among purposeful, or conscious, behaviors, interactive capacity solves two problems. First, it serves as a more *discrete* criterion than the current diagnostic criteria for consciousness. The current criteria require clinicians to determine where any given behavior falls along the spectrum between reflexive and purposeful behaviors. As such, a patient whose behavior falls only within this indeterminate spectrum will be difficult to categorize as either conscious or unconscious. In contrast, because interactive behaviors are more readily identifiable, deciding whether interactive capacity is present or absent, rather than whether a behavior is reflexive or purposeful, represents a more discrete diagnostic determination. As such, it can more effectively guide discrete decisions.

Second, the intuitiveness with which interactive capacity signifies consciousness reflects the *value* we place on interaction. For example, for many, a patient's ability to interact is crucial when deciding whether or not to withdraw LST. Fins writes that:

> Most decisions to withhold or withdraw care are made when a patient loses consciousness and the capacity to interact and participate in the decision-making process ... It is this loss of consciousness, the ability to interact with the other, that becomes the moral prompt for the conversation (Fins 2015, p.30)

Given that we often cannot definitively know whether consciousness has been lost, it is the loss of the "ability to interact" that prompts the conversation to withdraw LST, highlighting the value we place on interactive capacity when deciding if someone is conscious. Fins also relays the thoughts of someone whose loved one suffered from a disorder of consciousness:

> "First of all I wanted her alive and I wanted her to recover, but I didn't know what I wanted her to recover the most. And it's the ability to speak and communicate ... " The restoration of functional communication was " ... The world." (Fins 2015, p.197)

Surely the speaker would have wanted the patient to recover awareness, but it is telling that the speaker chose communication as the most important function to recover. For many, among the various empirical characteristics that might signify consciousness, it is interaction that is the most meaningful. It is also worth noting that the current behavioral criteria for the MCS (e.g., following commands, answering yes-or-no questions) already do center on interactive behaviors, another indication that interactive capacity intuitively matters to us when determining if another is conscious.

There are additional, more practical advantages to adopting interactive capacity as a criterion for consciousness. First, whereas the current diagnostic criteria for consciousness pertain specifically to behavior, interactive capacity is more generalizable, and thus can apply to technologically mediated forms of interaction. This is an important consideration in the context of emerging technologies, such as the fMRI tasks described earlier, which can demonstrate interactive capacity in patients who are behaviorally unresponsive. Second, selecting a criterion such as interactive capacity that falls safely among purposeful behaviors prevents tests for consciousness from becoming overly reductive (Fischer & Truog 2013). After fMRI studies revealed that behaviorally unresponsive patients could interact through neuroimaging, people began developing even more sensitive technological biomarkers for consciousness. However, as these tests began to stray from intuitive conceptions of consciousness (e.g., EEG responses to auditory stimuli (Faugeras et al. 2011), positron emission tomography responses to painful stimuli (Boly et al. 2008), fMRI responses to voice (Di et al. 2007), and correlations between spontaneous fluctuations in fMRI signals (Vanhaudenhuyse et al. 2010)), they risked over-attributing consciousness to subconscious neural phenomena. Interactive capacity is anchored in an intuitive conception of consciousness, thereby minimizing the risk of overly reductive tests.

Further considerations when incorporating values into decisions

Given that we cannot distinguish whether every patient is alive or dead, or conscious or unconscious, we have argued that the best alternative is to rely on what we know—that is, the characteristics of patients that we can observe—in order to guide decisions. As such, we have suggested that these empirical characteristics should be selected in accordance with our values. In other words, we should make these decisions based on the empirical characteristics that matter most to us for those decisions. However, if we are to incorporate values into our decisions, there are further issues to consider.

First, it is important to account for individual differences in values. For example, in the case of organ procurement, a survey conducted by Nair-Collins and colleagues in 2013 showed that most believe that the brain death criteria (e.g., coma and apnea) justify organ procurement, regardless of what they signify about death (Nair-Collins et al. 2015). However, approximately 30% of participants in the survey believed that the current diagnostic criteria for brain death do *not* justify organ procurement. This group may place a higher value on the certainty of a patient's death when procuring organs, and thus may favor more stringent criteria for organ procurement. Yet others might favor less stringent criteria, feeling that coma alone justifies organ procurement, regardless of apnea or the absence of brainstem reflexes. This view was exemplified in the case of a young girl named Jaiden, who became comatose after a strangulation accident (Sanghavi 2009; Truog 2015). Her parents wished to donate her organs, but because she maintained a few brainstem reflexes, she could not be considered brain dead. As a result, her organs could not be properly procured, a source of significant distress for Jaiden's parents who felt that donating her organs would lend an element of meaning to an otherwise tragic situation. Many others share the view of Jaiden's parents, that coma may justify organ procurement (Halevy & Brody 1993; Veatch 1993). Bioethicist and brain-injury survivor, William Winslade, has written that:

> Merely the organic functioning of our bodies doesn't constitute being human. Being persons requires having a personality, being aware of our selves and our surroundings, and possessing human capacities, such as memory, emotions, and the ability to communicate and interact with other people ... [I]n the case [of] ... the permanently unconscious, we are no longer human beings; hence society no longer has a moral responsibility to sustain our lives. A physician who orders a stop to artificial feeding and hydration in such a case ends a life, but only an organic life. He or she doesn't commit murder. (Winslade 1998, p.121)

Thus, whether based on values of autonomy, nonmaleficence, or personhood, many feel that coma alone (along with the permission of the patient or surrogate) may justify organ procurement. Ultimately, these examples illustrate the range in values that might apply to organ procurement.

The same variation in values likely pertains to other decisions that hinge on a patient's consciousness. As described, there are reasons to believe that interactive capacity reflects commonly held values about what "matters" for decisions about consciousness. However, this criterion will likely not match everyone's values. Some might place greater value on lesser degrees of consciousness—for example, intact perception of visual stimuli but no cognitive capacity to interact—when making decisions. Though we have no way of detecting these lesser forms of consciousness, some may so value these states as to select different criteria for their decisions. For example, perhaps this group might only be convinced that a patient is unconscious if a patient is comatose (i.e., lacks signs of arousal in additional to lacking signs of awareness). While selecting interactive capacity as a criterion might risk producing "false negatives" (i.e., may misclassify some conscious but non-interactive patients as unconscious), this alternative approach would risk "false positives" (i.e., may misclassify unconscious patients as conscious). Determining which of these consequences is more troubling, and therefore which criterion to choose, will depend upon one's values.

Beyond variation in values between people, a second consideration is that values may vary depending on the decision. For example, one might feel that the empirical characteristics that justify organ procurement are different from those that relieve clinicians of their obligation to treat; one might like the criteria for the latter to be more stringent (i.e., more certainly signify death) than for the former. Currently, the diagnostic category of brain death, by virtue of the assumption that it signifies death, is thought to justify both of these decisions. However, if we cannot be certain of such assumptions, we can no longer rely on one set of diagnostic criteria to guide different types of decisions. Thus, there may be added nuance to how different empirical characteristics are used in the context of different decisions.

Incorporating values into disorders of consciousness

Here, we argue that the distinctions between disorders of consciousness are not discrete, and thus cannot reliably guide discrete decisions as they currently do. Each disorder of consciousness has a conceptual definition, which is recognized based on diagnostic criteria. The conceptual distinctions between these disorders correspond to fundamental divisions—between life and death, and between consciousness and unconsciousness—and thus have significant implications for decisions. However, these conditions cannot be discretely distinguished, due to a continuum between these conceptual states: the distinctions between life and death, and between consciousness and unconsciousness, are inherently continuous and/or cannot be empirically recognized. Thus, as the current diagnostic criteria do not reliably distinguish between these conceptual states (in fact, perhaps no empirical criteria could), we cannot rely upon these conceptual distinctions to guide decisions as we currently do. If we are to continue making these decisions, we must instead rely upon what we have left: what we can empirically observe about patients. We argue that these empirical characteristics should be selected in accordance with our values.

This conclusion contradicts common convictions about how diagnostic criteria for death and unconsciousness should be defined. Many diagnoses are based on objective data (e.g., exam findings, laboratory results, or imaging), and given the profound implications of disorders of consciousness, we have preferred to view these diagnoses in this way. In 2004, Fins and influential neurologist Fred Plum published an article titled "Neurological diagnosis is more than a state of mind: diagnostic clarity and impaired consciousness" (Fins & Plum 2004), in which they advocate for the objectivity of the diagnostic criteria for disorders of consciousness. As Fins has argued, "Values should not distort the clinical reality" (Fins 2015, p.253). This stance has virtue; as mentioned, the variability in values between people will complicate how we think about disorders of consciousness. However, given that the "clinical reality" is deeply obscured in disorders of consciousness, values may indeed be necessary when managing these patients in the face of significant uncertainty. Indeed, other diagnoses that reflect complex and unknown pathophysiology have similarly incorporated values. For example, the diagnostic criteria for many psychiatric disorders, such as personality disorders, include significantly impaired function; what constitutes "function," let alone what it means to be significantly impaired, will often depend upon an individual's values.

Yet, perhaps we should not be thinking about disorders of consciousness in terms of discrete diagnostic categories at all. If we cannot reliably distinguish between the different conceptual states—between life and death, consciousness and unconsciousness—and if the empirical characteristics we value to make decisions will vary depending on the person and decision, then any attempt to standardize these criteria will inevitably conflict with the values of some. Indeed, the current standardized diagnostic criteria for brain death have already conflicted with the values of certain religious and cultural groups, leading some states to accommodate objections to brain death testing (Kimura 1991; Posner et al. 2007). Thus, perhaps instead of focusing on distinguishing between discrete diagnoses, we should focus on the range of empirical characteristics we can observe about patients. From there, clinicians may collaborate with families and surrogates, or ideally may refer to advance directives specified by the patients themselves, to determine which empirical characteristics matter most for the decisions to be made. This approach will certainly add complexity and nuance to our understanding of these disorders, but will also assure that our decisions are made in an intellectually honest and ethical manner.

Disorders of consciousness represent an enigmatic corner of the medical world, challenging physicians, scientists, philosophers, and ethicists alike. Unlike most other medical conditions, these disorders degrade our most fundamental attributes—those of life and consciousness. As such, they are as perplexing as they are critical to understand. These conditions present profound philosophical and scientific challenges, while also devastating the lives of patients, families, and loved ones. We have a responsibility to ensure that these patients are offered the most ethical treatment possible, which will require careful consideration of these complex philosophical problems. While we would like to base our management of these patients on whether they are alive or dead, or conscious or unconscious, we often cannot make these distinctions. By accepting the limitations in our knowledge, we can develop a more honest and thoughtful approach to these disorders and improve the ethical management of these patients.

Notes

1. Dreaming during rapid eye movement (REM) sleep might appear to violate this relationship, but in fact REM sleep shares many neuronal characteristics with wakefulness (Steriade et al. 1993).

2. Of note, wholism as applied to life is sometimes called vitalism. However, the term vitalism connotes a nonmaterial "life force" that enters an organism once whole, a notion largely rejected by the scientific and medical community.

References

Ad Hoc Committee of the Harvard Medical School to Examine the Definition of Brain Death (1968). A definition of irreversible coma. *JAMA*, **205**(6), 337–340.

Barnard, C. (1967). A human cardiac transplant: an interim report of a successful operation performed at Groote Schuur Hospital, Cape Town. *South African Medical Journal*, 1271–1274.

Becker, K.J., Baxter, A.B., Cohen, W.A., et al. (2001). Withdrawal of support in intracerebral hemorrhage may lead to self-fulfilling prophecies. *Neurology*, **56**(6), 766–772.

Bernat, J. (2006). Chronic disorders of consciousness. *The Lancet*, **367**(9517), 1181–1192.

Bernat, J., Culver, C., and Gert, B. (1981). On the definition and criterion of death. *Annals of Internal Medicine*, **94**(3), 389–394.

Boly, M., Faymonville, M.E., Schnakers, C., et al. (2008). Perception of pain in the minimally conscious state with PET activation: an observational study. *The Lancet Neurology*, **7**(11), 1013–20.

Di, H.B., Yu, S.M., Weng, X.C., et al. (2007). Cerebral response to patient's own name in the vegetative and minimally conscious states. *Neurology*, **68**(12), 895–9.

Ecker, J. (2014). Death in pregnancy—an American tragedy. *The New England Journal of Medicine*, **370**(10), 889–891.

Edelman, G.M. (1999). Building a picture of the brain. *Annals of the New York Academy of Sciences*, **882**(2), 68–89.

Executive Board of the American Academy of Neurology (1989). Position of the American Academy of Neurology on certain aspects of the care and management of the persistent vegetative state patient. *Neurology*, **39**, 125–126.

Faugeras, F., Rohaut, B., Weiss, N., et al. (2011). Probing consciousness with event-related potentials in the vegetative state. *Neurology*, **77**(3), 264–268.

Fins, J.J. (2015). *Rights Come to Mind: Brain Injury, Ethics, and the Struggle for Consciousness*. New York: Cambridge University Press.

Fins, J.J. and Plum, F. (2004). Neurological diagnosis is more than a state of mind: diagnostic clarity and impaired consciousness. *Archives of Neurology*, **61**(9), 1354–1355.

Fins, J.J., Miller, F.G., Acres, C.A., Bacchetta, M.D., Huzzard, L.L., and Rapkin, B.D. (1999). End-of-life decision-making in the hospital: current practice and future prospects. *Journal of Pain and Symptom Management*, **17**(1), 6–15.

Fischer, D.B. and Truog, R.D. (2013). Conscientious of the conscious: interactive capacity as a threshold marker for consciousness. *AJOB Neuroscience*, **4**(4), 26–33.

Fischer, D.B. and Truog, R.D. (2015). What is a reflex? A guide for understanding disorders of consciousness. *Neurology*, **85**(6), 543–548.

French, R. (2000). The Turing test: the first 50 years. *Trends in Cognitive Sciences*, **4**(3), 115–122.

Giacino, J. (2004). The vegetative and minimally conscious states: consensus-based criteria for establishing diagnosis and prognosis. *NeuroRehabilitation*, **19**, 293–298.

Giacino, J.T., Ashwal, S., Childs, N., et al. (2002). The minimally conscious state: definition and diagnostic criteria. *Neurology*, **58**(3), 349–353.

Gilbert, S.F. and Sarkar, S. (2000). Embracing complexity: organicism for the 21st century. *Developmental Dynamics*, **219**(1), 1–9.

Gramm, H.-J., Meinhold, H., Bickel, U., et al. (1992). Acute endocrine failure after brain death? *Transplantation*, **54**(5), 851–857.

Grigg, M.M., Kelly, M.A., Celesia, G.G., Ghobrial, M.W., and Ross, E.R. (1987). Electroencephalographic activity after brain death. *Archives of Neurology*, **44**(9), 948–954.

Halevy, A. and Brody, B. (1993). Brain death: reconciling definitions, criteria, and tests. *Annals of Internal Medicine*, **119**(6), 519–525.

Hopkin, M. (2006). "Vegetative" patient shows signs of conscious thought. *Nature*, **443**(7108), 132–133.

Jennett, B. and Plum, F. (1972). Persistent vegetative state after brain damage: a syndrome in search of a name. *The Lancet*, **299**(7753), 734–737.

Kimura, R. (1991). Japan's dilemma with the definition of death. *Kennedy Institute of Ethics Journal*, **1**(2), 123–131.

Lane, A., Westbrook, A., Grady, D., et al. (2004). Maternal brain death: medical, ethical and legal issues. *Intensive Care Medicine*, **30**, 1484–1486.

Laureys, S., Owen, A.M., and Schiff, N.D. (2004). Brain function in coma, vegetative state, and related disorders. *Lancet Neurology*, **3**(9), 537–546.

Laureys, S., Boly, M., Moonen, G., and Masquet, P. (2009). Two dimensions of consciousness: arousal and awareness. *Encyclopedia of Neuroscience*, **2**, 1133–1142.

Laureys, S., Celesia, G.G., Cohadon, F., et al. (2010). Unresponsive wakefulness syndrome: a new name for the vegetative state or apallic syndrome. *BMC Medicine*, **8**(68), 1–4.

Lock, M. (2002). *Twice Dead: Organ Transplants and the Reinvention of Death* (Vol. 1). Berkeley, CA: University of California Press.

Medical Consultants on the Diagnosis of Death to the President's Commission for the Study of Ethical Problems in Medicine and Biomedical and Behavioral Research (1981). Guidelines for the determination of death: report of the medical consultants on the diagnosis of death to the President's Commission for the Study of Ethical Problems in Medicine and Biomedical and Behavioral Research. *JAMA*, **246**, 2184–2186.

Miller, F.G. and Truog, R.D. (2009). The incoherence of determining death by neurological criteria: a commentary on "Controversies in the determination of death", a white paper by the President's Council on Bioethics. *Kennedy Institute of Ethics journal*, **19**(2), 185–193.

Monti, M., Vanhaudenhuyse, A., Coleman, M.R., et al. (2010). Willful modulation of brain activity in disorders of consciousness. *New England Journal of Medicine*, **362**(7), 579–589.

Nair-Collins, M., Green, S.R., and Sutin, A.R. (2015). Abandoning the dead donor rule? A national survey of public views on death and organ donation. *Journal of Medical Ethics*, **41**, 297–302.

Owen, A., Coleman, M.R., Boly, M., Davis, M.H., Laureys, S., and Pickard, J.D. (2006). Detecting awareness in the vegetative state. *Science*, **313**, 1402.

Pallis, C. (1983). ABC of brainstem death: the arguments about the EEG. *British Medical Journal*, **286**, 209–210.

Pennefather, S.H., Dark, J.H., and Bullock, R.E. (1993). Haemodynamic responses to surgery in brain-dead organ donors. *Anaesthesia*, **48**(12), 1034–1038.

Posner, J., Saper, C.B., Schiff, N., and Plum, F. (2007). *Plum and Posner's Diagnosis of Stupor and Coma* (4th edn.). Oxford: Oxford University Press.

Powner, D.J., Ackerman, B.M., and Grenvik, A. (1996). Medical diagnosis of death in adults: historical contributions to current controversies. *The Lancet*, **348**(9036), 1219–1223.

Repertinger, S., Fitzgibbons, W.P., Omojola, M.F., and Brumback, R.A. (2006). Long survival following bacterial meningitis-associated brain destruction. *Journal of Child Neurology*, **21**(7), 591–596.

Report of the Quality Standards Subcommittee of the American Academy of Neurology (1995). Practice parameters for determining brain death in adults (summary statement). *Neurology*, **45**, 1012–1014.

Robertson, J.A. (1999). The dead donor rule. *Hastings Center Report*, **29**(6), 6–14.

Rodin, E., Tahir, S., Austin, D., and Andaya, L. (1985). Brainstem death. *Clinical Electroencephalography*, **16**(2), 63–71.

Rubin, E.B. and Bernat, J.L. (2011). Ethical aspects of disordered states of consciousness. *Neurologic Clinics*, **29**(4), 1055–71.

Sanghavi, D. (2009). When does death start? *The New York Times Magazine*, December 16.

Schrader, H., Krogness, K., Aakvaag, A., Sortland, O., and Purvis, K. (1980). Changes of pituitary hormones in brain death. *Acta Neurochirurgica*, **52**(3–4), 239–248.

Searle, J.R. (1992). *The Rediscovery of the Mind*. Cambridge, MA: MIT Press.

Shah, S.K. and Miller, F.G. (2010). Can we handle the truth? Legal fictions in the determination of death. *American Journal of Law and Medicine*, **36**(4), 540–585.

Shah, S.K., Truog, R.D., and Miller, F.G. (2011). Death and legal fictions. *Journal of Medical Ethics*, **37**(12), 719–22.

Shewmon, D. (1998). Chronic "brain-death": meta-analysis and conceptual consequences. *Neurology*, **51**, 1538–1545.

Steriade, M., McCormick, D.A., and Sejnowski, T.J. (1993). Thalamocortical oscillations in the sleeping and aroused brain. *Science*, **262**(5134), 679–685.

Stins, J.F. (2009). Establishing consciousness in non-communicative patients: a modern-day version of the Turing test. *Consciousness and Cognition*, **18**, 187–192.

The Multi-Society Task Force on PVS (1994a). Medical aspects of the persistent vegetative state (1). *New England Journal of Medicine*, **330**(21), 1499–1508.

The Multi-Society Task Force on PVS (1994b). Medical aspects of the persistent vegetative state (2). *New England Journal of Medicine*, **330**(21), 1572–1579.

The President's Council on Bioethics (2008). *Controversies in the Determination of Death: A White Paper by the President's Council on Bioethics*. Washington, DC: The President's Council on Bioethics.

Truog, R.D. (1997). Is it time to abandon brain death? *The Hastings Center Report*, **27**(1), 29–37.

Truog, R.D. (2007). Brain death—too flawed to endure, too ingrained to abandon. *Journal of Law, Medicine and Ethics*, **35**(2), 273–281.

Truog, R.D. (2015). Defining death: getting it wrong for all the right reasons. *Texas Law Review*, **93**, 1885–1914.

US Supreme Court (1893). *Nix v. Hedden* 149 U.S. 304.

Vanhaudenhuyse, A. et al. (2010). Default network connectivity reflects the level of consciousness in non-communicative brain-damaged patients. *Brain*, **133**, 161–171.

Veatch, R.M. (1993). The impending collapse of the whole-brain definition of death. *Hastings Center Report*, **23**(4), 18–24.

Wijdicks, E.F.M., Varelas, P.N., Gronseth, G.S., Greer, D.M., American Academy of Neurology. (2010). Evidence-based guideline update: determining brain death in adults: report of the Quality Standards Subcommittee of the American Academy of Neurology. *Neurology*, **74**(23), 1911–1918.

Winslade, W.J. (1998). *Confronting Traumatic Brain Injury: Devastation, Hope and Healing*, New Haven, CT: Yale University Press.

Chapter 18

Brain death and the definition of death

James L. Bernat

Introduction

Brain death remains a controversial subject even now, 60 years after the publication of its original cases. The idea that a person whose brain functions have ceased irreversibly is dead (despite the continuation of heartbeat, circulation, and visceral organ function because of respiratory support) is the oldest and most enduring debate in neuroethics. Notwithstanding its growing acceptance among societies in the developed and developing world, a trend accompanied by its parallel incorporation into medical practices and laws, brain death remains inadequately understood by both medical and nursing professionals as well as by the general public (Bernat 2006). Within the academy, scholars continue to debate its conceptual validity (Truog 1997), a dispute amplified by a chorus of new opponents (Joffe 2010; Nair-Collins 2015). Yet while opponents cite certain valid conceptual inadequacies of brain death, their critiques for over four decades have not gained sufficient traction or public support to change prevailing medical practices or laws (Bernat 2014a).

In the first part of this chapter, I trace the history of brain death by explaining the technological, medical, and societal factors stimulating its origin and acceptance; discuss its medical, legal, religious, and social recognition; highlight a few of its controversies,; examine a recent commission report; and discuss two highly publicized cases that have reignited debates. In the second part, I provide the conceptual basis for brain death by analyzing the definition and criterion of death. I offer an analytic framework for a biophilosophical account of death that justifies the practice of brain death, pinpoints the areas of contention, and compares competing concepts of death. I conclude with thoughts about brain death as a contemporary neuroethical issue and predict the future landscape of the debate over brain death and the definition of death.

Brain death

Prior to the development of the mechanical ventilator and endotracheal intubation, human death was a unitary phenomenon. When illness or injury led to the cessation of heartbeat or breathing, brain function necessarily ceased and all organs lost functioning irreversibly. It did not matter which of the three vital functions (heartbeat/circulation, lungs/breathing, and brain) ceased first, their mutual interdependence required that the others inevitably also ceased within minutes. Physicians declaring death showed that circulation, respiration, and brain functions had ceased by simply examining the patient for the absence of heartbeat, breathing, responsiveness, and pupillary reactivity to light. Thus, prior to the 1950s, death always was a unitary phenomenon and its operational definition was the straightforward cessation of the three vital functions.

With the development of endotracheal intubation, positive-pressure ventilation, and cardiopulmonary resuscitation in the 1950s, for the first time, the cessation of heartbeat and breathing could be reversed or supported. It then became possible for a person to have sustained brain damage so widespread and severe that all brain functions had ceased irreversibly. Yet, with technological support of breathing and circulation, the person could be physiologically maintained for a short time, permitting the continuation of spontaneous heartbeat and systemic circulation to visceral organs though not to the brain. French neurologists, who offered the first detailed descriptions of such cases, labeled these patients *le coma dépassé* to indicate that they were in a state beyond coma (Mollaret & Goulon 1959).[1] Over the next decade, physicians began to assert that these patients were not merely deeply comatose from profound brain damage: they were dead.

The life status of these patients was ambiguous in the context of the prevailing unitary understanding of death because they had lost some vital functions but not others. The traditional unitary concept of death was not equipped to categorize the life status of such patients because, prior to technological advances in physiological support, such cases were impossible. A quick survey of their physical examination findings showed that they had some characteristics traditionally associated with death but not others. Favoring their classification as dead, they were utterly unresponsive to any stimuli, had no voluntary or reflex movement, and did not breathe. Favoring their classification as alive, they had heartbeat and systemic circulation, with intact visceral organ functioning to digest food introduced into the stomach and to make urine. Achieving consensus on whether such patients were alive or dead was impossible because they shared features of both states. Nor could physicians determine whether they were alive or dead until it was first agreed what death actually meant in the brave new technological world in which respiratory and circulatory support was possible after these functions had ceased.

In the 1960s, a growing consensus developed among physicians and scholars that these patients were, in fact, dead. In 1968, a committee of the Harvard Medical School faculty proposed the first set of clinical criteria to diagnose the emerging concept that they called *brain death* (Ad Hoc Committee 1968).[2] Although their pioneering work both established a new criterion of death and influentially fostered its widespread legal and public acceptance, their unfortunate choice of the term *brain death* promoted misunderstanding and confusion that has persisted to the present.[3] The term *brain death* now has become universal; it can be found in medical, scientific, scholarly, and legal contexts, has been translated into many languages, and has entered the public vernacular.

Factors stimulating the acceptance of brain death

Three factors accelerated the societal acceptance of brain death in the 1960s: technological developments in physiological support, the need to lawfully permit withdrawal of physiological support in hopeless cases, and the demand for donors for organ transplantation. Although these factors by themselves cannot justify the validity of the equivalence of brain death and human death, appreciating them offers insight into how the concept of brain death became accepted so quickly by the medical and legal communities. But notwithstanding brain death's obvious instrumental value, its validity as a concept of human death must stand or fall not on its utilitarian value but on its biophilosophical and scientific coherence.

Life-support technologies were developed in the 1940s to sustain respiration in patients with acute poliomyelitis whose respiratory failure was caused by weakness of the respiratory muscles. Negative-pressure ventilators ("iron lungs") could assist breathing when it became weak but could not replace breathing in the presence of complete apnea. In the early 1950s, endotracheal intubation and positive-pressure mechanical ventilation (tracheal positive-pressure ventilation (TPPV)) became available, technologies that were sufficient to maintain the lives of patients with complete respiratory failure by fully breathing for them. Although the intended usage of these technologies was by conscious patients or those who would later recover consciousness, they were also applied to support patients with deep coma from profound brain damage, including those with irreversible cessation of all brain functions. Physicians who examined these "brain dead" patients intuited that there was something essentially different about them from all comatose patients previously reported. Several authors asserted that, because of their complete absence of brain functions, they were dead.

Cardiopulmonary resuscitation (CPR) was first described in 1960, a technique allowing restarting heartbeat in patients during cardiac arrest from ventricular fibrillation or asystole, all of whom previously would have died. Although the goal of CPR was to restore circulation to permit complete recovery, some patients were belatedly resuscitated after the brain already had been destroyed by lack of oxygen and blood flow during the prolonged cardiac arrest. Many of the patients with the newly described condition of brain death acquired it through this mechanism. For them, CPR to restored circulation was not a successful life-saving intervention but rather was an ironic and tragic failed application of medical technology.

In the 1960s when TPPV was initiated to support respiration in a patient who had stopped breathing, there was no lawful (and, some argued, ethical) means to discontinue this therapy even when the presence of profound brain damage made the prognosis hopeless. There was general agreement that if the physician ordered discontinuation of TPPV and the patient died as a result of tracheal extubation, the physician would be liable for criminal homicide given the obvious causality of the physician's act. At that time, no laws had been enacted or high court cases decided that would legally permit physicians to take such an action. If, however, such a patient previously had been declared dead on the grounds of brain death, there would be no ethical or legal prohibition to discontinuing treatment. By the mid 1970s, the first legal rulings had been issued in the United States that permitted physicians to withdraw mechanical ventilator therapy under certain conditions. Once physicians had a lawful recourse to discontinue supportive therapy in hopeless cases, declaring brain death no longer had this instrumental benefit.

The role of organ transplantation in the development of brain death remains controversial but most medical historians of this era concluded that it played an important but often sub rosa role (Belkin 2003). The organ transplantation surgeon, Joseph Murray, presumably was chosen as a member of the Harvard Medical School Ad Hoc Committee because of the perceived relevance of organ donation to the new determination of death. Yet, the Ad Hoc Committee report offered little discussion of organ procurement from the brain dead donor. Although later, Ad Hoc Committee member and historian, Everett Mendelsohn, commented that he did not recall much discussion about organ procurement during committee meetings, an analysis of the papers of the committee chairman, Henry Beecher, reveals considerable concern over this issue (Pernick 1999). Because of the concern that the utility of organ transplantation might influence death determination standards, recent committees empaneled to draft uniform guidelines for circulatory death determination in organ donors purposely restricted any participation or influence by organ transplantation surgeons to prevent this conflict of interest.[4]

Societal acceptance of brain death

Legal acceptance

One measure of the degree of societal acceptance of a new idea or practice is the extent to which it becomes enshrined in public law. The emerging idea that brain death was equivalent to human death quickly became established in law in the United States. In retrospect, the rapid legalization of this new medical practice appears to have been more the result of the legal recognition of an advance in medical practice than it was a product of reasoned debate over its coherence as a biophilosophical concept. Only later did scholars seriously debate the essential conceptual question of whether and why brain dead patients were truly dead.

In 1970, Kansas became the first state to statutorily recognize brain death, only 2 years following publication of the Harvard Medical School Ad Hoc Committee report.[5] This development was followed shortly by the enactment of other state death statutes permitting physicians to use brain death as a determination of death. In 1981, the United States President's Commission for the Study of Ethical Problems in Medicine and Biomedical and Behavioral Research published their inaugural report *Defining Death* in which they provided a philosophical and physiological rationale for the new concept of brain death and in which they proposed and promoted the Uniform Determination of Death Act (UDDA) as a model statute of death that incorporated brain death. Working with the United States National Council of Commissioners on Uniform State Laws, the President's Commission urged all states to adopt the UDDA as their statute of death. Since then, nearly all states have adopted the UDDA or a close variant of it. The statute is drafted so simply and clearly that all attempted legal challenges to brain death statutes in the United States have failed (Burkle et al. 2011). The UDDA provides:

> An individual who has sustained either (1) irreversible cessation of circulatory and respiratory functions, or (2) irreversible cessation of all functions of the entire brain, including the brain stem, is dead. A determination of death must be made in accordance with accepted medical standards. (President's Commission 1981)

Although in its concise text, the UDDA does not clarify the relationship between the circulatory–respiratory and brain criteria, the discussion in *Defining Death* implies that the brain criterion is the fundamental criterion of death. In model death statutes proposed by other countries, most notably Canada, the brain criterion is offered within the statute as the fundamental criterion of death but one that may be tested by physicians at the bedside in two different ways: by showing the prolonged cessation of circulation and respiration in patients who are not on ventilators or who will receive CPR, and by showing the irreversible cessation of brain functions in patients who are maintained on ventilators (Law Reform Commission of Canada 1981).[6]

A few studies have examined the extent of international acceptance of brain death. In a 2002 survey study by Wijdicks, 80 countries reported medical practices or laws permitting brain death determination (Wijdicks 2002). In a more comprehensive study of 91 countries in 2015, Wahlster and colleagues found that brain death practices were prevalent worldwide but with a great variability among testing guidelines and laws (Wahlster et al. 2015). They found that 70% of countries had established legal provisions for brain death declaration and that the most powerful predictor of the presence of a lawful protocol was a functioning organ donation network. A smaller survey of developed countries also reported a high rate of brain death acceptance (Gardiner et al. 2012). Thus, despite academic debates during the past several decades over the conceptual and scientific validity of equating brain death with human death, during this same time, more countries in the developed and developing world have permitted or promoted brain death practices. Thaddeus Pope (2014) comprehensively summarized the current state-by-state legal status of brain death in the United States.

Religious acceptance

A second measure of social acceptance is the extent to which a new practice is consistent with prevailing religious traditions. Whether death determination using a brain criterion was consistent with the tenets, teachings, and beliefs of the world's principal religions has been a documented concern since the first writings on brain death. In early influential articles, Veith and colleagues asserted that brain death practice was compatible with the world's major religious beliefs (Veith et al. 1977a, 1977b). Although their claim was based on a selective reading of religious opinion, since then, their generalization has been shown to be mostly true although the actual situation is more complex.

There has been relatively little debate over brain death within Christianity. In Protestantism, brain death acceptance is essentially universal, even among fundamentalist sects (Campbell 1999). In Roman Catholicism, acceptance is official but not universal. Four Vatican Pontifical Councils and Academies charged with studying this issue, beginning in the 1980s, concluded that brain death was consistent with Roman Catholic beliefs and teachings (Furton 2002). The Roman Catholic Church Magisterium first took an official position in 2000, when, in an address to the International Congress of the Transplantation Society in Rome, Pope John Paul II formally endorsed brain death as fully consistent with Roman Catholic teachings (Pope John Paul 2000). In response to questions within the Vatican hierarchy after the death of Pope John Paul, the Pontifical Academy of Sciences later reaffirmed this position (Pontifical Academy 2007).[7]

Judaism lacks Roman Catholicism's top-down ecclesiastical hierarchy that dictates rulings on religious law; instead it relies on learned rabbis and other Talmudic scholars to interpret contemporary questions in light of ancient Jewish law and teachings. Given the opportunity for disagreement intrinsic to this method of interpretation, it is not surprising that a rabbinic debate over brain death persists. In general, Reform and Conservative Jews accept brain death but the debate over it continues to rage within Orthodox Judaism. One group of Orthodox Jewish scholars holds that brain death is consistent with Jewish tradition, citing the writings of Maimonides and Talmudic passages emphasizing the equivalence of breathing (arising from the brainstem) with life. Opponents counter that Talmudists correlated the loss of breathing with the loss of heartbeat and therefore brain dead patients with intact heartbeat remain alive.[8]

The acceptance of brain death within Islam remains partial and dependent on local Islamic rulings. Following a fatwah approving brain death determination, the Ullamah Council in Saudi Arabia confirmed the practice of brain death and permitted it in the kingdom of Saudi Arabia (Yaqub & Al-Deeb 1996). Yet brain death is not permitted in several other Islamic societies. The religious basis and varying interpretations of the acceptability of brain death and the laws governing it among various Islamic countries have been reviewed recently (Padela et al. 2013; Miller et al. 2014).

Other religious traditions also have addressed brain death. Hindu acceptance of brain death was the topic of conferences held in Bombay and Madras. In 1993, the Indian legislature enacted the Transplantation of Human Organs Act that contains provisions for determining brain death (Jain & Maheshawari 1995). In Japan, traditional Shinto and Buddhist concepts of death requiring cessation of heartbeat and breathing have collided with technological attempts to Westernize Japanese medicine including the introduction of brain death and organ transplantation (Lock 1995). Scholarly works, medical and scientific society advocacy, and public debates over the past two decades have led Japanese society to gradually accept the practices of brain death and deceased donor organ procurement (Lock 2002) but this practice was first supported explicitly in Japanese law only in 2009 (Aita 2009).

One area in which the prerogatives of religion, law, and public policy overlap is in the enactment of state laws incorporating religious exemptions for declaring brain death. In 1991, New Jersey became the first American jurisdiction to enact an exemption from brain death determination when family members of a patient could show that its declaration violated the religious beliefs of the patient (Olick 1991). New York initially provided administrative regulations accomplishing the same goal that later became a statutory exemption. Since then, California and Illinois have enacted similar religious exemptions. New York's and California's religious exemption admonishes physicians also to make reasonable efforts to accommodate moral objections to brain death. In these states, when the allowable circumstances are verified, physicians can declare death only by using circulatory–respiratory tests. Given the prevailing variation in state laws, an important public policy question remains over the extent of the elasticity of death statutes to accommodate personal religious and moral preferences before they become incoherent or unmanageable (Brock 1999; Miles 1999; Fry-Revere et al. 2010).

Professional and public acceptance

A third indicator of societal approval of brain death is the extent to which the general public and professionals accept it. Medical professional societies have uniformly accepted brain death since the first guidelines of the Harvard Ad Hoc Committee were published in 1968 (Wijdicks 2011). Medical expert groups continue to refine and publish evidence-based testing guidelines that have been accepted for adults (Wijdicks et al. 2010) and for infants and children (Nakagawa et al. 2011). In fact, the debates raging about brain death are largely restricted to the academy and generally are absent from medical practice where it has been uniformly accepted. Disagreements about brain death among practicing physicians usually involve only the technical issues of clinical and ancillary testing (Bernat 2013c).

One exception to this rule is the contention made by a few physicians and scholars that brain death represents a "legal fiction" to permit organ donation (Taylor 1997; Shah and Miller 2010). A legal fiction is a societal rule that compromises biological reality to accomplish a social good. Legal blindness is a good example. Many countries provide people with severe visual impairment the same social benefits as is given to those who are completely blind under the rubric "legal blindness" despite acknowledging that they are not truly blind. Scholars making this analogy claim that we all know that brain dead patients are not really dead but society creates the legal fiction of brain death to allow them to be declared dead for the societal benefits of organ donation.

Despite the strong endorsement of the practice of brain death from medical societies around the world, one generally accepted conclusion that has resulted from a number of surveys is a persistently broad level of misunderstanding and confusion over the meaning of brain death by both medical professionals (Joffe et al. 2011) and the public (Siminoff et al. 2004). This problem has been stubbornly immune to decades of professional educational interventions by the American Academy of Neurology and other groups.

Misunderstanding of brain death by laymen is even more widespread and has been chronically exacerbated by popular press accounts that describe and explain it inaccurately and misleadingly. Analyses of popular press accounts of brain death cases reveal the use of contradictory words like "life support" to describe ongoing treatment and commenting that the brain dead person "died" once the ventilator was disconnected (Racine et al. 2008; Daoust & Racine 2014).

The depth of lay and professional understanding of brain death varies widely (Bernat 2014a). The most rudimentary and common level is intuitive. Many people find it logical that patients whose brain functions have totally and irreversibly ceased occupy a qualitatively different state than those who continue to possess brain functions because the brain is the master organ and the seat of consciousness and human behavior. This superficial, intuitive level constitutes a sufficient degree of understanding to allow many people to accept brain death even though their concept remains only vague and inchoate.

Medical acceptance constitutes a higher level of understanding. Physicians and others with scientific sophistication may accept brain death because of their greater understanding of the role of the brain in the functioning of the human organism. They may regard the absence of consciousness, cognition, breathing, circulatory control, and centrally controlled homeostasis as equivalent to death because the remaining bodily functions in a brain dead patient are only those supported by mechanical devices. They may consider such patients suitable for organ donation and "as good as dead" even if they remain undecided over whether they are biologically dead.

The most profound level of understanding requires justifying the equivalence of brain death and human death as a result of a rigorous biophilosophical analysis. This level may be achieved by conducting an analysis that studies the concept of the human organism and the meaning of death in our technological era in which organs and organ subsystems, particularly circulation and respiration, can be maintained in modern intensive care units. This understanding requires appreciating the distinction between the life status of the human organism and of its component parts. In the next section of the chapter on the definition and criterion of death, I discuss the biophilosophical justification of the equivalence of brain death and human death and offer my response to criticisms.

Within the practice of death determination, brain death is not unique in generating misunderstanding or controversy. There remains a raging debate over the exact moment of death when physicians apply the circulatory–respiratory criterion. This controversy centers on whether death occurs once circulatory and respiratory functions have ceased irreversibly (i.e., they *cannot* return) or when these functions have ceased permanently (i.e., they *will not* return) as discussed in the second part of this chapter (Bernat 2010a, 2013a). A recent survey of medical and nursing professionals yielded the surprising finding that those surveyed felt greater confidence in the accuracy of a brain death determination than in the accuracy of a circulatory–respiratory death determination (Rodriguez-Arias et al. 2013).

Several high-profile councils have weighed in on the brain death debate. The most recent effort, in 2008, by the United States President's Council on Bioethics yielded a white paper entitled *Controversies in the Determination of Death*. Their work was stimulated by their earlier unpublished study of ethical questions in organ transplantation. The Council heard testimony from brain death opponents who explained why they believed that the concept was invalid. Despite acknowledging the shortcoming of relying on the loss of integration of bodily subsystems as the conceptual grounds for accepting the equivalency of brain death as human death, the Council "concluded that the neurological standard remains valid" (President's Council 2008, p.x).

Two tragic and highly publicized cases of brain death in the United States refocused public attention in 2014–2015 and led to controversy. The teenager Jahi McMath was diagnosed as brain dead following complications after throat surgery but her mother refused to accept the diagnosis and insisted upon continued treatment even following a second opinion that agreed with the finding (Magnus et al. 2014). The pregnant Marlise Muñoz was diagnosed as brain dead but the Texas hospital in which she was admitted insisted on continuing ventilator treatment against the wishes of her family by claiming that regulations in the Texas Health and Safety Code mandated treatment in pregnant women for the welfare of the fetus (Ecker 2014). These cases stimulated much controversy (and confusion) about the definition, conceptual validity, religious relevance, and legality of brain death (Gostin 2014). Much of the reporting of these cases by the popular press compounded the controversy by publishing often incorrect and misleading descriptions of the medical condition, treatment, and prognosis of both women (Bernat 2014a).[9]

Even within communities with laws and religions accepting brain death, family members of brain dead patients sometimes reject the diagnosis and insist that their loved one is not really dead and therefore requires continued treatment (Lewis et al. 2016). This was the tragic situation with McMath even before her mother cited her religious opposition to brain death. This poignant situation creates a dilemma for physicians caring for such patients and which requires compassionate and clear communication (Olick et al. 2009). It is essential for physicians to carefully explain the medical and legal rationale for brain death, to emphasize its hopelessness, and to clearly distinguish it from coma and other potentially reversible brain states.

Nevertheless, some families insist on continued treatment, hoping for improvement. Physicians managing such cases should use their judgment to decide whether or for how long to acquiesce to this request. Second consultations may be helpful as well as frequent, empathetic counseling sessions. Valid religious exemption claims should be accommodated to the extent consistent with maintaining good medical practice (Spike & Greenlaw 1995; Orr & Genesen 1997). If the principal barrier to acceptance is emotional, assistance from chaplains, nurses, social workers, or the hospital bioethics committee may be beneficial (Bernat 2008a). After reviewing several instructive cases, Flamm and colleagues provided guidelines for physicians to compassionately handle such difficult situations (Flamm et al. 2014).

The definition and criterion of death

The debate over whether brain death is equivalent to human death centers on the meaning of death in a technological era in which organ subsystems other than the brain can be restored and maintained after they have failed from illness or injury. Scholars advocating the equivalence of brain death and human death believe that once the brain is destroyed, the artificial maintenance of respiration and circulation allows portions of the body to remain alive but that the human organism ("as a whole") is dead. Those scholars who advocate that the brain dead person remains alive counter that the brain enjoys no special status over other organs in a determination of death and that death as a biological phenomenon requires that all organ function must cease as a result of absence of circulation (Shewmon 2010). Defining what it means to be dead in our technological era, therefore, is a prerequisite for settling the brain death–human death equivalency argument and accepting or rejecting the biological foundation of brain death.

During the early era of brain death acceptance, scholars began to offer systematic analyses of death to show that brain death people were dead. These brain death defense arguments ultimately progressed to a definition of death argument. In this section, I offer analytic approaches to the definition and criterion of death and discuss critiques of the conceptual basis of brain death.

A biophilosophical analysis of death

The analyses of death by Capron and Kass (1972) and Bernat et al. (1981) asserted that physicians cannot develop and validate tests for death until biologically sophisticated philosophers settle the conceptual definitional question of what death is. In an influential legal-philosophical analysis targeted at developing a model statute of death, Capron and Kass (1972) proposed four sequential levels for an analytic framework: definition, standard, criterion, and test. In a similar vein, Bernat et al. (1981) chose preconditions, definition, criterion, and test as the essential conceptual categories. Each analysis recognized that the definitional question of what do we mean when we say "death" had to be agreed upon before criteria or tests to determine death could be formulated. Each of these biophilosophical frameworks had the virtue of logically progressing from the theoretical/conceptual to the tangible/measureable and thereby created an analytic landscape permitting areas of disagreement to be pinpointed.[10]

The paradigm of death

The first analytic stage is to determine the preconditions of the argument or "paradigm" of death: that set of assumptions that frame the analysis by clarifying its goals and boundaries and stipulating the nature of what is being analyzed. Agreement on preconditions is a prerequisite for further discussion. Many of the disagreements among scholars results from their opposition to accepting one or more of the seven conditions that comprise the death paradigm (Bernat 2013b):

1. "Death" is a nontechnical word that we use correctly in ordinary conversation to refer to the cessation of the life of a previously living human being. The goal in an analysis should be to make explicit the implicit meaning of "death" that we all accept in our spoken and written usage but that has been made ambiguous by advances in life-support technology. The goal should not be to redefine "death" by contriving a new and different sense than its consensual meaning. In this effort, "death" should not be

overanalyzed to such a metaphysical depth that it is rendered devoid of its ordinary meaning.

2. Death is a biological phenomenon. We all agree that life is a biological phenomenon; thus its cessation also is fundamentally biological. Death therefore is an immutable, objective, and inevitable biological fact and is not a social contrivance. An appropriate biophilosophical analysis of the definition and criterion of death therefore solely considers the ontology of death and not its normative aspects.

3. We restrict the analysis to the death of higher vertebrate species for whom death is univocal. That is, we refer to the same phenomenon of "death" when we say our cousin died as we do when we say our dog died. There is no biological justification for defining death idiosyncratically only for humans. Of course, lower species, unicellular organisms, and organ, tissue, and cellular component parts of organisms also can die but those phenomena are not our focus here.

4. The term "death" should be applied directly and categorically only to organisms. All living organisms must die and only living organisms can die. When we say "a person died," we refer to the death of the living organism that embodied the person, and do not make the claim that their organism continues to live but no longer has the attributes of personhood. Personhood is a psychological, religious, moral, and legal concept that arguably may be lost in some cases of severe brain damage but that cannot die except metaphorically, such as when the word is used in the expression "death of a culture."

5. A higher organism can reside in only one of two states, alive or dead: no organism can be in both states simultaneously or in neither state. But because we lack the technology to always correctly determine an organism's life state at any given time, we may only know with confidence in retrospect that death has occurred. Viewed as sets in a Venn diagram, the only two possible states of an organism—alive and dead—are mutually exclusive (nonoverlapping) and jointly exhaustive (no intervening or other states).

6. Death is most accurately conceptualized as an event and not as a process. Because there are only two mutually exclusive underlying states of an organism—alive and dead—the transition from one state to the other, at least in theory, must be sudden and discontinuous because there is no intervening state (Shewmon & Shewmon 2004). However, because of technical limitations, the event of death that marks the transition from one state to the other may be determinable with confidence only in retrospect. Death is conceptualized most accurately as the event separating the biological process of dying and the process of bodily disintegration (Bernat et al. 1981).

7. Death is irreversible. All plausible concepts of death require its irreversibility. If the event of death were reversible it would not be death but rather incipient dying that was interrupted and reversed.

Critiques of the paradigm of death

Most scholars concur that the goal of an analysis of death is to make explicit its consensual meaning and not to contrive a new meaning and most analyses strive to accomplish that goal. One analysis that does not is the "higher brain formulation" of death, advocated by several scholars (Youngner & Bartlett 1983; Gervais 1986; Veatch 1993) which defines death as the loss of the uniquely human attributes of consciousness and cognition. This formulation leads to diagnosing as dead patients in a vegetative state who breathe spontaneously and have spontaneous eye and limb movements. Rather than making explicit the consensual meaning of death, this approach contrives a new definition because no society on earth regards such people as dead or allows breathing people to be buried. This effort also fails to respect the biological univocality of death across related species by creating a unique standard applicable only to *Homo sapiens*. Such a formulation erects boundaries where none exists in biology.

Some scholars reject the claim that death is primarily a biological phenomenon and instead assert that death is a socially determined phenomenon that varies among cultures (Veatch 1999). This claim seems to confound the ontological and normative issues surround death. Obviously there are innumerable rich traditions within cultures and religions encompassing behaviors surrounding death and dying. But the full respect of these traditions does not extend to the ontological question of when someone is dead. The normative aspects of death, including the subject of personal identity and morality, have been thoroughly reviewed (DeGrazia 2011). Some scholars who champion this idea have made the additional libertarian claim that respect for personal liberties should encompass the rights of patients to be declared dead according to their personal beliefs about when one is dead (Veatch 1999). As a public policy applied to clinical practice, however, it is hard to imagine how this idea could be feasible.

A more substantive disagreement centers on whether a unitary concept of death for a human being is possible or whether there must be two types of death: death of a human organism and death of a person. Jeff McMahan (1995, 2002) argues that because human beings are our persons and not simply our organisms, what counts most in a concept of death is the death of the person. He advocates the creation of separate, dual accounts of death for persons and for their embodied human organisms, and acknowledges that this dichotomy represents a form of Cartesian mind–body dualism. John Lizza (1999, 2005) has made a similar argument.

McMahan and Lizza reached their dualistic conclusions because their concept of person is more expansive than merely one of a human organism endowed with certain attributes, such as the concept of person and personhood that Gert, Culver, and I endorse (Bernat et al. 1981). McMahan's and Lizza's idea of a person incorporates a soul or a spiritual element that exists separately but in parallel with the bodily organism and therefore requires a dual account of death.[11]

Lee and George (2008) and Shewmon (2010) have further developed the concept that a person has both "animal and mental" components ("body–self dualism"), and assert that the mental component cannot be produced merely from the animal component. Shewmon (2010) elucidated this concept more explicitly: "Reflective self-awareness, universal concept formation, abstract reasoning, and free will all have properties that transcend spatiality and cannot in principle emerge from a complex electrochemical network. They therefore derive from an immaterial principle, but nevertheless, one profoundly oriented to operate in and through a body." These scholars therefore deny the "reductive" claim accepted by most cognitive neuroscientists that the mind is solely the product of the brain without the necessity of an animating spiritual element (Pinker 1997; Koch 2012).

For our paradigm, Gert, Culver, and I require that human beings are only our human organisms, that self-awareness and other human behaviors are solely emergent functions of the brain that do not derive from an immaterial principle, and therefore, human death is the death of the human organism. This approach is consistent with three other elements of the paradigm: (1) death is fundamentally a biological phenomenon; (2) only organisms can die; and (3) the biologically univocal usage of the word *death* requires that it mean the same thing for humans as for other higher animals such as dogs unimbued with souls. We regard personhood as a set of psychosocial, legal, moral, and religious attributes of human beings that arguably may be lost by severe brain injury or illness but cannot die except metaphorically. In his monograph, Eric Olson (1997) more rigorously defends the position that human beings are our organisms.

Linda Emanuel (1995) relies too heavily on metaphysics with her claim "there is no state of death ... to say 'she is dead' is meaningless because 'she' is not compatible with 'dead.'" The depth of Emanuel's metaphysical abstraction of death cannot identify a definition or criterion because it offers nothing to clarify the common usage of the term *death*. And as an experienced physician, she obviously does not truly believe that there is no state of death.

Amir Halevy and Baruch Brody (1993) claim that defining death is impossible because an organism can reside in a transitional state between life and death that possesses features of both states but is congruent with neither state. Using the mathematical model of fuzzy logic, they postulate that whereas no organism can fully belong to both the sets of living and dead organisms, because the sets represent mutually exclusive states, some organisms can reside in a transitional state in which they have some features of each state but do not fully belong to either the set of living or dead organisms.

Our argument against this position emphasizes the distinction between the ontology of the life-state of an organism and our ability to accurately determine that state. Because, as a consequence of technical limitations, we may not be able to conclusively determine at all times whether a given organism is alive or dead does not mean that it must reside in a transitional state between alive and dead. The paradigm provides that all organisms are either dead or alive, but, because of these technical limitations, we may not be able to make the accurate determination of its life-state in real time but only in retrospect. Future technological advances will improve the accuracy and timeliness of this determination.

In their mathematical analysis of state discontinuities, Alan and Elisabeth Shewmon (2004) settled the longstanding debate over whether death is best understood as an event or as a process.[12] The Shewmons showed that death must be an event because of the suddenness and discontinuity of the transition from the states of alive to dead, given the absence of an intervening state. They agreed that we may not always be able to identify the precise time of the event, and we may be able to identify it only in retrospect because of technical limitations. Everyone agrees that the biological phenomena of dying before death and of bodily disintegration after death are processes. Most scholars, including the Shewmons, now agree with us that death is best viewed as the event separating the process of dying from the process of bodily disintegration (Bernat et al. 1981).

The neurologist-philosopher Winston Chiong rejected the idea that death must be defined before analyzing criteria to measure it (Chiong 2005, 2014). He cited Wittgenstein's argument that some common terms, such as *games*, cannot have uniform definitions that are based on the possession of an essential meaning shared by all members of the set because all members of the set in question do not share an essential characteristic. Rather than communally sharing an essential characteristic, members of the set are related to each other in various other ways. Chiong claims that *death* is such a word. He argues that searching for the essence of the meaning of the word *death* from which to establish its definition is futile because there are no conditions that are both necessary and sufficient for death. He further maintained that defining death is an unnecessary step to provide a coherent argument supporting the whole-brain criterion of death. He concluded that our paradigm–definition–criterion–test sequential analytic method therefore should be rejected.

My late colleague, Bernard Gert, refuted Chiong's criticism, arguing that Chiong misunderstood the correct meaning of a definition by accepting a discredited essentialist concept of a definition (Gert 2006). The effort to choose a definition of death is not to make explicit the implicit essence of the concept of death but rather to make explicit the meaning implicit in our consensual and ordinary use of the nontechnical word *death*. Gert further pointed out that, given this proper intent of definition, Chiong, perhaps unknowingly, also relied on a definition of death—though one more diffuse than ours—in his defense of the whole-brain criterion of death. Therefore, Chiong's argument rejecting definition constituted an invalid criticism of our paradigm–definition–criterion–test method of analysis that requires starting with what is ordinarily meant by the term *death* before choosing its criterion.[13]

A number of commentators have reported cases of "near-death" experiences to support their claim that death is not irreversible because, as their experience shows, some people can return from being dead. The physiological explanation of these personal narratives is the presence of acute encephalopathies that occur during critical illness from marked disturbances in brain metabolism (Parnia & Fenwick 2002). Patients with "near-death" experiences were rescued from death while incipiently dying but were not dead and therefore did not return from the dead. Despite the poignancy and sincerity of these subjective accounts (Alexander 2012), they do not negate the axiom that death is irreversible.

The definition of death

The goal of divining the definition of death is to make explicit the meaning of death that is implicit in the way we use the word *death* in our everyday conversation but that has been obscured by technological advances of physiological organ support. Several putative definitions are unacceptable. The most conservative definition that requires the cessation of all cellular functioning is unnecessarily conservative because some cells can be kept alive in tissue culture and entire organs can be transplanted into other people allowing these cells to continue living long after a person has died. Some religious people regard death as the moment during dying at which the soul departs from the body. This definition of death is unacceptable because, as it constitutes a religious belief, is not shared universally. But more importantly, this definition does not represent the ordinary meaning of the word death implicit in our consensual usage of the term. Finally, it cannot yield a measurable criterion.

One plausible and intuitively acceptable definition of death is the irreversible cessation of all organ functions, usually as a consequence of the irreversible cessation of systemic circulation. The irreversible cessation of systemic circulation is an established criterion of death, featured in death statutes such as the UDDA as discussed previously. No one questions whether the irreversible cessation of all organ functions is sufficient for death; the essential question is whether it is necessary. Scholars supporting a definition focusing on the organism's unity and the interrelatedness of its component parts conclude that the irreversible cessation of all organ functions is unnecessary for death therefore this definition is unnecessarily conservative.

Brain-based approaches to the definition of death center on the concepts of the loss of the interrelatedness of the component parts of the organism and the loss of wholeness and coherent unity of the organism. They emphasize the critical distinction between the continued life of parts of the organism as a consequence of technological support and the continued life of the organism itself. It addresses the relationship between the whole and its parts and their ontological meaning.[14] The definition encompassing these ideas that has enjoyed the greatest staying power is that death is the cessation of functioning of the organism as a whole.

The organism as a whole, as first proposed a century ago by the biologist Jacques Loeb (1916), is not congruent with the whole organism, that is, the sum of the parts of the organism. Rather, the organism as a whole refers to those functions that are greater than the sum of the parts of organism, that reflect its unity and wholeness, and that serve the continued life and health of the organism, even at the expense of some of its parts. These features now are referred to as the emergent functions of the organism.

Emergent functions are characteristics of a whole that cannot be reduced or localized to any of its component parts but emerge spontaneously when the parts operate in their normal ensemble (Mahner & Bunge 1997, pp.29–30). Emergent functions are hierarchical. Tissues have emergent functions not reducible to their constituent cells and organs have emergent functions not reducible to their constituent tissues. In the context of the definition of death, the emergent functions of the organism are those that comprise its unity and wholeness. Biological theorists have shown that an organism's emergent functions are complex phenomena that are difficult to model or predict merely by studying its isolated parts (Clayton & Kauffman 2006).

Bonelli and colleagues offered a conceptual analysis of the role of the cessation of the organism as a whole in death (Bonelli et al. 2009). They stated that all life forms possess an intrinsic unity characterized by four criteria: (1) dynamics, or signs of life, such as metabolism, regeneration, growth, and propagation; (2) integration, the requirement that the life process derives from the mutual interaction of its component parts; (3) coordination, the requirement that the interaction of the component parts is maintained within a certain order; and (4) immanency, the requirement that the preceding characteristics originate from and are intrinsic to the life form. They identified four criteria that make a life form an integrated, unified, whole organism: (1) completion, the requirement that an organism is not a component part of another living entity but is itself an intrinsically independent and completed whole; (2) indivisibility, the condition of intrinsic unity that no organism can be divided into more than one living organism; and, if such a division occurs and the organism survives, the completed organism must reside in one of the divided parts; (3) self-reference or auto-finality, the characteristic that the observable life processes and functions of the component parts serve the self-preservation of the whole, even at the expense of the survival of its parts, because the health and survival of the living whole is the primary end in itself; and (4) identity, the circumstance that, despite incremental changes in form and the loss or gain of certain component parts (that even could eventually result in the exchange of all component atoms), the living being remains one and the same throughout life (Bonelli et al. 2009). My Dartmouth colleague, Andrew Huang, and I currently are critiquing and expanding upon Bonelli and colleagues' formulation.

The criterion of death

The criterion of death must satisfy the definition of death as the cessation of functioning of the organism as a whole by being both necessary and sufficient for death. The principal scholarly disagreement over the criterion of death is between those who hold that the functions of the organism as a whole are operated solely by the brain and those who hold that these functions are operated jointly by the brain and other body parts. Scholars who hold the first position consider irreversible cessation of brain function as a criterion of death whereas scholars who hold the second position require the irreversible cessation of function of all organs, a condition that can be satisfied only by the irreversible cessation of circulation and respiration.

Some of the numerous functions of the organism as a whole are more critical than others for the organism's life and health. The exact set of functions whose loss is both necessary and sufficient for death remains a topic of debate. Nearly all scholars agree that conscious awareness is an ineffable, emergent function of higher animals that counts as a critical function of the organism as a whole. Similarly, the systems controlling circulation, respiration, fluid–electrolyte balance, movement, and temperature regulation count as critical functions because the organism's health and life depend on them. Most of the critical functions of the organism as a whole can be classified as control and executive functions.

Advocates of "brain death" hold that because most of the critical functions of the organism as a whole are executed by the brain, irreversible cessation of brain functions eliminates these critical functions and is death (Korein 1978, 1997). Bonelli and colleagues stated this position most rigorously, arguing that with irreversible cessation of clinical brain functions ("brain death"), the organism has lost (1) immanency, because its life processes no longer arise from itself but result from external intensive care support; (2) auto-finality, because whatever control over the component organ subsystem parts that remains now is directed at the level of the surviving parts and no longer at the whole; (3) self-reference, because the continued functioning of its parts no longer supports to the function of the whole; and (4) completeness and indivisibility, because its separate component parts and subsystems no longer belong to each other and no longer constitute a whole. Once an organism has irreversibly lost its totality, completion, indivisibility, self-reference, and identity, it no longer functions as a whole and is dead (Bonelli et al. 2009).

Brain criterion advocates disagree to some extent on the question of exactly which elements of the brain's functioning must cease to be sufficient for death. Higher-brain criterion advocates require the cessation only of the functions of the cerebral hemispheres and thalami because those structures underlie conscious awareness and the cognitive functions that are the unique characteristics of the human being and whose cessation is death (Veatch 1975; Gervais 1986). The higher-brain criterion remains popular in the academy but has not been accepted by any medical society or jurisdiction in the world. The failure of its acceptance despite decades of argument is because, rather than trying to make explicit the implicit definition of death that has been obscured by technology, the higher-brain criterion is a radical redefinition of death. The evidence for this conclusion is that applying it wrongly labels many patients in coma or vegetative states as dead, despite the fact that they are considered as alive by societies throughout the world.

Brainstem formulation advocates require only the cessation of functions of the brainstem, arguing that the brainstem controls respiration, circulation, and consciousness, and serves as a through-station for nearly all hemispheric input and output. Further, the clinical tests determining "brain death" examine mostly for brainstem functions (Pallis 1995). But by not requiring the absence of hemispheric functioning, it permits the unlikely but possible case of false-positive death determination if awareness were retained despite the loss of other brainstem functions. The brainstem criterion and the whole-brain criterion have similar sets of bedside tests but it has been accepted as a criterion of death only by the United Kingdom and a few other countries.

The prevailing whole-brain criterion of death has been accepted in nearly all jurisdictions except those that accept the brainstem criterion. The whole-brain criterion acknowledges that the hemispheres, thalami, hypothalami, and brainstem all contribute to the critical functions of the organism as a whole, hence the clinical functions controlled by each of these parts of the brain must be abolished for death. But, despite its categorical name, the whole-brain criterion does not require the cessation of all brain neurons, only a critical number and array (currently unknown) that conduct the clinical functions of the brain underlying the critical functions of the organism as a whole (Bernat 1998, 2002).

An assessment of the physiological state of the "brain dead" patient shows the contrast between the technologically maintained and continued life of many component parts of the organism despite the death of the human organism itself. "Brain death" advocates do not deny that much of the human organism remains alive because of technological support, but assert that the body on the ventilator has died because of irreversible cessation of the clinical brain functions serving the organism as a whole.

Scholars who reject any brain criterion embrace the circulation–respiration criterion. Alan Shewmon, the intellectual leader of this group, accepts the definition of death as the cessation of function of the organism as a whole (Shewmon 2010). However, he rejects any brain criterion because he argues that many functions of the organism as a whole are operated by the spinal cord and other organs and tissues outside the brain (Shewmon 2001, 2004). Further, he cites cases of "chronic brain death" in whom people declared brain dead have been treated with circulatory and respiratory support of ventilators and other therapies to allow their visceral organs to be perfused and oxygenated successfully for months (Shewmon 1998) or, in one remarkable case, for 16 years (Repertinger et al. 2006). He argues that it is simply counterintuitive to the concept of death that people could have circulation and visceral organ function maintained for long periods, go through puberty, or gestate fetuses during this treatment, as have been reported. Shewmon and other advocates of the circulatory–respiratory criterion claim that the definition of death therefore is not fulfilled in typical brain death, and instead, death requires the irreversible cessation of circulation of oxygenated blood to all organs (Shewmon 2001, 2009). Brain death advocates counter that the definition of death is fulfilled in "brain death" and that the criterion of irreversible cessation of circulation and respiration, while sufficient for death, is not necessary. In a paper in preparation, my colleague Andrew Huang and I rebut Shewmon's claim.

The circulatory criterion of death

The alternative to the whole-brain criterion of death is the circulatory–respiratory criterion of death, often shortened to the circulatory criterion. Cessation of systemic circulation was the traditional criterion of death prior to the 1950s when technology rendered death no longer a unitary phenomenon. It remains a valid criterion of death in the overwhelming number of cases in which cardiopulmonary resuscitation will not be attempted (or had been attempted and failed) and in which mechanical ventilation will not be performed. Cessation of systemic circulation also causes cessation of brain circulation that leads to fulfillment of the brain criterion of death. If circulation ceases by cardiac arrest or respiration ceases by respiratory arrest and medical intervention is omitted, the mutual interdependence of vital functions leads immediately to their complete cessation. Thus, the traditional unitary concept of death remains valid in cases in which no resuscitative intervention takes place.

Although everyone agrees that the irreversible cessation of systemic circulation is sufficient for death, the current controversy over the circulatory criterion surrounds the question of how long physicians must wait to declare death once circulation (and respiration) have ceased. In cases of circulatory–respiratory death determination in which organ donation is not being performed, this question is usually inconsequential. Physicians examine the patient, determine the absence of breathing, heartbeat, responsiveness, and pupillary light reflexes, and simply declare death at that moment. There is no need to hurry and physicians usually conduct a leisurely examination without concern about timing. Death is declared at the completion of the examination.

In cases in which the patient is an organ donor, however, circulatory–respiratory death determination is an entirely different situation. Here, death determination must be performed as rapidly as possible to minimize warm ischemic damage to the organs to be transplanted. Conducting a timely death determination in the organ transplantation donor raises a knotty question: what is the shortest time interval after circulation and respiration have ceased after which physicians can declare death? This seemingly straightforward question has led to considerable controversy and, although a few guidelines have emerged, remains unresolved.

Many hospitals around the world have established programs of organ donation after the circulatory determination of death (DCDD), known formerly as nonbeating organ donation or donation after cardiac death (Bernat et al. 2006). In "controlled" DCDD as practiced in the United States and Canada, in which the donor is usually a patient with profound brain damage on a ventilator in an intensive care unit, family members or other lawful surrogates order withdrawal of life-sustaining therapy (the mechanical ventilator) to allow the patient to die and also consent for organ donation after death has been declared, presumably representing the wishes of the patient. Once respiration and circulation stop following endotracheal extubation, after a mandatory observation period to guarantee that breathing and circulation will not restart spontaneously ("auto-resuscitation"), the patient is declared dead and rushed to the surgical suite for organ donation. An analogous program of "uncontrolled" DCDD is practiced in some European countries, in which the often previously healthy donor suffers a sudden cardiac arrest, undergoes CPR but cannot be resuscitated, and is declared dead prior to organ donation (Munjal et al. 2013).

In death declaration of the controlled DCDD donor, the essential question centers on how long respiration and circulation must be absent before death can be declared. This interval varies from 2 minutes in the pioneering DCDD protocol at the University of Pittsburgh Medical Center (DeVita & Snyder 1993), to 5 minutes as recommended by the US Institute of Medicine (2000) which has been instituted by most US hospitals, to up to 20 minutes in some European countries. The principal requirement is that the duration chosen must exceed the interval in which auto-resuscitation can occur to be sure that circulation cessation is permanent. This empirical question was addressed in a study of published reports of auto-resuscitation after circulatory cessation (Hornby et al. 2010). The investigators found no cases at all of auto-resuscitation to restored circulation after withdrawal of mechanical ventilator therapy in the intensive care unit (a situation analogous to controlled DCDD) but many cases of auto-resuscitation to restored circulation for up to 7 minutes after CPR was stopped when it failed to restore heartbeat (a situation analogous to uncontrolled DCDD). Therefore, the problem of auto-resuscitation represents a greater concern in uncontrolled than in controlled DCDD (Bernat 2010b).

Even if commentators were to agree on the auto-resuscitation issue, however, there remains disagreement over whether the controlled DCDD donor is even dead at the moment of death declaration in most DCDD protocols. Critics claim that at 2–5 minutes after circulatory cessation, there remains the possibility that CPR could succeed in restoring heartbeat and circulation—even though it will not be performed because of a preexisting do-not-resuscitate (DNR) order—and that this potential circulatory reversibility proves that the donor was not dead at the moment death was declared because it is not manifestly irreversible (Miller and Truog 2008; Marquis 2010). Analyzing this criticism requires making two related distinctions: (1) between the biological concept of death and the medical standards for death determination, and (2) between the *irreversible* cessation of circulatory and respiratory functions and the *permanent* cessation of those functions (Bernat 2010a).[15]

Irreversible cessation of a function means that, once stopped, it *cannot* be restored using available technology. Permanent cessation of a function means that it *will not* be restored because, once stopped, because it will neither restart spontaneously (auto-resuscitate) nor will physicians attempt to restart it because of a DNR order. The prospective DCDD organ donor on whom death is declared after 5 minutes of ceased circulation and breathing, and for whom resuscitative effort will not be attempted because of a DNR order, has achieved permanent but not yet necessarily irreversible cessation of circulation and respiration (Bernat 2010a). But is the DCDD donor dead at that moment?

The medical practice of death determination using the circulatory criterion has traditionally followed the permanent cessation standard. When a physician declares a patient dead who has lost heartbeat, circulation, and respiration, the physician requires only the permanent cessation of those vital functions. In the setting of a DNR order, no attempt is made to resuscitate and there is no requirement for physicians to prove that the cessation is irreversible because permanent cessation is sufficient. For example, patients who are expected to die of an untreatable terminal disease who then stop breathing and lose heartbeat are immediately declared dead after a brief examination. There is no requirement for physicians declaring death to first prove that the patient's cessation of circulation and respiration is irreversible or to postpone declaring death for a sufficient time to assure that it is irreversible. In non-donation instances of circulatory death determination in whom CPR will not be performed, the medical consultants to the United States President's Commission, in *Defining Death* (1981, p.162), required "persistent cessation of functions during an appropriate period of observation" that is equivalent to the permanent cessation concept.

However, because all plausible biological concepts of death require that it is an irreversible state, there exists noncongruence between the biological concept of death requiring irreversible cessation of the circulatory–respiratory functions and the medical determination of death requiring only the permanent cessation of those functions. Thus, the answer to the question of whether the DCDD donor is dead at the moment death is ordinarily declared in DCDD protocols is "yes" by the medical standard of permanent cessation but "no" by the biological standard that requires irreversibility. The permanence standard to determining death has been accepted implicitly in all controlled DCDD protocols, explicitly by one medical society (American Academy of Pediatrics 2013), and was recommended by an expert panel advising DCDD programs on standardizing policy for circulatory death determination in organ donors (Bernat et al. 2010, 2014). Further scholarly and public policy debate among organ donation stakeholders should clarify if the clinical standard of permanent cessation of circulation and respiration also should pertain to death determination in the setting of organ donation (Bernat 2013a).

Future directions

The futures of brain death and the definition of death remain intertwined. The societal consensus accepting brain death as human death, although described by one critic as "superficial and fragile" (Youngner 1992), has endured for nearly half a century. One knowledgeable commentator accurately summarized the state of brain death in the developed world as "well settled but still unresolved" (Capron 2001). It has been settled to the extent that societies have enacted more or less uniform laws permitting physicians to use brain death as a determination of death. But it remains unresolved to the extent that its opponents validly identify weaknesses in its rationale causing them to reject it and requiring a stronger conceptual defense. As an advocate of brain death, I believe that its most coherent biophilosophical justification is that the definition of death should be based on the cessation of functioning of the organism as a whole. But more work needs to be done to fill in conceptual gaps in the present foundation—particularly in the criteria of the organism as a whole—that are necessary to more rigorously establish this rationale. My Dartmouth colleague, Andrew Huang, and I are presently constructing such an argument.

In any analysis of the definition of death or the determination of brain or circulatory death, it is essential to separate the social desirability and instrumental value of organ donation (Citerio et al. 2016) from conceptual and policy questions about death determination. Independent of its utilitarian value, the determination of death must be biologically accurate and the definition of death must be philosophically plausible and coherent. Public policies may permit certain agreed upon practical compromises to assure that medical practices involving organ donation and death determination are workable and socially acceptable. But these adjustments must be understood by physicians, accepted by the public, and operationalized into optimal practices of informed consent for the determinations of both brain death and circulatory death, particularly in the context of organ donation. Current informed consent practices in organ donation appear to be suboptimal for both brain death (Iltis 2015) and circulatory death (Overby et al. 2015).

It is in this regard that the noncongruence between the medical standards for death determination and the biological concept of death must be fully appreciated. Medical standards permit physicians to declare death at an earlier moment than the biological concept would otherwise allow. Primarily for social reasons, physicians routinely declare death when a person's vital functions have ceased permanently but at a time before these functions have ceased irreversibly (Bernat 2013a). As public policy, it is desirable for our society to rely on accepted medical standards for death determination in both organ donors and nondonors even though medical standards may not achieve the biological standard of death. Stakeholders in organ donation and death determination need to debate and accept a standard that they conclude is optimal for society. I advocate the permanent cessation standard that is implicit in all DCDD protocols and therefore appears to have achieved widespread acceptance.

An unresolved public policy question remains the extent to which moral or religious beliefs and personal preferences should influence the medical standards that physicians will apply in one's own death determination. This issue has arisen most frequently in the context of brain death but also could be raised in circulatory death determination for organ donors. The tension present here is between respect for personal liberties and the authority of medical practices and societal laws. As is true in most such controversies, differing opinions about the location of the equilibrium point of this tension result from individual differences in political philosophies, for example, between communitarianism and libertarianism. A successful public policy needs to identify an equilibrium point that adequately provides individual liberty rights but does not lead to social chaos.

Most people intuit the desirability of having uniform standards of brain and circulatory death determination—whatever they may be—across countries and societies (Choi et al. 2008). Yet, variability in medical practices remains the norm, even in the United States where expert guidelines on brain death determination have been in effect for over two decades (Greer et al. 2016). To some extent, this variability results from physicians who remain unaware of or who refuse to comply with national standards and from hospitals not enforcing them (Bernat 2008b). Achieving international standardization will be even more difficult because it must accommodate varying societal mores, religious beliefs, and medical practice variations (Wahlster et al. 2015). A few medical societies and the World Health Organization are currently working to try to develop such uniform standards but it presents a formidable undertaking.

Given the survey data documenting a stubbornly persistent and widespread level of ignorance about brain death (Siminoff et al. 2004), most people agree that there is a need to improve public and professional education on concepts and practices of death determination. Medical professionals need to increase their sophistication about the definition of death and their knowledge of the concept and determination of brain death. Among other goals, greater sophistication and knowledge will clarify the true level of support for the concept of brain death in the medical community that has been hard to assess accurately with current surveys. Better public education should include journalists who write articles on brain death in the popular press that educate (or mislead) the public and mold public opinion.

Finally, there remain gaps in the scientific understanding of incipient dying and death which, once filled, can inform the development of medical standards for death determination. For example, we know relatively little about the natural history and precise time course of how cessation of systemic circulation leads to hypoxic-ischemic neuronal damage and brain death, particularly, at what moment all brain functions cease and the point when brain death occurs. Although preliminary observational studies have made an important contribution to understanding this common phenomenon (Munshi et al. 2015), more such studies need to be conducted.

Notes

1. In their landmark 1959 article from the Claude Bernard Hospital in Paris, Mollaret and Goulon reported 23 cases of patients with utter coma, apnea, EEG electrocerebral silence, and total brain necrosis at autopsy (Mollaret & Goulon 1959). They coined the term *le coma dépassé* (which Prof. Goluon translated into English as "irreversible or irretrievable coma") to describe their unprecedented deep state of coma with no cognitive or vegetative functions (Wijdicks 2011). The article included the clinical, electroencephalographic, and pathological features of what later would be called "brain death." Mollaret and Goulon also discussed ethical and philosophical issues including the boundaries between life and death and the ethical duty of physicians to continue to support patients in that hopeless state. Because of the comprehensiveness of their description and discussion, Mollaret and Goulon are generally credited with describing the first cases of brain death although Lofstedt and von Reis had done so briefly 3 years earlier (Diringer & Wijdicks 2001).

2. The report of the Ad Hoc Committee of the Harvard Medical School to Examine the Definition of Brain Death, misleadingly subtitled "A definition of irreversible coma," first made the medical and legal communities and the general public aware of the use of the absence of brain functions to determine human death. Harvard anesthesiologist and research ethicist, Henry K. Beecher, chaired the committee. Other members included neurologists Raymond D. Adams, Derek Denny-Brown, and Robert Schwab; neurosurgeon William Sweet; transplant surgeon Joseph Murray; law professor William Curran; historian of medicine Everett Mendelsohn; and five others (Wijdicks 2003). The purpose of the Committee's report was "to define irreversible coma as a new criterion of death" (Ad Hoc Committee 1968). The Ad Hoc Committee proposed the first examination criteria to certify brain death. These tests have withstood the test of time and have not been altered significantly in the 49 years since they were proposed. The four cardinal characteristics of the brain dead patient are: (1) unreceptiveness and unresponsiveness; (2) no movements or breathing; (3) no reflexes, especially cranial nerve innervated reflexes; and (4) flat EEG. The tests needed to be repeated after 24 hours to insure irreversibility. They incorrectly predicted that if the medical community embraced this concept, no change in the law on death statutes would be required. They correctly asserted that considering the permanent cessation of all brain functions as death was implicit in the traditional concept of death.

3. The Harvard Medical School Ad Hoc Committee popularized the term "brain death." This was an unfortunate choice of terms because it misleadingly implied that only the brain was dead and that there were two types of death: brain death and ordinary death. The misleading term "brain death" has contributed to both public and professional persisting misunderstanding (Molinari 1982). Committee member Murray anticipated this problem and wrote a memo to committee chairman Beecher in which he pleaded: "The term 'brain death' should be eliminated. Death is what we are talking about, and adding the adjective 'brain' implies some restriction on the term as if it were an incomplete type of death...our charge is to define death..." (Giacomini 1997). But Beecher chose to retain the term "brain death."

4. When the Health Resources & Services Administration Division of Transplantation of the United States Department of Health and Human Services empaneled an expert committee to determine standards for the circulatory determination of death in DCDD organ donors, it explicitly and firmly barred all membership to and influence from organ transplantation surgeons although these surgeons could have contributed useful technical information to the process. See Bernat et al. (2010) and Bernat et al. (2014).

5. That Kansas became the first state in the United States to amend its death statute to incorporate brain death has been attributed to the outsize influence of a Kansas neurosurgeon who was friendly with state legislators and used his powers of personal persuasion to push through the bill (Kennedy 1971).

6. Ironically, Canadian provincial governments chose to ignore their federal council recommendation to adopt the model statute and, despite its clarity, did not incorporate the model statute into provincial laws.

7. Despite the papal pronouncement, opponents of brain death remain within Roman Catholicism who continue to assert that it is inconsistent with Roman Catholic, and even Christian, belief systems. Most persistent in this area over the past 40 years has been Paul Byrne (Byrne et al. 1979). Yet, even before the Pope's pronouncement, it was clear that Byrne's opinion was marginalized within Roman Catholic circles. The conservative Roman Catholic bioethics institute, the National Catholic Bioethics Center (formerly called the Pope John Center), held that Byrne was incorrect and that brain death was fully consistent with the teachings of Roman Catholicism (Furton 1999). Following Pope John Paul's endorsement of brain death in 2000, the Center predicted that the papal pronouncement should quell the disagreement by Byrne and other critics on Roman Catholic religious grounds (Furton 2002; Haas 2011).

8. The Orthodox rabbi-scientist Moshe Tendler argues that brain death is compatible with Halacha (historical-traditional Jewish law) because it is the physiological equivalent of decapitation. In the twelfth century, Maimonides asserted that a decapitated person was immediately dead and that transient muscle twitches observed in the decapitated body were not signs of life. Tendler and the physician-talmudist Fred Rosner cite the Talmudic tract, discussed by Rashi, in which it is stated that breathing, not heartbeat, is the primary sign of life. Therefore, brain death is human death according to Jewish law (Rosner & Tendler 1989; Rosner 1999). But Orthodox rabbis and Talmudic scholars David Bleich (1989) and Ahron Soloveichik (1979) reached the opposite conclusion. In emphasizing the distinction between a sign of death and death itself, they explained that ancient rabbis considered the cessation of respiration as the cardinal sign of death because it implied a prior cessation of heartbeat. They concluded that, according to Jewish law, physicians cannot determine death in the presence of spontaneous heartbeat. Although only the strictest Orthodox sects consistently embrace this position, it remains the official position in Israel though changes to the Brain Death Act in 2012 permitted brain death determination in some situations and thereafter, brain dead organ donation increased substantially (Cohen et al. 2012). For learned commentaries on this continuing religious dispute, see Halachic Organ Donor Society (https://www.hods.org/).

9. The follow-up stories of McMath and Muñoz are instructive. As of February 2017, McMath apparently remains supported on a ventilator somewhere in New Jersey as her mother and her attorneys try to convince a California judge to reverse her death declaration citing affidavits submitted by physicians who have examined her stating she is not dead (Winkfield v. Rosen 2015). It is difficult to objectively know her true condition because, other than these affidavits, her current medical findings are not accessible in the public record. In the Muñoz case, in response to a lawsuit filed by her husband, a Texas judge ruled that she should be removed from the ventilator because the Texas law requiring continued life-sustaining therapy to brain-damaged pregnant women for the benefit of the fetus did not apply to dead women. Her ventilator treatment was then discontinued.

10. Portions of this section were adapted from Bernat (2013b), Bernat (2014b), and Bernat (in press).

11. See John Lizza's (2009) anthology of articles relating personhood to death.

12. The classic debate over whether death is a process or an event was conducted in the pages of *Science* by Morison (1971) and Kass (1971).

13. Chiong thereafter countered that Gert misunderstood and misrepresented his position. For the latest chapter on this debate, see the point-counterpoint in (Bernat 2014b) and Chiong (2014). My colleague, Andrew Huang, and I respond further to Chiong in our manuscript in preparation.

14. Mereology is the branch of philosophy that analyzes the differences between the properties of a whole and its parts and explores their ontological relationship. The principles of biological mereology hold the key to understanding the concept of the organism as a whole. See Hovda (2009).

15. Portions of this section were adapted from Bernat (2014b).

References

Ad Hoc Committee (1968). A definition of irreversible coma. Report of the Ad Hoc Committee of the Harvard Medical School to Examine the Definition of Brain Death. *JAMA*, **205**, 337–340.

Aita, K. (2009). Japan approves brain death to increase organ donors: will it work? *The Lancet*, **374**, 1403–1404.

Alexander, E. (2012). *Proof of Heaven: A Neurosurgeon's Journey into the Afterlife*. New York: Simon & Schuster.

American Academy of Pediatrics Committee on Bioethics (2013). Policy statement: ethical controversies in organ donation after circulatory death. *Pediatrics*, **131**, 1021–1026.

Belkin, G.S. (2003). Brain death and the historical understanding of bioethics. *Journal of the History of Medicine and Allied Sciences*, **58**, 325–361.

Bernat, J.L. (1998). A defense of the whole-brain concept of death. *Hastings Center Report*, **28**(2), 14–23.

Bernat, J.L. (2002). The biophilosophical basis of whole-brain death. *Social Philosophy & Policy*, **19**, 324–342.

Bernat, J.L. (2006). The whole-brain concept of death remains optimum public policy. *Journal of Law, Medicine and Ethics*, **34**, 35–43.

Bernat, J.L. (2008a). *Ethical Issues in Neurology* (3rd edn.). Philadelphia, PA: Lippincott Williams & Wilkins.

Bernat, J.L. (2008b). How can we achieve uniformity in brain death determinations? *Neurology*, **70**, 252–253.

Bernat, J.L. (2010a). How the distinction between "irreversible" and "permanent" illuminates circulatory-respiratory death determination. *Journal of Medicine and Philosophy*, **35**(3), 242–255.

Bernat, J.L. (2010b). How autoresuscitation impacts death determination in organ donors. *Critical Care Medicine*, **38**, 1377–1378.

Bernat, J.L. (2013a). On noncongruence between the concept and determination of death. *Hastings Center Report*, **43**(6), 25–33.

Bernat, J.L. (2013b). The definition of death. In: Bernat, J.L. and Beresford, H.R. (Eds.) *Ethical and Legal Issues in Neurology. Handbook of Clinical Neurology* (3rd Series, Volume 118), pp. 419–435. (Series Eds. Aminoff, M.J., Boller, F., and Swaab, D.F.). Edinburgh: Elsevier.

Bernat, J.L. (2013c). Controversies in defining and determining death in critical care. *Nature Reviews Neurology*, **9**, 164–173.

Bernat, J.L. (2014a). Whither brain death? *American Journal of Bioethics*, **14**(8): 3–8.

Bernat, J.L. (2014b). There can be agreement as to what constitutes human death. In: Caplan, A.L. and Arp, R. (Eds.) *Contemporary Debates in Bioethics*, pp.377–387, 397–398. Chichester, UK: Wiley-Blackwell.

Bernat, J.L. (in press). Defining death. In: Moreman, C. (Ed.) *Routledge Companion to Death and Dying*. London: Taylor & Francis/Routledge.

Bernat, J.L., Culver, C.M., and Gert, B. (1981). On the definition and criterion of death. *Annals of Internal Medicine*, **94**, 389–394.

Bernat, J.L., D'Alessandro, A.M., Port, F.K., et al. (2006). Report of a national conference on donation after cardiac death. *American Journal of Transplantation*, **6**, 281–291.

Bernat, J.L., Capron, A.M., Bleck, T.P., et al. (2010). The circulatory-respiratory determination of death in organ donation. *Critical Care Medicine*, **38**, 963–970.

Bernat, J.L., Bleck, T.P., Blosser, S., et al. (2014). Circulatory death determination in uncontrolled organ donors: a panel viewpoint. *Annals of Emergency Medicine*, **63**, 384–390.

Bleich, J.D. (1989). Of cerebral, respiratory and cardiac death. *Tradition*, **24**(3), 44–66.

Bonelli, R.M., Prat, E.H., and Bonelli, J. (2009). Philosophical considerations on brain death and the concept of the organism as a whole. *Psychiatria Danubina*, **21**, 3–8.

Brock, D.W. (1999). The role of the public in public policy on the definition of death. In: Youngner, S.J., Arnold, R.M., and Schapiro, R. (Eds.) *The Definition of Death: Contemporary Controversies*, pp.293–307. Baltimore, MD: Johns Hopkins University Press.

Burkle, C.M., Schipper, A.M., and Wijdicks, E.F. (2011). Brain death and the courts. *Neurology*, **76**, 837–841.

Byrne, P.A., O'Reilly, S., and Quay, P.M. (1979). Brain death—an opposing viewpoint. *JAMA*, **242**, 1985–1990.

Campbell, C.S. (1999). Fundamentals of life and death: Christian fundamentalism and medical science. In: Youngner, S.J., Arnold, R.M., and Schapiro, R. (Eds.) *The Definition of Death: Contemporary Controversies*, pp.194–209. Baltimore, MD: John Hopkins University Press.

Capron, A.M. (2001). Brain death—well settled yet still unresolved. *New England Journal of Medicine*, **344**, 1244–1246.

Capron, A.M. and Kass, L.R. (1972). A statutory definition of the standards for determining human death: an appraisal and a proposal. *University of Pennsylvania Law Review*, **121**, 87–118.

Chiong, W. (2005). Brain death without definitions. *Hastings Center Report*, **35**(6), 20–30.

Chiong, W. (2014). There cannot be agreement as to what constitutes human death: Against definitions, necessary and sufficient conditions, and determinate boundaries. In: Caplan, A.L. and Arp, R. (Eds.) *Contemporary Debates in Bioethics*, pp.388–396, 399–400. Chichester, UK: Wiley Blackwell.

Choi, E.K., Fredland, V., Zachodni, C., Lammers, J.E., Bledsoe, P., and Helft, P.R. (2008). Brain death revisited: the case for a national standard. *Journal of Law, Medicine and Ethics*, **36**, 824–836.

Citerio, G., Cypel, M., Dobb, D.J., et al. (2016). Organ donation in adults: a critical care perspective. *Intensive Care Medicine*, **42**, 305–315.

Clayton, P. and Kauffman, S.A. (2006). On emergence, agency, and organization. *Biology and Philosophy*, **21**, 501–521.

Cohen, J., Ashkenazi, T., Katvan, E., and Singer, P. (2012). Brain death determination in Israel: the first two years experience following changes to the brain death law—opportunities and challenges. *American Journal of Transplantation*, **12**, 2514–2518.

Daoust, A. and Racine, E. (2014). Depictions of "brain death" in the media: medical and ethical implications. *Journal of Medical Ethics*, **40**, 253–259.

DeGrazia, D. (2011). The definition of death. In: *Stanford Encyclopedia of Philosophy*. Available from: http://plato.stanford.edu/entries/death-definition/ [accessed February 11, 2015].

DeVita, M.A. and Snyder, J.V. (1993). Development of the University of Pittsburgh Medical Center policy for the care of terminally ill patients who may become organ donors after death following the removal of life support. *Kennedy Institute of Ethics Journal,* 3, 131–143.

Diringer, M.N. and Wijdicks, E.F.M. (2001). Brain death in historical perspective. In: Widjicks, E.F.M. (Ed.) *Brain Death,* pp.5–27. New York: Oxford University Press.

Ecker, J.L. (2014). Death in pregnancy—an American tragedy. *New England Journal of Medicine,* 370, 889–891.

Emanuel, L.L. (1995). Re-examining death: the asymptotic model and a bounded zone definition. *Hastings Center Report,* 25(4), 27–35.

Flamm, A.L., Smith, M.L., and Mayer, P.A. (2014). Family members' requests to extend physiologic support after declaration of brain death: a case series analysis and proposed guidelines for management. *Journal of Clinical Ethics,* 25, 222–237.

Fry-Revere, S., Reher, T., and Ray, M. (2010). Death: a new legal perspective. *Journal of Contemporary Health Law and Policy,* 27(1), 1–75.

Furton, E.J. (1999). Reflections on the status of brain death. *Ethics and Medics,* 24(10), 2–4.

Furton, E.J. (2002). Brain death, the soul, and organic life. *National Catholic Bioethics Quarterly,* 2, 455–470.

Gardiner, D., Shemie, S., Manara, A., and Opdam, H. (2012). International perspective on the diagnosis of death. *British Journal of Anesthesia,* 108 (S1), i14-i28.

Gert, B. (2006). Matters of life and death. *Hastings Center Report,* 36(3), 4.

Gervais, K.G. (1986). *Redefining Death.* New Haven, CT: Yale University Press.

Giacomini, M. (1997). A change of heart and a change of mind? Technology and the redefinition of death in 1968. *Social Sciences in Medicine,* 44, 1465–1482.

Gostin, L.O. (2014). Legal and ethical responsibilities following brain death: the McMath and Muñoz cases. *JAMA,* 311, 903–904.

Greer, D.M., Wang, H.W., Robinson, J.D., Varelas, P.N., Henderson, G.V., and Wijdicks, E.F.M. (2016). Variability of brain death policies in the United States. *Neurology,* 73, 213–218.

Haas, J.M. (2011). Catholic teaching regarding the legitimacy of neurological criteria for the determination of death. *National Catholic Bioethics Quarterly,* 11(2), 279–299.

Halevy, A. and Brody, B. (1993). Brain death: reconciling definitions, criteria, and tests. *Annals of Internal Medicine,* 119, 519–525.

Hornby, K., Hornby, L., and Shemie, S.D. (2010). A systematic review of autoresuscitation after cardiac arrest. *Critical Care Medicine,* 38, 1246–1253.

Hovda, P. (2009). What is classical mereology? *Journal of Philosophical Logic,* 38(1), 55–82.

Iltis, A.S. (2015). Organ donation, brain death and the family: valid informed consent. *Journal of Law, Medicine and Ethics,* 43(2), 369–382.

Institute of Medicine (2000). *Non-Heart-Beating Organ Transplantation: Practice and Protocols.* Washington, DC: National Academy Press.

Jain, S. and Maheshawari, M.C. (1995). Brain death—the Indian perspective. In: Machado, C. (Ed.) *Brain Death,* pp.261–263. Amsterdam: Elsevier.

Joffe, A. (2010). Are recent defences of the brain death concept adequate? *Bioethics,* 24(2), 47–53.

Joffe, A., Carcillo, J., Anton, N., et al. (2011). Donation after cardiocirculatory death: a call for a moratorium pending full public disclosure and fully informed consent. *Philosophy, Ethics, and Humanities in Medicine,* 6, 17.

Kass, L. (1971). Death as an event: a commentary on Robert Morison. *Science.* 173. 698–702.

Kennedy, I.M. (1971). The Kansas statute on death—an appraisal. *New England Journal of Medicine,* 285, 946–950.

Koch, C. (2012). *Consciousness: Confessions of a Romantic Reductionist*. Cambridge, MA: MIT Press.

Korein, J. (1978). The problem of brain death: development and history. *Annals of the New York Academy of Sciences*, **315**, 19–38.

Korein, J. (1997). Ontogenesis of the brain in the human organism: definitions of life and death of the human being and person. *Advances in Bioethics* 2: 1–74.

Law Reform Commission of Canada (1981). *Criteria for the Determination of Death*. Ottawa, ON: Law Reform Commission of Canada.

Lee, P. and George, R.P. (2008). *Body–Self Dualism in Contemporary Ethics and Politics*. New York: Cambridge University Press.

Lewis, A., Varelas, P., and Greer, D. (2016). Prolonging support after brain death: when families ask for more. *Neurocritical Care*, **24**(3), 481–487.

Lizza, J.P. (1999). Defining death for persons and human organisms. *Theoretical Medicine and Bioethics*, **20**, 439–453.

Lizza, J.P. (2005). Potentiality, irreversibility, and death. *Journal of Medicine and Philosophy*, **30**, 45–64.

Lizza, J.P. (2009). *Defining the Beginning and End of Life. Readings on Personal Identity and Bioethics*. Baltimore, MD: Johns Hopkins University Press.

Lock, M. (1995). Contesting the natural in Japan: moral dilemmas and technologies of dying. *Culture, Medicine and Psychiatry*, **19**, 1–38.

Lock, M. (2002). *Twice Dead: Organ Transplants and the Reinvention of Death*. Berkeley, CA: University of California Press.

Loeb, J. (1916). *The Organism as a Whole*. New York: GP Putnam's Sons.

McMahan, J. (1995). The metaphysics of brain death. *Bioethics*, **9**(2), 91–126.

McMahan, J. (2002). *The Ethics of Killing: Problems at the Margins of Life*. New York: Oxford University Press.

Magnus, D.C., Wilfond, B.S., and Caplan, A.L. (2014). Accepting brain death. *New England Journal of Medicine*, **370**, 891–894.

Mahner, M. and Bunge, M. (1997). *Foundations of Biophilosophy*. Berlin: Springer-Verlag.

Marquis, D. (2010). Are DCD donors dead? *Hastings Center Report*, **40**(3), 24–31.

Miles, S. (1999). Death in a technological and pluralistic culture. In: Youngner, S.J., Arnold, R.M., and Schapiro, R. (Eds.) *The Definition of Death: Contemporary Controversies*, pp.311–318. Baltimore, MD: Johns Hopkins University Press.

Miller, A.C., Ziad-Miller, A., and Elamin, E.M. (2014). Brain death and Islam: the interface of religion, culture, history, law, and modern medicine. *Chest*, **146**, 1092–1101.

Miller, F.G. and Truog, R.D. (2008). Rethinking the ethics of vital organ donations. *Hastings Center Report*, **38**(6), 38–46.

Molinari, G.F. (1982). Brain death, irreversible coma, and words doctors use. *Neurology*, **32**, 400–402.

Mollaret, P. and Goulon, M. (1959). Le coma dépassé (mémoire préliminaire). *Révue Neurologique*, **101**, 3–15.

Morison, R.S. (1971). Death: process or event? *Science*, **173**, 694–698.

Munjal, K.G., Wall, S.P., Goldfrank, L.R., Gilbert, A., Kaufman, B.J., and Dubler, N.N. (2013). A rationale in support of uncontrolled donation after circulatory determination of death. *Hastings Center Report*, **43**(1), 19–26.

Munshi, L., Dhanini, S., Shemie, S.D., Hornby, L., Gore, G., and Shahin, J. (2015). Predicting time to death after withdrawal of life-sustaining therapy. *Intensive Care Medicine*, **41**, 1014–1028.

Nair-Collins, M. (2015). Taking science seriously in the debate on death and organ transplantation. *Hastings Center Report*, **45**(6), 38–48.

Nakagawa, T.A., Ashwal, S., Mathur, M., et al. (2011). Guidelines for the determination of brain death in infants and children: an update of the 1987 Task Force recommendations. *Critical Care Medicine*, **39**, 2139–2155.

Olick, R.S. (1991). Brain death, religious freedom, and public policy: New Jersey's landmark legislative initiative. *Kennedy Institute of Ethics Journal*, **4**, 275–288.

Olick, R.S., Braun, E.A., and Potash, J. (2009). Accommodating religious and moral objections to neurological death. *Journal of Clinical Ethics*, **20**, 183–191.

Olson, E.T. (1997). The identity of organisms. In: *The Human Animal: Personal Identity Without Psychology*, pp.131–135. New York: Oxford University Press.

Orr, R.D. and Genesen, L.B. (1997). Requests for inappropriate treatment based on religious beliefs. *Journal of Medical Ethics*, **23**, 142–147.

Overby, K.J., Weinstein, M.S., and Fiester, A. (2015). Addressing consent issues in donation after circulatory determination of death. *American Journal of Bioethics*, **15**(8), 3–9.

Padela, A.I., Arozullah, A., and Moosa, E. (2013). Brain death in Islamic ethico-legal deliberation: challenges for applied Islamic bioethics. *Bioethics*, **27**, 132–139.

Pallis, C. (1995). *ABC of Brainstem Death* (2nd edn.). London: British Medical Journal Publishers.

Parnia, S. and Fenwick, P. (2002). Near death experiences in cardiac arrest: visions of a dying brain or visions of a new science of consciousness. *Resuscitation*, **52**, 9–11.

Pernick, M.S. (1999). Brain death in a cultural context: the reconstruction of death 1967–1981. In: Youngner, S.J., Arnold, R.M., and Schapiro, R. (Eds.) *The Definition of Death: Contemporary Controversies*, pp.3–33. Baltimore, MD: Johns Hopkins University Press.

Pinker, S. (1997). *How the Mind Works*. New York: W.W. Norton & Co.

Pontifical Academy of Sciences (2007). *The Signs of Death*. Scripta Varia 110. Vatican City: Pontifical Academy of Sciences.

Pope John Paul II (2000, August 29). *Address of the Holy Father John Paul II to the 18th International Congress of the Transplantation Society, August 29, Rome*. http://www.vatican.va/holy_father/john_paul_ii/speeches/2000/jul-sep/documents/hf_jp-ii_spe_20000829_transplants_en.html [accessed June 4, 2012].

Pope, T.M. (2014). Legal briefing: brain death and total brain failure. *Journal of Clinical Ethics*, **25**(3), 245–257.

President's Commission for the Study of Ethical Problems in Medicine and Biomedical and Behavioral Research (1981). *Defining Death. Medical, Ethical, and Legal Issues in the Determination of Death*. Washington, DC: US Government Printing Office.

President's Council on Bioethics (2008). *Controversies in the Determination of Death: A White Paper by the President's Council on Bioethics*. Washington, DC: US Government Printing Office.

Racine, E., Amaram, R., Seidler, M., Karczuska, M., and Illes, J. (2008). Media coverage of the persistent vegetative state and end-of-life decision-making. *Neurology*, **71**, 1027–1032.

Repertinger, S., Fitzgibbons, W.P., Omojola, M.F., and Brumback, R.A. (2006). Long survival following bacterial meningitis-associated brain destruction. *Journal of Child Neurology*, **21**, 591–595.

Rodriguez-Arias, D., Tortosa, J.C., Burant, C.J., Aubert, P., Aulisio, M.P., and Youngner, S.J. (2013). One or two types of death? Attitudes of health professionals towards brain death and donation after circulatory death in three countries. *Medicine, Health Care and Philosophy*, **16**, 457–467.

Rosner, F. (1999). The definition of death in Jewish law. In: Youngner, S.J., Arnold, R.M., and Schapiro, R. (Eds.) *The Definition of Death: Contemporary Controversies*, pp.210–221. Baltimore, MD: Johns Hopkins University Press.

Rosner, F. and Tendler, M.D. (1989). Definition of death in Judaism. *Journal of Halacha and Contemporary Society*, **17**, 14–31.

Shah, S.K. and Miller, F.G. (2010). Can we handle the truth? Legal fictions in the determination of death. *American Journal of Law & Medicine*, **36**, 540–585.

Shewmon, D.A. (1998). Chronic "brain death:" meta-analysis and conceptual consequences. *Neurology*, **51**, 1538–1545.

Shewmon, D.A. (2001). The brain and somatic integration: insights into the standard biological rationale for equating "brain death" with death. *Journal of Medicine and Philosophy*, **26**, 457–478.

Shewmon, D.A. (2004). The "critical organ" for the organism as a whole: lessons from the lowly spinal cord. *Advances in Experimental Medicine and Biology*, **550**, 23–42.

Shewmon, D.A. (2009). Brain death: can it be resuscitated? *Hastings Center Report*, **39**(2), 18–24.

Shewmon, D.A. (2010). Constructing the death elephant: a synthetic paradigm shift for the definition, criteria, and tests for death. *Journal of Medicine and Philosophy*, **35**, 256–298.

Shewmon, D.A. and Shewmon, E.S. (2004). The semiotics of death and its medical implications. *Advances in Experimental Medicine and Biology*, **550**, 89–114.

Siminoff, L.A., Burant, C., and Youngner, S.J. (2004). Death and organ procurement: public beliefs and attitudes. *Kennedy Institute for Ethics Journal*, **14**, 217–234.

Soloveichik, A. (1979). The Halakhic definition of death. In: Rosner, F. and Bleich, J.D. (Eds.) *Jewish Bioethics*, pp.296–302. New York: Sanhedrin Press.

Spike, J. and Greenlaw, J. (1995). Ethics consultation: persistent brain death and religion: must a person believe in death to die? *Journal of Law, Medicine and Ethics*, **23**, 291–294.

Taylor, R.M. (1997). Reexamining the definition and criterion of death. *Seminars in Neurology*, **17**, 265–270.

Truog, R.D. (1997). Is it time to abandon brain death? *Hastings Center Report*, **27**(1), 29–37.

Veatch, R.M. (1975). The whole brain-oriented concept of death: an outmoded philosophical formulation. *Journal of Thanatology*, **3**, 13–30.

Veatch, R.M. (1993). The impending collapse of the whole-brain definition of death. *Hastings Center Report*, **23**(4), 18–24.

Veatch, R.M. (1999). The conscience clause: how much individual choice in defining death can our society tolerate? In: Youngner SJ, Arnold RM and Schapiro R (eds). *The Definition of Death: Contemporary Controversies*. Baltimore: Johns Hopkins University Press: 137–160.

Veith, F.J., Fein, J.M., Tendler, M.D., et al. (1977a). Brain death I. A status report of medical and ethical considerations. *JAMA*, **238**, 1651–1655.

Veith, F.J., Fein, J.M., Tendler, M.D., et al. (1977b). Brain death II. A status report of legal considerations. *JAMA*, **238**, 1744–1748.

Wahlster, S., Wijdicks, E.F., Patel, P.V., et al. (2015). Brain death declaration: practices and perceptions worldwide. *Neurology*, **84**, 1870–1879.

Wijdicks, E.F.M. (2002). Brain death worldwide. Accepted fact but no global consensus in diagnostic criteria. *Neurology*, **58**, 20–25.

Wijdicks, E.F.M. (2003). The neurologist and the Harvard criteria for brain death. *Neurology*, **61**, 970–976.

Wijdicks, E.F.M. (2011). *Brain Death* (2nd edn.). New York: Oxford University Press.

Wijdicks, E.F., Varelas, P.N., Gronseth, G.S., and Greer, D.M. (2010). American Academy of Neurology. Evidence-based guideline update: determining brain death in adults: report of the Quality Standards Subcommittee of the American Academy of Neurology. *Neurology*, **74**, 1911–1918.

Winkfield v. Rosen. First Amended Complaint for Damages for Medical Malpractice, Case No. Rg 15760730, Superior Court of the State of California for the County of Alameda, filed November 4, 2015.

Yaqub, B.A. and **Al-Deeb, S.M.** (1996). Brain death: current status in Saudi Arabia. *Saudi Medical Journal*, **17**, 5–10.

Youngner, S.J. (1992). Defining death: a superficial and fragile consensus. *Archives of Neurology*, **49**, 570–572.

Youngner, S.J. and **Bartlett, E.T.** (1983). Human death and high technology: the failure of the whole brain formulations. *Annals of Internal Medicine*, **99**, 252–258.

Social, legal, and regulatory frameworks: Lessons of the past guide policy for the future

Chapter 19

Minors and incompetent adults: A tale of two populations

Vasiliki Rahimzadeh, Karine Sénécal,
Erika Kleiderman, and Bartha M. Knoppers

Introduction

Biomedical research enables improvements in diagnosis and treatment of human diseases, and participation in research is the cornerstone of such medical progress. Indeed, the scholarly beginnings of the bioethics field are often attributed to human rights questions concerning the ethics of human participation in research, and which have since influenced every biomedical field from pediatrics (Diekema 2006) to aging (Kim et al. 2001), and genomics (Knoppers 2013) to neurology (Choudhury et al. 2014). The completion of the Human Genome Project in 2003 brought about paradigmatic shifts in the nature and conduct of biomedical research. This shift toward data-intensive science is evident in the ways that analysis, exchange, and reporting now occur increasingly in virtual (e.g., the cloud commons (Stein et al. 2015)) rather than physical spaces.

The participation of humans in research remains nevertheless a central consideration in bioethics, and an increasingly complex area for policy development as research becomes more data intensive and data driven. This is particularly true for the participation of categorically vulnerable populations in biomedical research—such as minors and incompetent adults—who warrant special protections against potential rights violations and exposure to undue risk or, to undue exclusion and deprivation of the benefits of research. In this chapter, we provide both a retrospective and prospective analysis of research involving these two populations with a special focus on the data-intensive sciences such as genomics and its related "omics" disciplines. In doing so, our analysis adopts what we term "reverse vulnerability" as one lens through which to examine the ethical intersectionalities between both populations in an effort to better complement governance strategies to the contemporary realities of data-intensive science and data sharing.

The first and second parts of this chapter provide a policy overview of research participation and the protection of minors and incompetent adults living with dementia. We comment on the practical and theoretical implications of reverse vulnerability to an emerging area of contemporary policy development: international data sharing. A reinvigorated discussion of research participation involving minors and patients living with dementia necessarily precedes, in our view, policy-making for sharing research data.

Historically, the *parens patriae* doctrine was the first to legitimize the legal status of vulnerable persons and the State's obligation to protect them. This legal doctrine stipulates that the government acts as a guardian of all persons legally incapable of acting on their own, even in the absence of specific legislation (Griffith 1991). Such State powers are protective of both property and personal interests, and are usually exercised by the courts. Both children and incompetent adults are legally presumed to be unable to make decisions concerning their health, welfare, or involvement in research. It was not until the *Nuremberg Code* in 1947, and later the *Declaration of Helsinki* in 1964, that the interests of minors and incompetent adults were specifically addressed in medical research.

Ethical principles outlined in the *Nuremberg Code* emphasized protection through exclusion, while the *Declaration* endorsed their inclusion albeit with special protections. Most countries recognize parents or family members as primary decision-making authorities for minors and incompetent adults. Surprisingly, however, neither the 1989 United Nations (UN) Convention on the Rights of the Child, nor the 2006 UN Convention on the Rights of Persons with Disabilities (the latter encompassing incompetent adults such as "those living" with dementia) explicitly addresses their inclusion in research.

From these historically protective stances toward vulnerable populations in research, international ethics norms evolved to adopt more promotional approaches. The European Clinical Trials Directive is but one example that testifies to this evolution, which positively mandates the inclusion of children and incompetent adults in clinical trials in Europe (European Parliament & the Council of the European Union 2014, s. 32(1)).

Table 19.1 summarizes the Directive, as well as other international ethics guidelines with specific mention of research involving vulnerable persons.

International ethics guidelines promote the inclusion of incompetent adults in research as for minors, provided certain special protections. The benefits, either direct or indirect, as a result of their participation justify such inclusion in part, to say nothing of the fact that certain diseases belong to these groups exclusively. Furthermore, improved standards of care for conditions earlier or later in life may not otherwise emerge without the participation of these populations. Data-intensive research involving vulnerable groups, we argue, must reconcile protective and promotional stances to make way for realistic and proportional risk analysis and governance. Only then can we facilitate, rather than obstruct, a future of open science required to advance (personalized) medicine for minors and patients living with dementia, among others.

Children and minors

The history of children in biomedical research has been (paradoxically) marked by both grave human rights abuses as well as groundbreaking clinical progress. Their participation in research has, in turn, long raised ethical concerns. An era of over-protectionism ensued in the wake of research abuses involving children such as those at the Willowbrook State

Table 19.1 International ethics guidelines on the participation of vulnerable persons in medical research

International convention/guidelines/policy	Participation of legally incompetent persons in research	Ethical rationale
World Medical Association (WMA) Declaration of Helsinki (2013) Ethical Principles for Medical Research Involving Human Subjects	**Article 13:** All vulnerable groups should receive specifically considered protection **Article 28:** For a potential research subject who is incapable of giving informed consent, the physician must seek informed consent from the legally authorised representative. These individuals must not be included in a research study that has no likelihood of benefit for them unless it is intended to promote the health of the group represented by the potential subject, the research cannot instead be performed with persons capable of providing informed consent, and the research entails only minimal risk and minimal burden.	**Article 20:** Medical research with a vulnerable group is only justified if the research is responsive to the health needs or priorities of this group and the research cannot be carried out in a non-vulnerable group. In addition, this group should stand to benefit from the knowledge, practices or interventions that result from the research.
Council for International Organizations of Medical Sciences (CIOMS) (2016) International Ethical Guidelines for Health-related Research Involving Humans	**Guideline 15:** Research involving vulnerable persons When vulnerable individuals and groups are considered for recruitment in research, researchers and research ethics committees must ensure that specific protections are in place to safeguard the rights and welfare of these individuals and groups in the conduct of the research.	**Commentary on Guideline 15:** …It is important to recognize that vulnerability involves not only the ability to provide initial consent to participate in research, but also aspects of the ongoing participation in research studies. In some cases, persons are vulnerable because they are relatively (or absolutely) incapable of protecting their own interests. This may occur when persons have relative or absolute impairments in decisional capacity, education, resources, strength, or other attributes needed to protect their own interests. In other cases, persons can also be vulnerable because some feature of the circumstances (temporary or permanent) in which they live makes it less likely that others will be vigilant about, or sensitive to, their interests.

(continued)

Table 19.1 Continued

International convention/guidelines/policy	Participation of legally incompetent persons in research	Ethical rationale
United Nations Declaration on Bioethics and Human Rights (2005) [Articles 7–8]	**Article 7:** Persons without the capacity to consent In accordance with domestic law, special protection is to be given to persons who do not have the capacity to consent: (a) authorization for research and medical practice should be obtained in accordance with the best interest of the person concerned and in accordance with domestic law. However, the person concerned should be involved to the greatest extent possible in the decision-making process of consent, as well as that of withdrawing consent; (b) research should only be carried out for his or her direct health benefit, subject to the authorization and the protective conditions prescribed by law, and if there is no research alternative of comparable effectiveness with research participants able to consent. Research which does not have potential direct health benefit should only be undertaken by way of exception, with the utmost restraint, exposing the person only to a minimal risk and minimal burden and, if the research is expected to contribute to the health benefit of other persons in the same category, subject to the conditions prescribed by law and compatible with the protection of the individual's human rights. Refusal of such persons to take part in research should be respected.	**Article 8: Respect for human vulnerability and personal integrity** In applying and advancing scientific knowledge, medical practice and associated technologies, human vulnerability should be taken into account. Individuals and groups of special vulnerability should be protected and the personal integrity of such individuals respected.

School, Staten Island, New York City, between the late 1950s and early 1970s (Diekema 2006). While the motivation for a protectionist approach was well intentioned, such policies resulted in children's near exclusion from biomedical research generally. The consequences of which resulted in a dearth of pediatric-specific therapies (Fernandez et al. 2003), felt even today as standards of care derived from clinical trial findings are often extrapolated from studies in adults.

Children and adolescents are not miniature adults, but differ both physiologically and psychologically. Thus, pediatric research is essential to developing treatments that are safe and effective for children and adolescents, specifically. Advances in pediatric health research improve the way we understand child and adolescent health, disease, and development, and how these are influenced by factors such as genetics and the environment. Classical tensions related to the involvement of children as vulnerable participants in research are synthesized in Table 19.2, and include how researchers determine the appropriate level of protection, the extent of parental authority and surrogate decision-making, and gauge the developing autonomy of minors. Transformative biotechnologies such as next-generation sequencing instantiate these classic tensions, but also shape new challenges around their responsible deployment and applications in the clinic.

Inclusion

Today, the need to include children and adolescents in research is recognized by international guidelines such as the United Nations Educational, Scientific and Cultural Organisation (UNESCO) *Declaration on the Human Genome and Human Rights* published in 1997 (s. 5(e)) and 2005 (s. 7), the Council for International Organizations of Medical Sciences (CIOMS) *International Ethical Guidelines for Health-Related Research Involving Humans* (2016, s. 3, 17), and the Council of Europe *Convention on Human Rights and Biomedicine* (1997, s. 17(1)(ii)(iii)) and its *Additional Protocol* (2005, s. 15(1)(i)(ii)). Indeed, such guidelines generally indicate that vulnerable persons, such as minors, should be included in research when it is justifiable, when their rights are protected, and when their safety and well-being have been considered.

Overall, minors can be involved in pediatric research when the research cannot be carried out on adults (World Medical Association (WMA) 2013, s. 20); parental consent, as well as the minor's assent (when possible) has been obtained; and the research involves minimal risk. There is a stronger justification for their inclusion when direct clinical benefit is anticipated (World Medical Association (WMA) 1964, s. 17; Council of Europe 1997, s. 17(1)(ii)(2), 2005, s. 15(2); UNESCO 1997, s. 5(e), 2005, s. 7(b); CIOMS 2002, s. 8–9; CIOMS 2016, s. 17). Other guidelines formulate differently their position that "the interventions and procedures should be studied in adults first [. . .], unless the necessary data cannot be obtained without participation of children or adolescents; and the risks must be minimized and no more than minimal." However, "when the social value of the studies with such research interventions and procedures is compelling, and these studies cannot be conducted in adults, a research ethics committee may permit a minor increase above minimal risk" (CIOMS, 2016, s. 17).

Table 19.2 International guidelines for the participation of minors in biomedical research

Guideline	Relevant clause

Council for International Organizations of Medical Sciences (CIOMS) (2016)
 	
International Ethical Guidelines for Health-related Research Involving Humans
	Guideline 17
	
	Children and adolescents must be included in health-related research unless a good scientific reason justifies their exclusion. [T]heir distinctive physiologies and emotional development may also place children and adolescents at increased risk of being harmed in the conduct of research. Moreover, without appropriate support, they may not be able to protect their own interests due to their evolving capacity to give informed consent. Specific protections to safeguard children's rights and welfare in the research are therefore necessary.
	
	◆ Before undertaking research involving children and adolescents, the researcher and the research ethics committee must ensure:
	  • A parent or a legally authorized representative of the child or adolescent has given permission; and
	  • The agreement (assent) of the child or adolescent has been obtained in keeping with the child's or adolescent's capacity, after having been provided with adequate information about the research tailored to the child's or adolescent's level of maturity.
	◆ If children reach the legal age of maturity during the research, their consent to continued participation should be obtained.
	◆ In general, the refusal of a child or adolescent to participate or continue in the research must be respected, unless, in exceptional circumstances, research participation is considered the best medical option for a child or adolescent.
	◆ For research interventions or procedures that have the potential to benefit children or adolescents, the risks must be minimized and outweighed by the prospect of potential individual benefit.
	◆ For research interventions or procedures that have no potential individual benefits for participants, two conditions apply:
	  • The interventions and procedures should be studied in adults first, when these interventions and procedures target conditions that affect adults as well as children and adolescents, unless the necessary data cannot be obtained without participation of children or adolescents; and
	  • The risks must be minimized and no more than minimal.
	◆ When the social value of the studies with such research interventions and procedures is compelling, and these studies cannot be conducted in adults, a research ethics committee may permit a minor increase above minimal risk.

United Nations Human Rights Office of the High Commissioner (1989) United Nations Convention on the Rights of the Child	**Article 12** 1. State Parties shall assure to the child who is capable of forming his or her own views the right to express those views freely in all matters affecting the child, the views of the child being given due weight in accordance with the age and maturity of the child. 2. For this purpose, the child shall in particular be provided the opportunity to be heard in any judicial and administrative proceedings affecting the child, either directly, or through a representative or an appropriate body, in a manner consistent with the procedural rules of national law.

Consent

As children do not have the legal capacity to consent to their own participation in research, international norms generally state that parental consent or permission of the authorized legal representative is required (WMA 2013, s. 28; Council of Europe 1997, s. 6.2, 17(1)(iv), 2005, s. 15(1)(ii); CIOMS 2016, s. 17; UNESCO 2005, s. 7), and that the best interests of the child should be considered in this decision (UNESCO 2003, 2005). They furthermore stipulate what information should appear in the consent form so as to ensure parental consent is fully informed (e.g., the goal and nature of the research, the potential risks and benefits, the right to withdraw, the protection of privacy and confidentiality, the compensation for participation) (Council of Europe 1997, s. 5, 2005, s. 13(2); UNESCO 2005, s. 6(2)). Some guidelines also set more technical requirements, such as adapting consent language in line with the capacity of parents (Council of Europe 2005, s. 13(1); UNESCO 2005, s. 6(2)). Although required, the scope of parental consent can be controversial in particular research contexts. Longitudinal cohort studies or pediatric biobanking testify to this, where consent to research participation is a continuous process that may span a lifetime.

Not only should consent be ongoing throughout the research project, but it should be renewed if significant changes are made to the research protocol (Council of Europe 2005, s. 24(2)). In addition, the CIOMS also states that, "if children reach the legal age of maturity during the research, their consent to continued participation should be obtained" (CIOMS 2016, s. 17). The likelihood requiring reconsent are perhaps greatest for longitudinal studies, where child participants eventually reach the age of majority while still enrolled in the study. A minor's capacity to consent can thus evolve over time, and during the course of the research. As a result, this may necessitate the re-contact of minors once they reach the age of majority, or once they become legally capable of deciding for themselves (Knoppers et al. 2016).

Assent of the child

In addition to parental or legal representative consent, applicable norms and policies surrounding the involvement of children in research consider the emerging maturity of a child, even if children do not have the legal capacity to consent. The 1998 international *Convention on the Rights of the Child* recognizes a child's right to be heard in decision-making despite their inability to consent. Specifically, "[a] child who is capable of forming his or her own views [has] the right to express those views freely in all matters affecting the child, the views of the child being given due weight in accordance with the age and maturity of the child" (United Nations General Assembly 1989, s. 12(1)). Thus, it is equally important for the researcher to obtain the assent of a child who has the capacity to participate at this level prior to inclusion in research (WMA 2013, s. 28–29; Council of Europe 1997, s. 6(2), 2005, s. 15(1)(iv); CIOMS 2016, s. 17, 2008, s. 13–14; UNESCO 2003, s. 8(b)(c), 2005, s. 7(a)). Since 1964, the WMA has adopted this position in the *Declaration of Helsinki* (2013, s. 29), as well as the CIOMS (2016, s. 17), the Council of Europe in the *Convention on Biomedicine* (1997, s. 6(2)) and its *Additional Protocol* (2005).

Despite recognizing the child's assent dependent on age and maturity, this concept is not uniformly defined or determined across clinical contexts. The UNESCO *Universal Declaration on Bioethics and Human Rights* defines assent as the duty to involve a person who is unable to express consent "[…] to the greatest extent possible in the decision-making process of consent, as well as that of withdrawing consent" (2005, s. 7(a)). The International Bioethics Committee of UNESCO identifies the circumstances in which this involvement should occur in their *Report on Consent*: individuals unable to consent "[…] should be involved in the decision-making process according to their age, maturity, and/or degree of capacity to consent" (2008, s. 164).

In addition to the absence of strict criteria, determining a child's capacity can also vary according to the quality or quantity of information given, the research environment, or the relationship with the researcher or the research team. A change in any of these contextual factors could have a significant effect on the child's capacity as a result, which may occur at any point throughout the course of a research project. As with consent, assent is a continuous process that may need to be reconfirmed throughout the duration of the research, especially in the case of longitudinal studies and, more increasingly with genomic data sharing that spans years or decades.

Dissent of the child

All norms governing pediatric research state that the opposition of a child to participate in research (dissent of the child) should be respected (WMA 2013, s. 29; Council of Europe 1997, s. 6(2), 2005, s. 15(1)(iv); CIOMS 2016, s. 17, 2008, s. 13–14; UNESCO 2003, s. 8(b)(c)), even if parental consent has been obtained. Most international norms, however, do not provide further guidance on how to formally acknowledge the child's dissent. Dissent typically requires a minor to possess the same level of capacity that is needed for assent. Thus, if the child is too young, too immature, or unable to understand the nature of the research, his or her dissent may be overridden. The CIOMS (2002), for example, outlines that the dissent of a minor must be respected, unless, in exceptional circumstances, research participation is considered the best medical option for a child or adolescent (s. 17). The return of genetic/genomic results and incidental findings with next-generation sequencing puts these issues of consent/assent and dissent into sharp relief, and will be discussed in depth in the following section ("Return of results and incidental findings").

Return of results and incidental findings

The focus on the return of results and incidental findings in the pediatric context dovetail on the increasing use of next-generation sequencing in the clinic. At the international level, the 2014 P³G (Public Population Project in Genomics and Society) international *Statement on the Return of Whole-Genome Sequencing Results in Paediatric Research,* in consideration of the child's best interests, holds that the potential to return inciden-tal findings should be addressed at the time that informed consent is obtained (i.e., the decision to return results or not should be agreed upon in advance) (Knoppers et al. 2014a). Incidental findings that are "scientifically valid, clinically useful, and reveal conditions that are preventable and actionable during childhood should be offered" (Knoppers et al. 2014a, p.5), while those that relate to an adult-onset disorder should not be returned, so as to preserve the child's future autonomy and decision-making ability. The *Statement* also confirms that the views of the child or adolescent should be consid-ered at the time of consent/assent based on his or her age and maturity (Knoppers et al. 2014a), reinforcing the need to respect the child's evolving decision-making capacities.

In sum, a framework addressing pediatric research exists at the international level. The principles and guidance, however, typically stem from research ethics norms applicable in the clinical setting, and which are not always applicable in the data-intensive research context typified by genomics. The norms included here underscore the need to include children in research, but do not provide specific guidance on how to manage the evolving maturity of children and their capacity to consent to research. While it has been estab-lished that the pediatric population should neither be excluded from medical research nor considered therapeutic orphans (Shirkey 1968, 1999; Rieder & Hazardous Substances Committee 2011), amending these international norms to address contemporary ethical uncertainties, namely in the data-intensive sciences and genomics, is warranted.

Guidance is furthermore lacking with respect to how these uncertainties in the data-intensive sciences relate to incompetent adults living with dementia. Such persons' inabil-ity to consent to research participation is an ethical intersectionality they share with children and minors. We explore this intersectionality, which we define as "reverse vul-nerability" in further depth in the next section ("Incompetent adults").

Incompetent adults

As the global population ages, so too has there been a steady increase in dementia and dementia-related diseases across high-, middle-, and low-income countries (Brookmeyer et al. 2007). With nearly 7.7 million new cases per year (World Health Organization 2012, 2016), there is considerable clinical demand (Fox & Petersen 2013; Ngandu et al. 2015) and political pressure for innovative research with curative goals (Department of Health & Prime Minister's Office 2013; Canadian Institutes of Health Research 2014). Timely diagnosis, health-related quality of life, and innovation in new therapies for patients living with dementia, however, are markedly lacking. Advances in genomics hold great promise toward improving patient outcomes through drug discovery, eluci-dating risk reduction strategies, and slowing disease progression. To realize this prom-ise, research priorities, data governance mechanisms, and alternative frameworks for consent are needed. In particular, collaboration between dementia research and care, and improvement in the accessibility of genomic and health data should be promoted across borders. Respect for persons and data protections for patients living with dementia have both emerged as ethical priorities in turn. Perhaps more acute than in the pediatric context, the scope of decision-making authority among legally authorized representatives is increasingly becoming a barrier to the sharing of incompetent adults' research data.

Inclusion

Similar to minors, the inclusion of incompetent adults living with dementia in research is protected under international conventions and guidelines summarized in Table 19.3. While these conventions and guidelines acknowledge that incompetent adults warrant special protections in research—and some explicitly identify living with dementia as a hallmark scenario of vulnerability in adults—the guidelines differ in their management of vulnerability and types of research permissible as determined by level of risk. For example, the *Declaration of Helsinki* stipulates that research with vulnerable populations is "only justified if the research is responsive to the health needs or priorities of this group" (WMA 2013, s. 20) while the CIOMS guidelines outline a set of criteria for determining ethically appropriate participation (2016, s. 17). The UNESCO *Declaration on Bioethics and Human Rights* (2005) is unique in this regard in two ways. First, it invokes concepts of both minimal risk as well as minimal burden in rationalizing the participation of vulnerable groups. Second, the human rights orientation of these guidelines marry the concepts of personal integrity and best interests standards upon which the decision to participate in research should be based for groups such as incompetent adults. Legally authorized representatives (LARs) are the primary shareholders of these best interests on behalf of incompetent adults, as well as for minors.

Legally authorized representatives

One emerging area of comparative policy interest relates to the scope of substitute decision-making authority to share research data derived from research with incompetent adults living with dementia. Markedly lacking in the consensus guidelines compared here is which individuals are eligible to serve as appropriate substitute decision-makers for incompetent adults in the research context. The limited guidance available has drawn primarily from the clinical context to date. For the purposes of care decisions, the substitute is often legally determined through the appointment of a LAR in most jurisdictions. One study found, however, that court-recognized LARs may in fact impede participation in dementia research in some European countries (Galeotti et al. 2012). Similar debates as to who may serve as LARs and the extent of their authority are underway in the United States (Derse & Spellecy 2015; Yarborough 2015) and Canada (Wildeman et al. 2013), where scholars are interrogating whether policies should "expand the concept of durable power of attorney for health care to include research participation to facilitate substituted judgments" (Taylor et al. 2015, p.64).

Table 19.3 International guidelines for the participation of incompetent adults living with dementia in biomedical research

Guideline	Relevant clause
Council for International Organizations of Medical Sciences (CIOMS) (2016) International Ethical Guidelines for Health-related Research Involving Humans	**Guideline 16: Research involving adults incapable of giving informed consent** Adults who are not capable of giving informed consent must be included in health-related research unless a good scientific reason justifies their exclusion. As adults who are not capable of giving informed consent have distinctive physiologies and health needs, they merit special consideration by researchers and research ethics committees. At the same time, they may not be able to protect their own interests due to their lack of capacity to provide informed consent. Specific protections to safeguard the rights and welfare of these persons in research are therefore necessary. Before undertaking research with adults who are not capable of giving informed consent, the researcher and the research ethics committee must ensure that: ♦ A legally authorized representative of the person who is incapable of giving informed consent has given permission and this permission takes account of the participant's previously formed preferences and values (if any); and ♦ Assent of the subject has been obtained to the extent of that person's capacity, after having been provided with adequate information about the research at the level of the subject's capacity for understanding this information. If participants become capable of giving informed consent during the research, their consent to continued participation must be obtained. In general, a potential participant's refusal to enrol in the research must be respected, unless, in exceptional circumstances, research participation is considered the best available medical option for an individual who is incapable of giving informed consent. If participants have made advance directives for participation in research while fully capable of giving informed consent, the directives should be respected. For research interventions or procedures that have the potential to benefit adults who are incapable of giving informed consent, the risks must be minimized and outweighed by the prospect of potential individual benefit. For research interventions or procedures that have no potential individual benefits for participants, two conditions apply: ♦ The interventions and procedures should be studied first in persons who can give consent when these interventions and procedures target conditions that affect persons who are not capable of giving informed consent as well as those who are capable, unless the necessary data cannot be obtained without participation of persons who are incapable of giving informed consent; and ♦ The risks must be minimized and no more than minimal When the social value of the studies with such research interventions and procedures is compelling, and these studies cannot be conducted in persons who can give informed consent, a research ethics committee may permit a minor increase above minimal risk.

(continued)

Table 19.3 Continued

Guideline	Relevant clause
	Commentary on Guideline 16 [...]A person may be incapable to give informed consent for a variety of reasons (for example, dementia, some psychiatric conditions and accidents). Persons can become capable of giving informed consent after a certain period, or they can be incapable to decide whether they should be treated for a certain disease but capable to decide whether they want to enjoy a meal. This illustrates that a lack of decisional capacity is time-, task- and context-specific
Political Declaration and Madrid International Plan of Action on Ageing (United Nations Second World Assembly on Ageing 2002)	12(j) Harnessing of scientific research and expertise and realizing the potential of technology to focus on, inter alia, the individual, social and health implications of ageing, in particular in developing countries; 75(e) Encourage, at all levels, arrangements and incentives to mobilize commercial enterprises, especially pharmaceutical enterprises, to invest in research aimed at finding remedies that can be provided at affordable prices for diseases that particularly afflict older persons in developing countries and invite the World Health Organization to consider improving partnerships between the public and private sectors in the area of health research 86(b) Develop, where appropriate, effective strategies to increase the level of quality assessment and diagnosis of Alzheimer's and related disorders at an early stage. Research on these disorders should be undertaken on a multidisciplinary basis that meets the needs of the patient, health professionals and carers;
United Nations Principles for Older Persons (United Nations General Assembly 1991)	**Article 7** Older persons should remain integrated in society, participate actively in the formulation and implementation of policies that directly affect their well-being and share their knowledge and skills with younger generations. **Article 8** Older persons should be able to seek and develop opportunities for service to the community and to serve as volunteers in positions appropriate to their interests and capabilities.

Consent, assent, and dissent

The ethical significance of assent is, as in the case of children, transferable to research with incompetent adults (Black et al. 2010). Mild to moderate cognitive and memory-impaired adults have been shown to meaningfully engage in discussions of research participation with researchers and their substitute decision-makers (Kim et al. 2004, 2011; Karlawish 2008). These studies importantly substantiate the involvement of such persons in research participation decisions as their abilities will allow. In incompetent adults living with dementia, their cognitive decline can be gradual, sudden, or episodic in accordance with the severity of their disease. Decision-making capacity may therefore be task, time, and context dependent.

Taken together, these decisional capacities should be evaluated on a continuous basis, and a LAR identified early in the dementia trajectory. According to the conventions and guidelines outlined in Table 19.3, adults living with dementia who are deemed incompetent should be granted the opportunity to assent or dissent to participation in research before becoming incompetent. A LAR may override a decision in cases where the participant has not stated an express wish to be involved in research, or when their participation would constitute greater than minimal risk. Like for children, the best interests standard varies considerably by jurisdiction and in how it is invoked to justify one's inclusion or exclusion from research, particularly if no direct clinical benefit is anticipated. *The United Nations Convention on the Rights of Persons with Disabilities* (2008) safeguards the right to engage in political, social, and cultural life, which could be interpreted to include participation in research as an exercise of civic engagement. The *Convention* protects the rights to dignity, autonomy, independence, and participation in society. We argue that decision-making associated with the type(s) of research participation, as well as the sharing of research data should be considered extensions of the latter participatory right protected under the *Convention*.

Examining ethical intersectionality in practice: Return of results and incidental findings in Canada

The Canadian College of Medical Genetics (CCMG) implicitly adopted the ethical intersectionalities of context, time, and task described herein between minors and incompetent adults in its position on the return of results (Boycott et al. 2015). Although the CCMG guidelines were drafted for application in clinical settings, they are useful for consideration in the research context as well. This is increasingly true as genomics/genetic research informs evidence-based practices, for example, and more firmly integrates into routine clinical care, diagnostics, and personalized therapies.

The CCMG *Professional and Ethical Guidelines* (Boycott et al. 2015) establish two responsibilities for the return of actionable results when they involve incompetent adults:

◆ Ensure the patient's best interest and appropriate level of understanding when conveying information to patients.

◆ Disclose all clinically relevant information to patients unless specifically instructed not to do so by the patient.

The decision to return results and incidental findings is often tailored to one of five circumstances that typify the involvement of vulnerable persons in genetic research. Figure 19.1 provides an overview of these five circumstances, and the corresponding guideline concerning the return of results. Based on the CCMG guidelines, results should be returned when incompetent adults have expressed wishes to this effect, or if the results are medically actionable for the patient in line with their best interest as is the case with children and minors.

Researchers are also implicated in the fulfillment of a participant's best interest, particularly when they express prior wishes related to the return of results/incidental findings or participation in research generally. As Figure 19.1 illustrates, participants' wishes may be unknown. In this case, researchers along with substitute decision-makers are charged with determining the appropriateness of returning results based on the clinical actionability of the result and the participant's best interest(s).

CCMG

guidelines for return of results and incidental findings for children and incompetent adults in research.

RESEARCHER

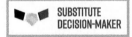

SUBSTITUTE DECISION-MAKER

Incidental findings that reveal highly penetrant, medically actionable results during childhood:

 RESEARCHER SHOULD RETURN RESULTS.

A child's risk for *adult-onset* genetic conditions should <u>not</u> be communicated unless:

1) The parents request disclosure; and

2) Disclosure of the information could prevent serious harm to the health of a parent or family member.

Participant is *competent* both at the time of the signing of the consent and when the medically actionable results are found, and has expressed a wish to receive such results:

 RESEARCHER SHOULD RETURN MEDICALLY ACTIONABLE RESULTS.

Participant is *competent* both at the time of the signing of the consent and when the medically actionable results are found, and has expressed a wish <u>not</u> to receive such results:

RESEARCHER SHOULD NOT RETURN MEDICALLY ACTIONABLE RESULTS.

Participant expressed <u>no preferences</u> to receiving medically actionable results when *competent*:

RESEARCHER IS OBLIGATED TO RETURN MEDICALLY ACTIONABLE RESULTS TO THE SUBSTITUTE DECISION-MAKER;

SUBSTITUTE DECISION-MAKER IS OBLIGATED TO ACT AS PROXY FOR THE BEST INTERESTS OF THE PARTICIPANT.

Participant is *incompetent*, but expressed an explicit refusal to be re-contacted with medically actionable results when competent:

SUBSTITUTE DECISION-MAKER AND RESEARCHER ARE BOUND BY THE PARTICIPANT'S WISHES, AND THE RESEARCHER SHOULD NOT RETURN MEDICALLY ACTIONABLE RESULTS.

Participant was *incompetent* at the time when the consent was signed; the researcher in a position to decide whether to return highly-penetrant, medically actionable results:

SUBSTITUTE DECISION-MAKER SHOULD AGREE TO RECEIVE MEDICALLY ACTIONABLE RESULTS ACCORDING TO PARTICIPANT'S BEST INTERESTS.

Figure 19.1 CCMG guidelines for return of results and incidental findings for children and incompetent adults in research.

Challenges related to feasibility of re-contact for the disclosure of medically action-able incidental findings remain a challenge, as is true for disclosure to child participants and their parents, yet for different reasons. Incompetent adults lose "task" capacity that prevents them from altering previously stated preferences. In contrast, the feasibility challenges associated with re-contacting minors centers on the fact that minors *gain* task capacity to make decisions to participate in research, which may be considerably different from their parents' original consent. In addition to the familiar (or sometimes legal) chal-lenges associated with identifying a LAR, Canadian guidelines are furthermore compli-cated by provincial differences in LAR policies (see, for example, Thorogood et al. 2016).

Summary

As illustrated in this chapter, minors and incompetent adults are considered situationally vulnerable within a research context. It is as a result of this vulnerability that special pro-tections in research are justified. The nuances of such vulnerability, however, have impor-tant implications for laying a responsible governance framework for sharing research data involving both minors and incompetent adults living with dementia. Lange and col-leagues have introduced the "situationality concept" for nuancing the typology of vulner-ability that patients with dementia, and other vulnerable groups with limited cognitive capacity may experience (Lange et al. 2013). It aligns furthermore with the framing of vulnerability in the CIOMS (2016) guidelines: "[o]ne widely accepted criterion of vulner-ability is limited capacity to consent or decline to consent to research participation" (com-mentary, s. 15). The limitations to capacity can be task, time, or context specific. In this way, the situational vulnerability that emerges from an inability to consent to research is a useful ethical intersectionality upon which a policy for genomic data sharing involving minors and incompetent adults living with dementia can build. Both populations share task-oriented deficiencies in their capacity to make research participation decisions, but differ in time- and context-oriented capacities. The decisional capacities of these two pop-ulations and their substitute decision-makers, as well as the special ethical protections the research community affords them, differ in their temporal and contextual specificities. Each component of capacity—time, task, and context—will be next compared for minors and incompetent adults living with dementia, respectively.

First, the situational vulnerability of minors with respect to decision-making in research is inversely related to their burgeoning autonomy. The convergence point heralding the end of minors' situational vulnerability and societal recognition of their decision-making ability is legally benchmarked upon reaching the age of majority. Whereas minors mature *out of* their situational vulnerability that arises from a tem-porary inability to consent to research (and in many cases clinical care), incompetent adults living with dementia can regress *into* situational vulnerability commensurate with their cognitive decline. The authors define this phenomenon "reverse vulnerability," a compelling ethical consideration when charting the ethics of responsible genomic and health-related data sharing.

Second, in the case of minors, parents are charged with deciding whether, and to what extent, to share their child's genomic data. Research suggests parents make this decision in line with their understanding of informational risks that data sharing poses for their children today, as they mature into adults, as well as their altruistic intentions to support further research through secondary data use.

Third, the ethical significance of task-related capacity can be directly compared between persons living with dementia and minors. That is, the cognitive tasks associated with providing informed consent to research require the same facilities and reasoning regardless of whether an incompetent adult person with dementia or a minor makes this decision. To make a decision to participate in research, a number of task-specific capacities are required. These include that prospective participants understand the nature and purpose of the investigation, they acknowledge the roles and activities necessary to participate, and they appreciate the immediate and future implications of their participation. Singh (2007), however, proposes a shift in the ethical weight researchers should attribute the task-specific capacities of vulnerable groups such as children and incompetent adults. She emphasizes the communicative process, rather than a strictly comprehension-oriented definition of task-specific capacity is relevant both to clinical and research decision-making:

> Children's task-specific capacities defy their characterization as vulnerable, incapacitated patients who are fully dependent on surrogate decision-makers ... children must have the capacity to discuss their understanding of diagnosis and treatment. This understanding need not be correct to be interesting and informative; therefore the capacity is specific to the task of communication, as opposed to understanding. (Singh 2007, p.S36)

Finally, surrogate decision-makers charged with ensuring the best interests of minors and incompetent adults are motivated by different decisional outcomes. These differences can provide some clarity to the contextual distinction in decision-making capacities between the two populations. Parents make decisions on behalf of their child temporarily, and with an overarching motivation to protect as a means of fostering independency in the future. Care providers or other legally authorized representatives also make decisions with a chiefly protective aim, yet do so as an exercise of managing the incompetent adult's gradual dependency on others.

Sharing research data: A dual imperative framework

Ethics governance of sharing research data involving minors and incompetent adults can draw from the ethical intersectionalities that arise from their shared sources of vulnerability to fulfill dual imperatives in the biomedical research endeavor. The first imperative is a scientific one, in which the sharing of research data is required to adequately power statistical associations between genetics and human etiologies of disease, and to provide care for particular health needs at all stages of life. The second imperative is an ethical one, in which only through sharing research data can the benefits of new scientific knowledge offset the anticipated risks. Knoppers and others have written elsewhere extensively on the human rights underpinnings of this second ethical imperative (Knoppers et al. 2014b). International support for data sharing resulted in the founding of which the Global Alliance for Genomics and Health, and adoption of its *Framework for the Responsible Sharing of Genomic and Health-Related Data* (Knoppers 2014). Founded on the right of all citizens to benefit from scientific progress (General Assembly of the United Nations 1948, s. 47), the *Framework* promotes a positive obligation on governments to act in respect of this right. For minors and incompetent adults who require LARs to act on their behalf, data sharing for research that is in their interest should be encouraged in light of their specific needs. A proportionate risk analysis that is based on the possible occurrence of real risks and the actual probability of their occurrence argues in favor of their inclusion in international data sharing initiatives. In line with the human rights foundations upon which the Global Alliance was founded, this chapter contends that the ethical intersectionalities between children and incompetent adults helps orient research and data governance that is better tailored to the contemporary challenges facing them in the post-genomic era. Policy collaboration and empirical policy research are needed if vulnerable populations are to continue benefiting from the fruits of scientific progress made possible through concerted data sharing efforts.

Acknowledgments

The authors wish to thank the Vanier Canada Graduate Scholarship (CIHR), the Canada Research Chair in Law and Medicine, Ministère de l'Économie, de la Science et de l'Innovation du Québec, PSR-SIIRI-850 (Canada), the Fond de Recherche du Québec-Santé, and the Québec Network of Applied Genetic Medicine (RMGA) for their funding support. The excerpt from Article 12 was reproduced from "Convention on the Rights of the Child," by United Nations Human Rights Office of the High Commissioner, © 1989/1990 United Nations. Reprinted with the permission of the United Nations.

References

Black, B.S., Rabins, P.V., Sugarman, J., and Karlawish, J.H. (2010). Seeking assent and respecting dissent in dementia research. *American Journal of Geriatric Psychiatry*, 18(1), 77–85.

Boycott, K., Hartley, T., Adam, S., et al. (2015). The clinical application of genome-wide sequencing for monogenic diseases in Canada: position statement of the Canadian College of Medical Geneticists. *Journal of Medical Genetics*, 52(7), 431–7.

Brookmeyer, R., Johnson, E., Ziegler-Graham, K., and Arrighi, H.M. (2007). Forecasting the global burden of Alzheimer's disease: forecasting the global burden. *Alzheimer's & Dementia*, 3(3), 186–191.

Canadian Institutes of Health Research (2014). *Second Global Dementia Legacy Event Harnessing the Power of Discoveries: Maximizing Academia-Industry Synergies*. Ottawa, ON: Canadian Institutes of Health Research.

Choudhury, S., Fishman, J.R., McGowan, M.L., and Juengst, E.T. (2014). Big data, open science and the brain: lessons learned from genomics. *Frontiers in Human Neuroscience*, 8, 239.

Council for International Organizations of Medical Sciences (CIOMS) and the World Health Organization (WHO) (2016). *International Ethical Guidelines for Health-Related Research Involving Humans*. Geneva, Switzerland: COMS.

Council of Europe (1997). *Convention on Human Rights and Biomedicine.* Oviedo, Spain: Council of Europe.

Council of Europe (2005). *Additional Protocol to the Convention on Human Rights and Biomedicine, concerning Biomedical Research.* Strasbourg, France: Council of Europe.

Department of Health and Prime Minister's Office (2013). *G8 Dementia Summit Agreement.* London: Department of Health.

Derse, A. and Spellecy, R. (2015). Ethical and regulatory considerations regarding enrollment of incompetent adults in more than minimal risk research as compared with children. *American Journal of Bioethics,* 15(10), 68–69.

Diekema, D.S. (2006). Conducting ethical research in pediatrics: a brief historical overview and review of pediatric regulations. *Journal of Pediatrics,* 149(Suppl. 1), S3–11.

European Parliament and the Council of the European Union (2014). *EU Clinical Trials Directive 2001/20/EC., L 269(September 2000),* 1–15.

Fernandez, C., Kodish, E., and Weijer, C. (2003). Informing study participants of research results: an ethical imperative. *IRB: Ethics & Human Research,* 25(3),12–9.

Fox, N.C. and Petersen, R.C. (2013). The G8 Dementia Research Summit – a starter for eight? *The Lancet,* 382(9909), 1968–1969.

Galeotti, F., Vanacore, N., Gainotti, S., et al. (2012). How legislation on decisional capacity can negatively affect the feasibility of clinical trials in patients with dementia. *Drugs & Aging,* 29(8), 607–14.

General Assembly of the United Nations (1948). *Universal Declaration of Human Rights.* Paris: UN General Assembly.

Griffith, D.B. (1991). The best interests standard: a comparison of the state's parens patriae authority and judicial oversight in best interests determinations for children and incompetent patients. *Issues in Law and Medicine,* 7(3), 283–338.

Karlawish, J. (2008). Measuring decision-making capacity in cognitively impaired individuals. *Neurosignals,* 16(1), 91–98.

Kim, S.Y.H., Caine, E.D., Currier, G.W., Leibovici, A., and Ryan, J.M. (2001). Assessing the competence of persons with Alzheimer's disease in providing informed consent for participation in research. *American Journal of Psychiatry,* 158(5), 712–717.

Kim, S.Y.H., Appelbaum, P.S., Jeste, D.V., and Olin, J.T. (2004). Proxy and surrogate consent in geriatric neuropsychiatric research: update and recommendations. *American Journal of Psychiatry,* 161(5), 797–806.

Kim, S.Y.H., Karlawish, J.H., Kim, H.M., Wall, I.F., Bozoki, A.C., and Appelbaum, P.S. (2011). Preservation of the capacity to appoint a proxy decision maker: implications for dementia research. *Archives of General Psychiatry,* 68(2), 214–20.

Knoppers, B.M. (2013). Genomics: from persons to populations and back again. *Genome,* 56(10), 537–9.

Knoppers, B.M. (2014a). Framework for responsible sharing of genomic and health-related data. *The HUGO Journal,* 8(1), 3.

Knoppers, B.M., Avard, D., Sénécal, K., and Zawati, M.H. (2014b). Return of whole-genome sequencing results in paediatric research: a statement of the P3G international paediatrics platform. *European Journal of Human Genetics,* 22(1), 3–5.

Knoppers, B.M., Harris, J.R., Budin-Ljøsne, I., and Dove, E.S. (2014c). A human rights approach to an international code of conduct for genomic and clinical data sharing. *Human Genetics,* 133, 895–903.

Knoppers, B.M., Sénécal, K., Boisjoli, J., et al. (2016). Recontacting pediatric research participants for consent when they reach the age of majority. *IRB: Ethics & Human Research,* 38(6)

Lange, M.M., Rogers, W., and Dodds, S. (2013). Vulnerability in research ethics: a way forward. *Bioethics*, **27**(6), 333–40.

Ngandu, T., Lehtisalo, J., Solomon, A., et al. (2015). A 2 year multidomain intervention of diet, exercise, cognitive training, and vascular risk monitoring versus control to prevent cognitive decline in at-risk elderly people (FINGER): a randomised controlled trial. *The Lancet*, **385**(9984), 2255–2263.

Rieder, M.J. and Hazardous Substances Committee (2011). Drug research and treatment for children in Canada: a challenge. *Paediatrics & Child Health*, **16**(9), 560–561.

Shirkey, H. (1968). Editorial comment: therapeutic orphans. *Journal of Pediatrics*, **72**(1), 119–120.

Shirkey, H. (1999). Editorial comment: therapeutic orphans. *Pediatrics*, **104**(Supplement 3), 583–584.

Singh, I. (2007). Capacity and competence in children as research participants. *EMBO Reports*, **8**(Special Issue), S35–S39.

Stein, L.D., Knoppers, B.M., Campbell, P., Getz, G., and Korbel, J.O. (2015). Data analysis: create a cloud commons. *Nature*, **523**(7559), 149–151.

Taylor, H.A., Kuwana, E., and Wilfond, B.S. (2015). Is it ethical to enroll cognitively impaired adults in research that is more than minimal risk with no prospect of benefit? *American Journal of Bioethics*, **15**(10), 64–65.

Thorogood, A., Deschênes St-Pierre, C., and Knoppers, B. M. (2016). Substitute consent to data sharing: a way forward for international dementia research? *Oxford Journal of Law and Biosciences*, Available from: 1–26. http://doi.org/10.1093/jlb/lsw063.

United Nations Educational, Scientific and Cultural Organisation (UNESCO) (1997). *Universal Declaration on the Human Genome and Human Rights*. Paris, France:UNESCO.

United Nations Educational, Scientific and Cultural Organisation (UNESCO) (2003). *International Declaration on Human Genetic Data*. Paris, France:UNESCO.

United Nations Educational, Scientific and Cultural Organisation (UNESCO) (2005). *Universal Declaration on Bioethics and Human Rights*. Paris, France:UNESCO.

United Nations Educational, Scientific and Cultural Organisation (UNESCO) (2008). *International Bioethics Committee, Report of the International Bioethics Committee of UNESCO (IBC) on Consent*. Paris, France:UNESCO.

United Nations General Assembly (1989). *United Nations Convention on the Rights of the Child* (UN Document A/RES/44/25). New York: United Nations.

United Nations General Assembly (1991). *United Nations Principles for Older Persons, Resolution 46/91* (UN Document A/RES/46/91). New York: United Nations.

United Nations Second World Assembly on Ageing (2002). *Political Declaration and Madrid International Plan of Action on Ageing*. New York: United Nations.

Wildeman, S., Dunn, L.B., and Onyemelukwe, C. (2013). Incapacity in Canada: review of laws and policies on research involving decisionally impaired adults. *American Journal of Geriatric Psychiatry*, **21**(4), 314–325.

World Health Organization (2012). *Dementia: A Public Health Priority*. Geneva: World Health Organization.

World Health Organization (2016). *Dementia Fact Sheet*. Available from: http://www.who.int/mediacentre/factsheets/fs362/en/ [accessed February 21, 2017].

World Medical Association (2013). *Declaration of Helsinki: Ethical Principles for Medical Research Involving Human Subjects. Amended by the 64 General Assembly, Fotaleza, Brazil*. World Medical Association.

Yarborough, M. (2015). Inconsistent approaches to research involving cognitively impaired adults: why the broad view of substituted judgment is our best guide. *American Journal of Bioethics*, **15**(10), 66–67.

Chapter 20

Behavioral and brain-based research on free moral agency: Threatening or empowering?

Eric Racine and Veljko Dubljević

Ipsa scientia potestas est [Knowledge is power].

(Francis Bacon 1597)

Introduction

The belief that people are free human beings, able to undertake actions based on will and desires, is central to much explanation of human behavior and to a broad set of social practices such as law, ethics, and politics. Indeed, threats to human freedom[1] and the ability to act according to individual beliefs and preferences can lead to serious conflict and is still the object of struggles of oppressed linguistic, ethnic, religious, and racial minorities throughout the world. It is often stated that there is nothing more precious than a person's freedom and this is well reflected in national constitutions where political freedoms and rights serve as the foundations of social and political orders. One of the iconic figures of the fight for human freedom in the twentieth century, Nelson Mandela, put forth in his famous discourse "No Easy Walk to Freedom" in 1953 to the African National Congress, in the context of the fight against the regime of apartheid that "[t]o overthrow oppression has been sanctioned by humanity and is the highest aspiration of every free man" (Mandela 2011). In a similar mind-frame, the influential moral and political thinking of the Enlightenment stressed that freedom was intrinsic to the definition of humans as social and political beings such that, in the words of Rousseau, humans had an intrinsic motivation to safeguard freedom because freedom defines human beings to the point of representing a duty to uphold this freedom for themselves and for others (Rousseau 1992).[2] In a nutshell, freedom has long been considered intrinsic to what people value in individual and collective life and constitutes a precondition for the enjoyment of many goods. Accordingly, the desire to be free has propelled democratic revolutions and social movements throughout history.

In contrast to this deep commitment to human freedom in the humanities and in society, and its omnipresence in individual life, domestic affairs, and international relationships, neuroscience has been heralded as a game-changing field which will radically change how individuals view themselves, including their beliefs in human freedom. Such changes have been captured under the concept of a neuroscience revolution (Wolpe 2002), the idea that neuroscience could reveal another view of humankind which potentially contradicts ordinary, non-scientifically informed views of ourselves. The claim is enormous, as the reader can easily guess. If freedom is a kind of illusion and human beings are ready to fight and even die in its name, are they then lured and wronging themselves in pursuing such a quasi-sacred good?

If the interpretation of some famous studies and neuroscientists is taken seriously, it would seem to be the case. Starting with the research of Libet in the 1980s (Libet et al. 1982, 1983), a stream of studies has been interpreted to suggest that free will—a basic descriptor of human freedom that we will come back to in "Behavioral and brain-based research as a threat to free moral agency"—does not exist. Perhaps human beings have at best some veto power allowing them to stop an action or thought already initiated by the brain but nothing much more (Libet et al. 1983). Others have argued that perhaps free will is simply an illusion (Wegner 2002), albeit a useful illusion which allows us to make sense of agency, but nevertheless an ill-founded concept, a concept that does not carve nature at its joints. Perhaps human beings just simply do not have any kind of freedom, and beliefs in freedom are supported by quirky metaphysics (Haggard, cited in Smith 2011). All these claims merit closer scrutiny, but they set an enormous challenge to the vast array of social practices that rely on some fundamental beliefs in human freedom. Adding to the importance of these interpretations of neuroscience research is the fact that the disqualification of human freedom by neuroscience has become lore in as much as, for example, the media has conveyed repeatedly that the existence of human freedom has been invalidated based, for example, on Libet and colleagues' famous study (Racine et al. 2016). Greene and Cohen have proposed that social practices such as ethics and law be extensively revised to reflect challenges to human freedom (Greene & Cohen 2004).

In this chapter, we examine some of the basic claims that challenge beliefs in free moral agency,[3] an expression we propose to encompass different abilities[4] of the moral agent often captured with different concepts such as free will, freedom, and autonomy. (These concepts are often included in the single term of freedom, broadly understood, as already suggested.) We strive to provide a generous understanding of the underlying points of views suggesting, for example, that decisions are not made consciously and voluntarily but are actually predetermined by prior brain activity, as revealed by, for example, readiness potentials. Other research has challenged how conscious and accurate human beings actually are in making choices and decisions which are believed to be free but are actually influenced by a host of implicit, nonconscious mechanisms (Bargh & Chartrand 1999). Although there are methodological caveats to what has been published in neuroscience with respect to this issue (e.g., about Libet's research which we and others have scrutinized (Dubljević 2013; Saigle et al. 2015)), this chapter looks at this problem from a conceptual and normative standpoint without engaging in a detailed methodological review of such claims. It assumes that some of the results of this research hold as true or at least can be taken as a basis for a discussion about different aspects of free moral agency.

In this respect, we review threats to different aspects of free moral agency and from various angles: some neuroscience experiments seem to threaten basic tenets about the ability to choose; social psychology seems to dispute freedom by providing environmental constraints in which a person is enacting choices; and research into microbiota and unconscious bias could be considered as an attack to autonomy. However, by having a second look at some of the interpretations of these results taken as a threat to free moral agency, we find that they have relied on ill-founded assumptions about moral agency and its relationship to empirical knowledge, such as the knowledge yielded by the behavioral and brain sciences. Accordingly, in the second part of the chapter—based on pragmatist theories of voluntary action and the concept of "contextualized autonomy" (Racine & Dubljević 2016; Racine et al. 2017) as well as recent work in social psychology which both stress the dynamic nature of moral agency—we reconsider a common objectivist and essential stance toward free moral agency and an equally common dichotomist interpretation of the fact–value (is–ought) tension which are at the center of these problematic interpretations. In contrast, we propose that knowledge can empower moral agents granted that there is, first, a genuine first-person experience and intersubjective recognition of the abilities captured within the concept of free moral agency. Second, the value of the experience of free moral agency is not jeopardized by the existence of third-person, behavioral, and brain-based research about free moral agency, but the understanding of this experience can actually be retooled and enriched by such knowledge provided that the fact–value dichotomy is reconsidered to reflect a more moderate tension or continuum.

Behavioral and brain-based research as a threat to free moral agency

The contribution of behavioral and brain research to the understanding of moral agency (e.g., moral reasoning and behavior from the point of view of the protagonist) has brought to light several domains of activity where moral agents are less free than they are thought to be. As can be imagined, disentangling the actual perception of this capacity for agency versus how it works in reality, with all cognitive biases and biases in self-interpretation at stake, is a complex undertaking and beyond the scope of this chapter. Hence, it is important to first introduce, even if only in broad strokes, what we have in mind when referring to the related concepts of free will, freedom, and autonomy (see Figure 20.1).

For the purposes of this paper, free will describes a basic ability to envision different options and choose among them. More often than not, the actual physical ability of enacting these choices is not part of the analysis of free will, although we do think that an analysis of free will without these considerations neglects an important component of the actual complexity of making choices in real-world contexts. Freedom is a more substantive term that describes and alludes to more concrete possibilities of enacting one's

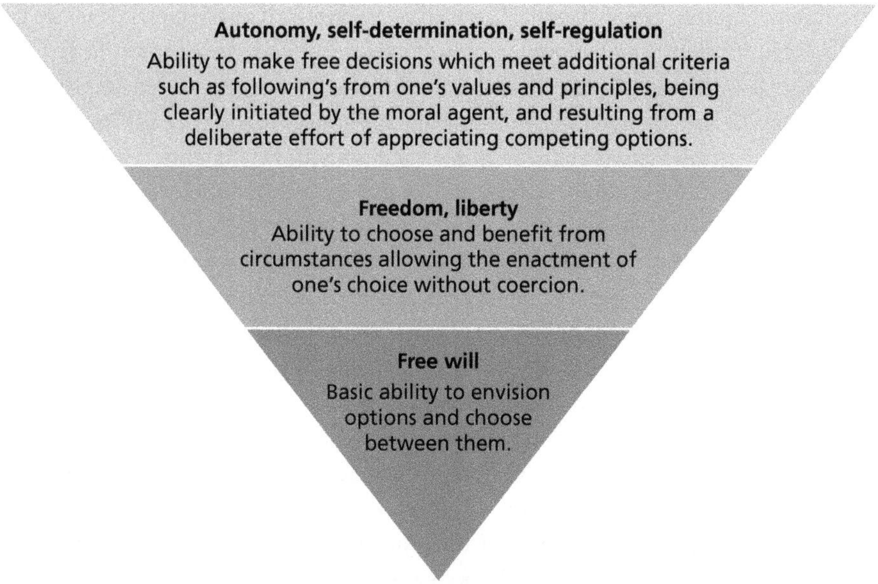

Figure 20.1 Key components of free moral agency: free will, freedom, autonomy.*
* This figure presents three sets of concepts commonly used to describe different components of free moral agency, but does not put forth a model of free moral agency per se. The hierarchical relationship reflects that free will is usually implied by freedom and freedom is usually implied by autonomy.

choices and preferences, often protected through specific rights in legal parlance such as freedom of association, freedom of speech, and freedom of religion. For example, whereas free will may be considered paradoxically to exist even if a choice cannot be enacted, this usually does not hold true of freedom that entails the actual circumstances allowing the exercise of one's freedom. Otherwise said, free will alludes to a more basic ability than freedom that captures more concrete aspects of choices and decisions. Now the concept of autonomy, and related concepts like self-determination and self-regulation, designates an even higher-level or more specific ability to choose or decide as a result of deliberation and reflection. Usually, there is an emphasis on the fact that the individual came to this or that choice or decision on his or her own, or that this choice reflects deeply held principles and values, although this criterion of authenticity is not univocal. According to these distinctions, the mere act of voting under circumstances where this act is not hindered or coerced could be an act of freedom, but it may not be autonomous if the choice to vote for such-and-such candidate has not been the result of a certain kind of deliberate appreciation of the options available, their implications and consequences, and their alignment with one's convictions. And simply having a basic ability to envision options and choose (having free will) would not entail that a person has the freedom to actually carry out this choice. For instance, women have the same basic ability to choose according to their preference among options of government as men, but historically did not have the freedom to actually carry out this choice. Women have only relatively recently been empowered with the right to vote, and there are still places where women do not enjoy political freedom.

Autonomy requires freedom, which itself requires free will, but free will does not necessarily entail freedom, and freedom does not necessarily lead to autonomy. In our view, these concepts—distinguished for the sake of clarity here—are interwoven in discussions about the impact and relevance of behavioral and brain sciences on agency.[5] Furthermore, moral agency is much more than the free or autonomous aspects of attitudes and behavior of the moral agent. For example, personality and responsibility represent other central aspects of moral agency although we do not focus on them in this chapter.

Again, as readers may have already envisioned, there are different kinds of claims that can be made about the impact of behavioral and brain sciences on free moral agency. A moderate claim could be that what is learned about behavior, decision-making processes, including non-conscious processes influencing human psychology, can inform how behavior is understood, including the kinds of actions described as being free. For example, an understanding of voting behavior that would exhibit an engagement of the amygdala, a structure potentially considered "the perceptual or evaluative tool by which the perceivers form their impressions" (Rule et al. 2010, p.354), does not mean that all voting behavior is under the control of the amygdala. Different accounts of relational autonomy have always integrated such moderate claims to revisit standard accounts of autonomy that originally paid little attention to determinants of choice (Mackenzie & Stoljar 2000). However, a stronger type of claim could capture statements and interpretations that suggest that revealing the existence of neurobiological mechanisms underlying a behavior, a decision-making process or the like, would lead to the conclusion that the phenomenon is outside the purview of free moral agency. In other words, the behavior or decision could be considered un-free, determined, lacking any control by the agent, being completely implicit and inaccessible to conscious awareness in ways that the agent cannot do anything about how he or she behaves, thinks, or reacts. By and large, the first, moderate claims are common in most everyday interpretations of what behavioral and brain sciences mean for human action. However, stronger claims are not uncommon and are often debated in the media because of their incredible (if true) and sensational implications (Racine et al. 2016). In the following sections, we illustrate how findings from the behavioral and brain sciences have been interpreted as challenging different aspects of free moral agency, notably (1) self-determined action integral to autonomy, (2) conscious volition involved in free will, (3) conscious control of action required for freedom, and (4) the sense of self required for autonomous decision making. These are by no means the only challenges that behavioral and brain sciences have brought to free moral agency or agency more generally, and we acknowledge that we can only provide an overview to the reader here.

Self-determined action versus action determined by others

The ability to direct one's attitudes and behaviors and orient one's conduct is instrumental to the concept of autonomy; autonomy entails an ability to come to decide for oneself what is the best course of action or the right decision. However, before the advent of contemporary discussions of the behavioral and brain sciences and moral agency, social psychology formulated a threat to the understanding of free moral agency because it exhibited serious social constraints on decisions and behaviors. Otherwise said, it put into question the putative prerequisites for free action and showed that human beings are less free than they are thought to be and more dependent on others, thus challenging the common understandings of free moral agency. Some key findings challenged autonomy, notably the observation of operative deference to the wishes of other individuals (e.g., obedience to authority—see Milgram (1974)) or of the group (e.g., conformity—see Bond (2005)). These are then the types of social influences (see Cialdini & Goldstein 2004) that we briefly review here.

In Milgram's classical experiment on obedience, subjects were made to believe that they are about to take part in a study of memory and learning, and each of them has been assigned the role of a teacher (Milgram 1974). The subject would then be paired with a learner, who is actually aware of the true nature of the experiment and is, in fact, a collaborator of the experimenter. For every error of the learner, the teacher was instructed to administer increasingly higher electric shocks, supposedly to facilitate and motivate learning (Nissani 1990). Once the presumed shock level reached a certain point, the strapped learner appeared to suffer pain and demanded to be set free, while the experimenter, if asked, insisted that the experiment was safe and that the teacher should continue (Milgram 1974). Contrary to expectations, which were that ordinary people would reject authority when confronted with a choice between hurting others and being obedient, 65% of subjects continued with the delivery of electric shocks up to the very highest levels (Nissani 1990). This experiment, and the research that followed it, had profound implications for the understanding of voluntary action and moral responsibility. In Milgram's words, "[t]he essence of obedience consists in the fact that a person comes to view himself as the instrument for carrying out another person's wishes, and he therefore no longer sees himself as responsible for his actions" (Milgram 1974, p.7).

However, actions and beliefs are not only influenced by powerful individuals, but they are also fundamentally shaped by more common group pressures. Psychologists have long been interested in the effects of group pressure on changes in the beliefs, actions, and even perceptual experiences of individuals. Starting from classical studies of Asch (1955) on effects of group pressure on judgments of perception of the length of drawn lines, psychologists have convincingly demonstrated the effects of conformity and pinpointed different social influence processes (see Cialdini & Goldstein 2004). Normative influences, for instance, may stem from the motive to please others and avoid negative sanctions, and they appear to be stronger when participants make public responses and are face-to-face with the majority (Bond 2005). Informational influences, on the other hand, may stem from the motive to reach accurate judgments, and appear to be stronger when individuals are faced with a difficult task and are less confident in their initial judgment (Baron et al. 1996). Belonging to a collectivist or individualist culture further affects the degree of conformity. Bond and Smith compared 134 studies in a meta-analysis and concluded that residents of collectivist countries are more likely to conform than people from individualistic countries (Bond & Smith 1996). This effect might be caused by the difference in meaning that different cultures assign to the concept of nonconformity—it might be interpreted alternatively as representing deviance or uniqueness (Cialdini & Goldstein 2004).

Neuroscience has recently contributed to the research on conformity with functional and structural imaging. On the functional side, evidence has been found for the involvement of the posterior medial frontal cortex in conformity (Izuma 2013). In terms of structural differences, greater gray matter volume is observed in the orbitofrontal cortex of people with high scores on tendency to conform (Campbell-Meiklejohn et al. 2012).[6] All in all, susceptibility to outside social influences has been firmly established by empirical data, and along with the psychological literature on moral development (e.g., Gilligan 1993), it has served as a backdrop for philosophers who reject the underlying assumptions (nonrelationality, value-neutrality, and hierarchical self-control) of traditional conceptions of autonomy and free moral agency (Benson 1990).

Conscious volition versus free and determined volition

In spite of the challenge from the behavioral sciences, the full-fledged debate on free moral agency, in particular on free will, later intersected with neuroscience with the emergence of studies that looked into the neural basis of volition. A series of experiments conducted by Benjamin Libet in the 1980s attempted to examine the physiological basis of spontaneous choice. Libet and colleagues investigated the readiness potential, that is, brain activation prior to the initiation of a conscious and deliberate action. By using electroencephalography, which measures electric activity of neurons, they established the existence of a readiness potential prior to the subjects' own consciousness of having made a decision. In other words, the brain was shown to be active before the subjects were aware of their desire to initiate an action. It is beyond this chapter to review the methodological and epistemological issues that challenge Libet and colleagues' initial conclusions (Mele 2009), but these experiments led them to propose that

> [O]ther relatively "spontaneous" voluntary acts, performed without conscious deliberation or planning, may also be initiated by cerebral activities proceeding unconsciously. These considerations would appear to introduce certain constraints on the potential of the individual to exert conscious initiation and control over his [this person's] voluntary acts. (Libet et al. 1983, p.641)

Libet did grant that there:

> [W]ould remain at least two types of conditions in which conscious control could be operative. (1) There could be a conscious 'veto' that aborts the performance even of the type of 'spontaneous' self-initiated act undress study here [...]. (2) In those voluntary actions that are not 'spontaneous' and quickly performed, that is, in those in which conscious deliberation (or whether to act or of what alternative choice of action to take) precedes the act, the possibilities for conscious initiation and control would not be excluded by the present evidence. (Libet et al. 1983, p.641)

Libet's experiments led to diametrically opposed interpretations, with few siding with his own original middle-ground interpretation (Saigle et al. 2015). These conclusions and interpretations are often taken to have significant implications. Based on Libet's initial findings and subsequent research, researchers have claimed that free will is simply an illusion (Wegner 2002). Others have followed Libet's lead to undertake a series of studies on the automaticity and unconscious components of behavior (Haggard et al. 2004; Haggard 2005, 2011; Smith 2011) and have concluded that free will does not exist because human behavior is essentially determined. A commentary in *Nature Neuroscience* (about a paper which has supposedly shown that the brain activity was initiated seconds before a decision) well reflects this strong eliminativist interpretation:

> The conscious decision to push the button was made about a second before the actual act but the team discovered that a pattern of brain activity seemed to predict that decision by as many as seven seconds. Long before the subjects were even aware of making a choice, it seems, their brains had already decided. As humans, we like to think that decisions are under conscious control — that we have free will. Philosophers have debated that concept for centuries, and now Haynes and other experimental neuroscientists are raising a new challenge. They argue that consciousness of a decision may be a mere biochemical afterthought, with no influence whatsoever on a person's actions. According to this logic, they say, free will is an illusion. "We feel we choose, but we don't," says Patrick Haggard, a neuroscientist at University College London. (Smith 2011, p.24)

Conscious control of action versus unconscious influence on action

Another aspect of free moral agency which has been questioned in behavioral and brain research is the ability for control over action, notably conscious control. The growing literature on implicit biases—a type of negative social attitudes that seem to guide most overt behavior—has been described as challenging traditional notions of freedom because it directly puts to light influences that appear to be beyond conscious control of action, and this kind of control is presumed in most understandings of freedom.

Negative social attitudes can be implicit and subtle (e.g., unconscious prejudice) or explicit and blatant (e.g., overt discrimination). Subtle or implicit social attitudes are not displayed openly—the individual appears to be unbiased and even might consider herself to be unbiased, whereas blatant biases are not denied—they are displayed for the world to see, and announce an individual's displeasure with someone or something (Keene 2011). Hate crimes are a perfect example of explicit bias—harm is caused to an individual due to the fact that this other individual is of a different race, gender, sexual orientation, or social status. The dislike or prejudice against the victim is openly displayed and often motivated by the belief that it is justified, with no regard for the victim (Fiske 2010). Implicit biases on the other hand, are an example of subtle and particularly insidious social attitudes. They describe unconscious and automatic prejudiced judgments, which often lead to social behavior that the individual might not be aware of (Brownstein 2016).

Many categories of implicit bias have been empirically studied, including race, age, gender, and ethnicity and other categories that fit more explicit negative social attitudes. However, the most important and well-known research has focused on attitudes toward members of racial and ethnic minorities or socially stigmatized groups, or both (see Brownstein 2016). Many studies have showed that implicit biases can affect social judgment and behavior independently from the subject's explicit attitudes. These unconscious and automatic influences, such as implicit racial bias, may be revealed on indirect measures of attitudes (e.g., sequential priming or the Implicit Association Test—see Brownstein (2016)). The growing research on implicit biases has many upshots, from chilling (e.g., biases in police shootings of unarmed men of African origin), very practical (e.g., the well-documented propensity of employers to prefer in-group candidates in job searches), to the more philosophical (e.g., how can a judgment be trusted once implicit biased are acknowledged?).

The most important implications of implicit biases in the context of our discussion stems from the fact that they stand in stark contrast to controlled cognitive processes. Namely, mental causation is an important theme in many conceptions of free moral agency—freely willed actions are viewed as actions that are derived from a person's conscious thoughts, whereas actions derived from the unconscious are viewed are unfree (Baumeister & Monroe 2014). Thus, research into implicit attitudes and the range of effects they have on judgment and behavior put the understanding of free moral agency into question. This is exacerbated by data from social psychology and neuroscience, which shows that prejudices and biases toward out-group members are widespread and persist in contemporary society (see Amodio 2014). The issue is whether automaticized behaviors related to, say, sexism, should be thought of as under control of the person in question. If they unfold in the absence of explicit reasoning, should individuals holding such attitudes be responsible for them, and for the judgments and behaviors influenced by their biases? Since there can be unawareness of the impact of a great many cognitive states on behavior, this may lead to a global skepticism about moral responsibility (Holroyd 2012). However, beyond the global issues of denying moral responsibility, the fact that implicit biases are pervasive introduces additional constraints to free moral agency and even forces us to consider ongoing, previously unreflected responsibility in two ways: responsibility for *having* implicit attitudes and responsibility for the behavioral expression of implicit attitudes (Faucher 2016).

Sense of self versus conglomerate of divergent wills

Another challenge to the understanding of free moral agency has come from surprising observations that the microorganisms living in the abdomen seem to affect many decisions considered to be those of the moral agent. Ownership of action has a complex relationship with free moral agency (e.g., a crime for which someone could be responsible could then be disowned), but there is at least a minimal sense by which a moral agent is someone who is the cause and the initiator of his or her actions. Recent research has called into question this sense of being a free and autonomous self.

In the last decade or so, there has been a growing recognition of the synergetic relationships between the gut microbiota comprising 10^{13}–10^{14} microorganisms and the nervous system, that is, the gut–brain axis (Neufeld & Foster 2009; Cryan & Dinan 2012). The importance of the commensal microbiota for the development of the immune system and mucosal tissues has long been known to immunologists and even before to some vision-ary scientists like Bernard, Darwin, James, Pavlov, and Lange (Cryan & Dinan 2012). However, the behavioral and brain sciences have witnessed a plethora of findings sug-gesting that not only disease states but also, so to speak, ordinary behavior is under the influence of the microbiota (Neufeld & Foster 2009). The mechanisms through which this action occurs are being explored (Cryan & Dinan 2012) but include the possible contri-bution of the gut commensals in the synthesis of neurotransmitters and other microbial mediators. These mediators then impact homeostasis in the central nervous system, acti-vation of the hypothalamic–pituitary–adrenal axis (HPA) or stress axis, neuroimmune response, and neurogenesis by affecting the vagus nerve, involved in the parasympathetic regulation of many organs, or crossing directly the blood–brain barrier (Sherwin et al. 2016), although other mechanisms could be at stake (Cryan & Dinan 2012).

One of the most intensively investigated areas of brain–gut interaction concerns stress response and the activity of the HPA axis in which several key observations have showed clear interactions between the central nervous system and the gut. For example, maternal separation (a stressor) in rhesus monkeys is associated with changes in the HPA axis but also affects temporarily the composition of the gut microbiota (Cryan & Dinan 2012). Long-term effects have also been found in rats and physical stressors (physical restraint) lead to changes in gut microbiota composition and reciprocally, stress response also appears to be modulated by microbiota profile (Cryan & Dinan 2012). A few years ago, a set of landmark studies suggested that the development of the immune system in mice led to the normalization of the HPA axis. For example, adult germ-free mice showed an exaggerated stress response in comparison to control animals (Sudo et al. 2004). The same study showed that colonization of the mice at 6 weeks (early adulthood) reversed this effect but not in adulthood (8 weeks of life). Additionally, the function of the intes-tinal barrier can be changed subsequent to stress exposure and can be reversed by the use of some probiotics (Cryan & Dinan 2012). In addition to effects on the HPA axis, a series of observations have suggested implications of gut composition and function on memory, mood, and visceral pain perception (Cryan & Dinan 2012). Of great potential clinical significance, neurodevelopmental disorders such as autism are associated with gut dysfunction, although the relationship (causal or correlational) is unclear given that diet and food intake (as well as use of antibiotics) can differ in individuals affected by autism (Cryan & Dinan 2012). An association between certain psychiatric conditions (e.g., mood disorders) and gastrointestinal disorders (e.g., irritable bowel syndrome) has also long been observed (Neufeld & Foster 2009).

In different headquarters, microbiome[7] research is now considered a genuine chal-lenge to the existence of the self, let alone the self which is free to choose. Kramer and Bressan (2015), describe human beings as superorganisms where the organisms constituting the microbiota would have interests of their own, "which need not coin-cide with ours" (p.464). For example, some behaviors of mammals could be dictated by the interests of microorganisms to use their hosts to perpetuate their existence. Rats inflected with toxoplasma lose their fear of cats; the toxoplasma is then able to reproduce sexually instead of asexually in the cat's intestine, thereby suggesting that the behaviors of these mammals are instrumentalized by toxoplasma. Similarly, food cravings and other behaviors and choices of humans could be directed by microorganisms, and could potentially run counter to long-term health and autonomously deliberated interests. Morar and Beever reviewed such microbiome research and conclude that it is "producing a radically new understanding of 'individual' organisms" (Morar & Beever 2016, p.39). In a nutshell, the human individual should be more aptly viewed as a "community, but also in recognizing the ways that agency and autonomy are additionally impacted by microbial interactions" (Morar & Beever 2016, p.40). Accordingly, they criticize standard accounts of autonomy (e.g., moral agents as having a capacity for self-sufficient, self-determined decisions, i.e., a capacity defined in abstract terms without much understanding or appreciation of determinants of choice) for not taking into account evidence from the biological and environmental sciences indicating that autonomy is influenced by a host of factors. They even fault the relational account of autonomy, which was developed explicitly to incorporate insights on social determinants of choice, as incomplete "to the extent that it overlooks this deep biological interconnectedness between human individuals and other organisms" (Morar & Beever 2016, p.41). However, the authors go further and claim that there is "no reason not to jettison the concept" [of autonomy] "if a suitable normative replacement for the value that autonomy has traditionally upheld can be identified" (Morar & Beever 2016, p.42). Otherwise said, microbiome research and brain–gut interaction radically put into question the understanding of free moral agency.

To conclude our analysis of threats stemming from behavioral and brain sciences, different claims have been made to the effect that they bear significant implications for the understanding of free (moral) agency. If different strands of research are to be believed, individuals are not free in the sense that they are considered to be. On the contrary, certain decisions would be heavily influenced by social determinants of choice, preparatory brain activity such as the readiness potential and neuronal activity preceding action, implicit biases, and as discovered more recently, the activity of the microbiota. These insights have been interpreted as threats to free moral agency and its different facets such as the capacities for self-determined action, conscious volition, conscious control of action, and the sense of self. In the next section of this chapter, we discuss two challenges to these claims. First, we highlight their objectivist and essentialist assumption toward abilities relevant for free moral agency. Second, we point to their misplaced devaluation of the normative principle of autonomy and human freedom, which stresses the value of free moral agency. We propose that a concept of "contextualized autonomy" (e.g., Racine et al. 2017)—which relies on a recognition that autonomy is a set of agential abilities (see also Dubljević 2013) exercised in context and originating in the experience of the agent—can integrate the useful insights they describe while avoiding these two pitfalls.

Behavioral and brain-based research as empowerment

Strong interpretations of research in different domains of the behavioral and brain sciences suggest that they explain away autonomy, freedom, and free will, and/or suggest that the insights provided by those concepts are wrong-headed and misleading, perhaps even simply useless from a scientific perspective. But, keeping in mind the importance of free moral agency in human affairs, can these concepts be dismissed as some propose (Wegner 2002)? Interestingly, some of those who more staunchly oppose such eliminativism have tended to argue that nothing could be actually learned from behavioral and brain science. For example, the behaviorist and legal scholar, Steven Morse, argues that the strong eliminativist interpretations rely on a "brain overclaim syndrome". According to this view, the behavioral and brain sciences bear no impact for the law because of the genuine value of beliefs in responsibility and autonomous decisions. Morse advances that the law is unconcerned by metaphysical discussions about free will, this concept being entirely abstract and barely relevant for law since the law defaults to folk psychological attribution of presumed autonomy and responsibility (Morse 2006, 2008).

More promising are middle-ground views that tend to give some credence to the importance of the behavioral and brain sciences but nevertheless frame the contribution of the behavioral and brain sciences as perhaps not a threat, but a challenge to free moral agency. For example, Felsen and Reiner have proposed that a standard model of autonomy—defined on the three criteria of (1) having a hierarchy of desires (reflexivity), (2) rationality, and (3) freedom from external influences—is to some extent inconsistent with evidence emerging from behavioral and brain research, notably because of the implicit mechanisms and biases to which human decision making is prone (Felsen & Reiner 2011). Regarding the third criterion, which is key to our discussion, the authors identify both internally generated influences to decisions and external influences to decisions, and focus on the latter. Rightfully, they stress that influences on decisions could be both detrimental and beneficial to the moral agent. Interestingly, they also note the possible interactions between external and internal influences, notably that covert external factors may become internalized and "indistinguishable from authentic internal sources" (Felsen & Reiner 2011, p.10). Their conclusion is that "at least some of our decisions do not appear to be as free from undue influence as the standard model of autonomy requires" (Felsen & Reiner 2011, p.10). Otherwise said, the commonly assumed ability of moral agents to truly deliberate and be completely free is unrealistic. Felsen and Reiner resist concluding that moral agents cannot undertake autonomous actions and should not be held responsible for their actions, and they thus open the door to more nuanced positions between the extremes found in the literature. However, they do not elaborate further on these promising ideas, including the synergies between internal and external factors as well as how influences can actually empower moral agents to achieve their goals, including those they choose voluntarily and autonomously.

In our eyes, middle-ground positions like Felsen and Reiner's are hinting at the genuine promises of the behavioral and brain sciences but fall short from overcoming explicitly the conceptual barriers to the meaningful integration of behavioral and brain science knowledge to the theoretical and practical understanding of free moral agency. Indeed, the discussion remains burdened—we argue—by two common fallacies within the view of abstracted free moral agency (Figure 20.2a): first, concerning an objectivist and essentialist stance (dualist metaphysics separating starkly the realms of the subjective and the objective) toward abilities relevant for free moral agency (e.g., free will, freedom, and autonomy) and second, a misplaced devaluation of the normative principles that stress the value of free moral agency (e.g., respect for autonomy, respect for individual rights, and responsibilities) based on behavioral and brain science. We explain these two tension points and provide a different view informed notably by pragmatist theory and the concept of "contextualized autonomy" (Dubljević 2013; Racine & Dubljević 2016; Racine et al. 2017). The first fallacy is resolved by the adoption of a nonessentialist (nondualistic) metaphysics; the second by taking a distance from postulating a strong fact–value (is–ought) dichotomy and siding with a description of this dichotomy as a tension which needs to be resolved in actual moral judgment (see Figure 20.2 for an overview).

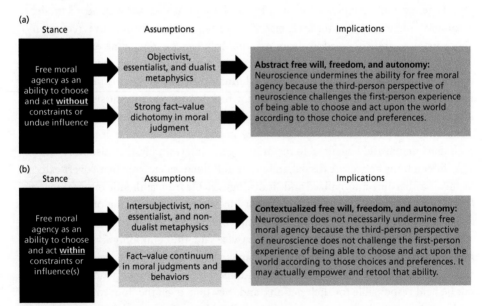

Figure 20.2 Two interpretations of the contribution of behavioral and brain science to the understanding of free moral agency. (a) Dualist metaphysical interpretation of behavioral and brain research on free moral agency. (b) Pragmatist interpretation of behavioral and brain research on free moral agency.

Assumption 1: Objectivist, essentialist, dualist metaphysics versus intersubjectivist, nonessentialist, and nondualist metaphysics

A large proportion of the debates about the existence of free moral agency, notably about free will but also autonomy, have been framed within an all-or-nothing discussion (Gert & Duggan 1979; Müller & Walter 2010; Baumeister & Monroe 2014). Either a person has free moral agency or not. The idea that there could be degrees of free will, freedom, or autonomy is often evacuated to the profit of such a polarized debate (Müller & Walter 2010; Baumeister & Monroe 2014; Racine 2017). Moreover, the corollary assumption is that concepts capturing free moral agency are considered essences that exist somewhat independently of the subject such that their existence can be established for all human beings in all different contexts. Yet this tendency seems to be a challenge for any moderate interpretation of behavioral and brain sciences' contribution to this discussion because empirical evidence from the behavioral and brain sciences enters a debate where the only two defendable options are extreme and without consideration for middle-ground positions. For example, Morar and Beever (2016, p.14), when reviewing findings from microbiome research come to the troubled conclusion that the status of autonomy, as an ethical principle, is jeopardized. However, this interpretation needlessly pits knowledge about social and biological determinants of choice against the abilities of free moral agency. Unfortunately, there is a long tradition of dualist metaphysics where the capacity to act freely or autonomously is salvageable only if considered to be outside the realm of causally described social, biological, and so on phenomena (see Figure 20.2a). Consequently, more knowledge about such phenomena appears to jeopardize the ability to be a free moral agent; the more is learned about such determinants of choice, the less freedom or autonomy there seems to remain. Even the more nuanced position of Felsen and Reiner (2011) describes greater knowledge about determinants of choice to be a moderate threat to beliefs about autonomy and freedom. However, it is not the actual scientific knowledge, but the outdated dualist metaphysics that is implicit in sweeping interpretations which is the source of the problem and needs to be considered and criticized.

Another starting point, consistent with pragmatist theory and the concept of contextualized autonomy, is offered by first granting that the concept of free moral agency originates in experience and describes, in folk psychological parlance, the ability to choose between actions and to reflect upon desirable outcomes to then orient them according to interests (Figure 20.1). This was well described philosophically by Gert and Duggan who called for a reformulation of free will as the ability to will or as a volitional ability (Gert & Duggan 1979). They posit a positive account of free moral agency as an ability to perform certain desired actions thereby overcoming the common fallacy that the more that is known about the causes of an action, the less that action seems free.[8]

Interestingly, some recent qualitative and social psychology research has shown that, contrary to scholarship claiming that lay views rely on substance dualism or exemption from the law of causality (Bargh 2008; Montague 2008), ordinary perspectives on free will are consistent with a positive definition of freedom as a fluctuating ability to make a choice, acting consistently with one's desires, and being reasonably free of constraints (Baumeister & Monroe 2014). The challenge of determinism to free will is unclear in lay understandings of free will because being free is viewed as something positive rather than simply as acting outside the realm of causes. These findings have been supported by other studies that have looked at ordinary views on free will (Baumeister & Monroe 2014). Otherwise said, the concepts describing free moral agency designate a genuine first-person experience of free moral agency and serve as an interpretation of these abilities. This sense of free and autonomous agency could be somewhat imperfectly described and captured by folk terminology, and it could be tricked in different ways as described by Wegner (2002), but it alludes to a real ability that humans have and robustly attribute to each other through different established social practices reflected in everyday interactions and also in social and legal norms.[9] As Baumeister (2008, p. 15) writes:

> Freedom and choice are woven deeply into the fabric of human relations and activities. If freedom and choice are completely illusions—if the outcome of every choice was inevitable all along—why must people agonize so over decisions? Why do they argue and strive so much for the right to decide (that is, for power and liberty)? Why has so much political, economic, and social struggle been aimed at increasing freedom if freedom is just an illusion?

This positive account of free moral agency as a set of abilities—and which actually is silent on the issue of its compatibility or incompatibly with determinism—maps well with the functions attributed to this concept in daily life. Every time we are able to fulfill basic or complex actions, we are comforted in our sense of being capable of acting both freely or autonomously, and, likewise, when we are frustrated in doing so, we experience resistance to our will and intents. As William James proposed, the feeling of effort and goal-focused activity "play a role in strengthening the internal sense of the will's capabilities" (Gray 2007, p.5). If this first-person experience and existence of free moral agency is not first recognized, then either a person falls into the trap of explaining away free moral agency (e.g., hard determinism) or reverts to "panicky metaphysics" (Strawson 1962), in order to account for the basic conditions of free moral agency such as free will. Clearly, the ability to act freely or autonomously can be jeopardized by a host of factors, but knowledge about those factors can also facilitate the realization of intents and desires such that the ability of human beings to act upon the world is not undermined by greater knowledge of how choices are made. The fact that free moral agency has a subjective and experiential dimension does not make it less real; and understanding and explaining it through social, psychological, or biological perspectives does not explain it away. As Dewey, the precursor of the concept of contextualized autonomy and free moral agency more broadly, wrote: "we need freedom in and among actual events, not apart from them"[10] (Dewey 1922, p.303)—our ability to act upon the world is best understood as a capacity which uses knowledge of personal, relational, social, biological, and environmental conditions to actually enact goals and projects (Figure 20.1). The traditional opposition (based on Kant's work (Kant 1994)) between freedom and causality is spuriously generated by a metaphysical view, which simply does not take into account seriously the basic subjective and intersubjective experience of making free and autonomous, and sometimes less free and autonomous, decisions. Acknowledging that this experience is the first foundation for the actual existence of free moral agency prevents an interpretation that more knowledge about its conditions actually somehow undermines this experience because the all-or-none and essentialist objective nature of free moral agency is not posited; rather, free moral agency describes agent-dependent characteristics which can be diminished or empowered. This is in line with the veto power interpretation of Libet: some decisions might be unfree, but the power to choose at least some things cannot be reasonably denied. Thus, empirical data gained about how different contexts can impede free moral agency becomes valuable knowledge for the moral agent to try to overcome these barriers or, if they are very difficult to surmount, to identify alternative remediation strategies (e.g., changing the context instead of the agent). In discussions of contextual influences on free moral agencies, studies similar to those that show that collectivist societies tend to foster conformity whereas individualist societies tend to reinforce the value of personal freedom and foster nonconformity, can be factored into the analysis of free moral agency.

Assumption 2: Strong fact–value dichotomy (is–ought fallacy) versus the fact–value continuum

One of the corollaries of strong interpretations of the behavioral and brain sciences is that the ethical principles relying on the existence of the concepts of autonomy and freedom (e.g., respect for autonomy, protection of individual rights, and freedom) would be imperiled because they rely on, at best, illusions or, perhaps worst, on flawed concepts which are deficient. If, for example, free will or freedom does not exist, then they are, in a sense, truly chimeras. And if this is the case, then such concepts should be expelled from moral thinking to be replaced, likely, by concepts and principles compatible with a scientific understanding of the world. An example of this move is Morar and Beever's claim that the normative value and force of autonomy are imperiled because the integrity of autonomy is challenged by microbiome research which undermines both the traditional and relational accounts of autonomy (Morar & Beever 2016). Similarly, Green and Cohen (2004) contend that the legal system needs to be realigned with consequentialism based on a deterministic view, allegedly supported by neuroscience. Otherwise said, behavioral and brain science knowledge about different aspects of free moral agency would not only alter how this ability is conceived of (epistemological revisionism) but actually challenge its existence (ontological revisionism). It is true that a positivist or dichotomist interpretation of the contribution of the behavioral and brain sciences does set the stage for such a strong opposition between fact and value (Figure 20.2a), but it is by no means the only possible epistemological stance (Figure 20.2b).

The claim that concepts describing free moral agency as free will, freedom, or autonomy should be expelled from the discourse of ethics is based on a fallacious understanding of the relationship between normative discourse (i.e., values, the "ought") and descriptive discourse (i.e., facts, the "is"). Claiming that knowing more about the factual aspects undermines the normative force of autonomy is profoundly erroneous because it implies that knowing more about the objective conditions of autonomy imperils not only the *existence* of free moral agency but also its *value* and *relevance*.[11] But, this would mean, somehow, that what is experienced and valued would be qualitatively distinct from what is known; experience could be invalidated by a scientific understanding of that world. However, the capacity for human agency would be worthless if it would not have some clear causality and logically, be describable according to the lenses of the behavioral and brain sciences but also those of sociology, psychology, and biology, not to say of personal experience. For instance, much has been made of the fact that 65% of people follow orders even when these orders are harmful and immoral. But what of the 35% that refuse to do so? Also, it has to be noted that normative values empower certain behaviors, specifically because they are reinforced by social influence. As Dewey commented about liberty, a term he used to describe a richer account of free moral agency, "[w]hat men have esteemed and fought for in the name of liberty is varied and complex—but [it] certainly has never been a *metaphysical* freedom of will" (Dewey 1922, p.304). Instead, Dewey proposed that what matters with respect to free moral agency is the ability to carry out plans, the ability to change them, and the power of the individual to be an actor in the course of events. There is actually much evidence pointing to the psychological existence and relevance of free moral agency.

Freedom and autonomy are valued because they describe a genuine ability to undertake actions, ponder about options, and imagine actionable scenarios—abilities which can generate an immense sense of achievement when they lead to outcomes that help individuals meet the challenges of daily lives and accomplish their intents (Peterson et al. 1995; Baumeister et al. 1998; Baumeister 2008). Baumeister writes, rightfully in our opinion, that "[l]aypersons may not understand the concept of free will in the same way as philosophers and scientists, but they use 'freedom' to denote some psychological phenomena that are powerful and important" (Baumeister 2008, p.15). Likewise, when the ability for free moral agency is compromised or diminished, it generates alienation (Habermas 1968), depressive feelings (Wegener et al. 2007), and apathy even in nonhuman animals (Seligman 1975). Furthermore, the presence of choice (freedom and autonomy versus no choice) has demonstrated impact on coping and stress (Glass et al. 1969, cited in Baumeister 2008), as well as self-motivation and healthy psychological development (Ryan & Deci 2000, cited in Baumeister 2008). Self-determination theory and evidence supporting it, suggest that the ability to have intrinsic motivations and be able to carry them out contributes to a sense of growth and fulfillment that foster psychological well-being and mental health (Ryan & Deci 2000). Likewise, when extrinsic motivations are shared by the agents and can be meaningfully internalized, individuals experience greater freedom or autonomy in action, even though the fulfillment of intrinsic motivations contributes more to well-being than extrinsic motivation (Ryan & Deci 2000). However, feelings of alienation (i.e., not being able to enact one's ability to will) can have a negative impact on psychological well-being. Accordingly, the existence of free moral agency has clear support in common-sense understandings of individual and social life as well as those of fields such as social psychology and sociology (where it is more customary to speak of "empowerment") (Perkins & Zimmerman 1995). Freedom presupposes having mature powers of volition and the capacity of self-control, but not that it is exercised all the time (Dubljević 2013). Self-control can be depleted and imperfect, but empirical research points toward the conclusion that it is undoubtedly present in most adults (Baumeister et al. 1998; Moller et al. 2006; Vohs et al. 2008).[12]

Autonomy and free will find their value in experience. And better factual or scientific understanding of the different aspects of free moral agency does not fundamentally undermine the value attributed to them because knowledge does not change the experience and exercise of free moral agency and its importance. Those claiming that the behavioral and brain sciences could bring new insights to the understanding of free moral agency are right, but they are wrong to conclude that ontological revisionism *will* be a dominant trend or even *should* be the dominant trend. It is true that the abilities described by the concepts of autonomy, freedom, and free will could be re-described in more accurate normative language. Pragmatist theory does grant and even encourage the scientific reconstruction of the understanding of physical and psychological realities (Racine 2013). However, as Dewey warned, contrary to the sweeping reductionism of positivist and eliminativist neurophilosophy, "it will be a long time before anything of this sort will be accomplished for human beings [in contrast to physics]. To expel traditional meanings and replace them by ideas that are products of controlled inquiries is a slow and painful process" (Dewey 1937, in Gouinlock 1994, p.48). Accordingly, we caution against the desire to reduce the experience of free moral agency to more rigorous or philosophical language if it leads to expelling important dimensions of the concept, as recent research does suggest it would be the case if the thinking of the eliminativists is followed (Monroe & Malle 2010). Meaningful contributions to the understanding of free moral agency should be understandable through the lenses of different disciplines. Contribution of diverse disciplines will likely provide enrichments and revisions to self-understanding and the understanding of other's behaviors. It would indeed be strange if, like proposed by Churchland (1981), the behavioral and brain sciences would systematically explain away the phenomena initially triggering the scientific inquiries. There are cases of ontological revisions (e.g., phlogiston in chemistry), but it might be more fruitful to expect that most scientific discoveries in neuroscience and psychology will enrich the understanding of common-sense (folk psychological) concepts that currently have some traction in individual and social life (e.g., law, ethics, and politics).

Conclusion

Neuroscience has been heralded as a game-changing field for its impact on the understanding of free human agency. Research in diverse fields examining, for example, social determinants of choice, brain activity preceding action, implicit biases, and more recently, the microbiome and its impact on brain function and behavior have all been interpreted as radically jeopardizing the existence of free moral agency. In this chapter we did not scrutinize the scientific and methodological foundations for such claims, which we have done elsewhere (Dubljević 2013; Saigle et al. 2015). We tackled another set of issues related to the conceptual and metaphysical assumptions underlying common interpretations of behavioral and brain science research. Much possible uptake of the behavioral and brain sciences with respect to the understanding of free moral agency could be hindered by strong dualistic metaphysical assumptions as well as dichotomous interpretations of the fact–value continuum. A resolution can be found in pragmatist theory and recent research in social psychology. These support the concept of contextualized autonomy that recognizes explicitly that different aspects of free moral agency originate in a first-person experience of agency. Moreover, the value placed on the abilities captured by the concept of free moral agency is not necessarily jeopardized by a factual understanding from the behavioral and brain sciences. In fact, knowledge can also empower moral agency. Behavioral and brain sciences provide us with important contributions and information on how free moral agency is limited and could even be seen as a depletable resource. However, scientific evidence can empower choices by teaching when to conserve limited resources, and when to undertake or delay decisions.

Acknowledgments

Support for this work comes from the Social Sciences and Humanities Research Council (Racine), the Fonds de recherche du Québec–Santé for career awards (Racine), and the Banting Postdoctoral Fellowships Programme (Dubljević). We extend our thanks to members of the Neuroethics Research Unit for feedback on previous versions of this manuscript. Special thanks to Kaylee Sohng, Simon Rousseau-Lesage, and Roxanne Caron for editorial assistance in the preparation of this manuscript. Some parts of this chapter have appeared previously in a short paper by the authors (Racine & Dubljević 2016).

Notes

1. We will introduce a finer-grained distinction between free will, freedom, and autonomy (and their synonyms), although "freedom" is often used as a generic term to describe these three different aspects of free moral agency.

2. "Renoncer à sa liberté c'est renoncer à sa qualité d'homme, aux droits de l'humanité, même à ses devoirs" (Rousseau 1992, p.34).

3. We also thereby acknowledge that there are many different aspects of agency (e.g., authenticity and responsibility) which we do not tackle since our discussion deals with the issue of freedom in a general sense and in the context of ethics, hence the qualifier "moral" used throughout even though some of our points about free agency could be valid beyond the context of ethics.

4. We frequently use the term "abilities" to stress the active dimension of these concepts; the term "capacities" has a more static connotation and reflects less clearly the context-dependent and agent-dependent nature of these concepts. However, when the abilities are not engaged, we use "capacities."

5. The concepts of autonomy, liberty, and free will have been used in different ways. Our conceptualization does not obviate other uses. It is simply meant to clarify how we use these terms in this chapter. The most important idea that is conveyed is the gradual expansion in the substantive commitments from free will to autonomy. One of us (Dubljević 2013) has used a different designation of the three

distinct levels (basic autonomy, free will, ideal of autonomy), but in this chapter we opted not to use autonomy twice, so as not to confuse the reader.

6. The tendency to conform is a stable trait with observable neuroanatomical correlates. The upshot is that not only do some people conform more than others, but that the neuroscience evidence may actually predict who is more likely to conform and who is more likely to resist such external influences.

7. In this chapter, we do not draw a strong distinction between the concepts of microbiome and microbiota.

8. Gert and Duggan (1979, p.197) write: "We are thus in agreement with Moore that the question whether man has Free Will is best viewed as a question about whether he has a certain power (we prefer to talk about ability) not as a question about whether what a man has willed to do was the result of coercion."

9. The fact that our abilities can be tricked or manipulated does not imply that we do not have them. Otherwise, visual illusions would suffice to debunk the fact that we have the ability to see, and rely on sight accurately, at least most of the time.

10. Dewey did not adopt the tripartite distinction (Figure 20.1) but stressed nevertheless how free will was a limited ability (in contrast to actual freedom).

11. Consider an analogy with slavery: the lack of freedom experienced by slaves does not mean that they lack appreciation or value of freedom, or that the right to be free has no meaning.

12. Vohs and colleagues equate self-control and self-regulation (which is the literal translation of the word autonomy) and define it as "the self-exerting control to override a prepotent response, with the assumption that replacing a response with another is done to attain goals and conform to standards" (Vohs et al. 2008, p.884). Moller and colleagues, in their self-determination theory make the distinction between self-control, which is ego-depleting, and self-regulation, which is not ego-depleting (Moller et al. 2006). "Ego depletion" refers to the phenomenon of diminished ability to enact self-regulation with repeated efforts, and neuroscience evidence supports the view that relevant capacities for exertion of control exist (Berkman & Miller-Ziegler 2013; Wagner & Heatherton 2013).

References

Amodio, D.M. (2014). The neuroscience of prejudice and stereotyping. *Nature Reviews Neuroscience,* **15**(10), 670–682.

Asch, S.E. (1955). Opinions and social pressure. *Scientific American,* **193**(5), 33–35.

Bargh, J.A. (2008). Free will is unnatural. In: Baer, J., Kaufman, J.C., and Baumeister, R.F. (Eds.) *Are We Free? Psychology and Free Will,* pp.128–154. New York: Oxford University Press.

Bargh, J.A. and Chartrand, T.L. (1999). On the unbearable automaticity of being. *American Psychologist,* **54**(7), 462–479.

Baron, R.S., Vandello, J.A., and Brunsman, B. (1996). The forgotten variable in conformity research: impact of task importance on social influence. *Journal of Personality and Social Psychology,* **71**(5), 915–927.

Baumeister, R.F. (2008). Free will in scientific psychology. *Perspectives on Psychological Science,* **3**(1), 14–19.

Baumeister, R.F. and Monroe, A.E. (2014). Recent research on free will: Conceptualizations, beliefs, and processes. In: Zanna, M.P. and Olson, J.M. (Eds.) *Advances in Experimental Social Psychology* (Vol. 50), pp. 1–52. Burlington, MA: Academic Press.

Baumeister, R.F., Bratslavsky, E., Muraven, M., and Tice, D.M. (1998). Ego depletion: is the active self a limited resource? *Journal of Personality and Social Psychology,* **74**(5), 1252–1265.

Benson, P. (1990). Feminist second thoughts about free agency. *Hypatia*, 5(3), 47–64.

Berkman, E.T. and Miller-Ziegler, J.S. (2013). Imaging depletion: fMRI provides new insights into the processes underlying ego depletion. *Social Cognitive and Affective Neuroscience*, 8(4), 359–361.

Bond, R. (2005). Group size and conformity. *Group Processes & Intergroup Relations*, 8(4), 331–354.

Bond, R. and Smith, P.B. (1996). Culture and conformity: a meta-analysis of studies using Asch's (1952b, 1956) line judgement task. *Psychological Bulletin*, 119(1), 111–137.

Brownstein, M. (2016). Implicit bias. *The Stanford Encyclopedia of Philosophy*. Available from: http://plato.stanford.edu/archives/spr2016/entries/implicit-bias/ [accessed March 30, 2016].

Campbell-Meiklejohn, D.K., Kanai, R., Bahrami, B., et al. (2012). Structure of orbitofrontal cortex predicts social influence. *Current Biology*, 22(4), R123–R124.

Churchland, P.M. (1981). Eliminative materialism and the propositional attitudes. *Journal of Philosophy*, 77(2), 67–90.

Cialdini, R.B. and Goldstein, N.J. (2004). Social influence: compliance and conformity. *Annual Review of Psychology*, 55, 591–621.

Cryan, J.F. and Dinan, T.G. (2012). Mind-altering microorganisms: the impact of the gut microbiota on brain and behaviour. *Nature Reviews Neuroscience*, 13(10), 701–712.

Dewey, J. (1922). *Human Nature and Conduct: An Introduction to Social Psychology*. New York: Holt.

Dewey, J. (1937). An address delivered before the College of Physicians in St. Louis, April 21, 1937, republished in Gouinlock, J. (Ed.) *The Moral Writings of John Dewey, Revised Edition* (Great Books in Philosophy), pp.47–55. Amhert, NY: Prometheus Books.

Dubljević, V. (2013). Autonomy in neuroethics: political and not metaphysical. *American Journal of Bioethics Neuroscience*, 4(4), 44–51.

Faucher, L. (2016). Revisionism and moral responsibility for implicit attitudes. In: Brownstein, M. and Saul, J. (Eds.) *Implicit Bias & Philosophy: Volume 2, Moral Responsibility, Structural Injustice, and Ethics*, pp.115–144. Oxford: Oxford University Press.

Felsen, G. and Reiner, P.B. (2011). How the neuroscience of decision making informs our conception of autonomy. *American Journal of Bioethics Neuroscience*, 2(3), 3–14.

Fiske, S.T. (2010). *Social Beings: Core Motives in Social Psychology*. Hoboken, NJ: Wiley.

Gert, B. and Duggan, T.J. (1979). Free will as the ability to will. *Noûs*, 13(2), 197–217.

Gilligan, C. (1993). *In a Different Voice: Psychological Theory and Women's Development*. Cambridge, MA: Harvard University Press.

Glass, D.C., Siger, J.E., and Friedman, L.N. (1969). Psychic cost of adaptation to an environmental stressor. *Journal of Personality and Social Psychology*, 12(3), 200–210.

Gray, M.T. (2007). Freedom and resistance: the phenomenal will in addiction. *Nursing Philosophy*, 8(1), 3–15.

Greene, J.D. and Cohen, J. (2004). For the law, neuroscience changes nothing and everything. *Philosophical Transactions of the Royal Society B: Biological Sciences*, 357(1451), 1117–1785.

Habermas, J. (1968). *Technik und Wissenschaft als „Ideologie."* Frankfurt am Main: Suhrkamp.

Haggard, P. (2005). Conscious intention and motor cognition. *Trends in Cognitive Sciences*, 9(6), 290–295.

Haggard, P. (2011). Decision time for free will. *Neuron*, 69(3), 404–406.

Haggard, P., Cartledge, P., Dafydd, M., and Oakley, D.A. (2004). Anomalous control: when 'free-will' is not conscious. *Consciousness and Cognition*, 13(3), 646–654.

Holroyd, J. (2012). Responsibility for implicit bias. *Journal of Social Philosophy*, 43(3), 274–306.

Izuma, K. (2013). The neural basis of social influence and attitude change. *Current Opinion in Neurobiology*, 23(3), 456-462.

Kant, E. (1994). *Fondements de la métaphysique des mœurs*. Paris: Delagrave.

Keene, S. (2011). Social bias: prejudice, stereotyping, and discrimination. *The Journal of Law Enforcement*, **1**(3), 1–5.

Kramer, P. and Bressan, P. (2015). Humans as superorganisms: how microbes, viruses, imprinted genes, and other selfish entities shape our behavior. *Perspectives on Psychological Science*, **10**(4), 464–481.

Libet, B., Wright, E.W., and Gleason, C.A. (1982). Readiness-potentials preceding unrestricted 'spontaneous' vs. pre-planned voluntary acts. *Electroencephalography and Clinical Neurophysiology*, **54**(3), 322–335.

Libet, B., Gleason, C.A., Wright, E.W., and Pearl, D.K. (1983). Time of conscious intention to act in relation to onset of cerebral activity (readiness-potential). The unconscious initiation of a freely voluntary act. *Brain*, **106**(3), 623–642.

MacKenzie, C. and Stoljar, N. (2000). *Relational Autonomy: Feminist Perspectives on Autonomy, Agency, and the Social Self*. Oxford: Oxford University Press.

Mandela, N. (2011). *Nelson Mandela by Himself: The Authorized Book of Quotations*. London: MacMillan.

Mele, A. (2009). *Effective Intentions: The Power of Conscious Will*. Oxford: Oxford University Press.

Milgram, S. (1974). *Obedience to Authority*. New York: Harper & Row.

Moller, A.C., Deci, E.L., and Ryan, R.M. (2006). Choice and ego-depletion: the moderating role of autonomy. *Personality and Social Psychology Bulletin*, **32**(8), 1024–1036.

Monroe, A.E. and Malle, B.F. (2010). From uncaused will to conscious choice: the need to study, not speculate about people's folk concept of free will. *Review of Philosophy and Psychology*, **1**(2), 211–224.

Montague, P.R. (2008). Free will. *Current Biology*, **18**(14), R584–R585.

Morar, N. and Beever, J. (2016). The porosity of autonomy: social and biological constitution of the patient in biomedicine. *American Journal of Bioethics*, **16**(2), 34–45.

Morse, S. (2006). Brain overclaim syndrome and criminal responsibility: a diagnostic note. *Ohio State Journal of Criminal Law*, **3**, 397–412.

Morse, S. (2008). Psychopathy and criminal responsibility. *Neuroethics*, **1**(3), 205–212.

Müller, S. and Walter, H. (2010). Reviewing autonomy: implications of the neurosciences and the free will debate for the principle of respect for the patient's autonomy. *Cambridge Quarterly of Healthcare Ethics*, **19**(2), 205–217.

Neufeld, K.A. and Foster, J.A. (2009). Effects of gut microbiota on the brain: implications for psychiatry. *Journal of Psychiatry & Neuroscience*, **34**(3), 230–231.

Nissani, M. (1990). A cognitive reinterpretation of Stanley Milgram's observations on obedience to authority. *American Psychologist*, **45**(12), 1384–1385.

Perkins, D.D. and Zimmerman, M.A. (1995). Empowerment theory, research, and application. *American Journal of Community Psychology*, **23**(5), 569–579.

Peterson, C., Maier, S.F., and Seligman, M.E.P. (1995). *Learned Helplessness: A Theory for the Age of Personal Control*. New York: Oxford University Press.

Racine, E. (2013). Pragmatism and the contribution of neuroscience to ethics. *Essays in the Philosophy of Humanism*, **21**(1), 13–20.

Racine, E. (2017). A proposal for a scientifically-informed and instrumentalist account of free will and voluntary action. Frontiers in Psychology, in press.

Racine, E. and Dubljević, V. (2016). Porous or contextualized autonomy? Knowledge can empower autonomous moral agents. *American Journal of Bioethics*, **16**(2), 48–50.

Racine, E., Nguyen, V., Saigle, V., and Dubljević, V. (2016). Media portrayal of a landmark neuroscience experiment on free will. *Science and Engineering Ethics*. Advance online publication. doi:10.1007/s11948-016-9845-3

Racine, E., Aspler J., Forlini, C., and Chandler, J. (2017). Broadening the lenses on complementary and alternative medicines in preclinical Alzheimer's disease: Tensions between autonomy and contextual determinants of choice. *Kennedy Institute of Ethics Journal*, 27(1), 1–41.

Rousseau, J.-J. (1992). *Du Contrat Social*. Paris: GF-Flammarion.

Rule, N.O., Freeman, J.B., Moran, J.M., Gabrieli, J.D., Adams, R.B., Jr., and Ambady, N. (2010). Voting behavior is reflected in amygdala response across cultures. *Social Cognitive and Affective Neuroscience*, 5(2–3), 349–355.

Ryan, R.M. and Deci, E.L. (2000). Self-determination theory and the facilitation of intrinsic motivation, social development, and well-being. *American Psychologist*, 55(1), 68–78.

Saigle, V., Dubljević, V., and Racine, E. (2015). The impact of a landmark paper on the concept of free will: reconsidering the legacy of the Libet et al. EEG experiments. *American Journal of Bioethics Neuroscience*, 6(4), 91.

Seligman, M.E.P. (1975). *Helplessness: On Depression, Development, and Death*. San Francisco, CA: W.H. Freeman.

Sherwin, E., Rea, K., Dinan, T.G., and Cryan, J.F. (2016). A gut (microbiome) feeling about the brain. *Current Opinion in Gastroenterology*, 32(2), 96–102.

Smith, K. (2011). Neuroscience vs philosophy: taking aim at free will. *Nature*, 477(7362), 23–25.

Strawson, P.F. (1962). Freedom and resentment. *Proceedings of the British Academy*, 48, 1–25.

Sudo, N., Chida, Y., Aiba, Y., et al. (2004). Postnatal microbial colonization programs the hypothalamic-pituitary-adrenal system for stress response in mice. *The Journal of Physiology*, 558(1), 263–275.

Vohs, K.D., Baumeister, R.F., Schmeichel, B.J., Twenge, J.M., Nelson, N.M., and Tice, D.M. (2008). Making choices impairs subsequent self-control: a limited-resource account of decision making, self-regulation, and active initiative. *Journal of Personality and Social Psychology*, 94(5), 883–898.

Wagner, D.D. and Heatherton, T.F. (2013). Self-regulatory depletion increases emotional reactivity in the amygdala. *Social Cognitive and Affective Neuroscience*, 8(4), 410–417.

Wegener, J.R., Ludlow, C.E., Olsen, A.J., Tortosa, M., and Wintch, P.H. (2007). Ego depletion: a contributing factor of hopelessness depression. *Intuition*, 3, 12–17.

Wegner, D. (2002). *The Illusion of Conscious Will*. Cambridge, MA: MIT Press.

Wolpe, P.R. (2002). The neuroscience revolution. *The Hastings Center Report*, 32(4), 8.

Chapter 21

Cognitive enhancement of today may be the normal of tomorrow

Fabrice Jotterand

Introduction

Contemporary debates on human enhancement often characterize the notion of enhancement as morally troubling because it undermines some deeply held beliefs concerning humanity and challenges what some deem as the appropriate use of biotechnologies to serve human ends. Most techniques and procedures beyond therapeutic aims raise the eyebrows of many concerned with the potential harmful implications of enhancement technologies. Fewer people might oppose the enhancement of human physiological and mental capacities if these techniques address specific diseases and disorders. If the goal of any intervention is therapeutic, it falls under the purview of medicine. Beyond therapy, where the aims of enhancement are not well defined, the moral landscape becomes less clear. To say that human enhancement strives to the betterment of the human condition or personal fulfillment does not make any normative claim robust enough to justify why we ought to embrace any type of enhancement. In addition, enhancement is often gauged against therapy to provide the backdrop necessary to discriminate what belongs to the domain of medicine and healthcare as well as to provide conceptual clarity. However, the conceptual distinction between therapy and enhancement has limitations that restrain our ability to demarcate the types of enhancements that could indeed benefit some individuals.

In light of these considerations, this chapter considers the use of cognitive enhancers in healthy individuals with cognitive impairment caused by mental disability, that is, disabled people with limited functioning but who are otherwise healthy and in a stable condition. The use of cognitive enhancers in healthy individuals with cognitive impairment illustrates the necessity to make more nuanced distinctions when examining the issue of cognitive enhancement technologies in order to harness their potential clinical benefits (Jotterand et al. 2015). The main objectives of this analysis are twofold: (1) to outline some of the problems associated with the attempt to distinguish the concept of enhancement from therapy. Both concepts assume that an understanding of health and normality is difficult to establish. I argue that these two concepts cannot be the basis for the assessment of the use of cognitive enhancers in people with mental impairment because the restoration of health or the attainment of normality cannot be realized in the case, for instance, of people with cerebral palsy (permanent state); and (2) to show the relevance of the distinction between two types of enhancement in the attempt to demonstrate why the notion of human enhancement, if it aims at the improvement of the quality of life of individuals with mental impairments, might become part of the therapeutic language of tomorrow. To this end, I introduce the concept of the clinical ideal to justify my claim about the use of cognitive enhancers in the mentally impaired.

The first section of this chapter examines the various conceptualizations of enhancement found in the literature, especially as outlined by Ruth Chadwick and Nicolas Agar. These two scholars present helpful categorizations of enhancement that will provide the basis for the development of the concept of the clinical ideal. Chadwick suggests four approaches to enhancement helpful to understand how the concept is used in current debates whereas Agar offers a valuable distinction between the objective ideal and the anthropocentric ideal. In the second section, building on the work of Agar, I argue that both ideals have limited clinical relevance and favor the clinical ideal, which allows for the evaluation of the concept of enhancement in the context of clinical interventions. The final section looks at the implications of the clinical ideal in relation to the use of cognitive enhancers in the mentally impaired. I contend that the moral evaluation of enhancement must take place within the context of the patient's clinical condition and in relation to the notion of quality of life.

Therapy and enhancement: Two evolving concepts

The pursuit of human enhancement through the transcending of human biological capabilities can be traced back to the 17th century in the writings of philosophers such as Francis Bacon (*The New Atlantis*, 1627), René Descartes (*Discours de la méthode pour bien conduire sa raison et chercher la vérité dans les sciences*, 1637), and Marquis de Condorcet (*Esquisse d'un tableau historique des progrès de l'esprit humain*, 1795) who suggested that human nature can be improved. Descartes, for instance, in his *Discourse on Method* (1637), advanced the idea that the human species can become master and possessor of nature whereas de Condorcet considered the possibility of human enhancement asking rhetorically:

> [M]ay it not be expected that the human race will be meliorated by new discoveries in the sciences and the arts, and, as an unavoidable consequence, in the means of individual and general prosperity; by farther progress in the principles of conduct, and in moral practice; and lastly, by the real improvement of our faculties, moral, intellectual and physical...? (de Condorcet 1795)

Contrary to the theoretical nature of these early reflections by Bacon, Descartes, and de Condorcet, some of the contemporary conceptualizations of human enhancement can be validated through use of specific technologies, either under development or already implemented. In current discussions, enhancement usually "characterize[s] interventions designed to improve human form or functioning beyond what is necessary to sustain or restore good health" (Juengst 1998, p.29). Cognitive enhancement encompasses "the amplification or extension of core capacities of the mind, using augmentation or improvements of our information processing systems" (Sandberg 2011). These core capacities include information acquisition (perception), active processing of information (attention), the ability to comprehend concepts and their meaning (understanding), and the capacity to store and retrieve information (memory). Cognitive enhancement comprises two types of technologies: (1) psychopharmaceutical enhancers (modafinil (Provigil'), dextroamphetamine, methylphenidate (Ritalin'), fluoxetine (Prozac'), donepezil (Aricept'), and propranolol (Inderal')); and (2) brain stimulation technologies (deep brain stimulation, transcranial direct current stimulation, and transcranial magnetic stimulation). In both instances, these modalities are used for therapeutic purposes, but when used in healthy subjects they may have enhancement effects such as increased focus and concentration; enhanced mood, memory, and learning; procedural learning tasks; motor learning; and visuomotor coordination tasks (Morein-Zamir et al. 2007, 2008; Sandberg 2011; Houdsen et al. 2011; Smith & Farah 2011; Gehring et al. 2012).

A closer look at how the concept of enhancement is used in the literature, however, reveals a different interpretation of what it entails. Two analyses are worth considering, the first by Ruth Chadwick and the second by Nicolas Agar. In her examination of the various interpretations of enhancement, Chadwick (2008) suggests four categories: (1) the "beyond therapy" view (President's Council on Bioethics 2003) holds the position that any procedures beyond therapeutic interventions are considered a type of enhancement. This interpretation, however, has deficiencies since some interventions might be viewed as enhancement but might be partly therapeutic too. In other words, the challenge is not only to define enhancement in relation to therapy. Therapy itself is likewise a difficult concept to describe and to suggest that the intentions behind any intervention can determine whether that particular intervention is therapeutic does not provide the necessary criteria to explain the difference between therapy and enhancement. (2) The additionality view holds that enhancement is understood quantitatively and signifies that "to enhance x is to add to, or exaggerate, or increase x in some respect." But in this case, without being "characteristic-specific," it is not possible to establish what counts as therapy or enhancement. To enhance cannot be understood in generic terms but requires focusing on particular capacities for moral evaluation (Chadwick 2008, pp.28–29). (3) The improvement view (enhancement understood qualitatively) means that to enhance is to improve or to make better. This view is not very useful because it does not provide the criteria necessary to determine what is considered an improvement. Any knowledge of the goals and intentions behind any request for improvement (i.e., enhancement) is contingent upon qualitative judgments and therefore makes it difficult to assess from a moral standpoint. (4) The umbrella view. Considering the challenges of the previous three approaches, this view offers a venue to assess enhancement case by case since it considers various potential changes: enhancement may be therapeutic (contra the "beyond therapy" view), enhancement may not add anything (contra the additionality view); and enhancement may not be an improvement (contra the improvement view) (See Table 21.1 for a side-by-side comparison of different definitions of enhancement.)

Table 21.1 Comparative table for the definition of enhancement

Concepts	Definitions	Key features and challenges
Views (Chadwick)		
Beyond therapy view	Any procedures beyond therapeutic interventions are considered a type of enhancement	Some interventions might be viewed partly as enhancement and partly as therapy
Additionality view	Enhancement is understood quantitatively: To enhance x is to add to, or exaggerate, or increase x in some respect	Requires focus on specific capacities to establish what defines therapy or enhancement Requires focus on particular capacities for moral evaluation
Improvement view	Enhancement is understood qualitatively: To enhance is to improve or to make better	Does not provide the criteria necessary to determine what is considered an improvement Contingent upon qualitative judgments Difficult to assess from a moral standpoint
Umbrella view	Enhancement is assessed case by case to determine its nature: ◆ Enhancement may be therapeutic (contra the beyond therapy view) ◆ Enhancement may not add anything (contra the additionality view) ◆ Enhancement may not be an improvement (contra the improvement view)	Too generic and does not provide a proper framework for conceptual clarity
Ideals		
Objective ideal (Agar)	All things being equal, technologies that produce greater enhancements are intrinsically more valuable than other technologies that generate fewer enhancement outcomes	Prudential value of enhancement is based on the degree to which it objectively enhances human capacity Requires setting objective standards to evaluate levels of enhanced capacities
Anthropocentric ideal (Agar)	Enhancement is evaluated based on a balance between the instrumental and intrinsic value of the enhanced capacity	Assigns value to enhancement relative to human standards and whether or not an enhancement is beneficial to the recipient based on standardized norms Requires a normative framework to determine how enhancement advances the well-being of the human species within the normal range of human capacities.

Table 21.1 Continued

Concepts	Definitions	Key features and challenges
Clinical ideal (Jotterand)	Enhancement is evaluated based on whether it enhances the physical, mental, and social capacities and the overall quality of life of an individual with mental impairment, where the baseline is disabled, and in the context of life (contextual standards)	Requires setting contextual standards Necessitates the development of criteria to evaluate degrees of enhancement in relation to quality of life

While Chadwick's categorization of enhancement is helpful in many respects (especially the umbrella view), Nicholas Agar presents a more fruitful classification to define human enhancement with regard to the focus of this chapter: the objective ideal and the anthropocentric ideal. The objective ideal states "an enhancement has prudential value commensurate with the degree to which it objectively enhances a human capacity." In other words, this approach holds that, all things being equal, technologies that produce greater enhancements are intrinsically more valuable than other technologies that generate less enhancement outcomes. This position is favored by transhumanists since they seek transcending our biological limitations, hence the maximization of human transformation. On the other hand, the anthropocentric ideal states: "some enhancements of greater objective magnitude are more prudentially valuable than enhancements of lesser magnitude ... or [it] assigns value to enhancement relative to human standards" (Agar 2014, pp.17, 27). This view asserts that the normal range of human capacities can be determined and that an enhancement can be evaluated based on a balancing between the instrumental and intrinsic value of the enhanced capacity. The enhancement of a human capacity has more intrinsic value if "instantiating more valuable internal goods" (i.e., an individual can directly benefit from enhancement only if that person enhances her/himself) whereas it has more instrumental value if it produces qualitative and quantitative superior external goods (goods that depends on social circumstances) (Agar 2014, p.29).

The objective ideal approach is a vision of enhancement supported by transhumanism but is somewhat at the margin of reflections pertaining to the clinical setting since it focuses on the radical enhancement of the human species. Transhumanists do not seek to address disorders or illnesses as such but rather improve the condition of the human species through technological means (posthumanity). Efforts to develop enhancement technologies, on their account, should be geared toward transcending the limitations of our brains and bodies as the survival of our species is closely linked to a technological future.

The anthropocentric ideal approach represents a more promising conceptualization with regard to the issue addressed in this chapter since enhancement is understood within the context of personal benefits (internal goods) and social benefits (external goods). Accordingly, the value of enhancement is not contingent upon an evaluative framework that fosters the maximization of the enhancement of human capacities but rather on the enhancement of the capacities of a particular individual based on his or her unique personal goals and values. However, according to Agar, the value of an enhancement is relative to the degree of prudential value in relation to the objective degree of enhancement within the normal range of human capacities. Beyond a certain point, he contends, enhancement becomes highly hypothetical, if not unrealistic, because it lacks objectivity with regard to its feasibility, and therefore its degree of prudential value decreases due to the uncertainty of how individuals could potentially be affected negatively by specific enhancements.

The anthropocentric ideal offers a potential way to evaluate the use of cognitive enhancers in people with mental impairments. Contrary to the objective ideal approach, the anthropocentric ideal specifically focuses on the established range of normal human capacities to evaluate the prudential value of particular enhancing technologies. As Agar puts it: "enhancement beyond human norms encompasses interventions whose purpose is to boost levels of functioning beyond biological norms" (Agar 2014, p.19). Thus, he assumes that the normal range of human capacities can be determined based on biological attributes of normal functioning, a claim I will challenge in the next section.

While Agar's distinction is valuable in many ways, I propose a third way of evaluating enhancement, especially since the objective ideal and the anthropocentric ideal, in my estimation, have limited clinical relevance. I am proposing what I call the clinical ideal based on the conceptualization of diseases as clinical problems (Engelhardt 1984).

The clinical ideal and enhancements

Diseases as clinical problems

In order to explain what I mean by clinical ideal, I will turn first to the concept of disease. Elsewhere, in relation to prison medicine, I outline why describing diseases as clinical problems (Engelhardt 1984) represents a more satisfactory account than descriptive explanations such as the ones adopted by Christopher Boorse (1975, 1977) and Leon Kass (1985) (Jotterand & Wangmo 2014).

Usually the term disease refers to a condition that is outside a set of functional standards within human biology. Boorse, for instance, states that the concept of disease can be understood in relation to the notion of species-typical levels of species-typical functions (Boorse 1975, 1977). Living organisms have an organization of biological functions that Boorse defines according to the following four criteria:

> 1) the *reference class* is a natural class of organisms of uniform functional design; specially, an age group of a sex of a species; 2) normal function of a part or process within members of the reference class is a statistically typical contribution by it to their individual survival and reproduction; 3) a disease is a type of internal state which is either an impairment of normal functional ability, i.e., a reduction of one or more functional abilities below typical efficiency, or a limitation on functional ability caused by environmental agents; 4) health is the absence of disease. (Boorse 1997, pp.7–8)

These criteria characterize a framework based on biostatistical analysis that provides standards of normal functioning (or healthy states) for which any deviation constitutes a state of disease. Kass likewise suggests that health, or the absence thereof, is based on biological standards. Each species displays specific bodily functionalities that can be recognized or determined by the specificities of its organism (Kass 1985).

Engelhardt, however, rejects the idea that diseased states can be established uniquely in terms of species-typical levels of species-typical functions. The indeterminateness of the forces of nature does not allow establishing a taxonomy of disease in relation to the functional organization of the body (Engelhard 1996). In addition, a species, the human species for instance, displays a range of characteristic polymorphisms that demands a specific understanding of what we mean by human nature, its purpose, and its values, which in turn justify the normality or abnormality of biological attributes (Engelhardt 1996). These values operate in a proscriptive manner since they inform what behaviors and habits (e.g., smoking, unhealthy diet) affect a person's health and therefore are considered detrimental to a person's health. In short, conceptualizations of health and disease are the result of a complex interaction between values and norms, in addition to biological standards, that qualify specific understandings of well-being and human flourishing (Engelhardt 1996).

In contrast to a descriptive account of disease, Engelhardt suggests considering diseases as clinical problems. The focus of clinical medicine is to address questions of pain, expectations concerning human form and grace, physical aptitudes, and mental capacities insofar as they affect a person's ability to achieve specific human goals and ends (Engelhardt 1984). This means that diseases are not considered as abnormalities outside the scope of what is deemed species-typical levels of species-typical functions. Diseases, as clinical problems, are contextualized based on the goals, ends, and notions of human flourishing held by individuals suffering from incapacitating ailments.

Based on the above analysis, I would argue that enhancement cannot be understood outside a normative framework and therefore without contextualizing enhancement in relation to notions of human flourishing, goals, and ends debates will remain somewhat entrenched in ideological positions rather than careful critical enquiry. In addition, as I stated earlier, my goal is to examine the concept of enhancement within the clinical context. To this end, I suggest that describing the notion of enhancement as a clinical ideal allows disentangling enhancement from extreme applications such as transhumanism, and from a narrow explanation based on a normal range of human capacities. In so doing, I consider the possibility of enhancement as a means to address clinical problems rather than as a way to promote the agenda of transcending the boundaries of human biology to achieve a posthuman state. But in order to justify this claim, an important distinction between two types of enhancement must be made as I show in what follows.

Enhancement #1 versus enhancement #2

Elsewhere I make an important conceptual distinction between enhancement #1 and enhancement #2 (Jotterand et al. 2015). Enhancement #1 is associated with the enhancement of healthy people with no cognitive impairments. This means that these individuals at baseline are nondisabled (normal function) and interventions in unhealthy individuals within that group of people imply therapy if the goal is to restore baseline function whereas in healthy individuals any intervention resulting in the enhancement of cognitive capacities is equated with enhancement #1. This is usually how people will find the distinction between therapy and enhancement in the bioethics literature. Enhancement #2, however, starts with a baseline disabled (limited function). These individuals are healthy but have cognitive impairments that limit their ability to function normally in major life activities (e.g., speaking, learning, communicating, interacting with others, caring, walking, sitting, and standing). This patient population is affected by genetic disorders, illnesses, or injuries that result in conditions such as cerebral palsy with accompanying developmental delay. Therapeutic interventions are warranted if these people have other underlying conditions that make them unhealthy but individuals who do not suffer from any ailment and have cognitive impairments might benefit from cognitive enhancers (enhancement #2).

The distinction between enhancement #1 and enhancement #2 is critical to the main argument presented in this chapter. Conceptually, it demonstrates that enhancement ought not to be limited to issues related to a posthuman agenda but could, in principle, become part of the therapeutic language of tomorrow under specific conditions. When enhancement is associated with the idea of transcending the limitations of our brains and bodies, it implies particular values and a vision of the human future that does not capture the realities and frailties of *this* life in contrast to a potential enhanced existence (posthuman condition). Such discourse relegates discussion about the clinical implications of enhancement to concerns about the current human condition, which is exactly what the strongest proponents of transhumanism want to transcend. In their view, human nature is improvable and can legitimately justify "reform[ing] ourselves and our nature in accordance with human values and personal aspirations" (Bostrom 2005 p.205). This perspective on the enhancement project does not include carefully any clinical aspects. It polarizes the debate over enhancement in ways that do not consider how some enhancements could actually be beneficial for the improvement of the quality of life of individuals whose levels of functioning cannot be regained or achieved in comparison to the nondisabled.

Clinical ideal

To reiterate Agar's earlier distinction, the objective ideal argues that the prudential value of enhancement is based on the degree to which it objectively enhances human capacity and therefore adopts a position that demands a set of objective standards to evaluate levels of enhanced capacities. On the other hand, the anthropocentric ideal assigns value to enhancement relative to human standards, that is, whether or not it benefits the recipient of an enhancement based on standardized norms. This approach assumes a normative framework that determines how enhancement advances the well-being of the human species.

In contrast to these two ideals, the clinical ideal assigned value to enhancement if it enhances the physical, mental, and social capacities and the overall quality of life of an individual with mental impairment (baseline disabled) in the context of life (contextual standards). The clinical ideal facilitates the introduction of the notion of enhancement in the clinical language without the need to contrast it or oppose it to the notion of therapy. It limits its scope of considerations to the life of a particular individual whose quality of life and well-being could be improved by cognitive enhancers. Ordinarily, therapy refers to the restoration, partial or complete, of the functions of an organism affected by a disease, a disability, or an impairment using appropriate treatment. Therapy is a useful concept that helps clinicians determine the scope of medical practice based on established baseline criteria. In addition, the language of therapy implies an understanding of what a healthy organism is, and the scope of normal functioning. But providing clear definitions and conceptual clarity to notions such as health, normality, impairment, and abnormality has been a notoriously difficult task. Notions of health and normality are concepts that have evolved, and continue to do, as progress in the biomedical sciences constantly reshapes the boundaries of what health is or what normal functioning for a biological organism entails. Health determinants are a complex combination of environmental, social, educational, and biological factors and therefore the expectation to achieve a healthy status is in constant flux. That said, the practice of medicine requires the establishment of standards that will allow practitioners to help their patients achieve the optimization of functioning established on standards of functionality within the boundaries of human biology.

On the other hand, enhancement #1 refers to the use of the power of biotechnologies to transcend the normal scope of species-typical human capacities to improve biological abilities. Therapy is limited to the scope of medicine whereas enhancement #1 applies to a different domain since its application focuses on healthy individuals who want to enhance some capacities. Yet, the line of demarcation between what constitutes therapy and enhancement respectively is increasingly fuzzy because some therapeutic interventions, devices, or drugs have dual effects or, as I point out in this chapter, enhancement #2 could be employed in the context of individuals whose levels of capacities are not consistent with being nondisabled.

Clinical ideal and mental impairment

Now that a general framework has been outlined as to the nature of the *clinical ideal*, we need to delve further into its implications for the mentally impaired. The overall purpose of the use of cognitive enhancers in individuals with cognitive impairment is to increase their quality of life. Various factors should be considered in the evaluation of quality of life in patients: (1) the nature of the illness and the side effects of treatment; (2) the patient's ability to perform basic everyday life activities such as walking, and preparing meals; (3) the patient's levels of independence, privacy, and dignity; and (4) the patient's experience of happiness, pleasure, pain, and suffering (Lo 2000 p.30). Typically, competent patients are able to communicate how they experience their own quality of life and to make decisions with regard to treatment options, their length, and their possible cessation. However, there are challenges to evaluate the quality of life of others either when these individuals are not competent, if they cannot voice their option, or when clinicians and family members make claims about the patient's quality of life. Studies have shown that there are discrepancies between how patients perceive their quality of life and what others say about it, especially since patients learn how to cope with illness, find support elsewhere, and in many instances remain positive about life in general and still find pleasure despite their situation (Lo 2000 p.30). Consequently, as Lo rightly states: "quality of life judgments by others might be inaccurate and biased unless they reflect the patient's own assessment of quality of life" (Lo 2000 p.30).

Another important issue to consider is the quality of life of individuals with cognitive impairments. Some authors argue that below a certain level of functionality considered essential for being a person, individuals cannot experience a meaningful quality of life and consequently clinical interventions are futile. Others are concerned that such evaluation of quality of life could lead to the contention that people who do not exhibit attributes of personhood will be "eliminated" because such lives are deemed "not worth living" (Schneiderman et al. 1990). The deep disagreement about the nature of the quality of life of the mentally impaired epitomizes the challenge for its assessment in those who are baseline disabled. So we need to be cognizant that ultimately the determination of the quality of life by others remains problematic.

With these considerations in mind, I turn now specifically to the criteria outlining the boundaries of the quality of life of the mentally impaired when potentially undergoing cognitive enhancement. The specific aims of the use of cognitive enhancers in people with mental disabilities include "increase[ing] functioning with a decrease in the need for health care and daily assistance" and "increase[ing] the individual's social capacity" (Jotterand et al. 2015 p.123). Social capacity refers to the ability to increase a person's ability to socialize and enter into relationships with others, which includes a broadening of personal choices, self-expression, and an optimization of how to engage in the world (Jotterand et al. 2015). Disabled people with mild cognitive impairments who exhibit self-awareness could benefit from cognitive enhancers with regard to a stronger social lifestyle and increased independence (Kittelsaa 2014). The case of the severely disabled is more complex and requires a more nuanced examination.

To make a stronger argument in my analysis, I will use the case of Ashley (often referred as the Pillow Angel Case). As far as I know, Ashley never underwent interventions involving cognitive enhancers. This case exemplifies how the use of cognitive enhancers could improve the quality of life of mentally impaired individuals such as children in a "permanently unabled state" for which there is no treatment currently available but who might benefit from enhancement technologies. Of course the assumption here is that these techniques are safe and efficacious—which I do not intend to assess in this chapter. My goal is simply to look at enhancement #2 and the implications of its applications in a real-life case.

Case scenario: Ashley, born in 1997, was diagnosed at the age of 6½ with static encephalopathy, a condition that keeps the brain of this patient population in a "permanent and unchanging state" and

for which there is no treatment. This severe brain disorder kept Ashley "permanently unabled" men-
tally (age of an infant) while her body continued to grow normally. The parents, concerned with the
care of Ashley, made three requests to the Children's Hospital and Regional Medical Center in Seattle,
Washington. They asked doctors to have Ashley undergo the following three procedures: (1) a hyster-
ectomy (to remove the uterus), (2) the removal of breast buds, and (3) a height attenuation to control
height and weight through hormone therapy. While alert and responsive to her surroundings Ashley
could not walk, sit up, grasp objects, use language or eat.

I will not analyze the case in terms of whether the doctors should proceed with the request of the parents as other scholars have already examined this case in depth (Gunther & Diekema 2006). Rather, my intention is to use the case of Ashley to raise the following question: Does the (hypothetical) improvement of the quality of life of individuals such as Ashley justify the use of cognitive enhancers?

To answer this question, it is essential to stress that context is crucial in our analysis. The clinical ideal asserts that any enhancement that improves the quality of life of an individual with a mental impairment is justified if it takes into account the particular situation of the person involved. This means that no general claims can be asserted as to the widespread usage of cognitive enhancers and as to the moral obligation of their use in the mentally impaired even if it has been proven that enhancers improve their overall quality of life. The prudential value of enhancement is determined by how clinical procedures—the degree of enhancement—meet realistic expectations set by the person caring for the (incompetent) cognitively impaired individual proportionally to the improvement of the quality of life, to the enhanced physical and mental functionality, and to the increased social capacity. Such a framework does not require setting any threshold since its goal is to promote the well-being of the individual subjected to these enhancements.

There is always a danger in using graphics to represent questions of enhancement and quality of life due to the difficulty in quantifying these concepts. However, the representation in Figure 21.1 might help visualize the concept of the clinical ideal.

The point here is to make a normative statement about the nature of enhancement with regard to quality of life. Consider the y-axis as representing quality of life on a scale of 0–10 and the x-axis as representing degrees of enhancement on a scale of 0–10. In addition, consider this graphic as a visual aid used by the person deciding whether to allow the use of cognitive enhancers in the mentally impaired under his/her responsibility. While I recognize that quantification of enhancement and quality of life are difficult to establish, Figure 21.1 is a graphic representation of the relationship between degrees of enhancement and degrees of quality of life. In an ideal scenario, the clinical ideal suggests that the increased degree of quality of life must be equal or superior to the degree of increased enhancement as indicated by the gray arrow. This means that quality of life should take priority over enhancement to avoid an instrumentation of patients through technological interventions. Unless enhancement technologies are safe, the focus should be on the improvement of quality of life and not on increased capacities. In a nonideal scenario (degree of enhancement higher than degree of quality of life—see Figure 21.1) where the improvement of quality of life of an individual requires extensive enhancements, each

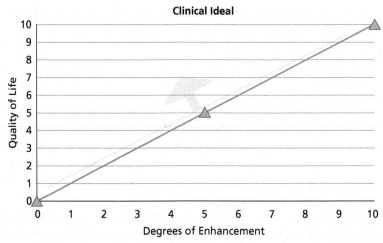

Figure 21.1 Graphic representation of the relationship between degrees of enhancement and degrees of quality of life. In an ideal scenario, the clinical ideal suggests that the increased degree of quality of life must be equal or superior to the degree of increased enhancement as indicated by the gray arrow.

case must be evaluated on its own facts using ethical guidelines that prevent the use of enhancement technologies that objectify the patient and promote his or her welfare and safety.

The clinical ideal does not necessitate setting a normal range of human capacities because the patient population concerned has a baseline disabled that is outside the normal range of human abilities. Therefore, any moral assessment as to the acceptability of an enhancement must be established on the subjective degree of improvement of quality of life. Furthermore, to protect and maximize the well-being of the recipient of the enhancement procedure, the degree of enhancement must be equal or superior to the degree of improved quality of life. In other words, if the degree of enhancement in relation to a baseline disabled is 4, ideally the degree of quality of life should be deemed to be 4 or higher.

The individualist approach of the clinical ideal is represented in Figure 21.2. It shows each individual starting at a different stage with regard to how quality of life is experienced or expressed and that the degree of enhancement does not necessarily improve well-being.

In Figure 21.2, we have three individuals with mental impairment whose quality of life has been determined to 2 at baseline disabled for person ◆, to 3 at baseline disabled for person ■, and to 4 at baseline disabled for person ✗. The clinical ideal does not provide a normative framework within specific parameters other than the fact that enhancements that do not increase quality of life above baseline line should be dismissed. In addition, quality of life is closely related to the condition of a particular person, how it is experienced and interpreted. Figure 21.2 reveals that there are different scenarios to

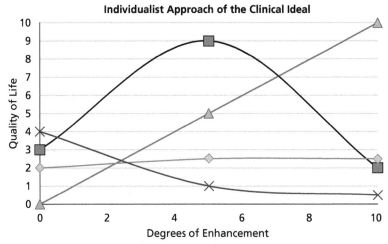

Figure 21.2 Graphic representation of various potential enhancement scenarios: ▲ represents the clinical ideal where increased degree of quality of life must be equal or superior to the degree of increased enhancement; ◆ represents the case of a disabled person where increased degree of enhancement minimally influences quality of life; ◆ represents the case of a disabled person where increased degree of enhancement decreases quality of life beyond a certain point; ✕ represents the case of a disabled person whereas increased degree of enhancement negatively impacts quality of life.

consider: in certain cases cognitive enhancement might indeed not improve quality of life (person ✕), in other cases the magnitude of enhancement only slightly changes the quality of life (person ◆), whereas in other instances there might be a dramatic improvement of quality of life up to a certain level of enhancement and then a drastic drop (person ◆). This is not a comprehensive list of scenarios nor is it based on empirical data, but it illustrates that each individual should be treated as a unique case, with specific needs and a particular experience and expression of well-being.

To summarize, the clinical ideal avoids the challenge of determining the normal range of human capacities, even in the case of the mentally impaired, or human standards. It simply states that, with regard to the mentally impaired, the moral evaluation of enhancement #2 can take place within the context of the patient's clinical condition and in relation to the notion of quality of life. There is a subjective and evaluative dimension that escapes any attempt to standardize enhancement #2. The alteration of the physical, mental, and social capacities of the mentally impaired is justified if it improves the overall well-being of the patient.

Ethical considerations

It is important to keep in mind that no study on the long-term effects of cognitive enhancers on healthy subjects has been completed so far and that no accepted ethical and legal framework has been fully developed for the use of cognitive enhancers.

Consequently, we ought to adopt a prudent approach in establishing standards for the use of cognitive enhancers in the mentally impaired. The following key points might provide a general framework for the development of more fine-tuned ethical approaches and guidelines (Jotterand et al. 2015). First, we need to keep in mind that the mentally impaired are a vulnerable population with specific needs and, therefore, there is always a danger of taking advantage of their condition. For this reason, extra layers of protection should be determined and implemented considering the experimental nature of the use of cognitive enhancers in the mentally impaired (enhancement #2). Second, the best-interest standard must guide any decision concerning the use of cognitive enhancers. Clinicians should always act in a way that promotes the well-being of their patients in accordance to the patient's being and wishes and avoid harm, even more so since these procedures would be experimental at this stage and no data is available as to their safety. Finally, there is a broader concern about the perception of the use of cognitive enhancers in the mentally impaired. Disability advocates stress the importance of allowing people with disabilities to have an adequate level of protection with regard to choice, empowerment, self-determination, and self-expression when possible (Fujiura & RRT Expert Panel on Health Measurement 2012). There is a danger that mentally impaired people may be used and abused for experimental purposes.

Conclusion

This chapter provides a philosophical justification of the potential clinical use of cognitive enhancers in the mentally impaired. The concept of the clinical ideal constitutes a possible approach that takes into account the complexities of clinical work in determining the overall experience of a patient with mental impairment in terms of expressed and perceived quality of life, and the interpretation of the notion of enhancement in medicine. While a philosophical justification is a first step toward a hypothetical acceptance case-by-case of cognitive enhancers in the clinical context, a more robust analysis needs to take place for the development of ethical guidelines and policies based on sound science. Without such analysis, public opinion might be influenced by various stakeholders concerned with the protection of the targeted patient population but might miss the opportunity to support an endeavor with potential clinical benefits.

References

Agar, N. (2014). *Truly Human Enhancement: A Philosophical Defense of Limits*. Cambridge, MA: The MIT Press.

Boorse, C. (1975). On the distinction between disease and illness. *Philosophy & Public Affairs*, 5, 49–68.

Boorse, C. (1977). Health as a theoretical concept. *Philosophy of Science*, 44, 542–573.

Boorse, C. (1997). A rebuttal on health. In: Humber, J.M. and Almeder, R.F. (Eds.) *What is Disease?*, pp.1–134. Totowa, NJ: Humana Press.

Bostrom, N. (2005). In defense of posthuman dignity. *Bioethics*, 19, 202–214.

Chadwick, R. (2008). Therapy, enhancement and improvement. In: Gordijn, B. and Chadwick, R. (Eds.) *Medical Enhancement and Posthumanity*, pp.25–37. Dordrecht: Springer Science + Business Media B.V.

Engelhardt, H.T., Jr. (1984). Clinical problems and the concept of disease. In: Nordenfelt, L. and Lindahl, B.I.B. (Eds.) *Health, Disease and Causal Explanations in Medicine*. Dordrecht: Kluwer.

Engelhardt, H.T., Jr (1996). *The Foundations of Bioethics*. Oxford: Oxford University Press.

Fujiura, G.T. and **RRT Expert Panel on Health Measurement** (2012). Self-reported health of people with intellectual disability. *Intellectual and Developmental Disabilities*, **50**(4), 352–369.

Gehring, K., Patwardhan, S.Y., Collins, R., et al. (2012). A randomized trial on the efficacy of methylphenidate and modafinil for improving cognitive functioning and symptoms in patients with a primary brain tumor. *Journal of Neurooncology*, **107**(1), 165–174.

Gunther, D.F. and Diekema, D.S. (2006). Attenuating growth in children with profound developmental disability. *Archives of Pediatrics and Adolescent Medicine*, **160**(10), 1013–1017.

Houdsen, C.R., Morein-Zamir, S., and Sahakian, B.J. (2011). Cognitive enhancing drugs: neuroscience and society. In: Savulescu, J., Meulen, T.R., and Kahane, G. (Eds.) *Enhancing Human Capacities*, pp.113–126. Oxford: Blackwell Publishing.

Jotterand, F. and Wangmo, T. (2014). The principle of equivalence reconsidered: assessing the relevance of the principle of equivalence in prison medicine. *American Journal of Bioethics*, **14**(7), 4–12.

Jotterand, F., McCurdy, J., and Elger, B. (2015). Cognitive enhancers and mental impairment: emerging ethical issues. In: Rosenberg, R.N. and Pascual, J.M. (Eds.), *Rosenberg's Molecular and Genetic Basis of Neurological and Psychiatric Disease* (5th edn.), pp.119–126. Philadelphia, PA: Elsevier, Inc.

Juengst, E.T. (1998). What does enhancement mean? In: Parens, E. (Ed.) *Enhancing Human Traits: Ethical and Social Implications*, pp.29–47. Washington, DC: Georgetown University Press.

Kass, L.R. (1985). *Toward a More Natural Science: Biology and Human Affairs*. New York: The Free Press.

Kittelsaa, A.M. (2014). Self-presentations and intellectual disability. *Scandinavian Journal of Disability Research*, **16**(1), 29–44.

Lo, B. (2000). *Revolving Ethical Dilemmas*. Philadelphia, PA: Lippincott Williams & Wilkins.

Morein-Zamir, S., Turner, D.C., and Sahakian, B.J. (2007). A review of the effects of modafinil on cognition in schizophrenia. *Schizophrenia Bulletin*, **33**, 1298–1306.

Morein-Zamir, S., Robbins, T.W., Turner, D., and Sahakian, B.J. (2008). State-of-science review: SR-E9: pharmacological cognitive enhancement. In: UK Government Foresight Mental Capital and Mental Wellbeing Project *Mental Capital and Wellbeing: Making the Most of Ourselves in the 21st Century*, pp.–16. London: The Government Office for Science.

Sandberg, A. (2011). Cognition enhancement: upgrading the brain. In: Savulescu, J, Meulen, T.R., and Kahane, G. (Eds.) *Enhancing Human Capacities*, pp.71–91. Oxford: Blackwell Publishing.

Schneiderman, L.J., Jecker, N.S., and Jonsen, A.R. (1990). Medical futility: its meaning and ethical implications. *Annals of Internal Medicine*, **112**, 949–954.

Smith, M.E. and Farah, M.J. (2011). Are prescription stimulants "smart pills"? The epidemiology and cognitive neuroscience of prescription stimulant use by normal health individuals. *Psychological Bulletin*, **137**(5), 717–741.

Chapter 22

Environmental neuroethics: Setting the foundations

Laura Y. Cabrera

At the intersection of brain and mental health, ethics, and environment

Humans have altered their environments in pursuit of self-improvement and for enhanced opportunities since ancient times, but the scope and impact of current changes are unprecedented. While human-initiated environmental changes have generally improved standards of living and provided new ways of protecting human health, paradoxically such changes have also contributed to global climate change, suburban sprawl, ecosystem loss, and increased health risks from toxic exposures and stress.

The various ways in which humans affect and are affected by their environments has been the focus of study across disciplines. For example, environmental ethics deals with questions concerning the value and moral standing of the natural environment, including whether the environment has instrumental or intrinsic value (Sagoff 2008) and whether present humans have particular moral obligations to protect and preserve the environment for future generations, and equally, if there should be limits imposed on the amount and rate of change that humans inflict in their environments; epigenetics considers the effect of present and past environments in gene expression (such as DNA methylation and hydroxymethylation, in addition to a plethora of histone modifications) (Crews 2010; Petronis & Mill 2011); and finally, environmental health focuses on those aspects of human health and disease affected by factors in the environment.

Questions emerging at the intersection of health, ethics, and environment, are complex and diverse. To properly address the associated breadth and depth, collaboration among environmental scientists, physicians, public health professionals, ethicists, lawyers, and policymakers is required. The variety of ethical issues at the nexus of environmental concerns and human health include issues of solid and hazardous waste disposal, regulation of pesticide use, environmental justice, and climate change. Those ethical issues explored by environmental health ethics commonly involve conflicts among fundamental values, including human-centered concerns, such as public health and human rights, as well as environmentally oriented concerns, such as protection of habitats and biodiversity (Shrader-Frechette 2006; Cranor 2011; Elliott 2011; Resnik 2009, 2012). Yet despite the wide range of disciplinary approaches examining the relationship between the environment and human health, and the attendant ethical implications, insufficient attention has been paid to the ethical, social, and legal implications that arise from environmental impacts and deleterious exposure in terms of brain and mental health effects.

In this chapter, I introduce environmental neuroethics as an emerging approach at the intersection of environmental studies, biomedical and social understandings of brain and mental health, and ethics. In order to make the case that environmental neuroethics is an important and needed approach, I provide in the next section examples of the various ways in which environmental exposures impact brain and mental health, both in positive and negative ways, and explore why those related impacts are a matter of moral concern. In the last sections I lay out the foundations for a first generation of environmental neuroethics and discuss key challenges ahead.

Environmental influences on brain and mental health

Humans have long recognized the interdependent relationship between environmental factors and health and disease. Today, it is widely accepted that health is determined by complex interactions between individual genetic predispositions, social and economic factors, behaviors, and perception of, as well as interactions with, the environment (Blas et al. 2011; Public Health Agency of Canada 2011). This chapter focuses on the interactions between the environment and brain health. Beyond neurological and physiological effects, environmental influences also impact psychological well-being, including interference both with the capacity to cope with normal stresses of life and to work productively and make useful contributions to community life. It is the latter group that I refer to as mental health impacts. I use brain health to refer to both brain and mental health. It is also important to clarify that while the environment encompasses all the physical, chemical, biological, and social forces that are external to a person (Smith et al. 1999), the focus here largely will be on the first three factors. However, I will touch on social aspects of the environment as this important area merits ongoing attention. As Marmot (2005) correctly observes, only in the most remote wilderness can a distinction be drawn between the social and other aspects of the environment.

Prior to the 1950s, some cases pointed to an association between neurological function and environmental changes; however, it was with the availability of more systematic epidemiological data that scientific consensus on that association was achieved. At present, there are some areas in which there is sufficient evidence linking brain and mental health impacts to environmental exposures, as in the case of chronic lead exposure and cognitive impairment, but in other areas the evidence is still limited as in the case of pesticide exposure and attention deficit hyperactivity disorder (ADHD). Among the barriers to identifying strong relational evidence is the fact that detrimental environmental exposures vary greatly depending on pollutant type, exposed population vulnerability, and dose and length of the exposure. In light of ongoing studies on environmental factors and their impact on brain health, in this section I review the evidence for environmental factors to be considered as developmental neurotoxicants and neurotoxic agents or as protective and enablers of brain health.

Developmental neurotoxicants and neurotoxic agents

Neurotoxic agents interfere with the functioning of the brain. Developmental neurotoxicants are agents that have the capacity to disrupt brain and nervous system development.

Metals, pesticides, and solvents

Based on robust animal data and large prospective epidemiological studies, there is well-documented evidence for the neurotoxic properties of certain metals (e.g., lead, methyl-mercury, arsenic, and manganese) as well as for certain industrial chemicals and solvents (e.g., organophosphate and organochlorine pesticides, polychlorinated biphenyls (PCBs), and toluene). For other substances such as cadmium, bisphenol A (BPA), and perfluoro-chemicals (PCFs) there is only limited animal data and a few, small epidemiological studies that would allow them to be categorized as suspected neurotoxicants.

Lead

United States medical journals began reporting on cases of paint-related lead poisoning in 1914. Patterson's influential paper "Contaminated and natural lead environments of man" (Patterson 1965), prompted investigations of lead as a pollution problem with human health impacts. A few years after the release of Patterson's findings, epidemiological studies demonstrated the inverse relationship between neurobehavioral functioning and blood-lead concentrations (Needleman et al. 1979, 1990). Higher levels of lead in childhood were found to be significantly associated with lower class standing in high school, increased absenteeism, lower vocabulary and grammatical-reasoning scores, poorer hand–eye coordination, and longer reaction times (Needleman et al. 1979). Recent studies are more worrisome, suggesting that chronic low-level lead exposure, even at prenatal stages, can adversely interfere with optimal brain development and functioning (Hu et al. 2006; Grandjean 2013), adversely affect cognitive skills later in life (Schwartz et al. 2000; Shih et al. 2007; Weisskopf et al. 2007; Grashow et al. 2013), and may even be linked to schizophrenia (Opler et al. 2008). Moreover, it is now widely recognized that there simply is no safe threshold that would avert deleterious effects of blood lead on neurobehavioral function (Lanphear et al. 2005).

Mercury

A well-known case documenting the neurological impact of a heavy metal comes from Minamata, Japan, where in the 1950s, hundreds of children were born with cerebral palsy, blindness, and seizures as a consequence of mercury exposure through eating contaminated local fish (Grandjean et al. 2010; Grandjean & Herz 2011). Other epidemiological studies have pointed to subclinical neurotoxicity in children from prenatal methylmercury exposure and its effects on attention, motor function, language, visual-spatial abilities, and verbal memory of those children (Grandjean et al. 1997; Debes et al. 2006).

Other metals

Arsenic is yet another neurotoxic heavy metal (Miodovnik 2011). In Japan, in 1955, arsenic-induced developmental neurotoxicity was reported as a result of consumption of contaminated dried milk powder. In follow-up studies, infants who survived the poisoning had lower IQs and higher rates of severe mental retardation when compared with control groups (Dakeishi et al. 2006). Other studies have found that arsenic exposure is associated with declines in verbal learning and memory (Wright et al. 2006), in attention (Tsai et al. 2003), and in intelligence in school-age children (Rosado et al. 2007).

There is also evidence that manganese (Lucchini et al. 2014) and cadmium (de Burbure et al. 2006; Rodríguez-Barranco et al. 2014) affect neurological function, either by acting directly on the brain or alternatively by adversely impacting hormones necessary to maintain healthy neurological function (Grandjean 2013; Grossman 2014). When accumulated in excess, copper and manganese can lead to liver failure, which in turn can trigger hepatocerebral disorders with morphologic changes similar to those seen in Alzheimer's disease, and neurological symptoms including parkinsonism[1] and cognitive dysfunction (Butterworth 2010).

Industrial chemicals and pesticides

A number of chemical compounds, such as pesticides (e.g., paraquat, maneb, DDT, and hexachlorobenzene) and solvents (e.g., toluene and trichloroethylene) are used widely worldwide and up to 201 compounds have been confirmed to be neurotoxic to humans with 8 specifically damaging to the developing brain (Miodovnik 2011). Long-term exposure to solvents and pesticides can affect verbal memory, attention, and spatial skills (Dick 2006; Tang et al. 2011).

Pesticides constitute the largest group of substances shown to produce neurological signs and symptoms (Grandjean & Landrigan 2006; Grandjean 2013). Research indicates that children born to mothers exposed to pesticides during pregnancy lag an average of 2 years behind in motor and spatial development when compared to children of mothers without exposure (Grandjean et al. 2006; Harari et al. 2010). Other studies have found an association between residential proximity to agricultural fields where pesticides were being applied during pregnancy and an increased risk of having children with neurodevelopmental disorders (Roberts et al. 2007; Shelton et al. 2014). However, few laboratory studies have addressed issues of concurrent exposure to multiple pesticides to consider the impact of synergistic or cumulative effects (Grandjean 2013). There is evidence connecting long-term exposure to certain metals, pesticides, and solvents with increased risk of developing Parkinson's disease and amyotrophic lateral sclerosis (Hertzman et al. 1990; Alavanja et al. 2004; Kamel & Hoppin 2004; Elbaz et al. 2009). Perrin and colleagues reported that paternal occupation as dry cleaners, working with the dry cleaning industry solvent tetrachloroethylene, was associated with a threefold increased risk of schizophrenia in offspring (Perrin et al. 2007). This study suggests that the spread of environmental contaminants not only directly affects individuals, but also increases adverse brain and mental health effects for future generations.

Other industrial chemicals with sufficient evidence to suggest neurotoxicity are PCBs. While PCBs were banned in the 1970s, they remain present in many products and materials produced before 1979. Various longitudinal studies have examined neurodevelopmental outcomes associated with exposure to PCBs, supporting modest associations with poorer attention (Jacobson & Jacobson 2003), IQ (Stewart et al. 2008), impulse control (Korrick & Sagiv 2008), and learning skills (Grandjean et al. 2001).

Industrialization and urban development

There are a myriad of ways in which industrialization and urban development impact brain and mental health.

Air and water pollution

Air pollution has a mixed exposure impact; in limited doses it is associated with headaches but in chronic exposure it is linked to neurodevelopmental deficits, including depression (Lim et al. 2012), fatigue, and reduced mental function (Calderon-Garcidueñas et al. 2002, 2004, 2008; Peters et al. 2006; Field et al. 2014). A study linked severe air pollution to lesions and growth abnormalities in the prefrontal cortex of children and dogs growing up in Mexico City (Calderon-Garcidueñas et al. 2008). Other studies have found that higher maternal exposure to particulate matter up to 2.5 micrometers in size during pregnancy is associated with greater odds of a child having autism spectrum disorders (Raz et al. 2014) and there are also potential associations between autism and estimated metal concentration (Windham et al. 2006). In the case of elderly adults, Lim and colleagues found that increases in particulate matter with an aerodynamic diameter less than 10 micrometers (PM10), such as nitrogen dioxide and ozone, are strongly associated in this population with emotional symptoms (Lim et al. 2012).

Water pollution represents yet another area of concern (Villanueva et al. 2013). More than 100 million people worldwide are chronically exposed to arsenic concentrations in drinking water at levels exceeding the World Health Organization guideline value (Miodovnik 2011). Ocean contamination represents another example of water pollution, in particular as bioaccumulation of heavy metals in the food chain or other chemical compounds found in water affect both brain and mental health.

Urban exposures

Another way in which brain and mental health are affected by the environment is through urbanization which introduces sources of urban pollution as well as toxic social circumstances (Bentall & Fernyhough 2008), including stress, fleeting social relationships, technology-driven remote interactions, higher crime rates, increased social competition, and wider socioeconomic disparities (Tost et al. 2015). Schizophrenia has the notorious recognition of being the mental health disorder with the best-examined link regarding urban birth, urban upbringing, and urban residence (Kelly et al. 2010; van Os et al. 2010). When compared with rural areas, psychiatric disorders are on average 34% more frequent in urbanized areas (Peen et al. 2010). A prospective study in Ireland found an incidence rate ratio of 1.92 for males and 1.34 for females in cities when compared to their rural counterparts (Kelly et al. 2010). Magnetic resonance imaging studies support the idea that on the functional level the degree of urbanization has implications for the neural social stress response system, and that the degree of exposure to urban environments is associated with para-anterior cingulate cortex activation and decreased gray matter volume in the prefrontal cortex (Lederbogen et al. 2011; Haddad et al. 2015). That finding plausibly relates to the high incidence of schizophrenia in the cities.

Proximity to noxious facilities

In addition to industry-generated chemicals polluting air and water supplies, simply living near to noxious facilities and waste repositories such as mining waste, nuclear plants, or natural resources extraction sites, can be harmful to brain health (Heiervang et al. 2010; Rabinowitz et al. 2015). A 2003 study noted that: "wastes from oil refineries can create health risks to facility workers and surrounding communities... including exposure to heat, polluted air, noise, and hazardous materials, including asphalt, asbestos, aromatic hydrocarbons, arsenic, hexavalent chromium, nickel, carbon monoxide, coke dust, hydrogen sulfide, lead alkyls, natural gases, petroleum, phenol, and silica" (O'Rourke & Connolly 2003). Brain health impacts from exposure to such material can include headaches, psychosis, and peripheral neuropathies (O'Rourke & Connolly 2003; Colborn et al. 2011). There are associated mental health problems as well. Two studies documented concerns from residents living in boom town communities in rural Northeastern Pennsylvania and Texas (Perry 2012; Stedman et al. 2012). They observed feelings of powerlessness, shock, and disgust among community members as a result of the rapid landscape change from agricultural and forested to industrial. They also noted stress associated with living within the vicinity of industrial facilities. Similar findings have been reported for communities who have experienced oil spills (Gay et al. 2010). There is a condition aptly labeled "eco-anxiety" which describes a chronic fear of living with the consequences of potentially unsound environmental practices (Watson 2008).

Extreme weather and changes in landscape

There is considerable evidence that human activities, such as those involved in industrialization and urbanization, are primarily responsible for climate change and the related side effects of global warming, including extreme weather fluctuations leading to more frequent droughts, deadly heat waves, and more damaging rainstorms, as well as landscape changes precipitated by the melting of glaciers, coastal flooding due to rising sea levels, and wildfires (APA Task Force 2011; Resnik 2012).

Climate change impacts on public health have primarily focused on physical health outcomes, linking changes in availability of food and water quality and quantity to malnutrition and vector-borne diseases, examining increasingly evident and oftentimes more visible health issues such as respiratory, cardiovascular, and kidney diseases. However, literature on the impact of climate change on mental health is now emerging (APA Task Force 2011). For example, research by Cunsolo and colleagues documents the impact of climate change on the mental health and well-being of Inuit people over time (Cunsolo Willox et al. 2013). Cunsolo and colleagues demonstrate how temperature increases and the concomitant thinning of the ice layer has prevented the Nunuvut Inuit from engaging in cultural food procurement practices, such as hunting, as frequently as they had done so in the past (Cunsolo et al. 2013). This cultural shift was found to be associated with mental health consequences including increased family stress and amplification of previous traumas and mental health stressors.[2] The term *solastalgia* (Albrecht et al. 2007) describes a particular distress experienced as a consequence of being deprived of the solace one derives from contact with one's native environment. Other research has focused on mental health impacts such as acute and post-traumatic stress disorder, depression, anxiety disorders, drug and alcohol abuse; higher suicide attempts and completions; and the general increased vulnerability of those with preexisting severe mental health issues suffering from climate change associated loss, disruption, and displacement (Fritze et al. 2008; Berry et al. 2010, 2011). Clearly, changes in the environment can have important direct impacts on the mental health and well-being of residents as well as disrupting underlying determinants of health.

Other factors: Diet and social support

There are other environmental exposures, such as diet and social support that can also negatively affect brain and mental health.

Diet

Over the past decade there has been good evidence for the influence of dietary factors on specific molecular systems and mechanisms involved in mental function. Dietary factors can negatively affect brain and mental health (for how diet can improve mental health see 'Diet' in the 'Protective/health-enabling environmental factors' section). The dramatic increase over the past 100 years in the intake of saturated fatty acids, linoleic acid, and *trans* fatty acids, together with a decreased consumption in omega-3 fatty acids might well be related to the elevated incidence of major depression in Western countries (Gómez-Pinilla 2008). Diets high in saturated fat are associated with reducing molecular substrates that support cognitive processing and increase the risk of neurological dysfunction (Greenwood & Winocur 2005). Studies have also found that higher blood glucose levels exert a negative influence on cognition (Kerti et al. 2013). A systematic review by O'Neil and colleagues found a significant relationship between unhealthy dietary patterns and compromised mental health in children and adolescents (O'Neil et al. 2014).

Social support

Among all possible stressors, negative parenting behaviors have the most profound effect on development, including increased risks for learning disabilities, emotional and behavioral abnormalities, and a broad range of disorders including depression, post-traumatic stress disorder, borderline personality disorder, anxiety, and schizophrenia (Tost et al. 2015). Individuals with a history of childhood maltreatment (including physical and emotional abuse and neglect) are likely to suffer from depression, anxiety disorders, and post-traumatic stress disorder (Teicher & Samson 2013), and to exhibit learning disabilities. The quality and extent of social support, or lack thereof, is also important in adulthood.

Ethnic minority status is a well-recognized social–environmental risk factor for schizophrenia (van Os et al. 2010). Perceived discrimination and social isolation from the larger community are likely psychological mechanisms linking increased mental health risks with minority status (Tost et al. 2015). Finally, the current explosion of digital technology and the ubiquitous exposure to it (e.g., computers, smart phones, search engines, and apps) is not only changing the way we communicate and live but is also altering our brains in profound ways (Small & Vorgan 2008), gradually strengthening new neural pathways while simultaneously weakening old ones. Emerging research links these changes to increases in social isolation, Internet addiction, ADHD, and related mental health risks.

Protective/health-enabling environmental factors

Research related to environmental influence and brain health has typically focused on the identification of adversity-related factors and their neural underpinnings. Less attention has been paid to those environmental exposures that in fact might be protective and enablers of brain and mental health; the following are examples of emerging evidence.

Experience with nature

A deep-seated connection between mental health and the natural world would, from an evolutionary perspective, be quite plausible. Several cross-cultural studies reinforce this view. Certain indigenous worldviews see identity, conceptions of the self, and mental wellness as linked directly and intimately to the environment, and one's ability to hunt, trap, fish, forage, and travel on the land, as well as to practice cultural traditions related to being "on the land" (Kirmayer & Valaskakis 2008; Vukic et al. 2011). This understanding, of the importance of land and land-based activities, highlights connections between individuals and their environments on well-being and mental health.

A growing body of literature demonstrates that exposure to natural landscapes, or to their composite features such as flora and fauna, has beneficial effects for child development, well-being, mental health, mood, mortality, and recovery from illness (Frumkin 2001; Maas 2006; Haluza et al. 2014). This connection is what David Cumes describes as "wilderness rapture" (Cumes 1998). That rapture includes feelings of awe, wonder, and humility, a sense of comfort and connection to nature, and feelings of renewal and vigor. Although the topic is under-researched, there is evidence associating exposure to natural environments with restoration from attention fatigue, recovery from stress, feelings of relaxation, as well as enhanced mental alertness, attention, and cognitive performance (Bowler et al. 2010; Tost et al. 2015). A large-scale epidemiological study revealed a relationship between the abundance of green space and human health (Maas 2006). This finding can be inversely related to the ill-health effects observed in urban settings, as a key difference between urban and rural settings is the available green space. Although the neurobiological effects of experiencing nature require further study, the current data suggests that human contact with nature is more than an aesthetic luxury. This knowledge has recently been integrated in horticultural therapy and wilderness therapy (Frumkin

2001), as well as in animal-facilitated therapy for treatment of psychiatric conditions (Matuszek 2010).

Further research is needed into the specific neural mechanisms linking nature exposure to mental health benefits, as well as to variation based on past experiences, social context, age, and gender.

Enriched environments

Seminal observations in the 1960s on the effect of environments rich in sensory stimuli to gross brain structures stimulated further research on the influence of enriched environments to brain function and behavior (van Praag et al. 2000). Nowadays it is recognized that environmental enrichment profoundly and positively affects the central nervous system at the functional, anatomical, and molecular level (Mora et al. 2007; Baroncelli et al. 2010), during critical developmental periods and in adulthood. In turn, this promotes various plastic responses in the brain, ranging from biochemical parameters to the arborization of dendrites and neurogenesis (van Praag et al. 2000).

Environmental enrichment improves quality of life by providing a "combination of multisensory/cognitive stimulation, increased physical activity, enhanced social interactions and by eliciting natural explorative behaviors" (Baroncelli et al. 2010, p.1093). While environmental enrichment studies are largely based on rodent research, they provide insights for application to humans. The admittedly minimal data from human studies and environmental enrichment suggests that a higher level of mental and physical activity reduces disease risk, including those associated with compromised brain function and cognition (Kramer et al. 2006; Baroncelli et al. 2010). A study with human subjects revealed that exercise promotes neurogenesis in the brain, suggesting that new proliferating neurons might possibly contribute to the effects of exercise on memory and learning (Pereira et al. 2007). Another study with children found environmental enrichment to be somewhat effective in ameliorating autism symptoms (Woo & Leon 2013). Considering evidence linking sustained stress with impaired immunity, disease, and neurological changes characteristic of major depressive illness and chronic anxiety disorders, it stands to reason that stress-reduced environments can positively impact brain health (Crews 2010). These findings corroborate the need for expanded attention to environmental enrichment paradigms both for therapy as well as to avert neurological disorders.

Other factors: Diet and social support

Dietary factors and social support factors were mentioned in the section on "Developmental neurotoxicants and neurotoxic agents" in relation to adverse health effects. Here I highlight the potential of dietary factors and social support for promoting brain and mental health.

Diet

Several studies point to the relationship between nutritional factors and mental health (O'Neil et al. 2014). By regulating neurotransmitter pathways, membrane fluidity, synaptic transmission, and signal transduction pathways, dietary factors can affect multiple brain processes (Gómez-Pinilla 2008). Lai and colleagues systematically reviewed the literature to conduct a meta-analysis of the association between dietary patterns and depression. They found that a healthy diet—one high in fruit, vegetables, fish, and whole grains—was significantly associated with reduced likelihood of depression (Lai et al. 2014). Another meta-analysis examined the association between adherence to a Mediterranean diet and the risks of stroke, cognitive impairment, depression, and Parkinson disease. High adherence to this diet was found to be associated with reduced risks for ischemic stroke, depression, and mild cognitive impairment (Psaltopoulou et al. 2013). Diets rich in antioxidants have become increasingly popular as they are linked with positive effects on neural function and might lower risks of developing age-related neurodegenerative diseases (Joseph et al. 2007). Finally, studies have shown that probiotics can improve anxiety levels, perceptions of stress, and mental outlook (Bested et al. 2013). Not surprisingly the production of neurotransmitters such as serotonin, as well as the function of various neurons, is highly influenced by the billions of "good" bacteria comprising the intestinal microbiota. Although these studies underscore the important effect of dietary factors on the brain, further work is necessary to determine mechanisms of action and how to best enhance brain function and mental health.

Social support
The extent and the quality of human social bonds influence positive affect, self-esteem, recovery, and risk for mental illness (Tost et al. 2015). Seeman and McEwen's study provides support for the hypothesis that the positive influence of social environment features reduces neuroendocrine reactivity (Seeman & McEwen 1996) which in turn decreases the odds for exaggerated physiological stress responses and negative emotional states (Tost et al. 2015). Laboratory experiments have corroborated the benefits of social support in managing psychological stress (Chen et al. 2011). As mentioned in the section on enriched environments, dynamic social interactions are critical, in particular during childhood between caregivers (particularly parents) and infants. Parenting behavior has demonstrable long-lasting effects on development. Stable, supportive, and loving caregiver behavior promotes one's ability to form trusting and empathetic social relationships as well as mitigating the detrimental effects of adverse life effects (Shonkoff 2012). Similarly, when older adults are satisfied with their available social support, they report a better health status than those who are dissatisfied (White et al. 2009).

Despite the available evidence, knowledge and understanding as to the effects of environmental changes and toxic exposures for brain and mental health is still in its infancy. Further research is required to gain a more complete understanding of the current and possible brain health impacts stemming from environmental influences. Given the current rapid rate of environmental changes, ongoing study both of risks and resilience mechanisms offers hope for developing preemptive and therapeutic approaches for brain and mental health disorders.

Setting the foundations for environmental neuroethics

The idea of combining neuroethics with the study of the effects of environmental exposures stems from cross-cultural research with a First Nations family, undertaken by researchers at the University of British Columbia. That collaboration examined experiences and understandings of early-onset familial Alzheimer's disease (EOFAD). The EOFAD study was a reminder about the need to consider how cultural differences mediate one's perceiving and valuing of the environment. Equally, it emphasized the increased vulnerability of certain populations to environmental changes and exposures. A closer look at the literature prompted recognition that ethical and social implications from environmental exposures and changes for brain and mental health were underexplored areas (Cabrera et al. 2016). This set the foundation for environmental neuroethics. This novel neuroethics approach aims to better understand and respond to those ethical and societal issues raised by the impact of environmental exposures on brain and mental health. The initial environmental neuroethics work largely focused on brain and mental health impacts from anthropogenic environmental changes—such as those brought forward by technological change, new industries, economic expansion, and population growth (Illes et al. 2014). As research moves forward, other areas such as the impact of natural environments, enriched environments, and social and technological environments are included within the scope of environmental neuroethics.

Why environmental neuroethics matters: A neuroethics perspective

Several neuroethics reasons support the claim that environmental neuroethics matters.

Brains matters

Environmental neuroethics matters because brain and mental health matter. The brain is arguably the most complex organ crucial to human well-being. It orchestrates all bodily functions, is in charge of affective and cognitive capacities of reasoning and decision-making, and is considered the seat of self-identity, autonomy, and self-narrative. The complicated development of the brain, as well as its plasticity, make it uniquely sensitive to environmental exposures, especially during development (Grandjean 2013). Such exposures can trigger either adverse or positive effects that can be manifested over a lifetime. A growing body of evidence suggests that neurological function in the elderly can be manifestations of exposures from earlier years (Landrigan et al. 2005). Likewise, evidence suggests that the brains of fetuses, infants, and children are uniquely sensitive to neurotoxicants at levels far below those known to harm adults.

Public health

Global impact studies have identified brain and mental health disorders as key contributors to disabilities and morbidity worldwide. Those disorders have a critical public health, societal, and economic impact (Prüss-Üstün & Corvalán 2006). Developmental disabilities or developmental delays (including mental retardation, autism, ADHD, and learning disabilities) occur in about one in six children in the United States (Centers for Disease Control and Prevention 2015). While the impact of the environment on brain and mental health cannot easily be measured and assessed with certainty, cumulative interactive effects are likely to be profound. Considering the role of the brain in defining a person, even small deficits can negatively impact mental health.

Policy implications

It is estimated that over 25% of the burden of human illness worldwide can be attributed to modifiable environmental conditions (Prüss-Üstün & Corvalán 2006). While urbanization and industry-precipitated environmental changes are unwittingly creating toxic environments for brain and mental health, we potentially could modify our lived environments to foster brain and mental health. Certain scholars have suggested that shaping our environments to optimize brain and mental health is likely to be more effective and less ethically problematic than genetic or pharmaceutical modifications of our capacities (Levy 2012; Cabrera 2015). In terms of exposures, approximately 200 chemicals have been found to be neurotoxic in humans (Grandjean & Landrigan 2006). However, thousands of chemicals commonly found in the environment and in consumer products have not undergone extensive neurotoxicity testing. Since even low levels of exposure can have far-reaching brain and mental health effects, prior to entering the market there should be more rigorous regulation in assessing the neurotoxic profile of chemicals and other substances (Cranor 2011).

Social justice and human rights

People live and work in diverse environments that vary in type and amount of chemical exposure. Equally, living experiences vary according to factors such as occupation, education, income, and ethnicity, leading to environmental risk and disease inequalities. Living in a healthy environment, one that optimizes well-being and mental health, should be considered a human right. Yet exactly what factors constitute a healthy environment requires further elaboration, as environments can vary dramatically from those that are truly awful to those that promote positive brain development.

What features are requisite for a given environment to be regarded as "healthy?" Indeed, while conceptually this might be difficult to define, the conversation opens the door for discussion of procedural rights to have a say over changes to one's environment that might affect brain and mental health. From that perspective, governments have a duty to promote healthy environments for all their citizens, environments that promote positive brain development and function, and to protect their citizens from the consequences of toxic environments.

Generational justice and social responsibility

After decades of research, evidence suggests that many chemicals in the environment and in consumer products not only affect those currently exposed, but also can damage the next generation's brain functioning. If nothing is done to mitigate environmental risks, future generations will bear extraordinarily disproportionate risks. Not only is the developing brain more vulnerable to environmental exposures, there are epigenetic mechanisms through which the dire consequences of environmental exposures are realized in the next generation. In his video "Little Things Matter," Bruce Lanphear argues that the widespread and chronic exposure to brain-damaging toxins can lead to an increase in the number of US children who are mentally challenged, from 6 million to more than 11 million (Canadian Environmental Health Atlas 2014). Thus in terms of environmental impacts for brain and mental health, the related concerns are intergenerational, and problematic both from a temporal distributive justice perspective as well as participative justice's grounds of equal rights to self-determination in societal decision-making (Shrader-Frechette 2006). Future generations are, in principle, unlikely to consent to environmental pollution depletion of natural resources, and other environmental changes that might pose a risk to their brain and mental health. To this concern, certain scholars make the argument that our choices will cause a different person to exist—the identity problem—so that future individuals have no moral grounds for lodging complaints against choices made by the present generation (Parfit 1971). Yet in the ethical assessment of acts that will impact future generations it does not in fact matter who the members of the future generation turn out to be. What matters is that future generations will suffer the consequences of our actions, and as such have serious grounds for complaint. John Rawls' *veil of ignorance* concept suggests that any reasonable person, not knowing to which generation, social class, or intelligence bracket they would belong, would accept as fair the principle of equal apportionment of risks, resources, and goods (Rawls 1971).

Principles of equality in the distribution of opportunity such as this as well as those suggested by the advocates of the capability approach seem intuitively fair (Nussbaum & Sen 1993). Other arguments touch on the notion of a social contract among all generations. According to this view, as beneficiaries of ancestral legacies, members of present generations have similar duties to future persons—regardless of whether or not those future persons can reciprocate their giving; regardless of whether or not the future persons are asked if they wish to receive benefits; and finally regardless of the degree to which the present persons might know the needs of future generations (Shrader-Frechette 2006, p.102). Minimally, we know that they will require clean water and air to flourish as human beings. Thus, at the simplest level, regardless of which generation they happen to belong to, each and every person will have the same *opportunity* as any other to use and benefit from resources. There is then a moral duty to provide future generations with the same level of opportunity as they would have had if the present generation had not depleted some resources or polluted their environment (Shrader-Frechette 2006, p.103). This moral duty is recognized in national and international documents, such as the 1972 preamble to the Stockholm Declaration on the Human Environment, which affirms that humans have "a solemn responsibility to protect and improve the environment for present and future generations" (United Nations Environment Programme 1972). In the United States, the first stated goal of the National Environmental Policy Act (NEPA) is to "fulfill the responsibilities of each generation as trustee of the environment for succeeding generations" (NEPA 1969). NEPA also states that present generations should not impose undue risks on "a future generation ... greater than those acceptable to the current generation" (NEPA 1969). There are approximately 50 federal statutes in the United States that contain explicit reference to future generations, many of which aim at preserving some current benefit for future generations (Shrader-Frechette 2006).

Considering the sensitivity and vulnerability of the brain to environmental exposures, environmental neuroethics reflects a timely, necessary research trajectory. The focus and investigation methods put forth by neuroethics provide a meaningful foundation to address the brain and mental health impacts of environmental change. Neuroethics-informed perspectives on environmental change and exposure impacts provide a theoretically and empirically based understanding of the brain and its relationship with the environment at the individual, community, and societal level. A neuroethics perspective also can help to prioritize and rank important environmental change causes, consequences, or responses as well as to make connections to research and concepts from other social, engineering, and natural science fields. In sum, a neuroethics lens is crucial to understand the ethical and social consequences of environmental change, to help reduce the adverse brain and mental health impacts from environmental change, and to enable effective individual and social adaptation strategies.

Useful principles and frameworks

The interaction between brain health and the environment puts a new twist on how we might understand and address those ethical and societal issues at stake. However, there are issues similar to those already explored in bioethics and environmental ethics, including conflicts between individual well-being and the common good, as well as different perspectives on justice (e.g., social or generational). Environmental health ethics scholars have provided guiding principles (Resnik 2012), some of which are relevant to environmental neuroethics, in particular those that appeal to the ethical principle of avoiding harm (nonmaleficence) and ensuring human welfare (beneficence), fairness and justice, utility, responsibility, and respect for rights. Determining the proper role of such principles in environmental neuroethics will require further attention, yet they currently have value in drawing attention to the environment's moral significance for its contribution to brain and mental health.

Currently, we face difficult environmental neuroethics questions involving conflicts among values, such as the value of human health, the intrinsic and instrumental value of the environment, and the value of economic development for societies. Human health and particularly brain and mental health, can be seen as a distinct human value. Brain health helps people to achieve their goals, promotes equality of opportunity (Daniels 2008), and prevents suffering and disability. From a social perspective, brain health deters money and productivity costs, associated with brain disease and mental disorders. There are, however, other values that modern society cherishes including economic growth. Environmental changes undertaken in the pursuit of economic growth might unwittingly create toxic environments for brain and mental health. In addition to the aforementioned value conflicts, issues at the intersection of brain health and the environment involve a

> ## Box 22.1 Useful frameworks for environmental neuroethics
>
> **Normative:** relational, global bioethics.
> **Justice:** social and environmental justice, human rights.
> **Vulnerability and resilience:** vulnerability as layers, individual vs. social
> vulnerability.
> **Social constructionism:** sociology of knowledge, traditional ecological knowledge.

clash of different understandings about the environment and its role in promoting or hindering brain health. Given those complexities, it is a crucial part of environmental neuroethics to bring together conceptual and methodological frameworks from diverse disciplines that do not often directly engage, including neuroscience, sociology, environmental science, and ethics (Illes et al. 2014). This drawing together of multidisciplinary insights launches the field of neuroethics and environmental ethics into broader interactive biomedical and social understandings of environmental change.

There are a variety of frameworks that are relevant to environmental neuroethics, as illustrated in Box 22.1.

Ethics frameworks

Normative frameworks focused on examining general concepts and principles of ethics, as well as empirical ethics frameworks focused on moral reasoning, judgment, behavior, development, and learning, are broad frameworks to explore environmental neuroethics issues. Van Rensselaer Potter's global bioethics is another useful normative framework. Potter's *Global Bioethics* (Potter 1988) focuses on integrating biological knowledge with diverse humanistic knowledge to systematically establish medical and environmental priorities for acceptable survival at a global scale. Rather than viewing neuroethics and environmental ethics as separate and distinct entities, within this strategy those two branches are harmonized to a consensual point, stressing the importance of brain and mental health consequences of environmental change, as well as the intimate human health and a healthy environment relationship. Care ethics is another useful framework, capturing the importance of the environment and human health and well-being relationship. The distinct relational perspective incorporated by care ethics and feminist approaches acknowledges that an individual's health and well-being are inextricably linked to the flourishing of others, of communities, and the natural environment (Held 2006; Slote 2007). The conceptual contribution frames human beings as "constituted [simultaneously] both as organisms within systems of ecological relations, and as persons within systems of social relations" (Ingold 2000, p.3). Such a view is intrinsic to certain indigenous considerations of an individual's relationship to their community and ecosystem. A quotation from a First Nation's member captures this well: "I am part of an environment and if my environment suffers I suffer as well." A relational framework rejects privileging individualistic values and instead promotes communal values, viewing the individual and community as vitally interconnected. From an ethical perspective, a relational framework is helpful to address individual and community interdependence with the environment, broadly construed to include natural, built, social, or cultural.

Social justice

A social justice framework considers unequal distribution, in the particular case of environmental neuroethics, distribution of exposure to neurotoxicants and environmental changes differentially impacting various populations in relationship to sociodemographic determinants, such as income, occupation, education, and ethnicity (Commission on Social Determinants of Health 2008). Those with lower socioeconomic status generally have experienced greater exposure to detrimental environmental conditions both within their homes and as part of their occupations. Minority groups, such as indigenous communities, might suffer greater burdens of exposure if the natural resources they hunt and fish are contaminated by industries operating near their communities. A social justice framework additionally takes into consideration vulnerable subpopulations, such as children, pregnant women, and the elderly, who due to age, genetics, or health status are especially susceptible to the effects of environmental exposures. From a social justice perspective, when making environmental planning decisions, such as choosing where to build an industrial site, locate a waste dump, or where to establish parks and green areas, communities and nations have a societal obligation to minimize and mitigate environmental injustices. There is an obligation to better regulate and ensure safety in consumer products and in the workplace. These decisions should be transparent, democratic, and fair, and should solicit and take into account in deliberation from those especially vulnerable to environmental risks (Resnik & Portier 2008).

Environmental justice

A related and important framework is that of environmental justice, which is based upon the idea that the health and well-being of the members of a community, both individually and collectively, are a product of the relationship between four interrelated environments: natural, built, social, and cultural/spiritual (Lee 2002). It represents a transformative public discourse on what constitutes truly healthy, livable, sustainable, and vital communities. It expands the discourse concerning public health and environmental risks to include issues of multiple, cumulative, and synergistic risk. Environmental injustice occurs whenever some individual or group bears disproportionate environmental risk, has unequal access to environmental goods, such as clean air, or has less opportunity to participate in the environmental decision-making (Shrader-Frechette 2006). In Western societies, poor and minority people who are disadvantaged in terms of education, income, and occupation have been especially affected by environmental injustices, as they are more likely than affluent Caucasians to live near polluting facilities, eat contaminated fish, and be employed at risky occupations. Environmental injustice is wrong because it presents multiple threats to well-being and health (both for present and future people), it violates the principle of prima facie political equity, including its components of distributed and participative justice (Shrader-Frechette 2006), including aspects of human rights to equal protection, due process, consent, and compensation. An environmental justice framework sets the foundations for conducting community-driven science and holistic, placed-based, systems-wide environmental protection (Lee 2002). It strives for a more equitable distribution of environmental goods and burdens, while at the same time it fosters greater public participation in evaluating and apportioning these goods and burdens (Shrader-Frechette 2006).

Vulnerabilities and resilience

Vulnerability has to do with a lack of capacity within individuals and communities to cope with, recover from, and adapt to external stresses placed on their livelihoods and well-being. While vulnerability is inherent in all human beings, it comes in different degrees. For example, children and elderly adults are considered more vulnerable to environmental exposures than young adults because their brains are either in development or because aging has mitigated the capacity for their brains to adapt to environmental stressors. However, vulnerability and resilience are not just a biological matter, broader issues of social, economic, and political inequality also affect them. Thus, economically disadvantaged, and racial and ethnic minorities are also vulnerable populations.

Florencia Luna nicely captures that the concept of vulnerability as a label is not sufficient to describe the underlying complexities. A better approach is to see vulnerabilities as layers. Different layers in operation can render a given individual more vulnerable than another from the same age group but with less layers of vulnerability operating (Luna 2009). A research framework focused on vulnerabilities and resilience from this point of view can help identify, address, and mitigate associated environmental factors. In addition, while considerable research attention has examined vulnerability from an individual perspective, not much attention has been paid to social vulnerability, which is "the product of social factors that influence or shape the susceptibility of various groups to harm and that also govern their ability to respond", as well as "those characteristics of communities and the built environment, such as the level of urbanization, growth rates, and economic vitality, that contribute to the social vulnerability of places" (Cutter et al. 2003). This is an important aspect of vulnerability that needs to be taken into account.

Human rights

After World War II, there was international consensus on the need to identify those activities, conditions, and freedoms that all human beings are entitled to enjoy by virtue of being human. These are regarded as human rights, and they include political, civil, social, economic, and cultural rights. Human rights cannot be granted or taken away, the enjoyment of one right affects the enjoyment of others and they must all be respected. A human rights framework focuses on ways to establish mechanisms for addressing serious breaches and promoting national governments' adherence regarding human rights obligations. In relation to environmental neuroethics, key human rights to be discussed would include the right to brain and mental health, to healthy environments, to access to healthy food and clean water, and equal protection from environmental risks.

Social constructionism

According to this theoretical framework, reality cannot be accessed directly; the world can only be perceived through our understandings. Thus what we know about the world is socially constructed, and this knowledge is developed, transmitted, and maintained in social situations. Social construction as a process refers then to how people collectively and through social interaction perceive and interpret, impose meaning and order on their world, and shape their shared reality (Berger & Luckmann 1966; Gergen 2001). Our perception of the world, including the environment, is influenced by the underlying culture, values, and local knowledge, by political and governance processes, by economic opportunities and costs, as well as by understandings of past experience, the present situation, and future goals together with appraisals of risk, environmental threat, and environmental change (see Box 22.1).

Finally, adapting previous work in neuroethics (Marcus 2002), there are five foundational areas (see Box 22.2) in which ethical issues that arise from environmental changes impacts on brain and mental health can be conceptualized and problematized (Cabrera et al. 2016).

Considering that environmental neuroethics is a relatively new approach within neuroethics, the principles and frameworks suggested here require further elaboration and analysis; however, they lay a strong foundation to continue interdisciplinary work and research in this important area.

Environmental neuroethics and future research

Investigation and discussion of the ethical, social, and policy implications of environmental change and exposures on brain and mental health is, both from a consequentialist and prudential perspective, needed in order to clarify ethical concepts and values at stake, address prioritization strategies, and make informed compromises at the interface of environment, brain, individuals, and society. Nonetheless, we should bear in mind that future research on environmental neuroethics is not without challenges. This section introduces key challenges to be addressed in moving forward with environmental neuroethics.

Conceptual

There are at least two conceptual issues that need to be addressed. The first one deals with the environment being too broad a concept or too complex as to be the object of discussion for mitigating brain disorders or fostering brain health. While it is true that the environment can include many variables, there are ways in which the discussion of how the environment affects brain and mental health is valuable. One way is to be specific about the particular aspect of the environment (e.g., physical or social, or air quality). Another way is to step back from a highly individualized perspective to a relational and interrelated one, embracing the underlying complexity and acknowledging how interconnected brain and mental health is to the environments we grow and live in.

Box 22.2 Environmental neuroethics: Top areas of contribution

Brain science and the environment: assessment of neuroscience methods and practices involved in studying the impact of environmental change and exposures; measurement and evaluation of factors that affect the way individuals, communities, and society adapt and cope with real or perceived environmental threats and unfolding impacts of environmental change on brain and mental health.

The relational self, the brain, and the environment: examination of the interdependence between the environment and brain and mental health, including relations underlying vulnerability and resilience in brain and mental health within the context of changing environments and environmental stressors; assessment of mechanisms by which exposures at key points in life may contribute to different brain and mental effects.

Cross-cultural factors, the brain, and the environment: exploration of the role of culture in the relationship and impact that environmental change has on brain and mental health; assessment of interactions between Traditional Ecological Knowledge and neuroscience knowledge and evidence.

Social policy, the brain, and the environment: consideration of priorities and allocation of resources to deal with environmental impacts on brain and mental health; deliberation and development of social policy related to environmental injustices affecting brain and mental health.

Public discourse, the brain, and the environment: facilitation of international, cross-disciplinary, transdisciplinary collaborations; engagement of diverse professional disciplines and communities in multidirectional communication and discourse about neurological, psychological, sociological and ethical dimensions of environmental change and exposures; creation of effective outreach programs that promote public understanding about the impact of environmental change and exposures on brain and mental health.

A second conceptual challenge has to do with how to distinguish between harm versus benefit with respect to environmental exposures. These concepts are relative to what an individual would have been like in the absence of a particular environmental factor. There is not a straightforward baseline level that would enable classification of a given environmental factor as being beneficial if it increases brain health above that level and harmful if it decreases brain health below that level. For many of the environmental factors already mentioned there seems to be a continuum, and there is not an obvious point on the continuum that counts as the presumed normal amount of that environmental factor. What counts as beneficial or harmful environmental exposure, or even what is a healthy environment, is open to debate and is in need of both more empirical evidence and normative discussion.

Knowledge

There are epistemic challenges. The rigid boundaries among disciplines of knowledge along with semiautonomous development of knowledge in each field have been long-standing obstacles to interdisciplinary thinking and practice necessary for integrated environmental impact assessment on brain and mental health. This raises questions such as "What knowledge should be counted in the assessment and how to assess diverse sources of knowledge?" There is also a relative lack of expertise bridging the different areas that are needed to obtain a better understanding of the issues involved, and to adequately project their expertise into the state and national debates (Goldstein et al. 2012). Political obstacles and time constraints are also a challenge for knowledge creation and dissemination. Policymakers typically use quantitative risk assessment and benefit–cost analysis in ways that are not sensitive to social and environmental justice issues.

Methodologies

The comparative dearth in understanding environmental impacts on brain and mental health has been partly due to methodological challenges in satisfying the scientific rigors of cause and effect in epidemiology. Useful methodologies, such as population-based studies that measure subtle effects on neurobehavioral outcomes, are often times challenging to interpret and costly to conduct (Miodovnik 2011). Evidence linking a particular exposure to brain health conditions relies on observational epidemiological studies or toxicity testing using animal models, or both. For obvious ethical reasons, experimental studies with purposeful dosing of substances with potential neurotoxic effects are not conducted in humans. Therefore, while cause and effect is not proven with any one epidemiology study, several well-designed studies in different populations, alone or combined with inferential evidence from animal exposures, can strongly support the likelihood that a given association is in fact causal in nature. There are also challenges in terms of accounting for the long timeframes of onset for many neurological illnesses, distinguishing environmental impacts from other influences on brain and mental health, and the fact that research on the brain is a relative latecomer to environmental and health impact studies.

The employment of locally accepted terminology and assessment strategies, study of the synergistic or additive effects from exposure to mixture compounds, the inefficiency of traditional research methods to obtain documentation and continuous monitoring of impacts, for example, in relation to the cumulative effects of contaminants, and the difficulties in analyzing biomarkers of exposure in human tissue (Cranor 2011) are other examples of the difficulties ahead.

Evidence

A related challenge is connected to evidence. Different environmental factors may not appear by themselves to cause any obvious or serious risk for brain and mental health, and damage might only be detectable from the combination of various chemicals often after continued exposure, which makes it difficult to assess causality (Grandjean 2013;

Grossman 2014). Likewise, while there is widespread agreement that exposing organisms to environmental toxins at high dose levels constitutes a significant threat to environmental and public health, low-dose effects continue to be a matter of intense disagreement among scientists, policymakers, and activist groups (Elliott 2011). There is also the issue of how to weigh scientific evidence and the role it has in decision-making. Some argue, for example, that we already have plenty of evidence to support actions aimed at protecting our brain and mental health, while others disagree, adding that the benefits of certain practices (e.g., use of pesticides) clearly outweigh the possible harms on brain and mental health. Others have argued that society cannot afford to wait for complete evidence, as the consequences could be catastrophic and irreversible. An alternative approach endorsed by a number of commentators and organizations is based on a precautionary principle, where the idea is that society should take reasonable steps to prevent or mitigate significant harm even when scientific evidence is incomplete or contested (Steel 2014). This leads to a related challenge, that of conflicts of interest.

Conflict of interest and bias

The debates around scientific evidence and the ethical issues at stake are exacerbated not only by uncertainty regarding risks and potential consequences, but also by vested interests, including profits from industry, as well as organized efforts by interest groups to strategically design studies, and spread confusing and misleading information in an effort to support their preferred positions (Oreskes & Conway 2010; Elliott 2013).

There are also issues connected to bias. An example of government bias is that of the lead-contaminated water scandal in Flint, MI. Water contamination continued for 18 months after citizens first voiced complaints about water quality. Presumably, action would have been taken more quickly, were Flint a more affluent community. Even more evident than government biases are the vested interests of industry. In cases where public safety and environmental protection get in the way of profit, these factors are often not considered (Shrader-Frechette 2006). An example of this relates to the fact that in the United States, industrial chemicals, with the exception of pesticides, have no restrictions prior to marketing (Cranor 2011). Furthermore, military and industry funding can be regarded as introducing bias in academic work related to the environment and its impact on brain and mental health.

Risks

Different social actors, based on their different worldviews and values, perceive and handle risks differently (Slimak & Dietz 2006). Risks can be overestimated or underestimated based on how they are perceived and communicated. Brain and mental health, for example, often lack the clearly detectable experience characteristic of other more observable bodily reactions such as coughing or red eyes resulting from high levels of air pollution, which could lead to an underestimation of the associated risks.

There are different dimensions of risk: "dread risk" captures emotional reactions to hazards such as nuclear reactor accidents where people become anxious because of a perceived lack of control over exposure to the risks and due to their catastrophic consequences. An example of this comes from the Ojibway who referred to mercury pollution as *pijibowin*, or poison, arguing that: "You can't see it or smell it, you can't taste it or feel it, but you know it's there. You know it can hurt you, make your spirit sick" (quoted in Grandjean 2013). Another dimension is that of "unknown risk", which refers to the degree to which a risk is new, with unforeseeable consequences. There are also challenges regarding the impacts on future generations and compensating them for environmental changes and pollution that affected their brain and mental health. Finally, by addressing a particular risk for a given group of individuals, we can end up creating new problems for a different group of individuals (the solution turns into a new problem).

While there are clearly various challenges ahead, environmental neuroethics provides a needed approach to start a discussion on priorities and strategies that we as a society can reasonably agree on regarding the type of environments that should be promoted without becoming a burden to innovation and progress.

Conclusion

Environmental neuroethics aims to identify pragmatic starting points and alternatives to resolve difficult ethical challenges presented at the intersection of environment and brain, from strategies to reduce the adverse brain and mental health impacts from environmental change and toxic exposures, to policies and strategies to foster healthy environments. It provides a platform to discuss different ethical principles and perspectives at the intersection of environmental change and its impact for brain and mental health, including issues of beneficence and nonmaleficence (avoiding harm and ensuring welfare for individuals and communities), social justice, and responsibility. It offers a novel conceptual, normative, and empirical approach, bringing various disciplines together into an interaction with broader ethical, biomedical, and social understandings of environmental change and exposures and their role for brain and mental health. The various frameworks and principles discussed here set the foundation to move forward the first generation of environmental neuroethics scholarship.

While many challenges lay ahead in the future of environmental neuroethics, it is clear that the unprecedented consequences for both present and future generations in terms of environmental impacts on brain and mental health make this new trajectory a needed and timely one.

Acknowledgments

The author would like to thank Kevin Elliot, Len Fleck, and Libby Bogdan-Lovis for their insightful comments and for helping with proofreading this chapter.

Notes

1. This is the clinical definition of a variety of different underlying pathologies that can cause Parkinson's-like symptoms.

2. There is also emerging evidence that chronic stress contributes to depression, post-traumatic stress disorder, and the development of certain pathologies by accelerating and/or exacerbating preexisting vulnerabilities.

References

Alavanja, M.C.R., Hoppin, J.A., and Kamel, F. (2004). Health effects of chronic pesticide exposure: cancer and neurotoxicity. *Annual Review of Public Health*, **25**, 155–197.

Albrecht, G., Sartore, G.M., Connor, L., et al. (2007). Solastalgia: the distress caused by environmental change. *Australasian Psychiatry*, **15**(Suppl. 1), S95–S98.

APA Task Force (2011). *Psychology and Global Climate Change*. American Psychological Association. Available from: http://www.apa.org/science/about/publications/climate-change.aspx [accessed January 18, 2016].

Baroncelli, L., Braschi, C., Spolidoro, M., Begenisic, T., Sale, A., and Maffei, L. (2010). Nurturing brain plasticity: impact of environmental enrichment. *Cell Death and Differentiation*, **17**(7), 1092–1103.

Bentall, R.P. and Fernyhough, C. (2008). Social predictors of psychotic experiences: specificity and psychological mechanisms. *Schizophrenia Bulletin*, **34**(6), 1012–1020.

Berger, P.L. and Luckmann, T. (1966). *The Social Construction of Reality: A Treatise in the Sociology of Knowledge*. London: Penguin Books.

Berry, H.L., Hogan, A., Owen, J., Rickwood, D., and Fragar, L. (2011). Climate change and farmers' mental health: risks and responses. *Asia-Pacific Journal of Public Health*, **23**(Suppl. 2), 119S–132S.

Berry, H.L., Bowen, K., and Kjellstrom, T. (2010). Climate change and mental health: a causal pathways framework. *International Journal of Public Health*, **55**(2), 123–132.

Bested, A.C., Logan, A.C., and Selhub, E.M. (2013). Intestinal microbiota, probiotics and mental health: from Metchnikoff to modern advances: part III—convergence toward clinical trials. *Gut Pathogens*, **5**(1), 4.

Blas, E., Sommerfeld, J., and Kurup, A.S. (2011). *Social Determinants Approaches to Public Health*. Geneva: World Health Organization.

Bowler, D.E., Buyung-Ali, L.M., Knight, T.M., and Pullin, A.S. (2010). A systematic review of evidence for the added benefits to health of exposure to natural environments. *BMC Public Health*, **10**(456), 1–10.

Butterworth, R.F. (2010). Metal toxicity, liver disease and neurodegeneration. *Neurotoxicity Research*, **18**(1), 100–105.

Cabrera, L.Y. (2015). *Rethinking Human Enhancement*. London: Palgrave Macmillan.

Cabrera, L.Y., Tesluk, J., Chakraborti, M., Matthews, R., and Illes, J. (2016). Brain matters: from environmental ethics to environmental neuroethics. *Environmental Health*, **15**(20), 1–5.

Calderon-Garcidueñas, L., Azzarelli, B., Acuna, H., et al. (2002). Air pollution and brain damage. *Toxicologic Pathology*, **30**(3), 372–389.

Calderon-Garcidueñas, L., Reed, W., Maronpot, R.R., et al. (2004). Brain inflammation and Alzheimer's-like pathology in individuals exposed to severe air pollution. *Toxicologic Pathology*, **32**(6), 650–658.

Calderon-Garcidueñas, L., Mora-Tiscareño, A., Ontiveros, E., et al. (2008). Air pollution, cognitive deficits and brain abnormalities: a pilot study with children and dogs. *Brain and Cognition*, **68**(2), 117–127.

Canadian Environmental Health Atlas (2014, November 11). *Little Things Matter: The Impact of Toxins on the Developing Brain.* [Online video.] https://youtu.be/E6KoMAbz1Bw [accessed January 29, 2016].

Centers for Disease Control and Prevention (2015). *Developmental Disabilities.* Available from: http://www.cdc.gov/ncbddd/developmentaldisabilities/index.html [accessed January 29, 2016].

Chen, F.S., Kumsta, R., von Dawans, B., Monakhov, M., Ebstein, R.P., and Heinrichs, M. (2011). Common oxytocin receptor gene (OXTR) polymorphism and social support interact to reduce stress in humans. *Proceedings of the National Academy of Sciences of the United States of America*, **108**(50), 19937–19942.

Colborn, T., Kwiatkowski, C., Schultz, K., and Bachran, M. (2011). Natural gas operations from a public health perspective. *Human and Ecological Risk Assessment: An International Journal*, **17**(5), 1039–1056.

Commission on Social Determinants of Health (2008). *Closing the Gap in a Generation: Health Equity Through Action on the Social Determinants of Health.* Geneva: World Health Organization.

Cranor, C. (2011). *Legally Poisoned: How the Law Puts Us at Risk from Toxicants.* Cambridge, MA: Harvard University Press.

Crews, D. (2010). Epigenetics, brain, behavior, and the environment. *Hormones*, **9**(1), 41–50.

Cumes, D. (1998). Nature as medicine: the healing power of the wilderness. *Alternative Therapies in Health and Medicine Journal*, **4**(2), 1–8.

Cunsolo Willox, A., Harper, S.L., Ford, J.D., et al. (2013). Climate change and mental health: an exploratory case study from Rigolet, Nunatsiavut, Canada. *Climatic Change*, **121**(2), 255–270.

Cutter, S.L., Boruff, B.J., and Shirley, W.L. (2003). Social vulnerability to environmental hazards. *Social Science Quarterly*, **84**, 1–20.

Dakeishi, M., Murata, K., and Grandjean, P. (2006). Long-term consequences of arsenic poisoning during infancy due to contaminated milk powder. *Environmental Health*, **5**(31), 1–7.

Daniels, N. (2008). *Just Health.* New York: Cambridge University Press.

de Burbure, C., Buchet, J.P., Leroyer, A., et al. (2006). Renal and neurologic effects of cadmium, lead, mercury, and arsenic in children: evidence of early effects and multiple interactions at environmental exposure levels. *Environmental Health Perspectives*, **114**(4), 584–590.

Debes, F., Budtz-Jørgensen, E., Weihe, P., White, R.F., and Grandjean, P. (2006). Impact of prenatal methylmercury exposure on neurobehavioral function at age 14 years. *Neurotoxicology and Teratology*, **28**(3), 363–375.

Dick, F.D. (2006). Solvent neurotoxicity. *Occupational and Environmental Medicine*, **63**(3), 221–226.

Elbaz, A., Clavel, J., Rathouz, P.J., et al. (2009). Professional exposure to pesticides and Parkinson disease. *Annals of Neurology*, **66**(4), 494–504.

Elliott, K.C. (2011). *Is a Little Pollution Good for You?: Incorporating Societal Values in Environmental Research.* Oxford: Oxford University Press.

Elliott, K.C. (2013). Financial conflicts of interest and criteria for research credibility. *Erkenntnis*, **79**(S5), 917–937.

Field, R.A., Soltis, J., and Murphy, S. (2014). Air quality concerns of unconventional oil and natural gas production. *Environmental Science: Processes & Impacts*, **16**(5), 954–969.

Fritze, J.G., Blashki, G.A., Burke, S., and Wiseman, J. (2008). Hope, despair and transformation: climate change and the promotion of mental health and wellbeing. *International Journal of Mental Health Systems*, **2**(1), 13.

Frumkin, H. (2001). Beyond toxicity: human health and the natural environment. *American Journal of Preventive Medicine*, **20**(3), 234–240.

Gay, J., Shepherd, O., Thyden, M., and Whitman, M. (2010). *The Health Effects of Oil Contamination: A Compilation of Research.* Worcester, MA: Worcester Polytechnic Institute.

Gergen, K.J. (2001). *Social Construction in Context*. London: SAGE.

Goldstein, B.D., Kriesky, J., and Pavliakova, B. (2012). Missing from the table: role of the environmental public health community in governmental advisory commissions related to Marcellus shale drilling. *Environmental Health Perspectives*, **120**(4), 483–486.

Gómez-Pinilla, F. (2008). Brain foods: the effects of nutrients on brain function. *Nature Reviews Neuroscience*, **9**(7), 568–578.

Grandjean, P. (2013). *Only One Chance: How Environmental Pollution Impairs Brain Development—and How to Protect the Brains of the Next Generation*. New York: Oxford University Press.

Grandjean, P. and Herz, K.T. (2011). Methylmercury and brain development: Imprecision and underestimation of developmental neurotoxicity in humans. *Mount Sinai Journal of Medicine*, **78**(1), 107–118.

Grandjean, P. and Landrigan, P.J. (2006). Developmental neurotoxicity of industrial chemicals. *The Lancet*, **368**, 2167–2178.

Grandjean, P., Weihe, P., White, R.F., et al. (1997). Cognitive deficit in 7-year-old children with prenatal exposure to methylmercury. *Neurotoxicology and Teratology*, **19**(6), 417–428.

Grandjean, P., Weihe, P., Burse, V.W., et al. (2001). Neurobehavioral deficits associated with PCB in 7-year-old children prenatally exposed to seafood neurotoxicants. *Neurotoxicology and Teratology*, **23**, 305–317.

Grandjean, P., Harari, R., Barr, D.B., and Debes, F. (2006). Pesticide exposure and stunting as independent predictors of neurobehavioral deficits in Ecuadorian school children. *Pediatrics*, **117**(3), e546–e556.

Grandjean, P., Satoh, H., Murata, K., and Eto, K. (2010). Adverse effects of methylmercury: environmental health research implications. *Environmental Health Perspectives*, **118**(8), 1137–1145.

Grashow, R., Spiro, A., Taylor, K.M., et al. (2013). Cumulative lead exposure in community-dwelling adults and fine motor function: Comparing standard and novel tasks in the VA Normative Aging Study. *NeuroToxicology*, **35**, 154–161.

Greenwood, C.E. and Winocur, G. (2005). High-fat diets, insulin resistance and declining cognitive function. *Neurobiology of Aging*, **26**(1), 42–45.

Grossman, E. (2014). Time after time: environmental influences on the aging brain. *Environmental Health Perspectives*, **122**(9), A238–A243.

Haddad, L., Schäfer, A., Streit, F., et al. (2015). Brain structure correlates of urban upbringing, an environmental risk factor for schizophrenia. *Schizophrenia Bulletin*, **41**(1), 115–122.

Haluza, D., Schönbauer, R., and Cervinka, R. (2014). Green perspectives for public health: a narrative review on the physiological effects of experiencing outdoor nature. *International Journal of Environmental Research and Public Health*, **11**(5), 5445–5461.

Harari, R., Julvez, J., Murata, K., et al. (2010). Neurobehavioral deficits and increased blood pressure in school-age children prenatally exposed to pesticides. *Environmental Health Perspectives*, **118**(6), 890–896.

Heiervang, K.S., Mednick, S., Sundet, K., and Rund, B.R. (2010). Effect of low dose ionizing radiation exposure in utero on cognitive function in adolescence. *Scandinavian Journal of Psychology*, **51**(3), 210–215.

Held, V. (2006). *The Ethics of Care*. New York: Oxford University Press.

Hertzman, C., Wiens, M., Bowering, D., Snow, B., and Calne, D. (1990). Parkinson's disease: a case-control study of occupational and environmental risk factors. *American Journal of Industrial Medicine*, **17**(3), 349–355.

Hu, H.W., Téllez-Rojo, M.M., Bellinger, D., et al. (2006). Fetal lead exposure at each stage of pregnancy as a predictor of infant mental development. *Environmental Health Perspectives*, **114**, 1730–1735.

Illes, J., Davidson, J., and **Matthews, R.** (2014). Environmental neuroethics: changing the environment – changing the brain. Recommendations submitted to the Presidential Commission for the Study of Bioethical Issues. *Journal of Law and the Biosciences*, 1(2), 221–223.

Ingold, T. (2000). *The Perception of the Environment*. London: Psychology Press.

Jacobson, J.L. and Jacobson, S.W. (2003). Prenatal exposure to polychlorinated biphenyls and attention at school age. *Journal of Pediatrics*, **143**(6), 780–788.

Joseph, J.A., Shukitt-Hale, B., and **Lau, F.C.** (2007). Fruit polyphenols and their effects on neuronal signaling and behavior in senescence. *Annals of the New York Academy of Sciences of the United States of America*, **1100**(1), 470–485.

Kamel, F. and **Hoppin, J.A.** (2004). Association of pesticide exposure with neurologic dysfunction and disease. *Environmental Health Perspectives*, **112**(9), 950–958.

Kelly, B.D., O'Callaghan, E., Waddington, J.L., et al. (2010). Schizophrenia and the city: a review of literature and prospective study of psychosis and urbanicity in Ireland. *Schizophrenia Research*, **116**(1), 75–89.

Kerti, L. Witte, A.V., Winkler, A., Grittner, U., Rujescu, D., and **Flöel, A.** (2013). Higher glucose levels associated with lower memory and reduced hippocampal microstructure. *Neurology*, **81**, 1746–1752.

Kirmayer, L.J. and **Valaskakis, G.G.** (2008). *Healing Traditions*. Vancouver, BC: UBC Press.

Korrick, S.A. and **Sagiv, S.K.** (2008). Polychlorinated biphenyls, organochlorine pesticides and neurodevelopment. *Current Opinion in Pediatrics*, **20**(2), 198–204.

Kramer, A.F., Erickson, K.I., and **Colcome, S.J.** (2006). Exercise, cognition, and the aging brain. *Journal of Applied Physiology*, **101**, 1237–1242.

Lai, J.S., Hiles, S., Bisquera, A., et al. (2014). A systematic review and meta-analysis of dietary patterns and depression in community-dwelling adults. *American Journal of Clinical Nutrition*, **99**(1), 181–197.

Landrigan, P.J., Sonawane, B., Butler, R.N., Trasande, L., Callan, R., Droller, D. (2005). Early environmental origins of neurodegenerative disease in later life. *Environmental Health Perspectives*, **113**(9), 1230–1233.

Lanphear, B.P., Hornung, R., Khoury, J., et al. (2005). Low-level environmental lead exposure and children's intellectual function: an international pooled analysis. *Environmental Health Perspectives*, **113**(7), 894–899.

Lederbogen, F., Kirsch, P., Haddad, L., et al. (2011). City living and urban upbringing affect neural social stress processing in humans. *Nature*, **474**(7352), 498–501.

Lee, C. (2002). Environmental justice: building a unified vision of health and the environment. *Environmental Health Perspectives*, **110**, 141–144.

Levy, N. (2012). Ecological engineering: reshaping our environments to achieve our goals. *Philosophy & Technology*, **25**(4), 589–604.

Lim, Y.-H., Kim, H., Kim, J.H., et al. (2012). Air pollution and symptoms of depression in elderly adults. *Environmental Health Perspectives*, **120**(7), 1023–1028.

Lucchini, R.G., Guazzetti, S., Zoni S., et al. (2014). Neurofunctional dopaminergic impairment in elderly after lifetime exposure to manganese. *NeuroToxicology*, **45**, 309–317.

Luna, F. (2009). Elucidating the concept of vulnerability: layers not labels. *International Journal of Feminist Approaches to Bioethics*, **2**(1), 121–139.

Maas, J. (2006). Green space, urbanity, and health: how strong is the relation? *Journal of Epidemiology & Community Health*, **60**(7), 587–592.

Marcus, S. (2002). *Neuroethics: Mapping the Field*. San Francisco, CA: Dana Press.

Marmot, M. (2005). Social determinants of health inequalities. *The Lancet*, **365**(9464), 1099–1104.

Matuszek, S. (2010). Animal-facilitated therapy in various patient populations. *Holistic Nursing Practice*, **24**(4), 187–203.

Miodovnik, A. (2011). Environment neurotoxicants and developing brain. *Mount Sinai Journal of Medicine*, **78**, 58–77.

Mora, F., Segovia, G. and del Arco, A. (2007). Aging, plasticity and environmental enrichment: structural changes and neurotransmitter dynamics in several areas of the brain. *Brain Research Reviews*, **55**(1), 78–88.

Needleman, H.L., Gunnoe, C., Leviton, A., et al. (1979). Deficits in psychologic and classroom performance of children with elevated dentine lead levels. *New England Journal of Medicine*, **300**(13), 689–695.

Needleman, H.L., Schell, A., Bellinger, D., Leviton, A., and Allred, E.N. (1990). The long-term effects of exposure to low doses of lead in childhood. An 11-year follow-up report. *New England Journal of Medicine*, **322**(2), 83–88.

NEPA (1969). *The National Environmental Policy Act*. Available from: https://ceq.doe.gov/laws_and_executive_orders/the_nepa_statute.html [accessed March 5, 2016].

Nussbaum, M. and Sen, A. (1993). *The Quality of Life*. Gloucestershire: Peterson's.

O'Neil, A., Quirk, S.E., Housden, S., et al. (2014). Relationship between diet and mental health in children and adolescents: a systematic review. *American Journal of Public Health*, **104**, e31–e42.

O'Rourke, D. and Connolly, S. (2003). Just oil? The distribution of environmental and social impacts of oil production and consumption. *Annual Review of Environment and Resources*, **28**(1), 587–617.

Opler, M.G.A., Buka, S.L., Groeger, J., et al. (2008). Prenatal exposure to lead, δ-aminolevulinic acid, and schizophrenia: further evidence. *Environmental Health Perspectives*, **116**(11), 1586–1590.

Oreskes, N. and Conway, E.M. (2010). *Merchants of Doubt*. New York: Bloomsbury Publishing USA.

Parfit, D. (1971). Personal identity. *The Philosophical Review*, **80**, 3–27.

Patterson, C.C. (1965). Contaminated and natural lead environments of man. *Archives of Environmental Health*, **11**, 344–360.

Peen, J., Schoevers, R.A., Beekman, A.T., and Dekker, J. (2010). The current status of urban-rural differences in psychiatric disorders. *Acta Psychiatrica Scandinavica*, **121**(2), 84–93.

Pereira, A.C., Huddleston, D.E., Brickman, A.M., et al. (2007). An in vivo correlate of exercise-induced neurogenesis in the adult dentate gyrus. *Proceedings of the National Academy of Sciences of the United States of America*, **104**(13), 5638–5643.

Perrin, M., Opler, M.G., Harlap, S., et al. (2007). Tetrachloroethylene exposure and risk of schizophrenia: offspring of dry cleaners in a population birth cohort, preliminary findings. *Schizophrenia Research*, **90**(1–3), 251–254.

Perry, S.L. (2012). Development, land use, and collective trauma: the Marcellus shale gas boom in rural Pennsylvania. *Culture, Agriculture, Food and Environment*, **34**(1), 81–92.

Peters, A., Veronesi, B., Calderón-Garcidueñas, L., et al. (2006). Translocation and potential neurological effects of fine and ultrafine particles a critical update. *Particle and Fibre Toxicology*, **3**(1), 1–13.

Petronis, A. and Mill, J. (2011). *Brain, Behavior and Epigenetics*. New York: Springer.

Potter, V.R. (1988). *Global Bioethics*. East Lansing, MI: MSU Press.

Prüss-Üstün, A. and Corvalán, C. (2006). *Preventing Disease Through Healthy Environments*. Geneva: World Health Organization.

Psaltopoulou, T., Sergentanis, T.N., Panagiotakos, D.B., Sergentanis, I.N., Kosti, R., and Scarmeas, N. (2013). Mediterranean diet, stroke, cognitive impairment, and depression: a meta-analysis. *Annals of Neurology*, **74**(4), 580–591.

Public Health Agency of Canada (2011). What determines health? *Public Health Agency of Canada*, 1–4. Available from: http://www.phac-aspc.gc.ca/ph-sp/determinants/index-eng.php [accessed January 18, 2016].

Rabinowitz, P.M., Slizovskiy, I.B., Lamers, V., et al. (2015). Proximity to natural gas wells and reported health status: results of a household survey in Washington County, Pennsylvania. *Environmental Health Perspectives*, **123**(1), 21–26.

Rawls, J. (1971). *A Theory of Justice*. Cambridge, MA: Harvard University Press.

Raz, R., Roberts, A.L., Lyall, K., et al. (2014). Autism spectrum disorder and particulate matter air pollution before, during, and after pregnancy: a nested case–control analysis within the nurses' health study II cohort. *Environmental Health Perspectives*, **123**(3), 264–270.

Resnik, D.B. (2009). Human health and the environment: in harmony or in conflict? *Health Care Analysis*, **17**(3), 261–276.

Resnik, D.B. (2012). *Environmental Health Ethics*. Cambridge: Cambridge University Press.

Resnik, D.B. and Portier, C.J. (2008). *Environment and Health*. Garrison, NY: The Hastings Center.

Roberts, E.M., English, P.B., Grether, J.K., Windham, G.C., Somberg, L., and Wolff, C. (2007). Maternal residence near agricultural pesticide applications and autism spectrum disorders among children in the California central valley. *Environmental Health Perspectives*, **115**, 1482–1489.

Rodríguez-Barranco, M., Lacasaña, M., Gil, F., et al. (2014). Environmental research. *Environmental Research*, **134**(C), 66–73.

Rosado, J.L., Ronquillo, D., Kordas, K., et al. (2007). Arsenic exposure and cognitive performance in Mexican schoolchildren. *Environmental Health Perspectives*, **115**(9), 1371–1375.

Sagoff, M. (2008). Zuckerman's dilemma a plea for environmental ethics. *Hastings Center Report*, **21**, 32–40.

Schwartz, B.S., Stewart, W.F., Bolla, K.I., et al. (2000). Past adult lead exposure is associated with longitudinal decline in cognitive function. *Neurology*, **55**, 1144–1150.

Seeman, T.E. and McEwen, B.S. (1996). Impact of social environment characteristics on neuroendocrine regulation. *Psychosomatic Medicine*, **58**, 459–471.

Shelton, J.F., Geraghty, E.M., Tancredi, D.J., et al. (2014). Neurodevelopmental disorders and prenatal residential proximity to agricultural pesticides: the CHARGE study. *Environmental Health Perspectives*, **122**, 1103–1109.

Shih, R.A., Hu, H., Weisskopf, M.G., and Schwartz, B.S. (2007). Cumulative lead dose and cognitive function in adults: a review of studies that measured both blood lead and bone lead. *Environmental Health Perspectives*, **115**(3), 483–492.

Shonkoff, J.P. (2012). Leveraging the biology of adversity to address the roots of disparities in health and development. *Proceedings of the National Academy of Sciences of the United States of America*, **109**(Suppl. 2), 17302–17307.

Shrader-Frechette, K.S. (2006). *Environmental Justice*. Oxford: Oxford University Press.

Slimak, M.W. and Dietz, T. (2006). Personal values, beliefs, and ecological risk perception. *Risk Analysis*, **26**(6), 1689–1705.

Slote, M. (2007). *The Ethics of Care and Empathy*. New York: Routledge.

Small, G. and Vorgan, G. (2008). *iBrain*. New York: Harper Collins.

Smith, K.R., Corvalan, C.F., and Kjellstrom, T. (1999). How much global ill health is attributable to environmental factors? *Epidemiology*, **10**(5), 573–584.

Stedman, R.C., Jacquet, J.B., and Filteau, M.R. (2012). Marcellus shale gas development and new boomtown research: wiews of New York and Pennsylvania residents. *Environmental Practice*, **14**, 382–393.

Steel, D. (2014). *Philosophy and the Precautionary Principle*. Cambridge: Cambridge University Press.

Stewart, P.W., Lonky, E., Reihman, J., et al. (2008). The relationship between prenatal PCB exposure and intelligence (IQ) in 9-year-old children. *Environmental Health Perspectives*, **116**(10), 1416–1422.

Tang, C.Y., Carpenter, D.M., Eaves, E.L., et al. (2011). Occupational solvent exposure and brain function: an fMRI study. *Environmental Health Perspectives*, **119**(7), 908–913.

Teicher, M.H. and Samson, J.A. (2013). Childhood maltreatment and psychopathology: a case for ecophenotypic variants as clinically and neurobiologically distinct subtypes. *American Journal of Psychiatry*, **170**, 1114–1133.

Tost, H., Champagne, F.A. and Meyer-Lindenberg, A. (2015). Environmental influence in the brain, human welfare and mental health. *Nature Neuroscience*, **18**(10), 1421–1431.

Tsai, S.-Y., Chou, H.Y., The, H.W., Chen, C.M., and Chen, C.J. (2003). The effects of chronic arsenic exposure from drinking water on the neurobehavioral development in adolescence. *NeuroToxicology*, **24**(4–5), 747–753.

United Nations Environment Programme (1972). *Report of the United Nations Conference on the Human Environment*. Available from: http://www.unep.org/Documents.multilingual/Default.asp?DocumentID=97&ArticleID=150 [accessed March 5, 2016].

van Os, J., Kenis, G., and Rutten, B.P.F. (2010). The environment and schizophrenia. *Nature*, **468**(7321), 203–212.

van Praag, H., Kempermann, G., and Gage, F.H., 2000. Neural consequences of environmental enrichment. *Nature Reviews Neuroscience*, 1(3), 191–198.

Villanueva, C.M., Kogevinas, M., Cordier, S., et al. (2013). Assessing exposure and health consequences of chemicals in drinking water: current state of knowledge and research needs. *Environmental Health Perspectives*, **122**, 213–221.

Vukic, A., Gregory, D., Martin-Misener, R., and Etowa, J. (2011). Aboriginal and Western conceptions of mental health and illness. *Pimatisiwin: A Journal of Aboriginal & Indigenous Community Health*, **9**(1), 65–86.

Watson, S. (2008, October 8). *How Eco-anxiety Works*. Available from: http://science.howstuffworks.com/environmental/green-science/eco-anxiety.htm [accessed January 23, 2015].

Weisskopf, M.G., Hu, H., Sparrow, D., Lenkinski, R.E., and Wright, R.O. (2007). Proton magnetic resonance spectroscopic evidence of glial effects of cumulative lead exposure in the adult human hippocampus. *Environmental Health Perspectives*, **115**(4), 519–523.

White, A.M., Philogene, G.S., Fine, L., and Sinha, S. (2009). Social support and self-reported health status of older adults in the United States. *American Journal of Public Health*, **99**(10), 1872–1878.

Windham, G.C., Zhang, L., Gunier, R., Croen, L.A., and Grether, J.K. (2006). Autism spectrum disorders in relation to distribution of hazardous air pollutants in the San Francisco Bay area. *Environmental Health Perspectives*, **114**(9), 1438–1444.

Woo, C.C. and Leon, M. (2013). Environmental enrichment as an effective treatment for autism: a randomized controlled trial. *Behavioral Neuroscience*, **127**(4), 487–497.

Wright, R.O., Amarasiriwardena, C., Woolf, A.D., Jim, R., and Bellinger, D.C. (2006). Neuropsychological correlates of hair arsenic, manganese, and cadmium levels in school-age children residing near a hazardous waste site. *NeuroToxicology*, **27**(2), 210–216.

Chapter 23

First Nations and environmental neuroethics: Perspectives on brain health from a world of change

Jordan Tesluk, Judy Illes, and Ralph Matthews

First Nations and environmental neuroethics: A call for understanding

This chapter calls for the bridging of knowledge between First Nations[1] and Western institutions of environmental and medical science to inform ethical debates related to environmental impacts on the brain. Amid changing environmental conditions and persisting health deficits, First Nations people in Canada face mounting risks to their neurological and mental health. Effective engagement between Western science and First Nations people requires an understanding of differences in the ontological and epistemological foundations of respective knowledge systems, so that the ethical issues related to environmental impacts on the brain can be brought into a space of common language and practice. This not only requires reconciling differences in understandings of the environment and the brain, but also requires reflection on the foundations of thought and reason that underpin First Nations' and settler society's respective systems of ethical decision-making.

Like other indigenous populations, First Nations people occupy a sensitive location where the consequences of industrial activity collide with traditional lifestyles, and where environmental change is experienced as a direct impact on communities and individuals. While the influence of environmental change on the human brain is a concern for all people, the ethical implications of such impacts are different for indigenous peoples who hold closer cultural and subsistence-based relationships with the environment. Animals, plants, and the elements of nature play a central role in the traditional mythology, artwork, social structure, economy, and belief systems of First Nations people. A majority of First Nations communities in Canada are located in isolated rural areas, where reliance upon the local environment for food and resources is high, and access to the amenities and services of urban life is often limited. The proximity of First Nations to the frontiers of development for mining, oil and gas development, and hydroelectric industries, together with their direct reliance upon the environment thus compounds the neurological risks of environmental change with complex social impacts that threaten the future of their culture and their communities.

It is for these reasons that the experience of First Nations has become an important research focus. Environmental change and the development of natural resources have important implications for the neurological and mental health of both First Nations and for indigenous people throughout Canada and the rest of the world. While we draw on findings from other Canadian aboriginal populations, we place First Nations at the center of this chapter with the understanding that the concerns they share with other populations are bound together by the global scope of environmental change that is occurring, and by common experiences under the institutions of Western science and medicine that have spread their influence across so much of the planet.

In this chapter, we examine past research that intersects studies of the environment, neurological and mental health, and First Nations experience. We begin with an examination of vulnerability to diseases of the brain and to environmental change in First Nations people, and situate these issues within the broader scope of change occurring in their societies. We examine some of the challenges that exist in conducting research on neurological and mental health with First Nations people, and identify examples of recent research that has helped create positive relationships between neurological researchers and First Nations people. We then explore how the corresponding progressions in classical Western thought and the study of brain and environment contrast with First Nations' paths of development. We use this analysis to confront the challenges inherent to the convergence of contrasting systems of knowledge, and we address the role that neuroethics can play in bringing First Nations and Western institutions of science together in protecting the health of the brain.[2]

This chapter focuses upon First Nations people, and their distinct experiences related to the environment and brain health. We remain cognizant of the related but distinct experiences of Metis and Inuit peoples in Canada, as well as aboriginal people in other regions. We acknowledge the need for exploration of the specific issues related to the intersections between their respective experiences and environmental neuroethics. We also recognize the diversity of experiences and perspectives that exists both between different groups of aboriginal people and among the 630 recognized First Nations governments and bands that are located across Canada.

Environment, brain, and knowledge

Environmental neuroethics takes as a central premise that the human brain is malleable, and that it changes in response to its environment (Rose & Abi-Rached 2013; Cabrera et al. 2016). Environmental factors can shape central nervous system health and psychopathology, and contribute to epigenetic processes that affect DNA and predispositions to neurological disease. Perceptions of the environment, and beliefs and knowledge about the environment, can directly influence mental health and shape behaviors that affect neurological health (Wilson 2003; Masuda & Garvin 2006; Jackson 2011) Perceptions of the environment are based in both information shared among individuals and in sensory experience. The environment is perceived through sensory systems (particularly sight, smell, and taste), which send signals to which the brain responds, and adapts, and which it ultimately translates into information and knowledge related to a person's surroundings and circumstances.

Both perceptions of the environment, and understandings of the brain in relation to human health and agency, are mediated through cultural frames of reference that define the boundary between the self and the natural environment, and that shape individuals' interactions with the environment. These frames of reference include systems of thought that situate humans as separate from nature (e.g., in traditional Western worldviews) and perspectives that see the environment and the self as intimately interconnected (such as traditional First Nations worldviews). Our most fundamental understandings of the relationship between environment and the brain are thus rooted in culturally situated systems of knowledge, and are shaped and reshaped through cumulative experience.

Thus, efforts to understand the impact of environmental factors on neurological and mental health among First Nations people must go beyond a mere acknowledgment of differences in vulnerability, perspectives, and experiences. Indeed, the fulsome investigation of this topic involves a broader endeavor. Thoughtful research must reach beyond the development of "culturally sensitive" programs and regulatory-based research ethics. Between the social dimensions of culture that link humans to the environment, and the neurological dimensions of environmental impacts on the central nervous system, is a broad space that spans biological, social, cultural, psychological, and genetic linkages between the brain and the environment. As a multidisciplinary field that examines environmental impacts on brain and mental health, environmental neuroethics seeks to develop a framework for understanding how new forms of knowledge are produced from the interaction between Western scientific worldviews and those of First Nations and other indigenous people.

First Nations and the context of change

The past four centuries of colonization have introduced many radical changes to First Nations people that have reshaped their environment and affected their neurological and mental health. The arrival of European settlers brought great troubles to First Nations people in the form of degradation of their environment, marked deficits in health and material privilege, and marginalization within the modern Western institutions that dominate the land around them (Stanbury 1973; Pinkerton 1987; Tennant 1990; High 1996; Anderson R.B. 1999). Among the challenges faced by First Nations people are high rates of suicide, depression, and substance abuse, and greater risks for brain injuries and neural tube defects among infants (Ray et al. 2004; Lehti et al. 2009; Northern Brain Injury Association 2015). The Public Health Agency of Canada (2014) acknowledges that First Nations and Metis people experience unique impacts as a result of neurological conditions, and illnesses of the brain compound disadvantage with existing hardship and inequality. Despite these concerns, data on neurological conditions among First Nations have not been gathered in Canadian census and national surveys, and neurological disorders such as cerebral palsy and autism that are disproportionately prevalent among other North American aboriginal people remain understudied among Canadian populations (Di Pietro and Illes 2014).

The need for a better understanding of the environmental impact on First Nations people's neurological and mental health is made more pressing by human-induced global climate change and ongoing natural resource development activities affecting territories of First Nations. Across Canada, First Nations are engaged in battles to assert control over the environment, and to deal with the impact of environmental change on their brains, bodies, and communities. In the north, for example, the Mikisew Cree and the Athabasca Chipewyan are confronting ongoing and unknown future health impacts resulting from oil and gas development in the Albertan tar sands projects (Kelly et al. 2009, 2010; Timoney & Lee 2009; Scarlett et al. 2012). In the east, the Grassy Narrows and Whitedog First Nations continue to fight the Canadian Government for the release of information and settlements related to mercury contamination that has negatively affected their neurological health over the past 70 years (Harada et al. 2011; McQuigge 2012; Crowe 2014). In the west, numerous First Nations are working to understand the potential future impacts of unconventional natural gas production (known as fracking). This includes the extraction of gas using hydraulic fracturing processes in the territories of the Treaty Six First Nations, and the compression and shipment of liquid natural gas through the territories of the Tsimshian, Haisla, and Squamish First Nations, among others (Goodine 2011; Benusic 2013; Jang 2015; Nikiforuk 2015).

These events are occurring within a broader context of change that includes historic transformation of First Nations traditional territories into industrialized landscapes, and the more dynamic and unpredictable impacts of global climate change. First Nations have experienced a more rapid and disruptive trajectory of change than other parts of the Canadian population with respect to the ways that environmental change has affected their health and their communities. The gradual industrialization of Western society that proceeded in step with the evolution of medical science and social adaptation has been compressed into a short and violent reorganization of life for First Nations people (Barron 1984; Tennant 1990; Young 1994; Furniss 1999; Yazzie 2000). Over the past four centuries of colonization, First Nations have witnessed the near extinction and recovery of their population, and the rise of resource industries that have reshaped the land and waters around them. Throughout this period, First Nations social needs and their traditional methods of health and healing have been marginalized by Western systems of science, education, and political control.

For some First Nations, these disruptions have been concentrated into a mere century of experience, with current generations bearing the direct impact of their lands being logged, mined, or flooded, and their communities forcibly relocated and reorganized by the state.[3] As these events have unfolded, many First Nations have been cut off from a life based on harvesting and hunting across vast territories, and forced into life within state-built modular housing accompanied by reliance upon limited healthcare services and unfamiliar Western food systems (Gladstone 1953; Tennant 1990; Parsons & Prest 2003;

Woolford & Thomas 2011). However, the central focus for this discussion is the implication that these radical transformations of environment and social organization hold for the neurological and mental health of First Nations people. Given the persisting deficits in the health of First Nations people, the need for research on the impact of environmental change on their brain and mental health has never been higher; yet significant obstacles remain to be overcome in creating new knowledge in this field.

Environmental impacts on First Nations neurological and mental health

Research on environmental impacts on First Nations neurological health occurs within a broader need for understanding of such issues for all populations. Assessments of human health risks posed by global climate change have focused more heavily on issues such as respiratory illness, skin cancer, malaria, and water-borne illnesses (McMichael et al. 2003, 2006; Patz et al. 2005; Haines et al. 2006) with little attention to the implications posed to neurological and mental health. Similarly, research on human health impacts from new forms of natural resource development, such as fracking, has moved more quickly with respect to diseases of the body, such as cancer and asthma, than with diseases specific to the brain (Illes et al. 2014). The comparative dearth in understanding of environmental impacts on brain and mental health compared to studies of impacts on general health has been partly due to methodological challenges in satisfying the scientific rigors of cause and effect in epidemiology. This includes accounting for the long timeframes of onset for many neurological illnesses, distinguishing environmental impacts from other influences on brain and mental health, and the fact that research on the brain is a relative latecomer to environmental and health impact studies.

As awareness of the impact of pollution on human health increased through the twentieth century, the earliest targets of scientific study were the impacts most readily corroborated by public experience. Examples include the effects of smog on respiratory illnesses, made self-evident by breathing difficulties experienced by people living in heavily industrialized areas (Schaefer 1907; Klotz 1914; Russell 1924; Mills 1943). However, neurological diseases often lack the clearly detectable experiential characteristics associated with afflictions affecting breathing, circulation, and other observable bodily processes.[4] The discovery of Minamata disease in 1959 due to methylmercury poisoning in Japanese villagers stands among the few known examples of a definitively established link between the environment and diseases of the brain (McAlpine & Araki 1958; Takeuchi et al. 1959). The convergence of an easily identifiable and previously suspected source of contamination with an unprecedented appearance of neurological illness in the people of Minamata provided an unusual recipe for direct causal attribution. However, extensive and clear linkages between environmental factors and other neurological illnesses have proven elusive until recent years.

Over the past two decades, increasing evidence has been found to support environmental change as a factor in neurological and mental illnesses. This includes research linking air pollution to neuroinflammation and autism spectrum disorder (Calderon-Garciduenas et al. 2004; Block & Calderon-Garciduenas 2009; Raz et al. 2015; Talbott et al. 2015), water pollution to sensory and neurodevelopmental deficits (Murata et al. 2004; Grandjean & Herz 2011; Grandjean & Landrigan 2014), and noise and light pollution to psychological illnesses (Passchier-Vermeer & Passchier 2000; De Kluizenaar et al. 2001; Chepesiuk 2009). In Canada, the vulnerability of First Nations and their extensive reliance upon natural resources has placed them at the center of several research initiatives. These include studies of methylmercury poisoning of First Nations people through water supplies and fisheries (Wheatley and Wheatley 2000; Gilbertson & Carpenter 2004), and contamination of marine food supplies with polychlorinated biphenyls, dioxins, and other chemicals (Kuhnlein et al. 1995; Van Oostdam et al. 1999; Wiseman & Gobas 2002). Social impact studies have shown that changes in land and natural resources have disrupted First Nations communities and interfered with the fundamental cultural practices that support community structures and practices that support positive mental health for First Nations people (Jackson 2011; Wilson 2003).

Stout and colleagues argue that the wide array of environmental impacts on the neurological and mental health of First Nations and other aboriginal people in Canada demands an expansion of research in this area (Stout et al. 2009). With climate change forming an increasing point of concern for society, it is unsurprising that research on environmental impacts on the brain have proliferated in Arctic Canada, where the most noticeable shifts in climate are occurring. The changing climate has affected ice cover and weather patterns, and is suspected as a key factor in the fluctuation of mercury levels that are implicated in the contamination of food sources for Inuit and northern First Nations (El-Hayek 2007). While signs of neurological deficits have been observed among these groups, there has been limited success in tracing these deficits to the environmental changes in question (El-Hayek 2007). However, with growing awareness of the potential impacts of climate change on mental health (Haines et al. 2006; Berry et al. 2008), attention has increasingly turned toward the broader web of environmental–health relationships that affect the lives of northern aboriginal people.

Like other aboriginal people in Canada, First Nations and Inuit people in the Arctic region experience high levels of mental illness (Lehti et al. 2009). Ample research has linked these problems to the impacts of colonization and the trauma of community and family disruptions (Sullivan & Brems 1997; Csonka & Schweitzer 2004; Hicks 2007). More recent work has focused on the role of climate change in undermining the mental health and coping capacity of Canadian aboriginal people in the arctic and subarctic regions. This includes climate change as a stressor affecting emotional wellness, anxiety, depression, chronic stress, mood disorders, and the ability to cope with other forms of mental illness (Macdonald et al. 2013; Willox et al. 2013a; Willox et al. 2015).

Researchers have found that negative impacts of climate change accrue as a result of disruptions to First Nations/Inuit seasonal practices, interruption of activities that support community and kin relations, and growth of physiological problems due to changes in traditional diet and difficulties hunting and gathering (Fritze et al. 2008; Willox et al.

2013b; Bourque & Willox 2014). The psychological distress experienced as a result of being deprived of the comfort derived from contact with one's native environment has been described as *solastalgia* (Albrecht et al. 2007).[5] While sharing a basis with nostalgia in the role of sensations of one's surroundings, solastalgia draws a focus on the role the environment plays in calming the mind, thus framing responses to drastic changes in the environment as a form of psychoterratic illnesses that can only be understood with reference to both environmental and psychological factors (Albrecht et al. 2007).[6]

Researchers working closely with Inuit people found the mental distress brought about by environmental change runs deeper than affective states, and is tied to changes in cognition and thinking. Willox and colleagues describe the *ecological affect* as a set of "pragmatic and site-specific tracing of infinitely complex ecological arrangements" (Willox et al. 2013a, p.17). These comprise patterns of thinking and feeling that shape everyday behaviors, recognition, and information processing that support human interactions with the environment. Emotions, rather than merely an affective state, embody a response to environmental stimuli that acts as a necessary and integral form of cognition for guiding instrumental and social behaviors among Inuit people such as navigation and dietary decision-making (Willox et al. 2013a). These intimate ties between environment and cognition reveal the blurring of lines between human and nature and between environment and the mind.

First Nations have been asserting the negative impacts of environmental change on their lives long before the most recent acknowledgments or discoveries of environmental impacts on neurological and mental health.[7] Meanwhile, researchers continue to work to understand the linkages between what is classified as affective experience and neurological health. There are many pieces to the puzzle that defines environmental impacts on neurological and mental health, but there is an ongoing lack of definitive cause-and-effect findings to bind these pieces together in a cohesive body of knowledge. However, if one accepts the premise that the neurological and mental health of First Nations people has historically been supported by a specific set of physiological, cultural, and cognitive relationships with the environment, then it stands to reason that upon the reshaping of the environment, these relationships will be disrupted and negative impacts to brain and mental health could follow.

Engaging First Nations in research of the brain

Engaging First Nations people on the subject of neurological and mental health is made complicated by the history of troubled relationships between First Nations people and medical researchers. The eugenics movement occupies the extreme end of this scale, including racist medical research and public policy that resulted in the involuntary sterilization of First Nations women in Canada as recently as the 1970s (Boyer Y. 2004). In some cases, research has been conducted with inadequate respect for First Nations and without securing consent. In a prominent case in western Canada, a genetic researcher used blood samples originally gathered for the purpose of studying rheumatism to conduct unrelated research on HIV without permission from the Nuu-chah-nulth people who participated in the original study (Wiwchar 2004). The same researcher later used the genetic materials in published research about human migration that contradicted the Nuu-Chah-Nulth's historical accounts, causing insult to the band and introducing theories that undermined their historical claims to their territories. This case has been cited in the fields of pharmacogenetics, sexual wellness, and genetic ancestry as part of the historical mistrust that must be overcome by researchers seeking to develop relationships with First Nations populations (Lee et al. 2009; Devries & Free 2010; Boyer B.B. et al. 2011).

In other cases, research has been criticized for stigmatizing and attaching essentialist judgments to First Nations populations. For example, the *thrifty gene hypothesis* attempted to link the high incidence of diabetes among First Nations in Manitoba to genetic predispositions among hunter-gatherer populations that equipped them to survive periods of hardship through retention of nutrients, but which resulted in maladaptive weight problems under modern Western diets. Like the *warrior gene* hypothesis that sought to link risk-taking behavior among Maori people to genetic variants that affect aggressive behavior, the *thrifty gene hypothesis* has been criticized for failing to adequately consider the disruptive impacts of colonization and the social and material disadvantages suffered by aboriginal people under Western institutions (Abraham 2011; Gillett & Tamatea 2012).

Even in cases where researchers are quite deliberate about approaching environmental health issues without imposing value judgments upon the study population, First Nations can be negatively impacted by the results. For example, research identifying the neurological health risks of consuming seafood contaminated with methylmercury and other chemicals can have negative consequences for First Nations. Proof of contamination can weaken First Nations people's relationships with the fisheries that form a critical basis of their culture and identity, and encourage increased reliance upon less healthy dietary choices in areas where food alternatives are scarce (Wiseman & Gobas 2002; Gilbertson & Carpenter 2004). The potential consequences of such research place scientific indicators of health at odds with cultural values, and raises ethical questions about the nature of research that must be explored before analysis even begins.

Bringing the brain into First Nations experience

First Nations have endured generalized stigma from their health being defined according to deficits and negative comparisons with the broader population (Hodge et al. 2002; Smylie et al. 2008; Allan & Smylie 2015). Too often, deficit-based explanations fail to account for the historical circumstances of First Nations, and fail to demonstrate respect for the pride of First Nations people as they seek to recover from a history of marginalization and mistreatment. As research on the neurological and mental health of First Nations proceeds, it is paramount that the brains of First Nations people are not treated as objects of study defined primarily by their flaws and vulnerabilities. In a study of health and wellness among American Indians, Hodge and colleagues advanced an alternative model of defining aboriginal health that centers on the practices that comprise positive health and wellness, and the storytelling techniques that are used to promote and define the wellness of community members (Hodge et al. 2002).[8] Indeed, to bring the study of the brain into the experience of First Nations people, efforts are needed to understand the dimensions of their culture and relationships with the environment that define positive neurological and mental health.

Researchers based in Edmonton, Alberta, focused upon First Nations practices that support healthy brains in their study of the Cree First Nations' traditional child-rearing practices and their impacts on neurological development (Pazderka et al. 2014). The researchers found correspondence between findings in neuroscience and Cree practices related to pre-birth nutrition, feeding, and handling of newborns. For example, they found that Cree practices of carrying infants in traditional leather pouches close to the body reflects recent findings in Western sciences that indicate such practices support maternal attachment and social integration, and improve the capacity for handling stress (Anderson G.C. et al. 2003; Feldman 2011; Pazderka et al. 2014). Such findings suggest that although First Nations may not commonly speak the technical language of neurology and medical science in their everyday lives, they have millennia of lived experience that informs the practices they follow and the relationships they hold with the environment in order to support the health of their brains.

Engaging First Nations on their own terms in matters of neurological and mental health requires moving beyond the limits of Western medical science and characterizations of the brain as merely an infinitely complex system of tissues, chemical reactions, electrical signals, and psychological processes. Native American geneticist Dr. Frank Dukapoo explains that biological materials that scientists view as merely tissue and personal property may be viewed by First Nations as the sacred essence of a person (Dukapoo in Arbour & Cook 2006, p.155).[9] This not only prompts important ethical considerations in relation to the handling and stewardship of bodily materials (Arbour & Cook 2006; Gillett & McKergow 2007), but also suggests that those who conduct research on the brain with First Nations and other indigenous people must question the very nature of that which they seek to study.

Health researchers working in Australia followed an innovative path to bring studies of neurological health into alignment with Aboriginal concepts of the brain.[10] Rather than focusing on the physiology of the brain or functioning of the central nervous system, researchers were able to advance their studies and serve Aborigine needs by focusing on representations of the brain according to its role in social and cultural processes (D'Abbs & MacLean 2000). A team of psychologists and psychiatrists collaborated with Aboriginal artists to create pictures of the brain that depicted different levels of cultural engagement focusing on stories, family, country, and body. These dimensions corresponded with distinct dimensions of functional cognition, including memory, socialization, identity, and motor function (Petrol Link-up 1994; Cohen & Stemmer 2011). The approach assisted in the development of resources to educate Aborigines about the impacts of specific drugs on neurological health.

Recent research on early-onset familial Alzheimer's disease with a northern First Nation in British Columbia provided a similar example of how beneficial partnerships can be formed by attending to the distinct needs of the study population and recognizing the complex relationship between traditional beliefs and medical science. Researchers facilitated the creation of a community advisory group, and engaged in dialogue with community members to explore their concerns about the stigmatization of dementia and to discuss the potential harms and benefits that may come from further examination of the issue (Stevenson et al. 2013). This engagement assisted in drawing attention to the wider range of ethical considerations, including concerns about their future ability to obtain medical insurance, potential stigmatization of families and individuals, and delicate matters of consent related to determining who speaks for a community when disclosing the existence of a health problem.

The work also yielded insight into the way that First Nations people integrate traditional beliefs and information gained through medical science into their understandings of brain disease. Participants in the research drew simultaneously from medical knowledge and traditional definitions of wellness to explain the illness, demonstrating the potential for different knowledge systems to converge with positive outcomes (Cabrera et al. 2015). In turn, the researchers drew on these hybrid understandings to create educational resources that combined the First Nations narrative and epidemiological data to assist community members in identifying and managing the disease.[11] These findings show how aboriginal and Western scientific worldviews can be united, and that First Nations people need not choose between exclusively traditional and scientific frames of reference. First Nations people do not reject medical science or empirical inquiry; they simply do not privilege it above all other ways of knowing, and seek to reconcile it with the traditional knowledge they have held since time immemorial. Researchers who study issues related to the environment and neurological health among First Nations must thus come to terms with their entry into a dynamic world of information exchange within which knowledge is continually being created and recreated.

First Nations ethics and environmental neuroethics

Bridging environmental neuroethics with First Nations experiences involves not only the percolation of medical science into the context of indigenous experiences, but also the introduction of indigenous knowledge systems into the realm of the researcher. This requires researchers to embrace alternative views on the environment, the mind, and the nature of the relationship between them, thus expanding the ontological bases of inquiry. First Nations people may view the environment as an autonomous sentient force, and their own bodily and neurological health as a reflection of the state or the will of the environment. Harm to the spirit, or to the relationship that joins the body, mind, and nature together may be viewed as more important than links between environmental factors and neurological disease that are apparent through reductive science.

Researchers must thus be prepared to discuss phenomena that may not otherwise enter their field of examination, and be willing to alter their research designs to accommodate dimensions that First Nations see as important. Definitions of both nature and medicine must be expanded to embrace First Nations perspectives that include views of the body as an integrated whole linked with its environment: "It [traditional medicine] looks at all of you. … Non-traditional looks at that cut on your hand or that ankle you keep spraining, whereas traditional medicine looks at the whole person" (First Nation Elder in interview dealing with well-being).

Cajete explains that while Western scientific methods revolve around establishing objective views to determine "a factual blueprint of the world," indigenous views emphasize experience and locate the center of thought and cognition within the interconnectedness of the body with the environment (Cajete 2000, pp.24–25). And, although the interaction of body, mind, and environment may not be directly observable through common scientific methods, Huntington reminds us that such experiences must nonetheless be taken into account when examining the impact of environmental change on indigenous people (Huntington in Krupnik & Jolly 2002, pp.xxii–xxiii).

The specifics of beliefs of First Nations people about the nature of the environment and their own bodies and brains may vary from nation to nation and even family to family; the epistemological foundations of indigenous knowledge systems are not based in centralized rational scientific authority, but instead within oral histories, traditional practices, and ceremonies (Battiste 2002). In light of these considerations, research on environmental brain impacts with First Nations necessarily becomes a multidisciplinary endeavor that draws from medical sciences, sociology, anthropology, and other potential contributing fields.

Beyond the considerations identified earlier in forming trusting relationships between researchers and First Nations, there are numerous conceptual and methodological challenges inherent in this field of study. Clearly there are many concepts that defy operationalization within pure medical research models. For example, the unmeasurable phenomena that comprise the spirit demands that qualitative inquiry form a central part of any research effort. In turn, the ability to establish cause–effect relationships becomes problematic when examining the way that environmental change affects the brain when the focus shifts to a wider definition of what the brain and body encompass, and when these parts of a person are viewed as being inseparable from the environment.

One of the most important considerations is that First Nations knowledge cannot be treated in the same manner as other forms of knowledge, particularly scientific knowledge. In this respect, traditional knowledge is not a currency that is exchanged within the public domain, and is not subject to the same processes of scrutiny and revision that characterize peer-reviewed scientific inquiry. Traditional knowledge is often bound in stories passed down between generations as sacred or private learnings. Knowledge of the environment and of medicine is often linked to the social and political structure of a nation, and is interwoven in the bonds of clan and family that give shape to a First Nation community. Researchers must recognize the importance of incorporating traditional knowledge into their work, but at the same time must understand the local conventions and customs that accompany engagement with this knowledge. Traditional medicine does not exist independently as a solution that is introduced to an afflicted body. Instead it is, like the body and the mind, part of a broader web of experience.

> If you look at cultural, even the medicinal parts of Cedar you know? You may get traditional medi-
> cine. It's not just what you swallow or rub on you, because I think the traditional medicine also may
> include that whole process of making yourself ready to even go get the bark. Because that's part
> of the healing. So, if you look at traditional healing it's not just a medicinal sense, you know, it's a
> whole mindset or way of thinking, the healing part of it. (First Nation Elder in interview dealing
> with issues of well-being)

First Nations beliefs about the environment and its relationship with health form the basis of many studies, and several volumes of books would be required to provide a comprehensive summary of the respective intricacies found among the many existing nations. The proliferation of research in this area indicates that Western science has at least begun to recognize the importance of respecting and integrating First Nations knowledge into studies of the environment and the brain. However, there remains the delicate and complex matter of moving beyond acknowledgments of differences in perspectives, to incorporating First Nations perspective into research models, and addressing the ethical implications of the knowledge that is produced in these endeavors. That is, how do we reconcile the belief systems of First Nations people with the normative foundations of decision-making that lie at the core of neuroethics?

Bridging ethical foundations

Scholars and researchers in the field of neuroethics occupy a space of debate within which they evaluate and apply various philosophical perspectives and normative theories for the purpose of developing acceptable and beneficial guidelines for action as related to matters of the brain. Our objectives do not necessarily focus on reaching definitive judgments of right or wrong, but instead support the articulation of decision-making practice and ethical tools. Within this space of debate, it is not yet clear where First Nations perspectives fit, or how the established normative theories that define the field of neuroethics and bioethics fit into First Nations people's own conceptualizations of right and wrong, harmony and balance, and other ethical notions. As one explores issues related to First Nations and environmental neuroethics, it may be tempting to locate First Nations perspectives among established normative theories or find consistencies between their beliefs and established normative theories. However, this consideration of normative ethics must be approached with an appreciation that the fundamental way in which ethical decision-making is conducted within Western institutions of science and medicine and within First Nations contexts flows from distinct and different, yet inextricably linked trajectories of historical, cultural, and legal development.

As Pinker (2003) observes, the contributions of classical ethical theorists continue to provide a basis for the systems of thought that shape the way society grapples with bioethical and moral issues, including those related to the human mind and the environment. Kant's assertion of human autonomy continues to form an influential reference point as scholars debate over the point at which the treatment of biological materials must include consideration of the personhood of the materials, or how to weigh the autonomy of the person against that of the public when forming policies dealing with neurological enhancements (Racine 2010, pp.131–133; Barker & Beaufort 2013). The friction between utilitarianism and the alternative methods of evaluating nature forms an enduring point of conflict as society deliberates over issues such as the genetic modification of food sources and the associated implications for human health (Faunce 2012). Questions regarding the extent to which the state should enforce the duty of care have arisen when the right to utilize traditional healing has been pitted against medical positions regarding the best interests of the patient (McLaren 2015).[12] These contemporary debates serve as a reminder that normative ethics not only express particular frameworks of reason, but also articulate responses to changes in the world around us and the evolution of society's relationship with the state and notions of the self. If we view Kantian deontology and utilitarian ethics along with their contemporary manifestations as part of a broader response to change in society and the world around us, how then might we view the expression of First Nations ethics? More specifically, how do we understand First Nations ethics in relation to complicated matters of neurological health as we confront the converging implications of advancing medical science and a changing environment? We must look more closely not only at the different traditions from which Western and First Nations respective notions of ethics arise, but also the conditions within which they have come to be expressed and defended.

When one seeks to identify the substance of what is referred to as First Nations ethics, some of the most commonly encountered ideas are respect, balance, interconnectedness, sustainability, reciprocity, community, and self-determination. Expressions of these ideas are pervasive within literature that examines environmental and health issues related to First Nations people (Browne 1995; Wilson 2003; McGregor 2004; Isaak & Marchessault 2008; King et al. 2009; Beckford et al. 2010; Jones et al. 2010). Of course, such values do not comprise a clear body of normative thought as commonly recognized within Western society, nor do they provide a clear means of determining how decisions should be made in relation to issues of environmental impacts on brain and mental health. Indeed, when one searches for a body of normative ethics within literature on First Nations people, there is no common body of knowledge to be found among the diversity that comprises this population. Certainly these concepts are not alien to Western society, and First Nations are not alone in their valuation of these ideals or their application of these concepts to ethical matters. However, the way in which these concepts are defined differently between cultures, and the way in which these concepts have emerged as the primary expressions of First Nations ethics provides important cues for engaging First Nations in discussions of environmental impacts on neurological health.

For example, the impact of neurological conditions must be communicated in more than physiological and cognitive terms. Discussing the impact of environmental change on the brain needs to attend to the way symptoms may be understood as impacts on the community. At the same time, the emphasis on community needs to be understood as an effort to reassert the importance of the bonds within and between First Nations families in response to the Western emphasis on personal autonomy and the individualization of medical practices as a process that occurs between doctor and patient. Discussing neurological conditions, environmental change, and the relationships between these issues with First Nations populations must also attend to the issue of balance. This not only includes remaining open to different ideas of how balance exists among mind, body, spirit, and nature, but also understanding the emphasis on balance as a response to Western systems of human and environmental care that have disrupted First Nations lives.

It is self-determination which provides the most powerful compass point in this endeavor. The ethics of First Nations are articulated within the context of the assertion of their aboriginal rights. While the autonomy of the individual forms a central reference point in ethical debates regarding Western studies of neurological health, it is the autonomy of each individual nation that is central to First Nations decision-making. First Nations people deserve to be able to choose their own path as they address impacts on their brains and bodies, and it is the internal and intimate pathways of decision-making that occur within their communities, families, and histories that shape the normative basis of their thought in regard to these matters. These positions are not always for outsiders to know, and the ethical values that are articulated (e.g., balance, interconnectedness) may only serve as course-corrections or a response to the dominant institutions of science and health that surround them.

Like that of most other indigenous people, the First Nations experience is forever altered by the impact of colonialism. When words such as interconnectedness and community are cited as First Nations ethics, they arise not only as an expression of their ethical foundations, but also as a response to the decisions that have shaped their current circumstances. These ethical compass points redirect scientific inquiry to alternative ontologies, and different targets for study and analysis. Western researchers must remain aware of how First Nations ethics are expressed in part as a response to the incursion of Western thought into explaining their lives. Even assumptions of time must be suspended: the ethic of interconnectedness implies that time may not be linear in their worldview, and the future and the past remain interconnected. The choices made today may have impacts on both the future and histories of a First Nation.

These ethical compass points direct researchers toward a broader appreciation of how First Nations people experience and perceive environmental impacts on the brain. When seeking guidance on decision-making related to the environment and the health of the brain, one must be prepared to submit to each First Nation's distinct flows of ethical processing. This may or may not be made visible or shared, and it may vary from group to group. Close collaboration is required with each individual nation along with a commitment to developing relationships of trust before such knowledge may be shared. Moreover, there is a need for intellectual flexibility, and a readiness to understand the process of change First Nations are continuing to experience. Only in this manner can we build bridges between the contrasting ontologies and normative ethics that define approaches of First Nations and Western institutions to understanding environmental impacts on the brain as a process of change.

Neuroethics represents a relatively recent shift in the way we consider the brain, and environmental neuroethics constitutes a new direction in this field. These studies not only embody advances in science, but also reflect changes in society's concerns over our relationships with nature and the way it shapes our lives and our minds. We must not lose sight of the ongoing changes within our societies that are shaping our ethical reasoning. In this sense, the West may have more to gain than First Nations in the bridging of our respective foundations of knowledge. Over the past 50 years, Western society has faced a reckoning over the impact of technology and industry on our planet, and we have collectively struggled to develop normative positions with sufficient power to provoke the changes necessary to protect our health and that of the environment we rely upon. While we cannot and should not view First Nations through a romantic lens that casts them nobly as one-with-the-land, we can benefit from observing the way that they integrate sciences of neurology and the environment with their own understandings of the brain as they articulate normative positions regarding these matters. The way forward for environmental neuroethics is to view our engagement with First Nations as a process of mutual learning, and the co-production of knowledge regarding environment–brain relationships.

The way forward

The pursuit of the goal of creating new understandings and new knowledge of environmental–brain relationships demands engagement with the specifics of local meanings. Researchers must continue and extend existing metaphysical lines of inquiry and engage First Nations in discussion of questions such as "What is the brain?" and "What is the environment?" In turn, this can lead to engaging First Nations on meta-ethical questions that are central to their ethical ontology, such as "What is community?" and "What is sustainability?" and "What is balance?"

The study of environmental neuroethics as related to First Nations people is confronted by a dilemma. The serious risks posed to the mental and neurological health of First Nations as a result of mounting environmental changes demand that immediate actions be taken to address these problems. Indeed, the downstream costs to healthcare systems, communities, and society require that steps be taken to mitigate these impacts. At the same time, production of new understandings related to these issues cannot proceed without addressing the fundamental gaps between First Nations and Western scientific worldviews, and their respective domains of ethical thought. However, this dilemma does not amount to an antimony based on one form of knowledge being used to define another. Indeed, we should not be paralyzed by the concern that the gap between First Nations and Western institutions of science are either too great to be overcome or that the history of their interaction has produced a state of irreconcilable differences. Environmental neuroethics acknowledges that studies with First Nations about environmental impacts on the brain involve co-creation of knowledge that is equally applicable to both First Nations and Western society; that our cultures and ways of viewing the world are indeed distinct, but that they are also frequently parallel and driving toward problems of mutual concern.

For Western scholars, the explicit emphasis on human–nature relationships within First Nations population creates an object of study, defined by its seemingly stark contrast with Western notions of separation between society and nature. In turn, this contrast provides reason to pause and to reconsider the impact of environmental change on the brain and ethics of environmental care for all. Through the past half century, during which Western society has come to confront the impacts of environmental change on human health, the focus has been primarily upon the body. The study of the environment has been shaped around these concerns. As we confront impacts on the brain, we enter new territory in our consideration of ethics related to the environment and our use of natural resources.

Notes

1. First Nations people, along with Inuit and Metis, comprise the groups referred to as aboriginal people in Canada. Inuit include the aboriginal people in the northern and arctic regions of Canada, and Metis include people with mixed origins that share First Nations and other (generally Western European) ancestry. First Nations were formerly referred to as Indians, and while this term is still utilized in the United States, the word Indian is no longer used in Canada to refer to aboriginal people.

2. The pronoun "we" is used extensively throughout this chapter, and its reference to Western society is by design. Indeed, it identifies a specific audience for this chapter with which not all readers may identify. However, the use of "we" is done self-consciously by the authors in acknowledgment of the social and cultural separation of First Nations from the society around them.

3. For many First Nations in British Columbia, for example, the wholesale industrialization of their forests and the flooding of their traditional lands occurred largely throughout the 1940s to 1960s. Examples include the Cheslatta and the Tsay Keh Dene, who were both displaced by large hydro-electric projects accompanied by the expansion of logging and mineral development in their territories.

4. Exceptions to this include severe central nervous system failure as a result of exposure to pollution at an extreme level, such as the events witnessed in Bhopal, India following the release of toxic gasses from a pesticide plant in 1984.

5. Solastalgia is taken from the words nostalgia and solace (Albrecht et al., 2007, p.96). Nostalgia, which was classified as a diagnosable illness until the late nineteenth century, is linked to the attachment of smells, tastes, and sounds to previous locational and geographic contexts (Hirsch 1992).

6. Albrecht (2011, 2012) defines psychoterratic illnesses as earth-based (terra) mental (psyche) health states or conditions that are caused by disruption of the conditions that shape people's feelings and emotions about nature and space.

7. First Nations have long expressed resistance to colonial impacts on their lands, but it is only in the past half-century that their voices have received proper legal recognition. Between 1927 and 1950, First Nations faced legal bans on their ability to mount protests latter half of the twentieth century. Between 1951 and 2000, there were an estimated 616 acts of Aboriginal protest in Canada (Clairmont & Potts 2006, p.20).

8. The term American Indian is the accepted terminology for indigenous people in the United States.

9. Indeed, it should be acknowledged that the infusion of the body with some intangible dimension of spirit is not exclusive to First Nations. As Pinker (2003) points out, the idea of a physical body indelibly imbued with sacred qualities or a predisposition toward a particular social nature also affects Western systems of thought and religious belief.

10. The word "Aboriginal" is capitalized when referring to the indigenous people of Australia. The term "Aborigine" is utilized when referring to a person of Aboriginal ancestry.

11. Resources included information folders that explained risk factors and described experiences of community members affected by Alzheimer's disease, and story books aimed at helping children understand brain functions and the way that the disease affects community and family relationships.

12. McLaren provides an account of a young First Nations girl from the Canadian province of Manitoba who was supported by her parents in refusing potentially life-saving chemotherapy in favour of traditional healing methods.

References

Abraham, C. (2011, February 25). How the diabetes-linked 'thrifty gene' triumphed with prejudice over proof. *The Globe and Mail*. Available from: http://www.theglobeandmail.com/news/national/how-the-diabetes-linked-thrifty-gene-triumphed-with-prejudice-over-proof/article569423/?page=all [accessed July 15, 2015].

Albrecht, G. (2011). Chronic environmental change: emerging "psychoterratic" syndromes. In: Weissbecker, I. (Ed.) *Climate Change and Human Well-Being*, pp.43–56. New York: Springer.

Albrecht, G. (2012). Psychoterratic conditions in a scientific and technological world. In: Kahn, P.H. and Hasbach, P. (Eds.) *Ecopsychology: Science, Totems, and the Technological Species*, pp.241–264. Cambridge, MA: MIT Press.

Albrecht, G., Sartore, G.M., Connor, L., et al. (2007). Solastalgia: the distress caused by environmental change. *Australasian Psychiatry*, **15**(Suppl. 1), S95–S98.

Allan, B. and Smylie, J. (2015). *First Peoples, Second Class Treatment: The Role of Racism in the Health and Well-Being of Indigenous Peoples in Canada*. Toronto, ON: The Wellesley Institute.

Anderson, G.C., Moore, E., Hepworth, J., and Bergman, N. (2003). Early skin-to-skin contact for mothers and their healthy newborn infants. *Cochrane Database of Systematic Reviews*, **2**, CD003519.

Anderson, R.B. (1999). *Economic Development Among the Aboriginal peoples of Canada: The Hope for the Future*. Toronto, ON: Captus Press.

Arbour, L. and Cook, D. (2006). DNA on loan: issues to consider when carrying out genetic research with aboriginal families and communities. *Public Health Genomics*, **9**(3), 153–160.

Barker, R.A. and de Beaufort, I. (2013). Scientific and ethical issues related to stem cell research and interventions in neurodegenerative disorders of the brain. *Progress in Neurobiology*, **110**, 63–73.

Barron, F.L. (1984). A summary of federal Indian policy in the Canadian west, 1867–1984. *Native Studies Review*, **1**(1), 28–39.

Battiste, M. (2002). *Indigenous Knowledge and Pedagogy in First Nations Education: A Literature Review with Recommendations*. Ottawa, ON: Apamuwek Institute.

Beckford, C.L., Jacobs, C., Williams, N., and Nahdee, R. (2010). Aboriginal environmental wisdom, stewardship, and sustainability: lessons from the Walpole Island First Nations, Ontario, Canada. *The Journal of Environmental Education*, **41**(4), 239–248.

Benusic, M. (2013). Fracking in BC: a public health concern. *British Columbia Medical Journal*, **55**(5), 238–239. Available from: http://www.bcmj.org/council-health-promotion/fracking-bc-public-health-concern [accessed August 17, 2015].

Berry, H.L., Kelly, B.J., Hanigan, I.C., et al. (2008). *Rural Mental Health Impacts of Climate Change. Commissioned Report for the Garnaut Climate Change Review.* Canberra, ACT: The Australian National University.

Block, M.L. and Calderón-Garcidueñas, L. (2009). Air pollution: mechanisms of neuroinflammation and CNS disease. *Trends in Neurosciences*, 32(9), 506–516.

Bourque, F. and Willox, AC. (2014). Climate change: the next challenge for public mental health? *International Review of Psychiatry*, 26(4), 415–422.

Boyer, B.B., Dillard, D., Woodahl, E.L., Whitener, R., Thummel, K.E., and Burke, W. (2011). Ethical issues in developing pharmacogenetic research partnerships with American Indigenous communities. *Clinical Pharmacology & Therapeutics*, 89(3), 343–345.

Boyer, Y. (2004). *First Nations, Métis and Inuit Health Care: The Crown's Fiduciary Obligation.* [Discussion Paper Series in Aboriginal Health: Legal Issues, No. 2.] Ottawa, ON: National Aboriginal Health Organization, Native Law Centre, University of Saskatchewan.

Browne, A.J. (1995). The meaning of respect: a First Nations perspective. *Canadian Journal of Nursing Research*, 27, 95–109.

Cabrera, L.Y., Beattie, B.L., Dwosh, E., and Illes, J. (2015). Converging approaches to understanding early onset familial Alzheimer disease: a First Nation study. *SAGE Open Medicine*, 3, 2050312115621766.

Cabrera, L.Y., Tesluk, J., Chakraborti, M., Matthews, R., and Illes, J. (2016). Brain matters: from environmental ethics to environmental neuroethics. *Environmental Health*, 15(1), 20.

Cajete, G. (2000). *Native Science: Natural Laws of Interdependence.* Santa Fe, NM: Clear Light Publishers.

Calderon-Garciduenas, L. Reed, W., Maronpot, R.R., et al. (2004). Brain inflammation and Alzheimer's-like pathology in individuals exposed to severe air pollution. *Toxicologic Pathology*, 32(6), 650–658.

Chepesiuk, R. (2009). Missing the dark: health effects of light pollution. *Environmental Health Perspectives*, 117(1), A20–A27.

Clairmont, D.H. and Potts, J. (2006). *For the Nonce: Policing and Aboriginal Occupations and Protests.* [A Background Paper Prepared for the Ipperwash Inquiry.] Ottawa, ON: Ipperwash Inquiry.

Cohen, H. and Stemmer, B. (Eds.) (2011). *Consciousness and Cognition: Fragments of Mind and Brain.* London: Academic Press.

Crowe, K. (2014, September 2). Grassy Narrows: why is Japan still studying the mercury poisoning when Canada isn't? *CBC News.* Available from: http://www.cbc.ca/news/health/grassy-narrows-why-is-japan-still-studying-the-mercury-poisoning-when-canada-isn-t-1.2752360 [accessed August 17, 2016].

Csonka, Y. and Schweitzer, P. (2004). Societies and cultures: change and persistence. In: Einarsson, N., Larsen, J.N., Nilsson, A., and Young, O.R. (Eds.) *Arctic Human Development Report*, pp.45–68. Akureyri, Iceland: Stefansson Arctic Institute.

D'Abbs, P. and MacLean, S. (2000). *Petrol Sniffing in Aboriginal Communities: A Review of Interventions.* Casuarina, NT: Cooperative Research Centre for Aboriginal and Tropical Health.

De Kluizenaar, Y., Passchier-Vermeer, W., and Miedema, H. (2001). *Adverse Effects of Noise Exposure to Health.* Report prepared for the EC Project UNITE by TNO PG. Leiden, the Netherlands: Division of Public Health.

Devries, K.M. and Free, C. (2010). "I told him not to use condoms": masculinities, femininities and sexual health of Aboriginal Canadian young people. *Sociology of Health & Illness*, 32(6), 827–842.

Di Pietro, N.C. and Illes, J. (2014). Disparities in Canadian indigenous health research on neurodevelopmental disorders. *Journal of Developmental & Behavioral Pediatrics*, 35(1), 74–81.

El-Hayek, Y.H. (2007). Mercury contamination in Arctic Canada: possible implications for aboriginal health. *Journal of Developmental Disabilities*, **13**, 67–89.

Faunce, T. (2012). Governing planetary nanomedicine: environmental sustainability and a UNESCO universal declaration on the bioethics and human rights of natural and artificial photosynthesis (global solar fuels and foods). *Nanoethics*, **6**(1), 15–27.

Feldman, R. (2011). Maternal touch and the developing infant. In: Hertenstein, M.J. and Weiss, S.J. (Eds.) *The Handbook of Touch: Neuroscience, Behavioral and Health Perspectives*, pp.373–407. New York: Springer Publishing Company.

Fritze, J.G., Blashki, G.A., Burke, S., and Wiseman, J. (2008). Hope, despair and transformation: climate change and the promotion of mental health and wellbeing. *International Journal of Mental Health Systems*, **2**, 13.

Furniss, E. (1999). *The Burden of History: Colonialism and the Frontier Myth in a Rural Canadian Community*. Vancouver, BC: UBC Press.

Gilbertson, M. and Carpenter, D.O. (2004). An ecosystem approach to the health effects of mercury in the Great Lakes basin ecosystem. *Environmental Research*, **95**(3), 240–246.

Gillett, G. and Tamatea, A.J. (2012). The warrior gene: epigenetic considerations. *New Genetics and Society*, **31**(1), 41–53.

Gillett, G. and McKergow, F. (2007). Genes, ownership, and indigenous reality. *Social Science & Medicine*, **65**(10), 2093–2104.

Gladstone, P. (1953). Native Indians and the fishing industry of British Columbia. *The Canadian Journal of Economics and Political Science/Revue canadienne d'Economique et de Science politique*, **19**(1), 20–34. Available from: http://www.jstor.org.ezproxy.library.ubc.ca/stable/10.2307/138471?origin=api [accessed July 9, 2016].

Goodine, C. (2011, October). Fracking controversy: rethinking the low-carbon label for natural gas. *Canadian Geographic*. Available from: https://www.canadiangeographic.ca/article/fracking-controversy [accessed July 9, 2016].

Grandjean, P. and Herz, K.T. (2011). Methylmercury and brain development: imprecision and underestimation of developmental neurotoxicity in humans. *Mount Sinai Journal of Medicine*, **78**(1), 107–118.

Grandjean, P. and Landrigan, P. J. (2014). Neurobehavioural effects of developmental toxicity. *Lancet Neurology*, **13**(3), 330–338.

Haines, A., Kovats, R.S., Campbell-Lendrum, D., and Corvalán, C. (2006). Climate change and human health: impacts, vulnerability and public health. *Public Health*, **120**(7), 585–596.

Harada, M., Hanada, M., Jajiri, M., et al. (2011). Mercury pollution in first nations groups in Ontario, Canada: 35 years of Canadian Minamata disease. *Journal of Minamata Studies*, **3**, 3–30.

Hicks, J. (2007). The social determinants of elevated rates of suicide among Inuit youth. *Indigenous Affairs*, **4**, 30–37.

High, S. (1996). Native wage labour and independent production during the "Era of Irrelevance." *Labour/le travail*, 243–264. Available from http://www.jstor.org.ezproxy.library.ubc.ca/stable/10.2307/25144044?origin=api [accessed August 17, 2016].

Hirsch, A.R. (1992). Nostalgia – a neuropsychiatric understanding. *Advances in Consumer Research*, **19**, 390–395.

Hodge, F.S., Pasqua, A., Marquez, C.A., and Geishirt-Cantrell, B. (2002). Utilizing traditional storytelling to promote wellness in American Indian communities. *Journal of Transcultural Nursing*, **13**(1), 6–11.

Illes, J., Davidson, J., and Matthews, R. (2014). Environmental neuroethics: changing the environment—changing the brain. Recommendations submitted to the Presidential Commission for the Study of Bioethical Issues. *Journal of Law and the Biosciences*, **1**(2), 221–223.

Isaak, C.A. and Marchessault, G. (2008). Meaning of health: the perspectives of Aboriginal adults and youth in a northern Manitoba First Nations community. *Canadian Journal of Diabetes*, **32**(2), 114–122.

Jackson, D.D. (2011). Scents of place: the dysplacement of a First Nations community in Canada. *American Anthropologist*, **113**(4), 606–618.

Jang, B. (2015, July 9). Five First Nations join forces to assess environmental impact of LNG exports. *The Globe and Mail*. Available from: http://www.theglobeandmail.com/report-on-business/industry-news/energy-and-resources/five-first-nations-join-forces-to-assess-impact-of-bc-lng-exports/article25388272 [accessed July 9, 2016].

Jones, R., Rigg, C., and Lee, L. (2010). Haida marine planning: First Nations as a partner in marine conservation. *Ecology and Society*, **15**(1), 12.

Kant, I. (1785). *Groundwork of the Metaphysics of Morals*.

Kelly, E.N., Short, J.W., Schindler, D.W., et al. (2009). Oil sands development contributes polycyclic aromatic compounds to the Athabasca River and its tributaries. *Proceedings of the National Academy of Sciences of the United States of America*, **106**(52), 22346–22351.

Kelly, E.N., Schindler, D.W., Hodson, P.V., Short, J.W., Radmanovich, R., and Nielsen, C.C. (2010). Oil sands development contributes elements toxic at low concentrations to the Athabasca River and its tributaries. *Proceedings of the National Academy of Sciences of the United States of America*, **107**(37), 16178–16183.

King, M., Smith, A., and Gracey, M. (2009). Indigenous health part 2: the underlying causes of the health gap. *The Lancet*, **374**(9683), 76–85.

Klotz, O. (1914). Pulmonary anthracosis—a community disease. *American Journal of Public Health*, **4**(10), 887–916.

Krupnik, I. and Jolly, D. (2002). *The Earth is Faster Now: Indigenous Observations of Arctic Environmental Change*. [Frontiers in Polar Social Science.] Fairbanks, AK: Arctic Research Consortium of the United States.

Kuhnlein, H.V., Receveur, O., Muir, D.C.G., Chan, H.M., and Soueida, R. (1995). Arctic indigenous women consume greater than acceptable levels of organochlorines. *The Journal of Nutrition*, **125**(10), 2501.

Lee, S.S.J., Bolnick, D.A., Duster, T., Ossorio, P., and TallBear, K. (2009). The illusive gold standard in genetic ancestry testing. *Science*, **325**(5936), 38.

Lehti, V., Niemelä, S., Hoven, C., Mandell, D., and Sourander, A. (2009). Mental health, substance use and suicidal behaviour among young indigenous people in the Arctic: a systematic review. *Social Science & Medicine*, **69**(8), 1194–1203.

MacDonald, J.P., Ford, J.D., Willox, A.C., and Ross, N.A. (2013). A review of protective factors and causal mechanisms that enhance the mental health of indigenous circumpolar youth. *International Journal of Circumpolar Health*, **72**, 21775.

McAlpine, D. and Araki, S. (1958). Minamata disease an unusual neurological disorder caused by contaminated fish. *The Lancet*, **272**(7047), 629–631.

McGregor, D. (2004). Coming full circle: indigenous knowledge, environment, and our future. *The American Indian Quarterly*, **28**(3), 385–410.

McLaren, L. (2015, January 21). Makayla Sault: whose rights are served when a little girl dies? *Special to The Globe and Mail*. Available from: http://www.theglobeandmail.com/life/parenting/whose-rights-are-served-when-a-little-girl-dies/article22562573/ [accessed August 18, 2015].

McMichael, A.J. (2003). *Climate Change and Human Health: Risks and Responses*. Geneva: World Health Organization.

McMichael, A.J., Woodruff, R.E., and Hales, S. (2006). Climate change and human health: present and future risks. *The Lancet*, **367**(9513), 859–869.

McQuigge, M. (2012, June 4). Two Ontario first nations still plagued by mercury poisoning: report. *The Globe and Mail.* Available from: http://www.theglobeandmail.com/news/politics/two-ontario-first-nations-still-plagued-by-mercury-poisoning-report/article4230507/ [accessed August 17, 2015].

Masuda, J.R. and Garvin, T. (2006). Place, culture, and the social amplification of risk. *Risk Analysis,* **26**(2), 437–454.

Mills, C.A. (1943). Urban air pollution and respiratory diseases. *American Journal of Epidemiology,* **37**(2), 131–141.

Murata, K., Weihe, P., Budtz-Jørgensen, E., Jørgensen, P.J., and Grandjean, P. (2004). Delayed brainstem auditory evoked potential latencies in 14-year-old children exposed to methylmercury. *Journal of Pediatrics,* **144**(2), 177–183.

Nikiforuk, A. (2015, January 10). Fracking industry shakes up northern BC with 231 tremors. *TheTyee. ca.* Available from: http://thetyee.ca/News/2015/01/10/Fracking_Industry_Shakes_Up_Northern_BC/ [accessed August 17, 2015].

Northern Brain Injury Association (2015). *First Nations.* http://nbia.ca/first-nations-brain-injury/ [accessed August 17, 2015].

Parsons, R. and Prest, G. (2003). Aboriginal forestry in Canada. *The Forestry Chronicle,* **79**(4), 779–784. Available from: http://pubs.cif-ifc.org.ezproxy.library.ubc.ca/loi/tfc?open=2003#id_2003 [accessed August 17, 2015].

Passchier-Vermeer, W. and Passchier, W. F. (2000). Noise exposure and public health. *Environmental Health Perspectives,* **108**(Suppl. 1), 123.

Patz, J.A., Campbell-Lendrum, D., Holloway, T., and Foley, J.A. (2005). Impact of regional climate change on human health. *Nature,* **438**(7066), 310–317.

Pazderka, H., Desjarlais, B., Makokis, L., et al. (2014). Nitsiyihkâson: the brain science behind Cree teachings of early childhood attachment. *First Peoples Child & Family Review,* **9**(1), 53–65.

Petrol Link-up. (1994). *Australian Government Department of Health and Aging, and Menzies School of Health Research.* Available from: http://resources.menzies.edu.au/download/Sniffing_and_the_Brain.pdf [accessed July 15, 2015].

Pinker, S. (2003). *The Blank Slate: The Modern Denial of Human Nature.* Chicago, IL: Penguin.

Pinkerton, E. (1987). The fishing-dependent community. In: Marchak, M.P., Guppy, N., and McMullan, J.L. (Eds.) *Uncommon Property: The Fishing and Fish-Processing Industries in British Columbia,* pp.293–325. Toronto, ON: Methuen Publications.

Racine, E. (2010). *Pragmatic Neuroethics: Improving Treatment and Understanding of the Mind–Brain.* Cambridge, MA: MIT Press.

Ray, J.G., Vermeulen, M.J., Meier, C., Cole, D.E., and Wyatt, P.R. (2004). Maternal ethnicity and risk of neural tube defects: a population-based study. *Canadian Medical Association Journal,* **171**(4), 343–345.

Raz, R., Roberts, A.L., Lyall, K., et al. (2015). Autism spectrum disorder and particulate matter air pollution before, during, and after pregnancy: a nested case–control analysis within the Nurses' Health Study II Cohort. *Environmental Health Perspectives,* **123**(3), 264–270.

Rose, N. and Abi-Rached, J.M. (2013). *Neuro: The New Brain Sciences and the Management of the Mind.* Princeton, NJ: Princeton University Press.

Russell, W.T. (1924). The influence of fog on mortality from respiratory diseases. *The Lancet,* **204**(5268), 335–339.

Scarlett, A.G., West, C.E., Jones, D., Galloway, T.S., and Rowland, S.J. (2012). Predicted toxicity of naphthenic acids present in oil sands process-affected waters to a range of environmental and human endpoints. *Science of the Total Environment,* **425**, 119–127.

Schaefer, T.W. (1907). The contamination of the air of our cities with sulphur dioxide, the cause of respiratory disease. *Boston Medical and Surgical Journal,* **157**(4), 106–110.

Smylie, J., Kaplan-Myrth, N., McShane, K., and Nation, P.F. (2008). Indigenous knowledge translation: baseline findings in a qualitative study of the pathways of health knowledge in three indigenous communities in Canada. *Health Promotion Practice*.

Stanbury, W.T. (1973). Indians in British Columbia: level of income, welfare dependency and poverty rate. *The British Columbian Quarterly*, **20**, 66–78. Available from: http://ojs.library.ubc.ca/index.php/bcstudies/issue/view/140 [accessed July 15, 2015].

Stevenson, S., Beattie, B.L., Vedan, R., Dwosh, E., Bruce, L., and Illes, J. (2013). Neuroethics, confidentiality, and a cultural imperative in early onset Alzheimer disease: a case study with a First Nation population. *Philosophy, Ethics, and Humanities in Medicine*, **8**(1), 15.

Stout, R., Stout, T.D., and Harp, R. (2009). *Maternal and Infant Health and the Physical Environment of First Nations and Inuit Communities: A Summary Review*. Winnipeg, MB: Prairie Women's Health Centre of Excellence.

Sullivan, A. and Brems, C. (1997). The psychological repercussions of the sociocultural oppression of Alaska native peoples. *Genetic, Social, and General Psychology Monographs*, **123**(4), 411–440.

Takeuchi, T., Kambara, T., Morikawa, N., Matsumoto, H., Shiraishi, Y., and Ito, H. (1959). Pathologic observations of the Minamata disease. *Pathology International*, **9**(S1), 769–783.

Talbott, E.O., Arena, V.C., Rager, J.R., et al. (2015). Fine particulate matter and the risk of autism spectrum disorder. *Environmental Research*, **140**, 414–420.

Tennant, P. (1990). *Aboriginal Peoples and Politics: The Indian Land Question in British Columbia, 1849–1989*. Vancouver, BC: UBC Press.

Timoney, K.P. and Lee, P. (2009). Does the Alberta tar sands industry pollute? The scientific evidence. *The Open Conservation Biology Journal*, **3**, 65–81.

Van Oostdam, J., Gilman, A., Dewailly, E., et al. (1999). Human health implications of environmental contaminants in Arctic Canada: a review. *Science of the Total Environment*, **230**(1), 1–82.

Wheatley, B. and Wheatley, M.A. (2000). Methylmercury and the health of indigenous peoples: a risk management challenge for physical and social sciences and for public health policy. *Science of the Total Environment*, **259**(1), 23–29.

Willox, A.C., Harper, S.L., Edge, V.L., Landman, K., Houle, K., and Ford, J.D. (2013a). The land enriches the soul: on climatic and environmental change, affect, and emotional health and well-being in Rigolet, Nunatsiavut, Canada. *Emotion, Space and Society*, **6**, 14–24.

Willox, A.C., Harper, S.L., Ford, J.D., et al. (2013b). Climate change and mental health: an exploratory case study from Rigolet, Nunatsiavut, Canada. *Climatic Change*, **121**(2), 255–270.

Willox, A.C., Stephenson, E., Allen, J., et al. (2015). Examining relationships between climate change and mental health in the Circumpolar North. *Regional Environmental Change*, **15**(1), 169–182.

Wilson, K. (2003). Therapeutic landscapes and First Nations peoples: an exploration of culture, health and place. *Health & Place*, **9**(2), 83–93.

Wiseman, C.L. and Gobas, F.A. (2002). Balancing risks in the management of contaminated First Nations fisheries. *International Journal of Environmental Health Research*, **12**(4), 331–342.

Wiwchar, D. (2004, December 16). Nuu-chah-nulth blood returns to west coast. *Ha-Shilth-Sa. Canada's Oldest First Nations Newspaper*, **31**(25). Available from: http://www.igb.illinois.edu/sites/default/files/Wiwchar%202004%20Nuu-chah-nulth.pdf [accessed July 15, 2015].

Woolford, A. and Thomas, J. (2011). Genocide of Canadian First Nations. *Genocide of Indigenous Peoples*, **8**, 61–86.

Yazzie, R. (2000). Postcolonial colonialism. Reclaiming Indigenous voice and vision. In: Battiste, M. (Ed). *Reclaiming Indigenous Voice and Vision*, pp.39–49. Vancouver, BC: UBC Press.

Young, T.K. (1994). *The Health of Native Americans: Toward a Biocultural Epidemiology*. New York: Oxford University Press.

Chapter 24

The neurobiology of addiction as a window on voluntary control of behavior and moral responsibility

Steven E. Hyman

Introduction

Drug addiction is a severe problem for public health, systems of justice, and for the functioning of societies worldwide. Views of addicted people vary widely between and even within countries. Perspectives range from seeing them as medically ill to having the status of moral pariahs. Since possession of addictive drugs other than tobacco or alcohol is a criminal offense within many jurisdictions, addiction to many drugs is de facto a crime. In this chapter, I use insights that have emerged from neuroscience and psychology in the last two decades to examine the predominant medical model of drug addiction as a form of behavior that has escaped volitional control, that is, compulsive drug use despite significant negative consequences. This model is codified in the fifth edition of the *Diagnostic and Statistical Manual of Mental Disorders* (DSM-5) (American Psychiatric Association 2013) and similarly in the tenth edition of the *International Classification of Diseases* (ICD-10) (World Health Organization 1992). I contrast this view of addiction with what might be called a moral model, which sees the behavior of the addicted person as reflecting a series of bad choices that could be brought under volitional control with appropriate effort. These contrasting views have significant ethical and pragmatic implications for how societies hold addicted people responsible for their behavior and how they treat them.

Addiction is an apt subject for neuroethical analysis because (1) it problematizes the concepts of volition and responsibility, which are central concerns for neuroscience, philosophy, law, and ethics; and (2) because enough is known scientifically about addiction to ground the discussion in a reasonable, if still incomplete body of knowledge. Addiction is a topic about which too many ungrounded inferences and assertions are made. My brief scientific discussion is a byproduct of mechanistic neurobiological analyses performed for other reasons. The pursuit of neurobiological mechanisms is partly driven by a desire for greater understandings of nature as well as human self-understanding. However, as with essentially all other medically significant conditions, mechanistic investigations of addiction are also meant to yield knowledge that will ultimately prove actionable for prevention and treatment.

Compulsion

Short of being externally coerced or psychotic, what can it mean to say that a person cannot control his or her consciously planned behaviors? What does society make of a person who has suffered dire health consequences from smoking tobacco, understands the health risks perfectly well, states a credible intention to stop smoking, repeatedly tries to cut down, but repeatedly resumes smoking? How do we interpret the behavior of an opioid user who had entered a treatment facility under the threat of losing his job and spouse, who soon after release from the facility (where he initially suffered painful withdrawal symptoms, and later participated seriously in psychotherapy and rehabilitative services), steals money from his spouse, buys heroin, prepares it for self-injection, and consumes it? Are these individuals likely to be lying glibly about their intentions to abstain? Are they weak-willed, heedlessly seeking pleasure or oblivion rather than effectively confronting urges to smoke tobacco or use drugs? Or is there a convincing alternative explanation for their behavior that can be grounded in the concepts of compulsion and loss of control that dominate medical description of addiction?

Much is now known about neural mechanisms of decision-making and control of behavior and how these mechanisms can be altered long term by addictive drugs. Nonetheless, addicted individuals, such as those briefly described in the vignettes, often attract moral condemnation, and if caught breaking the law for possessing illegal drugs or committing illegal acts to obtain them, may face severe sanctions. Attribution of the continuing drug use of addicted people to weak will or some kind of moral failure remains a potent factor in social, policy, and legal spheres. For example, in 1988 the Supreme Court of the United States decided two related cases, *Traynor v. Turnage* and *McKelvey v. Turnage* (1988), upholding the denial of a petition of two veterans by the Veterans Administration (VA) in the United States. The veterans had asked the VA to extend the usual 10-year limit on educational benefits because they suffered from complications of alcoholism, which, they argued, was a medical illness. Extensions had previously been granted to veterans who, because of other illnesses, had been unable to use their benefits during the allotted 10 years' time. The Court, however, upheld the denial of an extension, with the majority finding that the VA was reasonable in defining alcoholism as willful misconduct.

In contrast to the Court's decision, the DSM-5 (American Psychiatric Association 2013) defines drug addiction as compulsive drug use despite significant negative consequences. Loss of control over drug use may be evidenced, inter alia, by multiple unsuccessful attempts to cut down. Those who doubt that addicted individuals have truly lost control of their behavior and, like the Court, see it instead as willful misconduct, at least implicitly construe human decision-making to be a conscious process of deliberation that weighs costs and benefits (Heyman 2010). Based on such a model, a decision to use drugs would represent a conscious choice of gratification or relief from distress over the risk of exacerbating drug-related harms and perhaps committing illegal acts. In this model, a decision to abstain from drug use, and thus to forego the pleasure or relief gained by drug taking, might result from the incorporation of salient new information or recent aversive experiences that tilt the decisional scales against continued drug taking (Heyman 2010).

Those skeptical of the idea that addicted people cannot control their substance use have pointed out that seeking and using drugs often requires meticulous planning, first to gain necessary resources and then drugs; may be played out over time, giving ample opportunity for users to inhibit prepotent drug-seeking impulses; and requires significant cognitive and behavioral flexibility given real-world obstacles to obtaining drugs, especially when they are illegal or must be used furtively. Moreover, addicted people, except when intoxicated or experiencing significant withdrawal symptoms, may act in a rational manner, might at times display insight into their condition, and can typically control most other aspects of their behavior—unlike individuals with florid psychosis, dementia, or delirium who would seem to be better candidates for the descriptor, lack of control. A salient example of individuals who for the most part exhibit rationality outside the sphere of their drug habit are tobacco smokers—at least in part because tobacco is legal and not intoxicating—but to varying degrees, users of other drugs have periods of rational control of behavior.

Opposition to medical models of addiction is also grounded in pragmatic concerns. Medical models can be construed as decreasing incentives for recovery by appearing to explain away drug use as uncontrollable and providing a rationalization for being unable to abstain, (Satel 1999; Satel & Lillienfeld 2013). Concerns have been raised, in addition, about a trend toward medicalizing all forms of deviance in society (Conrad & Schneider 1992). After all, criminality is a form of deviance that, like addiction, has identifiable familial (presumably partly genetic), developmental, social, and contextual risk factors, yet it is not included in the DSM-5. Those concerned about what they see as creeping medicalization of all deviance also express concern about the placement of compulsive gambling in the same chapter with drug addiction in the DSM-5, and about arguments being made to further broaden the concept of behavioral addiction (Robbins & Clark 2015) based on shared risk factors and shared neural mechanisms. In fact, some scientists and clinicians have been arguing for the inclusion as addictions in future DSM editions of compulsive sex, compulsive shopping, compulsive use of the internet, and perhaps excess eating associated with obesity (Volkow et al. 2013).

The idea that human decision-making, including the choices made by addicted people, represent conscious deliberations that yield freely chosen goals and actions, is consistent with ubiquitous human intuitions that are sometimes described as folk psychology (Morse 2008). Upon introspection, human beings intuit a self that has experiences, makes decisions, and causes bodily actions. Thus folk psychology and systems of justice hold that human beings act for reasons. This view is entirely contradicted by neurobiology, which finds no evidence for anything like a self that possesses decisional capacity and causes behavior. What have been found instead are diverse neural mechanisms—opaque to introspection—that contribute to different aspects of decision-making and that control behavior. It appears that one of the outputs of the circuitry that controls action is the experience—but not the fact—of a self that authors our behavior (Wegner 2002, 2003; Haggard 2008). Without resorting to evolutionary "just so" stories, it can safely be posited that cognitive and behavioral mechanisms were selected in evolution by their adaptiveness and survival value, not by their ability to capture objective truth.

Actions can be conceptualized as ranging from reflexes, which are the most stimulus bound, to volitional acts in which external stimuli act only indirectly and do not directly determine the timing or nature of the act (Gold & Shadlen 2007; Haggard 2008). A partially overlapping way of parsing actions (which excludes simple reflexes) is to rank at one extreme, actions that arise wholly from unconscious processing and at the other, actions that reflect conscious, goal-oriented, deliberative decisions. However, decision-making is more complex than might be implied by these extremes. For example, voluntary action that might, on the evidence of introspection, appear to spring entirely from conscious, deliberative decision-making, is also invariably dependent on rapid, unconscious, parallel processing by neural circuits with access to such information as previously stored values (Gallagher et al. 1999; Xie & Padoa-Schioppa 2016), social stereotypes (Dunham et al 2013; Stolier & Freeman 2016), and ingrained behavioral tendencies. The decision-maker cannot be directly aware of this processing, or of the neural mechanisms that control motor outputs (Dayan 2007; Glockner 2007; Gold & Shadlen 2007; Koechlin & Hyafil 2007; Kable & Glimcher 2009; Cisek & Kalaska 2010; Custers & Aarts 2010). While such work is focused on general aspects of mechanisms underlying decision-making, it has profound implications for addiction. The process of becoming addicted is one in which drug taking begins as a volitional pursuit of such goals as pleasure, facilitation of social behavior, or relief from distress, but mediated by processes that will be described, devolves into behavior in which drug seeking becomes automatic, stimulus bound, extremely difficult to inhibit, and divorced from goals.

Individual differences in risk of drug use, addiction, and recovery

Much neurobiological study focuses on discovering general principles governing the development and functioning of cells, synapses, and circuits of the brain, and how cognition and behavior emerge from computations within those neural circuits. In this spirit, much investigation into the neural basis of addiction has focused on identifying the architecture and both normal and pathological functioning of reward circuits that are activated by drugs such as nicotine, ethyl alcohol, stimulants such as cocaine and amphetamines, cannabinoids, opioids, and N-methyl-D-aspartate receptor glutamate receptor blockers such as phencyclidine and ketamine. Other addiction research focuses on neural pathways that exert cognitive control over reward seeking. Alongside such investigations of general mechanisms, however, there are important questions, concerning individual differences in responses to addictive drugs. Individual differences occur among virtually all characteristics of humans and other living things. Humans exhibit diverse responses to environmental factors and perturbations ranging from microbes to nutrients to drugs. Heterogeneity among humans afflicted with diverse diseases is currently a central preoccupation of medical research, with the aim of being able to target precise interventions to the right individuals based on specific markers—precision medicine.

Based on individual differences, humans have differing likelihoods of experimenting with drugs when made available, of getting "captured" by a particular drug or rewarding behavior, such as gambling, upon repeated exposure, and different risks of ill-health outcomes (e.g., cirrhosis or lung cancer) consequent to chronic use of particular drugs. Individual differences in patterns of drug taking or reward seeking, including strength of compulsion, are not unique to humans but also are observed in laboratory rodents and other animals (Belin et al. 2011; Moreno et al. 2013; Pascoli et al. 2015).

Human individual differences in addiction liability result from the genetic makeup of the individual (Tsuang et al. 1996; Merikangas et al. 1998), developmental processes (many of which are stochastic) and environmental influences; these latter influences might include stressors, availability of drugs at vulnerable times in the lifespan such as teen years, and local social norms. In short, not every human or laboratory animal exposed to drugs, whether tobacco, alcohol, cocaine, or heroin, ultimately becomes a regular user, or for laboratory animals, regularly self-administering. Not all people who are regular users become addicted, and of those who do become addicted, some can cease drug use without ever seeking medical help. Some who find their way to treatment respond well; yet others continue to relapse despite many attempts at treatment and remain compulsive users until death (Hser et al. 2001).

Risk factors both for initiating drug use and for becoming addicted include genetic risk, male sex (across countries and cultures), chronic pain, certain mental disorders including attention deficit hyperactivity disorder and the manic phase of bipolar disorder, and availability of drugs in the person's environment and culture. Twin and adoption studies consistently demonstrate a significant influence of a person's genes in risk. For example, twin studies show higher rates of concordance for heavy drug use and addiction within monozygotic twin pairs (who share 100% of their DNA sequences) than within dizygotic twin pairs (who, on average, share 50%; Tsuang et al. 1996; Merikangas et al. 1998). Like all common neuropsychiatric disorders (Sullivan et al. 2012), the genetic contributions to risk of addictive disorders is significant, but highly complex, with variation in many genetic loci influencing different aspects of these complex syndromes. Moreover, while the role of genetic influences is significant, genes are not fate. The importance of environmental influences in drug initiation and of environmental cues and context in maintaining addiction is powerfully illustrated by the experience of US soldiers during the Vietnam War. Given the stress on these individuals, the ready availability of cheap heroin, and at least implicit permission to use heroin under certain circumstances, many soldiers became regular users who met criteria for addiction. However, on return to the United States most regular users were able to stop with the exception of those who had been regular users before going to Vietnam and those who returned to neighborhoods where heroin use was endemic (Robins et al. 1975).

Some specific genetic loci associated with smoking and nicotine dependence have been identified in part because the behavior is legal, making research participation more straightforward, and in part because it is relatively easy to know how many cigarettes a person has smoked per day. Pointing to the complexity of addiction risk, different genes have been associated with smoking initiation, smoking dependence, cessation, and smoking-associated diseases such as lung cancer (Thorgeirsson et al. 2010). Significant efforts to discover specific genes associated with different stages of drug use among other drug classes are currently under way. This research is being pursued because identification, with certainty, of specific disorder-associated genomic variants provides solid biological clues to pathogenesis, and hopefully molecular targets for preventive or therapeutic interventions.

Reward circuitry

The drive to obtain natural rewards such as food, water, social interactions, and mating opportunities is required for the survival of free-living animals and their species (Kelley & Berridge 2002). Rewards can be operationally defined as stimuli that elicit approach and appetitive behaviors; they also increase the probability that behaviors aimed at obtaining them will be repeated, that is, rewards are positively reinforcing. A specialized brain reward circuit that is conserved across mammalian evolution plays the central role in the ability of an organism to successfully obtain and consume or consummate innately programmed rewards and learn about and pursue new ones. This circuitry is comprised of neurons that synthesize and release the neurotransmitter dopamine and that project from the ventral tegmental area (VTA) of the midbrain to forebrain regions involved in motivation, emotion, and important aspects of cognition, notably the nucleus accumbens (NAc), amygdala, hippocampus, and prefrontal cortex. A separate cluster of midbrain dopamine neurons that project from the substantia nigra to the caudate and putamen facilitate motor learning that results in efficient reward seeking and consummatory behaviors. Reward circuitry has been extensively characterized in the brains of multiple mammals including mice, rats, several nonhuman primate species, and based on noninvasive methods such as functional brain imaging, in humans as well (Breiter et al. 1997; Diederen et al. 2016). Reward circuitry, which leads to appetitive and approach behaviors and amygdala-based and other circuits that produce aversion and avoidance, confer valuation and valence on experiences encountered in the world, and undergird decision-making and control of behavior. Maladaptive functioning of reward circuitry plays a central role in many human ills. Excessive stimulation of reward circuitry is central to drug addiction (Hyman et al. 2006; Nestler et al. 2015; Pascoli et al. 2015), to compulsive gambling and other impulsive behaviors (Dagher & Robbins 2009; Robbins & Clark 2015), and perhaps to obesity (Kenny 2011). Underactivity of reward circuitry likely contributes to the anhedonia (inability to experience pleasure) and behavioral inertia that characterize depression (Ferenczi et al. 2016).

Dopamine release in reward circuitry does not serve as the internal representation of pleasure as once was hypothesized, but as a learning signal that shapes behavior to maximize further rewards and thus the likelihood of survival (Schultz et al. 1997; Schultz 2006). In their basal (unstimulated) state, dopamine neurons exhibit a pattern of slow tonic firing. When a reward is encountered unexpectedly, these neurons fire a brief phasic burst of action potentials, producing a transient increase in synaptic dopamine in the forebrain. For example, an encounter with natural rewards such as palatable foods or mating opportunities causes VTA neurons to fire phasically (Kelley & Berridge 2002). Important insights into the function of dopamine came from electrophysiologic recordings of dopamine-releasing neurons in nonhuman primates, and later from rats and mice undergoing classical (Pavlovian) conditioning (Schultz et al. 1997; Schultz 2006). Animals readily learn to associate a previously neutral stimulus (e.g., a light) with delivery of reward when the stimulus reliably precedes reward over several trials. In the monkey studies by Schultz et al. (1997, 2006), appearance of highly palatable juice or food (an unconditioned stimulus—a natural reward) elicits phasic dopamine neuron firing and thus a transient increase in dopamine release. Once the animal—or as was later established a human subject studied by noninvasive neuroimaging—learns that the new stimulus (e.g., light) predicts reward, the dopamine neurons no longer fire phasically when the reward (unconditioned stimulus) appears, but rather at the appearance of the light, which has become a conditioned stimulus (i.e., predictive of reward). If an new stimulus is introduced (e.g., a tone), that reliably precedes the light, which in turn predicts the juice, dopamine neurons fire phasically in response to the tone, but no longer to the light or to reward delivery (Schultz et al. 1997; Schultz 2006). If the reward is omitted at the time when it was predicted by learned cues, there is a pause in dopamine neuron firing (i.e., a punishment signal). There is convergent evidence from many experiments (together with computational models), which make a strong case that dopamine neurons respond to errors in prediction about the availability of reward (Schultz 2006). They fire phasically at the first and therefore unpredicted appearance of a conditioned stimulus or reward. They fire only at their basal tonic rate when a cue or reward is encountered that has been fully predicted (i.e., when there is nothing more to learn) and pause when an expected reward is less than expected or omitted. The effect of increased synaptic dopamine is to initiate a set of molecular and cellular processes that consolidate learning about the predictive significance of cues and produce long-lived alterations in behavioral responses to reward-associated cues. A critical mechanism by which dopamine alters functional outputs of the forebrain is by facilitating long-term changes in the strength of synaptic communication between target neurons with the result that neural circuitry is altered long term (Berke & Hyman 2000; Hyman et al. 2006; Luescher & Malenka 2011).

Acting in the NAc, dopamine binds reward-associated stimuli to motivation, thus imbuing the cue with what has been called incentive salience (Robinson et al. 2005). The result is that the cue gains the ability to motivate pursuit and consummatory behavior. Within orbital prefrontal cortex, dopamine updates the representation of rewards and their relative valuation; neural processes within the orbital prefrontal cortex establish a common currency that permits an organism to choose among possible goals that may differ both qualitatively and quantitatively (Gallagher et al. 1999; Montague et al. 2004; Schoenbaum et al. 2006; Xie & Padoa-Schioppa 2016). In the caudate and putamen, phasic dopamine release supports motor learning that binds cues predictive of reward to action. Motor learning involving the caudate and putamen requires repetition (i.e., practice), but eventually becomes highly efficient, smooth, automatic, and resistant to forgetting. Examples include learning to ride a bike, learning to touch-type, or a baseball player learning to throw a fastball. In the case of direct pursuit of rewards, such as obtaining food under conditions of scarcity or intense competition, automatic recognition of predictive cues and speedy, efficient engagement in hunting, or foraging might make the difference between survival and starvation. While this last description may call up visions of animals in the wild, they apply also to modern humans when, because of crop failure, conflict, or extreme economic mismanagement, shops are empty and food is unavailable.

Well-learned action repertoires typically remain goal directed, but become automatized, highly efficient, and largely independent of conscious control. For example, fluent delivery of a well-rehearsed lecture by a professor or a well-practiced gymnastic routine during an athletic competition are initiated in a goal-directed manner but, once begun, unfold without conscious consideration of each word, gesture, or bodily movement. Under pathological conditions, however, behaviors of varying complexity from tics to extended routines may be activated by environmental or internal cues independently of, and even in contravention of a person's goals. Such deeply ingrained habits, which may become quite stereotyped, are highly resistant to conscious control (Graybiel 2008). Examples of common, relatively simple, stereotyped habits include compulsive nail biting and compulsive hair pulling (trichotillomania). Automatized behaviors and even deeply ingrained habits can be interrupted by a salient, unexpected event or a failure such as losing one's place while delivering a lecture or speech or falling during a well-practiced gymnastic routine; this results in a marked increase in arousal and attention, and a transient gain in behavioral flexibility (e.g., to find one's place or right oneself) before returning to the routine. Thus the addicted person who sees a uniformed policeman standing where he expects to see his drug dealer does not zombie-like try to buy drugs from the policemen, but aborts the plan and seeks out drugs elsewhere. Under such circumstances, the addicted person often suffers intense drug craving (Tiffany 1990).

It may seem counterintuitive based on what we think we know about ourselves, but the actions of dopamine in tuning motivation, cognition, and behavior toward future rewards are unavailable to conscious introspection. The basis of consciousness remains a deep scientific mystery, but the properties of conscious cognition are well studied. Among its attributes, consciousness is a remarkable, but capacity limited, serial processor of information. Conscious deliberative decision-making is slow and effortful relative to unconscious decision-making processes. In a complex, rapidly moving world in which survival often depends on speed and efficiency, it is unsurprising that decision-making and control of behavior are served by rapid, massively parallel unconscious processing that has access to multiple sources of relevant information (Dayan 2007; Cisek & Kalaska 2010).

Addictive drugs

Addictive drugs are chemically diverse and interact with different molecular targets in the nervous system (Nestler et al. 2015). Based on their chemical differences, these drugs exert diverse physiological and behavioral effects. For example, amphetamines are stimulants, that is, they increase arousal. At lower doses, amphetamines enhance cognitive performance; and at higher doses they produce euphoria, but may also cause anxiety and insomnia. Alcohol, in contrast, is a depressant; it is anxiolytic at low doses, and degrades cognitive and motor performance with higher, intoxicating doses. Despite their differences in mode of action in the brain, all addictive drugs share the property of causing dopamine release from VTA neurons and thus activating reward circuitry. Psychotropic drugs such as selective serotonin reuptake inhibitor antidepressants that do not cause significant dopamine release are not addictive. Even drugs that produce physical withdrawal symptoms (physiological dependence) are not considered addictive unless they produce compulsive behavior.

In a metaphorical sense, addictive drugs can be likened to Trojan horses in the brain. All addictive drugs mimic one or another endogenous neurotransmitter that can directly or indirectly increase synaptic dopamine. Cocaine is structurally similar enough to dopamine itself that it binds the dopamine transporter that normally clears dopamine from synapses. However, cocaine is different enough structurally from dopamine that it blocks the transporter without entering it. The result of cocaine blockade of the dopamine transporter is a buildup of dopamine in synapses to very high levels. Opiates, nicotine, alcohol, and cannabinoids mimic different neurotransmitters and act on different receptors in the brain, but by diverse mechanisms, all of them indirectly cause dopamine release (and in the case of opiates also exert independent rewarding effects) (Nestler et al. 2015).

Why do addictive drugs have an advantage over natural rewards, and, in vulnerable individuals, crowd out other, more adaptive goals? It is because addictive drugs activate reward circuits directly by pharmacologic action, independently of experience and despite their lack of any homeostatic or reproductive benefit. Such drugs reliably cause significant dopamine release and thus a positive reward prediction signal. As drug use is repeated under the initial influence of pleasure seeking, relief of distress, or social forces, the resulting grossly abnormal dopamine signal potently consolidates the motivational power of drug-associated cues, pathologically excessive valuation of drug taking, and drug seeking and consummatory behavioral repertoires (Montague et al. 2004; Schoenbaum et al. 2006; Everitt & Robbins 2016).

An additional advantage that addictive drugs have over natural rewards is that pharmacologically induced dopamine release masks the results of disappointing or aversive experiences. Thus biting into a rotten fruit would create, among other effects, a dopamine pause, which might devalue that kind of fruit. This is not the case for drugs that directly release dopamine. For example, if the inhalation of tobacco smoke causes painful coughing or shortness of breath in a chronically ill smoker, nicotine still causes dopamine release as a pharmacologic action mediated by nicotinic acetylcholine receptors. Despite the smoker's actual aversive experience, forebrain circuits still receive a signal that reinforces tobacco use (Redish 2004).

As individuals progress from repetitive drug use to addiction, action patterns involved in drug seeking and drug taking become habitual: drug-associated cues gain the ability to activate automatized drug-seeking behaviors in a manner that is increasingly independent of the person's goals (Berke & Hyman 2000; Everitt & Robbins 2005, 2016). Cue-dependent activation of ingrained drug-seeking habits helps explain why an addicted person continues to take drugs, even when late in the course of addiction, tolerance to drugs and damage to health may have sapped all pleasure from the experience of using them and may even be aversive. When such habitual drug seeking is delayed or foiled (e.g., by unavailability or by attempts to cut down), intense drug craving often occurs, increasing the motivation to find and use drugs (Tiffany 1990). Craving is so characteristic of addicted individuals as to be listed among the DSM-5 diagnostic criteria (American Psychiatric Association 2013).

Control systems and their limitations

Given that pursuit of rewards requires investment of energy and entails potential risks (e.g., exposure of prey species to predation when they leave shelter to seek food or water), diverse mechanisms have evolved to regulate reward seeking. For example, the drive to seek certain rewards, such as food and water, is regulated by homeostatic needs (e.g., hunger or thirst), the value of the reward, its proximity, and the dangers involved in obtaining it. Regions of the prefrontal cortex, such as the dorsolateral prefrontal cortex, subserve cognitive control mechanisms (Miller & Cohen 2001) that can support or suppress reward seeking in accordance with circumstances. For homeostatic functions such as balancing caloric intake and energy utilization, other specialized circuits are also involved (Kenny 2011). Control of feeding and energy balance involves not only reward circuitry, but also complex hypothalamic projection systems and both hormonal and neural signals involving the gastrointestinal system. Other specialized regulatory circuits in the hypothalamus and peripheral hormonal systems are involved in thirst and in sexual behaviors.

Research that has compared chronic drug users with healthy comparison subjects has found impairments in the ability of drug users to perform tasks that demand significant cognitive control of thought or behavior. These performance failures correlate with failure to activate prefrontal cortex in functional neuroimaging studies compared with healthy subjects (Goldstein & Volkow 2011). The implication is that the ability to exert executive control over impulses is weakened just at the time when an addicted person is also experiencing powerful drives to seek and consume drugs. This research suggests that in addiction, not only is the drug wanting, valuation, and drug-seeking behavior extremely potent, as a result of neural plasticity that occurred in response to repeated, pathological dopamine signals (Berke & Hyman 2000; Everitt & Robbins 2005, 2016), but the ability to exert top-down control over behavior may also be undermined. The result is a person whose drug seeking is increasingly unrelated to goals and resistant to devaluation by negative consequences even when the person (as in the two opening vignettes) is well aware of the situation (Hyman 2007; Ersche et al. 2016).

The persistence of addiction

One of the most challenging features of drug addiction for families, clinicians, and society is the lifelong persistence of relapse risk that characterizes a significant subset of addicted individuals (Hser et al. 2001). Even among individuals who recover, with or without treatment, successful achievement of abstinence often follows after many prior attempts have failed as a consequence of cue-activated drug seeking or drug craving or, as has been well studied, in the face as a result of a significant stress (Shaham et al. 2000; Koob 2008).

The grudgingly persistent nature of relapse risk is thought to reflect several biological processes that are known to be long-lived (Hyman et al. 2006; Wolf 2016). Among the most persistent alterations in the structure and function of the nervous system are forms of synaptic plasticity that underlie formation of associative memories—in the case of addiction, associations that bind specific drug cues to drug seeking. Drug-induced synaptic plasticity such as long-term potentiation (LTP) and long-term depression (LTD) that produce persistent changes in information processing within neural circuits have been well documented in VTA, and NAc in response to administration of addictive drugs (Hyman et al. 2006; Luescher & Malenka 2011). Measurements of LTP and LTD are physiological proxies for long-lived, most often structural changes in synaptic connections that can include the number and size of synaptic spines on the dendrites of the principal neurons of the NAc, caudate and putamen, and cerebral cortex. Activity-dependent changes in synapses and in synaptic spines alter information processing in the circuits in which they participate. Dopamine acts in the forebrain to regulate plasticity in synapses relevant to motivation, valuation, and reward-related behavior (Nestler et al. 2015). The persistence of alterations in synaptic connections that may be measured experimentally as LTP and LTD, is mediated by such biological processes as changing patterns of gene expression and protein translation in relevant neurons which might result, for example, in structural alterations in synapses and spines. There are many mechanisms of gene regulation influenced by addictive drugs including epigenetic modification of chromatin, the protein scaffolding that makes regions of DNA more or less available to be transcribed (Walker et al. 2015).

The foregoing review of mechanisms that contribute to addiction, perforce has limitations both with respect to its brevity and with respect to the still advancing state of the science. There is, however, a reasonable consensus that in individuals made vulnerable by genetic, developmental, and environmental risk factors, repetitive use of drugs that release dopamine in reward circuitry—a survival system conserved across mammalian evolution—produces the severe behavioral syndrome of addiction. Because addictive drugs reliably cause dopamine release, independently of any benefit, they have an advantage over natural rewards, ultimately producing profound pathologies of decision-making and behavior. Over time, as drugs shape behavior and crowd out the rewards of family, social interactions, and work, the addicted person's life is progressively narrowed to obtaining, using, and recovering from drugs. Despite these and other profoundly negative consequences, drug seeking and drug use, which has become stimulus bound and unrelated to a person's goals, becomes extremely difficult to interrupt for any extended period of time.

Some implications of the scientific model for understanding the behavior of addicted people

I began this chapter with two vignettes, one of a smoker who does not stop despite his understanding of the risks, experience of negative health consequences, and an intention to quit. The other described a person addicted to opioids who relapses despite having made it through severe withdrawal in a rehabilitation facility, and despite putting his marriage and job at risk. The neural mechanisms of addiction provide a basis for understanding the maladaptive behavior of such individuals. Both vignettes describe individuals who exhibit compulsive drug use despite negative consequences, and would thus be diagnosed with addiction by current, widely accepted DSM-5 criteria (American Psychiatric Association 2013). Both vignettes illustrate the limitations of an addicted person's knowledge of harms and of goals of the person might hold for abstinence and recovery.

Even for healthy people attempting to cut down on calories or certain classes of foods, homeostatic processes that drive hunger and reward-associated cues that activate wanting typically foil long-term success. The conscious use of willpower (i.e., exertion of cognitive control to inhibit prepotent impulses) is effortful and depletes cognitive resources over time. For addicted individuals, for whom drug seeking has become stimulus bound and dissociated from goals, awareness of harms and consciously held intentions typically yield little progress toward abstinence. Indeed, drug craving is perversely likely to arise from attempts to cut down because such attempts are an impediment to automatized drug seeking. To make matters worse, as suggested by neuroimaging studies, prefrontal cortical circuits may be impaired in addicted people, thus diminishing their ability to exert cognitive control compared with healthy people (Goldstein & Volkow 2011).

Now in the second decade of the twenty-first century the United States is experiencing an epidemic of deaths from opiate overdoses; even short of death, many addicted people continue to relapse over their lifespan (Hser et al. 2001). The situation for addicted people is not hopeless. Despite the neural mechanisms of addiction reviewed here, humans exhibit significant differences in their responses to drugs and to the addicted state. Some stop drug use on their own—often succeeding after many attempts; in the remarkable case of US soldiers returning from Vietnam, a pervasive change in circumstance permitted recovery for many without formal treatment (Robins et al. 1975). Of the many addicted people who seek treatment, the results are variable in terms of overall outcome, the number of episodes of high-quality treatment that are needed, and the overall length of time to a sustained recovery (during which harms unfortunately accrue). This essay is not the place to discuss the strengths and weaknesses of existing treatments, but overall, existing evidence-based treatments have been shown to be better than no treatment. That said, more effective treatments are much in need. Currently, some individuals gain significant benefit from existing psychosocial, pharmacologic, or combined treatments, but large numbers obtain only modest benefits, if any. As in the case of many severe chronic and relapsing human ills, especially those that affect the brain, progress has been slow. In the case of addiction, the search for better treatments has been hampered by low levels of interest on the part of pharmaceutical companies based, in part, on concerns about retaining addicted people in clinical trials and in part about the likelihood of reimbursement given the often tenuous state of health insurance for this stigmatized population. The most important reason to investigate neurobiological mechanisms of addiction is to gain deeper understandings that will yield biomarkers and effective treatments. One early example of a treatment that has come from research on basic mechanisms is the use of long-acting opiate antagonists that diminish drug craving across multiple classes of addictive drugs (Lee et al. 2016). These are proving a useful tool in treatment, but as noted, much more is needed.

A challenging problem is how to convince addicted individuals to accept treatment—a problem that is exacerbated by the dearth of drug treatment slots in most countries and the limitations on insurance payments. The foregoing discussion is consistent with the experience of many families and many clinicians. Information alone is rarely adequate; rather, sustained efforts at persuasion and some level of coercion are often required. Given the power of cues and stress to elicit drug seeking, opportunities to succeed in getting an addicted person to treatment may be transient—thus the limitation of treatment slots and insurance coverage are particularly unfortunate. Challenging and frustrating in addition, as in the second of the opening vignettes, are relapses after treatment. Given what we know of addiction, relapses are not surprising, nor should they be the occasion to give up on the addicted person. Based on what we know of the power and persistence of the underlying neural mechanisms, many attempts are often needed before an addicted person can achieve a sustained remission.

Some implications of the scientific model for moral responsibility and legal culpability

In most modern systems of justice, legal culpability is based both on commission of a criminal act (*actus reus*) and criminal intent (*mens rea*). Possession of certain drugs is a crime in most jurisdictions, and drug dealing universally so. A moral model of addiction imputes to the addicted drug user the capacity for deliberative choice and the ability to inhibit drug use, and thus both moral and legal responsibility for such illegal acts. The medical model of addiction, in contrast, problematizes the application of *mens rea* to addicted individuals, most clearly for crimes such as possession of illegal drugs intended for the person's use—and perhaps even low-level drug dealing when it is employed solely as a mode of obtaining resources to maintain a drug habit. Understanding addiction as compulsive behavior refutes the simple attribution of free choice and action assumed by the moral model and thus calls into question the imputation of criminal intent to most addicted individuals charged with drug possession, and to some degree other nonviolent drug offenses that are committed directly in the service of obtaining drugs.

Application of the medical model does not mean, however, that all sanctions would be inappropriate. Rationales for punishment in most systems of justice include a combination of retributivist and consequentialist goals. Retribution expresses the moral outrage of society about a crime and is meant to assuage the moral emotions of victims and of society more broadly. Consequentialist goals include deterrence of future crime by the perpetrator and by setting an example, deterrence of other potential perpetrators; incapacitation of dangerous individuals; and rehabilitation of the person being punished. The medical model of addiction does not militate against consequentialist goals of punishment, but explains why punishment often fails to end drug use. Moreover, if one goal of punishment is to protect society from the considerable harms wrought by drug use and addiction (including the profoundly negative effects of black markets), it can be asked whether incarceration in prison, which has been so common in the United States in recent decades, is beneficial to society, or whether alternative sanctions—a complex matter that cannot be discussed within the limits of this chapter—might better serve society's pragmatic goals. The medical model suggests that retribution makes no sense with respect to addicted individuals given the constraints its underlying pathologies put on volitional control, and on the ability of addicted persons to respond to moral condemnation. The moral model in systems of criminal justice, combined with stigmatization of addicted people and outrage against them, has produced a period of harsh punishment, notably in the United States, which has a significant percentage of its population serving long prison sentences for drug-related crimes, even when nonviolent.

Some legal scholars argue that moral responsibility and legal culpability can almost always be determined at the level of psychological evidence and that neuroscience has nothing to add (Morse 2006). This is because the factors that matter for determinations of *mens rea* are the capacities and especially the rationality of accused persons at the time that they committed an illegal act, and, they argue, facts about brain anatomy or brain function do not directly address these issues (Morse 2006). Such views create something of an impasse between the law and neuroscience, since the neuroscience view would argue that the apparent rationality of the addicted person is illusory: the person can plan to gain and take drugs, but is not in control of his or her motives.

Significant challenges remain if neuroscience is to effectively influence systems of justice and social policy more broadly. The mechanistic picture of addiction, which I have reviewed here, has ultimately been constructed out of many hundreds of individual studies. The majority of these studies are based on animal models rather than humans because of the inviolability of the human brain to intrusive experimentation in life. In addition, given the serious limitations of current noninvasive imaging and neurophysiological technologies, neural correlates of addiction—assuming these were better defined from basic studies—cannot be convincingly demonstrated for any given individual to a court, parole officer, or medical personnel. Even if published neuroimaging studies were more convincingly replicated than at the present time, it would still not be possible to determine the degree to which any specific person had been deprived by addiction of volitional control—and courts must deal with individuals, not with diagnostic classes. Perhaps at some future time, there will be convincing biomarkers that will provide a clear measure of the severity of a given person's loss of volitional control, but none exist today.

That might be the end of the matter for now, except that many scientists share the view that the moral model of addiction has often tilted courts and especially their sentencing practices toward injustice by overstating the degree to which addicted people can control their actions in accordance with the law. Certain courts and legal scholars—those more anxious about letting the guilty go free than about punishing the innocent—may also be concerned that symptoms of addiction, like symptoms of all physically invisible disabilities and disorders, such as mental illnesses and early-stage neurodegenerative disorders, might be feigned and impairments of decision-making and behavioral control overstated by defendants. These concerns notwithstanding, and understanding full well that today the severity of compulsion in any individual person cannot be objectively determined, it is still the case that mainstream scientific understandings converge on the view that drug seeking and drug taking by addicted individuals reflects severely impaired volitional control and markedly diminished ability to conform behavior to rational goals. Indeed, the moral model of addiction, based in folk psychological intuition, is inconsistent with scientific understandings both of normal decision-making (Dayan 2007; Glockner 2007; Gold & Shadlen 2007; Koechlin & Hyafil 2007; Kable & Glimcher 2009; Cisek & Kalaska 2010; Custers & Aarts 2010) and the pathological decision-making and behavior of addicted people (Berke & Hyman 2000; Redish 2004; Montague et al. 2004; Everitt & Robbins 2005; Hyman et al. 2006; Hyman 2007; Ersche et al. 2016; Everitt & Robbins 2016). The only way out of this impasse at this point in history is for the scientific investigation of addiction to continue to advance, with greater focus on humans, as opposed to animal models, and to engage more effectively with judges, legal scholars, and policymakers—both to understand their concerns and constraints, and to provide honest views of the state of the science. The most to hope for and energetically pursue are effective treatments that interrupt the progression of addiction and provide realistic and reliable alternatives to systems of criminal justice.

If the addicted person has significantly impaired agency, might coerced treatment be warranted?

The neuroscience-based claim that addiction is characterized by impaired decision-making and diminished behavioral control argues against attribution of moral responsibility to addicted people and for mitigation of legal culpability for crimes that flow directly from the addicted state. The neurobiological model of reduced agency supports arguments against punishment for nonviolent crimes that have been driven by the need to obtain drugs. However, given the personal and societal harms of addiction, the neurobiological view does not imply that society should take no action. Indeed, an implication of reduced decisional capacity in critical matters of health, and behavior and impaired ability to control behavior, including the ability to obey the law, invites the question of whether and under what circumstances involuntary treatment, if safe and effective, is ethical and warranted. Coerced treatment, especially of addicted individuals found guilty of nonviolent crimes has already engendered debate (Caplan 2008; Hall et al. 2008; Carter & Hall, Chapter 25, this volume). In addition to ethical analysis, however, far more data is needed from well-designed studies of particular societal interventions, for example, supervision by effective drug courts and of coerced treatments, including medications such as long-acting opiate blockers. Without safe and effective interventions that can be implemented in real-world settings, ethical analyses will remain too abstract to have utility.

A more difficult question is posed for addicted individuals who have never been convicted of a crime, but who are suffering significant health-related and other harms. Implementation of paternalistic medical treatment for brain disorders has a very dark history, including the widespread use of prefrontal lobotomy in the 1950s (Valenstein 1986). Thus as both an ethical and pragmatic matter there would have to be a high bar for determining the point at which an addicted person had so little decisional capacity or agency as to be a candidate for paternalistic treatment. There would also have to be stringent standards for the safety of a proposed intervention as well as its efficacy, and provision for interventions to be closely monitored by a qualified, disinterested observer. The high value placed on autonomy in bioethics and a history of many well-meant but ultimately damaging involuntary treatments (Valenstein 1986) militates against paternalism. Yet an argument for relaxing punitive sanctions because of lost rationality and control strongly entails open-minded consideration of paternalism under tightly regulated circumstances.

Conclusion

In this chapter I have described a scientific model of addiction that, while still incomplete, captures evidence-based views of many neurobiologists and psychologists. Understandings of neurobiological mechanisms that underlie both healthy functioning and pathological conditions such as addiction will deepen over time, especially with recent technological advances that, inter alia, have made analysis of genetic risk factors feasible. I have probed the implications of a neurobiological model of addiction for attributions of moral responsibility and legal culpability, contrasting it with a widely influential moral model that is grounded in folk psychology. The moral model lacks a scientific basis, but given its origin in the potent, if misleading, human intuition of a self that is free to make decisions and exert agency, this model has enormous persuasive power. In contrast, scientific formulations that appeal to complex and invisible neural mechanisms do not automatically influence public views of addiction or other neural pathologies that affect behavior; nor do they necessarily change the views lawmakers or courts, who are often sensibly skeptical of moving too rapidly on such claims. If scientific understandings of behavior are to contribute to more just societies, it is not enough for the science to advance. More specifically, understandings that are grounded in animal models must be convincingly translated to the human situation, and the disease state must be made "visible" through well-validated and convincing markers. As noted, perhaps most convincing to important lay audiences would be significant advances in treatment. Nothing would destigmatize more effectively than an ability to successfully reverse the addicted state.

References

American Psychiatric Association (2013). *Diagnostic and Statistical Manual of Mental Disorders* (5th edn.). Arlington, VA: American Psychiatric Association.

Belin, D., Berson, N., Balado, E., Piazza, P.V., and **Deroche-Gamonet, V.** (2011). High-novelty preference rats are predisposed to compulsive cocaine self-administration. *Neuropsychopharmacology*, **36**, 569–579.

Berke J.D. and **Hyman, S.E.** (2000). Addiction, dopamine, and the molecular mechanisms of memory. *Neuron*, **25**, 515–532.

Breiter, H.C., Gollub, R.L., Weisskoff, R.M., et al. (1997). Acute effects of cocaine on human brain activity and emotion. *Neuron*, **19**, 591–611.

Caplan, A. (2008). Denying autonomy in order to create it: the paradox of forcing treatment upon addicts. *Addiction*, **103**, 1919–1921.

Cisek, P. and **Kalaska, J.F.** (2010). Neural mechanisms for interacting with a world full of action choices. *Annual Review of Neuroscience*, **33**, 269–298.

Conrad, P. and **Schneider, J.W.** (1992). *Deviance and Medicalization: From Badness to Sickness, Expanded Ed.* St. Louis, MO: Mosby.

Custers, R. and **Aarts, H.** (2010). The unconscious will: how the pursuit of goals operates outside of conscious awareness. *Science*, **329**, 47–50.

Dagher, A. and **Robbins T.W.** (2009). Personality, addiction, dopamine: insights from Parkinson's disease. *Neuron*, **61**, 505–510.

Dayan, P. (2007). The role of value systems in decision making. In: Engel, C. and Singer, W. (Eds.) *Better Than Conscious? Decision Making, the Human Mind, and Implications for Institutions*, pp.51–70. Cambridge, MA, The MIT Press.

Diederen, K.M., Spncer, T., Vestergaard, M.D., Fletcher, P.C., and **Schultz, W.** (2016). Adaptive prediction error decoding in human midbrain and striatum facilitates behavioral adaptation and learning efficiency. *Neuron*, **90**, 1127–1138.

Dunham, Y., Chen, E.E., and **Banaji, M.R.** (2013). Two signatures of implicit intergroup attitudes: developmental invariance and early enculturation. *Psychological Science*, **24**, 860–868.

Ersche, K.D., Gillian, C.M., Simon Jones, P., et al. (2016). Carrots and sticks fail to change behavior in cocaine addiction. *Science*, **352**, 1468–1471.

Everitt, B.J. and **Robbins, T.W.** (2005). Neural systems of reinforcement for drug addiction: from actions to habits to compulsion. *Nature Neuroscience*, **8**, 1481–1489.

Everitt, B.J. and **Robbins, T.W.** (2016). Drug addiction: updating actions to habits to compulsions ten years on. *Annual Review of Psychology*, **67**, 23–50.

Fenno, L., Yizhar, O., and **Deisseroth, K.** (2011). The development and application of optogenetics. *Annual Review of Neuroscience*, **34**, 389–412.

Ferenczi, E.A., Zalocusky, K.A., Liston, C., et al. (2016). Prefrontal cortical regulation of brainwide circuit dynamics and reward-related behavior. *Science* **351**, 42–54.

Gallagher, M., McMahan, R.W., and **Schoenbaum, G.** (1999). Orbitofrontal cortex and representation of incentive value in associative learning. *Journal of Neuroscience*, **19**, 6610–6614.

Glockner, A. (2007). How evolution outwits bounded rationality. The efficient interaction of automatic and deliberate processes in decision making and implications for institutions. In: Engel, C. and

Singer, W. (Eds.) *Better Than Conscious? Decision Making, the Human Mind, and Implications for Institutions*, pp 259–284. Cambridge, MA: The MIT Press.

Gold, J.I. and Shadlen, M.N. (2007). The neural basis of decision making. *Annual Review of Neuroscience*, 30, 535–574.

Goldstein, R.Z. and Volkow, N.D. (2011). Dysfunction of the prefrontal cortex in addiction: neuroimaging findings and clinical implications. *Nature Reviews Neuroscience*, 12, 652–669.

Graybiel, A.M. (2008). Habits, rituals, and the evaluative brain. *Annual Review of Neuroscience*, 31, 359–387.

Haggard, P. (2008). Human volition: towards a neuroscience of will. *Nature Reviews Neuroscience*, 9, 934–946.

Hall, W., Capps, B., and Carter, A. (2008). The use of depot naltrexone under legal coercion: the case for caution. *Addiction*, 103, 1922–1924.

Heyman, G.M. (2010). *Addiction: A Disorder of Choice*. Cambridge, MA: Harvard University Press.

Hser, Y.I., Hoffman, V., Grella, C.E., and Anglin, M.D. (2001). A 33-year follow-up of narcotics addicts. *Archives of General Psychiatry*, 58, 503–508.

Hyman, S.E. (2007). The neurobiology of addiction: implications for voluntary control of behavior. *American Journal of Bioethics*, 7, 8–11.

Hyman, S.E., Malenka, R.C., and Nestler, E.J. (2006). Neural mechanisms of addiction: the role of reward-related learning and memory. *Annual Review of Neuroscience*, 21, 565–598.

Kable, J.W. and Glimcher, P.W. (2009). The neurobiology of decision: consensus and controversy. *Neuron*, 24, 733–745.

Kelley, A.E. and Berridge, K.C. (2002). The neuroscience of natural rewards: relevance to addictive drugs. *Journal of Neuroscience*, 22, 3306–3311.

Kenny, P.J. (2011). Reward mechanisms in obesity: new insights and future directions. *Neuron*, 69, 664–679

Koechlin, E. and Hyafil A. (2007). Anterior prefrontal function and the limits of human decision-making. *Science*, 318, 594–598.

Koob G. (2008). A role for brain stress systems in addiction. *Neuron*, 59, 11–14.

Lee, J.D., Friedman, P.D., Kinlock, T.W., et al. (2016). Extended-release naltrexone to prevent opioid relapse in criminal justice offenders. *New England Journal of Medicine*, 374, 1232–1242.

Luescher, C. and Malenka RC. (2011). Drug-evoked synaptic plasticity in addiction: from molecular changes to circuit remodeling. *Neuron*, 69, 650–63.

Merikangas, K.R., Stolar, M., and Stevens, D.E., (1998). Familial transmission of substance use disorders. *Archives of General Psychiatry*, 55, 973–979.

Miller E.K. and Cohen, J.D. (2001). An integrative theory of prefrontal cortex function. *Annual Review of Neuroscience*, 24, 167–202.

Montague, P.R., Hyman, S.E., and Cohen, J.D. (2004). Computational roles for dopamine in behavioural control. *Nature*, 431, 760–767.

Moreno, M., Economidou, D., Mar, A.C., et al. (2013). Divergent effects of D2/3 receptor activation in the nucleus accumbens core and shell on impulsivity and locomotor activity in high and low impulsive rats. *Psychopharmacology (Berlin)*, 228, 19–30.

Morse, S.J. (2006). Brain overclaim syndrome. *Ohio State Journal of Criminal Law*, 3, 397–412.

Morse, S.J. (2008). Determinism and the death of folk psychology: two challenges to responsibility from neuroscience. *Minnesota Journal of Law, Science, and Technology*, 9, 1.

Nestler, E.J., Hyman, S.E., Holtzman, D., and **Malenka, R.J.** (2015). *Molecular Neuropharmacology: Foundation for Clinical Neuroscience.* New York: McGraw Hill.

Pascoli, V., **Terrier, J.,** Hiver, **A.** and **Luscher, C.** (2015). Sufficiency of mesolimbic dopamine neuron stimulation for the progression to addiction. *Neuron,* **88,** 1054–1066.

Redish, A.D. (2004). Addiction as a computational process gone awry. *Science,* **306,** 1944–1947.

Robbins, T.W. and **Clark, L.** (2015). Behavioral addictions. *Current Opinion in Neurobiology,* **30,** 66–72.

Robins, L.N., **Helzer, J.E.,** and **Davis, D.H.** (1975) Narcotic use in southeast Asia and afterward. An interview study of 898 Vietnam returnees. *Archives of General Psychiatry,* **32,** 955–961.

Robinson, S., **Sandstrom, S.M.,** Denenberg, V.H., and **Palmiter, R.D.** (2005). Distinguishing whether dopamine regulates liking, wanting, and/or learning about rewards. *Behavioral Neuroscience,* **199,** 336–341.

Satel, S.L. (1999). What should we expect from drug abusers? *Psychiatric Services,* **50,** 861.

Satel, S. and **Lillienfeld, S.O.** (2013). *Brainwashed: The Seductive Appeal of Mindless Neuroscience.* New York: Basic Books.

Schoenbaum, G., **Roesch, M.R.,** and **Stalnaker, T.A.** (2006). Orbitofrontal cortex, decision-making and drug addiction. *Trends in Neurosciences,* **29,** 116–124.

Schultz, W. (2006). Behavioral theories and the neurophysiology of reward. *Annual Review of Psychology,* **57,** 87–115.

Schultz W., **Dayan P.,** and **Montague, P.R.** (1997). A neural substrate of prediction and reward. *Science,* **275,** 1593–1599.

Shaham, Y., **Erb, S.,** and **Stewart, J.** (2000). Stress-induced relapse to heroin and cocaine seeking in rats: a review. *Brain Research. Brain Research Reviews,* **33,** 13–33.

Stolier, R.M. and **Freeman, J.B.** (2016). Neural pattern similarity reveals the inherent intersection of social categories. *Nature Neuroscience,* **19,** 795–797.

Sullivan, P.F., **Daly, M.J.,** and **O'Donovan, M.** (2012). Genetic architectures of psychiatric disorders: the emerging picture and its implications. *Nature Reviews Genetics,* **13,** 537–551.

Thorgeirsson, T.E., **Gudbjartsson, D.F.,** Surakka, **I.,** et al. (2010). Sequence variants at CHRNB3-CHRNA6 and CYP2A6 affect smoking behavior. *Nature Genetics,* **42,** 448–453.

Tiffany, S.T. (1990). A cognitive model of drug urges and drug-use behavior: role of automatic and nonautomatic processes. *Psychological Review,* **97,**147–168.

Traynor v. Turnage and *McKelvey v. Turnage* (1988). 485 US 535; 108 S. Ct. 1372.

Tsuang, M.T., **Lyons, M.J.,** Eisen, **S.A.,** et al. (1996). Genetic influences on DSM-III-R drug abuse and dependence: a study of 3,372 twin pairs. *American Journal of Medical Genetics,* **67,** 473–477.

Valenstein, E. (1986). *Great and Desperate Cures: The Rise and Decline of Psychosurgery and Other Radical Treatments for Mental Illness.* New York: Basic Books.

Volkow, N.D., **Wang, G.J.,** Tomasi, **D.,** and **Baler, R.D.** (2013). The addictive dimension of obesity. *Biological Psychiatry,* **73,** 811–818.

Walker, D.M., **Cates, H.M.,** Heller, **E.A.,** and **Nestler, E.J.** (2015). Regulation of chromatin states by drugs of abuse. *Current Opinion in Neurobiology,* **30,** 112–121.

Wallis, J.D. (2007). Orbitofrontal cortex and its contribution to decision-making. *Annual Review of Neuroscience,* **30,** 31–56.

Wegner, D.M. (2002). *The Illusion of Conscious Will.* Cambridge, MA: MIT Press.

Wegner, D.M. (2003). The mind's best trick: how we experience conscious will. *Trends in Cognitive Science,* **7,** 65–70.

Wolf, M.E. (2016). Synaptic mechanisms underlying persistent cocaine craving. *Nature Reviews Neuroscience*, **17**, 351–365.

World Health Organization (1992). *The ICD-10 Classification of Mental and Behavioural Disorders*. Geneva: World Health Organization.

Xie, J. and **Padoa-Schioppa, C.** (2016). Neuronal remapping and circuit persistence in economic decisions. *Nature Neuroscience*, **19**, 855–861.

Chapter 25

Looking to the future: Clinical and policy implications of the brain disease model of addiction

Adrian Carter and Wayne Hall

Introduction

It has been two decades since Alan Leshner, then director of the US National Institute on Drug Abuse (NIDA), published his seminal paper in *Science* declaring that "addiction is a brain disease, *and it matters*" (emphasis added) (Leshner 1997). According to Leshner, the chronic use of addictive substances flicked a neurochemical switch in the brain that made it very difficult for addicted people to stop using drugs. The brain disease model of addiction (BDMA) professed to explain the high rates of relapse among people treated for addiction.

Leshner was not simply outlining the empirical case for why drug addiction is a brain disease; he was laying the groundwork for a clinical and policy agenda to transform the way in which addiction was treated by healthcare professionals, policymakers and legislators, and the general public. He argued that the BDMA would deliver more effective medical treatments for addiction that would be covered by health insurance and thereby be more readily accessible to addicted people (Leshner 1997). The brain disease view was also intended to convince a skeptical public that addiction was a real psychiatric disorder and that addicted people suffered from a condition that warranted more humane medical treatment, rather than being morally weak or bad people who should be punished either via incarceration or informally through stigmatization and social discrimination. These changes in public understanding, he argued, would lead to a greater investment in addiction treatment services and less reliance on punitive approaches to drug use and addiction.

Over the last 20 years, Nora Volkow, current NIDA director, and others have expanded this agenda. They have appealed to a growing body of neuroimaging research that argues that the brains of people with an addiction have been hijacked by their drug (Dackis & O'Brien 2005; Baler & Volkow 2006). This view has been very influential, particularly in the United States, given that NIDA funds over 85% of addiction research. In 2014, Michael Botticelli, then acting and now director of the US Office on National Drug Control Policy, declared "decades of research have demonstrated that addiction is a brain disorder—one that can be prevented and treated" (Botticelli 2014). The BDMA has also been widely promoted in leading science journals (Volkow & Li 2004; Volkow et al. 2016a) and was endorsed in an editorial in the UK journal, *Nature* (Editorial 2014). The American Society of Addiction Medicine has also recently redefined addiction as a "primary, chronic disease of brain reward, motivation, memory, and related circuitry" (American Society of Addiction Medicine 2011).

The notion that the brains of addicted individuals have been so altered to prevent them from choosing not to use their drug of addiction has significant ethical implications. It has been used to:

- justify the use of greater coercion or mandated treatment to support those who lack the autonomy to make decisions to support themselves (Caplan 2008)

- suggest that addicted people lack the decision-making capacity to freely consent to research or treatment that involves the administration of their drug of addiction (Charland 2002; Cohen et al. 2002), or its agonist (Carter & Hall 2008a)

- support the use of invasive and high-risk interventions such as brain surgery (e.g., deep brain stimulation or ablative psychosurgery) to correct the brain dysfunction that is believed to underpin addiction (Krack et al. 2010; Luigjes et al. 2012).

We have addressed these important ethical implications of a brain disease model elsewhere (Carter & Hall 2008a, 2008b, 2011; Hall et al. 2008; Carter et al. 2011). In this chapter, we examine the evidence supporting the claim that addiction is a brain disease and we assess whether this view has lived up to its promises. We also examine recent research on the impact that a BDMA has had on (1) public views of people with an addiction, and specifically whether it has reduced addiction-related stigma; (2) the way that people with an addiction view their own behavior and drug use, or engage in treatment; and (3) public health policies to reduce the harms of drug use. We explore a reframing of neuroscience research of addiction that takes into account the impact that the repeated use of addictive drugs has on the brain while ensuring that we apply these insights in ways that optimize public health policies to reduce the harms of drug use and addiction.

Evidence for the brain disease model of addiction

Animal models of addiction have been central to the development of the BDMA (Koob & Le Moal 2006; Feltenstein & See 2008; Ahmed 2012). Electrical stimulation and lesioning studies in animals have enabled researchers to identify the neural circuitry through which major drugs of addiction act. This is commonly referred to as the brain reward pathway and includes the ventral striatum, nucleus accumbens, amygdala, and frontal cortices (see Figure 25.1) (Hyman et al. 2006). Studies of drug use in animals, including rats, mice, and nonhuman primates, have shown that animals will repeatedly self-administer (by pressing a lever) the same drugs that are addictive in humans (Koob & Le Moal 2006; Feltenstein & See 2008); they will continue to do so despite the presence of aversive stimuli (such as an electrical footshock); and that self-administration of drugs is reduced by electrically stimulating the brain's reward center (Feltenstein & See 2008; Ahmed 2012). The

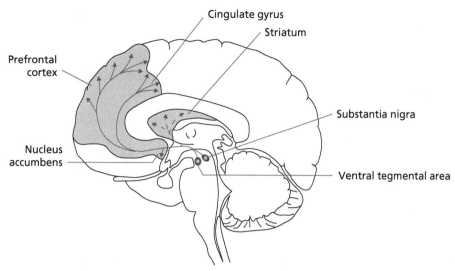

Figure 25.1 Projections from the midbrain to the nucleus accumbens (NAcc) and forebrain. Dopaminergic neurons from the ventral tegmental area (VTA) and substantia nigra project to the central reward area, the NAcc, and to the cortical areas primarily responsible for making decisions, such as whether to use drugs (e.g., the prefrontal cortex (PFC), and the anterior cingulate gyrus (aCG)).

Reproduced with permission of *Annual Review of Neuroscience*, Volume 29, Issue 1, Steven E. Hyman, Robert C. Malenka, Eric J. Nestler, 'Neural Mechanisms of Addiction: Role of Reward-Related Learning and Memory', pp. 565–598, DOI: 10.1146/annurev.neuro.29.051605.113009 © by Annual Reviews, http://www.annualreviews.org.

neurotransmitter, dopamine, plays a key role in addiction, although other neurotransmitters such as endogenous opioids and cannabinoids, glutamate, gamma-aminobutyric acid, corticotropin-releasing hormone, and the orexins are increasingly understood to play critical roles in the development and maintenance of different types of addiction (Feltenstein & See 2008).

Neuroimaging studies in normal and addicted human brains are broadly supportive of the findings from animal studies. These studies have identified changes in dopamine-regulated activity in key regions of the reward pathway, notably the nucleus accumbens and prefrontal cortices (Volkow et al. 2007, 2009, 2012). These changes in brain function correlate with impaired decision-making and poor impulse control (Volkow & Li 2004; Volkow & Baler 2014). The persistence of many of these changes after long periods of abstinence have been invoked to explain the high rates of relapse in people treated for addiction, even after years of abstinence (Reske & Paulus 2011).

The tendency for addictions to run in families is well known. Genetic factors are estimated to contribute between 40% and 60% of the risk of developing an addiction to a range of substances, including alcohol, nicotine, opioids, stimulants, and cannabis (Kendler et al. 2012). Genome-wide association studies have identified risk alleles that influence drug metabolism and the effects that drugs have on the brain, suggesting that chronic drug use acting on the brains of genetically vulnerable individuals produces addiction (Kendler et al. 2012).

How strong is the evidence for a brain disease model of addiction?

Volkow and Koob (2015) recently published an article where they again argue for the positive consequences of a BDMA. This article failed to outline prominent criticisms of the model (Buchman et al. 2010; Satel & Lilienfeld 2013; Satel & Lilienfeld 2014; Hall et al. 2015; Lewis 2015), although the authors did provide a brief response that was published as an online supplementary appendix (Volkow et al. 2016a, 2016b). We have provided a more detailed critique of the evidence against a BDMA elsewhere (Hall et al. 2015), and we summarize the argument here.

First, as a number of prominent social scientists have pointed out, the vast majority of people who have met the diagnostic criteria for dependence in epidemiological surveys are *not* drug dependent at the time of their interview. In contrast to the view that addiction is an inescapably chronic disease, the majority of individuals meeting a diagnosis of addiction mature out of their condition, stopping drug use often without seeking treatment (Robins 1993; Bachman et al. 1997; Heyman 2009; Satel & Lilienfeld 2014).

Second, the strong view of the BDMA where people with an addiction have been hijacked by their drug and unable to control their drug use is inconsistent with evidence that small financial rewards or avoiding 24 hours in jail for providing clean urine samples substantially reduces drug use in severely dependent individuals (Heyman 2009; Kleiman 2009; Kilmer et al. 2013; Satel & Lilienfeld 2013). The changes in drug use in response to such small incentives is hard to reconcile with the claim that such drug use is a compulsive behavior over which addicted people have little or no control (Hall et al. 2015).

Volkow and colleagues tacitly acknowledge these points in a rejoinder (Volkow et al. 2016b) to our critique of the BDMA (Hall et al. 2015). On our reading, they concede that the majority of individuals with an addiction do not fit with the BDMA and that the loss of control over their behavior is not complete, nor have they been hijacked by the drug (as NIDA colleagues have argued elsewhere, e.g., Dackis & O'Brien (2005)), and maintain some capacity to weigh up choices about whether to use a drug or not.

The BDMA may be more relevant if we acknowledge that addiction is a condition that varies significantly in severity with the majority of individuals with mild to moderate conditions not experiencing the chronic and relapsing version of the disorder who will overcome their condition without engaging in significant treatment (Hall et al. 2015). Volkow and colleagues acknowledge this point in their rejoinder. A chronic and relapsing account of addiction would only be applicable to a minority of addicted people who use drugs into their early thirties despite adverse health and social consequences and who are most likely to seek treatment after failing to control their drug use and most likely to show altered brain function. It is regrettable that these responses are not made more prominently and explicitly within the body of their numerous articles in support of the

BDMA (e.g., Volkow & Li 2004; Dackis and O'Brien 2005; Volkow & Koob 2015; Volkow & Morales 2015; Volkow et al. 2016a). In these articles they assert that addiction is a chronic, relapsing brain disease that hijacks behavior; any qualifications are relegated to supplementary material (Volkow et al. 2016b).

This acknowledgment has important implications for how clinicians, healthcare professionals, and policymakers discuss addiction and argue for its treatment and policies. We believe that advocates of the BDMA cannot continue to equate the lifetime prevalence of addictive disorders with the prevalence of the severe and chronic addictive disorders that appear to fit the BDMA. Nor can they argue for policy initiatives and treatment interventions that assume that the majority of addicted individuals are suffering from a chronic and relapsing disease that robs them of their ability to control their drug use. Although they state that the BDMA only applies to a minority of individuals with the most severe forms of addiction (Volkow et al. 2016b), Volkow and colleagues still cite estimates of the prevalence of lifetime and past year addiction from epidemiological surveys as if these were estimates of the prevalence of addiction of sufficient severity to make a brain disease a plausible hypothesis (Volkow et al. 2016a).

Does neurobiological evidence support a more modest version of the brain disease model of addiction?

Even if we accept that the BDMA may be relevant to a small fraction of addicted people with severe, chronic, and relapsing forms of addiction, a critical analysis of neurobiological research on addiction raises doubts about how compelling the evidence is for this more modest version of the BDMA.

Addictive patterns of behavior are not the invariable outcome of chronic drug self-administration in animals. Popular accounts of compulsive lever pressing and drug self-administration in animal studies of addiction diminish the context-dependent nature of these findings and the impact that the environment has on the behavior of animals. Animal studies often employ specifically bred strains of rats that have a heightened propensity to develop addictive behaviors. The results are also critically influenced by the very constrained conditions in which animal studies are conducted (Ahmed et al. 2013). Animals housed in more naturalistic environments with a range of additional activities to engage in and access to litter mates do not display the same extent of addictive behavior (Alexander et al. 1978; Xu et al. 2007; Ahmed 2010). This could be interpreted as evidence for the impact of impoverished environments and limited social opportunities on highly social animals, including humans.

Animal models of addiction also have little to say about the high rates of recovery in addiction in the absence of specific interventions (Kincaid & Sullivan 2010). Recovery is absent from Koob and LeMoal's analysis of the fit between animal models and the stages of human addiction (Koob & Le Moal 2006). Their assumption seems to be that an addicted animal or person will remain so unless treated. This is inconsistent with the epidemiological evidence reviewed in the previous section ("How strong is the evidence for a brain disease model of addiction?").

Neuroimaging studies of addiction are typically case–control studies that report statistically significant differences in brain function and structure between groups of severely addicted and nonaddicted people. These averaged differences do not reflect all addicted people—many addicted individuals will not display these deficits while some nonaddicted controls will. These studies also do not tell us whether addiction is a cause or a consequence of differences in brain structure (Ersche et al. 2013). Differences in patterns of activation in brain scans between addicted and nonaddicted people also do not prove that drug use in people with an addiction is compulsive (Hyman 2007). The fact that reduced activity in frontal brain regions is modestly correlated with self-reported craving does not demonstrate that drug use is driven by impulses that are irresistible.

Volkow and colleagues acknowledge that neuroimaging technologies have been unable to distinguish between addicted and nonaddicted individuals, but optimistically place this failure at the feet of current technological limitations, a lack of understanding of how the human brain works, and the complexity of the neurobiological changes triggered by drugs and the heterogeneity of substance use disorders (Volkow et al. 2016b). However, this failure may also reflect the fact that there are other important psychological and social factors at play in addiction. In the absence of acknowledgment of this failure, the role played by the brain in individual outcomes is overestimated.

While we know from twin and adoption studies that addiction runs in families and that genetics accounts for a significant portion of this risk, this pattern of risk is not explained by a small number of specific genetic or molecular pathways. Very large numbers of alleles are involved in the susceptibility to addiction and individually they are weak predictors of addiction risk (Kendler et al. 2012). Risk scores based on multiple risk alleles do not predict addiction risk any better than simple family history information (e.g., number of smoking parents) (Gartner et al. 2009). Risk involves a complex and unpredictable interaction between genetics and environmental events, as well as a possible constellation of gene–gene and epigenetic effects.

Our analysis suggests that the neuroscience research does not uncritically support the more modest version of the BDMA. We next examine whether the BDMA has delivered the clinical, social, and policy benefits promised and propose an alternative account of the relevance of neuroscience research to addiction.

Twenty years on: Has the brain disease model of addiction delivered?

When Leshner declared that addiction is a brain disease, he promised a range of therapeutic, social, and policy benefits. These benefits have continued to be promoted in support of treating and framing addiction as a brain disease (McLellan et al. 2000; Dackis & O'Brien 2005; Volkow & Li 2005; Volkow & Koob 2015). Given the important role that these outcomes have played in garnering support for the BDMA, it is worth asking whether the BDMA has lived up to its promise.

Improved medical treatments for addiction

Among the new treatments being trialed to treat addiction are novel drugs that aim to weaken the ability of cues associated with drug use or stress to trigger drug use; anti-craving medications; drugs to reverse epigenetic changes produced by chronic drug use; invasive and noninvasive brain stimulation devices to directly modulate aberrant brain activity; vaccines and implantable antagonists to protect against a relapse to drug use; and genomic and neurocognitive tests to match patients to the treatments that are most likely to be effective for them (Volkow & Li 2005; Nutt & Lingford-Hughes 2008). Despite significant investment by NIDA and other funders in the search for effective medical interventions, the promise of treatments that significantly improve upon the currently poor outcomes remain largely unfulfilled (Kalant 2010). The most widely used treatments for addiction are those that were developed prior to the considerable investment in neuroscience research and precede the BDMA by over three decades. These are primarily substitution treatments that attempt to replace commonly abused drugs of addiction, such as heroin and nicotine, with safer alternatives, such as methadone and nicotine replacement therapy.

There are ample claims that medications are among the most effective interventions for which they are available (Volkow et al. 2016b) but, in reality, few drugs derived from neurobiological research (e.g., naltrexone and varenicline) provide better results than older forms of treatment, such as disulfiram and nicotine replacement therapy (Lingford-Hughes et al. 2012). These represent a very small return from a large and sustained research investment in drug development in which the failures of a long list of promising new drugs have been quickly forgotten and attention and investment shifted to the next great hopes. Research on vaccines for nicotine and cocaine dependence have also been disappointing, with one of the leading developers of a nicotine vaccine withdrawing from clinical trials due to its limited effectiveness (Hall & Gartner 2011; Kosten et al. 2014).

There are significant challenges to the development of effective drug treatments for addiction, many of which are shared with development of pharmacological treatments in biomedicine more broadly (Koob et al. 2009). The development of drug treatments in addiction also raises additional challenges. Pharmaceutical companies have been reluctant to invest in addiction treatments due to doubts over the profitability of the medications when the majority of consumers will not be able to afford them. There is a concern that the stigma of addiction will discourage other potentially more profitable uses of these drugs, such as drugs to treat chronic pain.

Public views of addiction, stigma, and discrimination

Stigma can have serious deleterious effects on the health and well-being of individuals with mental illnesses such as addiction (Corrigan et al. 2014). Stigma is a complex construct that involves several elements. Structural stigma can lead to institutionalized discrimination that deprives individuals with certain disorders of effective healthcare. Social stigma involves the interpersonal victimization or discrimination against an individual that may discourage someone from seeking treatment. It can also exacerbate internalized or self-stigma, a collection of negative attitudes toward oneself that can lead to a loss of self-esteem and poorer health behaviors, such as continued or escalated drug use to escape negative thoughts and feelings.

A major benefit claimed for the BDMA is that its public acceptance will reduce stigma and discrimination against people with addictions. A similar hope has long been expressed in mental health more generally (Pescosolido et al. 2010; Sartorius 2010). The optimistic assumption is that the public will be more accepting of people with mental illnesses, such as depression and addiction, if they can be convinced that these are the result of biological, specifically neurobiological changes that produce the aberrant behavior, and hence that their sufferers have real diseases (Pescosolido et al. 2010).

Proponents of the BDMA do not provide sufficient evidence to support this claim and have ignored a growing body of social science survey evidence that suggests otherwise. Numerous social science surveys have found that increased acceptance of a biological basis for schizophrenia and alcohol dependence has had minimal impact on stigmatization, and may in fact increase rather than decrease stigma (Angermeyer & Matschinger 2005; Pescosolido et al. 2010; Reavley & Jorm 2011; Kvaale et al. 2013). One large study of the attitudes of US citizens over a 10-year period found that while acceptance of neurobiological and genetic causes of addiction and schizophrenia, two of the most highly stigmatized mental illnesses, increased substantially over this period, stigmatizing attitudes were unchanged. Some research suggests that addicted individuals often see the BDMA as adding to the stigma that they experience by reinforcing public fears that their behavior is an uncontrollable consequence of permanent changes in their brains. We have also found mixed support for the BDMA among Australian clinicians who treat addiction (Bell et al. 2014; Carter et al. 2014).

Reduced stigmatization and discrimination against people with an addiction is a desirable outcome, but we doubt that acceptance of a BDMA will achieve this. Stigma is a complex construct that involves a range of different stakeholders (e.g., clinicians, policymakers, the public, and the individual), each of whom will require a specific and targeted approach to reduce stigma and its negative consequences (Corrigan et al. 2014). There is a growing realization that alternative approaches are needed to reduce the stigma of mental illness, such as those employed by the nonprofit organization Bring Change 2 Mind, that target emotional processes to reduce social distance and fear (Corrigan et al. 2007; Corrigan & Shapiro 2010; Sartorius 2010). Future campaigns will need to identify evidence-based ways of reducing social distance and fear, rather than relying on plausible but misplaced suppositions that simply educating the public about the impact of drug use on the brain will reduce stigma.

Impact of brain disease explanations on addicted individuals

To date, the assertion that BDMA will increase treatment-seeking by people with drug addictions because it will increase acceptance of drug addiction as a persistent neurobiological change has been made in the absence of any research on its impact on addicted people. Satel, for example, believes that the BDMA rhetoric encourages addicted individuals to believe that they can never fully free themselves of their drug and alcohol problems (Satel 1998, 1999; Satel & Lilienfeld 2013). Davies (1998) argues that the BDMA exculpates the addicted person from taking responsibility for their condition: "it's not me, it's my disease" (Bell et al. 2014). Critics are concerned that biological understandings of addiction and other mental disorders may suggest that these disorders are incurable or untreatable, and will therefore reduce willingness to seek treatment (Phelan et al. 2006; Bell et al. 2014). Some critics suggest that the BDMA may in fact deter quit attempts by making quitting seem harder than it actually is (Chapman & MacKenzie 2010).

The empirical evidence, as limited as it is on these issues, is mixed. In a survey of public attitudes toward the treatability of mental illnesses, those who accepted psychological explanations were more likely to see mental illnesses as curable, less debilitating, and less likely to require professional assistance or hospitalization than those who supported biological explanations (Lam et al. 2005). Brain-based biological explanations also led people with these disorders to believe that their condition was harder to overcome than if they attributed them to psychological causes. If these findings are also true for addiction, the BDMA could discourage rather than encourage treatment-seeking and quit attempts.

By contrast, Walker (2010) has argued that if people accept that they have a disease that is beyond their control, they may be prompted to seek medical assistance rather than rely on willpower. Individuals who see their addiction as arising from a character flaw rather than a medical condition are less likely to seek treatment (Cunningham et al. 1993; Moyers & Miller 1993; Varney et al. 1995). Similarly, endorsement of genetic or neurochemical causes of mental illness, including alcohol dependence, has been associated with greater public support for medical treatment and the use of medication (Pescosolido et al. 2010).

Our own studies with addiction clinicians and people with various drug addictions have also produced mixed results (Meurk et al. 2013, 2016a, 2016b; Bell et al. 2014). Many interviewees were receptive to the findings of neuroscience research but their opinions varied on whether they believed that addiction constituted a brain disease and on what they thought the likely effects would be on addicted individuals. Many believed the BDMA could help addicted individuals to understand their own behavior and facilitate behavior change by reducing guilt about their personal responsibility for their condition (Meurk et al. 2013, 2016a, 2016b). Insofar as this happened, it could empower them to seek treatment and remain abstinent. The BDMA was also seen as potentially reducing what many interviewees believed to be the damaging effects of moralistic views of addiction (Meurk et al. 2013, 2016a, 2016b), and as a useful rhetorical device for improving treatment and reducing the punishment of individuals who need treatment for their addiction (Bell et al. 2014).

These positives were offset by views about the potentially adverse impacts of the BDMA on the treatment and drug use of a subpopulation of addicted individuals. The concern was that it could hinder addicted individuals' behavior change, reduce their willingness to enter treatment, undermine their ability to reduce drug use, and provide them with an excuse for not changing their behavior (Bell et al. 2014). These results are consistent with findings from a recent study of opioid-dependent individuals receiving opioid substitution treatment (Netherland 2011).

Policy consequences of the neuromedicalization of addiction

Another criticism of the BDMA is that it will focus on neurobiological solutions to treating addiction at the expense of more intensive psychosocial approaches. The majority of people with an addiction still do not receive minimal treatment or social support to improve their quality of life and reduce harmful drug use. The eminent pharmacologist, Harold Kalant, has argued that addiction neuroscientists have not critically reflected on their failure to deliver the promised advances in treatment of the addictions (Kalant 2010). We believe that the research focus of National Institutes of Health agencies on the neurobiology of severe addiction has deflected attention from population-based policies that are effective in reducing the harm from tobacco and alcohol use, namely, high rates of taxation, strategies to reduce drug availability, and smoke-free policies. This point was well made over a decade ago by Merikangas and Risch (2003).

In this context, the narrow focus on the neurobiological causes of severe cases of addiction has led to an over allocation of research and health resources toward neurobiological solutions that will benefit few addicted people. It also largely neglects the psychosocial drivers of drug use and addiction, such as poverty, homelessness, domestic violence, lack of access to quality education and employment, and social isolation. Indeed, the BDMA pays scant attention to social, cultural, and ethnic factors in addiction risk and severity or social obstacles to recovery. It gives little attention to the value of public health policies that have been the most successful in reducing cigarette smoking, gambling, and alcohol-related harm, such as increased taxation, reduced availability, and controls on the promotion of addictive products by the industries that profit from them. For example, taxes on cigarettes, advertising bans, and restrictions on where people can smoke have halved cigarette smoking rates in Australia (White et al. 2003) and the United States (Pierce et al. 1998) in the last 30 years. These strategies are much more efficient than high-risk strategies aimed at treating individual smokers or people at risk of smoking (Rose G. 1992; Hall et al. 2002). Similar evidence supports the greater efficiency of population-based strategies in reducing the societal harms of alcohol misuse (Doran et al. 2010). However, there is a real concern that an overemphasis on the BDMA may undermine population-level approaches, particularly when misused by the alcohol and tobacco industries to oppose public health policies and promote preventive policies that target at risk drug users or the treatment of addicted people (Miller et al. 2012).

The BDMA prioritizes treatment over public health policy. For example, of the 2014 NIDA and National Institute on Alcohol Abuse and Alcoholism $1065.24 million USD research budget, 41% was allocated to basic and clinical neuroscience and 17% to pharmacotherapies (NIDA 2014). Social and public health research forms a small part of the 24% allocated to epidemiological research (NIDA 2014). A National Academy of Science committee highlighted in 2001 that the National Institutes of Health had seriously underinvested in drug policy evaluation (National Research Council 2001). In a recent response to criticisms of the policy payoffs of the BDMA, Volkow and colleagues explicitly stated the greater need for even further investment in the search for medical treatments of addiction, despite acknowledging the failure to develop new effective treatments (Volkow et al. 2016b). No mention was made of increasing investment in public policy research.

It is important to emphasize that clinical treatment and population-level approaches are not mutually exclusive; we are not arguing for either one or the other. The observation that science should not ignore the very significant benefits of population-level approaches does not preclude the need to identify more effective treatments for addicted people. Society can, and should, look for more effective medical treatments of addiction. But this should not be at the expense of maintaining attention to more broadly effective population-level approaches for reducing drug use and preventing harm (Hall et al. 2015). We would like to see greater advocacy for increasing access to existing treatments rather than for a BDMA that is of uncertain utility and acceptability to the people who are supposedly its chief beneficiaries.

Volkow and colleagues claim that the BDMA has contributed to at least two policy milestones in recent years: (1) passing of the Mental Health Parity and Addiction Equity Act as part of the Patient Protection and Affordable Care Act that ensures that addiction treatment is covered by medical insurance; and (2) justification for increasing the drinking age to 21 years (Volkow et al. 2016b). It is difficult to assess whether the BDMA led to the Patient Protection and Affordable Care Act as part of US nationalized healthcare per se, however. A fair appreciation of the factors that contributed to the passage of this Act would require social and historical research on the drivers of this policy change.

Reframing neurobiological accounts of addiction

There is considerable scientific value in neurobiological and genetic research on addiction. Neuroscience research has uncovered the neurochemical and neurocognitive mechanisms that are associated with drug use. It helps to explain why those who have used addictive drugs over longer periods of time can find it difficult to refrain from drug use despite significant personal and social harms.

However, this research does not justify the simplified BDMA that dominates official discourse about addiction in the United States. Evidence that drugs alter brain function in a profound way is not new. The question is, what does neurobiology mean for the present or future of diagnosis and treatment? Our understanding of addiction, and the policies we adopt to treat and prevent problem drug use, should give biology its due, but no more than its due. Evidence from research in economics, epidemiology, and the social sciences shows that neurobiology is not the overriding factor when formulating policies toward drug use and addiction, and is not a major influence on our concern for people with an addiction (Meurk et al. 2016a).

Originators and current proponents of the BDMA uphold the forced choice of the 1990s, namely, that rejection of addiction as a chronic, relapsing brain disease equates to an acceptance of the moral view of addiction according to which addicted people can simply stop using (Leshner 1997; Volkow & Koob 2015). Critics can accept that there is value in neurobiological and genetic research on addiction but still reject the narrow view that addiction is a brain disease and be skeptical that promoting this view will overcome many of the limitations in the treatment of addicted individuals in society. As we have outlined already, we do not believe that telling people with an addiction that they have a brain disease that involves persistent rewiring of their brain that drives them to drug use will necessarily lead to better health outcomes.

It has recently been claimed that there is a consensus view among scientists and clinicians that addiction is a brain disease (American Society of Addiction Medicine 2011; Editorial 2014). There has been little empirical research to test this claim. In a small study of Australian neuroscientists and clinicians, less than one-third strongly endorsed the BDMA. A more reasonable and modified argument for the relevance of neurobiology to addiction would involve arguing that addiction can be, but is not always, a chronic, relapsing disorder; that people who develop chronic forms of addiction often require sustained treatment to become and remain abstinent; that this treatment often has to address the biological effects of sustained heavy drug use on brain and body; and that drug treatments can certainly assist in withdrawing from drugs and in maintaining abstinence after cessation of use. This view makes it clearer that the drug treatment of addiction is often only the beginning in addressing the many challenges that addicted people face in recovering and remaining abstinent, namely, securing useful employment, making friends who are not drug users, repairing damaged family relationships, and overcoming the stigma and social consequences of having a criminal record.

Neural plasticity: The forgotten science in addiction

Neuroplasticity is surprisingly absent from neurobiological accounts of addiction, particularly given the attention it has received in the academic and popular media. The BDMA rarely, if ever, discusses the relevance of neural plasticity to addiction treatment; it emphasizes instead the persistence of the changes produced by chronic drug use. One can only speculate that the view that chronic use of addictive drugs produces long-lasting or permanent changes in brain function provides a more morally and politically acceptable message, namely, that some forms of drug use damage the brain. While this message may deter some young people from experimenting with drugs of addiction, it may have a negative effect on those who develop an addiction, for example, by undermining their self-efficacy and treatment-seeking, thereby promoting fatalism and learned helplessness. A more effective, and accurate, application of neuroscience research on addictive drug use would take account of the ability of the brain to recover from drug-induced changes through abstinence and reduced drug use, therapy, and prosocial lifestyle choices, such as exercise, brain training, or cognitive exercises, to rewire and relearn more adaptive habits (Lewis 2015).

Conclusion

The chronic, relapsing brain disease model seems most plausible when used to describe the minority of severely addicted people who fail to quit using drugs and who present themselves, or are coerced by others, into treatment in middle adult life. These individuals comprise a minority of those who meet diagnostic criteria for addiction. The BDMA cannot be used to describe the vast majority of individuals who receive an addiction diagnosis at some point during their lives.

We conclude that the BDMA has also not delivered on its promises. We are yet to see substantially more effective treatments for drug addictions and the impact of the BDMA in reducing stigma and improving access and funding to appropriate treatment services has been modest. We share many of the hopes of those who advocate for the BDMA, such as the delivery of more effective treatment and less punitive responses to addicted people. But addiction is a complex biological, psychological, and social disorder that requires a combined clinical and public health approach (Carter & Hall 2012). The search for technological cures of addiction should not distract from increasing access to currently available psychosocial and drug treatments for addiction, which the majority of addicted individual are still unable to access, or developing simpler, cheaper, and more efficient population-based policies that discourage the whole population from smoking tobacco, drinking heavily, or abusing illicit drugs.

Our rejection of the BDMA should not be misinterpreted as a defense of the moral model of addiction (Carter & Hall 2012) or the denial of the value of neuroscience. Neuroscience research on addiction has provided useful insights into the neurobiology of decision-making, motivation, and behavioral control. These insights help to understand how chronic use of addictive drugs can impair cognitive and motivational processes and may partially explain why some people are more vulnerable than others to developing an addiction. The challenge for all addiction researchers is to incorporate the emerging insights of neuroscience research into those provided by economics, epidemiology, sociology, psychology, and political science so the harms caused by drug misuse and all forms of addiction can be reduced.

Neuroscience research can be extremely persuasive. It can influence expectations of the clinical effectiveness of treatments (Racine et al. 2007) and the understanding of the self in health and disease (Racine et al. 2010; Rose N.S. & Abi-Rached 2013). Care is needed in how clinicians, policymakers, and other relevant stakeholders communicate the results of neuroscience research on addiction (Satel & Lilienfeld 2013). It should not be assumed that messages about diseased brains will always lead to increased treatment-seeking and reduced drug use. Messages about neuroscience research must be tailored to individuals to foster better choices and health outcomes and avoid negative consequences. This requires evidence on how addicted individuals interpret this research. More work is needed to understand how addicted individuals understand and respond to neuroscience research of addiction, and how best to present this research to optimize treatment and reduce unintended adverse consequences.

References

Ahmed, S.H. (2010). Validation crisis in animal models of drug addiction: beyond non-disordered drug use toward drug addiction. *Neuroscience & Biobehavioral Reviews*, **35**, 172–184.

Ahmed, S.H. (2012). The science of making drug-addicted animals. *Neuroscience*, **211**, 107–125.

Ahmed, S.H., Lenoir, M., and Guillem, K. (2013). Neurobiology of addiction versus drug use driven by lack of choice. *Current Opinion in Neurobiology*, **23**, 1–7.

Alexander, B.K., Coambs, R.B., and Hadaway, P.F. (1978). The effect of housing and gender on morphine self-administration in rats. *Psychopharmacology*, **58**, 175–179.

American Society of Addiction Medicine (2011). *Public Policy Statement: Definition of Addiction (Long Version)*. Chevy Chase, MD: American Society of Addiction Medicine.

Angermeyer, M.C. and Matschinger, H. (2005). Labeling–stereotype–discrimination. An investigation of the stigma process. *Social Psychiatry and Psychiatric Epidemiology*, **40**, 391–395.

Bachman, J.G., Wadsworth, K.N., O'Malley, P.M., Johnston, L.D. and Schulenberg, J. (1997). *Smoking, Drinking, and Drug Use in Young Adulthood: The Impacts of New Freedoms and New Responsibilities*. Mahwah, NJ: Lawrence Erlbaum.

Baler, R.D. and Volkow, N.D. (2006). Drug addiction: the neurobiology of disrupted self-control. *Trends in Molecular Medicine*, **12**, 559–566.

Bell, S., Carter, A., Mathews, R., Gartner, C., Lucke, J., and Hall, W. (2014). Views of addiction neuroscientists and clinicians on the clinical impact of a "brain disease model of addiction". *Neuroethics*, **7**, 19–27.

Botticelli, M. (2014). *National Blueprint for Drug Policy Reform Released Today in Roanoke, VA*. Office of National Drug Control Policy Media Release.

Buchman, D.Z., Skinner, W., and Illes, J. (2010). Negotiating the relationship between addiction, ethics, and brain science. *AJOB Neuroscience*, **1**, 36–45.

Caplan, A. (2008). Denying autonomy in order to create it: the paradox of forcing treatment upon addicts. *Addiction*, **103**, 1919–1921.

Carter, A. and Hall, W. (2008a). Informed consent to opioid agonist maintenance treatment: recommended ethical guidelines. *International Journal of Drug Policy*, **19**, 79–89.

Carter, A. and Hall, W. (2008b). The issue of consent in research that administers drugs of addiction to addicted persons. *Accountability in Research*, **15**, 209–225.

Carter, A. and Hall, W. (2011). Proposals to trial deep brain stimulation to treat addiction are premature. *Addiction*, **106**, 235–237.

Carter, A. and Hall, W. (2012). *Addiction Neuroethics: The Promises and Perils of Neuroscience Research on Addiction*. London: Cambridge University Press.

Carter, A., Bell, E., Racine, E., and Hall, W. (2011). Ethical issues raised by proposals to treat addiction using deep brain stimulation. *Neuroethics*, **4**, 129–142.

Carter, A., Mathews, R., Bell, S., Lucke, J., and Hall, W. (2014). Control and Responsibility in Addicted Individuals: What do Addiction Neuroscientists and Clinicians Think? *Neuroethics*, **7**, 205–214.

Chapman, S. and MacKenzie, R. (2010). The global research neglect of unassisted smoking cessation: causes and consequences. *PLoS Medicine*, **7**, e1000216.

Charland, L.C. (2002). Cynthia's dilemma: consenting to heroin prescription. *American Journal of Bioethics*, **2**, 37–47.

Cohen, M.J.M., Jasser, S., Herron, P.D., and Margolis, C.G. (2002). Ethical perspectives: opioid treatment of chronic pain in the context of addiction. *Clinical Journal of Pain*, **18**, S99-S107.

Corrigan, P.W. and Shapiro, J.R. (2010). Measuring the impact of programs that challenge the public stigma of mental illness. *Clinical Psychology Review*, **30**, 907–922.

Corrigan, P.W., Larson, J., Sells, M., Niessen, N., and Watson, A. C. (2007). Will filmed presentations of education and contact diminish mental illness stigma? *Community Mental Health Journal*, **43**, 171–181.

Corrigan, P.W., Druss, B.G., and Perlick, D.A. (2014). The impact of mental illness stigma on seeking and participating in mental health care. *Psychological Science in the Public Interest*, **15**, 37–70.

Cunningham, J.A., Sobell, L.C., and Chow, V.M. (1993). What's in a label? The effects of substance types and labels on treatment considerations and stigma. *Journal of Studies on Alcohol*, **54**, 693–699.

Dackis, C. and O'Brien, C. (2005). Neurobiology of addiction: treatment and public policy ramifications. *Nature Neuroscience*, **8**, 1431–1436.

Davies, J.B. (1998). Pharmacology versus social process: competing or complementary views on the nature of addiction? *Pharmacology & Therapeutics*, **80**, 265–275.

Dejong, W. and Blanchette, J. (2014). Case closed: research evidence on the positive public health impact of the age 21 minimum legal drinking age in the United States. *Journal of Studies on Alcohol and Drugs, Supplement*, 108–115.

Doran, C.M., Hall, W.D., Shakeshaft, A.P., Vos, T., and Cobiac, L.J. (2010). Alcohol policy reform in Australia: what can we learn from the evidence. *Medical Journal of Australia*, **192**, 468–470.

Editorial (2014). Animal farm. *Nature*, **506**, 5.

Ersche, K.D., Williams, G.B., Robbins, T.W., and Bullmore, E.T. (2013). Meta-analysis of structural brain abnormalities associated with stimulant drug dependence and neuroimaging of addiction vulnerability and resilience. *Current Opinion in Neurobiology*, **23**, 615–624.

Feltenstein, M.W. and See, R.E. (2008). The neurocircuitry of addiction: an overview. *British Journal of Pharmacology*, **154**, 261–274.

Gartner, C.E., Barendregt, J.J., and Hall, W. (2009). Multiple genetic tests for susceptibility to smoking do not outperform simple family history. *Addiction*, **104**, 118–126.

Hall, W., Madden, P., and Lynskey, M. (2002). The genetics of tobacco use: methods, findings and policy implications. *Tobacco Control*, **11**, 119–124.

Hall, W., Capps, B., and Carter, A. (2008). The use of depot naltrexone under legal coercion: the case for caution. *Addiction*, **103**, 1922–1924.

Hall, W. and Gartner, C. (2011). Ethical and policy issues in using vaccines to treat and prevent cocaine and nicotine dependence. *Current Opinion in Psychiatry*, **24**, 191–196.

Hall, W., Carter, A., and Forlini, C. (2015). The brain disease model of addiction: is it supported by the evidence and has it delivered on its promises? *The Lancet Psychiatry*, **2**, 105–110.

Heyman, G. (2009). *Addiction: A Disorder of Choice*. Cambridge, MA: Harvard University Press.

Hyman, S.E. (2007). The neurobiology of addiction: implications for voluntary control of behavior. *American Journal of Bioethics*, **7**, 8–11.

Hyman, S.E., Malenka, R.C., and Nestler, E.J. (2006). Neural mechanisms of addiction: the role of reward-related learning and memory. *Annual Review of Neuroscience*, **29**, 565–598.

Kalant, H. (2010). What neurobiology cannot tell us about addiction. *Addiction*, **105**, 780–789.

Kendler, K.S., Chen, X., Dick, D., et al. (2012). Recent advances in the genetic epidemiology and molecular genetics of substance use disorders. *Nature Neuroscience*, **15**, 181–189.

Kilmer, B., Nicosia, N., Heaton, P., and Midgette, G. (2013). Efficacy of frequent monitoring with swift, certain, and modest sanctions for violations: insights from South Dakota's 24/7 sobriety project. *American Journal of Public Health*, **103**, e37–e43.

Kincaid, H. and Sullivan, J.A. (2010). Medical models of addiction. In: Ross, D., Kincaid, H., Spurrett, D., and Collins, P. (Eds.) *What is Addiction?*, pp.353–376. Cambridge, MA: MIT Press.

Kleiman, M. (2009). *When Brute Force Fails: How to Have Less Crime and Less Punishment*. Princeton, NJ: Princeton University Press.

Koob, G.F. and Le Moal, M. (2006). *Neurobiology of Addiction*. New York: Academic Press.

Koob, G F., Lloyd, G.K., and Mason, B.J. (2009). Development of pharmacotherapies for drug addiction: a Rosetta Stone approach. *Nature Reviews Drug Discovery*, **8**, 500–515.

Kosten, T.R., Domingo, C.B., Shorter, D., et al. (2014). Vaccine for cocaine dependence: a randomized double-blind placebo-controlled efficacy trial. *Drug and Alcohol Dependence*, **140**, 42–47.

Krack, P., Hariz, M.I., Baunez, C., Guridi, J., and Obeso, J.A. (2010). Deep brain stimulation: from neurology to psychiatry? *Trends in Neurosciences*, **33**, 474–484.

Kvaale, E.P., Haslam, N., and Gottdiener, W. H. (2013). The "side effects" of medicalization: a meta-analytic review of how biogenetic explanations affect stigma. *Clinical Psychology Review*, **33**, 782–794.

Lam, D.C.K., Salkovskis, P.M., and Warwick, H.M.C. (2005). An experimental investigation of the impact of biological versus psychological explanations of the cause of "mental illness". *Journal of Mental Health*, **14**, 453–464.

Leshner, A.I. (1997). Addiction is a brain disease, and it matters. *Science*, **278**, 45–47.

Lewis, M. (2015). *The Biology of Desire: Why Addiction is Not a Disease*. New York: Scribe Publications.

Lingford-Hughes, A.R., Welch, S., Peters, L., and Nutt, D.J. (2012). BAP updated guidelines: evidence-based guidelines for the pharmacological management of substance abuse, harmful use, addiction and comorbidity: recommendations from BAP. *Journal of Psychopharmacology*, **26**, 899–952.

Luigjes, J., van den Brink, W., Feenstra, M., et al. (2012). Deep brain stimulation in addiction: a review of potential brain targets. *Molecular Psychiatry*, **17**, 572–583.

McLellan, A.T., Lewis, D.C., O'Brien, C.P., and Kleber, H.D. (2000). Drug dependence, a chronic medical illness: implications for treatment, insurance, and outcomes evaluation. *JAMA*, **284**, 1689–1695.

Merikangas, K.R. and Risch, N. (2003). Genomic priorities and public health. *Science*, **302**, 599–601.

Meurk, C., Hall, W., Morphett, K., Carter, A., and Lucke, J. (2013). What does "acceptance" mean? Public reflections on the idea that addiction is a brain disease. *Biosocieties*, **8**, 491–506.

Meurk, C., Fraser, D., Weier, M., Lucke, J., Carter, A., and Hall, W. (2016a). Assessing the place of neurobiological explanations in accounts of a family member's addiction. *Drug and Alcohol Review*, **35**(4), 461–469.

Meurk, C., Morphett, K., Carter, A., Weier, M., Lucke, J., and Hall, W. (2016b). Scepticism and hope in a complex predicament: people with addictions deliberate about neuroscience. *International Journal of Drug Policy*, **32**, 34–43.

Miller, P., Carter, A., and De Groot, F. (2012). Investment and vested interests in neuroscience research of addiction: Why research ethics requires more than informed consent. In: Carter, A., Hall, W., and Illes, J. (Eds.) *Addiction Neuroethics: The Ethics of Addiction Research and Treatment*, pp. 278–301. New York: Elsevier.

Moyers, T.B. and Miller, W.R. (1993). Therapists' conceptualizations of alcoholism: Measurement and implications for treatment decisions. *Psychology of Addictive Behaviors*, **7**, 238.

National Institute on Drug Abuse (2014). *Fiscal Year 2015 Budget information—Congressional justification for National Institute on Drug Abuse*. Rockville, MD: National Institute on Drug Abuse.

National Research Council (2001). *Informing America's Policy on Illegal Drugs: What We Don't Know Keeps Hurting Us*. Washington, DC: National Academy Press.

Netherland, J. (2011). "We haven't sliced open anyone's brain yet": Neuroscience, embodiment and the governance of addiction. In: Pickersgill, M. and Van Keulen, I. (Eds.) *Sociological Reflections on the Neurosciences* (Advances in Medical Sociology, Volume 13), pp. 153–177. Bingley, UK: Emerald Group Publishing Limited.

Nutt, D. and **Lingford-Hughes, A.** (2008). Addiction: the clinical interface. *British Journal of Pharmacology*, **154**, 397–405.

Pescosolido, B.A., Martin, J.K., Long, J.S., Medina, T.R., Phelan, J., and **Link, B.** (2010). "A disease like any other?": a decade of change in public reactions to schizophrenia, depression, and alcohol dependence. *American Journal of Psychiatry*, **167**, 1321–1330.

Phelan, J.C., Yang, L.H., and **Cruz-Rojas, R.** (2006). Effects of attributing serious mental illnesses to genetic causes on orientations to treatment. *Psychiatric Services*, **57**, 382–387.

Pierce, J.P., Gilpin, E.A., Emery, S.L., White, M.M., Rosbrook, B., and **Berry, C. C.** (1998). Has the California tobacco control program reduced smoking? *JAMA*, **280**, 893.

Racine, E., Waldman, S., Palmour, N., Risse, D., and **Illes, J.** (2007). "Currents of hope": neurostimulation techniques in U.S. and U.K. print media. *Cambridge Quarterly of Healthcare Ethics*, **16**, 312–316.

Racine, E., Waldman, S., Rosenberg, J., and **Illes, J.** (2010). Contemporary neuroscience in the media. *Social Science and Medicine*, **71**, 725–733.

Reavley, N.J. and **Jorm, A.F.** (2011). Stigmatizing attitudes towards people with mental disorders: findings from an Australian National Survey of Mental Health Literacy and Stigma. *Australian and New Zealand Journal of Psychiatry*, **45**, 1086–1093.

Reske, M. and **Paulus, M.P.** (2011). A neuroscientific approach to addiction: ethical issues. In: Illes, J. and Sahakian, B. (Eds.) *Oxford Handbook of Neuroethics*, pp. 177–202. Oxford: Oxford University Press.

Robins, L.N. (1993). Vietnam veterans' rapid recovery from heroin addiction: a fluke or normal expectation? *Addiction*, **88**, 1041–1054.

Rose, G. (1992). *The Strategy of Preventive Medicine*. Oxford: Oxford University Press.

Rose, N.S. and **Abi-Rached, J.M.** (2013). *Neuro: The New Brain Sciences and the Management of the Mind*. Princeton: Princeton University Press.

Sartorius, N. (2010). Short-lived campaigns are not enough. *Nature*, **468**, 163–165.

Satel, S. (1998). For addicts, force is the best medicine. *Wall Street Journal*, January 6.

Satel, S. (1999). The fallacies of no-fault addiction. *Public Interest*, **99**, 52–67.

Satel, S. and **Lilienfeld, S.O.** (2013). *Brainwashed: The Seductive Appeal of Mindless Neuroscience*. New York: Perseus Books Group.

Satel, S. and **Lilienfeld, S.O.** (2014). Addiction and the brain-disease fallacy. *Frontiers in Psychiatry*, **4**, 141.

Varney, S.M., Rohsenow, D.J., Dey, A.N., Myers, M.G., Zwick, W.R., and **Monti, P.M.** (1995). Factors associated with help seeking and perceived dependence among cocaine users. *American Journal of Drug and Alcohol Abuse*, **21**, 81–91.

Volkow, N. and **Baler, R.** (2014). Addiction science: uncovering neurobiological complexity. *Neuropharmacology*, **76**, 235–249.

Volkow, N.D. and **Koob, G.** (2015). Brain disease model of addiction: why is it so controversial? *The Lancet Psychiatry*, **2**, 677–679.

Volkow, N.D. and Li, T.-K. (2004). Drug addiction: the neurobiology of behaviour gone awry. *Nature Reviews. Neuroscience*, **5**, 963–970.

Volkow, N.D. and Li, T.-K. (2005). Drugs and alcohol: treating and preventing abuse, addiction and their medical consequences. *Pharmacology & Therapeutics*, **108**, 3–17.

Volkow, N.D. and Morales, M. (2015). The brain on drugs: from reward to addiction. *Cell*, **162**, 712–725.

Volkow, N.D., Fowler, J.S., Wang, G.J., Swanson, J.M., and Telang, F. (2007). Dopamine in drug abuse and addiction: results of imaging studies and treatment implications. *Archives of Neurology*, **64**, 1575–1579.

Volkow, N.D., Fowler, J.S., Wang, G.J., Baler, R., and Telang, F. (2009). Imaging dopamine's role in drug abuse and addiction. *Neuropharmacology*, **56**(Suppl. 1), 3–8.

Volkow, N.D., Wang, G.-J., Fowler, J.S. and Tomasi, D. (2012). Addiction circuitry in the human brain. *Annual Review of Pharmacology and Toxicology*, **52**, 321.

Volkow, N.D., Koob, G.F., and McLellan, A.T. (2016a). Neurobiologic advances from the brain disease model of addiction. *New England Journal of Medicine*, **374**, 363–371.

Volkow, N.D., Koob, G.F., and McLellan, A.T. (2016b). Supplement to: Neurobiologic advances from the brain disease model of addiction. *New England Journal of Medicine*, **374**(4), 363–371. Available from: http://www.nejm.org/doi/suppl/10.1056/NEJMra1511480/suppl_file/nejmra1511480_appendix.pdf [accessed February 9, 2017].

Vrecko, S. (2010). Birth of a brain disease: science, the state and addiction neuropolitics. *History of the Human Sciences*, **23**, 52–67.

Walker, M.J. (2010). Addiction and self deception: a method for self control? *Journal of Applied Philosophy*, **27**, 305–319.

White, V., Hill, D., Siahpush, M. and Bobevski, I. (2003). How has the prevalence of cigarette smoking changed among Australian adults? Trends in smoking prevalence between 1980 and 2001. *Tobacco Control*, **12**, ii67.

Xu, Z., Hou, B., Gao, Y., He, F. and Zhang, C. (2007). Effects of enriched environment on morphine-induced reward in mice. *Experimental Neurology*, **204**, 714–719.

Chapter 26

Concussion, neuroethics, and sport: Policies of the past do not suffice for the future

Brad Partridge and Wayne Hall

Introduction

Concussion is a short-term disturbance in neurological function that is the result of a traumatic impact to the head. The term is sometimes used interchangeably with mild traumatic brain injury (MTBI). The Concussion in Sport Group has called concussion "a complex pathophysiological process affecting the brain, induced by traumatic biomechanical forces," which reflects a "functional" disruption to the brain rather than "structural" damage (McCrory et al. 2013).

The Centers for Disease Control has estimated that up to 3.8 million sports-related concussions occur in the United States alone each year, but these events are significantly under-reported (Daneshvar et al. 2011). It is a common injury in many body contact sports and its prevalence in this arena has recently garnered attention because of concerns about the possible long-term effects of these injuries. The focus of this chapter is on concussion injuries in body collision sports. It addresses ethical issues that arise in the diagnosis of concussion, debates about the longer-term consequences of repeated concussion injuries, and the design and implementation of policies that aim to prevent and manage concussion injuries that occur in the course of sporting matches.

Concussion in football codes

Some football codes exhibit high levels of risk of concussion because they involve repeated heavy collisions between participants who are running at speed directly at each other. The prevalence of concussion in amateur and professional Australian Rules football (Makdissi et al. 2009) and rugby league (Hinton-Bayre et al. 2004) are among the highest of any contact sports. At the high school and collegiate level in the United States, American football has the highest participation rate among sports and the highest rate of concussion per thousand playing hours (Daneshvar et al. 2011).

Concussion management policies have become a major priority worldwide for sports that involve frequent collisions between participants because repeated head trauma has been associated with long-term cognitive impairments (Guskiewicz et al. 2005), mental health problems (Guskiewicz et al. 2007), and some forms of neurological degeneration (Omalu et al. 2005; McKee et al. 2009, 2013, 2014). Traumatic brain injury (TBI) is the leading cause of death and disability among children and young adults worldwide, and approximately one-fifth of all TBI cases occur during sports participation.

Concussion management guidelines have been developed by a number of groups in an effort to either prevent concussion or to improve its management. These include the American Academy of Neurology (Giza et al. 2013) and Concussion in Sport Group (McCrory et al. 2009, 2013). Although procedures for identifying concussion have been formally implemented in many sports, clinicians are still confronted with conceptual, empirical, and ethical uncertainty in diagnosing and managing concussion (McNamee et al. 2016).

Challenges in diagnosis

Diagnosis can be a challenge because there is a long list of potential symptoms and signs of concussion that reflect disruptions to normal cognition, motor functioning, or affect. These include, for instance, amnesia, loss of balance, blurred vision, confusion, dizziness, feeling "in a fog," or being more emotionally labile. There is no consensus on when a diagnosis of concussion should be made and there are no specified necessary and sufficient symptoms for the diagnosis, not even the seemingly obvious loss of consciousness (McNamee et al. 2016).

Brief diagnostic tools have been developed to identify concussion but there is no definitive "concussion test" and so a concussion diagnosis always reflects a clinician's judgment. These conceptual uncertainties have the potential to undermine the reliability and validity of a diagnosis of the presence or absence of concussion. They also affect the validity of epidemiological data on the prevalence of concussion in different sports, with the likelihood that many cases of concussion go unrecognized.

Debates about the long-term effects of concussion

Most cases of concussion appear to resolve themselves without the need for medical intervention although there is individual variation in how long it may take for this resolution to occur. Some symptoms may be very brief and many of the symptoms of mild concussion will completely resolve within 7–10 days (McCrory et al. 2013). A minority of affected people may experience more lingering effects, such as headaches and mental fogginess that may last for weeks, months, or even longer. Deficits in functioning on cognitive tasks may also be evident in people with post-concussion syndrome.

Epidemiological studies in the general population have found that self-reported TBI in early life or midlife increases the risk of mild forms of cognitive impairment as well as dementia and Alzheimer's disease later in life (Graves et al. 1990; Molgaard et al. 1990; Salib and Hillier 1997; Guo et al. 2000; Lye & Shores 2000; Fleminger et al. 2003; Shively et al. 2012). A history of TBI has also been associated with an earlier onset of Alzheimer's dementia (Nemetz et al. 1999). Systematic reviews of these studies have concluded that there is sufficient evidence for a causal relationship between moderate to severe TBI and Alzheimer's and Parkinson's-related dementia. There is also an association between these conditions and mild TBI when there has been a loss of consciousness (Bazarian et al. 2009). This includes evidence from a large-scale prospective study which found that mild TBI was an independent risk factor for dementia (Lee et al. 2013). Prospective cohort studies also find that severe, moderate, and mild TBI are associated with an increased risk of developing depression (Fann et al. 2004).

There are fewer studies of the specific risks of concussion in athletes but epidemiological studies of players of collision sports have reported similar findings to the studies of TBI in the general population. Former American footballers who have suffered three or more concussions reported more mild cognitive impairment, memory problems, and depression than players without such a history. Retired football players also have an earlier age of onset of Alzheimer's disease than peers in the general population (Guskiewicz et al. 2005, 2007) and are three times more likely to die prematurely from neurodegenerative diseases than the general population (Lehman et al. 2012).

The available epidemiological evidence is sufficient to justify prudential policies to reduce the incidence of concussion in body collision sports. Instead, in the United States, these policies have been delayed by a narrow debate on whether the cumulative effects of repeated concussion cause a specific condition known as chronic traumatic encephalopathy (CTE)—a degenerative brain disease in which neurofibrillary tangles of tau protein are deposited in the brain—that was first described in boxers in the 1920s. Interest in CTE among American footballers increased after a 2005 postmortem neuropathological study of a former National Football League (NFL) player found evidence of the condition (Omalu et al. 2005). Since then, several groups have reported signs of CTE in postmortem studies of the brains of several dozen former American footballers and a smaller number of former rugby players and wrestlers (Omalu et al. 2010; McKee et al. 2014).

People who show neuropathological evidence of CTE often reported dementia-like symptoms, disturbed memory and mood, increased impulsiveness and aggression, and an increased risk of suicide (McKee et al. 2014). The prevalence of CTE in football players and other collision sport athletes is unknown but the seriousness of the condition and high public profile of some of its sufferers have placed the condition at the center of a contentious debate in the United States that has echoes in other countries such as Australia.

The evidence for the hypothesis that repeated head trauma causes CTE comes from postmortem evidence of the disorder in players with a history of repeated head trauma. Proponents of the CTE hypothesis assume that the disorder is so rare and so clearly related to repeated head trauma that these cases are sufficient to establish a causal relationship. Critics (Gardner et al. 2014) counter that there is no epidemiological evidence that CTE is more common among football players than in the general population and no evidence to rule out the role of factors other than concussion, such as genetic vulnerability to dementia, alcohol and other drug use, and risky behavior outside football. The most recent consensus statement about such injuries by the 4th International Conference on Concussion in Sport held in Zurich, November 2012 (McCrory et al. 2013) stated that: "a cause and effect relationship has not as yet been demonstrated between CTE and concussions or exposure to contact sports."

Well-designed, independent, and credible research studies are essential to decide whether CTE is caused by repeated concussions and head injuries in athletes. But this shortcoming in the literature on CTE should not detract attention from the larger epidemiological literature that strongly suggests that repeated head injuries increase the risk of cognitive impairment, depression, and dementia, whatever the underlying pathological mechanism.

Responses to concussion from professional sporting bodies

The governing bodies of many collision sports have struggled in how to appropriately manage concussion in the face of public controversy about the effects of repeated concussions and head injuries. In 2012, a class action lawsuit against the NFL was filed on behalf of more than 4500 former players which alleged that the NFL had:

◆ hidden and misrepresented evidence about the long-term neurological effects of football-related head injuries

◆ been aware of the neurological risks of playing football but deliberately failed to warn players

◆ funded a falsified body of scientific research that was conducted by its concussion advisory committee with the intention of raising doubts about independent evidence of the long-term effects of head injuries.

After failing to have the lawsuit dismissed, the NFL settled the case before trial. This unfortunately meant that the NFL was able to avoid admitting liability or that plaintiffs' injuries were caused by football and avoided a forensic examination of their alleged deception. The players also had good reasons to settle. The payout meant that players with "cognitive injuries" (yet to be defined) could access compensation and funding for medical examinations sooner than if the case had gone to trial. If the case had proceeded to a jury trial, the players may have received more money but they could also have lost and received nothing.

In Australia, the Australian Football League (AFL) and National Rugby League (NRL) claim that player safety and player welfare are key priorities in concussion management. The NRL has claimed that it will do "everything possible" to ensure the game is safe for all players (NRL 2012). The former Chief Medical Officer of the NRL has argued that "we have to protect players from themselves" (Crawley 2014). Taken together, these statements imply that these sporting bodies accept an ethical obligation to make their collision sports safer for participants and perhaps even paternalistically protect players from injuring themselves.

There is an alternative position that collision sports have not chosen to publicly advocate, namely, that adult athletes are autonomous agents who are free to participate in collision sports, if they choose to do so, and so should accept the risks of participation including concussion. If this stance were taken, then collision sports could argue that participants play at their own risk and therefore must accept responsibility for any future health problems that may arise (Partridge & Hall 2015).

If the "play at your own risk" stance were adopted then sporting bodies would need to ensure that players autonomously decide whether or not to play after being fully informed about the risks of participation. This would require (1) management of the potential for coerced consent and the conflicts of interest among players, coaches, trainers, and doctors that may lead them to circumvent concussion management; and, (2) full acknowledgment and disclosure of the risks of repeated head injuries by sports governing bodies.

The implicit assumption of current management policies seems to be that the risk of concussion cannot be eliminated from American football, rugby league, and Australian Rules football. The goal of these policies is therefore on managing specific instances of concussion. There are no policies to prevent players from suffering repeated head trauma during their playing careers. This stance is arguably at odds with the professed claims of the governing bodies of these sports that they give a high priority to protecting players.

These minimalist policies have often been justified by skepticism about the claim that concussions can cause serious long-term neurological harm in general and CTE in particular. Sporting officials have used uncertainty about the causal relationship to delay acting on their professed ethical obligation to protect players from repeated head injuries. For instance, some medical officers employed by, and researchers with professional links to, the AFL have described the evidence on the long-term risks of multiple concussions as "anecdotal" and overhyped by the media (Lane 2012, 2016). In 2012, for example, when the Chief Medical Officer of the International Rugby Board (IRB) was asked whether repeated concussions could cause long-term brain damage, he responded (McDermott and Hichens 2012): "The answer is: it is possible. But at the moment there is no clear evidence—there are some isolated cases suggesting it could—but there is no clear evidence that repeated concussions lead to long term brain damage."

In 2014 the Director of the AFL Medical Officers Association (Dr. Hugh Seward) also characterized evidence on the link between sports-related concussions and brain injury as circumstantial (Carlisle 2014).

If there was proof that repeated head trauma caused a degenerative condition such as CTE this would be a strong reason for collision sports to implement policies to prevent repeated head injuries. But uncertainty about whether a causal relationship between CTE and repeated concussion has been used to delay "major changes in the play or management of sports" (McKee et al. 2014). By avoiding acting until there is proof of causation for CTE, sports governing bodies have ignored the epidemiological evidence that justifies prudent measures to prevent repeated head injuries. In doing so, they have arguably failed to honor their professed commitment to protecting the health and safety of their players.

There are probably a number of reasons for these delaying tactics, namely, that concussion is seen as an inherent feature of many collision sports because many fans enjoy seeing fit athletes engage in bodily collisions; and that no sports administrator or fan would like to believe that repeated concussions can produce brain injuries because accepting this would require major changes in the way their sports are played. As with similarly motivated reasoning about the evidence on climate change, skepticism about the evidence for CTE has been used to justify inaction. This may be changing. In March 2016, an NFL executive was prepared to concede for the first time a "link between football and CTE" (Bieler 2016)—it remains to be seen how this change of mind will influence concussion management in collision sports around the world.

Proposed policies to manage concussion

It is common for players of collision sports to report repeated head trauma over their playing careers. An increasing number of players who have such a history have retired, or been advised to retire, because their medical advisors were concerned that additional head injuries could compromise their cognitive and mental health. Various policies have been proposed to deal with these concerns in the popular media, academic discourse, and by major sporting bodies in Australia and overseas.

Some examples of these proposals include:

◆ having concussion management in professional football codes overseen by independent doctors rather than team affiliated doctors

◆ legislating to prohibit junior athletes from returning to play after concussion without being given medical clearance

◆ the mandatory use of neuroimaging in football players who suffer a concussion

◆ genetic screening of athletes for genes implicated in the development of chronic neurological problems (e.g., *APOE4* allele) that may increase their risk of developing cognitive impairment from head injuries

◆ mandatory "sit-out" periods after concussion, or mandatory retirement after a specified number of concussions.

The effectiveness of these proposals is unknown and there has been little analysis of the ethical issues raised in implementing them.

Ethical issues in current policies to manage concussion

Several major position statements on concussion management have been published in the last 15 years by professional groups in the neurosciences (McCrory et al. 2009, 2013), neurology (Giza et al. 2013), and sports medicine (Harmon et al. 2013). Many of these guidelines on concussion management are regularly breached in sports at the professional and community levels (Price et al. 2012; Partridge 2014; White et al. 2014).

A popular recommendation, adopted by many sports organizations, is that concussed participants should be excluded from play on the day of injury—the "concussion exclusion rule" or "no same-day return-to-play" rule. The concussion exclusion rule represents a significant rule change that is specific to concussion because there are no analogous rules that prohibit players from continuing to play after receiving any other type of injury.

For example, neither the NRL nor the AFL prohibits same-day return to play for musculoskeletal injuries.

Previously in both leagues, team doctors were only encouraged not to allow concussed players to return to the field if they still had symptoms of concussion. It is clear that in the past, many players who suffered concussion continued to play for the remainder of the match even after being examined by a team doctor. Given recent concerns about the long-term implications of repeated head trauma, it is worrying that some former players have reported suffering multiple concussions in the same match on several occasions during their career (McDermott & Hichens 2012).

Efforts to improve compliance with this rule have focused on increasing awareness of the injury and educating coaching and medical staff on how to manage it. Important ethical and social issues pose significant barriers to the proper identification and management of concussion in many sports at the professional and community level (amateurs and juniors) (Partridge 2014; Partridge & Hall 2014). Lack of awareness may only be one reason for noncompliance with concussion guidelines. Conflicts of interest between doctors, players, and teams may be a more substantial obstacle to their adherence.

The decision to allow a concussed athlete to return to play is one of the most contentious in sports medicine (Kaye & McCrory 2012). Autonomy in decision-making, informed consent to participate in risky activities, coercion, and competing interests of key decision-makers all present difficulties when diagnosing and managing concussion—particularly in the case of junior athletes (Dunn et al. 2007). And there is little guidance on how key stakeholders should navigate these ethical issues.

In professional sports, the emergence of the "team doctor" (Polsky 1998) has transformed the doctor–patient relationship into a triad: doctor–patient–team (Dunn et al. 2007). A team's interest in winning may conflict with the welfare of an injured player when decisions are made about whether he is fit to play after injury. These issues are not confined to professional sport: they may present even greater difficulties in community level sport where other key stakeholders (e.g., coaches, parents) also have a duty of care toward junior athletes (Gilbert & Johnson 2011).

Severe concussion usually renders a player unable to continue playing but footballers may recover from many of the symptoms of milder cases during a match. Without a "concussion exclusion rule" these players have been allowed to stay on, or go back onto the field, and play at the risk of incurring further injury and complications. Some coaches may not want a player to continue after suffering a concussion, but there are examples of coaches allowing players to continue playing despite suspected concussion. In some cases, it may be in the team's interests of winning to prevent a potentially concussed player from being assessed (e.g., if there are no other substitute players available), if the match is very close, or if a particularly important team player is involved.

In surveys, sports physicians often report pressure from coaches (and injured players) to clear the injured players for a rapid return to play despite the risks to their welfare (Anderson & Gerrard 2005; Price et al. 2012). For example, Anderson and Gerrard (2005) interviewed sports physicians in New Zealand and found that conflicts of interest were one of the ethical issues team doctors had the most trouble managing.

Team doctors also report pressure from coaches to return concussed players to the field on the same day. The implementation of a concussion exclusion rule would reduce this pressure because it would require players to remain on the sideline for the remainder of the match after receiving a diagnosis of concussion. However, this rule could affect team doctors' clinical decisions as to whether a played has been concussed. A revealing insight into these tensions was given by a rugby league team doctor early in 2012 on the *British Journal of Sports Medicine* blog after the Australian NRL instigated a "concussion exclusion rule" (see Box 26.1) (Orchard 2012).

In this case, the team coaching staff prioritized team goals over player health by telling the doctor that if he "didn't examine the player, then the rules would allow him to continue." In practice, a team doctor could be prevented from examining a potentially concussed player if the sports trainer who initially assesses the player fails to refer him to a doctor at a coach's direction. The attitude of the coaching staff described in the blog post describes this situation.

The first point in Box 26.1 describes the tension created between the team doctors, the team, and the NRL by the concussion guideline. When the doctor says that "I am going to be pulling players out of the game who I have been comfortable letting continue for

---

Box 26.1 BJSM Blog post by a NRL team doctor

I told the coaching and training staff that the new official rule was that if I examined a player and determined that he had been concussed that day that, under the new rules, I couldn't let him return to the field and the club couldn't overrule me. However, it was quickly pointed out, if I didn't examine the player, then the rules would allow him to continue. I think everyone can see where this is heading. I am either going to be put in one of the 3 uncomfortable positions very soon:

1. That I am going to be pulling players out of the game who I have been comfortable letting continue for many years, and possibly hurting our team's chances of winning games.

2. That I am going to turn a blind eye and not examine or fully assess a player who looks as though he is fit to continue.

3. That I am going to re-name something I used to call "mild transient concussion" something different like "traumatic migraine" so the player can be allowed to continue, even though deep down I think that the player has probably had a very mild concussion that has quickly recovered.

many years," he refers to the previous policy of allowing concussed players to continue playing, if their symptoms had resolved. Under the most recent guidelines, the NRL has taken this clinical judgment out of the hands of team doctors. While this may accord with the views of some sports physicians, others may see it as an unwelcome intrusion into their clinical practice that reflects a lack of confidence in their management of their players. Concussion can sometimes be difficult to identify and there is still uncertainty about the threshold for diagnosing concussion, but the team doctor in Box 26.1 was aware that the concussion exclusion rule may "hurt our team's chances of winning" and so felt uncomfortable.

The second point raised in Box 26.1 is the worrying possibility that team doctors may avoid assessing players if a diagnosis of concussion may hurt their team's chances of winning. If this happens, then concussed players could be put at risk of further injuries—the very outcome the guidelines were intended to prevent.

The third point in Box 26.1 is that team doctors may re-label concussion as "being dazed" or "traumatic migraine" to enable a concussed player to continue playing. This runs the serious risk of delaying appropriate and timely treatment of concussion. Furthermore, because their injury would not be recorded as a concussion, the footballer would not be included in prospective epidemiological studies of how concussed players fare. Good data about the long-term consequences of concussion is essential for better management. Ironically, before the "concussion exclusion rule" there was less pressure to re-label concussion as something more benign because concussed players whose symptoms had resolved were permitted to continue (Orchard 2012).

Important questions remain about how to implement concussion management guidelines in ways that avoid nonadherence by team doctors. These questions include the following:

- What are the implications of concussion management guidelines for *autonomy* in decision-making by athletes, clinicians, parents, and coaches?
- How is decision-making about concussion affected by *coercive influences* to continue playing exerted on athletes by coaches, clinicians, parents and other stakeholders?
- How do we reconcile the *competing ethical obligations* among athletes, coaches, and clinicians when making decisions about the diagnosis and management of concussion?
- Does simply informing athletes (or parents of junior athletes) about the short- and long-term risks of suffering concussion make the athlete (or parent) *responsible for any future adverse health outcomes*?
- Is there still a moral obligation on team doctors or sports governing bodies to prescribe how concussions must be managed, and to *make paternalistic decisions for athletes* in their own best interests in some cases?
- Can athletes (or parents of junior athletes) give *full and informed consent in accepting the risks* of participating in sports with high rates of concussion when the evidence about the long-term harms of such injuries is incomplete?

- How can emerging neurotechnologies (e.g., neuro- and psychological diagnostic tests) be ethically incorporated into concussion management guidelines?

- What issues are raised by *financial links* between concussion experts (clinicians and researchers) and professional sporting bodies with high rates of concussion?

- How do *competing interests* among concussion policymakers and commercial entities affect the independence of concussion management guidelines?

Ethical, social, and policy implications of emerging neurotechnologies

Neuroscience promises to provide superior ways to diagnose and treat concussion that may be incorporated into future concussion management guidelines. This desirable goal will need to be achieved in ways that address the ethical, social, and policy impacts of proposed concussion management strategies.

Issues of conflict of interest are raised by a burgeoning market (what is now known as "Concussion Inc.") for neurogadgets, tests, devices, and other tools that claim to diagnose concussion, protect athletes from concussion, monitor the cognitive/neurological effects of concussion, and enable team doctors to make decisions about an athlete's fitness to return to play.

Over the last decade, a number of companies have marketed computerized products that they claim can assess the neuropsychological (NP) effects of concussion and assist physicians to manage these problems. These products include ImPACT, and CogState/ Axon. A series of global protocols for dealing with sports-related concussions—the Consensus Statements on Concussion in Sport (Aubry et al. 2002; McCrory et al. 2005, 2009, 2013)—have recommended that organized sports should make use of computerized NP tests to protect athletes.

In 2011, for example, the AFL and NRL adopted concussion policies that were based on the Consensus Statements. These policies made it compulsory for teams to use computerized NP tests when determining if a concussed athlete has recovered from their injury and can safely return to play. Players were required to undertake a pre-concussion assessment (e.g., at the start of a football season) so that post-concussion functioning could be compared to the athlete's "baseline" functioning. These tests were first sold to professional teams but are increasingly being used by amateur and youth participants in contact sport across the globe (often funded by sports sponsors). In Australia, the AFL and NRL both specifically mandated the use of the CogState product, and in a 2012 letter to the *Medical Journal of Australia*, the Chief Medical Officer of the NRL at the time stressed the central role that this test played: "[Concussed] players are not allowed to return to training until they are asymptomatic and their results have returned to baseline on CogState testing" (Muratore 2012).

The prominent role played by these tests in professional sports can be linked to the position statements on concussion—otherwise called the Consensus Statements—published by the self-appointed Concussion in Sport Group some of whose members have serious ethical issues of conflicts of interest.

First, the companies that market these tools, particularly ImPACT and CogState, have funded or undertaken many of the studies that have been published in the academic literature in support of the clinical utility of their products. Often these studies have been coauthored by owners or employees of the company, by researchers who have received research funding from the company, or by consultants to the company. Second, some of these researchers have been authors of the guidelines recommending the routine use of these tests. In the Consensus Statements on Concussion in Sport that have supported NP testing:

- only industry-funded studies were cited in support of the value of NP tests for concussion
- the industry funding of these studies was not disclosed in the consensus document
- some panel members have had direct links to the companies that own these NP tests but these links have not always been declared
- these documents do not cite research undertaken by independent researchers that raise doubts about the reliability and utility of these NP tests.

Some of the authors of the Consensus Statements who have links to NP testing companies have advised professional sporting leagues in the United States and Australia to adopt these tests. These leagues have subsequently mandated the use of these NP tests for all players—most notably, the NFL, the AFL, and, the NRL.

Major conflicts of interest arise when recommendations are made by a panel that includes members with financial and professional links to companies that sell computerized NP tests for concussion management. Without regulations to declare and manage conflicts of interest, it may be difficult to ascertain when special interests have improperly influenced a decision. Could we be confident that a panel without any links to the companies that market computerized NP tests would have made the same recommendations?

A major concern about their reliance on evidence from industry-linked research is that companies may simply not report studies that fail to support their products, as has happened in the past with the reporting of clinical trials by pharmaceutical companies. For example, systematic reviews that have compared neutrally sponsored drug trials with those sponsored by the pharmaceutical industry have found that the latter are much more likely to favor the company's product (Lexchin et al. 2003). Investigator bias may be even more likely when the investigators include owners and employees of the company.

Industry funding may also determine the questions that are asked and answered, often with the future marketing of the product in mind. This may bias the literature toward publication of results that are favorable to the funder's product, again as has been noted in the pharmaceutical field with selective publication of the results of clinical trials. The fact that the Consensus Statements relied on industry-sponsored evaluations of NP testing raises concerns that recommendations to use particular products have not been based on an unbiased appraisal of the evidence. The same concerns apply when sporting leagues establish their own concussion advisory groups whose members have links to computerized NP testing companies. How can sports governing bodies be sure that they received unbiased advice about the use of these products?

If independent assessment of these tests support the claims made by their designers, then no harm will have been done. But if these tests are not as useful in evaluating and managing concussion as their proponents claim, then athletes could be mistakenly diagnosed as "fit" to return to play and suffer additional concussions that produce longer-lasting forms of cognitive impairment.

In 2013, Resch and colleagues published a review of computerized NP tests for concussion including the ImPACT and CogState tools (Resch et al. 2013). They concluded that: "Although many of the studies reviewed demonstrate suboptimal reliability and validity, computerized testing is widely used in concussion management at all competitive levels … development, marketing and sales appear to have outpaced the clinical evidence base."

A high priority should be given to developing evidence-informed policies to manage concussions in sports. This should be done in ways that reduce the potential effects of conflicts of interest when those who design these tests make recommendations about their use and can by virtue of their positions also control the research that evaluates their effectiveness in reducing concussion.

Conclusion

Ethical issues can be important obstacles in the proper identification and management of concussion in sport. And yet despite the proliferation of published concussion management guidelines by professional groups over the last decade, there has been little guidance for stakeholders in how to navigate and resolve many of these issues. In the past, concussions in collision sports were often overlooked by players, coaches, team doctors, fans, and administrators. Describing a concussed player as having "gone to Disneyland" or "having their bell rung" reflects how concussion has often been spoken about in a nonserious way that downplays the potentially serious nature of the injury. The legacy of past policies and attitudes is a culture within many sports that sees concussion as a trivial injury, with participants who succumb to concussion being viewed as weak or not team players. In the past it was not uncommon for coaches, doctors, teammates, and the players themselves to continue playing after a concussion if it were physically possible—a stark contrast to the "no same-day return-to-play" policy adopted by many sports today. Despite increasing evidence that repeated concussions are associated with long-term cognitive and neurological sequelae, this legacy is not easily erased. Participants commonly encounter covert and overt forms of coercion to continue playing after a concussion at all levels of competition. What is best for the concussed player (sitting out) may not be what is best for the team (winning). In professional sports in particular, the lucrative rewards on offer mean that players, coaches, team doctors, and sports governing bodies may find that they have competing loyalties when it comes to concussion management. These conflicts of interest have often been over-looked or tolerated, and vigilance is required to ensure that conflicts of interest do not undermine the health and welfare of athletes. Sporting leagues would do well to explore managing some of these issues through the use of independent doctors who are not affiliated with particular teams.

Experts in sports medicine, neurology, and the neurosciences may also find they have conflicts of interest when it comes to concussion. The need to find better ways of identi-fying and managing concussion has presented a large commercial opportunity for clini-cians and researchers, and led to a preponderance of neurogadgets aiming to address these issues. But we need to ensure that in the pursuit of commercial opportunity, the marketing of these products does not outpace the evidence for their utility, reliability, and validity. Furthermore, there are professionally lucrative opportunities to consult for, or conduct research for, multimillion (or multibillion)-dollar sports leagues. Clinicians and researchers should ensure that they properly disclose these affiliations when publishing work in the area or treating athletes.

Acknowledgments

Sincere thanks to Sarah Yeates for her assistance in formatting this chapter.

References

Anderson, L.C. and Gerrard, D.F. (2005). Ethical issues concerning New Zealand sports doctors. *Journal of Medical Ethics*, **31**(2), 88–92.

Aubry, M., Cantu, R., Dvorak, J., et al. (2002). Summary and agreement statement of the first International Conference on Concussion in Sport, Vienna 2001. *British Journal of Sports Medicine*, **36**(1), 6–7.

Bazarian, J.J., Cernak, I., Noble-Haeusslein, L., Potolicchio, S., and Temkin, N. (2009). Long-term neurologic outcomes after traumatic brain injury. *Journal of Head Trauma Rehabilitation*, **24**(6), 439–451.

Bieler, D. (2016, March 14). In stunning admission, NFL official affirms link between football and CTE. *Washington Post*. Available from: https://www.washingtonpost.com/news/early-lead/wp/2016/03/14/in-stunning-admission-nfl-official-affirms-link-between-football-and-cte/ [accessed May 14, 2016].

Carlisle, W. (2014, June 22). Concussion games. *ABC Radio National Background Briefing*. Available from: http://www.abc.net.au/radionational/programs/backgroundbriefing/-v2/5497562#transcript [accessed May 14, 2016].

Crawley, P. (2014, March 13). Wake-up call: rugby league concussion culture "has to change" says head doctor. *The Daily Telegraph*. Available from: http://www.dailytelegraph.com.au/sport/nrl/wakeup-call-rugby-league-concussion-culture-has-to-change-says-head-doctor/story-fni3fbgz-1226852942110 [accessed May 14, 2016].

Daneshvar, D.H., Nowinski, C.J., McKee, A.C., and Cantu, R.C. (2011). The epidemiology of sport-related concussion. *Clinics in Sports Medicine*, **30**(1), 1–17.

Dunn, W.R., George, M.S., Churchill, L., and Spindler, K.P. (2007). Ethics in sports medicine. *American Journal of Sports Medicine*, **35**(5), 840–844.

Fann, J.R., Burington, B., Leonetti, A., Jaffe, K., Katon, W.J., and Thompson, R.S. (2004). Psychiatric illness following traumatic brain injury in an adult health maintenance organization population. *Archives of General Psychiatry*, **61**(1), 53–61.

Fleminger, S., Oliver, D.L., Lovestone, S., Rabe-Hesketh, S., and Giora, A. (2003). Head injury as a risk factor for Alzheimer's disease: the evidence 10 years on; a partial replication. *Journal of Neurology, Neurosurgery and Psychiatry*, **74**(7), 857–862.

Gardner, A., Iverson, G.L., and McCrory, P. (2014). Chronic traumatic encephalopathy in sport: a systematic review. *British Journal of Sports Medicine*, **48**(2), 84–90.

Gilbert, F. and Johnson, L.S.M. (2011). The impact of American tackle football-related concussion in youth athletes. *AJOB Neuroscience*, **2**(4), 48–59.

Giza, C.C., Kutcher, J.S., Ashwal, S., et al. (2013). Summary of evidence-based guideline update: evaluation and management of concussion in sports: report of the Guideline Development Subcommittee of the American Academy of Neurology. *Neurology*, **80**(24), 2250–2257.

Graves, A.B., White, E., Koepsell, T.D., et al. (1990). The association between head trauma and Alzheimer's disease. *American Journal of Epidemiology*, **131**(3), 491–501.

Guo, Z., Cupples, L.A., Kurz, A., et al. (2000). Head injury and the risk of AD in the MIRAGE study. *Neurology*, **54**(6), 1316–1323.

Guskiewicz, K.M., Marshall, S.W., Bailes, J., et al. (2005). Association between recurrent concussion and late-life cognitive impairment in retired professional football players. *Neurosurgery*, **57**(4), 719–726.

Guskiewicz, K.M., Marshall, S.W., Bailes, J., et al. (2007). Recurrent concussion and risk of depression in retired professional football players. *Medicine and Science in Sports and Exercise*, **39**(6), 903–909.

Harmon, K.G., Drezner, J.A., Gammons, M., et al. (2013). American Medical Society for Sports Medicine position statement: concussion in sport. *British Journal of Sports Medicine*, **47**(1), 15–26.

Hinton-Bayre, A.D., Geffen, G., and Friis, P. (2004). Presentation and mechanisms of concussion in professional rugby league football. *Journal of Science and Medicine in Sport*, **7**(3), 400–404.

Kaye, A.H. and McCrory, P. (2012). Does football cause brain damage? *Medical Journal of Australia*, **196**(9), 547–549.

Lane, S. (2012). *US Concussion Study May Not Apply to AFL: Doctor*. Available from: http://www.smh.com.au/afl/afl-news/us-concussion-study-may-not-apply-to-afl-doctor-20120603-1zq6h.html [accessed May 14, 2016].

Lane, S. (2016). *Concussion Problem Overblown: AFL Expert Paul McCrory*. Available from: http://www.theage.com.au/afl/afl-news/nfl-concussion-problem-overblown-afl-expert-paul-mccrory-20160406-go0ctx.html#ixzz45CM8W9VA [accessed May 14, 2016].

Lee, Y.K., Hou, S.W., Lee, C.C., Hsu, C.Y., Huang, Y.S., and Su, Y.C. (2013). Increased risk of dementia in patients with mild traumatic brain injury: a nationwide cohort study. *PloS One*, **8**(5), e62422.

Lehman, E.J., Hein, M.J., Baron, S.L., and Gersic, C.M. (2012). Neurodegenerative causes of death among retired National Football League players. *Neurology*, **79**(19), 1970–1974.

Lexchin, J., Bero, L.A., Djulbegovic, B., and Clark, O. (2003). Pharmaceutical industry sponsorship and research outcome and quality: systematic review. *BMJ (Clinical Research Ed.)*, **326**(7400), 1167–1170.

Lye, T.C. and Shores, E.A. (2000). Traumatic brain injury as a risk factor for Alzheimer's disease: a review. *Neuropsychology Review*, **10**(2), 115–129.

Makdissi, M., McCrory, P., Ugoni, A., Darby, D., and Brukner, P. (2009). A prospective study of postconcussive outcomes after return to play in Australian football. *American Journal of Sports Medicine*, **37**(5), 877–883.

McCrory, P., Johnston, K.M., Meeuwisse, W., et al. (2005). Summary and agreement statement of the 2nd International Conference on Concussion in Sport, Prague 2004. *British Journal of Sports Medicine*, **39**(4), 196–204.

McCrory, P., Meeuwisse, W., Johnston, K.M., Dvorak, J., Aubry, M., Molloy, M., and Cantu, R. (2009). Consensus Statement on Concussion in Sport: the 3rd International Conference on Concussion in Sport held in Zurich, November 2008. *British Journal of Sports Medicine*, **43**(Suppl 1), i76-i84.

McCrory, P., Meeuwisse, W.H., Aubry, M., et al. (2013). Consensus Statement on Concussion in Sport: the 4th International Conference on Concussion in Sport held in Zurich, November 2012. *British Journal of Sports Medicine*, **47**(5), 250–258.

McDermott, Q. and Hichens, C. (2012, May 14). Hard knocks. *ABC TV Four Corners*. Available from: http://www.abc.net.au/4corners/stories/2012/05/10/3499950.htm [accessed May 14, 2016].

McKee, A.C., Cantu, R.C., Nowinski, C.J., et al. (2009). Chronic traumatic encephalopathy in athletes: progressive tauopathy after repetitive head injury. *Journal of Neuropathology and Experimental Neurology*, **68**(7), 709–735.

McKee, A.C., Stein, T.D., Nowinski, C.J., et al. (2013). The spectrum of disease in chronic traumatic encephalopathy. *Brain*, **136**(1), 43–64.

McKee, A.C., Daneshvar, D.H., Alvarez, V.E., and Stein, T.D. (2014). The neuropathology of sport. *Acta Neuropathologica*, **127**(1), 29–51.

McNamee, M.J., Partridge, B., and Anderson, L. (2016). Concussion ethics and sports medicine. *Clinics in Sports Medicine*, **35**(2), 257–267.

Molgaard, C.A., Stanford, E.P., Morton, D.J., et al. (1990). Epidemiology of head trauma and neurocognitive impairment in a multi-ethnic population. *Neuroepidemiology*, **9**(5), 233–242.

Muratore, R. (2012). The need to tackle concussion in Australian football codes. *Medical Journal of Australia*, **197**(3), 146.

Nemetz, P.N., Leibson, C., Naessens, J.M., et al. (1999). Traumatic brain injury and time to onset of Alzheimer's disease: a population-based study. *American Journal of Epidemiology*, **149**(1), 32–40.

NRL (2012). *NRL Mailbox: Player Safety and Concussions*. Available from: http://www.nrl.com/nrl-mailbox-player-safety-and-concussions/tabid/10874/newsid/66477/default.aspx [accessed May 14, 2016].

Omalu, B.I., Dekosky, S.T., Minster, R.L., Kamboh, M.I., Hamilton, R.L., and Wecht, C.H. (2005). Chronic traumatic encephalopathy in a National Football League player. *Neurosurgery*, **57**(1), 128–134; discussion 128–134.

Omalu, B.I., Hamilton, R.L., Kamboh, M.I., Dekosky, S.T., and Bailes, J. (2010). Chronic traumatic encephalopathy (CTE) in a National Football League Player: Case report and emerging medicolegal practice questions. *Journal of Forensic Nursing*, **6**(1), 40–46.

Orchard, J. (2012). Concussion: how do we reconcile risk-averse policies with risk-taking sports? *BJSM Blog*. Available from: http://blogs.bmj.com/bjsm/2012/03/15/concussion-how-do-we-reconcile-risk-averse-policies-with-risk-taking-sports/ [accessed May 14, 2016].

Partridge, B. (2014). Dazed and confused: sports medicine, conflicts of interest, and concussion management. *Journal of Bioethical Inquiry*, **11**(1), 65–74.

Partridge, B. and Hall, W. (2014). Conflicts of interest in recommendations to use computerized neuropsychological tests to manage concussion in professional football codes. *Neuroethics*, **7**(1), 63–74.

Partridge, B. and Hall, W. (2015). Repeated head injuries in Australia's collision sports highlight ethical and evidential gaps in concussion management policies. *Neuroethics*, **8**(1), 39–45.

Polsky, S. (1998). Winning medicine: professional sports team doctors' conflicts of interest. *Journal of Contemporary Health Law and Policy*, **14**(2), 503–529.

Price, J., Malliaras, P., and Hudson, Z. (2012). Current practices in determining return to play following head injury in professional football in the UK. *British Journal of Sports Medicine*, **46**(14), 1000–1003.

Resch, J.E., McCrea, M.A., and Cullum, C.M. (2013). Computerized neurocognitive testing in the management of sport-related concussion: an update. *Neuropsychology Review*, **23**(4), 335–349.

Salib, E. and Hillier, V. (1997). Head injury and the risk of Alzheimer's disease: a case control study. *International Journal of Geriatric Psychiatry*, **12**(3), 363–368.

Shively, S., Scher, A.I., Perl, D.P., and Diaz-Arrastia, R. (2012). Dementia resulting from traumatic brain injury: what is the pathology? *Archives of Neurology*, **69**(10), 1245–1251.

White, P.E., Newton, J.D., Makdissi, M., et al. (2014). Knowledge about sports-related concussion: is the message getting through to coaches and trainers? *British Journal of Sports Medicine*, **48**(2), 119–124.

Chapter 27

Security threat versus aggregated truths: Ethical issues in the use of neuroscience and neurotechnology for national security

Michael N. Tennison, James Giordano, and Jonathan D. Moreno

History

Introduction

The historical relationship between the brain and military spans many cultures, continents, and centuries. From the age-old use of stimulants, depressants, intoxicants, and hallucinogens—including cannabis, coca, cocaine, and others—to today's pharmaceuticals and sophisticated brain–machine interfaces (BMIs), humankind has long studied and altered neurological functions to enhance mental and physical performance in warfare. Some of the same pharmacological and electronic neuroengineering breakthroughs that are being developed for use in medicine, including military medicine to treat neurological and psychiatric wounds of war, can also be employed within dual-use initiatives to facilitate or augment warfighter capabilities, or both. Additionally, in the United States at least, security agencies are increasingly addressing ways to leverage findings from neuroscience to enhance intelligence operations, including the acquisition, analysis, and use of diverse types and levels of information (Strategic Multilayer Assessment Group 2014, 2015; Giordano & Wurzman 2016).

Despite numerous success stories and persistently high hopes for the ways that brain science can be used in national security and intelligence operations, fundamental tensions between the goals of science and security riddle the history of US military and intelligence neuroscience research with case studies in ethics, law, and policy. Whereas national security relies on secrecy to maximize strategic, technological surprise, science is an inherently public enterprise characterized by transparency and peer review. Security agencies, moreover, focus on the ever-present immediacy of security threats, while science slowly tests and aggregates truths over time. History shows that an unfortunate result of these conflicts of values can manifest as unethical experimentation on human subjects and the deployment of technologies before being scientifically and ethically validated as safe and effective. And yet, as described by the Collingridge dilemma, the practical and ethical consequences of technological innovation may be unpredictable until widely deployed; by then, however, it may be too late to put the proverbial genie back in the bottle (Collingridge 1980).

The dual-use nature of national security neuroscience applications represents a fundamental tension in the science–security relationship. In the context of scientific research and the development and deployment of technology, scholars and policymakers use the term "dual-use" to refer to two separate, but interrelated, dichotomies. One meaning indicates that the research and/or use of a given technology may be intended to achieve either benevolent or malevolent ends, whereas the second refers to science and technology with both military and civilian uses (Evans 2014). Ongoing work, in both the United States and the European Union, addresses the historical uses of the term "dual-use" in order to clarify its meaning and implications.

Numerous science and technological developments for medical and other civilian uses have been adopted and incorporated for security and defense purposes (e.g., chemical and biological agents, among others). Moreover, under what is referred to as reverse dual-use, many innovations used and taken for granted in civilian medicine—and daily life—have trickled down from, or were developed in pursuit of, military applications (e.g., thermographic medical imaging and the Internet). With increasing momentum of the brain sciences, and growing interest in and dedication to the use of neurocognitive science for national security purposes, researchers, regardless of nationality, should be aware of both explicit dual-use research of concern (DURC), and the potential for any brain science to be viable and of value to security and defense agendas.

How should these tensions be managed and by whom? A 2003 *Nature* editorial admonished applied neuroscientists to consider the intentions of their military funders and to speak out (Editorial 2003). Does the "silence of the neuroengineers" that the editorial alleged persist? Recently, a concerned group of neuroscientists developed a document to establish a pledge to proscribe active participation in brain research that is explicitly to be used for military purposes (Bell 2014).

But can the goals that stand behind such a pledge be realized? After all, any published research could be utilized to develop information, techniques, and methods in military settings and operations. A study published in 2015 suggests that neuroscientists are far more likely to consider their colleagues' work of dual-use potential than their own (Kosal & Huang 2015). Ethical, legal, and social analysis of the science–security partnership receives scant attention in the academic world and government bureaucracy, allowing the military– and intelligence–industrial complex to churn along largely unfettered by the concerns that make national security and neuroscience a prime target for neuroethical critique. Perhaps a first question is: *How did we get here?*

Since the mid-nineteenth century, government agencies and advisory bodies have been periodically created to assess and direct science and technology policy issues as they relate to military operations and national security. In the United States, President Lincoln established the National Academy of Sciences to advise the Union on weapons purchasing, among other pressing technological issues. During World War I, the National Research Council was formed in an attempt to apply civilian science to the preparation for and waging of war. At that time, however, the military was reluctant to integrate findings from the academy. World War II marked the transition of US foreign policy from isolationism to a garrison state, a nation perennially militarized and engaged in conflict (Lasswell 1941). In 1945 the US presidential science advisor Vannevar Bush informed President Harry S. Truman that science represented an essential component of US national defense, one that the government must support, stating: "it has become clear beyond all doubt that scientific research is absolutely essential to national security" (Bush 1945). Bush went on to cite the critical advantages provided by radar and then-secret anti-rocket devices. Only a few months later, his point was bolstered by the results of the Manhattan Engineer District, popularly known as the Manhattan Project, that caused the Japanese Imperial Government to sue for peace. Implicit in Bush's description of science as "an endless frontier" was its dual-use nature, crucial for civilian as well as military purposes. Notably, in charging the former provost of the Massachusetts Institute of Technology (MIT) to prepare his report, US President Franklin Delano Roosevelt wrote in 1944 that "[n]ew frontiers of the mind are before us" (Roosevelt 1944).

Ever since the outset of the Cold War, US national security policy has called for technological superiority over all adversaries, not merely parity. In 1950, the National Security Council officially recognized the importance of the science–security partnership, declaring: "it is mandatory that in building up our strength, we enlarge upon our technical superiority by an accelerated exploitation of the scientific potential of the Unites States and our allies" (Nitze 1950). Following this recognition of the importance of innovation and technological surprise, a full-fledged military–academic complex emerged and remains an essential element of today's university funding portfolio. Reacting to the Soviet Union's launch of Sputnik, the first Earth-orbiting satellite, in 1958 President Dwight Eisenhower founded the Advanced Research Projects Agency (now known as the Defense Advanced Research Projects Agency (DARPA); one of a number of such ARPAs), tasked with the development of rockets and weather satellites among its first projects. DARPA's mission was and remains that of "keeping the United States out front when it comes to cultivating breakthrough technologies for national security rather than in a position of catching up to strategically important innovations and achievements of others" (DARPA 2016).

DARPA's most celebrated achievement, the development of the Internet, is proving useful in various BMI projects and integrating head-mounted displays to web-access for augmented reality. Since at least the early 2000s, DARPA has contracted with many university- and medical school-based neuroscientists to pursue projects of interest. According to one estimate, by fiscal year 2011 DARPA was funding at least $240 million USD in cognitive neuroscience studies (Kosal, quoted in Moreno 2012, p.53). In 2014, the presidential administration proposed to double, in fiscal year 2015, the $100 million USD federal investment in DARPA, the National Institutes of Health, and the National Science

Foundation, key participants in the federal Brain Research through Advancing Innovative Neurotechnologies (BRAIN) Initiative* (White House 2014).

Although DARPA has played a significant role in development and testing of innovative neuroscience techniques and technologies that can be translated into both military and civilian use, other US military agencies have also been, and are increasingly engaged in brain research. The Office of Naval Research (ONR) has sponsored a number of projects that have focused on enhancing warfighter performance. Through such work, ONR seeks to "[e]nhance individual and team decision-making and combat effectiveness" by optimizing biological efficiency and performance (Office of Naval Research: Science & Technology n.d. a). Additionally, the agency's Computational Neuroscience Program supports the development of neurologically inspired microcircuitry (Office of Naval Research: Science & Technology n.d. b). Replicating in silico the structural and functional bases of the algorithmic computations of the brain could enhance the scientific understanding of the relationship between brain and cognition. In addition, the US Department of Defense directly engages brain research through broad-agency announcements and contracts that solicit university-based projects. For example, in 2003, MIT received $500 million USD and Johns Hopkins University was awarded approximately $300 million USD in Department of Defense grants and contracts. Of course, not all of these efforts have proven to be fruitful. One especially colorful effort by the US Army in the 1980s involved an attempt to determine whether extrasensory perception and psychokinesis could aid in the training of "warrior monks" who could "remotely view" sites without ever having been physically present in them or seen them in images, and even cause animals to collapse at will (Druckhman & Swets 1988).

More sober efforts have been reflected in advisory reports from the National Research Council commissioned by agencies including the Army and the Defense Intelligence Agency during the early 2000s (National Research Council 2008; National Research Council (US) Committee on Opportunities in Neuroscience for Future Army Applications 2009). These reports included recommendations for the military and intelligence community to identify and pursue neurotechnologies that could be developed for operational use. This was prescient, for while a 2008 US National Academies Report, *Emerging Cognitive Neuroscience and Related Technologies*, was somewhat cautious in its view of the operational utility of brain science, subsequent reports, including a number of Pentagon white papers, have acknowledged that neuroscientific techniques and technologies have high potential for operational use in a variety of security, defense, and intelligence enterprises (Air Force Studies Board, 2008; Strategic Multilayer Assessment Group 2010, 2012a, 2012b, 2012c, 2013, 2015). These papers also advocated the need to address current and near-term ethical, legal, and social issues (ELSI) generated by such use. A subsequent report by the National Academies in 2014, *Emerging and Readily Available Technologies and National Security: A Framework for Addressing Ethical, Legal, and Societal Issues*, reflected this view, and the importance of ethical engagement (Chameau et al. 2014). At present, operationally viable products of brain science include microbiological agents, toxins, drugs, and devices (Wurzman & Giordano 2014). Certain microbiological agents, toxins, and chemicals are regulated and restricted by international policies, conventions, and treaties, such as DURC policies, and the Biological and Toxin and Weapons Convention (BTWC), and Chemical Weapons Convention (CWC). Therefore, this chapter will focus upon research and use of drugs and devices that are not constrained by these regulations, and which have been, and remain viable for use in defense and security initiatives on an international scale.

Enhancement/therapy

Drugs

Warfighters have long used myriad substances both to fortify performance of military tasks and to cope with the horrors of war. Alertness, wakefulness, and focus—key decision-making capacities of a warfighter—have been enhanced for centuries. An ephedrine-containing herb stimulated the senses of guards on China's Great Wall, just as coca leaves did for Incan fighters. Bavarian soldiers received cocaine from their officers, and amphetamines have been used by warfighters since World War II (Moreno 2012). Referred to as "go pills," amphetamines came to the attention of the public in 2003 when US Air National Guard pilots, allegedly enhanced by go pills, accidentally bombed Canadian troops in Afghanistan (Shanker & Duenwald 2003).

Warfighters have even used hallucinogens and intoxicating combinations of psychoactive herbs to enhance their combat effectiveness, or at least the appearance of ferocity. Turks reportedly used opium to enhance wartime bravery in the 1500s (Lewin 1998). Consumption of *Amanita muscaria*, a psychoactive and hallucinogenic mushroom, may have facilitated the "berserker" rage characteristic of Viking raids, though scholars remain divided on the veracity of this claim (Fabing 1956; Høyersten 2015). South African tribal warriors smoked dagga, a type of cannabis, in combination with the consumption of other herbs to enhance fearlessness and insensitivity to pain (Kamienski 2016, pp.29–30).

Not only have warfighters used substances to enhance capacities to engage in combat, the history of warfare is rich with examples of warriors using substances to disengage from combat. Homer's *Odyssey* famously described the consumption of nepenthe by warriors who quickly forgot their wartime sorrows (Kamienski 2016, pp.29–30). During the US Civil War, the term "soldier's disease" became synonymous with opiate addiction, likely as a result of the widespread use of opiate analgesics among warfighters (Miller 2013). Deployment to Vietnam correlated with an uptick in warfighter narcotic use and addiction, at least in part used for escapism (Robins et al. 1975). More recently, US warfighters have used "no go pills" to calm the mind and induce rest in preparation for, or in recovery from, combat (Golub & Bennett 2013). These interventions foreshadow ongoing research on drugs like propranolol, which could ultimately enable warfighters to disengage from combat without having formed traumatic memories (Donovan 2010).

Devices

The history of neuromodulation via electricity and magnetism also dates back centuries, if not millennia. Scribonius Largus, an ancient Roman physician, wrote the earliest known account of neurostimulation (Fregni & Pascual-Leone 2007). As a headache remedy he described the application of an electric fish to the scalp. Much more recently, eighteenth-century scientists demonstrated the therapeutic potential of transcranial electric current and the electrical stimulation of muscle contractions (Parent 2004). For the following two centuries, researchers attempted to treat various mental conditions with electric current, but their success varied. In the meantime, other electrical stimulation techniques gained traction. Scientists successfully used electroconvulsive therapy (ECT) to treat depression in the 1930s (UK ECT Review Group 2003). Another technique, deep brain stimulation (DBS), emerged in the 1980s as a treatment for Parkinson's and other movement disorders. Delivering pulses of electricity, DBS can activate or block action potentials in specific brain regions (Coffey 2008).

As the twentieth century came to a close, researchers rediscovered the potential of applying low-level electrical current through the skull to affect the brain and its function. Types of transcranial electrical stimulation (tES), including transcranial direct current stimulation (tDCS), have been used to modulate cortical excitability (Nitsche & Paulus 2000; Priori et al. 2008). In contrast to DBS, tDCS does not stimulate neurons by forcing or blocking their action potentials; rather, it modulates neurons by increasing or decreasing their threshold to fire. Studies have focused on the effects of tDCS on neuroplasticity as well as the neurological substrates of cognition and motor activity. Transcranial direct current stimulation has been shown to be safe, but the current understanding of its efficacy for enhancement is incomplete (Horvath et al. 2015). Some studies suggest that "tDCS can enhance cognitive processes occurring in targeted brain areas," but other scientists have failed to replicate this finding (Ukueberuwa & Wassermann 2010; Horvath et al. 2015). Recent analyses reveal that context matters, and the type(s) and extent of effects that can be elicited by tES strongly depend upon setting, and the neurocognitive state of the subject. Recreational tDCS devices are available on the consumer market and both clinical and direct-to-consumer tES technologies are of growing interest and potential utility to the military (Foc.us 2015; Kappenmann et al. 2016).

Magnetism is also used to manipulate neurological functions. Technologies include magnetic seizure therapy and transcranial magnetic stimulation (TMS). Approved to treat major depression, TMS may have additional applications for enhancement (Hamilton et al. 2011). In 2009, the US National Research Council identified TMS as a wakefulness enhancement for the US Army (Committee on Opportunities in Neuroscience for Future Army Applications, National Research Council of the National Academies 2009). Similarly, DARPA and the US Army funded studies of wearable, helmet-borne devices to affect neurological function through the delivery of patterned ultrasound pulses (Tyler 2010).

Brain–machine interfaces (BMIs, also known as brain–computer interfaces or BCIs) constitute another major area of military neurological device research. BMIs either translate neurological signals into inputs for computers or machines, or vice versa. BMIs have potential for therapeutic breakthroughs in civilian and military medicine. They also have a rich—though short—history of security applications.

In 2002, *Nature* published results of a DARPA study that created the "roborat," a remote-controlled rodent that utilized BMI technology. Scientists implanted electrodes in the rat's brain and in real time controlled its locomotion through complex mazes (Talwar et al. 2002). Scientists equivocate on whether such research actually advances the aims of medical prostheses, but DARPA's interest was different: the roborat could be sent into dangerous scenarios to detect mines, clear bombs, or identify survivors among rubble. More recently, scientists have created "cyborg beetles" whose flight characteristics can be remotely controlled (Sato et al. 2015). Such studies also raise ethical issues about the use and purpose of animals in such research. There is a nuanced general consensus that the public interest in animal studies performed for biomedical research may outweigh animal autonomy, but such sentiments are less clear when biomedical research is directed toward furthering military goals (Quigley 2007).

Other national security-funded animal BMI studies are more directly oriented toward therapeutic objectives. Around the same time as the roborat studies, scientists equipped monkeys with BMIs attached to robotic arms and successfully trained the primates to articulate and control the prostheses using neurological output from their brains (Lebedev 2006). This research could revolutionize prosthetics, both for warfighters and civilians. Since then, BMIs have also facilitated human neurological control of a mouse cursor, which could allow paraplegic patients to regain control and physical autonomy in their lives (Simeral et al. 2011). Current DARPA research seeks to complete the feedback loop between the brain and prostheses, not only granting the brain direct control over a robotic limb but also returning tactile feedback, such as pressure and temperature, from sensors in the prosthesis to the user's brain (DARPA 2015).

An early twenty-first century DARPA program, AugCog (short for "augmented cognition"), sought to fully integrate, at the neurological level, a warfighter's cognitive capacities and sensory perceptions with his or her combat vehicle environment (Cummings 2010). As computers monitor working memory, attention, executive function, and sensory input, the warfighter would be prompted in real time with information about cognitive load. This kind of biofeedback could help a warfighter manage his or her neurological resources; it might even bring into conscious awareness one's own subconscious recognition of danger, prompting the warfighter's attention to threats before they would have been identified naturally (Szondy 2012). Though in name AugCog no longer exists, similar research continues. Giordano and Dando provide current overviews of neuroscience and technology that is being studied, considered, developed, and used in national security, intelligence, and defense (Giordano 2014c; Dando 2015).

Ethical issues in historical context

The bioethics literature on optimizing human neurological function and performance tends to focus on cognitive enhancement via the use of drugs and devices and the implications for performance in school and in the workplace settings. Warfighter enhancement, however, can manifest more grave, if not lethal stakes and may entail additional enhancement modalities, including physical adaptions and modifications through wearable or implantable technologies, such as neurofeedback-equipped helmets and biointegrated BMIs, respectively. The ethics of civilian enhancement may be extrapolated and extended to address the implications of warfighter enhancement. When approaching these ethical issues, it becomes important to evaluate the actual capabilities and limitations of the technique(s) or technologies at hand, to assess if, and how any augmentation of function is imparted, and to evaluate the impact that any such effects will have (Shook & Giordano 2014; Shook et al. 2014; Giordano 2015). On a fundamental level, human enhancement raises ethical questions in at least two domains: (1) Does enhancement inherently transgress fundamental moral boundaries? (2) Do the known and unknown risks of enhancement interventions render them unethical?

In 2003, the then US President's Council on Bioethics released a report on the ethical issues associated with using ostensibly medical interventions to go beyond therapy into the realm of human enhancement (The President's Council on Bioethics & Kass 2003). The Council pointed out that enhancements could undermine human nature by obviating the natural connection between actions and accomplishments, which underlies human dignity. Enhancement may also represent a deeper transgression against what some view as the natural, given world: it could be a hubristic attempt to "play God" or master nature. These intuitions flag fundamental moral concerns about the normative implications of altering one's relationship to the world.

Assuming that performance enhancement is not intrinsically unethical, additional issues arise related to the risks of enhancement. Enhancing performance in a competitive endeavor is a zero-sum game. Performance optimization increases one's chance of winning, thereby directly decreasing one's opponent's chances. This underlies the sense of unfairness that, for example, gives rise to stringent drug testing in professional sports. And yet, athletes may feel coerced to enhance in order to remain competitive, risking not only getting caught but also potential short- and long-term health consequences as well as conceding a robust sense of autonomy. The military, however, does not necessarily seek a fair fight when entering combat. The entire purpose of the science–security partnership is to gain a technological advantage over the enemy, thereby enabling the accomplishment of an objective while minimizing loss on both sides. But here derivative ethical issues arise, including effects of enhancement on the health and status of an individual (and what occurs if such enhancements are discontinued), and the possibility of escalating development and use of neuroenhancements (and other applications of neuroscience and neurotechnology) within and for military operations (Giordano 2014a, 2016a; McCreight 2014; Shook & Giordano in press).

Current and future use of neurocognitive science in security operations

Although discussion thus far has focused upon efforts of the United States military and Department of Defense to operationalize or engage dual-use neuroscience, there is ongoing investment of academic, commercial, and governmental resources in developing and testing tools and techniques of the brain sciences in and for security and defense agendas around the world. As shown in Table 27.1, these include (but are not limited to) Australia, Canada, China, consolidated efforts of the North Atlantic Treaty Organization (NATO) member nations, India, and Russia.

In the main, brain science is primarily being used in international military and security contexts:

♦ for medical purposes

♦ to optimize operational training and performance in military and intelligence personnel

♦ to study (and perhaps develop) nonlethal and lethal weapons (e.g., chemical and microbial agents and toxins, and neurotechnologic devices) (Giordano 2014b).

The 2014 report of the National Academies asserted that the research, development, and use of brain science in international military and security scenarios represent a significant and growing concern. In the United States and most Western nations, governmentally funded neuroscience programs adhere to DURC policies, in keeping with the general constructs of the BTWC and CWC. But such control can also create a dilemma: on the one hand, it creates parameters for the conduct of brain science in participatory states; on the other, it can create opportunities for other nations or even nonstate actors to take advantage of these constraints to gain a competitive edge toward attaining power. What's more, international policies and treaties don't guarantee cooperation, and studies and applications of brain science don't have to be covert. As previously noted, the current BTWC and CWC do not restrict pharmaceutical formulations of neurotropic drugs for medical use, or neurotechnologies (e.g., neurostimulatory or modulatory devices); exemptions for biomedical experimental purposes and/or shields of commercial proprietary interests and intellectual property can subvert inquiry into the dual-use or military applications of brain science (Giordano 2016).

Addressing ethical issues

Military medicine can generate particular ethical circumstances and implications (e.g., conflicting military and medical duties; exigencies of battlefield conditions). However, brain research and the application of new techniques and technologies in military medicine evoke many of the same ethical issues, questions, and steps toward resolution as in civilian contexts. Also, several fundamental questions and issues about the type(s), extent, limit(s) and effects of neurological assessment and augmentation of military personnel can be viewed and addressed in much the same ways that these approaches are dealt with in civilian settings. There is continued discussion about the potentially coercive nature and non-consented conduct of biomedical research in the military, discussions that stem in large part from awareness of past ethical transgressions.

But, as matter of fact, illustration of such transgressions both in the military, and more widely in biomedical research, have fortified the stringency of responsible practices in the conduct of research in—and for—the military, at least in the United States, Canada,

Table 27.1 Representative international research programs in neuroscience and neurotechnology for national security

Country	Major research institutions and funding resources	Research themes	Example research projects and initiatives
Australia	Defense Science and Technology Group (previously DSTO, now DSTG) Land Human Services branch leads group on Cognition and Behavior Centre for Cognitive Work and Safety Analysis Office of the Chief Defense Scientist Chemical Warfare Agent Laboratory Network Defense Materials Technology Centre	Transcranial direct-current stimulation (tDCS) Artificial intelligence (AI) and machine cognition Cognitive enhancement/ performance Sleep deprivation studies	◆ Evaluating the index of cognitive activity as a measure of mental workload. ◆ Transferring training for battle management systems: From simulators to the real world. ◆ Evaluating longitudinal impacts of combat exposure: Neurocognitive, psychological and biological outcomes. ◆ Analyzing sleep characteristics of a small Australian Army population and the effect of sleep deprivation on their simulated driving performance. ◆ Defining usability of command and control technology while accumulating sleep loss. ◆ Reducing psychological dependency on ideological groups. ◆ Analyzing cognitive work beyond human factors and engineering: Application to military doctrine and strategy development. ◆ Evaluating non-invasive brain stimulation: Opportunities and implications for the Australian Defense Force (ADF). ◆ Determining the importance of cognitive neuroengineering and computational neuroscience for defense.
Canada	National Research Council Canada Biotechnology Research Institute (NRC-BRI) Defense Research and Development Canada (DRDC), particularly the Toronto, Valcartier and Suffield locations Defense and Security Research Institute Director General Military Personnel Research & Analysis	Sleep and circadian cycles Working memory Blood brain barrier-related topics Human factors, decision-making, and team/group behavior Performance enhancement Pharmacokinetics PTSD	◆ Novel ligands of prothrombin as inhibitors of blood and extra-vascular coagulation. ◆ Human factors engineering (HFE) support to Canadian soldier systems. ◆ Department of National Defense (DND) and SickKids working together to better diagnose post-traumatic stress disorder (PTSD) and mild traumatic brain injury (mTBI). ◆ The future Canadian soldier and enhancement of human performance: Research meets policy. ◆ Inflammatory cytokine and chemokine profiles are associated with patient outcome and the hyperadrenergic state following acute brain injury.

			◆ Hyperbaric stress in divers and non-divers: Neuroendocrine and psychomotor responses.
			◆ Thromboelastographic study of psychophysiological stress: A review.
			◆ Do framing effects reveal irrational choice?
China	National Natural Science Foundation of China	"Bio-chips" and biotechnology	◆ Determining how perceptual learning modifies the functional specializations of visual cortical areas.
	Ministry of Science and Technology (MOST)	Trauma	◆ Evaluating minocycline ameliorates hypoxia-induced blood-brain barrier damage by inhibition of HIF-1α through SIRT-3/PHD-2 degradation pathway.
	Institute of Neuroscience (ION) of the Chinese Academy of the Sciences (CAS)	Neurodegeneration	◆ Creating CGCG clinical practice guidelines for the management of adult diffuse gliomas.
	Chinese Society for Neuroscience	Tumor biology	◆ Evaluating meCP2 plays an analgesic role in pain transmission through regulating CREB/miR-132 pathway.
	Second Military Medical University	Pain and analgesia	◆ Learning to memorize: Shedding new light on prefrontal functions.
	Third Military Medical University	Drug abuse and addiction	◆ Analyzing medial prefrontal activity during delay period contributes to learning of a working memory task.
	Fourth Military Medical University in Xi'an		◆ Evaluating high-angular diffusion MRI in reward-based psychiatric disorders.
	Institute of Neurosciences		◆ Determining if PKMζ might be a potential target for the treatment of aging-related cognitive impairment, suggesting a potential therapeutic avenue.
	Zhujiang Hospital, Institute of Neuromedicine		◆ Analyzing functions and mechanisms of microglia/macrophages in neuroinflammation and neurogenesis during stroke.
	Beijing Society for Neuroscience		
	Neuroscience Research Institute, Peking University		
	IDG/McGovern Institute for Brain Research at Peking University		
	Beijing Normal University, National Key Laboratory of Cognitive Neuroscience and Learning		
	East China Normal University – School of Psychology and Cognitive Science		
	The Translational Neuroscience Center of West China Hospital of Sichuan University		

(continued)

Table 27.1 Continued

Country	Major research institutions and funding resources	Research themes	Example research projects and initiatives
Germany	Defense laboratories: Bundeswehr Institutes—part of a "Medical Academy" network Institute of Microbiology (InstMikroBioBw) Center for Aerospace Medicine of the Airforce German Society for Cognitive Science	Biological agent research PTSD Cognitive performance optimization	◆ Developing methods and measures to identify relevant characteristic parameters in military workplaces with high cognitive demands. ◆ Researching the pathophysiology of mild form of neurological decompression sickness. ◆ Analyzing spatial and temporal phylogenesis for the verification of biological agents using tick-borne encephalitis virus. ◆ Developing new therapies for nerve agent poisoning: From in vitro models to clinical use. ◆ Evaluating brain activity in soldiers with combat-related PTSD. ◆ Detecting molecular markers in the cerebellum after traumatic brain injuries (TBI).
India	Defense Research and Development Organisation (DRDO) Defense Institute of Psychological Research DIPAS –Defense Institute of Physiology and Allied Sciences DRDE –Defense Research and Development Establishment Institute of Nuclear Medicine and Allied Sciences (INMAS)	Effects of high altitude Effects of environmental stress Radiation neurobiology Biological/chemical agent research	◆ Differentiating neural activation for camouflage detection task in field-independent and field-dependent individuals: Evidence from fMRI ◆ Analyzing the neuroprotective role of L-NG-nitroarginine methyl ester (L-NAME) against chronic hypobaric hypoxia with crowding stress (CHC) induced depression-like behaviour. ◆ Analyzing the neuroprotective role of intermittent hypobaric hypoxia in unpredictable chronic mild stress-induced depression in rats. ◆ Evaluating the elevated pulmonary artery pressure and brain natriuretic peptide in high-altitude pulmonary edema-susceptible non-mountaineers. ◆ Evaluating neural activation patterns in self deceivers. ◆ Comparing (11)C-methionine and (18)F-fluorodeoxyglucose positron emission tomography-computed tomography scans in evaluation of patients with recurrent brain tumors. ◆ Identifying critical path in strategic domains using fuzzy cognitive maps.

Israel	Israel Institute for Biological Research (IIBR) Subsidiary: Life Science Research Israel Israel Ministry of Defense, Directorate of Defense R&D Affiliates Israel National Nanotechnology Initiative Ministry of Science, Technology, and Space Ben-Gurion University of the Negev, Zlotowski Center for Neuroscience	Therapeutics for CNS/PNS disorders Nanotechnology Neurodegeneration PTSD Medical diagnostic techniques Protein and enzyme synthesis and engineering Vaccines and pharmaceuticals	◆ Evaluating rapid brain MRI based on temporal sparsity exploitation. ◆ Analyzing novel bifunctional hybrid small molecule scavengers for mitigating nerve agent toxicity. ◆ Evaluating solvent effects on the reactions of the nerve agent VX with KF/Al2O3: Heterogeneous or homogeneous decontamination? ◆ Analyzing the attenuation of sarin-induced brain damage in rats by delayed administration of midazolam. ◆ Determining if rivastigmine is a safe pretreatment against nerve agents poisoning? A pharmacological, physiological and cognitive assessment in healthy young adult volunteers. ◆ Acute and long-term ocular effects of acrolein vapor on the eyes and potential therapies ◆ Improving functional recovery after experimental traumatic brain injury via inosine.
North Korea	Pyongyang Bio-Institute	Anthrax (hypothesized) Biological agents, neurotoxins	N/A
Russia	Russian Foundation for Advanced Research Projects Laboratory of Neurotechnology Perception and Recognition with focus areas Russian Academy of the Sciences Institute of Higher Nervous Activity 30th Central Scientific Research Institute, Ministry of Defense State Research Center of Virology and Biotechnology (VECTOR)	Neurotechnology Integrated biosystems Memory, perception, and recognition Public health and safety Neurotrauma	◆ Evaluating whether intracortical microinjections may cause spreading depression and suppress absence seizures. ◆ Analyzing a Bayesian classifier for brain–computer interface based on mental representation of movements. ◆ Activating brain structures by fMRI data when viewing the video clips and recall of shown actions. ◆ Determining whether two-faced nitric oxide is necessary for both erasure and consolidation of memory. ◆ Evaluating the influence of the working memory load on the spatial synchronization of prestimulus cortical electrical activity during recognition of facial expression. ◆ Analyzing the mechanisms of orientation sensitivity of human vision system. Part II: neural patterns of early processing of information about line orientation. ◆ Evaluating stress reactivity and stress-resilience in the pathogenesis of depressive disorders: Involvement of epigenetic mechanisms.

(continued)

Table 27.1 Continued

Country	Major research institutions and funding resources	Research themes	Example research projects and initiatives
United Kingdom	Defense Science and Technology Laboratory Porton Down Centre for Defense Enterprise University of Kent (academic partnership)	Next-generation neuro-pharmacology including nootropic agents Transcranial stimulation (DC and magnetic) Synthetic biology Machine-augmented cognition Neuro-prostheses Also interests in: neurotoxins, virtual reality, trauma, social and behavioral influence	◆ Analyzing R-Cloud human capability SOR—From safety net to augmented cognition: Using flexible autonomy levels for on-line cognitive assistance and automation. ◆ Initiatives on: • Adaptive technologies • Human factors • Understanding and influencing human behavior • Human performance and protection • Evaluating the cognitive systems project

United Kingdom, and European Union (Gross & Carrick 2013). There are still uncertainties regarding responsible conduct of research in those countries that employ differing codes of ethics, and what this portends both for research subjects and for the outcomes of such research and the capabilities of power incurred. Further questions arise if and when considering the use of brain science in settings and situations that are explicitly relevant to warfare, such as warfighter augmentation and the development and use of neuroweapons, and intelligence operations.

Casebeer has proposed a normative framework for neuroethical address of issues fostered by neuroscience research and its translation, which although primarily oriented toward military and civilian medicine, can be applied, and is nonetheless useful to national security, intelligence, and defense agendas (Casebeer 2014). This has been expanded to engage neuroethicolegal and social risk-assessment and mitigation paradigm, in which defined steps, queries, and framing constructs are employed to determine if and how specific uses of neuroscience and neurotechnologies give rise to ethical issues (Giordano et al. 2014; Giordano 2015). Critical to this approach is consideration of contexts of application. An undergirding question is whether neuroethical issues of brain science in contexts of national security and defense are best addressed in accordance with military ethics, ethics relevant to enterprises in the public domain, or some other system or ethical toolkit. It could be argued that ethics are ethics, but ethics provide systematic analysis and articulation of moral actions relevant to the specific goals, values, and tasks defined by and within communities of use even though the underlying principles are the same. This "telic" approach is not an attempt at veiled consequentialism, or mere ethical relativism, but rather, is ethical realism. It asks: What are the ends (i.e., the *telos*) of the profession, and how is the good to be attained and upheld in both end(s) and the means entailed toward their achievement?

On this realist view, the professional ethics of the military should define the focus, scope, and conduct of any and all its constituent scientific and technological enterprises. Just war theory, *jus ad bellum*, may substantiate using neuroscience and neurotechnology (or any science and technology) in national security, intelligence, and defense, in accordance with ethical precepts that define the need for aggressive actions. The premise is that war is a horror to be abhorred and avoided, but realistically may sometimes be necessary. Defined criteria attempt to discern if and how such conflict might be prevented, restrained, and made more humane.

In this latter regard, precepts for fair conduct of conflict (i.e., *jus in bello*) might define whether and ways that neuroscience and technologies may be employed within warfare, or to prevent warfare. But particular attention should be paid to the precept of *no means malum in se*, which proscribes the use of methods and weapons that violate some consensus construct of harm (e.g., chemical weapons), or that may be uncontrollable (e.g., biological and nuclear weapons), or both of these. This remains a gray zone: as noted previously, while some domains of brain science fall within the purview of the current BTWC, CWC, and DURC policy, others such as neurotechnologies do not. Thus, as brain science both continues to advance and to be assessed for use within national defense agendas, it will be important to address (1) whether extant and new neuroscientific techniques and technologies can incur harms relative to considerations of *means malum in se*, and (2) whether neurotechnologies fall under mandate constraints of biological weaponry.

An underlying issue is the ethics of power, specifically power exercised to preserve communal values and a way of life. The mission of national security and defense is to protect the ideals, objectives, and integrity of a state. Like any cutting-edge tool, brain science can afford an actor's defined advantage in leveraging relative balances of power in deterrence and influence. Furthermore, although in a free state the values of the polis are reflected in its politics, totalitarian regimes often inflict and enforce values of a power estate upon the polis. This gives rise to questions of relative good: what one state deems to be right and just may not necessarily be the same for other states. Formal, contemporary constructs of *jus ad bellum* define grave, public evil such as "aggression or massive violation of the basic human rights of ... populations" as substantive grounds to justify hostile action (i.e., war). If this is the case, should neuroscientific techniques and technologies be part of the armamentarium?

In light of these considerations, it is important to regard "the good" in a variety of dimensions, as security and defense operations represent state-level enterprises that are executed and exert influence in an international milieu. Such a view moves beyond the somewhat narrow confines of military ethics, and brings neuroethical and legal consideration of brain science in national defense agendas into a broader social context. Consistent with core precepts of other international deliberations upon the use of various implements in military and defense operations, such consideration would need to evaluate the ways that brain science should or should not be studied, developed, and employed. In many ways, these considerations reflect more general concerns about both unknown, unanticipated, and possibly uncontrollable effects of nascent science and technology, and the sociocultural manifestations and responses generated by the use or misuse of such techniques and technologies (Giordano & Benedikter 2012; Giordano 2014b). Key questions include whether the use of certain neuroscientific and technological approaches incurs greater or lesser risks and harms than other methods of intelligence, security, and defense, and what limits should be applied to any possible development and use of brain science in defense initiatives (Tractenberg et al. 2014).

Based upon the activities and results of other international conventions (e.g., the BTWC, CWC, and Geneva Convention) that have sought to govern military methods and weapons, we have speculated that such deliberation could lead to two possible outcomes. The first is a move to ban or restrict any research or use of brain science that is applicable to military operations. However, this would be difficult, if not impossible, in that existing brain research can be used for military purposes, and, as we have previously noted in this chapter, many countries already have directly subsidized military and defense enterprises of brain research. Thus, unless a ban was universal, restricted research and development by some could afford opportunities for others to exploit scientific and technological balances of power (Forsythe & Giordano 2011). The second, and we believe more plausible, possibility is to work to establish realistic criteria for development and use of specific types and extents of neuroscience and neurotechnological approaches within military and defense operations in accordance with strictly defined and implemented ethicolegal parameters, which would then require surveillance and enforcement on a variety of scales and levels.

Still, questions arise as to whether these criteria should be based upon or grounded to a particular philosophy, military ethics international law, or some other extant, new, or combined approach (Ferguson 2000; Forsythe & Giordano 2011; Benedikter & Giordano 2012; Dando 2012; Lanzilao et al. 2013; Abney et al. 2014; Farwell 2014; Shook & Giordano 2014). To be sure, any such deliberations must be articulated by dedicated groups of multidisciplinary professionals, from government and civilian sectors, with experience and expertise that is essential to making ethical decisions about the use, constraints, and outcomes of brain science in national security and defense initiatives on the global stage (Dando 2007; White 2008; Moreno 2012; Casebeer 2013). We have posited that these individuals and groups must be task-agile, scientifically and situationally knowledgeable, and ethically responsible, and have proposed methods for training and executing the process. Case-based analysis informed partly by historical information as presented in this chapter and elsewhere, provides a basis from which to assess the potential effects of current and emerging developments in brain science that can be employed in national security operations (Moreno 2012; Tabery 2014; Tractenberg et al. 2014). But ethical oversight is not simply a matter of retrospection; rather it must be forward-looking, descriptive, predictive, and not simply proscriptive, but rather preparatory for contingencies and exigencies that can occur as brain science and technology and global politics and military operations evolve. Moreover, such ethical engagement should not be a merely academic exercise: it must be conducted by groups that have credibility and capability to inform and influence formulation of international policies, treaties, and laws.

Infrastructure: Building for the challenges and opportunities ahead

This conclusion speaks to the need to develop infrastructure(s) capable of the tasks at hand, and to come. The 2014, the US National Research Council report on *Emerging and Readily Available Technologies and National Security* addressed ELSI relative to government agencies' work in disruptive technologies of potential interest to both state and nonstate actors. The committee's recommendations included a five-step process: initial screening of proposed research and development; further review of proposals that raise ELSI concerns; project monitoring and midcourse corrections as needed; public engagement; and periodic review of ELSI processes within an agency (Chameau et al. 2014).

But which agency, agencies, or organizations should be charged with these duties, how will their constituencies be decided, and what level and extent of interagency and public discourse can and should be engaged? Differing groups may have distinct views and goals, and, as with any approach to national defense, issues of security, operational readiness, and power will need to be evaluated and weighed in light of global humanitarian concerns. Axiomatically, defense and security operations require that some information remains classified, and so public discussion, while necessary, must be carefully engaged (Giordano et al. 2010). We believe that it is vital to inform the public about the reality and growing potential for brain science to be used in security, intelligence, and defense operations, so as to foster broad social awareness.

Indeed, it is the social impact of the use and misuse of brain science in military and security operations that gives rise to ethical and legal issues. Thus, we do not see military and civilian silos of ethicolegal deliberation and guidance as being wholly separate. Current intramural efforts at various national agencies dedicated to examining current and future ELSI are commendable. And while it is important to elucidate the ethical issues that arise in and from such research and its use, ethics alone are not sufficient. Ethics must inform and lead to the formulation of policies and regulations that guide and govern what aspects of brain science are studied and employed in these contexts (Dando 2007; Moreno 2012). The growing potential for the use of novel drugs and devices of brain science will prompt re-examination and revision of current categorizations, caveats, and constraints. As we have stated in prior work, it is undeniable that neuroscience is, and will be, employed in security and defense agendas (Moreno 2012; Wurzman & Giordano 2014). Our hope is that the knowledge and capabilities it confers will be used to prevent or mitigate violence, reduce conflict, and foster peace.

Acknowledgments

The authors gratefully acknowledge the research and editorial assistance of Celeste Chen and Kira Becker, and the ongoing collaboration of the Strategic Multilayer Assessment Group of the Joint Staff of the Pentagon. Aspects of this this work were supported in part by funding from the Lawrence Livermore National Laboratory (JG) and Office of Naval Research (JG).

Disclaimer

The work presented in this chapter is solely that of the authors and does not necessarily represent the perspectives of the United States Department of Defense, Defense Advanced Research Projects Agency (DARPA), or Strategic Multilayer Assessment Group of the Joint Staff of the Pentagon.

References

Abney, K., Lin, P., and Mehlman, M. (2014). Military neuroenhancement and risk assessment. In: Giordano, J. (Ed.) *Neurotechnology in National Security and Defense: Practical Considerations, Neuroethical Concerns*, pp.239–248. Boca Raton, FL: CRC Press.

Air Force Studies Board (2008). *Emerging Cognitive Neuroscience and Related Technologies*. Washington, DC: National Academies Press.

Bell, C. (2014). Why neuroscientists should take the pledge: a collective approach to the misuse of neuroscience. In: Giordano, J. (Ed.) *Neurotechnology in National Security and Defense: Practical Considerations, Neuroethical Concerns*, pp.227–238. Boca Raton, FL: CRC Press.

Benedikter, R. and **Giordano, J.** (2012). Neurotechnology: new frontiers for European policy. *Pan European Networks: Science & Technology*, **3**, 204–207.

Bush, V. (1945). Science: the endless frontier. *Transactions of the Kansas Academy of Science (1903–)*, **48**(3), 231–264.

Casebeer, W.D. (2013). *Plenary Testimonial Presented to US Presidential Commission for the Study of Bioethical Issues.* Available from: https://bioethicsarchive.georgetown.edu/pcsbi/node/2779.html [accessed February 19, 2017].

Casebeer, W.D. (2014). A neuroscience and national security normative framework for the twenty first century. In: Giordano, J. (Ed.) *Neurotechnology in National Security and Defense: Practical Considerations, Neuroethical Concerns*, pp.279–284. Boca Raton, FL: CRC Press.

Chameau, J.-L, Balhaus, W.F., and **Lin, H.S.** (2014). *Emerging and Readily Available Technologies and National Security: A Framework for Addressing Ethical, Legal, and Societal Issues.* Washington, DC: National Academies Press.

Coffey, RJ. (2008). Deep brain stimulation devices: a brief technical history and review. *Artificial Organs*, **33**(3), 208–220.

Collingridge, D. (1980). *The Social Control of Technology.* New York: St. Martin's Press.

Committee on Opportunities in Neuroscience for Future Army Applications, National Research Council of the National Academies (2009). *Opportunities in Neuroscience for Future Army Applications.* Washington, DC: National Academies Press.

Cummings, M.L. (2010). Technology impedances to augmented cognition. *Ergonomics in Design*, **18**(2), 25–27.

Dando, M. (2007). *Preventing the Future Military Misuse of Neuroscience.* New York: Palgrave-Macmillan.

Dando, M. (2012). *Bioterror and Biowarfare: A Beginner's Guide.* Oxford: OneWorld Publications.

Dando, M. (2015). *Neuroscience and the Future of Chemical-Biological Weapons.* New York: Palgrave-Macmillan.

DARPA (2015, April 23). *HAPTIX Starts Work to Provide Prosthetic Hands with Sense of Touch.* Available from: http://www.darpa.mil/news-events/2015-04-23 [accessed January 4, 2016] [accessed January 4, 2016].

DARPA (2016). *Where the Future Becomes Now.* Available from: http://www.darpa.mil/about-us/timeline/where-the-future-becomes-now [accessed January 4, 2016].

Donovan, E. (2010). Propranolol use in the prevention and treatment of posttraumatic stress disorder in military veterans: forgetting therapy revisited. *Perspectives in Biology and Medicine*, **53**(1), 61–74.

Druckman, D. and **Swets, J.A.** (Eds.) (1988). *Enhancing Human Performance: Issues, Theories, and Techniques.* Washington, DC: National Academies Press.

Editorial (2003). Silence of the neuroengineers. *Nature*, **423**(6942), 787. Available from: http://www.nature.com/nature/journal/v423/n6942/full/423787b.html [accessed January 11, 2016].

Evans, NG. (2014). Dual-use decision making: relational and positional issues. *Monash Bioethics Review*, **32**, 268–283.

Fabing, H.D. (1956). On going berserk: a neurochemical inquiry. *American Journal of Psychiatry*, **113**(5), 409–415.

Farwell, J. (2014). Issues of law raised by the developments and use of neuroscience and neurotechnology in national security and defense. In: Giordano, J. (Ed.) *Neurotechnology in National Security and Defense: Practical Considerations, Neuroethical Concerns*, pp.133–166. Boca Raton, FL: CRC Press.

Ferguson, J.R. (2000). Biological weapons and US law. In: Lederberg, J. (Ed.) *Biological Weapons: Limiting the Threat*, pp.81–91. Cambridge, MA: MIT Press.

Foc.us (2015). *Focus Take Charge®—Foc.us Electrical Brain Stimulators*. Available from: http://www.foc. us [accessed February 8, 2017].

Forsythe, C. and Giordano, J. (2011). On the need for neurotechnology in the national intelligence and defense agenda: scope and trajectory. *Synesis: A Journal of Science, Technology, Ethics and Policy*, 2(1), 5–8.

Fregni, F. and Pascual-Leone, A. (2007). Technology insight: noninvasive brain stimulation in neurology—perspectives on the therapeutic potential of rTMS and tDCS. *Nature Clinical Practice Neurology*, 3(7), 384.

Giordano, J. (Ed.) (2014a). *Neurotechnology in National Security and Defense: Practical Considerations, Neuroethical Concerns*. Boca Raton, FL: CRC Press.

Giordano, J. (2014b). Neurotechnology, global relations and national security: Shifting contexts and neurothical demands. In: Giordano, J. (Ed.) *Neurotechnology in National Security and Defense: Practical Considerations, Neuroethical Concerns*, pp.1–10. Boca Raton, FL: CRC Press.

Giordano, J. (2014c). The human prospect(s) of neuroscience and neurotechnology: domains of influence and the necessity—and questions—of neuroethics. *Human Prospect*, 4(1), 1–18.

Giordano, J. (2015). A preparatory neuroethical approach to assessing developments in neurotechnology. *AMA Journal of Ethics*, 17(1), 56–61.

Giordano, J. (2016, May). The neuroweapons threat. *Bulletin of the Atomic Scientists*. Available from: http://thebulletin.org/neuroweapons-threat9494 [accessed February 8, 2017].

Giordano, J. and Benedikter, R. (2012). An early—and necessary—flight of the Owl of Minerva: neuroscience, neurotechnology, human socio-cultural boundaries, and the importance of neuroethics. *Journal of Evolution and Technology*, 22(1), 14–25.

Giordano, J. and Wurzman, R. (2016). Integrative computational and neurocognitive science and technology for intelligence operations: horizons of potential viability, value and opportunity. *STEPS: Science, Technology and Engineering Policy Studies*, 4, 32–37.

Giordano, J., Casebeer, W., and Sanchez, J. (2014). *Assessing and managing risks in systems neuroscience research and its translation: A preparatory neuroethical approach*. Paper presented at 6th Annual Meeting of the International Neuroethics Society, Washington, DC.

Giordano, J., Forsythe, C., and Olds, J. (2010). Neuroscience, neurotechnology and national security: the need for preparedness and an ethics of responsible action. *AJOB Neuroscience*, 1(2), 1–3.

Golub, A. and Bennett, A.S. (2013). Introduction to the special issue: drugs, wars, military personnel, and veterans. *Substance Use & Misuse*, 48, 796.

Gross, M.L. and Carrick, D. (Eds.) (2013). *Military Medical Ethics for the 21st Century*. New York: Ashgate.

Hamilton, R., Messing, S., and Chatterjee, A. (2011). Rethinking the thinking cap: ethics of neural enhancement using noninvasive brain stimulation. *Neurology*, 76(2), 187–193.

Horvath, J.C., Forte, J.D., and Carter, O. (2015). Quantitative review finds no evidence of cognitive effects in healthy populations from single-session transcranial direct current stimulation (tDCS). *Brain Stimulation*, 8(3), 535–550.

Høyersten, J.G. (2015). Manifestations of psychiatric illness in texts from the medieval and Viking era. *Archives of Psychiatry and Psychotherapy*, 2, 57–60.

Kamienski, L. (2016). *Shooting Up: A Short History of Drugs and War*. New York: Oxford University Press.

Kappenmann, E., Bikson, M., and Giordano, J. (Eds.) (2016, March). *Mechanisms and Effects of Transcranial Direct Current Stimulation*. US Air Force Office of Scientific Research Report.

Kosal, M.E. and Huang, J.Y. (2015). Security implications and governance of cognitive neuroscience. *Politics and the Life Sciences*, 34(1), 93–108.

Lanzilao, E., Shook, J., Benedikter, R., and Giordano, J. (2013). Advancing neuroscience on the 21st century world stage: the need for—and proposed structure of—an internationally relevant neuroethics. *Ethics in Biology, Engineering and Medicine*, **4**(3), 211–229.

Lasswell, H.D. (1941). The Garrison State. *American Journal of Sociology*, **46**(4), 455–468.

Lebedev, M.A. and Nicolelis, M.A.L. (2006). Brain-machine interfaces: past, present and future. *Trends in Neurosciences*, **29**(9), 536–546.

Lewin, L. (1998). *Phantastica: A Classic Survey on the Use and Abuse of Mind-Altering Plants.* New York: Inner Traditions/Bear & Co.

McCreight, R. (2014). Brain brinksmanship: devising neuroweapons looking at battlespace, doctrine and strategy. In: Giordano, J. (Ed.) *Neurotechnology in National Security and Defense: Practical Considerations, Neuroethical Concerns*, pp.115–132. Boca Raton, FL: CRC Press.

Miller, R.J. (2013). *Drugged: The Science and Culture Behind Psychotropic Drugs.* New York: Oxford University Press.

Moreno, J.D. (2012). *Mind Wars.* New York: Bellevue Literary Press.

National Research Council. (2008). *Emerging Cognitive Neuroscience and Related Technologies.* Washington, DC: The National Academies Press.

National Research Council (US) Committee on Opportunities in Neuroscience for Future Army Applications (2009). *Opportunities in Neuroscience for Future Army Applications.* Washington, DC: National Academies Press. Available from: http://www.ncbi.nlm.nih.gov/books/NBK207975/.

Nitsche, M.A. and Paulus, W. (2000). Excitability changes induced in the human motor cortex by weak transcranial direct current stimulation. *Journal of Physiology*, **527**(3), 633–639.

Nitze, P. (1950). NSC 68: United States Objectives and Programs for National Security. In: May, E. (Ed.) *American Cold War Strategy: Interpreting NSC 68*, pp.31–32. Boston, MA: Bedford/St. Martins (1993).

Office of Naval Research: Science & Technology (n.d. a). *Code 34: Warfighter Performance Department.* Available from: http://www.onr.navy.mil/Science-Technology/Departments/Code-34.aspx [accessed April 15, 2016].

Office of Naval Research: Science & Technology (n.d. b). *Computational Neuroscience Program.* Available from: http://www.onr.navy.mil/Science-Technology/Departments/Code-34/All-Programs/human-bioengineered-systems-341/Computational-Neuroscience.aspx [accessed April 15, 2016].

Parent, A. (2004). Giovanni Aldini: from animal electricity to human brain stimulation. *Canadian Journal of Neurological Sciences*, **31**, 576–584.

Priori, A., Berardelli, A., Rona, S., Accornero, N., and Manfredi, M. (2008). Polarization of the human motor cortex through the scalp. *Neuroreport*, **9**(10), 2257–2260.

Quigley, M. (2007). Non-human primates: the appropriate subjects of biomedical research? *Journal of Medical Ethics*, **33**(11), 655–658.

Robins, L.N., Helzer, J.E., and Davis, D.H. (1975). Narcotic use in Southeast Asia and afterward. *Archives of General Psychiatry*, **32**(8), 955–961.

Roosevelt, F.D. (1944, November 17). *Letter to Vannevar Bush.* Available from: https://www.nsf.gov/about/history/vbush1945.htm#letter [accessed April 15, 2016.

Sato, H., Vo Doan, T.T., Kolev, S., et al. (2015). Deciphering the role of a coleopteran steering muscle via free flight stimulation. *Current Biology*, **25**(6), 798–803.

Shanker, T. and Duenwald, D. (2003, January 19). Bombing error puts a spotlight on pilots' pills. *The New York Times.* Available from: http://www.nytimes.com/2003/01/19/us/threats-and-responses-military-bombing-error-puts-a-spotlight-on-pilots-pills.html [accessed April 15, 2016].

Shook, J.R. and Giordano, J. (2014). A principled, cosmopolitan neuroethics: considerations for international relevance. *Philosophy, Ethics, and Humanities in Medicine*, **9**, 1.

Shook, J.R. and Giordano, J. (In press). Assessing performance enhancement before neuroethical evaluations. *Frontiers in Human Neuroscience*.

Shook, J.R., Galvagni, L., and Giordano, J. (2014). Cognitive enhancement kept within contexts: neuroethics and informed public policy. *Frontiers in Systems Neuroscience*, 8, 1–8.

Simeral, J.D., Kim, S.P., Black, M.J., Donoghue, J.P., and Hochberg, L.R. (2011). Neural control of cursor trajectory and click by a human with tetraplegia 1000 days after implant of an intracortical microelectrode array. *Journal of Neural Engineering*, 8(2), 1–24.

Strategic Multilayer Assessment Group, Joint Staff/J-3. (2010). *Neurobiology of Political Violence: New Tools, New Insights*. Washington, DC: The Pentagon.

Strategic Multilayer Assessment Group, Joint Staff/J-3. (2012a). *Neurobiological & Cognitive Science Insights on Radicalization and Mobilization to Violence: A Review*. Washington, DC: The Pentagon.

Strategic Multilayer Assessment Group, Joint Staff/J-3. (2012b). *Cyber on the Brain: The Effects of CyberNeurobiology & CyberPsychology on Political Extremism*. Washington, DC: The Pentagon.

Strategic Multilayer Assessment Group, Joint Staff/J-3. (2012c). *National Security Challenges: Insights from Social, Neurobiological, and Complexity Science*. Washington, DC: The Pentagon.

Strategic Multilayer Assessment Group, Joint Staff/J-3. (2013). *Topics in Operational Considerations on Insights from Neurobiology on Influence and Extremism*. Washington, DC: The Pentagon.

Strategic Multilayer Assessment Group, Joint Staff/J-3. (2014). *Leveraging Neuroscientific and Neurotechnological Developments with Focus on Influence and Deterrence in a Networked World (Revised)*. Washington, DC: The Pentagon.

Strategic Multilayer Assessment Group, Joint Staff/J-3. (2015). *Social and Cognitive Neuroscience Underpinnings of ISIL Behavior and Implications to Strategic Communication, Messaging, and Influence*. Washington, DC: The Pentagon.

Szondy, D. (2012, September 20). *DARPA's CT2WS Technology Uses "Mind Reading" to Identify Threats*. Available from: http://www.gizmag.com/tag-team-threat-recognition-technology/24208 [accessed April 15, 2016].

Tabery, J. (2014). Can (and should) we regulate neurosecurity? Lessons from history. In: Giordano, J. (Ed.) *Neurotechnology in National Security and Defense: Practical Considerations, Neuroethical Concerns*, pp.249–258. Boca Raton, FL: CRC Press.

Talwar, S. K., Xu, S., Hawley, E.S., Weiss, S.A., Moxon, K.A., and Chapin, J.K. (2002). Rat navigation guided by remote control. *Nature*, 417, 37–38.

The President's Council on Bioethics (US) and Kass, L. (2003). *Beyond Therapy: Biotechnology and the Pursuit of Happiness*. New York: Harper Perennial.

Tractenberg, R.E., FitzGerald, K.T., and Giordano, J. (2014). Engaging neuroethical issues generated by the use of neurotechnology in national security and defense: toward process, methods and paradigm. In: Giordano, J. (Ed.) *Neurotechnology in National Security and Defense: Practical Considerations, Neuroethical Concerns*, pp.259–278. Boca Raton, FL: CRC Press.

Tyler, W.J. (2010, September). Remote control of brain activity using ultrasound. *Armed with Science*. Available from: http://science.dodlive.mil/2010/09/01/remote-control-of-brain-activity-using-ultrasound/ [accessed April 15, 2016].

UK ECT Review Group. (2003). Efficacy and safety of electroconvulsive therapy in depressive disorders: a systematic review and meta-analysis. *The Lancet*, 361(9360), 799–808.

Ukueberuwa, D. and Wassermann, E.M. (2010). Direct current brain polarization: a simple, noninvasive technique for human neuromodulation. *Neuromodulation*, 13(3), 168–173.

White House (2014). *Fact Sheet: Over $300 Million in Support of the President's BRAIN Initiative.* Available from: https://obamawhitehouse.archives.gov/sites/default/files/microsites/ostp/brain_fact_ sheet_9_30_2014_final.pdf [accessed February 14, 2017].

White, S.E. (2008). Brave new world: neurowarfare and the limits of international humanitarian law. *Cornell International Law Journal*, **41**, 177–210.

Wurzman, R. and Giordano, J. (2014). NEURINT and neuroweapons: neurotechnologies in national intelligence and defense. In: Giordano, J. (Ed.) *Neurotechnology in National Security and Defense: Practical Considerations, Neuroethical Concerns*, pp.79–114. Boca Raton, FL: CRC Press.

Chapter 28

Communicating about the brain in the digital era

Julie M. Robillard and Emily Wight

Neuroscience communication: A priority for neuroethics

Brain science and public discourse, one of the four original pillars of neuroethics (Marcus 2002), explores the discourse that takes place around neuroscience research and supports the broader discussion about ethical, legal, and social issues arising from related advances. In recent years, progress in the development of methods for brain research and of treatment for conditions of the nervous system has generated much discussion at the intersection of ethics and neuroscience. For example, optogenetics, an emerging technique aimed at controlling genetically modified subsets of brain cells with light, has exploded in popularity in both academic research and on social media platforms such as Twitter. More closely related to human health, the regenerative properties of stem cells have been hailed as a miraculous stroke treatment by celebrities in traditional and new media (Rachul and Caulfield 2015). In parallel with advances in science, the development of Internet- and mobile-based communication channels have expanded and diversified how conversations about science and health take place. Interactive platforms now play an important role in shaping public perceptions of scientific and healthcare discoveries and provide new avenues for the public to interact with research on their terms, and to engage in multidirectional conversations about innovations in neuroscience.

In this chapter, we explore the history and evolving landscape of neuroscience communication, present work from our team in Canada, and look to the future to understand the ways that the evolution of technology have improved and challenged our ability to translate and communicate brain health research and discovery.

The evolution of science communication

The neuroscience communication aspect of neuroethics builds on the broader field of science communication. Well before the advent of dedicated empirical and theoretical research in the field of science communication, science discoveries were published in the mainstream press and were featured at fairs, exhibits, and in popular books (Jacques and Raichvarg 1991). By the 1700s, books such as Francesco Algarotti's *Il newtonianismo per le dame* ("Newtonism for ladies") were meant to translate science and increase its accessibility for the average person (Bucchi 2008). Over time, the process of science communication broadened and came to include multiple types of stakeholders. In his famous 1881 public experiment, French chemist Louis Pasteur demonstrated the concept of vaccination by injecting a group of livestock with an anthrax vaccine, while another group remained unvaccinated. Farmers, doctors, veterinarians, physiologists, politicians, and reporters all witnessed the inoculations. The science experiment was successful: all unvaccinated animals died and all inoculated animals survived. The public experiment was successful as well: the engagement powerfully demonstrated the value of research and discovery.

In more recent years, the dissemination of scientific knowledge has grown increasingly complex as once-broad fields of research have become specialized and academic journals have become the standard for publication of new research findings. In 1985, a report on the public's knowledge of science commissioned by the Royal Society urged scientists to "Learn to communicate with the public, be willing to do so, and consider it your duty to do so" (Royal Society 1985). However, many years later, scientist communicators are seen by their peers as less serious than their less-communicative counterparts: researchers understand the need to communicate broadly, but there is a stigma attached to those who actively pursue public engagement. A 2006 survey by the Royal Society on "factors affecting science communication by scientists and engineers" found that while 74% of Royal Society members surveyed had participated in "at least one science communication or public engagement activity in the past 12 months," in interviews researchers revealed that "public engagement activity was seen by peers as bad for their career," and that "public engagement was done by those who were 'not good enough' for an academic career." This view that public engagement is detrimental to a person's scientific career was especially damaging for junior and women researchers. In the same survey, 20% of respondents agreed that scientists who engage with the public are less well regarded by other scientists (Royal Society 2006); 64% of scientists surveyed that the need to spend more time on research was a barrier to engaging in public engagement activities. Known as "The Sagan Effect" for Dr. Carl Sagan who popularized science on television for lay audiences in the 1970s and 1980s, this belief that scientists who pursue media and public engagement opportunities are less serious and less focused on research persists in modern academia, despite being directly at odds with society's need to have an educated and informed populace who trusts and sees the value in scientific research.

Despite these real and perceived barriers in science communication, a 2014 *Public Attitudes to Science* Ipsos Mori poll found that 72% of citizens in the United Kingdom agree with the statement "it is important to know about science in my daily life"; 90% of those same people stated that they believe that scientists make an important contribution to society (Ipsos MORI 2014). The public desire for scientific knowledge is very high, even as overall science literacy appears to be threatened by the inscrutable, top-down nature of academic publishing. There is a clear disconnect between the goals of science communication and science literacy in the general population. According to the same poll, nearly 30% of respondents believe that science is speculative, and that "scientific research is never or only occasionally checked by other scientists before being published," illustrating the important gap in knowledge about the process of scientific discovery.

To increase public trust in science, and to improve science literacy, researchers and lay communicators alike have put forward models of knowledge translation. Earlier models, such as the deficit model, emphasized the transfer of knowledge in one-way and one-time processes, such as popularization of science through books. In recent decades, there has been much discussion and criticism of this top-down dissemination model (Gregory & Miller 2000; Wilsdon & Willis 2004). As the multidirectional nature of science communication emerged, the dialogue model was put forward, proposing a two-way, iterative process focused on consultation and the recognition and discussion of the implications of scientific discoveries (Bucchi 2008). In a shift driven by the urgency of fully engaging lay audiences in research, emerging models of science communication emphasize engagement between disciplines as well as with the general public. In the participation model, knowledge is co-produced; multidirectional communication takes place between all stakeholder groups with the goal of having all parties participate in shaping the agenda.

As these models evolve, so does the environment in which science communication takes place. The last century has seen an exponential increase in the specialization of science. Neuroscience, for example, now comprises dozens of specialties, each with their own methods, experimental designs, and terminologies. In addition to this specialization, neuroscience is also becoming increasingly multidisciplinary, with contributions from engineering, computer science, social sciences, psychology, and ethics, to mention just a few. As science broadly and neuroscience specifically become increasingly both specialized and multidisciplinary, new challenges arise, such as the lack of training in translating difficult concepts to lay audience (Illes et al. 2009) and the need to effectively translate across disciplines at the expert level.

The brain is uniquely challenging

Diseases of the brain are complex, and while the pace of discovery has accelerated in recent years with improved imaging and genomics technology, the origins of many neurological diseases are still poorly understood. Degenerative diseases of and injuries to the brain and nervous system—taken together—represent an urgent social and medical challenge and are an important cause of mortality, with the World Health Organization estimating that neurological disorders account for 12% of deaths globally every year (World Health Organization n.d.). The pace of brain health research challenges scientists and the public alike, and so engagement on public health issues, and clear communication of discoveries, challenges, timelines, and research needs are important priorities for the neuroscience community.

Beyond health and illness, the brain also represents our minds and hosts our sense of self. Unlike for other organs, understanding the brain and surrounding concepts such as emotions, states of consciousness, and behaviors has important implications for individuals and interpersonal and societal relationships (Illes 2006). Neuroscience communication thus presents challenges, distinct from other areas of science communication, in that many mysteries of the brain remain unsolved. Because so little is known about how the brain functions in health and illness, and because people are living longer and the population is aging, timely, accurate, and relevant communication about how to maximize the benefits of advances in neuroscience and neurology is an urgent global issue. While traditional media remains the primary source of science news for most people, online searches for and discussion of neuroscience topics are increasing.

In parallel with the shifting context of neuroscience communication, there has been significant growth and democratization of mass media in recent years, enabled by the development of dynamic platforms such as Internet and mobile applications. More than ever, the interactive web and mobile environments and the social media platforms they support are providing new opportunities to communicate and discuss neuroscience advances. The increase in the use of Internet-enabled applications has many potential and proven benefits over past approaches to reaching consumers of science and health information: the Internet allows for the rapid dissemination of information, promotes multidirectional engagement among different types of stakeholders, and reduces barriers to access information. For content with implications for health, online and mobile platforms may also promote autonomy and empowerment in health decision-making (Kim E.-H. et al. 2011), and may lead to increased social support for users interacting with others in discussion groups and forums (Idriss et al. 2009; van Uden-Kraan et al. 2009). However, the online environment as a venue for neuroscience communication is not without risks: many studies raise concerns over the quality of the information available online and in social media (Eysenbach et al. 2002; Scullard et al. 2010; Robillard et al. 2013), and the legitimacy of the information can be difficult to evaluate. Online discussion of discoveries tends to be trend based, following science news coverage in traditional media (Ipsos MORI 2014). Conflicts of interest are also frequent, such as when commercially driven websites promote content meant to educate web users and imply an association between specific products and healthy lifestyles (Illes et al. 2004). As the proportion of Internet users who seek health information grows, various industries hope to capitalize on this behavior, and the line between providing health information and direct-to-consumer advertising (DTCA) is increasingly blurred (Racine et al. 2007). This phenomenon predates the digital era: in 2004, investigators uncovered important concerns with print materials promoting self-referred imaging, such as a lack of balanced information necessary for the preservation of autonomy in health decision-making (Illes et al. 2004). Similar concerns were raised for web-based materials promoting DTCA for genetic services (Risk and Petersen 2002; Gollust et al. 2003) and for complementary and alternative medicine for dementia (Palmour et al. 2014), among other examples.

This changing media environment gives rise to new challenges and ethical considerations: how can the potential of new media platforms to enhance the delivery of engaging, high-quality, and multidirectional science communication be harnessed? Here we use dementia as a case study to answer this question.

Aging advice in a digital world

Recent survey data shows that older adults are increasingly online: over half of Canadians 65 years or older are Internet users, and of those, a majority use the Internet to search for health information and medical diagnoses (Cline and Haynes 2001; Song et al. 2002; Vance et al. 2009; Zulman et al. 2011). Online information and tools are reshaping health-care by offering powerful new ways for stakeholders to gather and share information (Hawn 2009). Online sources of information and platforms for exchange can potentially enhance the health of their users, if used in consultation with a physician who can provide context for how to apply this new information to their existing health or care regimen. Internet users can increase their social support and feelings of connectedness (Wangberg et al. 2008; Idriss et al. 2009) as well as feel empowered in managing their health (van Uden-Kraan et al. 2009). Interactive Internet applications can also increase the demo-cratic aspect of information-sharing, which in turn leads to a patient-centered experi-ence (Hawn 2009). Efforts to promote health through online platforms have a broader reach than traditional media and have been shown to have a significant impact for certain conditions that are tightly linked to healthy aging, such as smoking cessation and dietary interventions (Block et al. 2008; Norman et al. 2008; Thackeray et al. 2008). However, online health resources are not free of risk. Broad-reaching, easily accessible platforms can lead to the wide dissemination of misinformation (Kortum et al. 2008) and informa-tion of uneven quality. Confusion may arise when unregulated and commercially driven website content is represented as educational, as users may assume the information is free of bias (Wolfe 2002). Further, users have reported that they have difficulties in identifying and using appropriate information, and—without clarifying their findings with their fam-ily doctor—may experience anxiety as a result of consulting online resources (Eysenbach et al. 2002; Benigeri & Pluye 2003). These risks are particularly acute when the consumers of the information may suffer from possible cognitive impairment.

Unregulated online information introduces a new variable in health and quality of life and the patient–physician relationship. As dementia ranks among the most feared con-ditions by older adults (Corner & Bond 2004), online resources have emerged to meet their specific information demands. In this context, we explore the quality of three types of these resources about dementia: (1) social media posts, (2) websites containing health information about the prevention of dementia, and (3) self-assessments.

Social media

Social media platforms allow for content to be created and modified on an ongoing basis, and as such have transformed information-sharing from the passive consumption of top-down dissemination to active engagement with multidirectional, peer-based exchange. Extensive conversations about dementia are taking place on social media platforms.

From one 24-hour period in February 2012, for example, we captured 9200 unique tweets (Robillard et al. 2013). We used a customized content analysis methodology (Robillard et al. 2013) to characterize the following features of Twitter posts: (1) users contributing to this volume of activity on Twitter, (2) websites that are linked to in the tweets, and (3) thematic content of the tweets.

In answer to the first question, we collected and analyzed publically available data fields for freely contributed user information. We found that health professionals were the largest group contributing to the conversation. Of the sample, 17% were from this cohort of individuals operating in all realms of healthcare such as medicine, pharmacy, and nursing. News groups and commercial organizations also contributed important proportions of tweets (news: 10%; commercial: 10%). Care services facilities, journalists, patients and their families, and researchers represented the remaining groups of sources of tweets.

We answered the second question by analyzing the links to third-party websites contained within the text of tweets and found that 50% of the links redirected to an article hosted on a news site, such as BBC News or CBC News. This finding suggests that while new media provides unprecedented opportunity for engagement with information, at the present time popular platforms such as Twitter still rely heavily on traditional media to create and host content, at least for topics related to dementia.

To answer the third research question, we conducted a content analysis of the text of tweets. Although researchers themselves were relatively low contributors in the sample studied, their findings certainly made their way into the tweets: 48% referred to a specific research study, and of these, 76% were about peer-reviewed, original research. Especially intriguing was the way that different tweets from different users described the same research study: for example, an abstract from the 2012 American Academy of Neurology Annual Meeting entitled "Walking Speed, Handgrip Strength and Risk of Dementia and Stroke: The Framingham Offspring Study" was featured in a tweet that read "Fast walking speed and a strong grip in middle age may help predict dementia risk." Another tweet based on the same abstract read "Slow walking 'predicts dementia.'" Similarly, the research described in an article published in the *Journal of the American Medical Association Neurology* in 2012 entitled "Sleep Quality and Preclinical Alzheimer Disease" was tweeted both as "Disturbed sleep is associated with preclinical signs of Alzheimer's disease, researchers found" and as "Study shows sleep prevents Alzheimer's." In the case of both these examples about walking speed and sleep, the first tweet harnessed the 140 available characters to provide an accurate representation of the research and used cautious vocabulary designed to mitigate false hopes; the second tweet leans closer toward misleading and sensationalized oversimplifications.

Science discoveries and advances in biotechnologies that relate to the brain are often hyped in the media: their benefits exaggerated and their risks minimally discussed. These overly positive representations of neuroscience and neurology findings can lead to ethical conflict when they are responsible for the generation of false hopes in potentially vulnerable readers. Many different factors contribute to this problem along the entire process from scientific discovery to publication: on the one hand, scientists face pressures to appeal to funders, the industry, and to the public; on the other, journalists and media institutions must put forward newsworthy content to drive profit (Bubela et al. 2009). With the rise of Internet marketing, headline writers at traditional media outlets have been put into a position where they must compete with new media aggregators and digital sales channels for a limited share of Internet users' attention in an increasingly saturated online space. This has led to headlines tailored specifically to generate an emotional response in readers in order to entice them to click through to read more. In a 2016 study on science news headlines, researchers found that "although tentativeness is part and parcel of communicating science, popular science journalism employs epistemic modality strategically to mix reporting with predicting, recounting with speculating and evidentiality with possibility (Molek-Kozakowska 2016)". While the phenomenon of hype in neuroscience communication has been observed in traditional media (Caulfield and Rachul 2011; Caulfield and McGuire 2012), it can be further fueled by the format of social media platforms, for example, those that require very short descriptions of shared content, such as Twitter. For example, a recent analysis of the discussion about stem cells and Parkinson's disease or spinal cord injury on Twitter showed that research was at the forefront of the discussion but that the majority of tweets were either positive or neutral in tone and that there was very little discussion of risks (Robillard et al. 2015a). One concern with sensationalized neuroscience findings is that this would facilitate the early adoption, use, and promotion of unproven interventions (Ryan et al. 2010). Another consequence of overly simplified headline-driven science communication is that expertise is diluted to the point where researchers' authority over their subject matter appears questionable, reducing public trust in scientists and institutions (Fischoff 2013), or fueling the increase of trust in trends like homeopathy or natural health alternatives with no scientific validity or clinical endorsement.

Websites

Over half of the 65-years-and-older demographic uses the web specifically to seek health information. To examine the quality of the available information, we applied a quality evaluation tool to a sample of articles about the prevention of Alzheimer's disease. We retrieved 290 articles through a location-independent keyword search on Google, the most used search engine in North America. The evaluation tool is an original adaptation of previous work by Chumber and colleagues (Chumber et al. 2015), Sandvik (Sandvik 1999), and the *JAMA* benchmarks developed by Silberg and colleagues (Silberg et al. 1997). It consists of six weighted criteria for a maximum score of 28: (1) authorship, (2) attribution, (3) conflict of interest, (4) currency, (5) complementarity, and (6) tone.

The articles scored across the full range of quality, but we distinguished two clear tails of the distribution: articles that rated poorly (score: 4–11) and those that rated highly (score: 24–27). Three main characteristics separated the low- and high-quality articles: conflict of interest, attribution of claims, and tone. The presence of an endorsement or commercial advertisement of a product or service within an otherwise informational article about the prevention of Alzheimer's disease signaled conflict of interest. This important ethical concern was identified in 20% of the sample. None of the highest scoring articles contained this feature. We found that scientific claims were attributed to general research (broad, unverifiable claims using research words, such as "research studies show that … ," 60% of the articles), specific studies (58%), and quotes from doctors (30%); many articles contained more than one type of attribution. High-quality articles were more likely to cite a specific study that could be traced back by the researcher. Finally, the tone of recommendations differed across the sample. We found strong recommendations, unequivocal language, and little discussion of alternatives or limitations to the featured prevention strategies in 22% of the articles. Only 13% presented both balanced advice with cautious vocabulary, and discussed the limitations of the research supporting the claims.

As online behaviors have been shown to influence offline behaviors, poor communication about dementia but also about advances in neuroscience and neurology more broadly may lead to decision-making toward negative health outcomes. For example, a study by Walji and colleagues showed that one-quarter of websites about three popular herbal supplements contained recommendations that, if followed, would lead directly to physical harms (Walji et al. 2004). Of particular concern is the industry around complementary and alternative medicines for brain conditions ranging from depression to Alzheimer's disease. Dietary supplements and homeopathic remedies for a number of conditions are readily available for purchase on the Internet and promoted through social media. Palmour and colleagues examined websites about dietary supplements specifically for Alzheimer's disease and found that only 16% of these sites contained peer-reviewed evidence to support the claims about the benefits of the products (Palmour et al. 2014). While the notion of autonomy in health decision-making favors empowerment and accessibility to varied sources of information, an ethical conflict arises when the available resources are so greatly varied in quality.

Self-assessments

Online self-assessments are available for a wide variety of brain-related conditions including depression, anxiety, migraines, as well as neurodegenerative conditions such as Alzheimer's disease. They are also becoming increasingly popular as measured by the proportion of Internet users who report going online specifically to establish whether they or someone they know suffer from a medical condition. Today, that number is more than 30% (Fox 2013). No doubt, self-assessments for brain-related conditions may result in benefits for the target users: they may feel empowered, they can access information when traditional health services are difficult to reach, and they may become motivated to seek medical advice or self-monitor for preventive purposes (Trustram Eve & de Jager 2014). However, the limited oversight of technology-assisted self-assessments may also adversely lead to inaccurate self-diagnosis and treatment (Lovett et al. 2012). Users may misinterpret graphic displays of risk and associated terminology (Johnson & Shaw 2012), and select inappropriate actions that lead to negative health consequences (Waters et al.

2009; Lovett et al. 2012). These risks are exacerbated in the case of Alzheimer's disease for which no broadly effective treatments exist.

In 2015, we (Robillard and colleagues) published the results of the first study to our knowledge to evaluate online self-assessments for Alzheimer's disease. We studied 16 tests available on high-traffic websites (Robillard et al. 2015b) for criteria related to scientific validity such as test–retest reliability, ethical features such as conflict of interest, and human–computer interactions features such as usability. We found that most tests in our sample were not scientifically valid and failed to deliver meaningful information about Alzheimer's disease or other dementias in an ethically sound and user-friendly way, with no test scoring excellent across all criteria evaluated. Expert ratings for human–computer interaction items were variable, with ratings in the acceptable range for clarity of instructions and quality of the visuals (e.g., contract, font size) and lower ratings for the detecting of invalid performances and accommodations for varying levels of computer knowledge. Expert ratings for ethics-related factors were the lowest, and ranged from very poor to poor across most criteria. Despite the fact that the tests resembled legitimate medical interventions, the informed consent process was either lacking or overly dense. In response to the wealth of available information online and to concerns with both DTCA and informed consent in the digital age, Bal and Brenner ask "Has the Internet made informed consent obsolete?" (Bal & Brenner 2015). Indeed, the digital age has shifted how patients educate themselves about their conditions, the available conventional but also alternative treatments, and the risks and benefits associated with each option. While informed consent was initially designed in the context of a system whereby the physician is the primary source of health information, a recent survey found that 91% of physicians surveyed have seen patients who made inquiries about information they found online (Kim J. & Kim S. 2009). When writing about informed consent in the online environment more broadly, for example with regard to personalized advertising and behavior tracking, Friedman and colleagues state: "To date, informed consent is woefully understudied by the online community and underused as a means of cultivating trust online," and call for research into the development of principles to implement consent in the online environment (Friedman et al. 2000). As the medical community shifts toward new channels of communication about brain health, it becomes imperative to consider how information and misinformation acquired through online and mobile platforms will impact the ethical duty to obtain informed consent for both online interventions and those delivered in the clinic.

Similarly, for most tests, language around privacy and confidentiality of the data that were entered as part of the test-taking process was either absent or incorporated into lengthy terms of agreement. In Canada, the *Freedom of Information and Protection of Privacy Act* was enacted in the 1980s "to govern the collection, use and disclosure of personal information by organizations in a manner that recognizes both the right of individuals to protect their personal information and the need of organizations to collect, use or disclose personal information for purposes that a reasonable person would consider appropriate in the circumstances" (Office of the Information and Privacy Commissioner 1983). The Act protects the personal information of private citizens and limits disclosure of personal information by organizations or public bodies; however, this protection extends only to information stored on Canadian servers. Other jurisdictions may offer different or no protection for personal data. Without protective legislation in place, personal data may be vulnerable to third-party sale, subpoena, or theft.

In the United States, the *Genetic Information Nondiscrimination Act (GINA)* (United States 2008) protects private citizens from discrimination by employers or insurers based on genetic information, and builds upon the *Health Insurance Portability and Accountability Act (HIPPA)* (United States 2004) and related privacy rules. The intention of these privacy rules was meant to protect individuals' private health information, but there is no guideline or legislation in place to ensure privacy of personal health information given voluntarily via non-healthcare commercial entities, such as websites.

Terms of agreement may be vague in detailing how personal data may be used, and because this is relatively new legal terrain, it is unclear in many jurisdictions whether user agreements for web services and products are legally binding or if the terms themselves are ethically sound.

All 16 tests in our sample rated either very poorly or poorly with regard to conflicts of interest. For example, one of the self-assessments for dementia in the sample consisted of a number of questions about daily activities, behaviors, and risk factors. However, all possible combinations of answers yielded the same result: a strongly worded warning about a severe risk of developing Alzheimer's disease, followed by a recommendation to purchase three different supplements from the online store of the website. In this case, a predatory marketing strategy masqueraded as a self-test aimed at providing meaningful information regarding one's risk of developing dementia. Though not all conflicts of interests uncovered in our studies were so deeply problematic, they all had the potential to compromise the quality of the information offered to a vulnerable segment of the population, especially as older adults may experience difficulties with the identification of trustworthy information (Castle et al. 2012). In turn, information that is unbalanced, biased, or of poor quality may compromise personal autonomy and prevent meaningful decisions for brain health.

Looking to the future

"In the 16th century," writes Robert Lucky, "science progressed at the pace of the postal system. Often it would take six months for one scientist to learn of the ongoing results of another. It took even more time for scientists to build on one another's accomplishments" (Lucky 2000). As the ability of scientists to communicate more efficiently has improved, so has the ability to uncover the mysteries of the brain and behavior. In parallel, the rate of discovery has advanced together with technology, with publication rates in scientific journals increasing by nearly 5% per year (Larsen & von Ins 2010). The benefit of moving from a top-down to a participatory model of neuroscience communication in a process enabled by recent advances in technology results in a democratization of science, both for researchers and for the general population.

There are many advantages to improving scientific literacy, whether through traditional or novel mechanisms. In the book *Neuroethics: Mapping the Field* (Marcus 2002), Colin Blakemore, one of Britain's most influential science communicators, identifies three: an informed public is (1) better able to assess risks and benefits of decisions that are relevant for their lives; (2) empowered to participate in discussions and debates about the scientific and technology agendas; and (3) more likely to support evidence-based policies (Blakemore 2002). With these advantages in mind, educating scientists about how to communicate with the public, and educating the public on how to parse the language of science can only benefit both parties, both in terms of increased funding for research as public stakeholders understand the value of that funding, and improved access to essential information for those who seek it online or elsewhere. By understanding the contexts each group brings to framing neuroscience news, science communicators can move public thinking to a mutually beneficial and productive place.

While calls for more education for both scientists and science communicators have already been put forward (Illes et al. 2009), these initiatives should now consider incorporating lessons about new media. For example, as article and press release titles and portions of abstracts are often used verbatim in online social media, and particularly on platforms that limit the quantity of text that can be incorporated into a post, researchers should learn to phrase these in a way that, should they become widely accessible to the public, would not lead to false hopes or misunderstandings. Physicians and other healthcare providers, whose relationships with patients and their families are increasingly shaped by technology-enabled health resources, should also be provided with educational opportunities to better understand the evolving needs of the patient community. Neurologists and family physicians, for example, would benefit from leadership from the scientific community about how lay people are being directed in terms of resources and how these are incorporated into health decision-making, in order to better serve and treat highly informed patient populations.

In addition to efforts to educate knowledge producers, knowledge consumers and clinicians, more empirical research will be required to determine the impact of neuroscience communication in the digital age on health decision-making. The work described here and that of others has laid down a solid foundation of knowledge with regard to the quality and the ethics of available online resources for brain health. However, this is a dynamic and rapidly evolving landscape, and ongoing and collaborative efforts will be required to continue to monitor the available resources and their potential risks and benefits. In addition, providing ethical, high-quality online information is likely not sufficient to ensure that Internet users make decisions that lead to positive health outcomes. As such, it is imperative to evaluate the role that online information and services play in health decision-making and develop responsive and ethical resources that maximize the benefits to the users in this age of eHealth.

Over 30 years later, the recommendations of the Royal Society's 1985 report for scientists to consider science communication a duty are still very relevant. Looking to the future, all scholars in the life and social sciences must include in these recommendations to keep pace with the rapidly evolving communications landscape and embrace the conversation about brain, health, and ethics that is taking place on interactive platforms (Table 28.1). In the balance of the benefits and risks of new communication channels lies a significant opportunity for scientists to engage with the public about neuroscience and neuroethics, and to join forces in developing priorities for the future.

Table 28.1 Comparison between pre-Internet and digital era modes of neuroscience communication

	Pre-Internet era	Digital era
General		
Model of communication	Top-down Authoritative	Peer-to-peer Democratic
Direction	Unidirectional	Multidirectional
Media	Paper Television Radio	Paper Television Radio Internet Mobile technology
Culture	Siloed Expert-driven	Collaborative Participatory
Curation	News editors Science writers Academic journals	News editors Science writers Academic journals Bloggers Researchers Public
Vetting	Organizations	Organizations Individuals
Reach	Targeted	Global
Speed of information flow	Slow	Instant
Number of sources	Finite	Infinite
Information sharing	Original materials Paper photocopies, printouts	Hyperlinks Social sharing buttons Digital copies
Cost	Paid by subscription/membership	Free Paid by advertising Paid by subscription/membership

(continued)

Table 28.1 Continued

	Pre-Internet era	Digital era
Ethical considerations		
Privacy of personal information	Secure	At risk due to inconsistent regulations around data storage and server access, hacking
Commercial conflict of interest	Possible	Possible
Risks of self-diagnosis	Low	High
Sensationalism/hype	Possible, but limited to life cycle of publication	Possible, long-lasting as data is archived forever

Acknowledgments

J.M.R would like to acknowledge all co-authors who contributed to this published work, research assistant Tanya Feng at the National Core for Neuroethics, University of British Columbia, for her work on the analysis of websites about the prevention of Alzheimer's disease, and all members of the Robillard lab for their contributions to the development of the online health information quality evaluation tool.

References

Bal, B.S. and Brenner, L.H. (2015). Medicolegal sidebar: informed consent in the information age. *Clinical Orthopaedics and Related Research*, **473**, 2757–2761.

Benigeri, M. and Pluye, P. (2003). Shortcomings of health information on the Internet. *Health Promotion International*, **18**, 381–386.

Blakemore, C. (2002). From the "public understanding of science" to scientists' understanding of the public. In Marcus, S.J. (Ed.) *Neuroethics: Mapping the Field*, pp.211–221. New York: Dana Press.

Block, G., Sternfeld, B., Block, C.H., et al. (2008). Development of Alive! (A Lifestyle Intervention Via Email), and its effect on health-related quality of life, presenteeism, and other behavioral outcomes: randomized controlled trial. *Journal of Medical Internet Research*, **10**, e43.

Bubela, T., Nisbet, M.C., Borchelt, R., et al. (2009). Science communication reconsidered. *Nature Biotechnology*, **27**, 514–518.

Bucchi, M. (2008). *Handbook of Public Communication of Science and Technology*. Abingdon: Routledge.

Castle, E., Eisenberger, N.I., Seeman, T.E., et al. (2012). Neural and behavioral bases of age differences in perceptions of trust. *Proceedings of the National Academy of Sciences of the United States of America*, **109**, 20848–20852.

Caulfield, T. and McGuire, A. (2012). Athletes' use of unproven stem cell therapies: adding to inappropriate media hype? *Molecular Therapy*, **20**, 1656–1658.

Caulfield, T. and Rachul, C. (2011). Science spin: iPS cell research in the news. *Clinical Pharmacology & Therapeutics*, **89**, 644–646.

Chumber, S., Huber, J., and Ghezzi, P. (2015). A methodology to analyze the quality of health information on the internet the example of diabetic neuropathy. *The Diabetes Educator*, **41**, 95–105.

Cline, R.J.W. and Haynes, K.M. (2001). Consumer health information seeking on the internet: the state of the art. *Health Education Research*, **16**, 671–692.

Corner, L. and Bond, J. (2004). Being at risk of dementia: fears and anxieties of older adults. *Journal of Aging Studies*, **18**, 143–155.

Eysenbach, G., Powell, J., Kuss, O., and Sa, E.R. (2002). Empirical studies assessing the quality of health information for consumers on the world wide web: a systematic review. *JAMA*, **287**, 2691–2700. 1.

Fischhoff, B. and Scheufele, D.A. (2013). The science of science communication. *Proceedings of the National Academy of Sciences of the United States of America*, **110**(Suppl. 3), 14031–14032.

Fox, S. (2013). *Health Online 2013*. Washington, DC: Pew Research Center's Internet & American Life Project.

Friedman, B., Khan Jr, P.H., and Howe, D.C. (2000). Trust online. *Communications of the ACM*, **43**, 34–40.

Gollust, S.E., Wilfond, B.S., and Hull, S.C. (2003). Direct-to-consumer sales of genetic services on the Internet. *Genetics in Medicine*, **5**, 332–337.

Gregory, J. and Miller, S., 2000. *Science in Public: Communication, Culture, and Credibility*. Cambridge, MA: Basic Books.

Hawn, C. (2009). Take two aspirin and tweet me in the morning: how Twitter, Facebook, and other social media are reshaping health care. *Health Affairs (Millwood)*, **28**, 361–368.

Idriss, S.Z., Kvedar, J.C., and Watson, A.J. (2009). The role of online support communities: benefits of expanded social networks to patients with psoriasis. *Archives of Dermatology*, **145**, 46–51.

Illes, J. (Ed.) (2006). *Neuroethics: Defining the Issues in Theory, Practice, and Policy*. Oxford: Oxford University Press.

Illes, J., Kann, D., Karetsky, K., et al. (2004). Advertising, patient decision making, and self-referral for computed tomographic and magnetic resonance imaging. *Archives of Internal Medicine*, **164**, 2415–2419.

Illes, J., Moser, M.A., McCormick, J.B., et al. (2009). Neurotalk: improving the communication of neuroscience research. *Nature Reviews Neuroscience*, **11**, 61–69.

Ipsos MORI (2014, March). *Public Attitudes to Science 2014*. Available from: https://www.ipsos-mori.com/researchpublications/researcharchive/3357/Public-Attitudes-to-Science-2014.aspx [accessed April 16, 2016].

Jacques, J. and Raichvarg, D. (1991). *Savants et ignorants: Une histoire de la vulgarisation des sciences*. Paris: Seuil.

Johnson, C.M. and Shaw, R.J. (2012). A usability problem: conveying health risks to consumers on the Internet. *AMIA Annual Symposium Proceedings*, **2012**, 427–435.

Kim, E.-H., Coumar, A., Lober, W.B., and Kim, Y. (2011). Addressing mental health epidemic among university students via web-based, self-screening, and referral system: a preliminary study. *IEEE Transactions on Information Technology in Biomedicine*, **15**, 301–307.

Kim, J. and Kim, S. (2009). Physicians' perception of the effects of internet health information on the doctor–patient relationship. *Informatics for Health and Social Care*, **34**, 136–148.

Kortum, P., Edwards, C., and Richards-Kortum, R. (2008). The impact of inaccurate Internet health information in a secondary school learning environment. *Journal of Medical Internet Research*, **10**, e17.

Larsen, P.O. and von Ins, M. (2010). The rate of growth in scientific publication and the decline in coverage provided by Science Citation Index. *Scientometrics*, **84**, 575–603.

Lovett, K.M., Mackey, T.K., and Liang, B.A. (2012). Evaluating the evidence: direct-to-consumer screening tests advertised online. *Journal of Medical Screening*, **19**, 141–153. 1.

Lucky, R. (2000) The quickening of science communication. *Science*, **289**(5477), 259–64.

Marcus, S.J. (Ed.) (2002). *Neuroethics: Mapping the Field*. New York: Dana Press.

Molek-Kozakowska, K. (2016). Stylistic analysis of headlines in science journalism: a case study of New Scientist. *Public Understanding of Science*. Advance online publication. doi:10.1177/0963662516637321

Norman, C.D., McIntosh, S., Selby, P., and Eysenbach, G. (2008). Web-assisted tobacco interventions: empowering change in the global fight for the public's (e)Health. *Journal of Medical Internet Research*, 10, e48.

Office of the Information and Privacy Commissioner (1983). *Personal Information Protection Act*. Quebec, Canada: Office of the Information and Privacy Commissioner.

Palmour, N., Vanderbyl, B.L., Zimmerman, E., Gauthier, S., and Racine, E. (2014). Alzheimer's disease dietary supplements in websites. *HEC Forum*, 25, 361–82.

Rachul, C. and Caulfield, T. (2015). Gordie Howe's stem cell "miracle": a qualitative analysis of news coverage and readers' comments in newspapers and sports websites. *Stem Cell Reviews and Reports*, 11, 667–675.

Racine, E., van Der Loos, H.A., and Illes, J. (2007). Internet marketing of neuroproducts: new practices and healthcare policy challenges. *Cambridge Quarterly of Healthcare Ethics*, 16, 181–194.

Risk, A. and Petersen, C. (2002). Health information on the internet: quality issues and international initiatives. *JAMA*, 287, 2713–2715.

Robillard, J.M., Johnson, T.W., Hennessey, C., Beattie, B.L., and Illes, J. (2013). Aging 2.0: health information about dementia on Twitter. *PloS One*, 8, e69861.

Robillard, J.M., Cabral, E., Hennessey, C., Kwon, B.K., and Illes, J. (2015a). Fueling hope: stem cells in social media. *Stem Cell Reviews*, 11, 540–546.

Robillard, J.M., Illes, J., Arcand, M., et al. (2015b). Scientific and ethical features of English-language online tests for Alzheimer disease. *Alzheimer's & Dementia*, 1(3), 281–288.

Royal Society (1985). *The Public Understanding of Science*. Available from: https://royalsociety.org/~/media/Royal_Society_Content/policy/publications/1985/10700.pdf [accessed February 5, 2017].

Royal Society (2006, June). *Science Communication*. Available from: https://royalsociety.org/topics-policy/publications/2006/science-communication/ [accessed April 16, 2016].

Ryan, K.A., Sanders, A.N., Wang, D.D., and Levine, A.D. (2010). Tracking the rise of stem cell tourism. *Regenerative Medicine*, 5, 27–33.

Sandvik, H. (1999). Health information and interaction on the internet: a survey of female urinary incontinence. *BMJ*, 319, 29–32.

Scullard, P., Peacock, C., and Davies, P. (2010). Googling children's health: reliability of medical advice on the internet. *Archives of Disease in Childhood*, 95, 580–582.

Silberg, W.M., Lundberg, G.D., and Musacchio, R.A. (1997). Assessing, controlling, and assuring the quality of medical information on the internet: caveant lector et viewor – let the reader and viewer beware. *JAMA*, 277, 1244–1245.

Song, T.M., Park, E.J., and Lim, E.J. (2002). The survey of the demand for health information on the internet. *Journal of Korean Society of Medical Informatics*, 8, 17–24.

Thackeray, R., Neiger, B.L., Hanson, C.L., and McKenzie, J.F. (2008). Enhancing promotional strategies within social marketing programs: use of Web 2.0 social media. *Health Promotion Practice*, 9, 338–343.

Trustram Eve, C. and de Jager, C.A. (2014). Piloting and validation of a novel self-administered online cognitive screening tool in normal older persons: the Cognitive Function Test. *International Journal of Geriatric Psychiatry*, 29, 198–206.

United States (2004). *The Health Insurance Portability and Accountability Act (HIPAA)*. Washington, DC: US Department of Labor, Employee Benefits Security Administration.

United States (2009). *The Genetic Information Nondiscrimination Act of 2008 (GINA)*. Washington, DC: US Department of Labor, Employee Benefits Security Administration.

van Uden-Kraan, C.F., Drossaert, C.H.C., Taal, E., Seydel, E.R., and van de Laar, M.A. (2009). Participation in online patient support groups endorses patients' empowerment. *Patient Education and Counseling*, 74, 61–69.

Vance, K., Howe, W., and Dellavalle, R.P. (2009). Social internet sites as a source of public health information. *Dermatologic Clinics*, 27, 133–136, vi.

Walji, M., Sagaram, S., Sagaram, D., et al. (2004). Efficacy of quality criteria to identify potentially harmful information: a cross-sectional survey of complementary and alternative medicine web sites. *Journal of Medical Internet Research*, 6.

Wangberg, S.C., Andreassen, H.K., Prokosch, H.-U., Santana, S.M.V., Sørensen, T., and Chronaki, C.E. (2008). Relations between Internet use, socio-economic status (SES), social support and subjective health. *Health Promotion International*, 23, 70–77.

Waters, E.A., Sullivan, H.W., Nelson, W., and Hesse, B.W. (2009). What is my cancer risk? How internet-based cancer risk assessment tools communicate individualized risk estimates to the public: content analysis. *Journal of Medical Internet Research*, 11, e33.

World Health Organization (n.d.). *Neurology and Public Health*. Available from: http://www.who.int/mental_health/neurology/en/ [accessed April 16, 2016].

Wilsdon, J. and Willis, R. (2004). *See-Through Science: Why Public Engagement Needs to Move Upstream*. London: Demos.

Wolfe, S.M. (2002). Direct-to-consumer advertising—education or emotion promotion? *New England Journal of Medicine*, 346, 524–526.

Zulman, D.M., Kirch, M., Zheng, K., and An, L.C. (2011). Trust in the internet as a health resource among older adults: analysis of data from a nationally representative survey. *Journal of Medical Internet Research*, 13.

Chapter 29

The impact of neuroscience in the law: How perceptions of control and responsibility affect the definition of disability

Jennifer A. Chandler

Introduction

For better *and* worse, biomedical explanations of human behavior, including genetic and neuroscientific explanations, seem to affect public perceptions of a person's moral and legal responsibility for that behavior. This effect flows from the tendency to regard a person's biological constitution as given and largely fixed and, crucially, as a causal contributor to that person's behavior. In other words, the perception seems to be first that a person has no control over the origin and little control over the continuation of that essential biological nature, and second, that this biology determines or at least predisposes a person to a particular behavior. Because the person has little control over any of this, the person is perceived to have diminished responsibility for that behavior. As I will discuss, the extent to which this pattern of thought holds true, and the circumstances in which it does so, are both still matters of debate.

The purpose of this chapter is to consider the hypothesis that neurobiological accounts of behavior will affect the law, not just in the criminal context, but also in how disability is defined in the law. The chapter first presents a very short overview of some of the ways that neurobiology may affect judgments of criminal responsibility, in order to explore the importance of control in assessing responsibility as well as the countervailing impulse to control and supervise those who are judged to be dangerous and incapable of controlling themselves rather than unwilling to do so. The chapter then proceeds to explain how control and responsibility for one's condition affects the concept of disability and the social willingness to offer support and protection. Finally, examples drawn from the treatment in human rights law of contested conditions like addictions to gambling, sex or the Internet, and nicotine are used to illustrate the manner in which medicalization and, where possible, neurobiological explanations, have been used to try to suggest that the conditions constitute disabilities that ought to be accommodated rather than punished.

The positive consequences of this phenomenon lie in the reduction of blame for socially disfavored behavior and traits and increased willingness to forgive and to help. However, the negative consequences flow from the corollary of the idea that the person's behavior is explained by biological factors beyond his or her control, namely that the person is a faulty biological mechanism lacking self-control. In other words, the person is viewed as falling within a different biological category than other people, and furthermore, the person must be supervised or controlled if the disfavored behavior poses a threat. The particular mix of these reactions will likely vary from case to case, and the impulse to help may be swamped by the more negative reactions for behaviors that are perceived as particularly dangerous.

These are general reactions documented and discussed in the study of mental health stigma, biopolitics, and many other fields. In recent years, they have been increasingly discussed in the law. Since the questions of self-control, responsibility, and moral blameworthiness are of central importance in ascribing criminal responsibility and selecting the appropriate social response, a large literature has developed regarding the effect of behavioral genetics and neuroscience on criminal responsibility. The research has been both empirical, attempting to determine the likely or actual impact on moral reasoning and judgment, and also normative, arguing about whether, when, and how this evidence ought to affect moral reasoning and judgment in the context of criminal justice.

But these questions certainly do not exhaust the potential legal ramifications of genetic and neuroscientific explanations of human behavior. The questions of responsibility and self-control are important not just for affixing legal liability for harmful actions; they also turn out to be fundamental to social solidarity. People who are perceived to be in control of, and therefore responsible for their problems are much less likely to be perceived as having disabilities that are deserving of support in the form of financial assistance and protection from discrimination. This is particularly the case for psychiatric or behavioral conditions, as opposed to physical disabilities.

A review of the manner in which the brain disease concept has been used to repel blame and to support claims to solidarity and protection in relation to a multitude of different highly stigmatized behavioral conditions (e.g., schizophrenia, depression, addiction, and obesity) over the years reveals these patterns of thought and their political significance. Of course, the framing of these conditions as a biomedical disease is itself contested. The neurodiversity movement, for example, pushes back against the biomedical model of disability. It accepts biological essentialism and indeed crafts a group identity around the biological difference, but rejects the pejorative framing of the difference as a disease or disorder. The broader concept of the social construction of disability, which continues to battle the biomedical model of disability, locates the negative consequences of a physical or mental difference in the intolerant and inflexible social environment rather than within the person him- or herself. While these movements have succeeded to some extent, much of the law remains firmly within the biomedical model.

To some extent this is an old story. The pressure to medicalize various conditions in order to fend off blame and judgment and to elicit support is not new, and operates even where neither genetic nor neurobiological causal factors can be identified. Even a psychiatric diagnosis offers something in this ideological battle. However, there is evidence that the public tends to assign more control and therefore more blame to a person for his or her behavior when it is explained in psychiatric rather than organic terms. In other words, the brain-based account is likely to be a more potent force in reducing blame assigned to those with stigmatized and disfavored behaviors and conditions.

A final dimension in this exploration of the ideological impact of neurobiological explanations of behavior is the question of self-fulfilling prophecies. If people are pushed to claim a lack of self-control in order to ward off blame and judgment, or to claim a right to social support, are we in fact forcing them into that corner? The potentially dispiriting and anti-therapeutic effects of brain disease models of mental illness and addictions have been pointed out many times. In the legal context, the analogue to this concern is that biological deterministic explanations of criminal behavior may be demotivating and may undermine rehabilitation and reintegration. Similarly, a disability definition built on establishing a lack of control over the condition may itself be disabling because it undermines a person's motivation and sense of control.

Perceived control and criminal responsibility

The question of capacity is central to legal responsibility. We do not tend to blame people for their actions when they lack the capacity to act otherwise. We may, however, blame those who cause harm while in the grip of self-induced incapacity, according to what has been called the "tracing theory of moral responsibility" (Fischer and Ravizza 1998).

In Canadian law, the threshold for exoneration due to lack of capacity—the defense of mental disorder—is available only to those whose mental disorder renders them incapable of appreciating the nature and quality of their acts or of knowing they were wrong (*Criminal Code*, s.16). This test focuses on cognitive capacity as opposed to the capacity for self-control. Criminal responsibility also requires a voluntary act, so a person who commits an act while in the grip of a seizure or while unconscious will not be criminally responsible (Roach 2012, pp.117–119). Unlike some other jurisdictions such as Ireland, South Africa, and some jurisdictions in Australia and the United States, Canada does not recognize a defense of "irresistible impulse" although it is clear that people differ in their impulsiveness (Penney 2012).

Only cases of extreme cognitive incapacity or complete lack of self-control will deflect criminal responsibility altogether in Canada. Less severe degrees of impaired capacity for understanding or self-control are often considered in assessing the moral blameworthiness of an act at the stage of sentencing a convicted offender. The Canadian *Criminal Code* identifies as a fundamental principle that "a sentence must be proportionate to the gravity of the offence and the degree of responsibility of the offender" (*Criminal Code*, s.718.1; Roach 2012, pp.476–477).

At the sentencing hearing, convicted offenders frequently point to factors that suggest diminished capacity in order to try to reduce their degree of responsibility and thus the moral blameworthiness of their acts. For example, offenders may point to mental disorders that, while insufficient to render a person non-criminally responsible, may serve to reduce the moral blameworthiness of their actions and the severity of the sentence that is viewed as appropriate (Ruby 2004, para. 5.246; *R. v. Belcourt* 2010).

At the same time, causal explanations for the criminal behavior that suggest diminished capacity may increase the perception that an offender is out of control and a risk to the public. A sentence reflects many objectives, including not just retributive punishment proportionate to the degree of moral blameworthiness but also a range of consequentialist objectives including the protection of the public. Where prospects for rehabilitation are considered to be poor, a heavy sentence may be viewed as necessary for public safety even if moral blameworthiness is lowered due to diminished capacity. This double effect—the reduction of moral blameworthiness, which would reduce the length of the sentence, along with an increased perception of dangerousness, which might increase the length of the sentence—has been termed the "double-edged sword" of neurobiological causal explanations of criminal behavior (Aspinwall et al. 2012). To the extent that neurobiological explanations invite the conclusion of a "broken brain" that erodes self-control and is unlikely to be amenable to treatment, the outcome on sentence may depend upon the circumstances (e.g., the perceived riskiness of the offender or the type of offence in question).

An example of these countervailing impulses is the sentencing decision in the case of *R. v. Obed* (2006). Obed was sentenced to 10 years of imprisonment for a violent sexual assault. At the sentencing hearing, the judge considered evidence regarding Obed's chaotic and abusive childhood, as well as evidence of brain damage associated with prenatal alcohol exposure. The judge clearly felt that these factors meant that Obed had diminished responsibility for his crime. Yet, the judge also noted that there was neither a good treatment for Obed's brain damage nor good prospects for rehabilitation. The judge concluded that the predominant consideration in selecting the sentence was the protection of the public (*Obed* 2006, para. 54). At the conclusion of the sentence, the judge addressed the offender directly in a way that clearly shows the tension between retribution and consequentialism in a case of diminished capacity due to brain damage:

> Mr. Obed before you go. I don't know what more to say to you. I feel almost helpless in one way by the inability of the court in being able to do something that would offer you some form of rehabilitation. This is the first case I've ever had where I felt that rehabilitation was not a realistic factor. I hope that somehow the Federal or Provincial authorities will take a hard look at this FASD [fetal alcohol spectrum disorder] phenomenon. I hope, as well, that you become the person that they rally around and work with and see if somehow, some of this damage ... Dr. Rosales says that it's irreversible and that the brain no longer develops after a certain point. That's what we know right now ... Five years, two year from now maybe we'll know a whole lot of different things. Maybe in a couple of years from now we can actually go inside the brain and reorganize things. I don't know. I just hope in your case, because this is one of those cases that to me falls somewhere between total responsibility and diminished responsibility and I don't know really ... But I refuse to be without

hope and in your case I hope that something comes along that can focus on you and help you with this … You're not that old that further research can't have a profound effect on you and I would hope that you do benefit from such future research. In any event, in the meantime, I have to protect society from your violent behaviour. I have to protect people from you, even from behaviour that you don't know you're going to commit yet. (*R. v. Obed* 2006, para. 78)

With the development of behavioral genetics, particularly the identification of gene variants thought to be associated with an increased risk for impulsive violence, as well as neuroimaging that is looking for structural and functional correlates of particular behavioral abnormalities such as psychopathy, there has been an increase in interest in how such biological evidence might be used in the criminal justice context. Among the questions posed are the extent to which this evidence is being used in criminal cases, and the issue of how it is or should be affecting assessments of criminal responsibility and the selection of the appropriate responses to offending. Of course, evidence related to mental disorder has long been presented to criminal courts. The more novel question posed by these emerging forms of biologically focused accounts of behavior is whether they are more likely or not than the more typical psychiatric diagnoses to reduce attributions of control and blame.

While the research results are inconsistent so far, there are intriguing studies that do suggest that the effects of neurobiological explanations sometimes differ from those of psychiatric explanations of criminal behavior.

Aspinwall and colleagues surveyed a group of nearly 200 American state trial judges and found that evidence supporting a neurobiological cause of an offender's psychopathy reduced the sentence imposed as well as the extent to which the psychopathy was viewed as an aggravating factor (Aspinwall et al. 2012). The judges considered a vignette in which a psychiatrist provided expert testimony that a man who committed a violent assault was a psychopath. Half of the participants also received the expert testimony of a neurobiologist who presented a neurobiological explanation of the psychopathy. The participants were able to provide explanatory comments on their responses. One judge wrote that "[t]he evidence that psychopaths do not have the necessary neural connections to feel empathy is significant. It makes possible an argument that psychopaths are, in a sense, morally 'disabled' just as other people are physically disabled" (Aspinwall et al. 2012, p.847). The neurobiological explanation of psychopathy appears to have reduced perceived moral blame but at the same time may have increased the perception of increased dangerousness. Another judge wrote that "[p]sychopathy may make the defendant less morally culpable, but it increases his future dangerousness to society. In my mind, these factors balance out" (Aspinwall et al. 2012, p.848).

Fuss and colleagues repeated with a group of German judges a version of the study conducted by Aspinwall and colleagues (Aspinwall et al. 2012; Fuss et al. 2015). They presented the judges with a vignette in which an expert provided a psychiatric diagnosis of psychopathy. Half also received expert testimony from a neurobiologist explaining psychopathy as a genetically driven developmental condition and indicating that the offender had the gene in question. Fuss and colleagues found that the neurobiological explanation slightly reduced the judgment of legal responsibility, but it did not significantly affect the prison sentence imposed (Fuss et al. 2015). The presentation of the neurobiological evidence by the prosecution did greatly increase the decision to order involuntary commitment in a psychiatric hospital, which could lead to a longer detention time.

There have also been studies of public reactions to biological explanations of behavior. Monterosso and colleagues surveyed the public to see whether scientific explanations of undesirable behavior reduced perceived culpability (Monterosso et al. 2004). They looked at four types of undesirable behavior: setting fire to a building, killing a store clerk over a disagreement, overeating, and consistent failure to follow through on plans. They found that physiological explanations were more likely to lead to reduced judgments of culpability than socioenvironmental explanations in all four cases. For example, in the case of the man who killed the store clerk, attributions of culpability were lower where the explanation given for his behavior was a greatly elevated amount of a particular brain chemical versus the experience of severe and brutal abuse as a child. The physiological explanations also elicited lower scores on a composite measure, showing reduced judgments of voluntariness, greater sympathy, greater mitigation of blame, more positive treatment, and greater likelihood that the participants themselves would behave in the undesirable way if the explanatory condition were true of them (Monterosso et al. 2004, p.150).

Monterosso and colleagues speculated that the Western perception of the person is dualistic, and the mind (or soul) and body are perceived as two separate forces independently capable of bringing about behavior (Monterosso et al. 2004, p.154). Further, behavior tends to be perceived as voluntary or "owned by the self" only when it flows from the mind (or soul), and where one force is operating it is perceived as denying the causal efficacy of the other. They suggest that this explains "why, in our vignettes, when a physiological explanation was given, participants tended to view the body as the cause of the behaviour and motivations as less relevant, with the result that the behaviour was perceived as less voluntary" (Monterosso et al. 2004, p.155). They suggest that this manner of thought may lead to reduced space for intentional agency as the physiological behavioral sciences advance.

There is also research questioning whether biological explanations of criminal behavior will affect judgments of responsibility. Scurich and Appelbaum (2015) failed to find an effect of biological (genetic and neurobiological) explanations of wrongful behavior on public decisions about punishment. They suggest several possible explanations. First, ordinary people may regard biological predisposition as only one influence on behavior, and one that is not necessarily the prime determinant of the behavior in a given context. Second, they suggest that these explanations might activate countervailing concerns about dangerousness, which negate any mitigating impact on punishment.

In sum, the evidence is presently mixed on the question of whether or when biological explanations of criminal behavior produce different perceptions of control and responsibility than psychosocial explanations. Nonetheless, some of the experimental evidence suggests such an effect in both public and judicial decision-making. The ultimate effect on punishment decisions may be slight for serious offences, however, given that perceptions of dangerousness may increase at the same time as perceived culpability decreases.

Perceived control, responsibility, and social solidarity

Perceptions of control and therefore responsibility for a condition also affect the willingness to assist and protect people with disabilities or those in financial need. When a person is perceived to have the power to avoid or to mitigate a particular condition or situation, others are less likely to respond with sympathy and the desire to help.

Bernard Weiner has suggested that attributions about cause and control leading to judgments about personal responsibility are ubiquitous in human interactions, and they drive emotional reactions such as anger or sympathy as well as behavioral reactions like help or punishment (Weiner 1993). This model has come to be known as "attribution theory" in social psychology, and has been explored as an explanatory model in a wide range of contexts including mental health stigma (Weiner et al. 1988; Corrigan 2000; Corrigan et al. 2003) and support for social welfare programs (Van Oorschot 2000, p.36; Weiner et al. 2011).

In the model suggested by attribution theory, controllability refers to the degree of voluntary influence a person has over a cause. A further distinction in relation to controllability is made between "onset controllability" (control over developing a condition) and "offset responsibility" (capacity and responsibility for coping with or overcoming a condition) (Corrigan 2000 pp.52–53).

People who are perceived to be in control of a negative event are more likely to be held responsible and to elicit angry or punitive reactions. Those who are perceived to lack control are more likely to be pitied and helped. A third reaction, independent of attributions of control and responsibility, is driven by perceptions of dangerousness that produce fear and elicit efforts to avoid or to control a person (Corrigan et al. 2003).

Another body of research looks at whether causal explanations of behavior based on biology or physiology (as opposed to mental or psychological explanations) have different effects on attributions of control and responsibility for that behavior. This has been a topic of particular interest in the area of mental health stigma, where it was hoped that biological explanations of psychiatric illnesses would reduce perceived responsibility for behavior and symptoms and help to de-stigmatize these conditions. A meta-analysis by Kvaale and colleagues suggests that biological explanations may indeed diminish the tendency to blame people for psychological problems (Kvaale et al. 2013). However, these explanations increase perceived unpredictability or dangerousness, as well as prognostic pessimism. The impact of biological explanations may, however, vary according to the specific psychological problem in question, the type of biological causal explanation given (e.g., genetic versus neurochemical), and treatability, among other variables (Kvaale et al. 2013; Lee et al. 2014).

Finally, researchers have also explored the willingness of the public to recognize conditions as disabilities worthy of social assistance and protection. Mueller and colleagues explored public views on which of 57 different health conditions should be viewed as disabilities worthy of legal protection under the *Americans with Disabilities Act* (ADA) (Mueller et al. 2010). The conditions differed according to whether they were physical or psychological, visible or not, static or progressive, and whether there was personal control over onset. They found that respondents did not regard conditions that may result from poor decision-making or lifestyle choices as disabilities worthy of ADA protection. These included alcoholism, drug addiction, obesity, HIV/AIDS, emphysema, and eating disorders. Ultimately, "[v]isible physiological health conditions were accorded higher rates of protection under the provisions of the ADA than invisible, psychological, or behavioural/self-imposed health conditions" (Mueller et al. 2010, p.178).

Esses and Beaufoy (1994) explored attitudes to three types of disability: chronic depression, AIDS, and amputees. They found that attitudes to amputees were more favorable than toward people with AIDS or chronic depression. Participants attributed greater control over the occurrence of their conditions to people with AIDS and chronic depression and very little control to amputees. Furthermore, high attributions of control were correlated with particularly unfavorable attitudes to those with AIDS and chronic depression.

Together, these three lines of inquiry all provide some support for several ideas. First, a person who is perceived to be in control of their condition or situation is more likely to be considered responsible for it, increasing blame and reducing the impulse to assist. Second, a biological explanation of a condition or behavior may be a more powerful inducer of these reactions than a psychological explanation. This provides the foundation for the following exploration of how neurobiological explanations of behavioral conditions may serve to support claims that those conditions should be treated as disabilities entitled to social assistance and protection from discrimination.

The meaning and scope of disability in the law

The human rights of people with disabilities are now protected through international treaties such as the UN *Convention on the Rights of Persons with Disabilities* (United Nations General Assembly 2007) and, in Canada, through the *Charter of Rights and Freedoms* (1982) and provincial human rights codes. The *Charter* requires federal and provincial governments in Canada to respect the right to equality and equal benefit of the law without discrimination on the basis of physical or mental disability (s.15(1)).

In addition, provincial human rights legislation, such as Ontario's *Human Rights Code*, prohibits discrimination on the ground of disability by public and private actors in particular contexts such as access to services, employment, and residential accommodations. The *Code* requires only reasonable accommodation of disability, or accommodation up to the point of undue hardship, and there will be no violation of a person's rights under the *Code* if the obstacles cannot be remedied without undue hardship (s.17(1)). It is also important to note that the jurisprudence that has developed around the concept of reasonable accommodation also makes it clear that the claimant has a responsibility to make

reasonable efforts to facilitate the accommodation by, for example, participating in a rehabilitation program:

> When an alcoholic employee has failed to respond to multiple rehabilitation efforts and there is no
> objective evidence that further efforts at accommodation would be likely to succeed, it is generally
> concluded that the employee has been accommodated to the point of undue hardship. (*Kellogg
> Canada Inc v. Bakery, Confectionary, Tobacco Workers & Grain Millers* 2006)

Many of these legal instruments do not provide particularly precise definitions of disability, meaning that the list of specific impairments that fall within the scope of entitlements
to support or protection from discrimination evolves over time and remains uncertain in
some cases (Law Commission of Ontario 2009, pp.10–11). It is this area of uncertainty
that may ultimately be influenced by emerging neurobiological explanations of stigmatized behavior and conditions. As noted in the section "Perceived control, responsibility, and social solidarity," attribution theory predicts that conditions and behaviors that
are seen to fall outside personal control are more likely to generate reactions of social
solidarity—willingness to provide support and protection from discrimination.

The political and economic significance of the definition of disability is clear. As the
Law Commission of Ontario (2009, p.11) puts it, claimants seek "the label of 'disability'
in order to access benefits and supports, while institutions [withhold] the label in order
to maintain program standards." Other institutional actors may also use the definition of
disability strategically, inappropriately finding disabilities where this will generate higher
funding allocations (Law Commission of Ontario 2009, p.12). Others police the boundary of disability to ward off what they perceive as inappropriate attempts to appropriate
resources that should be made available to those whose impairments they consider more
genuinely disabling. For example, Gilbert and Majury cite the argument that a legal claim
lodged by an infertile couple that the exclusion of a form of infertility treatment from
provincially funded health care discriminated on the basis of disability (i.e., infertility)
"appropriated a disability rights discourse [in order] to gain access to resources and services forged by earlier efforts at advocating very different disability issues" (Gilbert &
Majury 2006, p.112). Of course the use of models that attract social sympathy and support will be taken up by others seeking a similar response, a pattern revealed by the use
of the concept of addiction to describe other compulsive behaviors unrelated to drugs or
alcohol, including compulsive behavior related to food, the Internet, and sex.

The medicalization of a condition tends to support its inclusion within the scope of disability for legal purposes, although this is not invariably the case. The biomedical model
of disability—the dominant understanding of disability for much of the twentieth century, and still the dominant public conception of disability—is reflected in many laws. For
example, access to disability benefits is largely defined in biomedical terms and eligibility for benefits is verified by medical professionals (Law Commission of Ontario 2009,
p.17). As a result, the development of a biomedical conception of a condition is often
a key requirement to accessing the protections available by law. It is not strictly necessary that this conception involve a known biological etiology to qualify as a disability for

legal purposes, because the human rights laws cover both physical and mental disabilities. However, the claim to inclusion as a disability and to social solidarity is strengthened where there is a lack of control over a disabling condition, which biological causal explanations of the condition suggest more persuasively than do psychological causal explanations.

Drug and alcohol dependency and addiction are considered disabilities under human rights legislation (*Canadian Human Rights Act* s.25; Ontario Human Rights Commission 2014). Thus, the exclusion of people with drug and alcohol dependency or addiction from eligibility for disability benefits constituted discrimination on the basis of disability (*Ontario v. Tranchemontagne* 2009). The human rights jurisprudence now commonly refers to drug and alcohol dependency and addiction as a disease of the brain (e.g., *Milazzo v. Autocar Connaisseur Inc.* 2003), a framing that has been in the ascendant since the early 1970s (Vrecko 2010, p.58).

The path forged by drug and alcohol addictions toward a neurobiological disease model is now being emulated in relation to a range of other compulsive behaviors. As Vrecko (2010, p.54) puts it, "[p]ersonal, social and legal issues associated not only with drugs, but with a range of compulsive and problematic behaviours—including criminal offending, spending and debt, gambling and obesity—are coming to be represented as, at least in part, problems of the molecular body and brain." Canadian legal cases reveal the manner in which neurobiology is being adopted in the legal context to reduce perceptions of control and moral blameworthiness for this broader range of stigmatized behavior.

Disability claims involving addictions to gambling, nicotine, sex, and the Internet are presented in the following sections. While nicotine addiction has been medicalized for some time, warranting its own section—"Nicotine-related disorders"—within the *Diagnostic and Statistical Manual of Mental Disorders*, fourth edition (DSM-IV) section on "Substance-related disorders" (American Psychiatric Association 2000, p.264), gambling addiction has recently been reclassified because of neurobiological evidence. Pathological gambling was formerly classified in the DSM-IV as an "impulse control disorder" (American Psychiatric Association 2000 p.312.31) and was reclassified in the DSM-5 within the section on "Substance-related and addictive disorders" (American Psychiatric Association 2013). The notes on the section indicate that gambling disorder has been added because of "evidence that gambling behaviors activate reward systems similar to those activated by drugs of abuse and produce some behavioral symptoms that appear comparable to those produced by the substance use disorders" (American Psychiatric Association 2013).

As for addictions to sex and the Internet, official psychiatric diagnoses are not yet recognized. As the notes on gambling disorder in the DSM-5 (American Psychiatric Association 2013) state:

> Other excessive behavioral patterns, such as Internet gaming, have also been described, but the research on these and other behavioral syndromes is less clear. Thus, groups of repetitive behaviors, which some term *behavioral addictions*, with such subcategories as "sex addiction," "exercise addiction," or "shopping addiction," are not included because at this time there is insufficient

peer-reviewed evidence to establish the diagnostic criteria and course descriptions needed to identify these behaviors as mental disorders.

Gambling addiction

Gambling addictions typically arise in legal cases in one of two contexts. First, they are raised as explanations for criminal behavior (e.g., stealing from an employer to finance gambling). The judicial reactions to this type of argument are mixed, perhaps reflecting the degree of perceived lack of self-control in specific cases. Second, they are raised to try to challenge dismissal from employment. In these employment grievances, the argument made is that the gambling addiction is a disability and so the wrongful behavior like stealing from the employer merits less severe disciplinary measures and in fact may even require an employer to accommodate the employee's disability by, for example, reassigning the employee to a position in which theft will not be possible. Courts and arbitrators have not been very receptive to these arguments. Despite attempts to frame gambling addiction as a brain disorder, they remain suspicious that the condition necessarily entails a sufficient lack of self-control to relieve a person from responsibility. Furthermore, the accommodations requested may be perceived as unreasonable.

The case of *R. v. Smith* (2015) involved a woman who pleaded guilty to stealing from her employer to finance her gambling. At the sentencing, the defense tendered evidence that pathological gambling was now classified as an addiction in the DSM-5 because of "[b]rain imaging studies and neurochemical tests [that] have provided compelling evidence that gambling activates reward systems in the brain much the same way that drugs and alcohols do," and that it generates "physiological cravings and the experience of being 'high' or in an altered state of consciousness similar to the experience of substance dependence" (*Smith* 2015, para. 48).

The defense expert also indicated that, as with substance addiction, gambling addicts also experienced craving and loss of control, and that although "[t]he neurobiology associated with behavioural and substance addictions overlap but are not identical ... there is also convincing research indicating common genetic factors including heritable phenotypes that underlie substance abuse and gambling disorders" (*Smith* 2015, para. 49).

The judge considered precedents in which gambling addictions were cited as mitigating factors at sentence, one of which was *R. v. Horvath* (1997), in which a distinction was drawn between pathological and compulsive gambling (*R. v. Horvath* 1997, para. 81). The court in *Horvath* had accepted the medical model of pathological gambling and concluded that it reduced moral blame:

> The offences were the products of a distorted mind—a mind seriously diseased by a disorder now recognized by the medical community as a mental disorder. The acts committed at the command of that mind were not acts of free choice in the same sense as are the acts of free choice of a normal mind. A pathological gambler does not have the same power of control over his or her acts as one who does not suffer from that complex disease. Accordingly, where those acts constitute criminal offences, the moral culpability—moral blameworthiness—and responsibility are not of the same order as they would be in those cases where the mind is not so affected.

However, the judge in *Smith* was not ultimately persuaded that Smith had a condition that undermined her self-control, and so rejected the argument that her gambling problem should mitigate personal responsibility:

> People with serious mental disorders cannot, through sheer willpower and resolve, rid themselves of their afflictions. In notable contrast, once Ms. Smith's crime was discovered and she resolved to stop gambling, she was able to do so without apparent difficulty and with lasting results. If only it were so easy for the mentally disordered. (*R. v. Smith* 2015, para. 92)

In the employment context, employees dismissed for theft sometimes bring grievances on the basis that the disciplinary measure was unfair because the behavior was caused by a disability—pathological gambling—that had subsequently been addressed. The role of demonstrable biological etiology in buttressing such a claim is revealed in an older case, in which the arbitrator considered the claim for reinstatement and drew a distinction between so-called easy cases in which there was a clear physical reason for the wrongful conduct, and the more common cases where there are no physical reasons for it (*Canada Safeway Ltd.* 1999):

> In the former category would fall cases where there is an identifiable physical reason for aberrant conduct. An easy example is where an employee had a brain tumor [that] caused improper conduct such as violence or theft. If it can be shown that the brain tumor was a direct contributor to the aberrant conduct and that it has now been cured, most employers and arbitrators would reinstate the employee. The most common cases that seem to arise do not deal with clearly identifiable physical problems such as a brain tumor, rather they most often deal with alcohol addiction, drug addiction and, more recently, gambling addiction. Even in these cases there is a tendency to characterize the "condition" as an illness. Often much of the evidence will be led for that purpose. (para. 62)

The arbitrator then went on to say it wasn't absolutely necessary to establish a "recognizable condition" or "true illness," but there must be a condition that leads to the improper behavior and it must be one that essentially removes self-control from the employee:

> Again using the brain tumor example, it may be concluded that the employee was acting as an automaton and was not responsible whatever for his or her conduct. On the other hand, even if a gambling or alcohol addiction is established, and it is established that *but for* the condition the aberrant conduct, such as theft, would not have occurred, it still may be concluded that the grievor possesses sufficient responsibility for his or her actions so that a substitution of penalty is not appropriate. This is precisely what occurred in the *SaskTel* case where it was accepted that the grievor had a pathological or compulsive gambling addiction and that it contributed to his acts of theft but it was concluded that he was still responsible because he had been fully aware of his problem and he was fully aware of the avenues open to him to have his problem dealt with. In other words, the mere existence of an addiction does not in itself explain or justify serious aberrant conduct. There are many people with alcohol, narcotic and gambling addictions but a very small number of those people steal money. (para. 66)

Similar reasoning has been subsequently followed in a range of other labor arbitrations involving pathological gambling. For example, in *Manitoba v. Manitoba Government and General Employees' Union* (2005), the arbitration panel doubted that gambling addiction made the decision to steal involuntary, despite evidence that gambling addictions "hijack the brain" (paras. 48, 59, 85).

In the US context, Tovino (2015) writes about another way in which the neurobiological conceptions of pathological gambling may affect benefits entitlements and protection from discrimination. While her example does not have to do directly with how biological causal explanations affect judgments of control and responsibility, it does suggest another way in which the legal treatment of disabilities may be affected by neurobiological conceptions of disability. Tovino (2015) observes that the neurobiologically driven reclassification of pathological gambling as a form of addictive behavior in the DSM-5 may support the argument that it should no longer fall within the category of "impulse control disorders" for the purposes of health insurance in Nevada, which excludes that category from coverage but includes coverage for substance-related conditions (drug and alcohol use disorders). She also predicts that neurobiological advances may lead to challenges to certain exclusions from the scope of anti-discrimination laws. For example, Title I of the ADA excludes people engaging in the use of illegal drugs from protection against discrimination in employment (Tovino 2015, p.723). However, those who have engaged in drug rehabilitation programs and are not currently using illegal drugs are protected. Title I of the ADA also specifically excludes from employment discrimination protection "compulsive gambling," and several other impulse control disorders, but does not include a similar proviso protecting those who have been successfully treated. Tovino (2015) suggests that the evolving understanding of compulsive gambling may permit a stakeholder to argue that gambling disorder should be treated more like alcohol or drug use disorders under antidiscrimination laws.

The example of gambling addictions sheds some light on the effect of neurobiological explanations of behavior on the concept of disability. Although brain-based evidence may help to support the idea that someone may become addicted to the act of gambling, this may not be sufficient where the disfavored behavior is the upstream activity of stealing in order to finance the compulsive behavior. Unlike the following examples of nicotine and Internet sex addiction where the disfavored behavior that is being punished is the compulsive behavior itself, in pathological gambling it is not the gambling but the stealing that is being punished. In these cases, decision-makers appear to doubt that the compulsion to gamble necessarily made the theft sufficiently involuntary to reduce blame. Nevertheless, the invocation of neurobiological evidence supports the idea that it is being used in an attempt to suggest lack of self-control and responsibility, and may also have other effects on disability rights protections through the shifting classification of various conditions.

Nicotine addiction

Although nicotine addiction is a medicalized condition, the case law on whether it is a disability is mixed. The focus of the dispute is the degree of loss of voluntary self-control.

In *McNeill v. Ontario Ministry of the Solicitor General and Correctional Services* (1998), an inmate challenged a smoking ban in a detention center on the basis that it discriminated against him on the basis of disability. He argued that addiction to drugs and alcohol were recognized as disabilities for the purposes of human rights protections and that addiction to tobacco should similarly be recognized as a disability. The Ontario Court of Justice ruled that smokers do not have a "mental or physical disability." The Court held that "[a]ddiction to nicotine is a temporary condition that many people voluntarily overcome, albeit with varying levels of difficulty related to the strength of their will to discontinue smoking," and that "smoking and the addiction that often accompanies it does not interfere with a person's effective physical, social and psychological functioning, the results that often characterize addiction to alcohol" (para. 29). The court also observed sarcastically that the claimant "was in a position to do something about the discrimination of which he complained. He could forego his frequent attendances at the Wellington Detention Centre" (para. 31).

Other courts have rejected the idea that nicotine addiction is a disability for the purposes of the antidiscrimination provisions of the *Charter of Rights and Freedoms* (*R. v. Ample Annie's Itty Bitty Roadhouse* 2001; *Yellowknife (City) v. Denny* 2004). The judge in *Yellowknife* wrote that an addiction to smoking:

> may be a mental or physical disability by somebody else's definition, but not pursuant to section 15 [of the *Charter*]… Smokers can successfully overcome their addiction. Help is close at hand. There is nothing discriminatory in the [non-smoking] by-law. If it has a purpose or effect that addresses the issue of addiction, it is a positive, rehabilitative by-law that should be welcomed by the public at large, including those so addicted. (paras. 56–60)

In a different context dealing with a union's challenge to a company's nonsmoking policy, an arbitration panel later found that nicotine addiction was a disability protected from discrimination for the purposes of human rights legislation (*Cominco Ltd.* 2000). The union claimed that the policy discriminated on the basis of disability, namely nicotine addiction, because addicted smokers cannot control their addiction (*Cominco Ltd.* 2000, para. 14). The company argued that addiction to nicotine or smoking was not a disability because it is a temporary condition that many people overcome voluntarily and it does not interfere with a person's physical, social, or psychological functioning (*Cominco Ltd.* 2000, para. 174).

The panel in *Cominco Ltd.* declined to follow the *McNeill* decision on the basis that the judge in that case had not considered the kind of scientific evidence made available to the panel (*Cominco Ltd.* 2000, para. 179). It cited evidence that that nicotine was as or more addictive than cocaine or heroin, that a puff of cigarette smoke was a more direct method of delivery to the brain than injecting into a vein in the arm (*Cominco Ltd.* 2000, para. 180), and that those who are heavily addicted are disabled as they are "unable to control their addiction" (*Cominco Ltd.* 2000, para. 203). Thus, the nonsmoking policy was discriminatory at least for heavily addicted smokers as it is essentially "tantamount to telling them that they will not continue to be employed at Cominco because the inevitable result is that they will be terminated" (para. 204).

This being said, another labor arbitrator rejected a union grievance on behalf of an employee terminated for smoking contrary to a clear nonsmoking policy in a workplace with flammable chemicals. The union argued that the employee's misconduct was not willful because he was addicted to nicotine. The arbitrator said no medical evidence had been provided, and even if he was addicted, he could have used nicotine-containing alternatives as a substitute to smoking (*Invista (Canada) Co.* 2013). The question of whether nicotine addiction will be accepted as a disability under human rights legislation continues to evolve, and a human rights complaint arguing that nicotine addiction constitutes a disability that must be reasonably accommodated is apparently scheduled to be heard in the summer of 2016 in a challenge to a condominium's nonsmoking bylaw (*Strata Plan NW 1815* 2016).

The issue of self-control and nicotine addiction is also raised outside human rights litigation. In a recent product liability class action lawsuit against tobacco companies in the province of Quebec, the experts engaged in a battle over whether nicotine dependence should be characterized as a "brain disease." The importance of this point had to do with whether the smokers ought to be viewed as solely or partly responsible for the health consequences if they persisted in smoking despite knowing it was harmful. As the companies put it, smokers "are entitled to take risks and ... they knew or could have known about the health risks associated with smoking" (*Letourneau* 2015, para. 34.) As a result, the questions of whether smokers retain control and, if so, how much, were important issues in dispute, and the experts on both sides cited neurobiological evidence.

One of the companies' experts rejected the medical model of drug use in favor of a socioenvironmental approach, but was strongly criticized by the judge for a perceived lack of objectivity (*Letourneau* 2015, para. 163):

> [H]is almost total dismissal of the pharmacological effects of nicotine on the brain is not supported by the experts in the field. He implicitly recognized this when, after much painful cross-examination, he admitted that nicotine does, in fact, have a pharmacological effect on the brain. He stated that nicotine binds to receptors in the brain, thus causing "brain changes." ... Such changes do not mean that the brain is damaged, in his view, because they are not permanent. He cited a study ... showing that the brains of people who quit smoking "return to normal" after twelve weeks. That this indicates that the smoker's brain was, therefore, not "normal" while he was smoking seems not to have been considered by him. (*Letourneau* 2015, paras. 161–162)

The plaintiffs' expert, on the other hand, described nicotine dependence as a brain disease, and characterized the user as a "slave" who had lost "freedom of action" (*Letourneau* 2015, paras. 172–173). The companies' experts energetically disputed this, arguing that nicotine use does not remove freedom of action, as is clearly demonstrated by the many people who have quit smoking (*Letourneau* 2015, para. 173).

The Court ultimately found that "nicotine affects the brain in a way that makes continued exposure to it strongly preferable to ceasing that exposure. In other words, although it can vary from individual to individual, nicotine creates dependence. That is the point" (*Letourneau* 2015, para. 179). The judge also noted that tobacco dependence was listed as a psychiatric disorder in the DSM-5, and the marketing of a dependence-inducing product could be the basis of a product liability claim (*Letourneau* 2015, paras. 183–184). As a result, those who started smoking at a time when the addictive and harmful properties of tobacco were known were held partly responsible, but the others who were already dependent when the harms became known were not at fault (*Letourneau* 2015, paras. 829–830).

Although the sample size does not permit firm conclusions about the role of neurobiol-ogy on the concept of disability, these cases suggest a number of points. First, medicaliza-tion is not in itself sufficient to guarantee that a condition will be recognized as a disability for legal purposes. In some of these cases, the effort to deflect blame and to claim accom-modation of a disability was unsuccessful because the claimant was perceived to retain some control over the condition. The invocation of neurobiological explanations of nico-tine addiction seems to help to produce a more sympathetic response to the difficulties of smokers, as the cases of *Cominco* and *Letourneau* reveal. Finally, the legal context appears to matter. There seems to be a greater willingness to support the claimant where the other side is profiting in some way from the claimant's alleged addiction (as in the product liability case in *Letourneau*), than there is in cases where a party is being asked to accom-modate the claimant's alleged addiction.

Internet sex addiction

The claim of Internet sex addiction is now being raised occasionally in cases concerning criminal activities involving child pornography or child luring, (e.g., *R. v. H.T.* 2010) as well as in labor arbitration cases where employees dispute disciplinary actions taken by employers arising out of the same behavior. In addition, the claim is also being raised in other cases that involve non-criminal behavior that is nonetheless contrary to workplace policies, such as compulsive viewing of pornography at work.

As with the examples of gambling and nicotine addictions, the focus remains on whether the person concerned has lost self-control. At present, the Canadian cases do not seem to refer to any neurobiological evidence to support the claim of loss of control in Internet sex addiction, which is surprising given that such evidence is available in the literature (e.g., Love et al. 2015). Although a diagnosis of "hypersexual disorder" (includ-ing compulsive use of pornography) was proposed for the DSM-5, it was not accepted (Love et al. 2015). On the other hand, "Internet gaming disorder" has been proposed as a condition warranting further study within the DSM-5 (American Psychiatric Association 2013, Section III). The current lack of a recognized medical diagnosis may explain the fact that legal arguments remain tentative and underdeveloped, for now. Several examples illustrate the current shape of the arguments about the compulsive use of the Internet for sexual purposes.

In *Manitoba Government and General Employees' Union v. Manitoba (M.W. Grievance)* (2013), the labor union filed a grievance challenging the termination of an employee who sent and received a large volume of pornography on his work computer. The arbitrator rejected the argument that sex addiction was a mitigating factor that made the termina-tion excessive because "sex addiction" was not a medically recognized diagnosis (paras. 94–96). Interestingly, the employer attempted to turn this argument against the employee, suggesting that "the grievor's continued reliance on a sex addiction as an explanation demonstrated that he was continuing to blame factors out of his control for his behaviour" (para. 102). The grievance was ultimately successful nonetheless for other reasons.

Another case involving an employee fired for pornography use at work contrary to acceptable use policies was *Interior Health Authority (South Similkameen Health Centre) v. Hospital Employee's Union (RP Grievance)* (2013). The employee argued that he had the condition of "sexually compulsive behaviour regarding viewing pornography," and that the employer was obligated to make reasonable accommodation of his disability. There was supporting evidence provided in this case by a clinical social worker, accepted as an expert for the purpose of making diagnoses under the DSM-IV. The expert acknowledged that sexual compulsivity was not a diagnosis under the DSM-IV and so no claim about a diagnosis was advanced, but he did proceed to provide evidence that RP ranked in the moderate to severe range of compulsive pornography use on a Sexual Addiction Screening Test and an Internet Sex Screening Test. The arbitrator rejected the claim saying that, in the absence of expert medical or psychiatric evidence supporting the existence of such a condition, she was "unable to conclude that there exists a disability of 'sexually compulsive behavior regarding viewing pornography'" (paras. 57–59). She went on further to note that the employee's behavior did not resemble an addiction since he did not appear to have "an irresistible drive to view pornography, or any loss of control" (para. 60). Instead, he timed his pornography consumption to times when work was slow.

In *Seneca College v. Ontario Public Service Employees Union (Discharge Grievance)* (2002), a college professor pleaded guilty to child pornography offences after he was caught using the student computer lab to download child pornography. When his employer fired him for this behavior, he argued that his pathological Internet use was the result of a mental disorder, and that there was an obligation under the Ontario Human Rights Code to accommodate this disability. He further argued that he had paid a penalty in the criminal courts, and that termination of his employment was too harsh a response by the employer in the circumstances. The arbitration panel rejected his suggestion that he had "a type of impulse control disorder that took the form of pathological attraction to internet pornography," because the evidence showed he did have control:

> The grievor through his own testimony indicated that his activity in searching out and viewing Internet pornography was both selective and controlled, suggesting that he could exercise self restraint if he chose to do so. (para. 16)

There was an intriguing dissent by the panel member nominated by the union. The member said that even though his condition did not fit the "accepted medical model of psychiatric disorders" in the DSM-IV, the medical community does recognize compulsive behavior. The member noted that this was a person who turned to the Internet to escape from reality due to difficult life circumstances, and asked "[h]ow is that different from those who turn to alcohol, drugs, gambling and other forms of escapism? The difference, this member submits, is our own personal view on the morality of the issue and of course, legality of the issue" (para. 26). Because other employees are given time off to "fight their own personal demons" the grievor also should have been given this opportunity and been awarded severance even if reinstatement was not possible due to the breach of the fundamental responsibilities of his teaching position (paras. 26–30).

The Internet sex addiction cases demonstrate the importance of an accepted medical diagnosis. Even where expert medical evidence is provided, its weight may be discounted by the lack of a generally accepted condition, as occurred in *Interior Health Authority* (2003) case. The cases also hint at the possibility of a backlash against those who are perceived to be in control of a condition yet are attempting to deny control and responsibility by invoking a disability. Where this claim is not believed, blame may be aggravated by the perception of failure to properly take responsibility and the lack of remorse.

Conclusion

A common theme in discussions of law and neuroscience is that behavioral, genetic, and neuroscientific explanations of behavior will affect judgments of causal control and responsibility. The suggestion is that people tend to interpret these explanations deterministically to a greater extent than other types of causal explanations of behavior, such as environmental or psychological factors. Of course, all behavior involves the brain, and pointing out this fact does not settle the question of whether society should hold someone legally responsible for it. Legal practice and social psychology seem to show that people are sensitive to a person's perceived control over disfavored behaviors or symptoms, and assign blame and punishment to those who are perceived to have the capacity to control them and are therefore responsible for them.

This effect has been most often explored in the criminal context, where the impact of behavioral genetics and neuroscience on retribution are being actively researched. However, this does not exhaust the possible legal effects of the manner in which people link brain explanations to a lack of control and to diminished responsibility. The willingness to recognize a behavioral condition as a disability, entitling a person to social assistance and protection from discrimination, also is sensitive to perceived control.

A review of cases in which emerging forms of behavioral addictions are claimed as disabilities reveals a complex picture. Brain-based explanations are sometimes used in an attempt to emulate the successful reframing of drug and alcohol addiction as a disease of the brain. However, this in itself does not seem to be enough to succeed in legal claims. Decision-makers often question whether self-control is sufficiently diminished to excuse the behavior in question. Furthermore, claimants may be punished for failing to take responsibility and to show remorse when they ascribe behavior to a condition over which they are perceived to retain at least some control.

As research continues to elucidate the underlying biological mechanisms involved in behavioral control, it seems likely that human rights lawyers and complainants will adopt this evidence in relation to stigmatized conditions not yet firmly accepted as disabilities. An additional way that neuroscience research may alter the legal response to disabilities is by developing novel treatments based on greater understanding of the biology of behavior. Since claimants are expected to participate in attempts to make reasonable accommodation of their disabilities under human rights law, claimants may find themselves under pressure to adopt these new treatments.

There are potential negative consequences, familiar from the literature on human rights theory and the sociological literature on medicalization, to the framing of behavioral problems as matters of biology or physiology. They deflect attention from other social or environmental conditions that are also causal contributors, and so may weaken attempts to make broader responses to behavioral conditions. Furthermore, while it is understandable why people may seek to attribute causal control for socially disfavored behavior to biological mechanisms beyond their control, might this not also undermine their actual self-control in a self-fulfilling way? Future research looking specifically at the law and legal institutions could fruitfully explore these and other themes to help societies incorporate neuroscientific explanations of human behavior.

Acknowledgments

The author gratefully acknowledges the research assistance of Tijana Potkonjak and Hilary Bond, as well as the financial support of the Bertram Loeb Research Chair and the Social Sciences and Humanities Research Council of Canada.

References

American Psychiatric Association (2000). *Diagnostic and Statistical Manual of Mental Disorders* (4th edn., text rev.). Washington, DC: American Psychiatric Association.

American Psychiatric Association (2013). *Diagnostic and Statistical Manual of Mental Disorders* (5th edn.). Washington, DC: American Psychiatric Association.

Aspinwall, L.G., Brown, T.R., and Tabery, J. (2012). *Science*, **337**, 846–849.

Canada Safeway Ltd. and RWDSU (MacNeill) Re [1999] S.L.A.A. No. 1 (Sask. Labour Arbitration).

Charter of Rights and Freedoms, Part I of the *Constitution Act, 1982*, being Schedule B to the *Canada Act 1982* (UK), 1982, c 11.

Cominco Ltd. v. United Steelworkers of America, Local 9705, [2000] B.C.C.A.A.A. No. 62 (QL) Para. 179.

Corrigan, P.W. (2000). Mental health stigma as social attribution: implications for research methods and attitude change. *Clinical Psychology: Science and Practice*, **7**(1), 48–67.

Corrigan, P., Markowitz, F.E., Watson, A., Rowan, D., and Kubiak, M.A. (2003). An attribution model of public discrimination toward persons with mental illness. *Journal of Health and Social Behavior*, **44**(2), 162–179.

Criminal Code. R.S.C. 1985 c. C-46.

Esses, V.M. and Beaufoy, S.L. (1994). Determinants of attitudes toward people with disabilities. *Journal of Social Behavior and Personality*, **9**(5), 43–64.

Fischer, J.M. and Ravizza, M. (1998). *Responsibility and Control: A Theory of Moral Responsibility*. Cambridge: Cambridge University Press.

Fuss, J., Dressing, H., and Briken, P. (2015). Neurogenetic evidence in the courtroom: a randomized controlled trial with German judges. *Journal of Medical Genetics*, **42**, 730–737.

Gilbert, D. and Majury, D. (2006). Infertility and the parameters of discrimination discourse. In: Pothier, D. and Devlin, R. (Eds.) *Critical Disability Theory: Essays in Philosophy, Politics, Policy, and Law*, pp.285–304. Vancouver, BC: UBC Press.

Human Rights Code, R.S.O. 1990, c. H.19.

Interior Health Authority (South Similkameen Health Centre) v. Hospital Employee's Union (RP Grievance) [2013] B.C.C.A.A.A. No. 44.

Invista (Canada) Co., Kingston Site v. Kingston Independent Nylon Workers Union (Asselstine Grievance) [2013] O.L.A.A. NO. 38.

Kellogg Canada Inc. v. Bakery, Confectionary, Tobacco Workers & Grain Millers, Local 154-G (Fickling Grievance), [2006] O.L.A.A. No. 375 at 60.

Kvaale, E.P., Haslam, N., and Gottdiener, W.H. (2013). The 'side effects' of medicalization: a meta-analytic review of how biogenetic explanations affect stigma. *Clinical Psychology Review*, 33, 782–794.

Law Commission of Ontario (2009). *The Law as it Affects Persons with Disabilities: Preliminary Consultation Paper, Approaches to Defining Disability*. Available from: http://www.lco-cdo.org/en/disabilities-threshold-paper [accessed February 11, 2017].

Lee, A.A., Laurent, S.M., Wykes, T.L., Kitchen Andre, K.A., Bourassa, K.A., and McKibbin, C.L. (2014). Genetic attributions and mental illness diagnosis: effects on perceptions of danger, social distance, and real helping decisions. *Social Psychiatry and Psychiatric Epidemiology*, 49, 781–789.

Letourneau v. JTI-MacDonald Corp. 2015 QCCS 2382.

Love, T., Laier, C., Brand, M., Hatch, L., and Hajela, R. (2015). Neuroscience of internet pornography addiction: a review and update. *Behavioral Sciences*, 5(3), 388–433.

Manitoba Government and General Employees' Union v. Manitoba (M.W. Grievance) [2013] M.G.A.D. No. 10 at para. 81.

Manitoba v. Manitoba Government and General Employees' Union 2005 MGAD No. 14.

McNeill v. Ontario Ministry of the Solicitor General and Correctional Services [1998] O.J. No. 2288 (Ont. Ct. J.—Gen Div.).

Monterosso, J., Royzman, E.B., and Schwartz, B. (2004). Explaining away responsibility: effects of scientific explanation on perceived culpability. *Ethics & Behavior*, 15(2), 139–158.

Milazzo v. Autocar Connaisseur Inc. 2003 CHRT 37.

Mueller, P.K., Houser, J.A., and Riddle, M.K. (2010). Regarding disability: perceptions of protection under the Americans with Disabilities Act. *Research in Social Science and Disability*, 5, 159–180.

Ontario (Director, Disability Support Program) v. Tranchemontagne (2009) O.R. (3d) 327, aff'd 2010 ONCA 593.

Ontario Human Rights Commission (2014). *Policy on Preventing Discrimination Based on Mental Health Disabilities and Addictions*. Available from: http://www.ohrc.on.ca/en/policy-preventing-discrimination-based-on-mental-health-disabilities-and-addictions [accessed February 11, 2017].

Penney, S. (2012). Impulse control and criminal responsibility: lessons from neuroscience. *International Journal of Clinical and Health Psychology*, 35, 99–103.

R. v. Ample Annie's Itty Bitty Roadhouse (2001) O.J. No. 5968 (Ont. C.J.).

R. v. Belcourt 2010 ABCA 319.

R. v. H.T. 2010 MBPC 8.

R. v. Obed 2006 NLTD 155.

R. v. Smith 2015 BCSC 1267.

R. v. Horvath [1997] S.J. No. 385 (Sask. C.A.).

Roach, K. (2012). *Criminal Law* (5th edn.). Toronto, ON: Irwin Law Inc.

Ruby, C.C. (2004). *Sentencing* (6th edn.). Markham, ON: Butterworths.

Seneca College v. Ontario Public Service Employees Union (Discharge Grievance [2002] O.L.A.A. No. 415.

Tovino, S.A. (2015). Will neuroscience redefine mental injury? Disability benefit law, mental health parity law, and disability discrimination law. *Indiana Health Law Review*, 12(2), 695–728.

United Nations General Assembly (2007). *Convention on the Rights of Persons with Disabilities: Resolution/Adopted by the General Assembly, 24 January 2007*. [UN Document A/RES/61/106.] New York: United Nations.

Van Oorschot, W. (2000). Who should get what, and why? On deservingness criteria and the conditionality of solidarity among the public. *Policy and Politics*, **28**(1), 33–49.

Vrecko, S. (2010). Birth of a brain disease: science, the state and addiction neuropolitics. *History of the Human Sciences*, **23**(4), 52–67.

Weiner, B. (1993). On sin versus sickness: a theory of perceived responsibility and social motivation. *American Psychologist*, **48**(9), 957–965.

Weiner, B., Perry, R.P., and Magnusson, J. (1988). An attributional analysis of reactions to stigmas. *Journal of Personality and Social Psychology*, **55**(5), 738–748.

Weiner, B., Osborne, D., and Rudolph, U. (2011). An attributional analysis of reactions to poverty: the political ideology of the giver and the perceived morality of the receiver. *Personality and Social Psychology Review*, **15**(2), 199–213.

Yellowknife (City) v. Denny 2004 NWTTC 2.

Chapter 30

Neuroethics and global mental health: Establishing a dialogue

Dan J. Stein and James Giordano

Introduction

Neuroethics, alongside the other neuro-disciplines, has largely been developed within high-income Western countries. This is not surprising, given that neuroscience is an expensive research endeavor that has mostly been undertaken in academic institutions based in these locations, and given that the conceptual framework of neuroethics is related to work in areas such as bioethics and neurophilosophy, which have also most often been pursued in this high-income Western context. Conversely, an electronic search of the research literature at the intersection of "neuroethics" and "low-income or middle-income countries" reveals vanishingly few publications, although there is certainly clear recognition of the importance of issues surrounding neuroethics and globalization (Lombera & Illes 2009; Giordano 2013).

Global mental health is another recently defined field of intellectual and practical endeavor, which has emphasized that although the majority of research on mental disorders takes place in high-income Western countries, the vast bulk of the world's people with mental disorders live in low- and middle-income countries (LMICs) (the so-called 10:90 research gap) (Becker & Kleinman 2012; Patel 2012). Researchers in global mental health have not only emphasized the research gap, but they have also brought attention to the treatment gap; in LMICs, there are significant barriers to accessing and scaling up effective treatments (Wang et al. 2007; Andrade et al. 2013). Conversely, an electronic search of "global mental health" and "neuroscience" yields relatively few publications at this intersection, although there is a growing emphasis on the need for both implementation and discovery science in LMIC contexts (Collins et al. 2011; Stein et al. 2015).

This chapter makes the argument that despite their different histories and trajectories, neuroethics and global mental health have a good deal in common, and there is much to be gained from an ongoing dialogue and interaction between these fields. There are a number of important areas of overlap between neuroethics and global mental health, including (1) a naturalist and empirical approach, (2) a concern with both disease and with well-being, (3) a concern with human rights and with patient empowerment—intersections which have previously been reviewed (Stein & Giordano 2015). Each of these areas of overlap is considered in more detail throughout this chapter, and an additional area, namely, (4) an appreciation of human diversity, including neurodiversity, is also addressed.

Naturalist and empirical approach

A central set of concerns for neuroethics includes work on the neuroscience of ethics and morality (Buniak et al. 2014; Darragh et al. 2015). Methodological advances in fields such as neuroimaging and electroencephalography have provided insights into the neurocircuitry of ethical and moral judgments and decision-making. Neuroethics is concerned with the implications of such new findings for age-old questions about the nature of the self and of agency, free will and weakness of will, moral culpability and responsibility, and so forth. Its approach is a naturalist and empirical one insofar as it emphasizes the value of both neuroscience and empirical neuroethics in providing more accurate and more comprehensive answers to such questions (Illes 2007; Racine 2008; Levy 2012).

Analogously, global mental health has advocated an evidence-based approach and has advanced in part by developing rigorous study designs and methods for addressing long-standing questions in psychiatry and clinical psychology (Patel & Thornicroft 2009). Rather than simply speculating about the relevance of psychotherapies developed in Western contexts for LMICs, for example, researchers have done careful research on how best to adapt such work to new contexts, and how best to build capacity in the delivery of such adapted interventions (Patel & Prince 2010; Murray et al. 2011; Remien et al. 2013). Such work has in turn arguably contributed to our understanding of the nature and mechanisms of psychotherapy (Murray et al. 2014).

Important concerns can certainly be raised about the sort of approach that both neuroethics and global mental health have undertaken. From a metaphilosophical perspective, for example, it might be strongly argued that no amount of science can simply reduce core conceptual questions in philosophy and ethics to empirical ones (Overgaard et al. 2013). And critics of global mental health have argued that the field has been overly accepting of Western diagnostic constructs and therapeutic approaches, failing to see how psychiatry is embedded in one particular conceptual mold, how it reflects the values of its own social origins and context, and how it imposes its theories and values in a neocolonialist way on the rest of the world (Summerfield 2013). Certainly, an absence of evidence does not imply evidence of absent efficacy (Altman & Bland 1995), simplistic extrapolation of the evidence based from one setting to another is not always appropriate (Stein et al. 2002), and there may well be a need to supplement an evidence-based approach with values-based medicine (Fulford 2011).

We would submit that a conversation and collaboration between neuroethics and global mental health may be useful in addressing some of these concerns. Empirical study of a broad range of ethical and moral issues, drawn from contexts around the world, is important in raising an awareness of—and appropriately addressing—the complexity of ethical and moral decision-making (Lanzilao et al. 2013). Conversely, a rigorous analysis of the theories and values underlying and underpinning global mental health is important to ensure that these are made fully explicit, that these are rigorously discussed and debated, and that adaptation of different constructs and approaches across different contexts is done in a sensitive and appropriate way.

Concern with both disease and wellness

While much work in neuroethics had been concerned with a range of neuropsychiatric disorders, the field has also had a particular concern with issues about human enhancement (Savulescu 2006; Sahakian & Morein-Zamir 2011; Stein 2012b; Shook & Giordano 2016). Neuroethics has raised questions about the nature of well-being, the importance of societal happiness, and issues of optimal neurodevelopment and aging, and has emphasized that countries need to capitalize on the cognitive resources of their citizens (Beddington et al. 2008). Again, these concerns emerge from advances in neuroscience and from new methodologies, which allow a more detailed understanding of both risk and resilience factors, which provide a range of tools for monitoring and measuring disease and wellness, and which have led to a range of novel interventions that may maximize the potential of individuals and of societies (Stein 1998; Farah et al. 2004; Greely et al. 2008).

Global mental health has emphasizing the spectrum of mental disease and health, ranging from mental disease, through to suffering and distress, and on to different forms of well-being (Jacob & Patel 2014). It has focused not only on understanding disease mechanisms and the treatment of disorders, but also on understanding the mechanisms underlying resilience and the prevention of disorders. It has argued that we should be careful not to pathologize different idioms of distress, that we also address the structural determinants of mental health, and that we aim not only for symptom reduction but also for recovery (Davidson & Strauss 1995; Kirmayer & Pedersen 2014). Whereas clinical neuroscience has focused on the endophenotypes, or intermediate phenotypes that underlie the spectrum of states and traits, a public health perspective has also focused on exophenotypes, or the range of social and cultural factors that impact such phenotypes (Stein et al. 2013).

An emphasis on both disease and well-being is not without controversy. A reasonable argument can be made for the need to focus mental health resources on the small proportion of the community with serious mental disorders, where appropriate human and health infrastructure can make an enormous difference. In contrast, it may be argued that a focus on the "worried well," where long-term outcomes are good without intervention, is an investment that has a relatively poor return. Certainly there is a relative paucity of data on the efficacy and effectiveness of interventions to improve well-being (Stein 2012a). Furthermore, while it may be relatively easy to obtain consensus about what constitutes severe mental disorders and to agree on universal interventions to address such conditions, there may be considerable disagreement about the nature of human well-being and on the best ways of attempting to achieve such well-being (Stein 2012a). In addition, many have written persuasively about the problems inherent in relying on technological approaches to resolving social ills (Parens 1998; Sandel 2004).

Both neuroscience and global mental health can usefully contribute to the debate about the nature and boundaries of mental disorder and mental well-being. Community survey data have emphasized, for example, that respondents who do not meet threshold criteria for presence of disorder, but who have subclinical symptoms, nevertheless suffer from a great deal of impairment (Kessler et al. 2003). While absenteeism is a major cost, the burden due to presenteeism also requires recognition (Beddington et al. 2008). Delineation of the boundaries between illness and disease partly reflects the availability and costs for treatment; thus the introduction of generic statins led to a lowering of definitions of hypercholesterolaemia (Kessler et al. 2003). Similarly, in the area of mental disorders and mental health, both conceptual and empirical work is needed to address optimally the delineation of disorders, and to answer fully the questions of when and how best to intervene in a global context (Stein 2008; Stein et al. 2013). We can expect that novel opportunities to improve individual and societal flourishing will become an increasingly prominent topic for debate and practice (Shook & Giordano 2014).

Concern with human rights and patient empowerment

Important aspects of neuroethics include work on the ethics of neuroscientific studies, as well as the intersection of neuroscience and social policy (Buniak et al. 2014). Advances in fields such as neurogenetics, neuroimaging, and neuropsychopharmacology have raised a new set of ethical issues for contemporary society. Neurogeneticists are asked about the implications of polygenic risk scores, neuroimaging scientists encounter incidentalomas, and neuropsychopharmacologists have introduced new chemical entities which raise the possibility of cognitive and emotional enhancement (Illes 2004; Sahakian et al. 2015). Neuroethicists are concerned that we address such questions in a way that maximizes beneficence, non-maleficence, autonomy, justice, and other key values, and that neurotechnologies are not used to fortify asymmetric relationships between individuals or communities (Giordano 2014; Shook & Giordano 2014). Some have put particular emphasis on the importance and value of autonomy in decisions about issues such as enhancement (Schaefer et al. 2014).

Analogously, global mental health is crucially driven by a concern for human rights and well-being (Patel et al. 2006; Becker & Kleinman 2013). The field has framed stigmatization of mental illness, and lack of parity for mental disorder treatments, as human rights issues (Silove 1999; Yamin & Rosenthal 2005). It has argued that mental health is central to the developmental agenda, and that there is "no health without mental health" (Prince et al. 2007). Moreover, authors writing from a global mental health perspective have emphasized that mental disorders represent only one end of a spectrum of human suffering, so efforts must be directed not only at reducing symptoms of disorders, but also at improving resilience and well-being. Global mental health has emphasized the importance of patient voices, consumer advocacy, destigmatization, and social inclusion, as depicted in the slogan "nothing for us, without us." Some have put particular emphasis on the importance of equitable relationships between researchers from the global North and South (Fricchione et al. 2012).

Important concerns can be raised about these foci of neuroethics and of global mental health. There is an argument which emphasizes that human rights should not be thought of as a natural kind but rather that such constructs are bound to specific times and places (Donnelly 2013); there is a view that to understand and evaluate an ethical issue, concepts of human rights need to be augmented with additional constructs such as "duty" and "virtue" (Benatar 2006); and there is a position that the right to mental healthcare may not always be the most appropriate approach to ensuring optimization of mental health services (Glover-Thomas & Chima 2015). Some critics have emphasized the risk of including patient perspectives merely for the sake of political correctness (Spitzer 2004), and there are certainly limits to the effectiveness of social inclusion as a sole treatment for those suffering from mental disorders. There is a need for careful conceptual and empirical work on issues such as the stigmatization of mental disorder, in order to ensure that responses to mental illness fully recognize its reality and its consequences, but at the same time, do not unfairly discriminate against those who suffer from these conditions.

Indeed, in response to such concerns, there is much room for neuroethics and global mental health to contribute in a collaborative way. Although they come from quite different perspectives, each emphasizes the growing role of the patient in mental healthcare, and each strongly values the autonomy of the individual in deciding on what healthcare to access and how. Each is also, however, aware of important limits to autonomy; sometimes those with serious mental illness are hospitalized against their will, and sometimes consumers use neurotechnologies, including psychotropic agents, in harmful ways. Rigorous theoretical and empirical work is needed in order to ensure the optimization and maximization of human rights and consumer empowerment in the area of clinical neuroscience. As the agenda of global mental health research, including work on discovery and implementation science, gives increasing impetus to studies of mental health and mental disorders in LMICs, so there will be a need to increase attention to neuroethical issues arising in these settings, and to the promotion of responsible policies (Shook and Giordano 2014).

Appreciation of human diversity and neurodiversity

Neuroethics is acutely aware that there is not simply one sort of human brain, but that each brain is unique. The more we learn about genetics and about the interaction of genes and environment, the more we are aware of the complexity and diversity of brain–mind development and of human phenotypes. New neurotechnologies, including neuroimaging, neurogenetics, neurotranscriptomics, and neurometabolomics, have become ever more reliant on, and engaged by, the incentives of "big data," which is challenging our fastest super-computers and most sophisticated data-processing algorithms (Monteith et al. 2015; DiEuliis & Giordano 2016).

Analogously, global mental health is very much aware that most psychology research is undertaken on Westernized, educated, industrialized, rich, and democratic (WEIRD) populations (Henrich et al. 2010). It is also crucially aware of how mental disorders present differently in different contexts, of how there are multiple different explanatory models of mental disorder, and of how these can impact the course of treatment (Kleinman 1991; Stein 1993; Kirmayer 2006). Whereas clinical neuroscience has focused on biological diversity, global mental health has emphasized the importance of cultural and social diversity.

Once again, some concerns can perhaps be raised in this area. Science, including the field of genetics, does not always have a good track record of celebrating diversity, and it is crucial to bear in mind those times when neuroscience and psychiatry have had an ignominious and deplorable past (Weiss & Lambert 2011; Zeidman 2012). At the same time, there are perhaps potential risks in not providing neuroscientific and psychiatric treatments when these can relieve suffering. In the current era, recognition of past excesses has led to a strong focus on the avoidance of over-medicalization. At the same time, in certain contexts, medicalization is entirely appropriate (Stein & Gureje 2004).

Neuroethics and global mental health can again perhaps develop a mutually enriching research agenda in this area. It is important, for example, to understand fully our genetic diversity by employing global populations in the study of normal and abnormal behavior. There is a strong argument that the inclusion of African populations is needed to delineate fully the genetic risk factors for mental disorders, and that the exclusion of African populations from such studies may potentially lead to exclusion from the positive benefits of certain kinds of research (e.g., not fully understanding pharmacogenetics of a particular psychotropic drug) (Dalvie et al. 2015). Indeed, the study of global populations may yield crucially important lessons for neuroscience and for psychiatry (Patel 2010; Patel & Saxena 2014).

Conclusion

Neuroethics and global mental health have strikingly different origins and trajectories. The former has focused mainly on new and specialized technologies being introduced into high-income countries, while the latter has focused on primary care interventions and their scale-up in LMICs. At the same time, these fields also share a number of important concerns and approaches. There has been relatively little conversation between neuroethics and global mental health to date; this chapter argues that collaboration between these fields may be useful in advancing a number of key areas of work and of debate. Our hope is that lessons from global neuroscience may impact positively on the development and articulation of an internationally sensitive, relevant, and valuable neuroethics.

Such a conversation and collaboration between neuroethics and global mental health may be important in ensuring ongoing progress in advancing a naturalist and empirical approach to mental disorders, in addressing both disease and wellness in the mental health context, in maintaining an ongoing focus on the importance of human rights and patient empowerment, and in appreciating human diversity, including neurodiversity. Such work will need to address a range of issues that are important for both fields, including neuroethical aspects of global mental health research, as well as the expansion of neuroethical discourse and research to address the introduction of neuroscientific developments and novel neurotechnologies into the low- and middle-income world (Stein & Giordano 2015).

References

Altman, D.G. and Bland, J.M. (1995). Absence of evidence is not evidence of absence. *BMJ*, **311**, 485.

Andrade, L.H., Alonso, J., Mneimneh, Z., et al. (2013). Barriers to mental health treatment: results from the WHO World Mental Health surveys. *Psychological Medicine*, 1–15.

Becker, A.E. and Kleinman, A. (2012). An agenda for closing resource gaps in global mental health: innovation, capacity building, and partnerships. *Harvard Review of Psychiatry*, **20**, 3–5.

Becker, A.E. and Kleinman, A. (2013). Mental health and the global agenda. *New England Journal of Medicine*, **369**, 1380–1381.

Beddington, J., Cooper, C.L., Field, J., et al. (2008). The mental wealth of nations. *Nature*, **455**, 1057–1060.

Benatar, D. (2006). Bioethics and health and human rights: a critical view. *Journal of Medical Ethics*, **32**, 17–20.

Buniak, L., Darragh, M., and Giordano, J. (2014). A four-part working bibliography of neuroethics: part 1: overview and reviews – defining and describing the field and its practices. *Philosophy, Ethics, and Humanities in Medicine*, **9**, 9.

Collins, P.Y., Patel, V., Joestl, S.S., et al. (2011). Grand challenges in global mental health. *Nature*, **475**, 27–30.

Dalvie, S., Koen, N., Duncan, L., et al. (2015). Large scale genetic research on neuropsychiatric disorders in African populations is needed. *EBioMedicine*, **2**, 1259–1261.

Darragh, M., Buniak, L., and Giordano, J. (2015). A four part working bibliography of neuroethics: part 2: neuroscientific studies of morality and ethics. *Philosophy, Ethics, and Humanities in Medicine*, **10**, 1.

Davidson, L. and Strauss, J.S. (1995). Beyond the biopsychosocial model: integrating disorder, health, and recovery. *Psychiatry*, **58**, 44–55.

DiEuliis, D. and Giordano, J. (2016). Neurotechnological convergence and big data: a "force multiplier" toward advancing neuroscience. In: Collmann, J. and Matei, S. (Eds.) *Ethical Reasoning in Big Data: An Exploratory Analysis*, pp.71–80. New York: Springer.

Donnelly, J. (2013). *Universal Human Rights in Theory and Practice* (3rd edn). Ithaca, NY: Cornell University Press.

Farah, M.J., Illes, J., Cook-Deegan, R., et al. (2004). Neurocognitive enhancement: what can we do and what should we do? *Nature Reviews Neuroscience*, **5**, 421–425.

Fricchione, G.L., Borba, C.P., Alem, A., Shibre, T., Carney, J.R., and Henderson, D.C. (2012). Capacity building in global mental health: professional training. *Harvard Review of Psychiatry*, **20**, 47–57.

Fulford, K.W. (2011). The value of evidence and evidence of values: bringing together values-based and evidence-based practice in policy and service development in mental health. *Journal of Evaluation in Clinical Practice*, **17**, 976–987.

Giordano, J. (2013). Ethical considerations in the globalization of medicine – an interview with James Giordano. *BMC Medicine*, **11**, 69.

Giordano, J. (2014). The human prospect(s) of neuroscience and neurotechnology: domains of influence and the necessity—and questions—of neuroethics. *Human Prospect*, 4(1), 1–18.

Glover-Thomas, N. and Chima, S.C. (2015). A legal "right" to mental health care? Impediments to a global vision of mental health care access. *Nigerian Journal of Clinical Practice*, 18(Suppl), S8–S14.

Greely, H., Sahakian, B., Harris, J., et al. (2008). Towards responsible use of cognitive-enhancing drugs by the healthy. *Nature*, 456, 702–705.

Henrich, J., Heine, S.J., and Norenzayan, A. (2010). Most people are not WEIRD. *Nature*, 466, 29.

Illes, J. (2004). Medical imaging: a hub for the new field of neuroethics. *Academic Radiology*, 11, 721–723.

Illes, J. (2007). Empirical neuroethics. Can brain imaging visualize human thought? Why is neuroethics interested in such a possibility? *EMBO Reports*, 8(Spec No), S57–60.

Jacob, K.S. and Patel, V. (2014). Classification of mental disorders: a global mental health perspective. *Lancet*, 383, 1433–1435.

Kessler, R.C., Merikangas, K.R., Berglund, P., Eaton, W.W., Koretz, D.S., and Walters, E.E. (2003). Mild disorders should not be eliminated from the DSM-V. *Archives of General Psychiatry*, 60, 1117–1122.

Kirmayer, L.J. (2006). Beyond the 'new cross-cultural psychiatry': cultural biology, discursive psychology and the ironies of globalization. *Transcultural Psychiatry*, 43, 126–144.

Kirmayer, L.J. and Pedersen, D. (2014). Toward a new architecture for global mental health. *Transcultural Psychiatry*, 51, 759–776.

Kleinman, A. (1991). *Rethinking Psychiatry: From Cultural Category to Personal Experience*. New York: Free Press.

Lanzilao, E., Shook, J., Benedikter, R., and Giordano, J. (2013). Advancing neuroscience on the 21st century world stage: the need for—and proposed structure of—an internationally relevant neuroethics. *Ethics in Biology, Engineering and Medicine*, 4(3), 211–229.

Levy, N. (2012). Neuroethics. *Wiley Interdisciplinary Reviews: Cognitive Science*, 3, 143–151.

Lombera, S. and Illes, J. (2009). The international dimensions of neuroethics. *Developing World Bioethics*, 9, 57–64.

Monteith, S., Glenn, T., Geddes, J., and Bauer, M. (2015). Big data are coming to psychiatry: a general introduction. *International Journal of Bipolar Disorders*, 3, 21.

Murray, L.K., Dorsey, S., Bolton, P., et al. (2011). Building capacity in mental health interventions in low resource countries: an apprenticeship model for training local providers. *International Journal of Mental Health Systems*, 5, 30.

Murray, L.K., Dorsey, S., Haroz, E., et al. (2014). A common elements treatment approach for adult mental health problems in low- and middle-income countries. *Cognitive and Behavioral Practice*, 21, 111–123.

Overgaard, S., Gilbert, P., and Burwood, S. (2013). *An Introduction to Metaphilosophy*. New York: Cambridge University Press.

Parens, E. (1998). Is better always good? The Enhancement Project. *Hastings Centre Reports*, 28, S1–S17.

Patel, V. (2010). Research priorities for Indian psychiatry. *Indian Journal of Psychiatry*, 52, S26–29.

Patel, V. (2012). Global mental health: from science to action. *Harvard Review of Psychiatry*, 20, 6–12.

Patel, V. and Prince, M. (2010). Global mental health: a new global health field comes of age. *JAMA*, 303, 1976–1977.

Patel, V. and Saxena, S. (2014). Transforming lives, enhancing communities—innovations in global mental health. *New England Journal of Medicine*, 370, 498–501.

Patel, V. and Thornicroft, G. (2009). Packages of care for mental, neurological, and substance use disorders in low- and middle-income countries: PLoS Medicine Series. *PLoS Medicine*, 6, e1000160.

Patel, V., Saraceno, B., and Kleinman, A. (2006). Beyond evidence: the moral case for international mental health. *American Journal of Psychiatry*, 163, 1312–1315.

Prince, M., Patel, V., Saxena, S., et al. (2007). No health without mental health. *Lancet*, 370, 859–877.

Racine, E. (2008). Which naturalism for bioethics? A defense of moderate (pragmatic) naturalism. *Bioethics*, 22, 92–100.

Remien, R.H., Mellins, C.A., Robbins, R.N., et al. (2013). Masivukeni: development of a multimedia based antiretroviral therapy adherence intervention for counselors and patients in South Africa. *AIDS and Behavior*, 17, 1979–1991.

Sahakian, B.J. and Morein-Zamir, S. (2011). Neuroethical issues in cognitive enhancement. *Journal of Psychopharmacology*, 25, 197–204.

Sahakian, B.J., Bruhl, A.B., Cook, J., et al. (2015). The impact of neuroscience on society: cognitive enhancement in neuropsychiatric disorders and in healthy people. *Philosophical Transactions of the Royal Society B: Biological Sciences*, 370, 20140214.

Sandel, M.J. (2004). The case against perfection: what's wrong with designer children, bionic athletes, and genetic engineering. *Atlantic Monthly*, 292, 50–54, 56–60, 62.

Savulescu, J. (2006). Justice, fairness, and enhancement. *Annals of the New York Academy of Sciences*, 1093, 321–338.

Schaefer, G.O., Kahane, G., and Savulescu, J. (2014). Autonomy and enhancement. *Neuroethics*, 7, 123–136.

Shook, J.R. and Giordano, J. (2014). A principled and cosmopolitan neuroethics: considerations for international relevance. *Philosophy, Ethics, and Humanities in Medicine*, 9, 1.

Shook, J.R. and Giordano, J. (2016). Neuroethics beyond normal: performance enablement and self-transformative technologies. *Cambridge Quarterly of Healthcare Ethics*, 25, 121–140.

Silove, D. (1999). The psychosocial effects of torture, mass human rights violations, and refugee trauma: toward an integrated conceptual framework. *Journal of Nervous and Mental Disease*, 187, 200–207.

Spitzer, R.L. (2004). Good idea or politically correct nonsense? *Psychiatric Services*, 55, 113.

Stein, D.J. (1993). Cross-cultural psychiatry and the DSM-IV. *Comprehensive Psychiatry*, 34, 322–329.

Stein, D.J. (1998). Philosophy of psychopharmacology. *Perspectives in Biology and Medicine*, 41, 200–211.

Stein, D.J. (2008). *Philosophy of Psychopharmacology*. Cambridge: Cambridge University Press.

Stein, D.J. (2012a). Positive mental health: a note of caution. *World Psychiatry*, 11, 107–109.

Stein, D.J. (2012b). Psychopharmacological enhancement: a conceptual framework. *Philosophy, Ethics, and Humanities in Medicine*, 7, 5.

Stein, D.J. and Giordano, J. (2015). Global mental health and neuroethics. *BMC Medicine*, 13, 274.

Stein, D.J. and Gureje, O. (2004). Depression and anxiety in the developing world: is it time to medicalise the suffering? *Lancet*, 364, 233–234.

Stein, D.J., Xin, Y., Osser, D., Li, X., and Jobson, K. (2002). Clinical psychopharmacology guidelines: different strokes for different folks. *World Journal of Biological Psychiatry*, 3, 64–67.

Stein, D.J., Lund, C., and Nesse, R.M. (2013). Classification systems in psychiatry: diagnosis and global mental health in the era of DSM-5 and ICD-11. *Current Opinion in Psychiatry*, 26, 493–497.

Stein, D.J., He, Y., Phillips, A., Sahakian, B.J., Williams, J., and Patel., V. (2015). Global mental health and neuroscience: potential synergies. *Lancet Psychiatry*, 2, 178–185.

Summerfield, D. (2013). "Global mental health" is an oxymoron and medical imperialism. *BMJ*, 346, f3509.

Wang, P.S., Angermeyer, M., Borges, G., et al. (2007). Delay and failure in treatment seeking after first onset of mental disorders in the World Health Organization's World Mental Health Survey Initiative. *World Psychiatry*, 6, 177–185.

Weiss, K.M. and Lambert, B.W. (2011). When the time seems ripe: eugenics, the annals, and the subtle persistence of typological thinking. *Annals of Human Genetics*, 75, 334–343.

Yamin, A.E. and Rosenthal, E. (2005). Out of the shadows: using human rights approaches to secure dignity and well-being for people with mental disabilities. *PLoS Medicine*, 2, e71.

Zeidman, L.A. (2012). Re: neuroscience in Nazi Europe part I: eugenics, human experimentation, and mass murder. Can J Neurol Sci. 2011; 38:696–703. *Canadian Journal of Neurological Sciences*, 39, 400.

Part IV

Epilogue

Chapter 31

Neuroethics and neurotechnology: Instrumentality and human rights

Joseph J. Fins

Neuroethics as an ethics of technology

Five years ago, in writing an epilogue for the *Oxford Handbook of Neuroethics* (Illes & Sahakian 2011), I questioned if neuroethics was distinct from mainstream medical ethics (Fins 2011). I was not sure, but asserted that if neuroethics could be differentiated it would be as an ethics heavily predicated upon a confluence of technological advances that opened up a heretofore unchartered problem space. So if we were to differentiate neuroethics, it would be fundamentally as an ethics of technology. I want to return to this question here for *Neuroethics: Anticipating the Future*.

Neuroethics and technology is a perennial theme, with its own eternal recurrence. Think of what one emerging technology, neuroimaging, has done for our rather naïve understanding of disorders of consciousness. Our faith in the classic bedside exam has been shaken. Behavioral manifestations of self, of consciousness used to assess brain states from coma to the vegetative and minimally conscious states have been augmented by a new reality made possible by brain scans, colorful flares on functional magnetic resonance imaging, which sometimes suggest that individuals once thought irretrievably lost, might still be there, albeit only in their head. Technological advance made apparent what once went unnoticed and thus unreflected upon for its normative significance. What was once beyond our gaze now looked us straight in the eye asking us to reconsider our moral obligations to others.

Neuroimaging has revealed a degree of complexity akin to what Mendel discerned from the colorful pea plants in his monastery's garden (Fins 2007). Some patients revealed cognitive motor dissociation, a discordance between overt action and implicit thought, much like the distinction between phenotype and genotype (Schiff et al. 2005; Owen et al. 2006; Monti et al. 2010; Bardin et al. 2011). What was seemingly obvious at the bedside was no longer so simple. As the technology of neuroimaging expanded our empirical reality, the normative universe in which people exist needed to expand (Fins 2008a). Indeed, it had to become more inclusive about who might retain covert consciousness.

Technology was a powerful normative force. Not only did neuroimaging alter the scientific status quo, it fundamentally disrupted the moral landscape (Fins 2005). Even though neuroimaging is actually less sensitive in detecting covert consciousness than the gold standard Coma Recovery Scale-Revised (Giacino et al. 2004), the colorful visuals of neuroimaging captured media attention and generated academic discourse. Although neuroimaging was inferior as a diagnostic tool, it was far more spectacular because only a nonbehavioral metric could reveal a dissociation between overt and covert signs of consciousness. This discordance, born of technological advance, prompted neuroethical reflection about what should be done for those who might have a life of the mind (Naccache 2006).

Halfway technologies

As neuroscience began to unravel what the great Canadian neurosurgeon Wilder Penfield (1976; Fins 2008b) called the "mystery of the mind", the revelations were accompanied by both angst and hope. People thought forever gone might literally be trapped in their heads, deprived of community and communication (Fins 2015), isolated and alone. And it also suggested the seemingly impossible. Might those stranded somehow be rescued?

Once again neuroethics becomes concerned with its recurring theme of technology. If technology reveals a problem, might it not also resolve it? That is, could neurotechnology help address the isolation of those recently identified as suffering from cognitive motor dissociation and reintegrate them into community through the restoration of functional communication? And of course, this effort has begun with the use of neuroprosthetics like deep brain stimulation (DBS) (Schiff et al. 2007), drugs (Giacino et al. 2012), and neuroimaging paradigms to communicate with the severely injured brain (Giacino et al. 2014).

It would seem to be a neat resolution to our problem with technology. One technology creates an ethical challenge only for another to solve it. But if it were that simple, it would be impoverishing. We would risk falling prey to Leon Kass's (2003) critique of modern bioethics with its penchant for a technological quick fix of complex problems, reducing human experience to engineering solutions that truncate the lived experiences of affected communities. He warns that:

> To a society armed with biotechnology, the activities of human life may come to be seen in purely technical terms, and more amenable to improvement than they really are. Or we may imagine ourselves wiser than we really are. Or we may get more easily what we asked for only to realize it is vastly less than what we really wanted. (Kass 2003)

While the use of neuroprosthetics is a brilliant scientific response to cognitive motor dissociation, it is hardly enough. Neuroscience's current therapeutic forays are primitive and cumbersome and what the American physician and essayist Lewis Thomas (1974) might have considered halfway technologies, medical innovations which improve but do not eliminate a condition. But it is more than merely a question of feasibility and utility: if our technologies are still halfway there, so too is an analysis which views the problem space too exclusively as a technological one. By doing so we discount our motivations for intervening and neglect the broader import of any effort to manipulate, heal, or restore the function of the human brain and mind. To view the challenge of brain science solely through the prism of technology is both clinically incomplete and normatively thin.

A self-critique

While neuroscience is nurtured by the scientist's curiosity, neuroethics should help ensure that the quest discovery is more than a science project. The stakes are higher than that and the work has to be understood in a broader humanistic context in which the purpose and telos of technological advance is well articulated. Short of that, neuroethics will be a hollow, self-justifying framework, one prone to misdirection, distraction, and even abuse.

This is not a trivial statement. To be more than just the handmaiden of technological advance, neuroethics needs to examine the uses of instrumentality. But asking such questions can be a challenge. It is all too easy to be seduced by an emergent ability to manipulate the chemistry of a synapse or alter current flow through a circuit. The explanatory power of these maneuvers can be overwhelming, especially when they achieve an elusive (and often dramatic) therapeutic need. Such success can be blinding, especially when ancient nihilism about brain-based disorders is replaced by a fledgling optimism. When this happens, the technological can easily become triumphant, obscuring important ethical considerations too easily pushed into the shadows caused by a glaring success, making it all the more difficult to question success.

Indeed, a thoughtful commentator can be thought a Philistine when even broaching concerns about the purpose and limits of emerging technologies in the wake of an important discovery. But question we must, lest we fall prey to what the bioethicist Daniel Callahan (1996) has termed "scientism" by which he suggests a "non-falsifiable faith in the capacity of science not simply to provide reliable knowledge but also to solve all or most human problems, social, political, and economic."

Callahan critiques science as a member of what he describes the "loyal opposition" (1996) committed to, but still questioning its goals. His skepticism about the sweep of science is wise and prudent, especially in an age of discovery that we live today. But I would add a friendly amendment to his critique. Such skepticism is essential to, and not counter to, good science itself.

This was known to Isadore I. Rabi, who was awarded the Nobel Prize for physics for his work on nuclear magnetic resonance, essential to modern neuroimaging. When asked why he became a scientist, he reportedly responded that he became a scientist because of his mother's influence upon him, even though that was not her intent. Instead of asking him, as all the other Jewish mothers in Brooklyn did, whether he had learned anything in school, Mrs. Rabi would query him about methods. She'd ask him if he had asked a "good question." Rabi attributes this focus to his becoming a scientist (Sheff 1988).

Rabi, better than most, understood that it was not certainty about our answers but our ability to question that leads us forward. So while discovery is cause for celebration, these moments also call for humility and careful deliberation. When we do this we are not disloyal to the scientific enterprise but rather squarely in its best traditions.

Sustaining success

Humility even in moments of scientific triumph is prudent because success is no guarantee of sustained progress. Despite the spectacular advances we have seen in the past decades, much of which is recounted in this volume, barriers to progress remain (Fins et al. 2012). One critical challenge is the proverbial "valley of death," a phrase used to describe the premature demise of promising proof-of-principle efforts for lack of funding that bridges the perilous terrain, typically between government funding and that of venture capital or pharma or device companies (Finkbeiner 2010). When this happens, good ideas linger on the sidelines.

This challenge is a profound one for two types of what one might call messy innovation. The first is for emerging drugs and devices aimed at conditions that do not constitute a ready market and thus do not attract capitalization. While the Orphan Drug Act of 1983 (Orphan Drug Act of 1983), has helped to ease some of the US Food and Drug Administration's (FDA's) regulatory barriers for such drugs (and to a lesser extent devices under the aegis of the "Orphan Products Board," now the FDA Office of Orphan Products Development (FDA 2016) and the more recently constituted Center for Devices and Radiological Health (FDA 2015)), there is no provision to address the fiscal challenges that can doom promising research lingering between basic science funding streams and underwriting from conventional market sources. Moreover, there are distinct challenges for device development for orphan conditions versus the promotion of pharmacological approaches to therapy (Fins et al. 2011a; Mokhtarzadeh et al. 2016).

The challenge of ongoing support is even more complicated when basic neuroscience research is predicated upon having access to devices and to devices of greater and greater sophistication needed for experimental work. This has been an ongoing challenge in the realm of neuromodulation where DBS may not always be readily available to investigators by their manufacturers. Device makers may withhold access to devices, or their rights of reference required for regulatory approval, because of concerns about liability when they are used in experimental trials, cost, or in order to maintain a competitive advantage of one sort or another (Mokhtarzadeh et al. 2016). This can be highly problematic because devices like DBS are not only therapeutic agents but probative agents (Fins 2012) that can help map circuits essential to the understanding of systems neurobiology, and ultimately the creation of a three-dimensional brain map or connectome (Fins and Shapiro 2013). When these devices are unavailable, tools of discovery are taken out of the hands of the scientist, disadvantaging all and imperiling access to the fruits of neuroscience.

In sum, while the future is bright for neuroscience, success is not assured. Economic forces, the commodification of new ideas through patenting (Fins 2010a; Fins and Schiff 2010a; Fins et al. 2011b), and the all too human quest for notoriety can distort the central ethos of scientific discovery.

The instrumentality of the humanities

The constraints I have just discussed suggest that whatever ethos governs emerging neu-rotechnologies, it needs to be grounded in something more robust than the technologi-cal imperative or unbridled faith in market forces as neither is sustainable nor reliable. Instead, neuroethics, as a field, needs to place this work into a broader context worthy of those who might depend upon its success.

In this we need to seek the wisdom of the humanities (Fins et al. 2013). In turning to the humanities for deeper insights into the sciences, and the biomedical sciences in particular, William Osler (1919) said it best. In an address, "The Old Humanities and the New Sciences," given to Oxford's Classical Association just before his death, Osler told those assembled, "Now, the men of your guild secrete materials which do for society at large what the thyroid gland does for the individual. The humanities are the hormones." He was forceful in his admonition asserting a divide between the sciences and the humanities, observing that "… the so-called Humanists have not enough Science, and Science sadly lacks the Humanities" suggesting as a remedy "liberal education" that would provide guidance on how to "… give the science school the leaven of an old philosophy, how to leaven the old philosophical school with the thoughts of science" (Osler 1919). Osler's admonition is as current today as it was nearly a century ago.

So how do we bring the "leaven of an old philosophy"—as Osler put it—to bear on the future of this still growing field? (Osler 1919). If neuroethics is indeed an ethics of tech-nology, what then should be the guiding ethos of its use? To what goals and objectives should these tools be directed? How do we broaden out the conversation so as not to fall prey to scientism or its neuroethics neologism, neuroscientism? And how do we engage the humanities and social sciences in this critical discussion?

As a physician, I am naturally drawn to the therapeutic possibilities posed by neuro-technology. I think that the amelioration and relief of human suffering caused by neu-ropsychiatric disorders should be the primary goal of our efforts. While some might criticize this focus as overly medicalizing these efforts, and a perspective expected from a doctor, I would maintain that such goals can generate a broad consensus and avoid the controversy that comes with more contentious efforts like enhancement or the uses of neurotechnology for law enforcement or national security purposes. These prohibi-tions were articulated by the National Commission for the Protection of Human Subjects of Biomedical and Behavioral Research in its 1977 report, "Use of Psychosurgery in Practice and Research" which proscribed nonclinical uses of that technology (National Commission 1977; Fins 2003).

While the line can be blurred between restoration and enhancement, and nonclinical applications are inevitable, I believe that it is normatively appropriate and politically useful to cast neurotechnology in the service of those who are sick, injured, or suffering from a neuropsychiatric condition. This will maintain the broad societal consensus and provide a platform for translational discovery that will be good for basic and clinical neuroscience. In short, by prioritizing the clinical instrumentality of emerging neuro-technologies, neuroethics can minimize contention, provide comfort to those afflicted with neuropsychiatric conditions, and foster scientific advance both at the bench and the bedside.

But mine is not solely a clinical appeal. It is a broader one grounded under the broad banner of human rights, specifically the civil and disability rights of individuals with neu-ropsychiatric conditions (Fins 2010b, 2015). As a physician, and as a humanist, I believe that such a prioritization is just and fair. Directing emerging neuro-instrumentality to the remediation of conditions that cause physical or mental distress, constrain liberty, or compromise a person's ability to participate in human community helps to equalize opportunity for those who suffer from life's most malignant conditions.

And beyond the effects of these therapies on the well-being of individuals, as a class, people with neuropsychiatric conditions have a long history of marginalization and dis-crimination. We only need turn to the history of medicine and consider the legacy of the asylum for those with psychiatric conditions (Rothman 1971) or the more modern neu-ronal segregation of those with severe brain injury that I have described in my work (Fins and Hersh 2011; Fins 2013, 2015), to appreciate that these populations have been under-served by medicine and society at large. Emerging neurotechnologies have the potential to even that score and restore functional status and equanimity to those burdened by these conditions, though of course technology can only be part of any solution to such complex social problems.

Toward a rights agenda

Viewing neuroethics through this prism of rights both distances it from Callahan's "sci-entism" (1996) and brings it closer to a sad, but important, truth about neuropsychiatric conditions and their treatment in society. Historically, those with neuropsychiatric condi-tions have been marginalized by society. Whether it is paralysis or a movement disorder that makes mobility a challenge or cognitive or psychiatric conditions that require insti-tutional care, diseases of the brain and mind can lead to the segregation of affected indi-viduals and their sequestration from others in society, much as racism of the Jim Crow, American South segregated individuals because of race.

Legal scholar Timothy Cook (1991) has written of the historic confluence between atti-tudes toward the disabled and racism. He observed that, "The Jim Crow system estab-lished after *Plessy* and the government-supported, systematic segregation of persons with disabilities during precisely the same time were no mere coincidence of historical events. The historical record abounds with evidence that disability discrimination emanated from the same attitudes and prejudices fomenting at the turn of the century regarding race." (Cook 1991).

Although the Americans with Disability Act (Americans with Disability Act 1990) and the United Nations Convention on the Rights of Persons with Disabilities (United Nations General Assembly 2007) both call for the maximal integration of individuals with disability into civil society, this legacy of segregation and marginalization has been difficult to overcome, much less acknowledge or recognize as a problem. Bias toward these populations remains, and individuals with neuropsychiatric conditions continue to suffer, as former US Surgeon General David Satcher (2000) noted, from profound health disparities and substandard care. More recently, Jaffe and Jimenez (2015) have written of health disparities in rehabilitation medicine.

My own research examining the treatment of individuals with disorders of consciousness reveals a class, of mostly young people, segregated in chronic care facilities designed for the aged with degenerative conditions. There they struggle for services, a credible diagnosis, and are denied needed rehabilitation (Fins 2013, 2015). Collectively these omissions can deprive these individuals of the possibility of returning to their homes and families, often robbing them of fundamental rights, such as the right to remain part of their communities.

Of course sometimes, the burden of injury or illness makes institutional care necessary. But increasingly, and expectantly, neurotechnology might make this less necessary, providing opportunities for people with disabilities to be more routinely integrated into society with the help of neuroprostheses. This is more than wishful thinking. If society takes stock of what neurotechnology has done and promises to achieve in the next decade, this aspiration can be viewed as more than a promissory note.

Neurotechnology is beginning to lift the burden of neuropsychiatric maladies and ameliorate many conditions. To name but a few notable examples . . . The work of engineers like P. Hunter Peckam (Moss et al. 2011), Miguel Nicolelis (2012), and others have provided mobility to those with paralysis. John Donoghue and Leigh Hochberg (Hochberg et al. 2012), and Andrew Schwartz (Collinger et al. 2013) have even begun to broach the locked-in-state giving individuals mobility with brain–computer interfaces. Our team's efforts with thalamic DBS helped an individual in the minimally conscious state speak for the first time in 6 years and eat food by mouth (Schiff et al. 2007). Bart Nuttin's (1999) pioneering use of DBS for obsessive–compulsive disorder and Helen Mayberg's for depression (2005) have begun to bring relief to those whose conditions were refractory to pharmacologic and conventional therapy. And beyond invasive technologies, neuroimaging—as exemplified by Monti et al. (2006)—has the potential to serve as a communication channel in the setting of what appeared to be the vegetative state but was in fact, a patient with apparent cognitive motor dissociation in the minimally conscious state (Fins and Schiff 2010b).

What all these technologies have in common is the possibility of reintegration, bringing people with motor, cognitive, or psychiatric disabilities back into the room, back into civil society by ending the sequestration caused by both their illness and society's treatment of their disability. When I consider how the active neuroimaging paradigm used by

Monti et al. (2006) liberated a conscious individual from cognitive motor dissociation, I am reminded of that famous image of Nelson Mandela (1994) when he returned to Robben Island and the place of his decades-long imprisonment and peered out of the bars that had kept him captive.

In the context of our efforts using thalamic DBS in the minimally conscious state, neuroprosthetics have helped to restore voice, and with the restoration of functional communication a person severed from his family was able to rejoin his community. Note the resonant cognates: communication makes the return to community possible (Schiff et al. 2007; Fins 2015).

Each of these efforts is a story of liberation and reintegration into civil society. While work with thalamic DBS in the minimally conscious state is not an established therapy and remains in the realm of proof of principle (Schiff et al. 2009), I believe that it, and the aforementioned examples of technological innovation, are emblematic of the constructive role that neurotechnology might come to play in securing rights for those left too long on the margins by neuropsychiatric conditions. It also suggests how neuroprosthetic technology might become the centerpiece of an emerging rights agenda for patients with neuropsychiatric disabilities (Schiff et al. 2007).

Beyond providing direct assistance to patients in need, this confluence of neuroethics and the law will also help expand neuro-law scholarship beyond its current purview with its focus on criminal law, culpability, and the credibility of evidentiary information (The MacArthur Foundation n.d.), to broader considerations of how a human rights perspective can help inform the future of neuroscience and neuroethics (Fins 2016; Wright & Fins 2016), and become an important tool for civil rights advocacy to advance health in the twenty-first century (McGowan et al. 2016).

Neurotechnology as heuristic

A refrain ... returning to my original argument that if neuroethics was a distinct ethics, it would be as an ethics of technology, we have seen here that not only has neurotechnology like neuroimaging pointed us to a technical solution, but that it also pointed to arguments beyond itself, making us appreciate that society was willfully disregarding the needs of members of its citizenry and ignoring their right to be more fully integrated into civil society. This is surely a question that transcends technology, and it would be a mistake to resort to technology for its solution. But we would also be mistaken not to be grateful to technology as a heuristic that allows society to better apprehend dimensions of our reality that heretofore had been obscure and unrecognized. While neuroscience does not provide an easy answer to our questions, it does help us pose new and sometimes deeply disturbing questions. For that, we should be grateful.

Acknowledgments

The author is grateful to Amy B. Ehrlich and Cathleen A. Acres for their comments, and Jennifer Hersh for editorial assistance, and acknowledges the Jerold B. Katz Foundation for its support of the Consortium for the Advanced Study of Brain Injury (CASBI) at Weill Cornell Medical College and Rockefeller University and an NIH Clinical & Translational Science Center Award (UL1-RR024966) to Weill Cornell Medical College.

References

Americans with Disability Act of 1990. Public Law 101-336. 104 stat. 328 (1990).

Bardin, J.C., Fins, J.J., Katz, D.I., et al. (2011). Dissociations between behavioural and functional magnetic resonance imaging-based evaluations of cognitive function. *Brain*, **134**(3), 769–782.

Callahan, D. (1996). Calling scientific ideology to account. *Society*, **33**(4), 14–19.

Collinger, J.L., Wodlinger, B., Downey, J.E., et al. (2013). High-performance neuroprosthetic control by an individual with tetraplegia. *The Lancet*, **381**(9866), 557–564.

Cook, T.M. (1991). The Americans with Disability Act: the move to integration. *Temple Law Review*, **64**, 393–470.

FDA Center for Devices and Radiological Health (2015). *2014–2015 Strategic Priorities*. Available from: http://www.fda.gov/downloads/AboutFDA/CentersOffices/OfficeofMedicalProductsandTobacco/CDRH/CDRHVisionandMission/UCM431016.pdf [accessed July 11, 2016].

FDA Office of Orphan Products Development (2016). *Developing Products for Rare Diseases & Conditions*. (Updated March 30, 2016.) Available from: http://www.fda.gov/ForIndustry/DevelopingProductsforRareDiseasesConditions/ucm2005525.htm [accessed July 11, 2016].

Finkbeiner, S. (2010). Bridging the Valley of Death of therapeutics for neurodegeneration. *Nature Medicine*, **16**(11), 1227–1232.

Fins, J.J. (2003). From psychosurgery to neuromodulation and palliation: history's lessons for the ethical conduct and regulation of neuropsychiatric research. *Neurosurgery Clinics of North America*, **14**(2), 303–319.

Fins, J.J. (2005). Rethinking disorders of consciousness: new research and its implications. *The Hastings Center Report*, **35**(2), 22–24.

Fins, J.J. (2007). Border zones of consciousness: another immigration debate? *American Journal of Bioethics-Neuroethics*, **7**(1), 51–54.

Fins, J.J. (2008a). Neuroethics & neuroimaging: moving towards transparency. *American Journal of Bioethics*, **8**(9), 46–52.

Fins, J.J. (2008b). A leg to stand on: Sir William Osler and Wilder Penfield's "Neuroethics." *American Journal of Bioethics*, **8**(1), 37–46.

Fins, J.J. (2010a). Deep brain stimulation, free markets and the scientific commons: is it time to revisit the Bayh-Dole Act of 1980? *Neuromodulation*, **13**, 153–159.

Fins, J.J. (2010b). Minds apart: severe brain injury. In: Freeman, M. (Ed.) *Law and Neuroscience, Current Legal Issues* (Vol. 13), pp.367–384. New York: Oxford University Press.

Fins, J.J. (2011). Neuroethics and the lure of technology. In: Illes, J. and Sahakian, B.J. (Eds.) *Handbook of Neuroethics*, pp.895–908. New York: Oxford University Press.

Fins, J.J. (2012). Deep brain stimulation as a probative biology: scientific inquiry & the mosaic device. *AJOB Neuroscience*, **3**(1), 4–8.

Fins, J.J. (2013). Disorders of consciousness and disordered care: families, caregivers and narratives of necessity. *Archives of Physical Medicine and Rehabilitation*, **94**(10), 1934–1939.

Fins, J.J. (2015). *Rights Come to Mind: Brain Injury, Ethics and the Struggle for Consciousness*. New York: Cambridge University Press.

Fins, J.J. (2016, May 10) Bring them back. *AEON*. Available from: https://aeon.co/essays/thousands-of-patients-diagnosed-as-vegetative-are-actually-aware [accessed 12 July 2016].

Fins, J.J. and Hersh, J. (2011). Solitary advocates: the severely brain injured and their surrogates. In: Hoffman, B., Tomes, N., Schlessinger, M., and Grob, R. (Eds.) *Transforming Health Care from Below: Patients as Actors in U.S. Health Policy*, pp.21–42. New Brunswick, NJ: Rutgers University Press.

Fins, J.J. and Schiff, N.D. (2010a). Conflicts of interest in deep brain stimulation research and the ethics of transparency. *Journal of Clinical Ethics*, **21**(2), 125–132.

Fins, J.J. and Schiff, N.D. (2010b). In the blink of the mind's eye. *The Hastings Center Report*, **40**(3), 21–23.

Fins, J.J. and Shapiro, Z.E. (2013). Deep brain stimulation, brain maps & personalized medicine: lessons from the human genome project. *Brain Topography*, **27**(1), 55–62.

Fins, J.J., Schlaepfer, T.E., Nuttin, B., et al. (2011a). Ethical guidance for the management of conflicts of interest for researchers, engineers and clinicians engaged in the development of therapeutic deep brain stimulation. *Journal of Neural Engineering*, **8**(3), 033001.

Fins, J.J., Mayberg, H.S., Nuttin, B., et al. (2011b). Neuropsychiatric deep brain stimulation research and the misuse of the humanitarian device exemption. *Health Affairs (Millwood)*, **30**(2), 302–311.

Fins, J.J., Dorfman, G.S., and Pancrazio, J.J. (2012). Challenges to deep brain stimulation: a pragmatic response to ethical, fiscal and regulatory concerns. *Annals of the New York Academy of Sciences*, **1265**, 80–90.

Fins, J.J., Pohl, B., and Doukas, D. (2013). In praise of the humanities in academic medicine: values, metrics and ethics in uncertain times. *Cambridge Quarterly of Health Care Ethics*, **22**, 355–364.

Giacino, J.T., Kalmar, K., and Whyte, J. (2004). The JFK Coma Recovery Scale-Revised: measurement characteristics and diagnostic utility. *Archives of Physical Medicine and Rehabilitation*, **85**(22), 2020–2029.

Giacino, J.T., Whyte, J., Bagiella, E., et al. (2012). Placebo-controlled trial of amantadine for severe traumatic brain injury. *New England Journal of Medicine*, **366**(9), 819–826.

Giacino, J.T., Fins, J.J., Laureys, S., and Schiff, N.D. (2014). Disorders of consciousness after acquired brain injury: the state of the science. *Nature Reviews Neurology*, **10**, 99–114.

Hochberg, L.R., Bacher, D., Jarosiewicz, B., et al. (2012). Reach and grasp by people with tetraplegia using a neurally controlled robotic arm. *Nature*, **485**(7398), 372–375.

Illes, J. and Sahakian, B.J. (Eds.) (2011). *Handbook of Neuroethics*. Oxford: Oxford University Press.

Jaffe, K.M. and Jimenez, N. (2015). Disparity in rehabilitation: another inconvenient truth. *Archives of Physical Medicine and Rehabilitation*, **96**: 1371–1374.

Kass, L.R. (2003). Foreword. In: President's Council on Bioethics, *Beyond Therapy: Biotechnology and the Pursuit of Happiness*, p.xvii. New York: Harper Perennial.

The MacArthur Foundation Research Network on Law and Neuroscience (n.d.). Available at: http://www.lawneuro.org [accessed July 11, 2016].

Mandela, N. (1994). *Long Walk to Freedom*. New York: Little, Brown and Company.

Mayberg, H.S., Lozano, A.M., Voon, V., et al. (2005). Deep brain stimulation for treatment-resistant depression. *Neuron*, **45**(5), 651–660.

McGowan, A.K., Lee, M.M., Meneses, C.M., Perkins, J., and Youdelman, M. (2016). Civil rights laws as tools to advance health in the 21st century. *Annual Review of Public Health*, **37**, 185–204.

Mokhtarzadeh, M., Eydelman, M. and Chen, E. (2016). Challenges and opportunities when developing devices for rare disease populations. *Expert Opinion on Orphan Drugs*, **4**(5), 457–459.

Monti, M.M., Vanhaudenhuyse, A., Coleman, M.R., et al. (2010). Willful modulation of brain activity in disorders of consciousness. *New England Journal of Medicine*, **362**(7), 579–589.

Moss, C.W., Kilgore, K.L., Peckham, P.H. (2011). Training to improve volitional muscle activity in clinically paralyzed muscles for neuroprosthesis control. *Conference Proceedings of the Annual International Conference of the IEEE Engineering in Medicine and Biology Society*, **2011**, 5794–5797.

Naccache, L. (2006). Is she conscious? *Science*, **313**(5792), 1395–1396.

National Commission for the Protection of Human Subjects of Biomedical and Behavioral Research (1977). Use of psychosurgery in practice and research: report and recommendations of National Commission for the Protection of Human Subjects of Biomedical and Behavioral Research. *Federal Register*, **42**(99), 26318–26332.

Nicolelis, M.A. (2012). Mind in motion. *Scientific American*, **307**(3), 58–63.

Nuttin, B., Cosyns, P., Demeulemeester, H., Gybels, J., and Meyerson, B. (1999). Electrical stimulation in anterior limbs of internal capsules in patients with obsessive-compulsive disorder. *The Lancet*, **354**(9189), 1526.

Orphan Drug Act of 1983. Public Law 97–414. 96 Stat. 2049 (1983).

Osler, W. (1919). *The Old Humanities and the New Science: An Address before the Classical Association, Oxford*. London: John Murray.

Owen, A.M., Coleman, M.R., Boly, M., et al. (2006). Willful modulation of brain activity in disorders of consciousness. Detecting awareness in the vegetative state. *Science*, **313**(5792), 1402.

Penfield, W. (1976). *Mystery of the Mind: A Critical Study of Consciousness and the Human Brain*. Princeton, NJ: Princeton University Press.

Rothman, D. (1971). *The Discovery of the Asylum: Social Order and Disorder in the New Republic*. New York: Little Brown.

Satcher, D. (2000). Executive summary: a report of the surgeon general on mental health. *Public Health Reports*, **115**(1), 89–101.

Schiff, N.D., Rodriguez-Moreno, D., Kamal, A., et al. (2005). fMRI reveals large scale network activation in minimally conscious patients. *Neurology*, **64**, 514–523.

Schiff, N.D., Giacino, J.T., Kalmar, K., et al. (2007). Behavioral improvements with thalamic stimulation after severe traumatic brain injury. *Nature*, **448**(7153), 600–603.

Schiff, N.D., Giacino, J.T., and Fins, J.J. (2009). Deep brain stimulation, neuroethics and the minimally conscious state: moving beyond proof of principle. *Archives of Neurology*, **66**(6), 697–702.

Sheff, D. (1988, January 19). Izzy, did you ask a good question today? *The New York Times*. Available from: http://www.nytimes.com/1988/01/19/opinion/l-izzy-did-you-ask-a-good-question-today-712388.html [accessed July 12, 2016].

Thomas, L. (1974). *The Lives of a Cell: Notes of a Biology Watcher*. New York: The Viking Press.

UN General Assembly (2007). *Convention on the Rights of Persons with Disabilities*. Available from: http://www.un.org/disabilities/convention/conventionfull.shtml [accessed July 12, 2016].

Wright, M.S. and Fins, J.J. (2016). Rehabilitation, education, and the integration of individuals with severe brain injury into civil society: towards an expanded rights agenda in response to new insights from translational neuroethics and neuroscience. *Yale Journal of Health Policy, Law, and Ethics*, **16**(2), 233–288.

Author Index

7 Cups of Tea 92–7
9Solutions 84

Aarts, H. 480, 491
Abend, N.S. 182, 184
Abi-Rached, J.M. 456, 509
Abney, K. 547
Abraham, C. 462
Achard, S. 187
Adam, H. 214
Adams, F. 109
Adams, R.D. 109
Ader, R. 228
Adesman, A. 73
Advisory Committee, National Institutes
   of Health Director 147
Agar, N. 415, 416
Agarwal, N. 56
Ahmed, S.H. 498, 501
Air Force Studies Board 534
Aita, K. 341
Aizawa, K. 109
Akcakaya, M. 126
Akkad, A. 90
Al-Deeb, S.M. 341
Alavanja, M.C.R. 429
Alba, D. 83, 87
Albert, M.S. 295
Albrecht, G. 432, 461, 470
Alexander, B.K. 501
Alexander, E. 349
Allan, B. 462
Alpert, S. 130
Alphonso, C. 269
Alphs, H.H. 57
Altman, D.G. 592
Alzheimer's Association 144
Amanzio, M. 222, 223
American Medical Association 206, 216
American Psychiatric Association 94, 488, 585
American Society of Addiction Medicine 508
Americans with Disability Act 609
Amodio, D.M. 396
Anderson, G.C. 456, 463
Anderson, J.A. 6
Anderson, K. 123
Anderson, K.D. 136
Anderson, L.C. 521
Anderson, R.B. 457
Andrade, L.H. 591
Andrade, P. 237
Andrews, K. 186
Angermeyer, M.C. 504

APA Task Force 431
Apold, V.S. 61
Appelbaum, P.S. 130
Apple Inc 81
Araki, S. 459
Arbour, L. 463
Arias, J.J. 7
Ariely, D. 202
Arrighi, H.M. 302
Arrowsmith, J.B. 209
Asch, S.E. 393
Ashwal, S. 188
Aspinwall, L.G. 573
Aston-Jones, G. 128
Atlas, L.Y. 199, 204
Au, P.K. 204
Aubry, M. 524

Bachman, J.G. 500
Baillet, S. 26, 35
Baker, A.P. 25
Baker, L.A. 165
Bal, B.S. 562
Baler, R. 497, 499
Baler, R.D. 497
Bandettini, P.A. 17
Banerjee, T. 84, 85, 87
Bardin, J.C. 603
Bargh, J.A. 389, 402
Barker, R.A. 467
Barnard, C. 324
Barnhill, A. 226
Baron, R.S. 393
Baroncelli, L. 434
Barron, F.L. 458
Barth, M. 15
Bartlett, E.T. 347
Basser, P.J. 20, 38
Bassett, D.S. 19
Bateman, R.J. 297
Battiste, M. 465
Baumeister, R.F. 396, 401, 402, 404, 405
Baumel, A. 93
Baxendale, S. 257
Bayer, T.L. 204
Baylis, F. 251
Bazarian, J.J. 517
Beaty, R.E. 17
Beauchamp, T.L. 112, 216, 250
Beaufoy, S.L. 577
Beaulieu, C. 16
Becker, A.E. 591, 594
Becker, K.J. 312

Beckford, C.L. 467
Beddington, J. 594
Bedo, N. 27, 28
Beever, J. 398, 403
Belin, D. 481
Belkin, G.S. 339
Belknap, R. 81, 85
Bell, C. 532
Bell, E. 131, 180, 250
Bell, S. 504, 505, 506
Bell, W.O. 235
Benatar, D. 594
Bender, A. 186
Bendtsen, L. 223
Benedetti, F. 219, 220, 221, 222, 223, 224, 228
Benedikter, R. 546, 547
Benigeri, M. 558
Benjaminy, S. 270
Bennett, A.S. 535
Bennett, J.D. 209
Benson, P. 394
Bentall, R.P. 430
Benusic, M. 458
Berendse, H.W. 26
Berger, H. 28
Berger, R.P. 183
Bergfeld, I.O. 241, 242
Bering, J. 111
Berke, J.D. 483, 486, 491
Berkman, E.T. 407
Bernat, J. 306, 308, 309, 312, 315, 319, 336–65, 343, 345, 346, 353, 355
Berridge, K.C. 481, 483
Berry, H.L. 432, 460
Bested, A.C. 435
Betzel, R.F. 20
Bieler, D. 520
Bilgic, B. 16
Bingel, U. 228
Birbaumer, N. 125, 126, 131
Bird, S.J. 3, 150
Bishop, F.L. 228
Bishop, S. 163
Black, B.S. 380
Blakemore, C. 564
Bland, J.M. 592
Blankfield, R.P. 198
Blas, E. 427
Blease, C. 198, 206, 216, 218
Bleich, D. 359
Block, G. 558
Block, M.L. 460
Bloom, D.E. 144
Boersma, M. 26
Bohn, D. 101
Boly, M. 329
Bond, J. 558
Bond, R. 393
Bonelli, R.M. 351, 352
Boorse, C. 416
Booth, B.G. 37, 38, 39, 40

Booth, T.C. 57
Boothman, N. 205
Borich, M.R. 16
Bostick, N.A. 216, 217
Bostrom, N. 418
Botticelli, M. 498
Boudway, I. 98
Bourdette, D. 129
Bourque, F. 461
Boutet, C. 62
Bowler, D.E. 433
Boycott, K. 381
Boyd, S.C. 266
Boyer, B.B. 461
Boyer, E.L. 271
Braak, E. 283
Braak, H. 283
Brackmann, D.E. 31
Brand, C. 119
Brandeis, L. 115
Breakthrough 85, 93
Breden, T.M. 250
Breiter, H.C. 481
Brems, C. 460
Brenner, L.H. 562
Brenninkmeijer, J. 119
Bressan, P. 397
Breton, E. 38
Brock, D.W. 169, 342
Brody, B. 314, 330, 348
Brody, H. 135, 215, 218
Bromet, E. 128
Bronstein, J.M. 252
Brookes, M.J. 24, 37
Brookmeyer, R. 377
Brown, C.J. 40
Brown, M.A. 63
Browne, A.J. 467
Brownstein, M. 395
Bubela, T. 560
Bucchi, M. 554, 556
Buccino, G. 18
Buchanan, A.E. 169
Buchman, D.Z. 500
Buckley, K.M. 31
Buckner, R.L. 289
Bullmore, E.T. 19
Bunge, M. 350
Buniak, L. 592, 594
Bunnik, E.M. 54
Burgstaller, J.M. 199
Burkle, C.M. 340
Burton, R. 215
Burwood, S. 119
Bush, V. 533
Butterworth, R.F. 429
Buzsáki, G. 26
Byrne, P. 359

Cabrera, L.Y. 175, 436, 437, 443, 456, 464
Cacioppo, J. 17

Cacioppo, J.T. 205
Cajete, G. 465
Calabro, R.S. 185, 186
Calderon-Garcidueñas, L. 430, 460
Callahan, D. 605
Callaway, E. 5
Campbell, A.T. 160
Campbell, C.S. 341
Campbell, K.B. 31
Campbell, S.B. 160
Campbell-Meiklejohn, D.K. 394
Canada Safeway Ltd. 581
Canadian Environmental Health Atlas 438
Canadian Institutes of Health Research 377
Caplan, A.L. 361, 492, 498
Capron, A.M. 345, 356
Carlisle, W. 519
Carpenter, D.O. 460, 462
Carroll, D.C. 289
Carter, A. 492, 497, 504, 509
Casebeer, W.D. 545, 547
Castle, E. 563
Castro, A.L. 76
Caulfield, T. 554, 560
Cavazos, J. 84
Cavuoto, J. 241, 242
Centers for Disease Control and Prevention 203, 437
Centers for Medicare and Medicaid Services 203
Chadwick, R. 413
Chafe, R. 269
Chalmers, D. 99, 108
Chameau, J.L. 9, 533, 547
Chandler, J.A. 570–613
Chapman, S. 505
Charland, L.C. 498
Chartrand, T.L. 389
Chaudhary, U. 126
Chen, F.S. 435
Chen, J. 200
Chen, L. 6
Chepesiuk, R. 460
Cheung, K. 56
Chew, C. 266
Chiappa, K.H. 31
Childress, J.F. 112, 216, 250
Childs, N.L. 186
Chima, S.C. 595
Chin, V.N. 7
Chiong, W. 349
Cho, M.K. 7, 62
Choa, D. 56
Chodakiewitz, Y. 235, 238
Choi, E.K. 357
Chooi, C.S. 199
Choudhury, S. 369
Christen, M. 238, 250, 257
Christman, J. 113
Chumber, S. 560
Church, G.M. 91
Churchland, P.M. 405
Cialdini, R.B. 393, 394

Cieszanowski, A. 56
Cipriani, J. 87
Cisek, P. 480, 484, 491
Citerio, G. 356
Claassen, J. 191
Clairmont, D.H. 470
Clark, A. 99, 108
Clark, L. 479, 481
Clausen, J. 123
Clayton, P. 351
Clifford, D.B. 135
Cline, R.J.W. 558
Clouser, K.D. 135, 136
Coffey, R.J. 536
Cohen, E.D. 127
Cohen, H. 463
Cohen, J. 111, 359, 389, 403
Cohen, J.D. 486
Cohen, M.J.M. 498
Colborn, T. 431
Cole, J. 58
Colletti, V. 127
Collinger, J.L. 125, 137, 609
Collingridge, D. 532
Collins, P.Y. 591
Colloca, L. 219, 220, 221, 222, 225
Coluccia, D. 17
Combs, H.L. 130
Cominco Ltd. 583, 585
Commission on Social Determinants of Health 441
Committee on Opportunities in Neuroscience for
    Future Army Applications, National Research
    Council of the National Academies 536
Conrad, P. 479
Conway, E.M. 446
Cook, D. 463
Cook, M. 235
Cook, T.M. 609
Corner, L. 558
Corrigan, P.W. 503, 504, 576
Corripio, I. 237
Corvalán, C. 436, 437
Council of Europe 373, 375, 376
Council for International Organizations of Medical
    Sciences (CIOMS) 373, 374, 379
Covi, L. 226
Cox, S.M. 64
Coyle, P.K. 135
Cramer, S.C. 60
Cranor, C. 426, 437, 445
Crawford, M.L. 18
Crawley, P. 518
Crews, D. 426, 434
Crow, James 8
Crowe, K. 458
Cruse, D. 30
Cryan, J.F. 396, 397
Csonka, Y. 460
Cumes, D. 433
Cummings, M.L. 537
Cunningham, J.A. 505

Cunsolo Wilcox, A. 431, 432
Custers, R. 480, 491
Cutter, S.L. 442
Cyna, M.A. 198

D'Abbs, P. 463
Dackis, C. 497, 500, 501, 502
Dagher, A. 481
Dakeishi, M. 429
Dalal, S. 36
Dalvie, S. 596
Damasio, A.R. 283
Dando, M. 537, 547
Daneshvar, D.H. 515
Daniels, N. 439
Daoust, A. 343
DARPA 533–7
Darragh, M. 592
Dasgupta, L. 5
Davidson, A. 287
Davidson, J.S. 593
Davies, J.B. 505
Dayan, P. 480, 484, 491
de Beaufort, I. 467
de Burbure, C. 429
de Calignon, A. 289
de Condorcet 412
de Jager, C.A. 561
De Kluizenaar, Y. 460
De La Fuente-Fernandez, R. 220, 223, 224
De Salles, A. 235, 241, 242
De Salles, A.A. 239, 253, 254
De Strooper, E. 298
De Vos, A. 300
De Zwaan, T.E. 256
Deary, I.J. 55
Debes, F. 428
Debette, S. 58
Decety, J. 17
Deci, E.I. 404
Deco, G. 27
DeGrazia, D. 347
deHaan, W. 26
Deisseroth, K. 236
DeLee, S.T. 204
DeLong, M.R. 146
Delorme, A. 27
Deng, Z.D. 146
Denning, T. 123, 131
Department of Health and Prime Minister's
    Office 377
DePasquale, F. 25
DeRidder, D. 251
Derse, A. 378
DeVita, M.A. 354
Devries, K.M. 462
Dewey, J. 402, 404, 405, 407
Di Blasi, Z. 219
Di, H. 188
Di, H.B. 329
Di Pietro, N.C. 458

Diamond, A. 159, 175
Dibiase, R. 137
Dick, F.D. 429
Diederen, K.M. 481, 482
Diederich, N.J. 224
Diekema, D.S. 162, 369, 421
Diering, S.L. 235
Dietz, T. 446
Dieuliis, D. 595
Dimitriadis, S.J. 26
Dinan, T.G. 396, 397
Ding, K. 19
Diringer, M.N. 358
Dodds, S. 250
Doepp, F. 268
Doesburg, S.M. 27, 28, 41
Donnelly, J. 594
Donovan, E. 535
Doran, C.M. 506
Dougherty, D.D. 128, 241, 242, 253
Downie, J. 61
Dresser, R. 275, 276
Druckhman, D. 533
Dubljevic, V. 389, 390, 400, 405, 406
Dubois, B. 295, 298, 299
Duenwald, D. 535
Duggan, T.J. 401, 407
Duncan, G.J. 167
Dunham, Y. 480
Dunkley, B.T. 25, 26
Dunn, W.R. 521
Dupré, D.D. 237
Dutt-Gupta, J. 206
Duyn, J.H. 16
Dworkin, G. 112
Dworkin, R.H. 223, 278, 284
Dyson, F. 145

Eberling, J.L. 23
Ecker, J. 315
Ecobee 101
Edelman, G.M. 321
Edidin, J.P. 76
Editorial (Nature) 508, 532
Edwards, L. 100
El-Hayek, Y.H. 460
Elbaz, A. 429
Elder, G.J. 127
Eljamel, S. 243
Elkins, G. 204
Elliott, K.C. 426, 446
Elshahabi, A. 26
Emanuel, E.J. 135, 136
Emanuel, L.L. 135, 136, 348
Enck, P. 224
Engel, A.K. 25
Engelhardt, H.T. 416, 417
Enserink, M. 7
Erickson-Davis, C. 250
Ersche, K.D. 486, 491, 502
Eskandary, H. 56

Espay, A.J. 223
Esses, V.M. 577
Etkin, J. 99
European Commission 90
European Union 90, 370
Evans, N.G. 532
Everett, J.J. 204
Everitt, B.J. 485, 486, 491
Evers, K. 214
Ewald, A. 32
Ewin, B.N. 204
Executive Board of the American Academy
    of Neurology 312
Eysenbach, G. 266, 558

Fabing, H.D. 535
Fabsitz, R.R. 5
Faden, R. 173
Fagan, A.M. 299
Fann, J.R. 517
Farah, M.J. 111, 115, 118, 150, 159, 413, 593
Farrar, J. 59
Farwell, J. 547
Fassler, M. 215, 216
Faucher, L. 396
Faugeras, F. 329
Faunce, T. 467
Favaro, A.E. 268
Faymonville, M.E. 204
FDA 87, 89, 606
Feinberg, J. 161
Feinberg, T.E. 119
Feldman, R. 463
Felsen, G. 399, 400, 401
Felt, A.P. 89
Feltenstein, M.W. 498, 499
Fent, R. 216
Fenton, A. 130
Fenwick, P. 349
Ferenczi, E.A. 481, 482
Ferguson, J.R. 547
Fernandez, C. 373
Fernyhough, C. 430
Feudtner, C. 186
Field, R.A. 430
Fink, S. 209
Finkbeiner, S. 606
Finley Caulfield, A. 190
Finniss, D.G. 215
Fins, J.J. 133, 150, 186, 187, 240, 250, 256, 257, 312,
    313, 323, 328, 329, 332, 603–10, 610
Fischer, D.B. 317, 318, 327, 328, 329
Fischer, J.M. 572
Fischoff, B. 560
Fisher, C.B. 75
Fisher, C.E. 250
Fiske, S.T. 395
Fitz, N.S. 108, 118
Fitzgerald, T.H.B. 41
Fitzgibbon, S.P. 28
Flamm, A.L. 343

Fleminger, S. 516
Fleurence, R. 227
Florin, E. 26
Focquaert, F. 251
Foc.us 536
Foddy, B. 228
Fogg, B.J. 113
Folkins, C.H. 84
Ford, P.J. 130, 132
Ford, P.L. 215
Forman, E.N. 163
Fornai, F. 267
Forsythe, C. 546
Fortnum, H. 31
Foster, J.A. 396, 397
Fox, M.D. 21, 290, 561
Fox, N.C. 377
Fox, S. 561
Franca, A. 5
Frankfurt, H.G. 112
Franssen, E. 283
Franzini, A. 237
Frayne, R. 16
Free, C. 462
Freedman, R. 228
Freeman, J.B. 480
Fregni, F. 536
French, R. 328
Freund, H.J. 127
Fricchione, G.L. 594
Friedman, B. 562
Friedman, M. 113
Friedman, T.L.
Friehs, G.M. 239, 243
Fries, P. 24, 26
Frisaldi, E. 224
Friston, K.J. 21, 28
Fritze, J.G. 432, 460
Frumkin, H. 433
Fry-Revere, S. 342
Fuchs, T. 119
Fujiura, G.T. 424
Fukushi, T. 132
Fulford, K.W. 592
Furniss, E. 458
Furton, E.J. 341, 359
Fuss, J. 574, 575

Galeotti, F. 378
Gallagher, M. 480, 484
Gallagher, S. 119
Garbarino, J. 91
Gardiner, D. 340
Gardner, A. 517
Gartner, C. 503
Gartner, C.E. 502
Garvin, T. 456
Gay, J. 431
Geddes, J. 85, 87
Gehring, K. 413
General Assembly of the United Nations 385

Genesen, L.B. 344
George, M.S. 128
George, R.P. 347
Gergen, K.J. 443
German National Cohort Consortium 54
Gerrard, D.F. 521
Gert, B. 349, 401, 407
Gervais, K.G. 347, 352
Geula, C. 282
Giacino, J. 127, 130, 186, 308, 309, 310, 604
Giacomini, M. 358
Giaschi, D.E. 18
Gigerenzer, G. 100
Gilbert, D. 578
Gilbert, E. 124, 521
Gilbert, F. 235, 245, 250, 251, 253
Gilbert, S.F. 320
Gilbertson, M. 460, 462
Gillett, G. 462, 463
Gilligan, C. 394
Giordano, J. 251, 531–53, 533, 538, 548, 591–614,
    594, 595
Girotra, S. 181, 183
Giza, C.C. 515, 520
Gladstone, P. 458
Glannon, W. 119, 124, 131, 251
Glass, D.C. 404
Glimcher, P.W. 480, 491
Globe Editorial 269
Glockner 480, 491
Glover-Thomas, N. 595
Goate, A.M. 299
Gobas, F.A. 460, 462
Goebel, M.U. 220
Goering, S. 123, 124
Goetz, C.G. 223, 224
Gold, J.I. 480, 491
Goldberg, D.S. 250
Goldberg, S. 76
Goldenberg, A.J. 5
Goldstein, R.D. 445, 488
Gollust, S.E. 557
Golub, A. 535
Gómez-Pinilla, F. 432
Goodine, C. 458
Goodman, M. 266
Goodyear, B.G. 18
Google Play 94–7
Gorgulho, A. 239, 253, 254
Gostin, L.O. 344
Goubert, L. 200
Goulon, M. 337, 358
Grafton, S.T. 18
Gramm, H.-J. 314
Grandjean, P. 428, 429, 430, 436, 437, 445, 460
Grant, S. 72
Grashow, R. 428
Graves, A.B. 516
Gray, J.R. 163, 164
Gray, M.T. 402
Graybiel, A.M. 484

Greely, H.T. 115, 118, 593
Green, J.J. 27
Greenberg, B.D. 243
Greene, J. 111, 389, 403
Greene, J.A. 88
Greenlaw, J. 343
Greenwood, C.E. 432
Greer, D.M. 183, 357
Gregory, J. 556
Griffith, D.B. 370
Grigg, M.M. 314
Grill-Spector, K. 18
Grisso, T. 130, 163
Groopman, J. 270
Gross, J. 36
Grossman, E. 429, 446
Guan, X. 16
Gunnoe, J. 137
Gunther, D.F. 162, 421
Guo, J.Y. 220
Guo, Z. 516
Gureje, O. 596
Guskiewicz, K.M. 515, 516, 517

Haas, J.M. 359
Habermas, J. 404
Haddad, L. 431
Hadjistavropoulos, T. 130
Haga, S.B. 5
Haggard, P. 389, 395, 480
Haines, A. 459, 460
Halevy, A. 314, 330, 348
Hall, K.T. 225, 500, 506
Hall, W. 492, 497, 503, 509, 519
Haluza, D. 433
Hamalainen, M.S. 32
Hamarneh, G. 38, 39
Hamilton, R. 18, 536
Hansson, S.O. 137
Harada, M. 458
Harari, R. 429
Hardy, J. 297, 300
Harmon, K.G. 520
Harvard Medical School: Ad Hoc Committee
    to Examine the Definition of Brain Death
    324, 358
Harwerth, R.S. 18
Haselager, P. 123, 131, 251
Hashmi, J.A. 199
Haslam, C. 84
Haslam, N. 119
Hastings Center 7–8
Hatfield, E. 205
Hatira, P. 204
Haubensak, W. 148, 155
Hawn, C. 558
Hayashi, K. 199
Haynes, J.D. 115
Haynes, K.M. 558
He, J.H. 186
Heatherton, T.F. 407

Hebb, A.O. 124
Heesen, C. 135
Hegenscheid, K. 55, 57, 62
Heidegger, M. 98
Heiervang, K.S. 431
Heiss, W.D. 187
Held, V. 440
Henderson, J.M. 132
Henrich, J. 596
Herdman, A. 36
Herfst, S. 7
Hermundstad, A.M. 20
Herron, J. 126
Hertzman, C. 429
Herz, K.T. 428, 460
Hester, D.M. 163, 170
Heyman, G.E. 478, 479
Heyman, G.M. 500
Hichens, C. 519, 521
Hickner, J. 215, 216, 218, 225
Hicks, J. 460
High, S. 457
Hill, K. 117
Hillier, V. 515
Hindle, E.M. 183
Hinton-Bayre, A.D. 515
Hinxton Group 9
Hipp, J.F. 24
Hirsch, A.R. 470
Ho, A. 270
Ho, C.S. 18
Hochberg, L.R. 123, 125, 609
Hodge, F.S. 462, 463
Hodson, H. 89
Hoggard, N. 59
Hohagen, F. 253
Holroyd, J. 396
Holsheimer, J. 132
Holz, E.M. 40
Holzer, M. 182
Holzkamp, K. 119
Hopkin, M. 318
Hoppin, J.A. 429
Hornby, K. 354
Horstkötter, D. 176
Horvath, J.C. 536
Houdsen, C.R. 413
Houston, S.M. 165
Hovda, P. 360
Howick, J. 216
Høyersten, J.G. 535
Hradsky, J. 38
Hrobjartsson, A. 216
Hser, Y.I. 481, 487, 488
Hu, H.W. 428
Huang, J.Y. 532
Huang, M.X. 25
Hubel, D.H. 18
Huggins, J.E. 137
Hui, H.B. 36
Hull, S.C. 215, 216, 226

Human Rights Code 577, 586
Humphrey, N. 225
Hutmacher, D.W. 6
Hyafil, A. 480, 491
Hyman, S. 483, 486, 487, 491
Hyman, S.E. 491

iClarified 101
Idriss, S.Z. 557, 558
Ikram, M.A. 55
Ilieva, L.P. 118
Illes, J. 3, 7, 9, 58, 115, 150, 151, 160, 264, 271, 436,
    440, 459, 556, 557, 564, 591, 603
Ilmoniemi, R.J. 32
Iltis, I.S. 356
Immordino-Yang, M.H. 17
Ingold, T. 440
Institute of Medicine and National Research
    Council 7, 354
Interior Health Authority (South Similkameen
    Health Centre) 586
Invista (Canada) Co. 589
IOM 164
Isaak, C.A. 467
iTunes App Store 94
Izuma, K. 394

Jack, C.R., Jr 297
Jackson, D.D. 456, 460
Jacob, K.S. 593
Jacobson, J.L. 430
Jacobson, R. 88
Jacobson, S.W. 430
Jacques, J. 554
Jaffe, K.M. 609
Jain, S. 341
Jang, B. 458
Jaremko, J.L. 56
Jaworska, A. 276
Jeffrey, C. 85
Jennett, B. 309
Jerbi, K. 25
Jian, B. 38
Jimenez, N. 609
Joffe, A. 336, 342
Johnson, C.M. 561
Johnson, L.S.M. 521
Johnston, J. 115
Joint Commission in Accreditation of Healthcare
    Organizations 203
Jolly, D. 465
Jonas, W.B. 219
Jones, D.K. 16
Jones, R. 467
Jorm, A.E. 504
Joseph, J.A. 435
Jotterand, F. 416
Jox, R.J. 189
Jozaghi, E. 266
Juengst, E.T. 412
Jurcoane, A. 19

Kable, J.W. 480, 491
Kahn, J. 8
Kaiser, E. 76
Kaiser, M. 20
Kalant, H. 503, 506
Kalaska, J.F. 480, 484, 491
Kamath, S. 56
Kam-Hansen, S. 206, 223
Kamel, F. 429
Kamienski, L. 535
Kang, J. 23
Kant, E. 402
Kant, I. 467
Kaplan, J.T. 17
Kappenmann, E. 536
Kaptchuk, T.J. 219, 226, 227
Karch, C.M. 299
Karlawish, J. 380
Karran, E. 298
Kass, L. 360, 416, 538, 608
Kass, L.R. 345
Katz, D.J. 185
Katzman, G.L. 57
Kauffman, S.A. 351
Kaye, A.H. 521
Kaye, K. 90
Keene, S. 395
Kelley, A.E. 481, 483
Kelly, B.D. 430
Kelly, E.N. 456
Keltner, J.R. 204
Kelty, C. 92
Kendell, R.E. 228
Kendler, K.S. 499, 500, 502
Kennedy, I.M. 359
Kenny, P.J. 481
Kermen, R. 215, 225
Kerti, L. 432
Kesey, K. 235
Kessler, R.C. 594
Kessler, S.K. 181
Khandanpour, N. 56
Kikuchi, A. 23, 26
Kilmer, B. 500
Kilwein, J.H. 200
Kim, E.-H. 557
Kim, J. 562
Kim, P. 167
Kim, S. 562
Kim, S.Y. 302, 369, 380
Kimmelman, J. 270
Kimura, R. 317, 332
Kincaid, H. 501
King, C. 125
King, M. 467
King, N.J. 168
Kinomura, S. 149
Kiorpes, L. 18
Kirmayer, L.J. 228, 432, 593, 596
Kirsch, I. 220, 222
Kirschen, M.P. 64, 182, 184, 186

Kittelsaa, A.M. 420
Kitzbichler, M.G. 26
Klaming, L. 251
Kleiderman, E. 5
Kleiman, M. 500
Klein, E. 123, 129, 137
Kleinman, A. 591, 594, 596
Klotz, O. 459
Klyuzhin, L.S. 23
Knight, H.M. 63
Knoppers, B.M. 7, 369, 377, 384, 385
Koch, C. 187, 348
Koechlin, E. 480, 491
Kohl, S. 237, 242, 244, 245, 252
Kohlberg, L. 159
Kolata, G. 131
Kolibree 101
Konrad, P. 127
Konrad, T.R. 229
Koob, G. 500, 502, 508
Koob, G.F. 487, 498, 501, 503
Koppelmans, V. 56
Korein, J. 352
Korrick, S.A. 430
Kortum, P. 558
Kosal, M.E. 532
Kosten, T.R. 503
Kowal, S.P. 266
Kowalski, R.G. 189
Krack, P. 498
Kraemer, F. 251
Kramer, A.F. 434
Kramer, P. 133, 397
Krames, E.S. 132, 133
Kringelbach, M.L. 27
Kroger, W.S. 204
Krupnik, I. 465
Kübler, A. 131, 137
Kubu, C.S. 130
Kuehn, B.M. 74
Kuhn, J. 127
Kuhnlein, H.V. 460
Kullo, I.J. 5
Kuo, M.H. 74
Kupers, R. 204
Kvaale, E.P. 504, 576
Kwong, K.K. 17

La Vaque, T. 228
Laakso, M.P. 282
Lachaux, J.P. 33, 36
Lacour, A. 299
Ladd, R.E. 163
Lai, J.S. 435
Lam, D.C.K. 505
Lamas, D. 85
Lambert, B.W. 596
Lamkin, P. 81
Landrigan, P.J. 429, 436, 437, 460
Lane, A. 315
Lane, S. 519

Lang, D.J. 16
Lang, E.V. 198, 201, 204, 205, 206
Lange, M.M. 383
Lanphear, B.P. 428, 438
Lanzilao, E. 547, 592
Lao, Y. 166
Larsen, P.O.M. 563
Larson-Prior, L.J. 26
Laser, E. 205
Lasswell, H.D. 533
Laule, C. 16
Laupacis, A. 268
Laureys, S. 185, 187, 308, 309, 310
Lavigne, K.M. 21, 22
Law Commission of Ontario 578
Law Reform Commission of Canada 340
Leavitt, N. 89
Lebedev, M.A. 537
LeBihan, D. 38
Lederbogen, F. 431
Lee, A.A. 576
Lee, C. 442
Lee, I.S. 220, 489, 576
Lee, J.D. 489
Lee, K. 175
Lee, P. 347, 456
Lee, S.S.J. 462
Lee, Y.K. 517
Legon, W. 149
Lehman, E.J. 517
Lehmkuhle, M. 81, 84, 87
Lehti, V. 456, 460
LeMoal, M. 498, 501
Leng, S. 203
Lenhart, A. 70
Leon, M. 434
Leonard, A. 83, 99
Lerner, Y. 18
Leshner, A. 264
Leshner, A.I. 497, 508
Lévèque, M. 235, 237, 242, 243, 245
Levitt, P. 172
Levy, N. 111, 118, 437, 592
Lewin, L. 535
Lewis, A. 343
Lewis, C.M. 28, 500
Lewis, M.H. 5
Lexchin, J. 525
Li, T.-K. 498, 499, 502, 503
Li, W. 16
Li, X. 19
Liang, X. 187
Liao, S.M. 162
Libet, B. 389, 394
Licht, D.J. 184
Lichtenberg, P. 228
Lidstone, S.C. 220, 221, 222, 223
Lidz, C.W. 250
Lilienfeld, S.O. 479, 500, 505, 509
Lim, Y.-H. 430
Lin, L.Y. 72

Lin, X. 19
Lindemann, H. 171
Lindquist, C. 253, 254
Lingford-Hughes, A.R. 503
Liossi, C. 204
Lipsman, N. 235, 244, 245, 251, 252
Lisanby, S.H. 146
Liu, J.C. 116
Liu, S. 16, 23, 154
Liu, X. 154
Lizza, J. 347, 360
Llinás, R. 40
Lo, B. 419, 420
Lobier, M. 28
Lock, M. 317, 341
Lodygensky, G.A. 16
Loeb, J. 350
Logothetis, N.K. 21
Lomax, G.P. 5
Lombera, S. 591
Loria, K. 83, 85
Lou, J.S. 221
Love, T. 585
Lovett, K.M. 561
Lubman, D.I. 56, 64
Luby, J. 167
Lucchini, R.G. 429
Luckhoo, H. 37
Luckmann, T. 443
Lucky, R. 563
Luescher, C. 483, 487
Lufkin, B. 94
Lui, F. 221
Luigjes, J. 235, 498
Luna, F. 442
Luparello, T. 220
Lupton, D. 58
Lupton, M.K. 81
Lutkenhoff, E.S. 189
Luxton, D.D. 83, 85, 88
Lye, T.C. 516
Lynch, M. 117
Lynoe, N. 215, 216, 226

Maas, J. 433
MacArthur Foundation Research Network on Law
   and Neuroscience 610
MacDonald, J.P. 460
MacDonald, M.E. 16
MacKay, A. 16
Mackenzie, C. 113, 392, 505
Mackenzie, R. 251
MacLean, S. 463
Magnus, D.C. 344
Magnuson, K. 167
Maheshawari, M.C. 341
Mahner, M. 350
Mainsah, B.O. 127
Majury, D. 578
Mak, J.N. 123, 125, 127, 131
Makdissi, M. 515

Malenka, R.C. 483, 487
Malle, B.F. 405
Malloy, D.C. 130
Malone, D.A. 128
Mandela, N. 388, 610
Manogaran, P. 16
Marchant, J. 205
Marchessault, G. 467
Marcin, J.P. 182
Marcus, S. 3, 264, 443, 554, 564
Markus, H.S. 58
Marmot, M. 427
Marquis, D. 355
Martínez-Álvarez, R. 243, 245
Mason, C.E. 91
Masuda, J.R. 456
Mathews, D.J.H. 5, 6, 8
Matschinger, H. 504
Matthews, P.M. 55
Matuszek, S. 434
Maxwell, A.W.P. 60
Mayberg, H.S. 128, 146, 224, 609
Mayer, C.A. 268
Mazanderani, F. 269, 271
Mazurek, M.O. 74
McAlpine, D. 459
McCaffery, K.J. 64
McCall, C. 17
McCombs, M.E. 265
McCreight, R. 538
McCrory, P. 517, 520, 521, 524
McDermott, Q. 519, 521
McDermott, V.G.M. 202
McEwen, B.S. 435
McGie, S.C. 129, 131
McGowan, A.K. 610
McGregor, D. 467
McGuire, A. 560
McKee, A.C. 515, 517, 519
McKee, S.P. 18
McKergow, F. 463
McLaren, L. 467
McLellan, A.T. 502
McLoyd, V.C. 167
McMahan, J. 347
McMichael, A.J. 459
McNamee, M.J. 515
McNeal, G.S. 89
McQuigge, M. 458
Medical Consultants on the Diagnosis of Death
    to the President's Commission for the Study of
    Ethical Problems in Medicine and Biomedical and
    Behavioral Research 308
Medical Research Council and Wellcome
    Trust 54, 59, 61
Medina, L.S. 31
Meissner, K. 216
Mele, A. 394
Melnyk, R.M. 88, 93
Menary, R. 109
Mendelsohn, D. 235, 252, 339

Mercado, R. 223
Merikangas, K.R. 481, 506
Merkel, R. 251
Meshi, D. 6
Meslin, E.M. 6
Metzak, P.D. 21
Meurisse, M. 204
Meurk, C. 505, 508
Meyers, D. 113
Mezzacappa, E. 175
Mfutso-Bengo, J. 65
Micheau, J. 282
Michiels, E. 184
Miles, S. 342
Milgram, S. 393
Mill, J. 426
Miller, A.C. 341
Miller, E.K. 486
Miller, F.G. 129, 218, 219, 220, 222, 225, 315, 325,
    341, 342, 355
Miller, G. 214
Miller, P. 506
Miller, R.J. 535
Miller, S. 556
Miller, W.R. 505
Miller-Ziegler, J.S. 407
Mills, C.A. 459
Milstein, A.C. 57
Minassian, A. 84
Miodovnik, A. 429, 430, 445
Miresco, M.J. 228
Mirzadeh, Z. 127, 237
Miskovic, V. 25
Mitchell, S.M. 73
Mitchell, T.M. 115
Moerman, D.E. 219
Mogil, J.S. 200
Mohamed, F.B. 145
Mohan, G. 168
Moiseev, A. 33, 36
Mokhtarzadeh, M. 606
Molek-Kozakowska, K. 560
Moler, F.W. 182
Molgaard, C.A. 516
Molinari, G.F. 358
Mollaret, P. 337, 358
Moller, A.C. 405, 407
Monroe, A.E. 396, 401, 402
Montague, P.R. 402, 484, 485, 491
Monteith, S. 595
Monterosso, J. 575
Montgomery, G.H. 204, 220
Monti, M.M. 188, 318, 610
Mora, F. 434
Morales, M. 501
Morar, N. 398, 403
Morein-Zamir, S. 413, 593, 594
Moreno, J.D. 531–53
Moreno, M. 481
Mori, S.P.C.M. 38
Morishita, T. 237, 240, 241, 242, 244, 245

Morison, R.S. 360
Morris, Z. 56, 150
Morse, S. 399, 479, 490
Moss, C.W. 609
Moyers, T.R. 505
Muckli, L. 18
Mueller, P.K. 577
Mullen, T. 27
Müller S. 236, 237, 243, 245, 246, 250, 251, 252, 253, 255, 256, 257, 258, 401
Müller, U.J. 237
Multiple Sclerosis International Federation 267
Multi-Society Task Force on PVS 310
Munjal, K.G. 354
Munshi, L. 358
Murata, K. 460
Muratore, R. 524
Murphy, E. 65
Murphy, M.D. 123
Murphy, N. 111
Murray, C.J. 144, 146
Murray, L.K. 592
Murray, T.H. 170
Mustanski, B. 75
Mustanski, R. 75

Na, Y.C. 235, 240, 243, 244, 246
Naccache, L. 604
Nagarajan, S. 31
Nagel, S.K. 108, 110, 119
Nair-Collins, M. 330, 336
Nakagawa, T.A. 342
Nand, K. 38
Nardone, R. 128
Naselaris, T. 115
Nasrallah, F.A. 16
National Commission for the Protection of Human Subjects of Biomedical and Behavioral Research 607
National Institute on Drug Abuse 497
National Institutes of Health Director, Advisory Committee 147
National Research Council 507, 540, 547
Nedelsky, J. 113
Needleman, H.L. 428
Nelson, H.L. 171
Nemetz, P.N. 516
NEPA 439
Nestler, E.J. 481, 485, 487
Netherland, J. 506
Neufeld, K.A. 396, 397
Ngandu, T. 377
Nicholas, C.R. 188
Nicolelis, M.A. 609
Nielsen, N. 182
Nikiforuk, A. 458
Niso, G. 27
Nissani, M. 393
Nitsche, M.A. 536
Nitze, P. 533
Noble, K.G. 167

Noë, A. 119
Nolan, J.P. 183
Nolan, T.A. 224
Norbash, A. 205, 208
Norman, C.D. 558
Northern Brain Injury Association 457
Northoff, G. 187, 251
Norup, M. 216
Nussbaum, M.C. 438
Nutt, D. 503
Nuttin, B. 252, 257, 609

O'Brien, C. 497, 500, 501, 502
O'Doherty, K. 266
Office of the Information and Privacy Commissioner 562-3
Office of Naval Research: Science & Technology 534
Office of Technology Assessment 8
Ogawa, S. 17
Olick, R.S. 342
Olson, E. 348
Omalu, B.I. 515, 517
O'Neil, A. 432
O'Neill, G.C. 25
O'Neill, O. 226
Ontario Human Rights Commission 579
Opler, M.G.A. 428
Orchard, J. 522, 523
Oreskes, N. 446
O'Rourke, S. 431
Orphan Drug Act 606
Orr, R.D. 344
Osler, W. 607
Ovadia, D. 250
Overby, K.J. 356
Overgaard, S. 592
Owen, A.M. 187, 188, 318, 603
Oz, G. 16

Padela, A.I. 341
Padoa-Schioppa, C. 480, 484
Pallis, C. 314, 352
Palmour, N. 557, 561
Palop, J.J. 22
Palva, J.M. 24, 36
Palva, S. 24, 36
Pannek, K. 26
Panofsky, A. 92
Pantic, I. 72
Papanikolaou, V. 56
Pardo, M.S. 119
Parens, E. 115, 118, 593
Parent, A. 536
Parfit, D. 438
Park, H.-J. 56
Park, L.C. 226
Parker, J. 56
Parnia, S. 349
Parsons, R. 458
Partridge, B. 519, 520, 521
Pascoli, V. 481, 482

Pascual-Leone, A. 536
Pasic, M.D. 151
Passchier, W.F. 460
Passchier-Vermeer, W. 460
Patel, V. 591, 592, 593, 594, 596
PatientsLikeMe 91
Patil, P.G. 127
Patterson, C.C. 428
Patterson, D. 119
Patz, J.A. 459
Paulus, M.P. 499
Paulus, W. 536
Pavlov, I.P. 219
Pazderka, H. 463
Peace, W.J. 162
Pedersen, D. 593
Peen, J. 430
Peerdeman, K.J. 199
Penfield, W. 604
Pennefather, S.H. 314
Penney, S. 572
Pepper, J. 238, 241, 242, 243, 244, 245, 246, 252
Pereira, A.C. 434
Perera, F.P. 167
Perkins, D.D. 405
Perman, S.M. 182, 190
Pernick, M.S. 364
Perova, Z. 148
Perrin, M. 429
Perry, S.L. 431
Pescosolido, B.A. 504, 505
Peters, A. 430
Peters, C. 160
Peters, K.R. 299, 302, 303, 430
Petersen, A. 270
Petersen, C. 557
Petersen, R.C. 377
Peterson, B.S. 166, 167
Peterson, C. 404
Peterson, T.O. 203
Petrol Link-up 463
Petronis, A. 426
Petrovic, P. 220
Pew Research Center 112
Peyron, R. 199
Pfurtscheller, G. 125
Pham, T. 73
Phelan, J.C. 505
Piaget, J. 159
Picton, T.W. 31
Pierce, J.P. 506
Pierce, R. 6
Pinker, S. 348, 466
Pinkerton, E. 456
Plum, F. 309, 332
Pluye, P. 558
Podd, J. 222
Poldrack, R.A. 17
Pollo, A. 220
Polsky, S. 521
Pontifical Academy of Sciences 341

Pope John Paul II 341, 359
Pope, T.M. 340
Portier, C.J. 441
Posner, J.B. 308, 309, 316, 317
Potter, V.R. 440
Potts, J. 470
Powell, H.F. 56
Powers, M. 173
Powner, D.J. 306
Prasloski, T. 16
Presidential Commission for the Study of Bioethical Issues 151
President's Commission for the Study of Ethical Problems in Medicine and Biomedical and Behavioral Research 7, 339
President's Council on Bioethics 320, 343, 413, 538
Prest, G. 458
Preston, J.L. 111
Price, J.L. 520, 521
Prince, M. 592
Prince, M.R. 61, 594
Priori, A. 536
Proctor, R.D. 56
Pruessmann, K.P. 15
Prüss-Üstün, A. 436, 437
Psaltopoulou, T. 435
Public Health Agency of Canada 427
Pugh, C. 101
Pugh, J. 215, 216
Pullman, D. 269, 270
Purtilo, R.B. 138, 225

Quattrocchi, C. 56, 57
Quigley, M. 537
Quimby, E.G. 76

Rabinowitz, P.M. 431
Rachels, J. 169
Rachul, C. 554, 560
Racine, E. 160, 180, 190, 192, 214, 250, 343, 389, 390, 392, 398, 400, 406, 467, 509, 557, 592
Raeburn, S. 61
Raffin, T.F. 160
Raichle, M.E. 24, 290
Raichvarg, D. 554
Raj, A. 289
Randall, B. 61
Ranney, M.L. 73
Ransohoff, R.M. 135
Rasminsky, M. 269
Raspe, H. 253
Rauscher, A. 16
Ravizza, M. 572
Rawls, J. 438
Ray, J.G. 456
Raz, R. 430
Reavley, N.J. 504
Redish, A.D. 485, 491
Reekers, J.A. 268
Rees, G. 115
Régis, J. 239

Reimer, J. 270
Reiner, P.B. 108, 110, 111, 118, 399, 400, 401
Remien, R.H. 592
Repertinger, S. 353
Resch, J.E. 526
Reske, M. 499
Resnik, D.B. 426, 431, 439, 441
Ribary, U. 40
Rice, E. 76
Richardson, H.S. 62
Riedel, G. 282
Rieder, M.J. and Hazardous Substances
    Committee 377
Risch, N. 506
Risk, A. 557
Rissman, J. 115
Rizvi, S.A. 135
Roach, K. 572
Robbins, T.W. 479, 481, 485, 486, 491
Roberts, E.M. 429
Roberts, R.M. 188
Robertson, J.A. 311
Robillard, J.M. 6, 557, 559
Robins, L.N. 488, 500, 535
Robinson, S. 36, 483
Rodin, E. 314
Rodriguez-Arias, D. 343
Rodríguez-Barranco, M. 429
Roepke, A.M. 85
Rogers, B.P. 19
Rommelfanger, K.S. 215, 218
Roosevelt, F.D. 533
Rosado, J.L. 429
Rose, G. 506
Rose, J.P. 227
Rose, N.S. 456, 509
Rose, S.P.R. 119
Rosenberg, C.E. 218
Rosenthal, E. 594
Rosenzweig, M.R. 167
Rosin, B. 236
Roskies, A.L. 111, 214
Rosner, F. 359
Rosner, G. 90
Ross, B.D. 17
Ross, L.F. 163, 170
Roth, B.L. 149
Rothman, D. 608
Rothstein, M.A. 91
Rousseau, J.J. 388, 406
Roy, C. 162
Royal College of Radiologists, The 60, 62, 63
Royal Society 555, 565
RRT Expert Panel on Health Measurement 424
Rubin, E.B. 312
Rubinov, M. 20
Ruby, C.C. 573
Rück, C. 245
Rule, N.O. 392
Rupert, R.D. 109
Russell, W.T. 459

Ryan, K.A. 560
Ryan, R.M. 404

Sabat, S.R. 281
Sacks, O. 274, 291
Sadatsafavi, M. 62
Saggar, M. 17
Sagiv, S.K. 430
Sagoff, M. 426
Sahakian, B.J. 150, 593, 594, 603
Saigle, V. 389, 394
Salgado-López, L. 237
Salib, E. 515
Salter, E.K. 170
Samson, J.A. 432
Samson, K. 83, 84, 269
Sandberg, A. 413
Sandel, M.J. 593
Sandeman, E.M. 62, 64
Sander, C.Y. 23
Sandler, A.D. 228
Sandroni, C. 183
Sandvik, H. 560
Sanghavi, D. 330
Sanicola, L. 89
Sankar, T. 128
Sarasohn-Kahn, J. 88, 92
Sarkar, S. 320
Sartorius, N. 504
Satcher, D. 609
Satel, S. 479, 500, 505, 509
Sato, H. 537
Sauseng, P. 40
Savill, J. 59
Savulescu, J. 593
Saxena, S. 596
Scarlett, A.G. 456
Schaefer, G.O. 594
Schaefer, T.W. 459
Schäfer, C.B. 25, 37, 215, 227
Schafer, S.M. 227
Schechtman, M. 251
Scheltens, P. 294, 299
Schermer, M. 133, 250
Scheufele, D.A. 265
Schiff, N.D. 127, 149, 603, 604, 607, 609, 610
Schläpfer, T.E. 240, 242, 250
Schmidt, C.O. 61
Schnakers, C. 186
Schneider, J.W. 479
Schneider, M.J. 123
Schneiderman, L.J. 420
Schoenbaum, G. 484, 485
Schoffelen, J.M. 36
Schonert-Reichl, K.A. 175
Schrader, H. 314
Schreiber, T. 28
Schüll, N.D. 98, 100
Schultz, W. 224, 483
Schwab, A. 270
Schwartz, B.S. 428

Schweitzer, P. 460
Schweser, F. 16
Scialabba, G. 119
Sclove, R. 9
Scott, D.J. 222
Scullard, P. 557
Searle, J.R. 321
Secen, J. 19
See, R.E. 498, 499
Seel, R.T. 127
Seeley, W.W. 289, 290
Seeman, T.E. 435
Sekihara, K. 31
Selemon, L.D. 165
Seligman, M.E.P. 404
Selkoe, D.J. 297, 300
Selters, W.A. 31
Sen, A. 438
Sénécal, K. 5, 6
Shadlen, M.N. 480, 491
Shafer, J.P. 37
Shah, N.J. 17
Shah, S.K. 325, 342
Shaham, Y. 487
Shanker, T. 535
Shanks, D.R. 222
Shanks, T. 127
Shapiro, J.R. 504
Shaw, D.L. 265
Shaw, R.J. 561
Sheff, D. 605
Shelton, J.F. 429
Shen, F.X. 115, 116
Sherman, R. 215, 216, 218, 225
Sherwin, E. 397
Shewmon, A. 346, 347, 348, 353
Shewmon, D. 315, 353
Shewmon, E. 346, 348
Shih, J.J. 123, 428
Shirkey, H. 377
Shively, S. 516
Shonkoff, J.P. 172, 435
Shook, J. 538, 594, 595
Shores, E.A. 516
Shrader-Frechette, K.S. 426, 438, 439, 441, 442, 446
Silberg, W.M. 560
Silove, D. 594
Silvers, A. 135
Sime, W.E. 84
Simeral, J.D. 537
Siminoff, L.A. 357
Singer, T. 200
Singh, I. 384
Singh, V. 17
Siu, A.L. 93
Sivashanker, K. 72
Sjögren, M.J.C. 127
Skloot, R. 5, 7
Skuban, T. 250
Slaughter, L.M. 153
Slimak, M.W. 446

Slote, M. 440
Small, G. 432
Smith, C.F. 99
Smith, E.E. 55
Smith, K. 395
Smith, K.R. 427, 580
Smith, M.E. 413
Smith, P.B. 393
Smith, S. 136
Smith, S.M. 38, 153
Smylie, J. 462
Snow, C.P. 8
Snyder, J. 269, 270
Snyder, J.V. 354
Soekadar, S.R. 123, 124, 125, 126, 127
Soloveichik, A. 359
Song, T.M. 558
Souren, L. 283
Sowell, E.R. 165
Spearman, C. 20
Spellecy, R. 378
Spiegel, D.P. 19
Spike, J. 343
Spiro, H.M. 223
Spitzer, R.L. 594
Sporns, O. 20, 33
Sproul, A.A. 6
Squire, L.R. 282
Stables, J. 84
Stam, C.J. 26, 32
Stanbury, W.T. 456
Stanford Medicine 91
Stanley, S.A. 149
Stanovich, K.E. 109
Starling, R.M. 181
Stedman, R.C. 431
Steel, D. 446
Stein, D.J. 591–614
Stein, L.D. 369
Steinsbekk, K.S. 62
Stejskal, E.O. 38
Stemmer, B. 463
Stephan, B.C.M. 55
Steriade, M. 333
Stevenson, S. 464
Stewart, P.W. 430
Stewart-Williams, S. 222
Stilgoe, J. 266
Stins, J.F. 328
Stockard, J.J. 31
Stolier, R.M. 480
Stolyar, N. 392
Stout, R. 460
St Philip, E. 268
Strategic Multilayer Assessment Group 534
Strauss, J.S. 593
Strawson, P.F. 402
Sturm, V.E. 290
Stutzman, F. 117
Sudlow, C. 55
Sudo, N. 397

Sullivan, A. 460
Sullivan, J.A. 501
Sullivan, P.F. 481
Summerfield, D. 592
Sun, B. 235, 241, 242, 243, 245
Sutner, S. 87
Swan, M. 100
Swets, J.A. 533
Synofzik, M. 133
Szondy, D. 537

Tabery, J. 547
Tahmasian, M. 23
Tairyan, K. 9
Takahashi, K. 5
Takeuchi, T. 459
Talbott, E.O. 460
Talwar, S.K. 537
Tamatea, A.J. 462
Tan, T.J. 250
Tang, C.Y. 429
Tanner, J.E. 38
Tass, P.A. 236
Taylor, H.A. 378
Taylor, J.P. 127
Taylor, R.M. 342
Teicher, M.H. 432
Tendler, M.D. 359
Tennant, P. 456, 458
Teoh, G.Z. 6
Thackeray, R. 558
Thomas, J. 459
Thompson, A. 65
Thompson, B. 18
Thompson, P.M. 163, 164
Thomson, B. 61
Thorgeirsson, T.E. 481
Thornicroft, G. 592
Tiffany, S.T. 484, 486
Tilburt, J.C. 215, 216, 218
Timoney, K.P. 456
Tipper, C.M. 17, 18
Tokmetzis, D. 116
Tomasi, D. 187
Tononi, G. 187
Topjian, A.A. 181, 182, 183
Torres, C.V. 237
Tost, H. 432, 433, 435
Tournier, J.D. 38
Tovino, S.A. 580
Townsend, A. 64
Traboulsee, A.L. 268
Tractenberg, R.E. 546, 547
Trufyn, J. 56
Truog, R.D. 311, 314, 315, 317, 318, 324, 327, 328,
    329, 336, 355
Trustram Eve, C. 561
Tsai, S.-Y. 429
Tsuang, M.T. 481
Tuch, D.S. 38
Turner, D.A. 127

Tuttle, A.H. 223
Tyler, W.J. 536

Uhlhaas, P.J. 40
UK ECT Review Group 536
Ukueberuwa, D. 536
United Nations Educational, Scientific and Cultural
    Organisation (UNESCO) 373, 375, 376, 378
United Nations Environment Programme 438
United Nations General Assembly 577
US Department of Health and Human
    Services 5, 96
US Genetic Information Nondiscrimination
    Act 156, 563
US Health Insurance Portability and Accountability
    Act 89, 563
US Preventive Services Task Force 93
US Supreme Court 325

Valaskakis, G.G. 432
Valenstein, E.S. 235, 492
van der Werf, Y.D. 149
Van Fleet, D.D. 203
Van Laarhoven, A.I. 220
Van Oorschot, W. 576
Van Oostdam, J. 460
van Os, J. 430, 433
van Praag, H. 434
van Uden-Kraan, C.F. 557, 558
Van Veen, B.D. 31, 36
van Zellem, L. 183
van Zijl, P.C.M. 38
Vance, K. 558
Vanhaudenhuyse, A. 329
Varela, F. 28, 40
Varela, F.J. 119
Varelmann, D. 206
Varma, G. 16
Varney, S.M. 505
Vavasour, I.M. 16
Veatch, R.M. 134, 330, 347, 352
Veith, F.J. 340
Vemuri, B.C. 38
Verbeek, P.-P. 113
Vernooij, M.W. 54, 56, 59
Vespa, P. 191
Vicente, R. 28
Vierling, L. 163
Villanueva, C.M. 430
Villemagne, V.L. 22, 23, 297
Vincent, N.A. 251
Vinck, M. 36
Visser-Vandewalle, V. 237
Vohs, K.D. 405, 407
Voida, S. 84
Volkmann, J. 133
Volkow, N.D. 479, 488, 497, 499, 500, 501, 502, 503,
    506, 507, 508
Vollmann, J. 250
von Ins, M. 563
Vorgan, G. 432

Vos, S.J. 299, 300
Vrba, J. 36
Vrecko, S. 579
Vukic, A. 432

Wager, T.D. 199, 204, 221
Wagner, B. 93
Wagner, D.D. 407
Wahlster, S. 340, 357
Wahlund, L.O. 56
Walji, M. 561
Walker, D.M. 487
Walker, J. 88, 89
Walker, M.J. 505
Walter, H. 236, 251, 401
Walter, J.K. 182, 184, 186
Wang, F. 188, 591
Wang, H.E. 20
Wang, P.S. 591
Wang, T. 19
Wang, X.J. 40
Wang, Y. 16
Wangberg, S.C. 558
Wangmo, T. 416
Ward, K.M. 28
Ward, L.M. 40
Warden, D. 128
Wardlaw, J.M. 62, 64
Warren, S. 115
Wassermann, E.M. 536
Waters, E.A. 561
Watson, B.O. 26
Watson, S. 431
Watt-Watson, J. 201
Wearable Technologies 81
Wegener, J.R. 404
Wegner, D. 389, 395, 399, 480
Weijer, C. 189
Weiner, B. 576
Weinstein, E.J. 204
Weir, R.F. 160
Weiss, K.M. 596
Weisskopf, M.G. 428
Weithorn, L.A. 160
Weldeselassie, Y. 38
Wellcome Trust 64
Wenar, L. 172
Westin, C.F. 38
Wheatley, B. 460
Wheatley, M.A. 460
Wheeler, M.A. 149
White, A.M. 435, 547
White House 534
White, P.E. 520
White, S.E. 547
White, V. 506
Whitman, J.C. 37
Whittle, A.J. 149
Wibral, M. 28
Wichmann, T. 146

Wicks, P. 267
Widge, A.S. 128
Wiggermann, V. 16
Wijdicks, E.F. 183, 308, 340, 342, 358
Wildeman, S. 378
Wilkinson, D. 160, 184, 189
Willis, R. 556
Willox, A.C. 164, 461
Wilsdon, J. 556
Wilson, F.C. 186
Wilson, K. 456, 460, 467
Wilson, S.J. 253
Wilson, T.D. 100
Winblad, B. 294, 297
Windham, G.C. 430
Winkfield v. Rosen 359
Winocur, G. 432
Winslade, W.J. 330
Winston, G.P. 56
Winter, C. 84
Wintermark, M. 189
Wiseman, C.L. 460, 462
Witt, K. 251
Wittes, B. 116
Wiwchar, D. 462
Wolf, G. 57
Wolf, M.E. 487
Wolf, S.M. 5, 6, 98
Wolfe, S.M. 558
Wollmer, M.A. 205
Wolpaw, E.W. 123, 139
Wolpaw, J.R. 123, 125, 127, 131, 139
Wolpe, P. 115, 389
Wong, A.M. 18
Woo, C.C. 434
Woodward, T.S. 19, 21
Woolford, A. 459
Woopen, C. 134, 250
World Health Organization 94, 377, 477, 556
World Medical Association 371, 373
Wright, A.D. 16
Wright, M.S. 610
Wright, R. 429
Wu, H. 250
Wu, X. 187
Wurzman, R. 548

Xie, J. 480, 484
Xu, Z. 501

Yagiela, J.A. 209
Yamanaka, S. 5
Yamin, A.E. 594
Yang, C.I. 19, 165
Yang, Q. 214
Yang, Y. 165
Yaqub, B.A. 341
Yarborough, M. 378
Yazzie, R. 458
Ybarra, M.L. 75

Ye, A.X. 26
Yoshinaga-Itano, C. 30
Young, P.A. 283
Young, P.H. 283
Young, T.K. 458
Youngner, S.J. 347, 356
Ystby, Y. 289

Zalesky, A. 24, 37
Zamboni, P. 267, 268
Zealley, L.A. 57

Zeidman, L.A. 596
Zhai, J. 19
Zhang, R.R. 220, 221
Zhao, J.Q. 5
Zheng, Y.L. 81
Zhou, J. 290
Zimmerman, M.A. 405
Zinko, C. 81
Zola-Morgan, S. 282
Zubieta, J.K. 223
Zulman, D.M. 558

Subject Index

Note: "*b*", "*f*" and "*t*" indicate box, figure, and table

ablative microsurgical procedures *see* neurosurgery, ablative microsurgical procedures
addiction
  animal models of 483, 498–9
    high rates of recovery 501
  concept, describing other compulsive behaviors 578–80
  drugs
    abnormal dopamine signal 485
    advantage over natural rewards 485–6
    dependency and 578–9
    individual differences in liability and risk of drug use 480–2
  gambling, neurobiology claimed/adopted in legal context 579, 580–2
  impaired agency 491–2
  medical model *see* brain disease model of addiction (BDMA)
  moral model 477, 489, 490, 491, 492, 509
  nicotine 582–5
  persistence of relapse risk 487–8
  public views of addiction, stigma, and discrimination 503–4
  punishment and 490, 492
  scientific model 477–93
    implications for moral responsibility and legal culpability 490–1
    implications for understanding behavior 488–90
  see also brain disease model of addiction (BDMA)
adolescents
  consent issues 73
  social media, cyberbullying 70–9
  vulnerability and homelessness 70–9
advance directives 258
aging, advice in a digital world 558
air and water pollution 430
Albertan tar sands projects 458, *see also* First Nations Canada
alpha-synuclein, Parkinson's disease 23
alpha-theta-gamma (ATG) switch, thalamocortical processing 41
Alzheimer's disease
  antiamyloid drugs and amyloid-related imaging abnormalities 300, 302
  case study 275
  competing theoretical perspectives 275–6
  costs of care, US 144
  genetic risk 302
  hippocampal function 289–91
  preclinical Alzheimer's disease diagnosis 294–304
  preventive clinical trials 301–3
  self-assessments 562
  see also dementia, demented patients
amblyopia
  event-related, parametric fMRI 18–19
  localizing neuropathology 18–19
Americans with Disabilities Act (ADA) 577
amphetamines, enhancement therapy 535
amyloid-related imaging abnormalities (ARIAs) 302
amyloidosis
  antiamyloid drugs 300, 302
  beta-amyloid protein 22, 295–8
  clinical trials, monoclonal antibodies 295
analytic framework sharing 8–11, 10*t*
anorexia nervosa
  ablative procedures 243
  deep brain stimulation (DBS) 242
anterior limb of the internal capsule (ALIC) 242
antiamyloid drugs 300, 302
antiviral drugs, compassionate access 266
Apple, *Health Kit* 81
arsenic intoxication 428
arterial spin labeling, estimates of cerebral blood flow rates 16
attention deficit hyperactivity disorder 167
  placebo therapy 226–7
attribution theory 576, 578
auditory brain stem implants, speech recognition 127
auditory brainstem response (ABR) 30–1
auditory event-related potentials 30–2
auditory steady-state response (ASSR) 30–1
AugCog (augmented cognition) 537
autism, gut dysfunction 397
autism spectrum disorders and social media 73–4
autonomy
  constrained parental autonomy 170
  deception of placebo therapy 225–7
  definition 407
  demented patients 284–9
  ethics of deep brain stimulation (DBS) 250
  extended mind hypothesis (EMH) 112–15, 124
  free moral agency and 392
  free will 169
  microbiome research and 398
  notes 291–2
  personality-changing interventions 251
  standard model 399

beamforming
  multiple constrained minimum-variance 33–7*f*
  source leakage 36
  traditional LCMV (single-source) 34*f*, 35*f*
behavior
  behavioral addictions 579–80
  behavioral genetics 574
  brain-based research on free moral agency
      and 388–410
  compulsive behaviors, concept of addiction to
      describe 578
  voluntary control of, addiction and 477–97
beta-amyloid protein 22–3, 295–8
  see also amyloidosis
bias
  explicit/implicit 395–6
  government biases 446
  publication bias 240–1
Biological and Toxin and Weapons Convention
      (BTWC) 535
biomarkers
  disclosure 301
  ethical debate 296–300
  prostate-specific antigen, test dilemma 301
Bits of Freedom, metadata 116
brain, computers and *see* brain–computer interface
      (BCI) medicine
Brain Canada 147
brain death
  definitions, flawed definition,
      elimination 314–15, 358
  definitions and diagnostic criteria 306–8*t*, 344–56
    circulatory criterion of death 353–6
    common definitions 350
    criterion of death 351–2
    paradigm 346–9
  factors stimulating acceptance 338–44
  future consensus 356–8
  history 336–7, 358
  religious exemptions for declaring 341–2
  vs coma 311, 313, 314–17
brain disease model of addiction (BDMA) 497–514
  assessment of evidence 498–501
    genetic factors 499–500
  effectiveness of public health policies vs individual
      treatments 506
  escaped volitional control 477, 480
  ethical implications 498–9
  failure to develop new effective treatments 507
  has model of addiction delivered 502–4
  impact of brain disease explanations on addicted
      individuals 505–7
  more modest version of model 501–3
  most widely used medical treatments 503–4
  neural plasticity relevance 508
  neuroimaging studies 499
  policy consequences of neuromedicalization of
      addiction 506–7
  reframing neurobiological accounts 507–9
  rejection of 501, 509
brain imaging

perfusion measurement techniques 16
  signal processing advances 35–7
  see also magnetic resonance imaging
BRAIN Initiative® (US National Institutes of
      Health) 144–56
  ethical issues 150–4
  history 146–8
  scientific focus 148*t*
brain injury in children
  acute phase 182–3
  advanced neuroimaging techniques 187–9
  ethical issues 189–90
  neuroprognostication 180–96
    European Resuscitation Council algorithm 183
    future directions 190–2
    key questions from parents 183*b*
    resuscitation phase 181–2
    subacute phase 184–5
BRAIN Multi-Council Working Group
      (MCWG) 152
brain networks
  connectivity analyses 35
  localization of specific functions 19
  magnetoencephalography /
      electroencephalography (MEG/EEG) analyses
      of brain signals 24–37
  number of neurons 144
  organization of brain networks 19
  personality and 214
  reversible lesion 237
  reward circuitry 482–4, 487, 498–9*f*
  spatial and temporal replication 21–2
  topological network analysis 19–20
  tractography 16
  water diffusion 16
brain-based research on free moral
      agency 388–410
brain–computer interface (BCI) medicine 123–43
  adverse effects on cognition 130
  affect and 128
  biocompatibility and durability 129–30
  clinical ethics 124–30
    communication 126
    consciousness and cognitive processes 127–30
    consent 128–30
    near-term horizon 124–6
    neuromodulation 132–3
    sensation 127
  as enabling technology 137–8
    motor impairment 125–6
  military research 536
  privacy and security 131–2
  progressive cognitive impairment 129–30
  research 123–4, 128
  safety 129
  unrealistic expectations 131
brain–gut microbiota interactions 396–8
brain–machine interface (BMI) 139
Braingate system 125
Brodmann areas 36
  Criminal Code, criminal responsibility and 572

Canadian College of Medical Genetics (CCMG),
  Professional and Ethical Guidelines
  (incompetent adults) 381, 382*f*
Canadian Prospective Urban Rural Epidemiological
  (PURE) study 55
capacity
  interactive capacity 328–9
  task-related capacity, children and incompetent
    adults 384
  to value 276–8
case registries 256
cerebral blood flow rates, arterial spin labeling 16
cerebrospinal venous insufficiency *see* chronic
    cerebrospinal venous insufficiency (CCSVI)
    research
chemical shift saturation transfer 16
Charter of Rights and Freedoms, Canada 577
Chemical Weapons Convention (CWC) 535
child/childhood 159–79, 180–96
  adolescents, social media, cyberbullying 70–9
  bioethics issues 159–79
    best interests standard (BIS) 169–70
    core issues 168–9
    neuroimaging in at-risk populations 164–6
    rights for self-determination and
      decision-making 168–71
  biomedical research, minors and incompetent
    adults 369–85
    consent, assent and dissent 375–6
    incidental findings 376–7
    inclusion in research 373, 374*t*
  brain injury 180–96
  case study 181, 182, 184
  developmental issues in identity and
    self-determination 163–5
  maltreatment, learning disabilities 432–3
  neurocognitive potential, altering 164
  neuroprognostication, brain injury 180–96
  open future 161–3
    competing models 169
    concept 161–3
    environmental pollutants and toxins 166–8, 175
    pragmatic interventions 175–6
    resource allocation 174
    social justice for, capabilities of 172–4
    task-specific capacities 384
  transparency of communication with 163–4
China, psychiatric neurosurgery 250
chronic cerebrospinal venous insufficiency
    (CCSVI) 265–71
chronic traumatic encephalopathy (CTE) 517–20
  sport-linked 517–20
climate change and global warming 431–2
clinical ethics, and consent, brain–computer
    interface (BCI) medicine 128–9
clinical neuroimaging, applications and
    challenges 30–1
closed-loop medical devices with embedded artificial
    intelligence 124
cognition, parity principle 109
cognitive enhancement *see* enhancement

cognitive motor dissociation 603, 604
  apparent 609
  prosthetics 604
coma 185, 308–9
  vs brain death 311, 313, 314–17
communication *see* science communication
community engagement *see* public participation in
    science
compassionate access, antiviral drugs 266
competence *see* incompetent adults
compulsion 479–81
  medicalization of deviance and 479
compulsive behaviors, concept of addiction to
    describe 578
concussion in sport 515–30
  challenges in diagnosis 516–18
  long-term effects 516–20
    chronic traumatic encephalopathy
      (CTE) 517–20
  neuropsychological test devices, conflicts of
    interest 524–5
  prevalence 515
  professional sporting bodies 518–27
    conflicts of interest 524–7
    ethical obligations and alternative
      position 518–20
conformity, functional and structural imaging and 394
conscious awareness 321
consciousness, definitions of, interactive
    capacity 328–9
consciousness, disorders of 185*b*, 305–35
  civil rights and 312–13
  clinical assessment 186
  definitions and diagnostic criteria 305–10
    using diagnostic criteria as a legal fiction 324–6
  epistemological continuity between 322–3
  importance of discrete distinctions 313–15
  incorporating values 331–3
  lack, withholding food/fluid 187
  minimally conscious state 310–11
  natural history of brain recovery 185
  ontological continuity between 319–20
  problematization of diagnostic boundaries 323–4
  significance of diagnostic categories 310–12
  unresponsive wakefulness syndrome 185
consent issues
  adolescents 73
  alternative model for neuromodulation,
    rehabilitation medicine 138–9
  brain–computer interface (BCI)
    medicine 128–30
  inclusion of children in research 375–6
  incompetent adults 380–3
control
  attribution theory 576, 578
  diminished/removed self-control 572, 573, 581
  onset controllability and offset responsibility 576
  perceived, responsibility and social
    solidarity 576–8
  of substance use 479
control systems and their limitations 486–7

Council of Europe Convention on Human Rights
and Biomedicine 373
Council for International Organizations of Medical
Sciences (CIOMS)
consent issues, research, minors 373
Guidelines for Health-related Research Involving
Humans 379t
vulnerable individuals 371t
vulnerable individuals 371t, criterion of
vulnerability 383
creatine/phosphocreatine pool 16
criminal offending, neurobiology claimed/adopted in
legal context 579
criminal responsibility/criminality 570
addictions 479, 489
diminished self-control 572, 573
perceived control and 572
see also moral responsibility

data, privacy and confidentiality 562
data sharing, Global Alliance for Genomics and
Health 385
death, definitions and diagnostic
criteria 306–8t, 344–56
debt and compulsive spending, neurobiology
claimed/adopted in legal context 579
deception
and placebo therapy 225–7
disclosure 226
patient autonomy 225–6
physician belief 225–7
see also placebo therapy
decision-making
conscious vs unconscious 484
contradiction by neurobiology 479–80
control and implications for addiction 479–80
volitional acts 480
Declaration of Helsinki, rights of the child 370, 375
deep brain stimulation (DBS)
advantages 236, 253–4
adverse effects 244–5
applications 237–8
brain–computer interface (BCI)
research 123–4, 128
claims for 252
closed-loop 128
comparison of techniques 247–8t
costs 238, 253
efficacy, publication bias 240–1
ethical issues 237
superiority questionable 252
experimental use, psychiatric disorders 235–6
mapping circuits 606
methodological and study heterogeneity 241–4
minimally conscious state 130, 185, 307,
603, 609–10
moral/legal responsibility of patient 251
as neuromodulatory technique 236
in Parkinson's disease 146
personality change and legal responsibility
following 251

personal identity changes 252
quick onset of effect 254
reversibility 238
reversible lesion 237
sham-controlled trials 256
Defense Advanced Research Projects Agency
(DARPA) 533–4, 537
delirium 185
dementia, demented patients
autonomy 284–9
brain–computer interface (BCI)
medicine 127, 127–8
experiential interests vs critical interests 276–9
incompetent adults 377–999
onset age 303
quality of three types of resources 558–60
self-assessments and 563
understanding values 279–84
depression
ablative procedures 243
brain–computer interface (BCI) medicine 128
electroconvulsive therapy 146
treatment-refractory major, deep brain
stimulation 242
Diagnostic and Statistical Manual of Mental
Disorders (DSM-5) 477
behavioral addictions 579–80
definition of drug addiction 478
dietary factors, brain and mental health 432, 434–5
diffusion magnetic resonance imaging
(dMRI) 37–40
diffusion tensor imaging (DTI) 16
digital culture, cyberbullying 72–9
digital literacy 72
direct-to-consumer advertising
(DTCA) 557–8, 562
disability
Americans with Disabilities Act (ADA) 577
attitudes to 576–7
concept, used to describe compulsive
behaviors 578
definition
meaning and scope in law 570, 577–9
political and economic significance 578
social assistance and protection 576–7
social construction 571
disease as clinical problem(s) 416–18
concept 416
species-typical functions and 417
see also enhancement
dopamine
abnormal signals from addictive drugs 485, 486
Parkinson's disease and 221
regulation of plasticity in synapses 487
release
positron emission tomography/magnetic
resonance imaging 23
reward circuitry 482–4, 487, 498–9
smoking 485
unavailability (conscious introspection) 484
dopamine agonists, placebo responses 220–1

drowning, long QT syndrome 184
drugs *see* addiction

ecological affect, environmental change 461
electroconvulsive therapy (ECT) 249, 536
  treatment-resistant depression 128, 146
electroencephalography (EEG)
  brain injury cases 182, 187
  connectivity maps 35*f*
  continuous EEG monitoring 84
  diagnosis of epilepsy foci 27
  dynamic EEG synchronization 29*f*
  functional and dynamic imaging 27–8
  independent component analysis (ICA) 27
  limitations to be overcome 28
  neural basis of volition 394
  with sonoelectric tomography 149
  source mixing of signals 32–3, 34*f*
  spontaneous magnetoencephalography
    synchrony 26
electromyography (EMG) 28
emerging technologies 3–14
  ethical, legal, and social implications
    (ELSI) 3
  multiple morally contested areas 4–5
  neuroimaging 15–53
  shared analytic framework 8–11, 10*t*
  siloed research and analysis 6–8
engineering model of medicine 135–7
  division of labor 135
  fixing frame 136–7
  moral responsibility, limitations 136–7
enhancement 411–25
  clinical implications of enhancement 418–19
  comparative table for the definition of
    enhancement 414–15*t*
  ethical considerations 423–5
  ideals 414*t*
    anthropocentric ideal 415
    clinical ideal 415*t*, 416, 418–23
    graphic representations 422–3*f*
    objective ideal 415, 418
  posthuman agenda 418
  therapy vs enhancement 412–13, 418, 535
  views
    additionality view 413
    beyond therapy view 413
    improvement view 413
    umbrella view 413
environmental neuroethics 426–54
  brain/mental health risks 427–33, 446–7
  burden of human illness, policy implications 437
  change, ecological affect 461
  conceptual issues 443–4
  conflict of interest and bias 446–7
  evidence 445–6
  future research 443–7
  generational justice and social
    responsibility 437–9
  methodologies 445–6
  principles and frameworks 439–43

protective/health-enabling environmental
    factors 433–4
  social support and 432–3, 435
ethical, legal, and social implications (ELSI)
    (of neurotechnology) 3
  associated with Human Genome Project 6
  conflict between clinician and patient 133
  debates and future challenges 54–69
  regenerative medicine and 7
European Clinical Trials Directive 370
event-related potential (ERP) 30
evidence-based medicine
  chronic cerebrospinal venous insufficiency
    experience 264–73
  comparisons, psychiatric neurosurgery 256
extended mind hypothesis (EMH) 108–23
  (cognitive) enhancement 118–19
  neuroethical issues 111–13
    autonomy 112–15, 124
    potential conflicts of interest 115
    privacy of thought 115–17
  Otto and Inga 108–9

Facebook users, privacy-seeking
    behavior 117
First Nations Canada 455–76
  Albertan tar sands projects 458
  Alzheimer's disease and 464
  bridging ethical foundations 466–9
  colonization, context of change 457–9
  engaging in research of the brain 461–3
  environmental impacts on neurological
    health 459–61
  ethics and neuroethics 464–5
  fracking 458
  hydro-electric projects 470
  mental illness 460
  mercury contamination and 458, 459, 460, 462
  mistrust (historical) 461–2
  proper legal recognition 470
  study of the brain 462–3
  terminology 470
  way forward 470
folk psychology 479
Food and Drug Administration, humanitarian
    device exemption 256
football, codes/rules 515
fracking, First Nations people and 458
free moral agency 388–410
  behavioral and brain-based research as
    empowerment 399–400
  conscious control of action vs unconscious
    influence on action 395–6
  conscious volition vs free and determined
    volition 394
  contribution of behavioral and brain science to
    understanding 400*f*
  fallacies: assumption 1, objectivist,
    essentialist, dualist metaphysics vs
    intersubjectivist, nonessentialist, nondualist
    metaphysics 401–3

free moral agency (*cont.*)
  fallacies: assumption 2, is–ought fallacy vs
    fact–value continuum 403–5
  free will an illusion 395
  key components 391*f*
  obedience and 393
  self-determined action vs action determined by
    others 392–4
  sense of self vs conglomerate of divergent
    wills 396–7
  shaping by group pressures 393
  standard model of autonomy 399
  threats to 390–8
  see also moral responsibility
functional magnetic resonance imaging
    (fMRI) 17–21
  amblyopia 18–19
  blood oxygenation level-dependent (BOLD)
    contrast 17, 21
  constrained principal component analysis
    (fMRI-CPCA) 19–21
  detecting lying and deception 145
  event-related fMRI analyses 17, 18
  hemodynamic response (HDR) shape 20–1
  injury outcome prediction 187
  repetition suppression/fMRI adaptation 18
  task-based fMRI signal reliability and
    validity 20–2

gambling addiction 580
  neurobiology claimed/adopted in legal
    context 579–82
Gamma Knife® radiosurgery 239–40, 243, 253–6
  adverse effects 245
  minimal invasiveness 255
  numbers 252
generational justice, and social responsibility 437–8
Genetic Information Nondiscrimination Act
    (GINA) 153, 155, 563
genetic risk
  Alzheimer's disease 302
  brain disease model of addiction (BDMA) 499–500
  behavioral genetics 574–6
German Association for Psychiatry, Psychotherapy
    and Psychosomatics, obsessive–compulsive
    disorder (OCD) and 253
German National Cohort 55
Global Alliance for Genomics and Health, data
    sharing 385
global mental health 591–600
  10:90 research gap 591
  disease and wellness 593–4
  human diversity and neurodiversity 596
  human rights and patient empowerment 594
  naturalist and empirical approach 592–3
government biases 446

Harvard Medical School, definition of brain
    death 324, 358
headline-driven science, reducing research
    authority 560

health insurance, discrimination based on genetic
    information 153
Health Insurance Portability and Accountability Act
    (HIPPA) 563
health technology, wearable and mobile 80–107
HeLa cell line 7
hippocampal function, Alzheimer's
    disease 289, 289–91
homelessness, youth 76–9
homeopathy, trust in 560, 561
Human Fertilization and Embryology Authority
    (UK) 10
Human Genome Project 6
human rights
  instrumentality and 603–8
  legislation, drug and alcohol dependency/
    addiction 578–9
  patient empowerment and 594, 594–5
  social justice and 437
  towards an agenda 608–9
humanitarian device exemptions (FDA) 256
humanities, instrumentality 607
hypnosis 197–213
  lay hypnosis 208
  procedure hypnosis 207–9
  self-hypnosis 201, 204, 205
  shaping experience at onset of interaction 205–7
  see also pain management
hypothalamic–pituitary–adrenal axis (HPA), stress
    response and 397

immunomodulatory therapies, tradeoffs 135
impulse control disorders (Nevada) 582
incidental findings (IFs) on neuroimaging 6, 54–69
  children included in research 376–7
  clinically significant IFs 57–8
  defined 54
  disclosure of medically actionable, re-contact 383
  future importance 65
  impact 61–3
  prevalence of IFs 59–60
  public expectations 63–5
  scale of problem 55–63
    large projects 55
  service use and cost of follow-up 61–2
  types of IFs 56–7
incompetent adults 377–87
  consent, assent, and dissent 380–3
  ethical intersectionality, results and incidental
    findings in Canada 381–2
  inclusion in research 378–83
industrial chemicals and pesticides 429–31, 446
information, obligation of clinician, engineering
    model 134–7
Institute of Medicine and National Research
    Council 5
institutional care 609
institutional review boards (IRBs) 75, 77–8
instrumentality and human rights 603–8
interactive capacity 328–9
International Classification of Diseases (ICD-10) 477

Internet information, tagging of Internet-connected algorithmic 117
Internet sex addiction 585–7
is–ought fallacy vs fact–value continuum, free moral agency 403–5

Jewish law 359

law
    perceptions of control and responsibility, disability 570
    see also legal cases
lead intoxication 428
legal cases
    American Civil Liberties Union v. James Clapper (2013) 116
    Interior Health Authority (South Similkameen Health Centre) v. Hospital Employee's Union (2013) 586
    Kellogg Canada Inc v. Bakery, Confectionary, Tobacco Workers & Grain Millers (2006) 578
    McKelvey v. Turnage (1988) 478
    McNeill v. Ontario Ministry of the Solicitor General and Correctional Services (1998) 582
    Manitoba Government and General Employees' Union v. Manitoba (2013) 585
    Manitoba v. Manitoba Government and General Employees' Union (2005) 581
    Milazzo v. Autocar Connaisseur Inc. (2003) 579
    Nix v. Hedden (1895) 325
    Ontario v. Tranchemontagne (2009) 579
    R. v. Ample Annie's Itty Bitty Roadhouse (2001) 583
    R. v. Belcourt (2010) 573
    R. v. Horvath (1997) 580
    R. v. H.T. (2010) 585
    R. v. Obed (2006) 573
    R. v. Smith (2015) 580, 581
    Seneca College v. Ontario Public Service Employees Union (2002) 586
    Traynor v. Turnage (1988) 478
    Winkfield v. Rosen (2015) 359
    Yellowknife (City) v. Denny (2004) 583
legal fiction, using diagnostic criteria as 324–5
legally authorized representatives (LARs) 378–80
lesbian, gay, bisexual, and transgender (LGBT), youth and social media 74–5
life-saving treatment (LST), withdrawal decisions 315, 327
life/death distinction see brain death
locked-in syndrome 185, 609
long QT syndrome 184
long-term depression (LTD) 487
long-term potentiation (LTP) 487
Lothian Birth Cohort 59
low- and middle-income countries (LMICs), global mental health 591–2
lumbar spine, magnetic resonance imaging examinations 56

MacArthur Competence Assessment Tool for Treatment 130
magnetic resonance imaging
    assessment of changes in tissue structure in neurodegenerative diseases 16
    basis 15
    with carbon-13 nuclei 17
    costs caused by patient anxiety 209
    diffusion (dMRI) 37–40
    evolving into a quantitative tool 15
    fMRI, intrinsic neurophysiological amplitude network correlations 24–5
    functional (fMRI) 17–21
    magnetic resonance spectroscopy, brain metabolites 16
    phase information 16
    positron emission tomography (PET/MRI) 22–4
    quantification of tissue properties 15–17
    sodium imaging 17
    structural (sMRI) 15–17
magnetic resonance-guided focused ultrasound (MRgFUS) 236, 237, 240–2
    adverse effects 246
    comparison of techniques 247–8t
    contraindications 246
    obsessive-compulsive disorder (OCD) 243
    reduction of anxiety and depression 244
magnetoencephalography (MEG) imaging 24
    default mode network (DMN) 25
    injury outcome prediction 187
    magnetoencephalography/ electroencephalography (MEG/EEG) analyses of brain signals 27–37
    quantitative translational magnetoencephalography (MEG) 26–7
    spontaneous phase synchrony 25–6
Mandela, on freedom 388
manganese intoxication 429
Manhattan Engineer District (Manhattan Project) 533
media
    conflicts of interest 557
    direct-to-consumer advertising (DTCA) 557–8, 562
    pressure, compassionate access to antiviral drugs 266
    public participation in science 265–6
    self-assessments 561–3
    websites 560–1
medicalization
    addiction concept, describes other compulsive behaviors 578
    of deviance in society 479, 572
mental impairment, clinical ideal 419–23
mercury contamination 428–9
    First Nations people 458
mereology 360
microbiome, research, existence of self and 397–8
microsurgical ablative procedures see ablative microsurgical procedures
military, professional ethics 545–6

mind
  as blend between brain and algorithm 111
  defined 109–10
  lying and deception 145
minimally conscious state (MCS) 187–8,
    307, 310–13
  behavioral criteria 329
  fMRI 187–8
  misdiagnosis 186
  vs vegetative state 311–13, 317–19, 321–2, 324,
    327, 603, 609
minors and incompetent adults 369–87
moral responsibility
  definition of disability and 570–90
  engineering model of medicine 135–7
  perceived control 576–8
  scientific model of addiction 490–1
  self-control 571
  see also criminal responsibility/criminality; free
    moral agency
multiple constrained minimum-variance (MCMV)
    beamforming 33–7, 34–5f
  source mixing 32–7
multiple sclerosis (MS), CCSVI experience 267–71
multiple-comparisons problem, and approaches
    (statistics) 37
myelin water imaging 16

N-acetylaspartate 16
naloxone, inhibition of placebo-induced
    analgesia 223
National Child Development Study (UK) 63–4
National Commission for the Protection of Human
    Subjects of Biomedical and Behavioral
    Research 607
National Environmental Policy Act (NEPA) 438
National Institute on Alcohol Abuse and Alcoholism,
    prioritization of brain disease model of
    addiction (BDMA) treatment over public
    health policy 506–7
National Institute on Drug Abuse (NIDA), brain
    disease model of addiction (BDMA) and
    funding 497–514
National Institutes of Health
  BRAIN Initiative* 144–56
  overemphasis on the brain disease model of
    addiction (BDMA) 506
National Research Council 533
  review of ethical, legal, and social implications
    (ELSI) 547
national security 531–53
  criteria for development and use of specific
    neurotechnology 547
  devices 535–6
  dual-use research of concern (DURC) 532
  ethical issues in historical context 537–40
  ethical, legal, and social issues (ELSI) 534, 547
    ethics of power 546
    telic approach 545
  neurocognitive science, current and future
    use 538–40

performance enhancement/therapy 535–9
representative international research
    programs 540–4t
tensions between goals of science and
    security 531–2
neonatal care, improvement in intensive care
    techniques 165–6
neuroimaging
  advanced neuroimaging techniques,
    children 187–90
  emerging technologies 15–53
neurological disorders
  12% of deaths globally 556
  degenerative disease 274–93
    amyloid and tau 295
    capacity to value 276–8
  psychiatric conditions, marginalization and
    discrimination 608
neurological, mental health, and substance abuse
    (NMS) disorders 144
neuromodulation
  defined 132
  engineering model of medicine 134–7
  role of the clinician 133
neuroprognostication
  advanced neuroimaging techniques 187–90
  severe brain injury in children 180–96
neuroprosthetics, communication and 610
neurosurgery, ablative microsurgical
    procedures 236–52
  adverse effects 245
  comparison of techniques 247–8t
  efficacy 242–4
    publication bias and 240–1
  ethical issues 252
  history 238
  see also psychiatric neurosurgery
neurotechnologies
  emerging 3–14
  global mental health and 591–613
  as heuristic 610
  invasive 152–3
  neuroethics instrumentality of the humanities 607
  neuroethics of 603–5
  research and development see also BRAIN
    Initiative*
  targeting brain circuitry 145–6
  see also technologies of the extended
    mind (TEMs)
neurotoxicants and neurotoxic agents 428–33
nicotine addiction 582–5
Northwestern University's Center for Behavior
    Intervention Technologies (CBITs) 77
Nuremberg Code 370

obedience, Milgram's classical experiment 393
obsessive-compulsive disorder (OCD)
  ablative procedures 238–43
  anterior capsulotomy 239, 244, 253
  comparisons of treatment 254
  deep brain stimulation (DBS) 237, 242

Food and Drug Administration humanitarian
device exemption 256
Gamma Knife® radiosurgery 243, 252
magnetic resonance-guided focused
ultrasound 243, 246
pathogenesis 235
quick onset of treatment effect 254
targets for treatment 238–9
Office of Technology Assessment 9
Ontario, Human Rights Code 577–8
opiate antagonists, long-acting 489
opiates, addiction (Civil War) 535
opioid crisis/epidemic 203, 208–9, 488
organ transplantation 324, 339
dead donor rule' 311, 326–7
failed 163
rapid death determination 354
Health Resources & Services Administration
Division of Transplantation 359
see also brain death
Orphan Drug Act (1983) 606

pain/pain management
advanced rapport skills 205–6
beneficence/nonmaleficence equation 199–200
drug amounts and 201–2
eliciting deleterious behaviors in care
providers 201
experience and expectation 199, 202, 204–5
healthcare provider-induced 198
hypnotic or imagery-focused techniques 204
inflicting 200–1
language 206–7
negative suggestions and 198
procedure hypnosis 207–8
societal influence 203
parens patriae doctrine 370
parity principle 109
parkinsonism
arsenic and 429, 448
dementia and 517
Parkinson's disease
deep brain stimulation (DBS) 126, 130, 132, 133,
146, 235, 252
diet and 435
dopamine and 221
Gamma Knife® radiosurgery 239
magnetoencephalography (MEG) 23, 26
pesticides 429
placebo pathway 221, 223–4
Pasteur, Louis, vaccination 555
Pediatric Imaging, Neurocognition, and Genetics
(PING) Study 167–8, 167–9
personal identity, recognition 171
personality-changing interventions 251
personalized diagnostics, prognosis, targeted
intervention, and ethical challenges 15–53
pesticides 429–31
physician belief, placebo therapy, deception 225–7
Pillow Angel case 420–1
placebo therapy 214–34

as an ethical way forward 227–9
as an evolved endogenous healthcare system 224
deception and 225–7
disease with established biochemical
pathways 223–5
history 215–16
mechanisms 219–23
opiate release and 223
potential gains in research 228–9
prejudicial terminology 218–19
reward pathways 224
subjective experience and 229
placebome 225
pollution, air and water 430
polycyclic aromatic hydrocarbons (PAH) 166–7
American Medical Association (AMA)
professional values 216–17
in practice 216–17
pornography, Internet sex addiction 585–7
positron emission tomography (PET/MRI)
imaging 22–4
$_{18}$F-fluorodeoxyglucose 23
preclinical detection of Alzheimer's disease 294–304
President's Commission for the Study of Ethical
Problems in Medicine Biomedical Behavioral
Research 7
President's Council on Bioethics 315, 320, 538
privacy
and confidentiality, wearable and mobile health
technology 88–90
health insurance discrimination, information
sources 153
privacy of thought
extended mind hypothesis (EMH) 115–18
online privacy 115–16
prostate-specific antigen, test dilemma 301
prosthetics
brain–machine interface (BMI) for neural
interfaces 139
brain–computer interface (BCI) medicine 125–6
cognitive motor dissociation 604
limbs
control 125, 537
handshakes and 137
stroke-related motor prostheses 127
voice and 610
psychiatric neurosurgery 235–63
ablative procedures 236–7
comparison of the different techniques 246–7t
efficacy, publication bias 241
ethical, legal, and social context 249, 252
ethical issues 236–7, 252–3
evidence-based comparisons 256
law enforcement and social control 251
recommendations 255–8
psychopathy, neurobiology vs genetics 574
psychotherapies
evidence-based and values-based approaches 592
low- and middle-income countries (LMICs) 592
Public Attitudes to Science, Ipsos Mori poll 555
public engagement activity, bad for career 555

Public Health Agency of Canada, First Nations
    people 457
public health policies
    prioritization of brain disease model of addiction
        (BDMA) treatment over 506–7
    value re smoking 506, 509, 558
    views of addiction, stigma, and discrimination 503–4
public participation in science 265–6
    success stories 266–7
publication bias, efficacy of psychiatric
    neurosurgery 241

quantitative susceptibility mapping (QSM) 16

radiofrequency ablation procedures 252
radiosurgery
    comparison of techniques 247–8t
    see also Gamma Knife® radiosurgery
Recombinant DNA Advisory Committee 7
reference class 416–17
regenerative medicine 5
    ethical, legal, and social implications (ELSI) and 7
rehabilitation medicine
    alternative model of consent for
        neuromodulation 138–9
    enabling technology 137–8
reward circuitry
    dopamine release 482–4, 487, 499
    placebo therapy 224
roborat (DARPA) 537
robotics, brain–computer interface (BCI)
    medicine 125–6
Roman Catholic bioethics institute 359
Royal Society, on public engagement activity 555

Sagan Effect 555
science communication 554–69
    headline-driven, reducing research authority 560
    pre-Internet vs digital era modes 565t
    progress 563–4
    science literacy in general population and 556
security see national security
self-assessments
    online 561
    predatory marketing strategy 563
self-control see control
sex addiction (Internet-based) 585–7
sleep, rapid eye movement (REM) 333
smart technology see wearable and mobile health
    technology
smoking
    autonomy and 162
    control 478
    genetic loci associated 482
    nicotine addiction 582–4
    value of public health policies 506, 509, 558
social influences
    conformity and 394
    explicit/implicit bias and 395–6
social justice, human rights and 437
social media

cyberbullying 70–9
dementia and 558–60
ethical, legal, and social context of psychiatric
    neurosurgery 250
headline-driven science reducing research
    authority 560
new 71
promoting autonomy and empowerment in health
    decision-making 557
social support 432–3, 435
sociocognitive phenotypes 18
sonoelectric tomography 149
speech recognition, auditory brain stem
    implants 127
spontaneous network phase synchrony 26
spontaneous neuromagnetic phase locking 25–6
static encephalopathy 420–1
statistics, multiple-comparisons problem and
    approaches 37
Stockholm Declaration on the Human
    Environment 438
stress response, hypothalamic–pituitary–adrenal axis
    (HPA) 397
stroke
    brain–computer interface (BCI) medicine 127
    European Co-operative Acute Stroke Study trial 60
structural magnetic resonance imaging see magnetic
    resonance imaging

tau aggregation 22–3
tau tracers 23
technologies of the extended mind
    (TEMs) 108, 109–11
    (cognitive) enhancement 118–19
    global positioning system (GPS) as 110–11
    intrusions upon privacy of thought 115–18
    new framework for neuroethics 118–20
    tagging of Internet-connected algorithmic devices
        as TEMs 117–18
thalamocortical processing, neurophysiological
    frameworks 40–1
theft, caused by gambling addiction 581–2
thermal ablation see magnetic resonance-guided
    focused ultrasound (MRgFUS)
thermocoagulation 252
Three City Study (France) 55
topological network analysis 19–20
transcranial electrical stimulation (TES) 536
transcranial magnetic stimulation (TMS) 127, 536
traumatic brain injury (TBI) 149
Turing test 328
twin studies 165–7
    developing brain structure and cognitive
        processing 165–6

UK Biobank 55, 62
ultrasound, sonoelectric tomography 149
Ulysses contracts 258
UN Convention on the Rights of the Child 370, 375
UN Convention on the Rights of Persons with
    Disabilities 370, 381, 577

UN Office of the High Commissioner 374*t*
UN Principles for Older Persons 380*t*
UN Second World Assembly on Ageing 380*t*
UNESCO Declaration on Bioethics and Human
    Rights 372*t*, 378
unresponsive wakefulness syndrome 185
urban pollution 430
US Department of Health & Human Services 5
US National Institute on Drug Abuse (NIDA) 497
US National Institutes of Health BRAIN *see* BRAIN
    Initiative® (US National Institutes of Health)

vaccination, Pasteur, Louis and 555
value
    capacity to 276–8
    understanding values 279–84
vegetative state (unresponsive wakefulness
    syndrome) 185, 309–10
vestibular schwannomas (aka acoustic neuromas) 31
volitional control 392, 491, 495
    brain–computer interface (BCI) medicine
        and 126
    conscious vs free and determined 394, 401, 405
    escaped 477, 480
    minimally conscious state (MCS) and 318
    neural basis 394

vulnerability, youth, and homelessness 70–9

water pollution 430
wearable and mobile health technology 80–107
    authentic living 98–100
    case study: mental health apps 92–7
    classifications 82–7
    consent 90–1
    democratization of healthcare 91–2
    ethical considerations 87–92
    examples 81
    functionalities 84–6
    privacy and confidentiality 88–90
    purposes 82–3
    usage profiles 86–7
    see also brain–computer interface (BCI) medicine
westernized, educated, industrialized, rich, and
    democratic (WEIRD) populations 596
white coat trance 204
whole genome sequencing 4–5
wholism 333
World Medical Association (WMA), vulnerable
    individuals, research 371*t*, 373
worried well 593

youth homelessness 76–9